RELATIONAL MATHEMATICS

Relational mathematics is to operations research and informatics what numerical mathematics is to engineering: it is intended to aid modelling, reasoning, and computing. Its applications are therefore diverse, ranging from psychology, linguistics, decision studies, and ranking to machine learning and spatial reasoning. Although many developments have been made in recent years, they have rarely been shared amongst this broad community of researchers.

This first comprehensive overview begins with an easy introduction to the topic, assuming a minimum of prerequisites; but it is nevertheless theoretically sound and up to date. It is suitable for applied scientists, explaining all the necessary mathematics from scratch with a multitude of examples which are visualized using matrices and graphs. It ends with graspable results at research level. The author illustrates the theory and demonstrates practical tasks in operations research, social sciences and the humanities.

T0302248

All the titles listed below can be obtained from good booksellers or from Cambridge University Press. For a complete series listing, visit

http://www.cambridge.org/uk/series/sSeries.asp?code=EOM

ENCYCLOPEDIA OF MATHEMATICS AND ITS APPLICATIONS

Relational Mathematics

GUNTHER SCHMIDT

Universität der Bundeswehr München

CAMBRIDGE
UNIVERSITY PRESS

CAMBRIDGE UNIVERSITY PRESS
Cambridge, New York, Melbourne, Madrid, Cape Town, Singapore,
São Paulo, Delhi, Dubai, Tokyo, Mexico City

Cambridge University Press
The Edinburgh Building, Cambridge CB2 8RU, UK

Published in the United States of America by Cambridge University Press, New York

www.cambridge.org
Information on this title: www.cambridge.org/9780521762687

First published 2011

A catalogue record for this publication is available from the British Library

ISBN 978-0-521-76268-7 Hardback

Contents

PART III   ALGEBRA

PART IV   APPLICATIONS

Notes on illustrations

The author and the publishers acknowledge the following sources of copyright material and are very grateful for the permissions granted.

Leibniz' portrait, Fig. D.3 on the left, is an engraving the author owns personally. The handwriting on the right is owned by the Gottfried Wilhelm Leibniz Library – Niedersächsische Landesbibliothek – Hannover (Signature LH 35, 3, B2, Bl. 1 r).

The nineteenth century portrait of Augustus De Morgan, Fig. D.6 on the left, is in the public domain because its copyright has expired; one may find the photograph, for example, in [47]. Figure D.6 on the right was featured in an article by Adrian Rice in the EMS (European Mathematical Society) Newsletter of June 2003. Adrian Rice mentioned that this *seems to be* public domain and granted permission to use it.

Figure D.7 on the left is from http://www.kerryr.net/pioneers/gallery/ns_boole2.htm, where it is remarked that it is *probably* public domain. George Boole's picture on the right is a frontispiece of his book.

Figure D.8 on the left is from the Deutsches Museum (BN 4208). Figure D.8 on the right shows Sir William Rowan Hamilton as president of the Royal Irish Academy in 1842.

The Seki painting, Fig. D.9 on the left, was produced for the mathematician Nobuyoshi Ishikuro ($\approx$ 1760–1836). It is widely accepted as representing Seki – although it was painted long after his lifetime. Koju-kai is the rightsholder of the portrait. Shinminato Museum, Imizu City, acts as caretaker for it and was so kind as to provide a better copy and to grant permission to include it. The Seki stamp on the right may be printed since there is no regulation on printing used stamps in Japan.

Figure D.11 on the left, J. J. Sylvester, was sent by the Deutsches Museum (R1247-1) as a scan. Figure D.11 on the right, showing Arthur Cayley, is the frontispiece of *The Collected Mathematical Papers of Arthur Cayley*, volume VII, Cambridge University Press, 1894.

For the pictures of Charles Sanders Peirce, Fig. D.12 left and right, Houghton Library at Harvard University (Signature MS CSP 1643) has sent a more recent scan.

Ernst Schröder, Fig. D.13, is shown with permission of the Generallandesarchiv Karlsruhe (Signature J-A-S Nr. 106) which holds the copyright.

Figure D.14 on the left shows a page from the second Cambridge University Press edition of *Principia Mathematica*, volume I. The photograph from 1907 on the right is a touched-up scan from the Bertrand Russell Archives at McMaster University.

Leopold Löwenheim as shown in Fig. D.15 may be found in many mathematics history contexts in the web, marked as *probably free*.

Figure D.16 shows Alfred Tarski during the Tarski Symposium 1971 in Berkeley. The photograph was taken by Steven Givant, who is the rightsholder.

The author was happy to receive Fig. D.17 from Jacques Riguet for the present purpose.

All the other photographs stem from the Wikimedia Commons, a freely licensed media file repository; in particular Aristotle in Fig. D.1 as a photograph by Eric Gaba.

Preface

This book addresses the broad community of researchers in various fields who use relations in their scientific work. Relations occur or are used in such diverse areas as psychology, pedagogy, social choice theory, linguistics, preference modelling, ranking, multicriteria decision studies, machine learning, voting theories, spatial reasoning, data base optimization, and many more. In all these fields, and of course in mathematics and computer science, relations are used to express, to model, to reason, and to compute with. Today, problems arising in applications are increasingly handled using relational means.

In some of the above mentioned areas it sometimes looks as if the wheel is being reinvented when standard results are rediscovered in a new specialized context. Some areas are highly mathematical, others require only a moderate use of mathematics. Not all researchers are aware of the developments that relational methods have enjoyed in recent years.

A coherent text on this topic has so far not been available, and it is intended to provide one with this book. Being an overview of the field it offers an easy start, that is nevertheless theoretically sound and up to date. It will be a help to scientists even if they are not overly versed in mathematics but have to apply it. The exposition does not stress the mathematical side too early. Instead, visualizations of ideas – mainly via matrices but also via graphs – are presented first, while proofs during the early introductory chapters are postponed to an appendix.

It has often been observed that researchers and even mathematicians frequently encounter problems when working with relations; in particular they often hesitate to use relations in an algebraic form. Instead, they write everything down using predicate logic, thus accepting notation that is – by a rough estimation – six times as voluminous as an algebraic handling of relations. Therefore, concepts may not be written down with the necessary care, and it may not be easy to see at which points algebraic rules might have been applied.

In former years, this reluctance was often attributed simply to insufficient familiarity with relations; later one began to look for additional reasons. It seemed that people often could not properly interpret relational tasks or their results, even when these were explained to them in detail. Grasping the concept, which by now is over a hundred years old, of a *set* turned out to be a main obstacle. Mathematicians had developed the concept of a set and computer scientists used it indiscriminately although it is not immediately recognizable as a *data structure*. Since a set is unordered, a relation may not be understood directly as a matrix.

Engineers on the other hand are happy to use matrices. So it has been decided to give the reader some help. As far as representation is concerned, a set is conceived as being linearly ordered; then a relation between two sets has a more or less uniquely determined matrix representation, i.e., with rows and columns ordered.

In the same way as for real or complex matrices, linearity is a great help in dealing with Boolean matrices, i.e., binary predicates or relations. In this book, relational theory is conceived much in the same way as linear algebra in the classic form – but now for logics. The book provides a diversity of hooks showing where in mathematics or its applications one should intensify the use of relational mathematics. It is, thus, at many places 'open-ended'.

Regrettably, a common way of denoting, formulating, proving, and programming around relations has not yet been agreed upon. To address this point, the multipurpose relational reference language TιτυRεl has been developed in parallel with the writing of this book and is often made use of.

First of all such a language must be capable of expressing whatever has shown up in relational methods and structures so far – otherwise it will not gain acceptance. Secondly, the language has to convince people that it is indeed useful and brings added value. This can best be done by designing it to be *immediately operational*.[1] To formulate in the evolving language means to write a program (in the new language, and thus in Haskell) which can be executed directly at least for moderately sized problems. One should, however, remember that this uses *interpretation*, not *compilation*. To make this efficient requires still a huge amount of work which may well be compared with all the work invested in numerical algorithms over decades. Here, however, the algebraic formulation will greatly enhance any conceivable attempt at program transformation. TιτυRεl is an appeal to the Haskell community to check whether it is worth the effort to engage in the many new concepts, not least in the domain construction aspect.

Gunther Schmidt

---

[1] We thus address what Chapter 4 of [22] ends with: *A topic that we shall not address in this book is that of executing relational expressions. Clearly, it would be desirable to do so, but it is as yet unclear what the model of computation should be.*

Remark. This book has been written by an author whose native language is German, not English – of course with considerable additional work by native speakers. The majority of potential readers are certainly non-German and the author is not willing to miss any of them. He expresses the hope that the reader will excuse any shortcomings in this text due to his insufficient command of the English language.

The reader will also recognize that one feature of the German way of expressing concepts has prevailed: Germans frequently use compound words and at several places this book adheres to this style. A good example is the *strict order* in English, which is definitely not an order that is strict. Germans would use *strictorder* with the traditional idea that the added particle in such a compound, *strict-* for example, serves the purpose of restricting the whole but also of shifting the meaning. So the reader is asked to accept that (s)he will find here a *weakorder*, a *semiorder*, etc.

Acknowledgments. Much of the cooperation and communication around the writing of this book took place during the European COST[2] Action 274: TARSKI,[3] which the author had the honor to chair from 2001 to 2005; in particular with Harrie de Swart, Ewa Orłowska, and Marc Roubens in co-editing the volumes [43, 44]. He owes much to the many friends and colleagues with whom he has had discussions, cooperation and joint papers over the years. Not all can, but some must be mentioned explicitly: Roger Maddux checked the genesis of unsharpness ideas. Franz Schmalhofer asked several times for concepts to be clarified. Britta Kehden made reasonable suggestions for shortening the text. Peter Höfner read parts of a nearly final version and gave a plethora of detailed suggestions of many kinds. Josef Stoer read an early version concentrating on analogies with numerical mathematics and sent numerous hints and comments. Michael Winter commented on an early complete version of the book and afterwards also went through it a second time giving very valuable advice.

The historical annotations in particular received support from various people: Ulf Hashagen gave hints and Brigitte Hoppe suggested substantial contributions. Kurt Meyberg as well as Lothar Schmitz commented in a detailed way on an early version.

It should also be mentioned that the author enjoyed perfect support by interlibrary loan via the library of Universität der Bundeswehr München concerning the many historically interesting sources he had to access.

The author is very grateful to David Tranah, to all the other staff of Cambridge University Press, and to the copy editor Siriol Jones for their efficient assistance and support.

---

[2] European Cooperation in Science and Technology.
[3] Theory and Application of Relational Structures as Knowledge Instruments.

1
Introduction

A comparison may help to describe the intention of this book: natural sciences and engineering sciences have their differential and integral calculi. Whenever practical work is to be done, one will easily find a numerical algebra package at the computing center which one will be able to use. This applies to solving linear equations or determining eigenvalues, for example, in connection with finite element methods.

The situation is different for various forms of information sciences as in the study of vagueness, fuzziness, spatial or temporal reasoning, handling of uncertain/rough/ qualitative knowledge in mathematical psychology, sociology, and computational linguistics, to mention a few areas. These also model theoretically with certain calculi, the calculi of logic, of sets, the calculus of relations, etc. However, for applications practitioners will usually apply PROLOG-like calculi. Hardly anybody confronted with practical problems knows how to apply relational calculi; there is almost no broadly available computer support. There is usually no package able to handle problems beyond toy size. One will have to approach theoreticians since there are not many practitioners in such fields. So it might seem that George Boole in 1854 [26, 28] was right in saying:

> *It would, perhaps, be premature to speculate here upon the question whether the methods of abstract science are likely at any future day to render service in the investigation of social problems at all commensurate with those which they have rendered in various departments of physical inquiry.*

We feel, however, that the situation is about to change dramatically as relational mathematics develops and computer power exceeds previous expectations. Already in [35] an increasingly optimistic attitude is evident, claiming for an approach with matrices that it '... permits an algorithmic rather than a postulational-deductive development of the calculus of relations'. There exists, however, Rudolf Berghammers's RELVIEW system for calculating with relations fairly beyond toy size. This is a tool with which many diverse applications have been handled successfully.

With this text we present an introduction to the field which is at the same time easily accessible and theoretically sound and which leads to (reasonably) efficient computer programs. It takes into account problems people have encountered earlier. Although mainly dealing with discrete mathematics, at many places the text will differ from what is presented in standard texts on that topic. Most importantly, complexity considerations will not be the center of our focus – notwithstanding the significance of these. The presentation favors rather basic principles that are broadly applicable.

In general, we pay attention to the process of delivering a problem to be handled by a computer. This means that we anticipate the diversity of representations of the basic constituents. It shall no longer occur that somebody arrives with a relation in set function representation and is confronted with a computer program where matrices are used. From the very beginning, we anticipate conversions to make this work together.

We aim to provide three forms of work with relations simultaneously, namely *modelling* with relations, *reasoning* about relations, transforming relational terms considered as program transformation, and, finally, *computing* the results of relational tasks. We are convinced that such support is necessary for an extremely broad area of applications.

PART I
REPRESENTATIONS OF RELATIONS

Part I starts by recalling more or less trivial facts on sets, their elements or subsets, and relations between them. It is rather sketchy and will probably be uninteresting for literate scientists such as mathematicians and/or logicians.

Sets, elements, subsets, and relations can be represented in different ways. We will give hints to the programmer how to work with concrete relations and put particular emphasis on methods for switching from one form to another. Such transitions may be achieved on the representation level; they may, however, also touch a higher relation-algebraic level which we hide at this early stage.

We are going to recall how a partition is presented, or a permutation. Permutations may lead to a different presentation of a set or of a relation on or between sets. Functions between sets may be given in various forms, as a table, as a list, or in some other form. A partition may reduce the problem size when factorizing according to it. We show how relations emerge. This may be simply by writing down a matrix, or by abstracting with a cut from a real-valued matrix. For testing purposes, the relation may be generated randomly.

There is a clash in attitudes and expectations between mathematicians and information engineers. While the first group is interested in *reasoning about properties*, the second aims at *computing and evaluating* around these properties and thus tries to have the relations in question in their hands. Logicians will be satisfied with assertions and sketches in theory fragments, while information scientists try to obtain Boolean matrices and operate on these with machine help.

A point deserving special mention is the language TITUREL, which is frequently referred to in this book. It has been developed in parallel with the writing of this book and is a thoroughgoing relational reference language, defined in some syntactical way. It may be used in proofs and transformations or for interpretation in some model and environment.

However, one cannot start a book on a not yet commonly known topic using a sophisticated supporting language. If one is not sufficiently acquainted with relations, one will simply not be able to assess the merits of the language. This enforces an unconventional approach. We cannot start with the language from scratch and have to say first what a relation, a subset, or a point is like in a rather naïve way. Also, one cannot start a book on relations and relational methods without saying what symmetry and transitivity mean, although this can only be formulated later in a theoretically more satisfactory way. We cannot maintain an overly puristic attitude: the potential reader has to accept learning about *concrete* relations first and switching to the intended pure and abstract form only later.

At several places, however, we will give hints when approaching the intended higher level. At such a level it will be possible to write down terms in the language mentioned and – when given a model and an environment – to evaluate and observe the results immediately (of course only for small or medium-sized finite examples). All of the figures and examples of this book emerged in this way.

To sum up: Part I does not resemble the ultimate aims of the present book; rather, it has been inserted as a sort of 'warm up' – with more advanced remarks included – for those who are not well acquainted with relations.

2
Sets, Subsets, and Elements

Usually, we are confronted with sets at a very early period of our education. Depending on the respective nationality, it is approximately at the age of 10 or 11 years. Thus we carry with us quite a burden of concepts concerning sets. At least in Germany, *Mengenlehre* as taught in elementary schools will raise bad memories on discussing it with parents of school children. All too often, one will be reminded of Georg Cantor, the inventor of set theory, who became mentally ill. At a more advanced level, we encounter a number of paradoxes making set theory problematic, when treated in a naïve way. One has to avoid colloquial formulations completely and should confine oneself to an adequately restricted formal treatment.

The situation does not improve when addressing logicians. Most of them think in just one universe of discourse containing numbers, letters, pairs of numbers, etc., altogether rendering themselves susceptible to numerous semantic problems. While these, in principle, can be overcome, ideally they should nevertheless be avoided from the beginning.

In our work with relations, we will mostly be restricted to finite situations, which are much easier to work with and to which most practical work is necessarily confined. A basic decision for this text is that a (finite) set is always introduced together with a *linear ordering* of its elements. Only then we will have a well-defined way of presenting a relation as a Boolean matrix. When we want to stress this aspect, the set will be called a baseset. Other basesets will be generated in a clear way from already given basesets. Furthermore, we distinguish such *basesets* from their *subsets*: they are handled completely differently when they are transferred to a computer. Subsets of basesets necessarily refer to their baseset. We will be able to denote elements of basesets explicitly and to represent basesets for presentation purposes in an accordingly permuted form. Changing the order of its elements means, however, switching to another baseset. Altogether, we will give a rather constructive approach to basic mathematics – as well as to theoretical computer science.

2.1 Set representation

The sets we start with are (hopefully sufficiently small) finite **ground sets** as we call them. To denote a ground set, we need a name for the set and a list of the different names for all the names of its elements as, for example

Politicians = {Clinton,Bush,Mitterand,Chirac,Schmidt,Kohl,Schröder,Thatcher,
                    Major,Blair,Zapatero}
Nationalities = {US,French,German,British,Spanish}
   Continents = {North-America,South-America,Asia,Africa,Australia,Antarctica,Europe}
       Months = {Jan,Feb,Mar,Apr,May,Jun,Jul,Aug,Sep,Oct,Nov,Dec}
     GermSocc = {Bayern München,Borussia Dortmund,Werder Bremen,Schalke 04,VfB Stuttgart}
       IntSocc = {Arsenal London,FC Chelsea,Manchester United,Bayern München,
                    Borussia Dortmund,Real Madrid,Juventus Turin,Olympique Lyon,
                    Ajax Amsterdam,FC Liverpool,Austria Wien,Sparta Prag,FC Porto}.

There is no need to discuss the nature of ground sets as we assume them to be given 'explicitly'. Since such a set is intimately combined with the order of its representation, we will call it a **baseset**. An easy form of representation in a computer language like HASKELL is possible. One will need a name for the set – the first string below – and a list of names for the elements – here delimited by brackets – giving a scheme for denoting a 'named baseset' as

BSN String [String].

Here BSN starts with an upper case letter – without double quotes – and, thus, denotes a 'constructor'. A constructor has been chosen so as to be able to match against it. We would have to write, for example,

BSN "Nationalities" ["US","French","German","British","Spanish"].

In this way, paradoxes such as the 'set of all sets that do not contain themselves as an element' cannot occur; these are possible only when defining sets 'descriptively' as in the preceding sentence.

A variant form of the ground set is, for example, the '10-element set Y' for which we tacitly assume the standard element notation and ordering to be given, namely[1]

$$Y = \{1, 2, 3, 4, 5, 6, 7, 8, 9, 10\}.$$

Ordering of a ground set

Normally, a set in mathematics is *not* equipped with an ordering of its elements. Working practically with sets, however, this level of abstraction cannot be maintained. Even when presenting a set on a sheet of paper or on a blackboard, we can hardly avoid some ordering. So we demand that ground sets correspond to *lists* and not just to *sets*. As this is the case, we take advantage of it in so far as we are allowed to choose a favorable ordering of elements of a set. This may depend

---

[1] When implementing this in some programming language, one will most certainly run into the problem that the elements of the set are integers – not strings.

on the context in which the set is presented. The necessary permutation will then somehow be deduced from that context.

As an example, consider the baseset MonthsS of short month names under the additional requirement that month names be presented alphabetically as in

{Apr,Aug,Dec,Feb,Jan,Jul,Jun,Mar,May,Nov,Oct,Sep}.

The necessary permutation shown as numbers to which position the respective original month name should be sent is

[5,4,8,1,9,7,6,2,12,11,10,3].

'Jan', for example, must for this purpose be sent to position 5. Occasionally, it will be necessary to convert such a permutation back, for which we use the inverse permutation

[4,8,12,2,1,7,6,3,5,11,10,9]

sending, for example, 'Apr' back to its original position 4.

Another example of a ground set is that of Bridge card denotations and suit denotations. The latter need a permutation so as to obtain the sequence suitable for the game of Skat[2]

$$\begin{aligned} \text{CardValues} &= \{\text{A,K,Q,J,10,9,8,7,6,5,4,3,2}\} \\ \text{BridgeColors} &= \{\spadesuit,\heartsuit,\diamondsuit,\clubsuit\} \\ \text{SkatColors} &= \{\clubsuit,\spadesuit,\heartsuit,\diamondsuit\}. \end{aligned}$$

A ground set is an object consisting of a name for the ground set – uniquely chosen in the respective situation – and a list of element names. Handling it in a computer requires, of course, the ability to ask for the name of the set, to ask for its cardinality, and to ask for the list of element names. At this point we do not elaborate this any further.

What should be mentioned is that our exposition here does not completely follow the sequence in which the concepts have to be introduced theoretically. When we show that sets may be permuted to facilitate some visualization, we are already using the concept of relations which is introduced later. We cannot avoid using it here in a naïve way.

Constructing new basesets

Starting from ground sets, further basesets will be obtained by construction, as pair sets, as variant sets, as powersets, or as the quotient of a baseset modulo some equivalence. Other constructions that are not so easily identified as such are

---

[2] Skat is a popular card game in Central Europe with the same suits as Bridge, but ordered differently as mentioned here; card values range from 7 to Ace only.

subset extrusion and also baseset permutation. The former serves the purpose of promoting a subset of a baseset to a baseset in its own right,[3] while the latter enables us, for example, to present sets in a nice way. For all these constructions we will give explanations and examples later.

2.2 Element representation

So far we have been concerned with the (base)set as a whole. Now we concentrate on its elements. There are several methods of identifying an element, and we will learn how to switch from one form to another. In every case, we assume that the baseset is known when we try to denote an element in one of these forms

— as an element number out of the baseset　　`NUMBElem BaseSet Int`
— marked **1/0** along the baseset　　　　　　`MARKElem BaseSet [Bool]`
— as a name out of the elements of the baseset　`NAMEElem BaseSet String`
— as a one-element mark in the diagonal　　　　`DIAGElem BaseSet [[Bool]]`.

First we might choose to indicate the element of a *ground* baseset giving the position in the enumeration of elements of the baseset, as in `Politicians`$_5$, `Colors`$_7$, `Nationalities`$_2$. Because our basesets, in addition to what is normal for mathematical sets, are endowed with the sequence of their elements, this is a perfect way of identifying elements. Of course, we should not try an index above the cardinality which will result in an error.

However, there is another form which is useful when using a computer. It is very similar to a bit sequence and may, thus, be helpful. We choose to represent such an element identification as in Fig. 2.1.

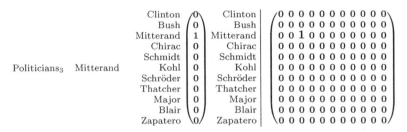

Fig. 2.1 Element as a marking vector or as a marking diagonal matrix

Again we see that combination with the baseset is needed to make the column vector of **0**s and **1**s meaningful. We may go even further and consider, in a fully

---

[3] The literate reader may identify basesets with objects in a category of sets. Category objects constructed generically as direct sums, direct products, and direct powers will afterwards be interpreted using natural projections, natural injections, and membership relations.

naïve way, a partial identity relation with just one entry **1**, or a partial diagonal matrix with just one entry **1** on the baseset.[4]

With these four attempts, we have demonstrated heavy notation for simply saying the name of the element, namely `Mitterand` ∈ `Politicians`. Using such a complicated notation is justified only when it brings added value. Mathematicians have a tendency of abstracting over all these representations, and they often have reason for doing so. On the other hand, those using applications cannot switch that easily between representations. Sometimes, this prevents them from seeing the possibilities of more efficient work or reasoning.

An element of a baseset in the mathematical sense is here assumed to be transferable to the other forms of representation via functions that might be called

   `elemAsNUMB`, `elemAsMARK`, `elemAsNAME`, `elemAsDIAG`,

as required. Each of these takes an element in whatever variant it is presented and delivers it in the variant indicated by the name of the function. All this works fine for ground sets. Later we will ask how to denote elements in generically constructed basesets such as direct products (pair sets), powersets, etc.

2.3 Subset representation

We now extend element notation slightly to the notation of subsets. Given a baseset, subsets may be defined in at least six different forms which may be used interchangeably and which are also indicated here in a computer usable form:

— as a list of element numbers of the baseset `LISTSet BaseSet [Int]`
— as a list of element names of the baseset `LINASet BaseSet [String]`
— as a predicate over the baseset   `PREDSet BaseSet (Int -> Bool)`
— as element in the powerset of the baseset `POWESet BaseSet [Bool]`
— marked **1 / 0** along the baseset   `MARKSet BaseSet [Bool]`
— as partial diagonal matrix on the baseset `DIAGSet BaseSet [[Bool]]`.

Again, the capitalized name is a constructor in the sense of a modern functional language; what follows is the type to indicate the set we draw the subset from. This methods works fine as long as the baseset is ground. In this case we have the possibility of giving either a set of numbers or a set of names of elements in the baseset. One has to invest more care for constructed basesets or for infinite basesets. For the latter, negation of a finite subset may well be infinite, and thus more difficult to represent.

---

[4] We quote from [77]: ... Moreover, the **0** and **1** reflects the philosophical thought embodied in symbols of dualism of the cosmos in *I Ching* (The Classic of Changes or The Book of Divination) of ancient China. That is, **0** represents
  Yin (negative force, earth, bad, passive, destruction, night, autumn, short, water, cold, etc.)
and **1** represents
  Yang (positive force, heaven, good, active, construction, day, spring, tall, fire, hot, etc.).

Given the name of the baseset, we may list numbers of the elements of the subset as in the first variant or we may give their names explicitly as in the second version of Fig. 2.2. We may also use marking along the baseset as in variant three, and we may, as already explained for elements, use a partial diagonal matrix over the baseset.

Politicians$_{\{1,2,6,10\}}$

$\{$Clinton,Bush,Kohl,Blair$\}$

Clinton	1
Bush	1
Mitterand	0
Chirac	0
Schmidt	0
Kohl	1
Schröder	0
Thatcher	0
Major	0
Blair	1
Zapatero	0

Clinton	1	0	0	0	0	0	0	0	0	0	0
Bush	0	1	0	0	0	0	0	0	0	0	0
Mitterand	0	0	0	0	0	0	0	0	0	0	0
Chirac	0	0	0	0	0	0	0	0	0	0	0
Schmidt	0	0	0	0	0	0	0	0	0	0	0
Kohl	0	0	0	0	0	1	0	0	0	0	0
Schröder	0	0	0	0	0	0	0	0	0	0	0
Thatcher	0	0	0	0	0	0	0	0	0	0	0
Major	0	0	0	0	0	0	0	0	0	0	0
Blair	0	0	0	0	0	0	0	0	0	1	0
Zapatero	0	0	0	0	0	0	0	0	0	0	0

Fig. 2.2 Different subset representations

In any case, given that

```
Politicians =  {Clinton,Bush,Mitterand,Chirac,Schmidt,Kohl,
                Schröder,Thatcher,Major,Blair,Zapatero}
```

what we intended to denote was simply something like

```
{Clinton,Bush,Kohl,Blair} ⊆ Politicians.
```

Subsets may be given in either of these variants; we assume, however, functions

```
setAsLIST, setAsLINA, setAsPRED, setAsPOWE, setAsMARK, setAsDIAG
```

to be available that convert to a prescribed representation of subsets. Of course, the integer lists must provide numbers in the range of 1 to the cardinality of the baseset.

Another technique should only be used for sets of minor cardinality, although a computer will handle even medium sized ones. We may identify the subset of X as an element in the powerset $\mathcal{P}(X)$ or $\mathbf{2}^X$. For reasons of space, it will only be presented for the subset $\{\heartsuit,\diamondsuit\}$ of the 4-element Bridge suit set $\{\spadesuit,\heartsuit,\diamondsuit,\clubsuit\}$.[5]

---

[5] Another well-known notation for the empty subset is $\emptyset$ instead of $\{\}$.

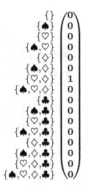

Fig. 2.3 Subset as powerset element

The main condition for the powerset definition requires that precisely one entry is 1 or True and all the others are 0 or False while enumerating all the subsets. Even more difficult to handle are predicate definitions. We explain them along with the set N_{20} of numbers from 1 to 20. In many programming languages, one may characterize the numbers that may be divided by 7 without remainder as p ‘rem‘ 7 == 0. The predicate form is now specific in as far as it is hardly possible to guarantee that the predicate representation will be found again when iterating cyclically as in

setAsPRED (setAsLIST ss)

where ss is given selecting arguments p by a predicate as p ‘rem‘ 7 == 0. How should a computer regenerate such nice formulations when given only a long list of multiples of 7?

One will certainly ask how subsets of complex constructed basesets can be expressed. This is postponed until some more algebraic background is available.

So far we have been able to denote only a few *definable* subsets. Over these basic subsets we build further subsets with operators. While in most mathematics texts union and intersection may be formed more or less arbitrarily, we restrict these operators to subsets of some common baseset. For complement formation this is required from the beginning.

Subset union and intersection

We will be very sketchy here, since we touch the lowest level of our exposition. Assume the baseset of numerals from 1 to 20. Then we may unite and intersect subsets in different representations as follows:

- $\{3, 6, 9, 12, 15, 18\} \cup \{2, 4, 6, 8, 10, 12, 14, 16, 18, 20\}$
  $= \{2, 3, 4, 6, 8, 9, 10, 12, 14, 15, 16, 18, 20\}$

$$\{3,6,9,12,15,18\} \cap \{2,4,6,8,10,12,14,16,18,20\} = \{6,12,18\}$$

- analogously in list form
- in predicate form we obtain

$$\{n \in N_{20} \mid 3|n\} \cup \{n \in N_{20} \mid 2|n\} = \{n \in N_{20} \mid 2|n \quad \text{or} \quad 3|n\}$$
$$\{n \in N_{20} \mid 3|n\} \cap \{n \in N_{20} \mid 2|n\} = \{n \in N_{20} \mid 2|n \quad \text{and} \quad 3|n\} = \{n \in N_{20} \mid 6|n\}$$

in general,

$$\{x \mid E_1(x)\} \cup \{x \mid E_2(x)\} = \{x \mid (E_1 \vee E_2)(x)\}$$
$$\{x \mid E_1(x)\} \cap \{x \mid E_2(x)\} = \{x \mid (E_1 \wedge E_2)(x)\}$$

- in vector form as shown in Fig. 2.4.

	A	∪	B	=	C		D	∩	E	=	F		G	⊆	H
1	0		0		0		0		1		0		0		0
2	0		1		1		0		1		0		0		1
3	1		0		1		1		0		0		0		0
4	0		1		1		0		1		0		0		1
5	0		0		0		0		0		0		0		0
6	1		1		1		1		1		1		1		1
7	0		0		0		0		0		0		0		0
8	0		1		1		0		1		0		0		1
9	1		0		1		1		0		0		0		0
10	0		1		1		0		1		0		0		1
11	0		0		0		0		0		0		0		0
12	1		1		1		1		1		1		1		1
13	0		0		0		0		0		0		0		0
14	0		1		1		0		1		0		0		1
15	1		0		1		1		0		0		0		0
16	0		1		1		0		1		0		0		1
17	0		0		0		0		0		0		0		0
18	1		1		1		1		1		1		1		1
19	0		0		0		0		0		0		0		0
20	0		1		1		0		1		0		0		1

Fig. 2.4 Subset union, intersection, and containment

In the last version with vectors, union means that the result is marked **1** if at least one of the arguments is marked **1**. Intersection of the result is marked **1** if both arguments are marked **1**.

Subset complement

Also in the various representations, complement forming over finite sets is more or less obvious:

- $\overline{\{3,6,9,12,15,18\}} = \{1,2,4,5,7,8,10,11,13,14,16,17,19,20\}$
- analogously in list form
- predicate form $\overline{\{n \in N_{20} \mid 6|n\}} = \{n \in N_{20} \mid 6 \nmid n\}$, in general,

$$\overline{\{n \mid p(n)\}} = \{n \mid \neg p(n)\},$$

- in vector form, exchange **0** and **1**.

So far, we have had a glimpse of *operations* on subsets, taking one or two subsets as arguments and delivering another subset as a result.

Subset containment

There is also an important binary *predicate* for subsets namely subset containment:

- $\{6, 12, 18\} \subseteq \{2, 4, 6, 8, 10, 12, 14, 16, 18, 20\}$
  holds, since the first subset contains only elements which are also contained in the second
- analogously in list form
- predicate form $\{n \in N_{20} \mid 6|n\} \subseteq \{n \in N_{20} \mid 2|n\}$, in general, $\{x \mid E_1(x)\} \subseteq \{x \mid E_2(x)\}$ holds if for all $x \in V$ we have $E_1(x) \rightarrow E_2(x)$, which may also be written as $\forall\, x \in V : E_1(x) \rightarrow E_2(x)$
- in vector form, if a **1** is on the left, then also on the right, as shown in Fig. 2.4.

We have presented the definition of such basic subsets in some detail, although many people know what the intersection of two sets, or union of these, actually means. The purpose of our detailed explanation is as follows. Mathematicians in everyday work are normally not concerned with basesets; they unite sets as they come, {red,green,blue} $\cup$ $\{1, 2, 3, 4\}$. This is also possible on a computer which works with a text representation of these sets, but it takes some time. However, when maximum efficiency of such algorithms is required with regard to space or time, one has to go back to sets represented as bits in a machine word. Then the position of the respective bit becomes important. This in turn is best dealt with in the concept of a set as an ordered list with every position meaning some element, i.e., with a baseset.

Permuting subset representations

One will know permutations from early school experience. They may be given as a function, decomposed into cycles, or as a permutation matrix as in Fig. 2.5. There is one 3-cycle as 1 will be sent to 4, 4 to 7, after which 7 is sent to the starting point 1. But there are also 2-cycles, as 3 and 5 for example will toggle when applying the permutation several times.

$$
\begin{array}{lll}
1 \mapsto 4 & & \\
2 \mapsto 6 & & \\
3 \mapsto 5 & & \\
4 \mapsto 7 & & \\
5 \mapsto 3 & \texttt{[[1,4,7],[3,5],[6,2]]} & \\
6 \mapsto 2 & & \\
7 \mapsto 1 & \text{or} & \\
\end{array}
\qquad
\begin{pmatrix}
0 & 0 & 0 & 1 & 0 & 0 & 0 \\
0 & 0 & 0 & 0 & 0 & 1 & 0 \\
0 & 0 & 0 & 0 & 1 & 0 & 0 \\
0 & 0 & 0 & 0 & 0 & 0 & 1 \\
0 & 0 & 1 & 0 & 0 & 0 & 0 \\
0 & 1 & 0 & 0 & 0 & 0 & 0 \\
1 & 0 & 0 & 0 & 0 & 0 & 0
\end{pmatrix}
$$

$$\texttt{[4,6,5,7,3,2,1]}$$

Fig. 2.5 Representing a permutation as a function, using cycles, or as a matrix

Each form has its specific merits.[6] Sometimes the inverted permutation is useful.

---

[6] In particular, we observe that permutations partly subsume to relations.

It is important also that subsets can be permuted. Permuting a subset means that the corresponding baseset is permuted followed by a permutation of the subset – conceived as a marked set – in reverse direction. Then the listed baseset will show a different sequence, but the marking vector will again identify the same set of elements.

When we apply a permutation to a subset representation, we insert it in between the row entry column and the marking vector. While the former is subjected to the permutation, the latter will undergo the reverse permutation. In effect we have applied p followed by $inverse(p)$, i.e., the identity. The subset has not changed, but its appearance has.

To make this claim clear, we consider the month names baseset and what is shown in the middle as the subset of 'J-months', namely *January, June, July*. This is then reconsidered with month names ordered alphabetically; cf. the permutation from page 7.

April		January	$\begin{pmatrix}1\\0\\0\\0\\0\\1\\1\\0\\0\\0\\0\\0\end{pmatrix}$
August		February	
December		March	
February	names sent to positions	April	
January	[4,8,12,2,1,7,6,3,5,11,10,9]	May	
July	i.e., permutation p	June	
June		July	
March		August	
May		September	
November		October	
October		November	
September		December	

January $\begin{pmatrix}1\\0\\0\\0\\0\\1\\1\\0\\0\\0\\0\\0\end{pmatrix}$ Boolean values sent to positions [5,4,8,1,9,7,6,2,12,11,10,3] i.e., permutation $inverse\{p\}$ $\begin{pmatrix}0\\0\\0\\0\\1\\1\\1\\0\\0\\0\\0\\0\end{pmatrix}$

Fig. 2.6 Permutation of a subset: the outermost describe the permuted subset

3

Relations

Already in the previous chapters, relations have shown up in a more or less naïve form, for example as permutation matrices or as (partial) identity relations. Here, we provide ideas for more stringent data types for relations. Not least, these will serve to model graph situations, like graphs on a set, bipartitioned graphs, or hypergraphs.

What is even more important at this point is the question of denotation. We have developed some scrutiny when denoting basesets, elements of these, and subsets; all the more will we now be careful in denoting relations. Since we restrict ourselves mostly to binary relations, this will mean denoting the source of the relation as well as its target and then denoting the relation proper. It is this seemingly trivial point which will be stressed here, namely from which set to which set the relation actually leads.

3.1 Relation representation

We aim mainly at relations over finite sets. Then a relation R *between* sets V, W is announced as $R : V \longrightarrow W$. It may be presented in at least one of the following forms:

— as a set of pairs $\{(x, y), \ldots\}$ with $x \in V, y \in W$
— as a list of pairs $[(x,y), \ldots]$ with x :: V, y :: W
— in predicate form $\{(x, y) \in V \times W \mid p(x, y)\}$ with a binary predicate p
— in matrix form, discriminating pairs over $V \times W$, the latter in rectangular presentation
— in vector form, discriminating pairs over $V \times W$, the latter presented as a linear list
— in vector form, indicating an element in the powerset $\mathbf{2}^{V \times W}$
— as a 'set-valued function', assigning to every element of V a set of *numbers* of elements from W

— as a 'set-valued function', assigning to every element of V a set of *names* of elements from W

— visualized as a bipartitioned[1] graph with vertices V on the left side and W on the right and an arrow $x \mapsto y$ in the case $(x, y) \in R$.

Yet another important representation for relations is possible using the very efficient reduced ordered binary decision diagrams (ROBDDs) as is used for the RELVIEW system [15, 85].

In many cases, V and W will be different sets. When we want to stress this possibility, we speak of a *heterogeneous* relation. A *homogeneous* relation in contrast will be a relation on one and the same set, i.e., $V = W$. When an operation or proposition concerns a relation of either form, we sometimes give a hint 'possibly heterogeneous'.

After these preparatory remarks we mention the diversity of variants possible in TITUREL. The intention is easy to decode from the denotation – at least for programmers. Only matrices with row or column lengths, or list lengths respectively, that correspond to the basesets mentioned before, will be accepted as correctly defined.

```
PALIRel BaseSet BaseSet [(Int,Int)]
PREDRel BaseSet BaseSet (Int -> Int -> Bool)
MATRRel BaseSet BaseSet [[Bool]]
VECTRel BaseSet BaseSet [Bool]
POWERel BaseSet BaseSet [Bool]
SETFRel BaseSet BaseSet (Int -> [Int])
SNAFRel BaseSet BaseSet (Int -> [String])
```

These variants are explained in the following examples that make the same relation look completely differently. In Fig. 3.1, we present the matrix form together with the set function that assigns sets given via lists of element numbers and then via element names.

---

[1] Be aware that a graph is *bipartite* if its point set *may be subdivided* in one way or another, but *bipartitioned* if the subdivision *has already taken place, and is, thus, fixed*.

	Kingdom	Very large	European	Strong industry	
USA	0	1	0	1	has $properties_{2,4}$
UK	1	0	1	1	has $properties_{1,3,4}$
Brasil	0	1	0	0	has $properties_2$
France	0	0	1	1	has $properties_{3,4}$
Germany	0	0	1	1	has $properties_{3,4}$
Hungary	0	0	1	0	has $properties_3$
Spain	1	0	1	0	has $properties_{1,3}$

(USA,Kingdom)	0
(UK,Kingdom)	1
(USA,Very large)	1
(Brasil,Kingdom)	0
(UK,Very large)	0
(USA,European)	0
(France,Kingdom)	0
(Brasil,Very large)	1
(UK,European)	1
(USA,Strong industry)	1
(Germany,Kingdom)	0
(France,Very large)	0
(Brasil,European)	0
(UK,Strong industry)	1
(Hungary,Kingdom)	0
(Germany,Very large)	0
(France,European)	1
(Brasil,Strong industry)	0
(Spain,Kingdom)	1
(Hungary,Very large)	0
(Germany,European)	1
(France,Strong industry)	1
(Spain,Very large)	0
(Hungary,European)	1
(Germany,Strong industry)	1
(Spain,European)	1
(Hungary,Strong industry)	0
(Spain,Strong industry)	0

USA ↦ {Very large,Strong industry}
UK ↦ {Kingdom,European,Strong industry}
Brasil ↦ {Very large}
France ↦ . {European,Strong industry}
Germany ↦ {European,Strong industry}
Hungary ↦ {European}
Spain ↦ {Kingdom,European}

Fig. 3.1 Different representations of the same relation

We will sometimes say (UK,European) $\in R$, but in other cases also use the matrix indexing R_{ik}, as in for example, $R_{\text{UK,European}}$. We will frequently use the matrix style to visualize relations.[2]

Then we show the relation with pairs of row and column numbers, which requires that one knows the source and target not just as sets but as element sequences, i.e., basesets:

$$\{(1,2),(1,4),(2,1),(2,3),(2,4),(3,2),(4,3),(4,4),(5,3),(5,4),(6,3),(7,1),(7,3)\}.$$

It is also possible to mark the way along the list of all pairs as on the right of Fig. 3.1.

Let us have a closer look at the way in which relations are denoted. If we have two finite basesets V, W, we are able – at least in principle – to denote every element in V and in W, and thus, every relation between V and W. In practice, however, we will never try to *denote an arbitrary* 300×5000 relation while we can easily work on a computer with a 300×5000 matrix as an operand. It is highly likely that our relations are composed in a simple way based on subsets of the two basesets. A simple calculation makes clear what a tiny percentage this is. There are

---

[2] Already in [35] L. M. Copilowish 'enjoyed the full benefits of the matrix approach' and regretted that this 'elegant machinery is apparently too little known'. We think these feelings can be shared today. Bednarek and Ulam in [7] motivated such investigations by their 'suitability to machine computations'.

$2^{300 \times 5000}$ relations – only finitely many, but a tremendous number. Relations that are composed of subsets of the first or second component set are far fewer, namely only $2^{300} \times 2^{5000}$, or $0.000\ldots$ % of the latter. If the set with 5000 elements is itself built as a product of a 50 and a 100 element set, it is highly likely also that the more interesting of the 2^{5000} subsets are built from smaller sets of components.

Representing relations is possible in various ways, as we are going to show in yet another example, starting with the version

```
MATRRel bsGenders bsColorsRedBlue [[False,True],[True,True]].
```

It is possible to convert this matrix to a set function, for example. It is also possible to list the pairs of indices – where the sets are understood to be known. However, one may also indicate the subset of pairs along the list of all pairs. Finally, for not too large sets, one may indicate the subset as an element of the powerset as in Fig. 3.2. Again, we assume that a relation can be given in either of these forms. Should we, for some reason, wish to have it presented in one specific variant, we assume functions

```
relAsMATRRel, relAsPREDRel, relAsSETFRel, relAsSNAFRel,
relAsVECTRel, relAsPALIRel, relAsPOWERel
```

to switch to this representation – as far as this is possible.

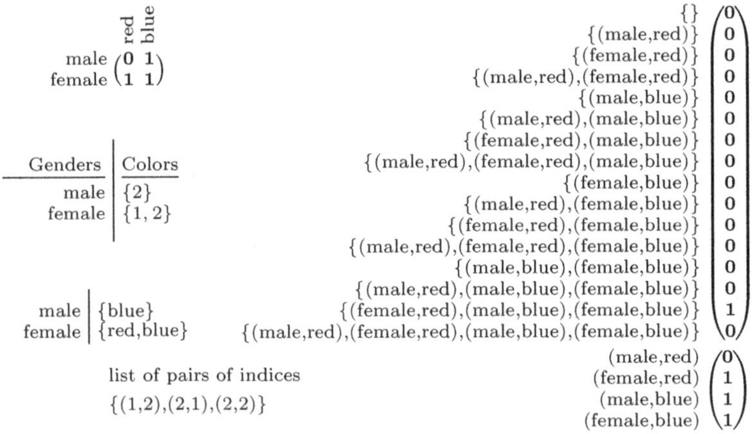

Fig. 3.2 The same relation represented in a variety of ways

Permuting relation representations

We now study how a relation representation may be varied using permutations. Two main techniques are possible, permuting simultaneously and permuting rows and columns independently.

$$A = \begin{array}{r} \\ 1\\2\\3\\4\\5\\6\\7\\8\\9\\10\\11\\12 \end{array} \begin{array}{c} \begin{array}{cccccccccccccc} 1&2&3&4&5&6&7&8&9&10&11&12&13&14 \end{array} \\ \begin{pmatrix} 1&0&1&0&1&0&0&1&1&0&1&1&0&1\\ 0&1&0&1&0&0&1&0&0&1&0&0&0&0\\ 0&0&0&0&0&1&0&0&0&0&0&0&1&0\\ 1&0&1&0&1&0&0&1&1&0&1&1&0&1\\ 0&1&0&1&0&0&1&0&0&1&0&0&0&0\\ 1&0&1&0&1&0&0&1&1&0&1&1&0&1\\ 0&1&0&1&0&0&1&0&0&1&0&0&0&0\\ 0&0&0&0&0&0&0&0&0&0&0&0&0&0\\ 0&0&0&0&0&1&0&0&0&0&0&0&1&0\\ 1&0&1&0&1&0&0&1&1&0&1&1&0&1\\ 0&0&0&0&0&1&0&0&0&0&0&0&1&0\\ 1&0&1&0&1&0&0&1&1&0&1&1&0&1 \end{pmatrix} \end{array} \qquad A_{\text{rearranged}} = \begin{array}{r} \\ 1\\4\\6\\10\\12\\2\\5\\7\\3\\9\\11\\8 \end{array} \begin{array}{c} \begin{array}{cccccccc|cccc|cc} 1&3&5&8&9&11&12&14&2&4&7&10&6&13 \end{array} \\ \left(\begin{array}{cccccccc|cccc|cc} 1&1&1&1&1&1&1&1&0&0&0&0&0&0\\ 1&1&1&1&1&1&1&1&0&0&0&0&0&0\\ 1&1&1&1&1&1&1&1&0&0&0&0&0&0\\ 1&1&1&1&1&1&1&1&0&0&0&0&0&0\\ 1&1&1&1&1&1&1&1&0&0&0&0&0&0\\ \hline 0&0&0&0&0&0&0&0&1&1&1&1&0&0\\ 0&0&0&0&0&0&0&0&1&1&1&1&0&0\\ 0&0&0&0&0&0&0&0&1&1&1&1&0&0\\ \hline 0&0&0&0&0&0&0&0&0&0&0&0&1&1\\ 0&0&0&0&0&0&0&0&0&0&0&0&1&1\\ 0&0&0&0&0&0&0&0&0&0&0&0&1&1\\ \hline 0&0&0&0&0&0&0&0&0&0&0&0&0&0 \end{array}\right) \end{array}$$

Fig. 3.3 A relation with many coinciding rows and columns, original and rearranged

In many cases, the results of some investigation automatically reveal information that might be used for a partition of the set of rows and the set of columns, respectively. In this case, a partition into groups of identical rows and columns is easily obtained. It is a good idea to permute rows and columns so as to have the identical rows of the groups side by side. This means permuting rows and columns independently as in Fig. 3.3.

There may, however, also occur a homogeneous relation, for which rows and columns should not be permuted independently, but simultaneously. Figure 3.4 shows how in this case a block form may be reached.

$$\Xi = \begin{array}{r} \\ 1\\2\\3\\4\\5 \end{array} \begin{array}{c} \begin{array}{ccccc} 1&2&3&4&5 \end{array} \\ \begin{pmatrix} 1&0&1&0&0\\ 0&1&0&0&1\\ 1&0&1&0&0\\ 0&0&0&1&0\\ 0&1&0&0&1 \end{pmatrix} \end{array} \quad P = \begin{array}{r} \\ 1\\2\\3\\4\\5 \end{array} \begin{array}{c} \begin{array}{ccccc} 1&3&2&5&4 \end{array} \\ \begin{pmatrix} 1&0&0&0&0\\ 0&0&1&0&0\\ 0&1&0&0&0\\ 0&0&0&0&1\\ 0&0&0&1&0 \end{pmatrix} \end{array} \quad \begin{array}{r} \\ 1\\3\\2\\5\\4 \end{array} \begin{array}{c} \begin{array}{ccccc} 1&3&2&5&4 \end{array} \\ \begin{pmatrix} 1&1&0&0&0\\ 1&1&0&0&0\\ 0&0&1&1&0\\ 0&0&1&1&0\\ 0&0&0&0&1 \end{pmatrix} \end{array} = P^{\mathsf{T}}; \Xi; P$$

Fig. 3.4 Equivalence Ξ with simultaneous permutation P to a block-diagonal form

When permuting only the rows of a homogeneous relation, it will become a heterogeneous relation, albeit with square matrix.

3.2 Relations describing graphs

Relational considerations may very often be visualized with graphs. On the other hand, several questions of graph theory are heavily related to matrix properties, i.e., relations. Usually, graph theory stresses other aspects of graphs than our relational approach. So we will present what we consider relational graph theory. This will mean, not least, making visible the differences between the various forms in which graphs are used with relations.

The following type of graphs will most directly resemble a relation between two sets.

Definition 3.1. Given a heterogeneous relation $M : P \longrightarrow V$, we speak of a **hypergraph**, interpreting

(i) P as a set of **hyperedges**,

(ii) V as a set of **vertices** or **points**,

(iii) M as an **incidence** relation.   □

The term *incidence* reminds us of a geometric origin, that may be found in the example of a so-called 'complete quadrangle' in Fig. 3.5. It consists of the four points P, Q, R, S with all six diagonals and the three additional diagonal cutting points D_1, D_2, D_3 added.

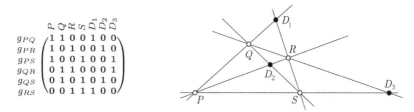

$$
\begin{array}{c}
\\
g_{PQ} \\
g_{PR} \\
g_{PS} \\
g_{QR} \\
g_{QS} \\
g_{RS}
\end{array}
\begin{array}{c}
\begin{array}{cccccccc}
P & Q & R & S & D_1 & D_2 & D_3
\end{array} \\
\left(\begin{array}{ccccccc}
1 & 1 & 0 & 0 & 1 & 0 & 0 \\
1 & 0 & 1 & 0 & 0 & 1 & 0 \\
1 & 0 & 0 & 1 & 0 & 0 & 1 \\
0 & 1 & 1 & 0 & 0 & 0 & 1 \\
0 & 1 & 0 & 1 & 0 & 1 & 0 \\
0 & 0 & 1 & 1 & 1 & 0 & 0
\end{array}\right)
\end{array}
$$

Fig. 3.5 Complete quadrangle as an example of a very special hypergraph

The next type is also intimately connected with a relation, in particular in the case when the second relation S is equal to $\perp\!\!\!\perp$.

Definition 3.2. Given any pair of heterogeneous relations $R : V \longrightarrow W$ and $S : W \longrightarrow V$, we may speak of a **bipartitioned graph**, interpreting

(i) V as a set of **vertices** on the left,

(ii) W as a set of **vertices** on the right,

(iii) R as a relation from the left to the right side,

(iv) S as a relation from the right to the left side.   □

The relation S will sometimes be the empty relation $\perp\!\!\!\perp$, so that all arrows will lead from the left to the right side.

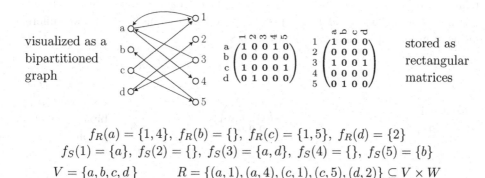

$$f_R(a) = \{1, 4\}, \quad f_R(b) = \{\}, \quad f_R(c) = \{1, 5\}, \quad f_R(d) = \{2\}$$
$$f_S(1) = \{a\}, \quad f_S(2) = \{\}, \quad f_S(3) = \{a, d\}, \quad f_S(4) = \{\}, \quad f_S(5) = \{b\}$$
$$V = \{a, b, c, d\} \qquad R = \{(a,1), (a,4), (c,1), (c,5), (d,2)\} \subseteq V \times W$$
$$W = \{1, 2, 3, 4, 5\} \qquad S = \{(1,a), (3,a), (3,d), (5,b)\} \subseteq W \times V$$

Fig. 3.6 Relation as a bipartitioned graph and as set-valued functions

A special variant is the relation *on a set*, i.e., with $V = W$, and thus a homogeneous relation, which more often will appear in a graph context.

Definition 3.3. Given any homogeneous relation $B : V \longrightarrow V$, we may speak of a **1-graph**, interpreting

  (i) V as a set of **vertices**,

 (ii) B as a relation with arrows between vertices of V, called the **associated relation**.   □

As an example we present

$$V = \{a, b, c, d\} \qquad B = \{(a,a), (d,a), (d,c), (c,d)\} \subseteq V \times V$$

which is also shown as a graph and as a matrix in Fig. 3.7.

Fig. 3.7 Homogeneous relation as a 1-graph

A little bit of **caution** is necessary when presenting a relation on some set without further context. The discussion may indeed be concerned with homogeneous relations; it may, however, be principally about heterogeneous relations holding between two sets that are identical just by coincidence. This leads to two different forms of graph representation, as in Fig. 3.8.

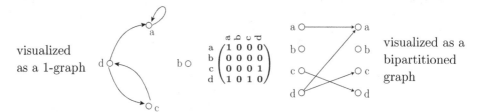

visualized as a 1-graph

$$\begin{array}{c} \\ a \\ b \\ c \\ d \end{array} \begin{pmatrix} 1 & 0 & 0 & 0 \\ 0 & 0 & 0 & 0 \\ 0 & 0 & 0 & 1 \\ 1 & 0 & 1 & 0 \end{pmatrix}$$

visualized as a bipartitioned graph

Fig. 3.8 Two different types of graph representing the same relation

Just looking at the relation and without knowing whether the homogeneous or the heterogeneous form was intended, one cannot decide readily on one or the other graph presentation.

Definition 3.4. Given any pair of heterogeneous relations $A : P \longrightarrow V$ and $E : P \longrightarrow V$ that are both mappings, we may speak of a **directed graph**, interpreting

(i) V as a set of **vertices**,
(ii) P as a set of **arrows**,
(iii) A as a relation assigning precisely one initial vertex to each arrow,
(iv) E as a relation assigning precisely one target vertex to each arrow. □

An example is provided with Fig. 3.9 in which one should look in particular at the parallel arrows s, t. They could not be distinguished by the concept of a 1-graph as presented in Def. 3.3.

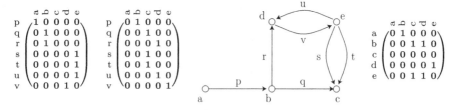

$$\begin{array}{c} \\ p \\ q \\ r \\ s \\ t \\ u \\ v \end{array} \begin{pmatrix} 1 & 0 & 0 & 0 & 0 \\ 0 & 1 & 0 & 0 & 0 \\ 0 & 1 & 0 & 0 & 0 \\ 0 & 0 & 0 & 0 & 1 \\ 0 & 0 & 0 & 0 & 1 \\ 0 & 0 & 0 & 0 & 1 \\ 0 & 0 & 0 & 1 & 0 \end{pmatrix} \qquad \begin{array}{c} \\ p \\ q \\ r \\ s \\ t \\ u \\ v \end{array} \begin{pmatrix} 0 & 1 & 0 & 0 & 0 \\ 0 & 0 & 1 & 0 & 0 \\ 0 & 0 & 0 & 1 & 0 \\ 0 & 0 & 1 & 0 & 0 \\ 0 & 0 & 1 & 0 & 0 \\ 0 & 0 & 0 & 1 & 0 \\ 0 & 0 & 0 & 0 & 1 \end{pmatrix} \qquad \begin{array}{c} \\ a \\ b \\ c \\ d \\ e \end{array} \begin{pmatrix} 0 & 1 & 0 & 0 & 0 \\ 0 & 0 & 1 & 1 & 0 \\ 0 & 0 & 0 & 0 & 0 \\ 0 & 0 & 0 & 0 & 1 \\ 0 & 0 & 1 & 1 & 0 \end{pmatrix}$$

Fig. 3.9 Directed graph representation with associated relation

When given just a drawing on the blackboard, it may not immediately be clear what sort of relation(s) is meant. The drawing on the left of Fig. 3.8, for example, may also be interpreted as showing a directed graph, i.e., providing us with two relations A, E. This would require an identification to be attached to the four arrows (say, arrow$_{aa}$, arrow$_{da}$, arrow$_{dc}$, arrow$_{cd}$), and the corresponding relations to be extracted from the drawing. We would immediately observe that arrows are uniquely given by their starting vertex determined via A and their target vertex determined via E. The latter condition says that from any vertex to another, there will be *at most one* arrow. This is the condition making a directed graph appear as a 1-graph.

Often one is not interested whether an arrow leads from a vertex x to the vertex y or the other way round and concentrates only on the question whether x and y are related somehow. To express this bidirectional situation, one will use a matrix that is symmetric when mirroring along its main diagonal. This is always combined with being uninterested in arrows from x to x, so that there should be only **0**s in this main diagonal.

Definition 3.5. Given any symmetric and irreflexive homogeneous relation Γ : $V \longrightarrow V$, we may speak of a **simple graph**, interpreting

(i) V as a set of **vertices**,
(ii) Γ as a symmetric and irreflexive relation of **adjacency** on vertices of V.   □

A tiny example of a simple graph is given in Fig. 3.10.

$$
\begin{array}{c}
 \\
a \\
b \\
c \\
d \\
e
\end{array}
\begin{array}{ccccc}
a & b & c & d & e \\
\left(\begin{array}{ccccc}
0 & 1 & 0 & 0 & 0 \\
1 & 0 & 1 & 1 & 0 \\
0 & 1 & 0 & 0 & 1 \\
0 & 1 & 0 & 0 & 1 \\
0 & 0 & 1 & 1 & 0
\end{array}\right)
\end{array}
$$

Fig. 3.10 Simple graph representation with its adjacency

3.3 Relations generated by cuts

Relations often originate from real-valued sources. An example of a real-valued matrix representing percentages is shown in Fig. 3.11. A closer look at this case

$$\begin{pmatrix}
38.28 & 9.91 & 28.42 & 36.11 & 25.17 & 11.67 & 87.10 & 84.73 & 81.53 & 35.64 & 34.36 & 11.92 & 99.73 \\
93.35 & 93.78 & 18.92 & 44.89 & 13.60 & 6.33 & 25.26 & 36.70 & 34.22 & 98.15 & 8.32 & 4.99 & 21.58 \\
5.69 & 94.43 & 47.17 & 95.23 & 86.50 & 80.26 & 41.56 & 86.84 & 47.93 & 40.38 & 3.75 & 19.76 & 12.00 \\
93.40 & 20.35 & 25.94 & 38.96 & 36.10 & 25.30 & 89.17 & 19.17 & 87.34 & 85.25 & 5.58 & 18.67 & 1.13 \\
6.37 & 83.89 & 23.16 & 41.64 & 35.56 & 36.77 & 21.71 & 37.20 & 43.61 & 18.30 & 97.67 & 27.67 & 42.59 \\
30.26 & 43.71 & 90.78 & 37.21 & 16.76 & 8.83 & 88.93 & 15.18 & 3.58 & 83.60 & 96.60 & 18.44 & 24.30 \\
29.85 & 14.76 & 82.21 & 35.70 & 43.34 & 99.82 & 99.30 & 88.85 & 46.29 & 24.73 & 47.90 & 92.62 & 46.65 \\
19.37 & 88.67 & 5.94 & 38.30 & 48.56 & 87.40 & 46.46 & 34.46 & 17.92 & 24.30 & 33.46 & 34.30 & 43.95 \\
97.89 & 96.70 & 4.13 & 44.50 & 23.23 & 81.56 & 95.75 & 34.30 & 41.59 & 47.39 & 39.29 & 86.14 & 22.98 \\
18.82 & 93.00 & 17.50 & 16.10 & 9.74 & 14.71 & 21.30 & 45.32 & 19.57 & 24.78 & 82.43 & 41.00 & 43.29 \\
5.38 & 36.85 & 4.38 & 28.10 & 17.30 & 45.30 & 33.14 & 81.20 & 13.24 & 33.39 & 23.42 & 18.33 & 83.87 \\
14.82 & 18.37 & 1.87 & 19.30 & 4.82 & 93.26 & 28.10 & 26.94 & 19.10 & 43.25 & 85.85 & 15.48 & 49.57 \\
7.63 & 28.80 & 10.40 & 89.81 & 17.14 & 7.33 & 96.57 & 16.19 & 35.96 & 8.96 & 47.42 & 39.82 & 8.16 \\
89.70 & 14.16 & 7.59 & 41.67 & 34.39 & 88.68 & 18.80 & 99.37 & 7.67 & 8.11 & 86.54 & 86.65 & 44.34 \\
31.55 & 13.16 & 86.23 & 45.45 & 92.92 & 33.75 & 43.64 & 46.74 & 27.75 & 89.96 & 37.71 & 84.79 & 86.32 \\
25.48 & 7.40 & 43.67 & 1.69 & 85.18 & 27.50 & 89.59 & 100.00 & 89.67 & 11.30 & 2.41 & 83.90 & 96.31 \\
48.32 & 93.23 & 14.16 & 17.75 & 14.60 & 90.90 & 3.81 & 41.30 & 4.12 & 3.87 & 2.83 & 95.35 & 81.12
\end{pmatrix}$$

Fig. 3.11 A real-valued 17×13-matrix

shows that the coefficients are clustered around 0–50 and around 80–100. So it will not be the deviation between 15 and 20, or 83 and 86, which is important but the huge gap between the two clusters. Constructing a histogram is, thus, a good idea, as shown in Fig. 3.12.

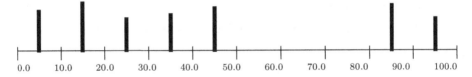

Fig. 3.12 Histogram for value frequency in Fig. 3.11

So one may be tempted to apply what is known as a **cut**, at 60, for example, considering entries below as `False` and entries above as `True` in order to arrive at the Boolean matrix of Fig. 3.13.

$$\begin{pmatrix}
0 & 0 & 0 & 0 & 0 & 0 & 1 & 1 & 1 & 0 & 0 & 0 & 1 \\
1 & 1 & 0 & 0 & 0 & 0 & 0 & 0 & 0 & 1 & 0 & 0 & 0 \\
0 & 1 & 0 & 1 & 1 & 1 & 0 & 1 & 0 & 0 & 0 & 0 & 0 \\
1 & 0 & 0 & 0 & 0 & 0 & 1 & 0 & 1 & 1 & 0 & 0 & 0 \\
0 & 1 & 0 & 0 & 0 & 0 & 0 & 0 & 0 & 0 & 1 & 0 & 0 \\
0 & 0 & 1 & 0 & 0 & 0 & 1 & 0 & 0 & 1 & 1 & 0 & 0 \\
0 & 0 & 1 & 0 & 0 & 1 & 1 & 1 & 0 & 0 & 0 & 1 & 0 \\
0 & 1 & 0 & 0 & 0 & 1 & 0 & 0 & 0 & 0 & 0 & 0 & 0 \\
1 & 1 & 0 & 0 & 0 & 1 & 1 & 0 & 0 & 0 & 0 & 1 & 0 \\
0 & 1 & 0 & 0 & 0 & 0 & 0 & 0 & 0 & 0 & 1 & 0 & 0 \\
0 & 0 & 0 & 0 & 0 & 0 & 0 & 1 & 0 & 0 & 0 & 0 & 1 \\
0 & 0 & 0 & 0 & 0 & 1 & 0 & 0 & 0 & 0 & 1 & 0 & 0 \\
0 & 0 & 0 & 1 & 0 & 0 & 1 & 0 & 0 & 0 & 0 & 0 & 0 \\
1 & 0 & 0 & 0 & 0 & 1 & 0 & 1 & 0 & 0 & 1 & 1 & 0 \\
0 & 0 & 1 & 0 & 1 & 0 & 0 & 0 & 0 & 1 & 0 & 1 & 1 \\
0 & 0 & 0 & 0 & 1 & 0 & 1 & 1 & 1 & 0 & 0 & 1 & 1 \\
0 & 1 & 0 & 0 & 0 & 1 & 0 & 0 & 0 & 0 & 0 & 1 & 1
\end{pmatrix}$$

Fig. 3.13 Boolean matrix corresponding to Fig. 3.11 according to a cut at 60

In principle, one may use any real number between 0 and 100 as a cut, but this will not make sense in all cases. In order to obtain qualitatively meaningful subdivisions, one should obey certain rules. Often one will introduce several cuts and test to what extent this results in orderings, for example.

There exist techniques to analyze real-valued matrices by investigating a selected cut at one or more levels. Typically, a cut is acceptable when the cut number can be moved up and down to a certain extent without affecting the relational structure. In the case of Fig. 3.11, one may shift the cut up and down between 50% and 80% without changing the relation in Fig. 3.13.

It may be the case, however, that there is one specific entry of the matrix which changes the structure of the matrix dramatically according to whether the entry is a **1** or a **0**. When this is just one entry, one has several options how to react. The first is to check whether it is a typographic error, or an error of the underlying test. If this is not the case, it is an effect to be mentioned, and may be important. It is not easy to execute a sensitivity test in order to find out whether the respective entry of the matrix has such a key property, but there exist graph-theoretic methods.

3.4 Relations generated randomly

To investigate programs working on relations, one will sometimes need test relations. The programming language HASKELL, for example, provides a mechanism to generate random numbers in a reproducible way. This also allows random relations to be generated. To this end one can convert any integer into a 'standard generator', which serves as a reproducible offset.

Because we are often interested in random matrices with some given degree of filling density, we further provide 0 and 100 as lower and upper bounds and assume a percentage parameter to be given as well as the desired row and column number,

```
randomMatrix startStdG perc r c.
```

For the realization of this function, random numbers between 0 and 100 produced from the offset are first generated infinitely, but afterwards only $r \times c$ are actually taken. Then they are filtered according to whether they are less than or equal to the prescribed percentage cut. Finally, they are grouped into r rows of c elements each.

It is much more complicated to generate random relations with prescribed properties such as being a univalent and injective relation. One may also wish to construct a random permutation matrix for n items or a random difunctional relation. Here, we cannot elaborate on this any further.

Figure 3.14 shows relations which are far from being random: a privately printable railway ticket and a stamp as used in recent times. Conceiving a black square as **1** and a white square as **0**, these are obviously relations in our sense – probably with a high degree of redundancy to make sure they can be read automatically even if they are partly damaged.

Fig. 3.14 Relations in practice: privately printable railway ticket and stamp in Germany

3.5 Function representation

A most important class of relations are, of course, mappings or totally defined functions. While these may easily be handled as relations with some additional properties, they are so central that it is often more appropriate to give them special treatment.

In TITUREL, we distinguish unary and binary functions as may be seen in the examples in Fig. 3.15 and Fig. 3.16. For both unary and binary mappings, a matrix representation as well as a list representation is provided. For unary mappings this looks like

```
MATRFunc BaseSet BaseSet [[Bool]]
LISTFunc BaseSet BaseSet [Int].
```

It is easy to see what this presentation means. The unary mapping matrix will only be accepted when row and column lengths meet the sizes of the basesets specified before and when every row contains besides **0**s precisely one **1**. In the list representation, the list must contain as many entries as the size of the first baseset specifies, with integers ranging over the interval from 1 to the number of entries in the second baseset. In addition, for binary mappings we have

```
TABUFct2 BaseSet BaseSet BaseSet [[Int]]
MATRFct2 BaseSet BaseSet BaseSet [[Bool]].
```

In a similar way, the matrix for the table will have as many rows as the size of the first baseset indicates and as many columns as the second baseset indicates, while the integer entries range between 1 and the size of the third baseset. The matrix form of a binary mapping has as many rows as the product of the first two baseset sizes and columns according to the third baseset, and it is again required that every row has among the **0**s precisely one **1**. When such mappings are input to start some computation, it is important that they are checked for consistency. Functions for switching between the two respective versions are assumed to exist as

funcAsMATR, funcAsLIST

fct2AsTABU, fct2AsMATR.

Let the nationalities of politicians be indicated by a function in list form

LISTFunc bsPoliticians bsNationalities [1,1,2,2,3,3,3,4,4,4,5].

It is rather immediate how this may be converted to the relation of Fig. 3.15. To actually program such a transition will probably be considered boring and may, thus, turn out to be error-prone.

$$
\texttt{funcAsMATR politiciansNationalities} =
\begin{array}{r}
\\
\\
\\
\text{Clinton} \\
\text{Bush} \\
\text{Mitterand} \\
\text{Chirac} \\
\text{Schmidt} \\
\text{Kohl} \\
\text{Schröder} \\
\text{Thatcher} \\
\text{Major} \\
\text{Blair} \\
\text{Zapatero}
\end{array}
\begin{array}{c}
\begin{array}{ccccc}
\text{US} & \text{French} & \text{German} & \text{British} & \text{Spanish}
\end{array} \\
\left(
\begin{array}{ccccc}
1 & 0 & 0 & 0 & 0 \\
1 & 0 & 0 & 0 & 0 \\
0 & 1 & 0 & 0 & 0 \\
0 & 1 & 0 & 0 & 0 \\
0 & 0 & 1 & 0 & 0 \\
0 & 0 & 1 & 0 & 0 \\
0 & 0 & 1 & 0 & 0 \\
0 & 0 & 0 & 1 & 0 \\
0 & 0 & 0 & 1 & 0 \\
0 & 0 & 0 & 1 & 0 \\
0 & 0 & 0 & 0 & 1
\end{array}
\right)
\end{array}
$$

Fig. 3.15 Function in matrix representation

As a further example we choose the operations in the famous Klein four-group.[3] Assume a rectangular playing card such as is used for Bridge or Skat and consider transforming it with the barycenter fixed in 3-dimensional space. The possible transformations are limited to identity ⬜, flipping vertically ⊟ or horizontally ⬜, and finally rotation by 180 degrees ⬜. Composing such transformations, one will easily observe the group table of Fig. 3.16. The middle representation obviously relies on the ordering {⬜,⊟,⬜,⬜} of the baseset and gives just the position number.

---

[3] In German: Kleinsche Vierergruppe.

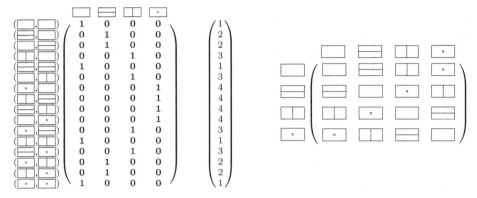

Fig. 3.16 Three representations of composition in the Klein four-group

3.6 Permutation representation

Permutations will often be used for presentation purposes, but are also interesting in their own right. They may be given as a matrix, a sequence, via cycles, or as a function. We provide mechanisms to convert between these forms and to apply permutations to some set.

$$
\begin{array}{c}
\begin{array}{c} \, 1\ 2\ 3\ 4\ 5\ 6\ 7 \end{array} \\
\begin{array}{c} 1 \\ 2 \\ 3 \\ 4 \\ 5 \\ 6 \\ 7 \end{array}
\begin{pmatrix}
0\ 0\ 1\ 0\ 0\ 0\ 0 \\
0\ 0\ 1\ 0\ 0\ 0\ 0 \\
0\ 0\ 0\ 0\ 1\ 0\ 0 \\
0\ 0\ 0\ 0\ 0\ 0\ 1 \\
0\ 0\ 0\ 0\ 0\ 1\ 0 \\
0\ 1\ 0\ 0\ 0\ 0\ 0 \\
1\ 0\ 0\ 0\ 0\ 0\ 0
\end{pmatrix}
\end{array}
\qquad
\begin{array}{c}
\begin{array}{c} \, 1\ 4\ 7\ 2\ 3\ 5\ 6 \end{array} \\
\begin{array}{c} 1 \\ 4 \\ 7 \\ 2 \\ 3 \\ 5 \\ 6 \end{array}
\left(\begin{array}{ccc|cccc}
0\ 1\ 0 & 0\ 0\ 0\ 0 \\
0\ 0\ 1 & 0\ 0\ 0\ 0 \\
1\ 0\ 0 & 0\ 0\ 0\ 0 \\
\hline
0\ 0\ 0 & 0\ 1\ 0\ 0 \\
0\ 0\ 0 & 0\ 0\ 1\ 0 \\
0\ 0\ 0 & 0\ 0\ 0\ 1 \\
0\ 0\ 0 & 1\ 0\ 0\ 0
\end{array}\right)
\end{array}
$$

Fig. 3.17 A permutation rearranged to cycle form

One will easily confirm that both matrices in Fig. 3.17 represent the same permutation. The second form, however, shows more easily how this permutation works in cycles. Specific interest arose around permutations with just one cycle of length 2, namely transpositions, and all other cycles of length 1, i.e., invariant. Every permutation may (in many ways) be generated by a series of transpositions.

Univalent and injective heterogeneous relations

Looking at such a permutation matrix, we easily observe that in every row and column among the **0**s precisely one **1** may be found; considered forward and backwards, a permutation is thus a mapping. Often, however, it is just a (partially defined) function, i.e., assigning at most one **1** per row (then called *univalent*) and column (then called *injective*).

With Fig. 3.18, we study an example showing how permutation helps visualization. First we assume a *heterogeneous* relation that is univalent and injective. This obviously gives a one-to-one correspondence where it is defined, that should somehow be made visible.

```
        1 2 3 4 5 6 7 8 9 10             8 2 5 4 3 6 9 7 1 10
    a  /0 0 0 0 0 0 0 1 0 0\        a  /1 0 0 0 0 0 0 0 0 0\
    b  | 0 1 0 0 0 0 0 0 0 0 |      b  | 0 1 0 0 0 0 0 0 0 0 |
    c  | 0 0 0 0 0 0 0 0 0 0 |      d  | 0 0 1 0 0 0 0 0 0 0 |
    d  | 0 0 0 0 1 0 0 0 0 0 |      e  | 0 0 0 1 0 0 0 0 0 0 |
    e  | 0 0 0 1 0 0 0 0 0 0 |      g  | 0 0 0 0 1 0 0 0 0 0 |
    f  | 0 0 0 0 0 0 0 0 0 0 |      h  | 0 0 0 0 0 1 0 0 0 0 |
    g  | 0 0 1 0 0 0 0 0 0 0 |      i  | 0 0 0 0 0 0 1 0 0 0 |
    h  | 0 0 0 0 0 1 0 0 0 0 |      k  | 0 0 0 0 0 0 0 1 0 0 |
    i  | 0 0 0 0 0 0 0 0 1 0 |      c  | 0 0 0 0 0 0 0 0 0 0 |
    j  | 0 0 0 0 0 0 0 0 0 0 |      f  | 0 0 0 0 0 0 0 0 0 0 |
    k  \0 0 0 0 0 0 1 0 0 0/        j  \0 0 0 0 0 0 0 0 0 0/
```

Fig. 3.18 A univalent and injective relation, rearranged to diagonal-shaped form; rows and columns permuted independently

Univalent and injective homogeneous relations

It is more complicated to visualize the corresponding property when satisfied by a *homogeneous* relation. One is then no longer allowed to permute independently. Nonetheless, an appealing form can always be reached.

```
     1 2 3 4 5 6 7 8 9 10 11 12 13 14 15 16 17 18 19                13 15 3 10 1 11 2 4 14 6 17 5 16 12 8 7 19 9 18
1   /0 0 0 0 0 0 0 0 0 1 0 0 0 0 0 0 0 0 0\              13   /0 1|0 0 0 0|0 0|0 0 0|0 0|0 0 0|0 0 0\
2   | 0 0 1 0 0 0 0 0 0 0 0 0 0 0 0 0 0 0 0 |            15   | 1 0|0 0 0 0|0 0|0 0 0|0 0|0 0 0|0 0 0 |
3   | 0 0 0 0 0 0 0 0 1 0 0 0 0 0 0 0 0 0 0 |             3   | 0 0|0 1 0 0|0 0|0 0 0|0 0|0 0 0|0 0 0 |
4   | 0 0 0 0 0 0 0 0 0 0 0 0 0 0 0 0 1 0 0 |            10   | 0 0|0 0 1 0|0 0|0 0 0|0 0|0 0 0|0 0 0 |
5   | 0 0 0 0 0 0 0 0 0 0 0 0 0 0 0 0 1 0 0 |             1   | 0 0|0 0 0 1|0 0|0 0 0|0 0|0 0 0|0 0 0 |
6   | 0 0 0 0 0 0 0 0 0 0 0 0 0 0 0 0 1 0 0 |            11   | 0 0|1 0 0 0|0 0|0 0 0|0 0|0 0 0|0 0 0 |
7   | 0 0 0 0 0 0 0 0 0 0 0 0 0 0 0 0 0 0 0 |             2   | 0 0|0 0 0 0|1 0|0 0 0|0 0|0 0 0|0 0 0 |
8   | 0 0 0 0 0 1 0 0 0 0 0 0 0 0 0 0 0 0 0 |             4   | 0 0|0 0 0 0|0 0|0 0 0|0 0|0 0 0|0 0 0 |
9   | 0 0 0 0 0 0 0 0 0 0 0 0 0 0 0 0 0 0 0 |            14   | 0 0|0 0 0 0|0 0|1 0 0|0 0|0 0 0|0 0 0 |
10  | 1 0 0 0 0 0 0 0 0 0 0 0 0 0 0 0 0 0 0 |             6   | 0 0|0 0 0 0|0 0|0 1 0|0 0|0 0 0|0 0 0 |
11  | 0 0 1 0 0 0 0 0 0 0 0 0 0 0 0 0 0 0 0 |            17   | 0 0|0 0 0 0|0 0|0 0 0|0 0|0 0 0|0 0 0 |
12  | 0 0 0 0 0 0 0 1 0 0 0 0 0 0 0 0 0 0 0 |             5   | 0 0|0 0 0 0|0 0|0 0 0|1 0|0 0 0|0 0 0 |
13  | 0 0 0 0 0 0 0 0 0 0 0 0 0 0 1 0 0 0 0 |            16   | 0 0|0 0 0 0|0 0|0 0 0|0 0|0 0 0|0 0 0 |
14  | 0 0 0 0 1 0 0 0 0 0 0 0 0 0 0 0 0 0 0 |            12   | 0 0|0 0 0 0|0 0|0 0 0|0 0|1 0 0|0 0 0 |
15  | 0 0 0 0 0 0 0 0 0 0 0 0 1 0 0 0 0 0 0 |             8   | 0 0|0 0 0 0|0 0|0 0 0|0 0|0 1 0|0 0 0 |
16  | 0 0 0 0 0 0 0 0 0 0 0 0 0 0 0 0 0 0 0 |             7   | 0 0|0 0 0 0|0 0|0 0 0|0 0|0 0 0|0 0 0 |
17  | 0 0 0 0 0 0 0 0 0 0 0 0 0 0 0 0 0 0 0 |            19   | 0 0|0 0 0 0|0 0|0 0 0|0 0|0 0 0|0 0 0 |
18  | 0 0 0 0 0 0 0 0 0 0 0 0 0 0 0 0 0 0 0 |             9   | 0 0|0 0 0 0|0 0|0 0 0|0 0|0 0 0|0 0 0 |
19  \0 0 0 0 0 0 0 0 0 0 0 0 0 0 0 0 0 0 0/             18   \0 0|0 0 0 0|0 0|0 0 0|0 0|0 0 0|0 0 0/
```

Fig. 3.19 Arranging a univalent and injective relation by simultaneous permutation

The principle is as follows. The univalent and injective relation clearly subdivides the set into cycles and/or linear strands as well as possibly a set of completely unrelated elements. These are obtained as classes of the symmetric reachability according to the corresponding matrix. (The unrelated elements should be collected in one class.) Now every class is arranged side by side. To follow a general principle, let cycles come first, then linear strands, and finally the group of completely unrelated elements.

Inside every group, one will easily arrange the elements so that the relation forms an upper neighbor of the diagonal, if a strand is given. When a cycle is presented, the lower left corner of the block will also carry a **1**.

It is easy to convince oneself that Fig. 3.19 shows two different representations of the same relation; the right form, however, facilitates an overview on the action of this relation: when applied repeatedly, 13 and 15 will toggle infinitely. A 4-cycle is made up by 3, 10, 1, 11. From 2, we will reach the end in just one step. The question arises of how such a decomposition can be found systematically. In Appendix C, we will give hints to how this may be achieved.

Arranging an ordering

The basic scheme to arrange an ordering (*is less than or equal to, divides for numbers*, or *is contained in for subsets*) is obviously to present it contained in the upper right triangle of the matrix. When given the relation on the left of Fig. 3.20, it is not at all clear whether it is an ordering, i.e., a reflexive, transitive, and antisymmetric relation; when arranged nicely, it immediately turns out to be. It is, thus, a specific task to identify the techniques according to which one may obtain a nice arrangement. For a given example, this will easily be achieved. But what about an arbitrary input? Can one design a general procedure?

```
       1 2 3 4 5 6 7 8 9 10 11 12 13 14 15 16 17 18 19            17 2 19 9 3 7 4 8 6 5 1 10 15 11 12 13 14 16 18
  1  / 1 0 0 0 0 0 0 0 0 1 0 0 0 0 0 0 0 0 0 \        17  / 1 1 1 1 1 1 1 1 1 1 1 1 1 1 1 1 1 1 1 \
  2  | 0 1 0 0 0 0 0 0 0 0 0 0 0 0 0 0 0 0 0 |         2  | 0 1 0 0 0 0 0 0 0 0 0 0 0 0 0 0 0 0 0 |
  3  | 1 0 1 1 1 1 1 0 0 1 1 0 0 1 0 0 0 0 0 |        19  | 0 0 1 1 0 1 1 0 0 0 0 0 0 0 0 1 1 0 0 |
  4  | 0 0 0 1 0 0 0 0 0 0 0 0 0 0 0 0 0 0 0 |         9  | 0 0 0 1 0 0 1 0 0 0 0 0 0 0 0 0 0 0 0 |
  5  | 1 0 0 0 1 0 0 0 0 1 0 0 0 1 0 0 0 0 0 |         3  | 0 0 0 0 1 1 1 0 1 1 1 1 1 0 1 0 0 1 0 |
  6  | 0 0 0 0 0 1 0 0 0 0 0 0 0 0 0 0 0 0 0 |         7  | 0 0 0 0 0 1 1 0 0 0 0 0 0 0 0 0 1 0 0 |
  7  | 0 0 0 1 0 0 1 0 0 0 0 0 0 1 0 0 0 0 0 |         4  | 0 0 0 0 0 0 1 0 0 0 0 0 0 0 0 0 0 0 0 |
  8  | 0 0 0 0 0 1 0 1 0 0 0 0 0 0 0 0 0 0 0 |         8  | 0 0 0 0 0 0 0 1 1 0 0 0 0 0 0 0 0 0 0 |
  9  | 0 0 0 1 0 0 0 0 1 0 0 0 0 0 0 0 0 0 0 |         6  | 0 0 0 0 0 0 0 0 1 0 0 0 0 0 0 0 0 0 0 |
 10  | 0 0 0 0 0 0 0 0 0 1 0 0 0 0 0 0 0 0 0 |         5  | 0 0 0 0 0 0 0 0 0 1 1 1 0 0 0 0 1 0 0 |
 11  | 0 0 0 0 0 0 0 0 0 1 1 0 0 0 0 0 0 0 0 |         1  | 0 0 0 0 0 0 0 0 0 0 1 1 0 0 0 0 0 0 0 |
 12  | 0 0 0 0 0 0 0 0 0 0 0 1 0 0 0 0 0 0 0 |        10  | 0 0 0 0 0 0 0 0 0 0 0 1 0 0 0 0 0 0 0 |
 13  | 0 0 0 0 0 0 0 0 0 0 0 0 1 0 0 0 0 0 0 |        15  | 0 0 0 0 0 0 0 0 0 0 0 0 1 1 0 1 0 0 0 |
 14  | 0 0 0 0 0 0 0 0 0 0 0 0 0 1 0 0 0 0 0 |        11  | 0 0 0 0 0 0 0 0 0 0 0 0 0 1 0 0 0 0 0 |
 15  | 0 0 0 0 0 0 0 0 0 0 1 0 1 0 1 0 0 0 0 |        12  | 0 0 0 0 0 0 0 0 0 0 0 0 0 0 1 0 0 0 0 |
 16  | 0 0 0 0 0 0 0 0 0 0 0 0 0 0 0 1 0 0 0 |        13  | 0 0 0 0 0 0 0 0 0 0 0 0 0 0 0 1 0 0 0 |
 17  | 1 1 1 1 1 1 1 1 1 1 1 1 1 1 1 1 1 1 1 |        14  | 0 0 0 0 0 0 0 0 0 0 0 0 0 0 0 0 1 0 0 |
 18  | 0 0 0 0 0 0 0 0 0 0 0 0 0 0 0 0 0 1 0 |        16  | 0 0 0 0 0 0 0 0 0 0 0 0 0 0 0 0 0 1 0 |
 19  \ 0 0 0 1 0 0 1 0 1 0 0 0 1 1 0 0 0 0 1 /        18  \ 0 0 0 0 0 0 0 0 0 0 0 0 0 0 0 0 0 0 1 /
```

Fig. 3.20 Division order on permuted numbers 1...19; arranged to upper right triangle

One will indeed find out that Fig. 3.20 shows an ordering, i.e., transitive, reflexive, and antisymmetric. But assume an attempt to draw a graph for it which is sufficiently easy to overview. This will certainly be possible when one is able to arrange the matrix so as to have the **1**-entries in the upper right triangle of the matrix as in Fig. 3.20. Even better when drawing this ordering is to use the so-called Hasse diagram based on it, as shown on the left in Fig. 3.21.

The relation on the right in Fig. 3.21 is an identity relation. It is, however, represented as a matrix with row marking different from column marking, which is possible because we have decided to work with basesets that are ordered – possibly in different ways. It represents the permutation for this transition: the permuted sequence appears when scanning it from top to bottom and identifying the **1**-entries.

```
        17 2 19 9 3 7 4 8 6 5 1 10 15 11 12 13 14 16 18        1 2 3 4 5 6 7 8 9 10 11 12 13 14 15 16 17 18 19
17     /0  1  1 0 1 0 0 1 0 0 0 0  1  0  1  0  0  1  1\    17  /0 0 0 0 0 0 0 0 0 0  0  0  0  0  0  0  1  0  0\
 2     |0  0  0 0 0 0 0 0 0 0 0 0  0  0  0  0  0  0  0|     2  |0 1 0 0 0 0 0 0 0 0  0  0  0  0  0  0  0  0  0|
19     |0  0  0 1 0 1 0 0 0 0 0 0  0  0  1  0  0  0  0|    19  |0 0 0 0 0 0 0 0 0 0  0  0  0  0  0  0  0  0  1|
 9     |0  0  0 0 0 0 1 0 0 0 0 0  0  0  0  0  0  0  0|     9  |0 0 0 0 0 0 0 0 1 0  0  0  0  0  0  0  0  0  0|
 3     |0  0  0 0 0 1 0 0 1 1 0 0  0  1  0  0  0  0  0|     3  |0 0 1 0 0 0 0 0 0 0  0  0  0  0  0  0  0  0  0|
 7     |0  0  0 0 0 0 1 0 0 0 0 0  0  0  0  1  0  0  0|     7  |0 0 0 0 0 0 1 0 0 0  0  0  0  0  0  0  0  0  0|
 4     |0  0  0 0 0 0 0 0 0 0 0 0  0  0  0  0  0  0  0|     4  |0 0 0 1 0 0 0 0 0 0  0  0  0  0  0  0  0  0  0|
 8     |0  0  0 0 0 0 0 0 1 0 0 0  0  0  0  0  0  0  0|     8  |0 0 0 0 0 0 0 1 0 0  0  0  0  0  0  0  0  0  0|
 6     |0  0  0 0 0 0 0 0 0 0 0 0  0  0  0  0  0  0  0|     6  |0 0 0 0 0 1 0 0 0 0  0  0  0  0  0  0  0  0  0|
 5     |0  0  0 0 0 0 0 0 0 1 0 0  0  0  0  1  0  0  0|     5  |0 0 0 0 1 0 0 0 0 0  0  0  0  0  0  0  0  0  0|
 1     |0  0  0 0 0 0 0 0 0 0 0 1  0  0  0  0  0  0  0|     1  |1 0 0 0 0 0 0 0 0 0  0  0  0  0  0  0  0  0  0|
10     |0  0  0 0 0 0 0 0 0 0 0 0  0  0  0  0  0  0  0|    10  |0 0 0 0 0 0 0 0 0 1  0  0  0  0  0  0  0  0  0|
15     |0  0  0 0 0 0 0 0 0 0 0 0  1  0  1  0  0  0  0|    15  |0 0 0 0 0 0 0 0 0 0  0  0  0  0  1  0  0  0  0|
11     |0  0  0 0 0 0 0 0 0 0 0 0  0  0  0  0  0  0  0|    11  |0 0 0 0 0 0 0 0 0 0  1  0  0  0  0  0  0  0  0|
12     |0  0  0 0 0 0 0 0 0 0 0 0  0  0  0  0  0  0  0|    12  |0 0 0 0 0 0 0 0 0 0  0  1  0  0  0  0  0  0  0|
13     |0  0  0 0 0 0 0 0 0 0 0 0  0  0  0  0  0  0  0|    13  |0 0 0 0 0 0 0 0 0 0  0  0  1  0  0  0  0  0  0|
14     |0  0  0 0 0 0 0 0 0 0 0 0  0  0  0  0  0  0  0|    14  |0 0 0 0 0 0 0 0 0 0  0  0  0  1  0  0  0  0  0|
16     \0  0  0 0 0 0 0 0 0 0 0 0  0  0  0  0  0  0  0/    16  \0 0 0 0 0 0 0 0 0 0  0  0  0  0  0  1  0  0  0/
18      0  0  0 0 0 0 0 0 0 0 0 0  0  0  0  0  0  0  0     18   0 0 0 0 0 0 0 0 0 0  0  0  0  0  0  0  0  1  0
```

Fig. 3.21 Hasse relation and permutation for Fig. 3.20

More detailed hints on how matrices are rearranged in this book may be found in Appendix C. It should be mentioned that the technique is based on algebraic considerations and not just prettiness or pulchritude.

3.7 Partition representation

Partitions are frequently used in mathematical modelling. We introduce them rather naïvely. They subdivide a baseset, i.e., they consist of a set of mutually disjoint non-empty subsets of the baseset.

Figure 3.22 shows an equivalence relation Ξ on the left. However, its elements are not sorted in a way that allows elements of an equivalence class to stay together. It is a rather simple programming task to arrive at the rearranged matrix $\Theta := P_{;}\Xi_{;}P^{\mathsf{T}}$ on the right, where P is the permutation used to rearrange rows and columns simultaneously. It is not so simple a task to define P in terms of the given Ξ and the ordering E of the baseset over which Ξ is defined.

$$
\Xi =
\begin{array}{c}
\\ 1 \\ 2 \\ 3 \\ 4 \\ 5 \\ 6 \\ 7 \\ 8 \\ 9 \\ 10 \\ 11 \\ 12 \\ 13
\end{array}
\begin{array}{c}
1\;2\;3\;4\;5\;6\;7\;8\;9\;10\;11\;12\;13 \\
\left(\begin{array}{ccccccccccccc}
1&1&0&0&0&0&0&1&0&0&0&0&1\\
1&1&0&0&0&0&0&1&0&0&0&0&1\\
0&0&1&0&1&0&1&0&1&0&0&1&0\\
0&0&0&1&0&1&0&0&0&1&1&0&0\\
0&0&1&0&1&0&1&0&1&0&0&1&0\\
0&0&0&1&0&1&0&0&0&1&1&0&0\\
0&0&1&0&1&0&1&0&1&0&0&1&0\\
1&1&0&0&0&0&0&1&0&0&0&0&1\\
0&0&1&0&1&0&1&0&1&0&0&1&0\\
0&0&0&1&0&1&0&0&0&1&1&0&0\\
0&0&0&1&0&1&0&0&0&1&1&0&0\\
0&0&1&0&1&0&1&0&1&0&0&1&0\\
1&1&0&0&0&0&0&1&0&0&0&0&1
\end{array}\right)
\end{array}
$$

$$
P =
\begin{array}{c}
1\;2\;8\;13\;3\;5\;7\;9\;12\;4\;6\;10\;11 \\
\left(\begin{array}{ccccccccccccc}
1&0&0&0&0&0&0&0&0&0&0&0&0\\
0&1&0&0&0&0&0&0&0&0&0&0&0\\
0&0&0&0&1&0&0&0&0&0&0&0&0\\
0&0&0&0&0&0&0&0&1&0&0&0&0\\
0&0&0&0&0&1&0&0&0&0&0&0&0\\
0&0&0&0&0&0&0&0&0&1&0&0&0\\
0&0&0&0&0&0&1&0&0&0&0&0&0\\
0&0&1&0&0&0&0&0&0&0&0&0&0\\
0&0&0&0&0&0&0&1&0&0&0&0&0\\
0&0&0&0&0&0&0&0&0&0&0&1&0\\
0&0&0&0&0&0&0&0&0&0&0&0&1\\
0&0&0&0&0&0&0&0&0&0&1&0&0\\
0&0&0&1&0&0&0&0&0&0&0&0&0
\end{array}\right)
\end{array}
\begin{array}{c}
\\ 1 \\ 2 \\ 8 \\ 13 \\ 3 \\ 5 \\ 7 \\ 9 \\ 12 \\ 4 \\ 6 \\ 10 \\ 11
\end{array}
$$

$$
\Theta =
\begin{array}{c}
1\;2\;8\;13\;3\;5\;7\;9\;12\;4\;6\;10\;11 \\
\left(\begin{array}{ccccccccccccc}
1&1&1&1&0&0&0&0&0&0&0&0&0\\
1&1&1&1&0&0&0&0&0&0&0&0&0\\
1&1&1&1&0&0&0&0&0&0&0&0&0\\
1&1&1&1&0&0&0&0&0&0&0&0&0\\
0&0&0&0&1&1&1&1&1&0&0&0&0\\
0&0&0&0&1&1&1&1&1&0&0&0&0\\
0&0&0&0&1&1&1&1&1&0&0&0&0\\
0&0&0&0&1&1&1&1&1&0&0&0&0\\
0&0&0&0&1&1&1&1&1&0&0&0&0\\
0&0&0&0&0&0&0&0&0&1&1&1&1\\
0&0&0&0&0&0&0&0&0&1&1&1&1\\
0&0&0&0&0&0&0&0&0&1&1&1&1\\
0&0&0&0&0&0&0&0&0&1&1&1&1
\end{array}\right)
\end{array}
$$

Fig. 3.22 Rearranging an equivalence relation to block-diagonal form: Ξ, P, Θ

Observe the first four columns of P. They regulate that 1,2,8,13 are sent to positions 1,2,3,4. It is possible to give a nested recursive definition for this procedure. More information on how P has been designed is given in Appendix C.

PART II

OPERATIONS AND CONSTRUCTIONS

At this point of the book, a major break in style may be observed. So far we have used free-hand formulations, not least in HASKELL, and have presented the basics of set theory stressing how to represent sets, subsets, elements, relations, and mappings. However, so far we have not used relations in an algebraic form.

From now on, we shall mainly concentrate on topics that inherently require some algebraic treatment. We cannot start immediately with formal algebraic proofs and with the relational language TITUREL in which all this has been tested. Rather, we first present Part II, which is full of examples that demonstrate the basics of algebraic rules. Whenever one of these early rules needs a proof, this proof will be very simple, but nevertheless omitted at this point. The postponed proofs can be found in Appendix B.

We will, however, often show how point-free versions, i.e., those hiding quantifiers and individual variables, are derived from a predicate-logic form. These deductions are *definitely not* an aim of this book. However, they seem necessary for many researchers who are not well trained in expressing themselves in a point-free algebraic manner. These deductions are by no means executed in a strictly formal way. Rather, they are included so as to convince the reader that there is a reason to use the respective point-free construct.

There is another point to mention concerning the domain constructions, or data structures, that we are going to present. Considered from one point of view, they are rather trivial. Historically, however, they did not seem to be so. The first theoretically reasonably sound programming language ALGOL 60 did *not* provide data structures beyond arrays. When improved with PASCAL, variant handling was *not* treated as a basic construct and was offered only in combination with tuple forming. Dependent types have long been considered theory laden and are *not yet* broadly offered in programming languages. When we offer generic domain construction methods at an early stage, we face the problem that there is not yet a sufficiently developed apparatus available to prove that the respective construct is unique up to isomorphism – of course, our constructs are. We present them here mentioning always that there is essentially one form, as one is in a position to provide a bijection based on the syntactic material already given. This is then shown via examples. For the literate reader these examples will be sufficiently detailed to replace a proof.

Several concepts may be expressed in either style, in a programming language (HASKELL) as well as in an algebraic form. So one or the other topic which we have already dealt with will now be rewritten in the new style. This will make the differences visible.

To sum up: we present here an introduction to relations that may fit the needs of many researchers but is not yet given in a fully formal style.

4

Algebraic Operations on Relations

A full account of relation algebra cannot be given here, just enough to enable us to work with it. This will be accompanied by examples showing the effects. When relations are studied in practice, they are conceived as subsets $R \subseteq X \times Y$ of the Cartesian product of the two sets X, Y between which they are defined to hold. Already in Chapter 3, we have given enough examples of relations between sets together with a diversity of ways to represent them. We will now present the operations on relations in a way that is both algebraically sound and sufficiently underpinned with examples.

With this algebraic approach we start in full contrast to such texts as, for example, [125], where relations are also presented in extensu. There, however, everything is based on point-wise reasoning.

4.1 Typing relations

In many cases, we are interested in solving just *one problem*. More often, however, we want to find out how a whole *class of problems* may be solved. While in the first case, we may use the given objects directly, we must work with constants and variables for relations in the latter case. These variables will be instantiated when the concrete problem is presented.

Let us consider timetabling as an example. It may be the case that a timetable has to be constructed at a given school. However, it would not be wise to write a timetabling program just for this school with the present set of teachers and classrooms. If it is intended to sell the program several times so that it may be applied also to other schools, one must allow for the possibility of different sets of teachers, different sets of classrooms, etc.

The standard technique in such cases is to introduce **types**. One will have to be able to handle the type *teacher*, the type *classroom*, etc. At one school, there may exist 15 teachers and 10 classrooms, while instantiation at another school will give

20 teachers and 15 classrooms. Although the concept of time is in a sense universal, one will also introduce the type *hour*; it may well be the case that at one school a scheme of hours or timeslots 8.00–9.00, 9.00–10.00 is used, while at another school the timeslots are 8.15–9.00, 9.00–9.45. It is therefore wise to offer an instantiable concept of 'hour'.[1]

In any case, a type variable is a denotation of a set of items that may afterwards be interpreted with a baseset. In this strict sense, the following relations are of different type – in spite of the fact that their constituent matrices are equal.

$$
R_1 = \begin{array}{c} \text{red} \\ \text{green} \\ \text{blue} \\ \text{orange} \end{array}
\begin{array}{c} \text{\scriptsize Mon Tue Wed Thu Fri Sat} \end{array}
\begin{pmatrix} 0 & 1 & 1 & 0 & 1 & 0 \\ 1 & 0 & 1 & 0 & 0 & 1 \\ 0 & 0 & 0 & 1 & 1 & 0 \\ 1 & 0 & 0 & 1 & 0 & 1 \end{pmatrix}
\qquad
R_2 = \begin{array}{c} \spadesuit \\ \heartsuit \\ \diamondsuit \\ \clubsuit \end{array}
\begin{array}{c} \text{\scriptsize 1 2 3 4 5 6} \end{array}
\begin{pmatrix} 0 & 1 & 1 & 0 & 1 & 0 \\ 1 & 0 & 1 & 0 & 0 & 1 \\ 0 & 0 & 0 & 1 & 1 & 0 \\ 1 & 0 & 0 & 1 & 0 & 1 \end{pmatrix}
$$

Fig. 4.1 Differently typed relations

While $R : X \longrightarrow Y$ is the type, the concrete instantiation of the first relation is

$R_1 \subseteq \{\text{red}, \text{green}, \text{blue}, \text{orange}\} \times \{\text{Mon}, \text{Tue}, \text{Wed}, \text{Thu}, \text{Fri}, \text{Sat}\}$.

In the other case, $R : X \longrightarrow Y$ is the type but

$R_2 \subseteq \{\spadesuit, \heartsuit, \diamondsuit, \clubsuit\} \times \{1, 2, 3, 4, 5, 6\}$

the instantiation. We see, that X is instantiated in two ways, namely as

$\{\text{red}, \text{green}, \text{blue}, \text{orange}\}$ or as $\{\spadesuit, \heartsuit, \diamondsuit, \clubsuit\}$.

When we do not ask for both, X, Y, we will frequently determine the **source** and **target** by src $R = X$ and tgt $R = Y$.

A word is in order concerning the concept of a **monotype** that is met frequently in the literature. This indicates a homogeneous setting, i.e., all items of discourse are assumed to reside in one huge domain. As far as typing in the sense developed here is necessary, people think of a partial diagonal relation selecting this set. The distinction between a vector and a partial diagonal is usually not made; expressed differently, it is abstracted over.

---

[1] To what extent this by now general practice was difficult to achieve may be estimated from George Boole's investigations on the laws of thought of 1854 [26, 28]: *In every discourse, whether of the mind conversing with its own thoughts, or of the individual in his intercourse with others, there is an assumed or expressed limit within which the subjects of its operation are confined. The most unfettered discourse is that in which the words we use are understood in the widest possible application, and for them the limits of discourse are co-extensive with those of the universe itself. But more usually we confine ourselves to a less spacious field. ... Furthermore, this universe of discourse is in the strictest sense the ultimate subject of the discourse. The office of any name or descriptive term employed under the limitations supposed is not to raise in the mind the conception of all the beings or objects to which that name or description is applicable, but only of those which exist within the supposed universe of discourse.*

4.2 Boolean operations

We recall in a more formal way what has already been presented in the examples of Part I. Relations with type $X \longrightarrow Y$, when interpreted, may be understood as subsets of a Cartesian product: $R \subseteq V \times W$. Being interpreted as a subset immediately implies that certain concepts are available for relations that we know from subsets.

Definition 4.1. Given two relations $R : X \longrightarrow Y$ and $S : X \longrightarrow Y$, i.e., relations *of the same type*, we define

union	$R \cup S$	$\{(x,y) \in X \times Y \mid (x,y) \in R \vee (x,y) \in S\}$
intersection	$R \cap S$	$\{(x,y) \in X \times Y \mid (x,y) \in R \wedge (x,y) \in S\}$
complementation	$\overline{R}$	$\{(x,y) \in X \times Y \mid (x,y) \notin R\}$
null relation	$\bot\!\!\!\bot$	$\{\}$ or $\emptyset$
universal relation	$\top\!\!\!\top$	$X \times Y$
identity relation	$\mathbb{I}$	$\{(x,x) \in X \times X \mid x \in X\}$

On the right we have indicated what we intend this to mean when the relations are interpreted as subsets of $X \times Y$ or, in the case of the identity, $X \times X$. □

There is also a tradition of denoting this differently,[2] which we do not follow. It makes no sense to unite, for example, relations of different types, and therefore this is not allowed. Concerning the top and bottom relations, we have been a bit sloppy here. The symbols for the empty or null and for the universal relation should have been $\bot\!\!\!\bot_{X,Y}$ and $\top\!\!\!\top_{X,Y}$, respectively. While we know the typing in cases of union, intersection, and negation from the operands R, S, we should provide this information explicitly for the null and the universal relation. It is evident that the relations of Fig. 4.2

$$\top\!\!\!\top_{\text{colors,weekdays}} = \begin{array}{c} \\ \text{red} \\ \text{green} \\ \text{blue} \\ \text{orange} \end{array} \begin{array}{c} \text{\scriptsize Mon Tue Wed Thu Fri Sat} \\ \begin{pmatrix} 1 & 1 & 1 & 1 & 1 & 1 \\ 1 & 1 & 1 & 1 & 1 & 1 \\ 1 & 1 & 1 & 1 & 1 & 1 \\ 1 & 1 & 1 & 1 & 1 & 1 \end{pmatrix} \end{array} \qquad \top\!\!\!\top_{\text{brigdesuits,numerals}} = \begin{array}{c} \spadesuit \\ \heartsuit \\ \diamondsuit \\ \clubsuit \end{array} \begin{array}{c} \text{\scriptsize 1 2 3 4 5 6} \\ \begin{pmatrix} 1 & 1 & 1 & 1 & 1 & 1 \\ 1 & 1 & 1 & 1 & 1 & 1 \\ 1 & 1 & 1 & 1 & 1 & 1 \\ 1 & 1 & 1 & 1 & 1 & 1 \end{pmatrix} \end{array}$$

Fig. 4.2 Universal relations of different type

should not be mixed up although the matrices proper seem equal. It is important to keep them separate when even row and column numbers differ. Often, however, the typing is so evident from the context that we will omit the type indices and write just $\bot\!\!\!\bot, \top\!\!\!\top$.

---

[2] In many classical, but also in modern, texts, usually in a homogeneous setting, 0 is used for $\bot\!\!\!\bot$, 1 for $\top\!\!\!\top$ and 1' for $\mathbb{I}$.

Since we have introduced union, intersection, negation, and the counterpart of the full and the empty set in a relational setting, it is clear that we are in a position to reason and calculate in the same way as we have learned from set theory, from predicate logic, or from Boolean algebra. This means that "$\cup, \cap$", when restricted to some given type, behave as follows

$$R \cup S = S \cup R \quad \text{commutative} \quad R \cap S = S \cap R,$$
$$R \cup (S \cup T) = (R \cup S) \cup T \quad \text{associative} \quad R \cap (S \cap T) = (R \cap S) \cap T,$$
$$R \cup (R \cap S) = R \quad \text{absorptive} \quad R \cap (R \cup S) = R,$$
$$R \cup (S \cap T) = (R \cup S) \cap (R \cup T) \quad \text{distributive} \quad R \cap (S \cup T) = (R \cap S) \cup (R \cap T).$$

Concerning negation or complementation, we have

$\overline{\overline{R}} = R$ is involutory,

$\overline{R \cup S} = \overline{R} \cap \overline{S}$ obeys the De Morgan rule,[3]

$R \cup \overline{R} = \mathbb{T}_{\text{src } R, \text{tgt } R}$ is complementary.

So far, we have handled union and intersection as binary operations. We may, however, form unions and intersections in a descriptive way over a set of relations of the same type. They are then called the supremum or the infimum of a set $\mathcal{S}$ of relations, respectively,

$\sup \mathcal{S} = \{(x, y) \mid \exists R \in \mathcal{S} : (x, y) \in R\},$

$\inf \mathcal{S} = \{(x, y) \mid \forall R \in \mathcal{S} : (x, y) \in R\}.$

The two definitions differ in just replacing "$\exists$" by "$\forall$". This turns out to be indeed a generalization of "$\cup, \cap$" because for $\mathcal{S} := \{R, S\}$ we find out that $\sup \mathcal{S} = R \cup S$ and $\inf \mathcal{S} = R \cap S$. One should recall the effect that

$$\sup \emptyset = \{(x, y) \mid \texttt{False} \} = \mathbb{L} \qquad \inf \emptyset = \{(x, y) \mid \texttt{True} \} = \mathbb{T}.$$

After having defined the operations, we now introduce predicates for relations:

$$R \subseteq S :\Longleftrightarrow \forall x \in X : \forall y \in Y : \big[(x, y) \in R \rightarrow (x, y) \in S\big] \quad \text{inclusion, containment.}$$

The latter definition could also have been expressed point-free as

$$R \subseteq S :\Longleftrightarrow \mathbb{T} = \overline{R} \cup S \iff \mathbb{L} = R \cap \overline{S} \iff R = R \cap S \iff S = R \cup S$$

resembling the traditional logical equivalence of $a \rightarrow b$ and $\texttt{True} = \neg a \vee b$.

4.3 Relational operations proper

The first operation on relations beyond the Boolean operations is conversion or transposition; it is characteristic for relations.

---

[3] According to [79], p. 295, these were known to medieval logicians and 'occur explictly' in William of Ockham's (1287–1347) *Summa Totius Logicae*.

Definition 4.2. Given a relation $R : X \longrightarrow Y$, its **converse** (or transpose) $R^\mathsf{T} :$ $Y \longrightarrow X$ is that relation in the opposite direction in which for all x, y containment $(y, x) \in R^\mathsf{T}$ holds precisely when $(x, y) \in R$. □

Colloquially, transposition is often expressed using separate wordings:

Bob *is_taller_than* Chris $\longleftrightarrow$ Chris *is_smaller_than* Bob

Dick *is_boss_of* Erol $\longleftrightarrow$ Erol *is_employee_of* Dick

Fred *sells_to* Gordon $\longleftrightarrow$ Gordon *buys_from* Fred

Here, as well as in the set function representation of Fig. 4.3, conversion is not easily recognized; even worse, it depends heavily on the language used, English, German, Japanese, for example.[4]

$$
R_1 = \begin{array}{c|l}
\spadesuit & \{\text{Tue,Wed,Fri}\} \\
\heartsuit & \{\text{Mon,Wed,Sat}\} \\
\diamondsuit & \{\text{Thu,Fri}\} \\
\clubsuit & \{\text{Mon,Thu,Sat}\}
\end{array}
\qquad
R_1^\mathsf{T} = \begin{array}{c|l}
\text{Mon} & \{\heartsuit,\clubsuit\} \\
\text{Tue} & \{\spadesuit\} \\
\text{Wed} & \{\spadesuit,\heartsuit\} \\
\text{Thu} & \{\diamondsuit,\clubsuit\} \\
\text{Fri} & \{\spadesuit,\diamondsuit\} \\
\text{Sat} & \{\heartsuit,\clubsuit\}
\end{array}
$$

Fig. 4.3 Transposition in set function representation

For matrix representations it is more schematic; it means exchanging row and column types and mirroring the matrix along the diagonal upper left to lower right.

$$
R_1 = \begin{array}{c}
\spadesuit \\ \heartsuit \\ \diamondsuit \\ \clubsuit
\end{array}
\begin{pmatrix}
0 & 1 & 1 & 0 & 1 & 0 \\
1 & 0 & 1 & 0 & 0 & 1 \\
0 & 0 & 0 & 1 & 1 & 0 \\
1 & 0 & 0 & 1 & 0 & 1
\end{pmatrix}
\qquad
R_1^\mathsf{T} = \begin{array}{c}
\text{Mon} \\ \text{Tue} \\ \text{Wed} \\ \text{Thu} \\ \text{Fri} \\ \text{Sat}
\end{array}
\begin{pmatrix}
0 & 1 & 0 & 1 \\
1 & 0 & 0 & 0 \\
1 & 1 & 0 & 0 \\
0 & 0 & 1 & 1 \\
1 & 0 & 1 & 0 \\
0 & 1 & 0 & 1
\end{pmatrix}
$$

Fig. 4.4 Transposition in matrix representation

Obviously, conversion is involutory, i.e., application twice results in the original argument again:

$$(R^\mathsf{T})^\mathsf{T} = R.$$

It is standard mathematics to check how a newly introduced operation behaves when applied together with those operations already available. Such algebraic rules are rather intuitive. It is immaterial whether one first negates a relation and then transposes or the other way round (Fig. 4.5):

$$\overline{R}^\mathsf{T} = \overline{R^\mathsf{T}}.$$

---

[4] The early authors who wrote texts on relations complained repeatedly how difficult it was to free themselves from standard use of the natural languages; see Appendix D.4.

$$
R = \begin{array}{c} \spadesuit \\ \heartsuit \\ \diamondsuit \\ \clubsuit \end{array}
\begin{pmatrix} 0 & 1 & 1 & 0 & 1 & 0 \\ 1 & 0 & 1 & 0 & 0 & 1 \\ 0 & 0 & 0 & 1 & 1 & 0 \\ 1 & 0 & 0 & 1 & 0 & 1 \end{pmatrix}
\qquad
\overline{R} = \begin{pmatrix} 1 & 0 & 0 & 1 & 0 & 1 \\ 0 & 1 & 0 & 1 & 1 & 0 \\ 1 & 1 & 1 & 0 & 0 & 1 \\ 0 & 1 & 1 & 0 & 1 & 0 \end{pmatrix}
$$

(columns: Mon Tue Wed Thu Fri Sat)

$$
R^\mathsf{T} = \begin{array}{c} \text{Mon} \\ \text{Tue} \\ \text{Wed} \\ \text{Thu} \\ \text{Fri} \\ \text{Sat} \end{array}
\begin{pmatrix} 0 & 1 & 0 & 1 \\ 1 & 0 & 0 & 0 \\ 1 & 1 & 0 & 0 \\ 0 & 0 & 1 & 1 \\ 1 & 0 & 1 & 0 \\ 0 & 1 & 0 & 1 \end{pmatrix}
\qquad
\overline{R}^\mathsf{T} = \overline{R^\mathsf{T}} = \begin{pmatrix} 1 & 0 & 1 & 0 \\ 0 & 1 & 1 & 1 \\ 0 & 0 & 1 & 1 \\ 1 & 1 & 0 & 0 \\ 0 & 1 & 0 & 1 \\ 1 & 0 & 1 & 0 \end{pmatrix}
$$

(columns: ♠ ◇ ◇ ♣)

Fig. 4.5 Transposition commutes with negation

In a similar way, transposition commutes with union and intersection of relations and with the relation constants

$$(R \cup S)^\mathsf{T} = R^\mathsf{T} \cup S^\mathsf{T} \qquad (R \cap S)^\mathsf{T} = R^\mathsf{T} \cap S^\mathsf{T}$$
$$\mathbb{\perp}^\mathsf{T}_{X,Y} = \mathbb{\perp}_{Y,X} \qquad \mathbb{T}^\mathsf{T}_{X,Y} = \mathbb{T}_{Y,X} \qquad \mathbb{I}^\mathsf{T}_X = \mathbb{I}_X.$$

When a relation is contained in another relation, then the same will be true for their converses

$$R \subseteq S \quad \Longleftrightarrow \quad R^\mathsf{T} \subseteq S^\mathsf{T}.$$

The next vital operation to be mentioned is relational composition. Whenever x and y are in relation R and y and z are in relation S, one says that x and z are in relation $R\,;S$. It is unimportant whether there is just one intermediate y or many. The point to stress is that *there exists*[5] an intermediate element.

Definition 4.3. Let $R : X \longrightarrow Y$ and $S : Y \longrightarrow Z$ be relations. Their (relational) **composition**, or their **multiplication**, or **product** $R\,;S : X \longrightarrow Z$ is defined as the relation

$$R\,;S := \{(x, z) \in X \times Z \mid \exists y \in Y : (x, y) \in R \wedge (y, z) \in S\}.$$

Concerning composition, there exist left unit elements and right unit elements, the so-called **identity relations**.   □

The identity relations vary over the sets: $\mathbb{I}_X, \mathbb{I}_Y, \mathbb{I}_Z$. For composition, a stronger binding power is assumed, $R\,S \cap Q = (R\,S) \cap Q$, than for union or intersection. We illustrate composition and identity relations with a completely fictitious example in which indeed three different types are involved.[6] Assume owners of several cars who travel and have to rent a car, but not all car rental companies offer every car type.

---

[5] Appendix D.3 shows how difficult it was to arrive at a sufficiently clear concept of quantification suitable for defining composition of relations.

[6] The author apologizes for blaming RentACar for offering no car type at all.

The following composition of the relation `ownsCar` with the relation `isOfferedBy`
results in a relation

`mayRentOwnedCarTypeFrom = ownsCar ; isOfferedBy.`

$$
\begin{array}{c}
\text{Arbuthnot}\\\text{Perez}\\\text{Dupont}\\\text{Botticelli}\\\text{Schmidt}\\\text{Larsen}
\end{array}
\begin{pmatrix}
1&0&1&1&1&1&0&1\\
0&0&1&0&0&0&0&1\\
0&1&1&0&0&1&0&1\\
1&0&0&1&1&1&0&0\\
1&0&0&0&1&0&0&0\\
0&1&1&1&0&1&0&1
\end{pmatrix}
=
\begin{array}{c}
\text{Arbuthnot}\\\text{Perez}\\\text{Dupont}\\\text{Botticelli}\\\text{Schmidt}\\\text{Larsen}
\end{array}
\begin{pmatrix}
0&1&0&1&0&0&1\\
0&1&0&0&0&0&0\\
0&1&1&0&0&1&0\\
1&0&0&0&0&0&1\\
0&0&0&1&0&0&0\\
0&1&1&0&1&0&0
\end{pmatrix}
\;;\;
\begin{array}{c}
\text{Dodge}\\\text{Audi}\\\text{Renault}\\\text{Bentley}\\\text{BMW}\\\text{Seat}\\\text{Ford}
\end{array}
\begin{pmatrix}
1&0&0&0&1&0&0&0\\
0&0&1&0&0&0&0&1\\
0&1&0&0&0&1&0&0\\
1&0&0&0&1&0&0&0\\
0&1&0&1&0&0&0&1\\
0&0&1&0&0&0&0&1\\
1&0&0&1&0&1&0&0
\end{pmatrix}
$$

Fig. 4.6 The chance of a car owner renting a car of one of the types he already owns

For the unit relations, we have again the choice of being fully precise or not

$$R ; \mathbb{I}_Y = R, \quad \mathbb{I}_X ; R = R \qquad \text{as opposed to} \qquad R ; \mathbb{I} = R, \quad \mathbb{I} ; R = R$$

see, for example, Fig. 4.7. We will also use notation with an exponent when composing a (homogeneous) relation with itself, so that $R^2 = R ; R$ and $R^0 = \mathbb{I}$.

$$
\mathbb{I}_{\text{BridgeSuits}} =
\begin{array}{c}
\spadesuit\\\heartsuit\\\diamondsuit\\\clubsuit
\end{array}
\begin{pmatrix}
1&0&0&0\\
0&1&0&0\\
0&0&1&0\\
0&0&0&1
\end{pmatrix}
\qquad
R =
\begin{array}{c}
\spadesuit\\\heartsuit\\\diamondsuit\\\clubsuit
\end{array}
\begin{pmatrix}
0&1&1&0&1&0\\
1&0&1&0&0&1\\
0&0&0&1&1&0\\
1&0&0&1&0&1
\end{pmatrix}
$$

$$
\begin{array}{c}
\text{Mon}\\\text{Tue}\\\text{Wed}\\\text{Thu}\\\text{Fri}\\\text{Sat}
\end{array}
\begin{pmatrix}
1&0&0&0&0&0\\
0&1&0&0&0&0\\
0&0&1&0&0&0\\
0&0&0&1&0&0\\
0&0&0&0&1&0\\
0&0&0&0&0&1
\end{pmatrix}
= \mathbb{I}_{\text{WeekDays}}
$$

Fig. 4.7 Left and right identity relations

Composition is very similar to multiplication of real matrices. Observe that "$\exists$"
stands as the possibly infinite version of "$\lor$" in much the same way as "$\sum$" does
for "$+$":

$$(R ; S)_{xz} = \exists i \in Y : R_{xi} \land S_{iz} \qquad (R \cdot S)_{xz} = \sum_{i \in Y} R_{xi} \cdot S_{iz}.$$

For the presentation of the whole text of this book it is important to have some
formulae available which in the strict line of development would only be presented
later. We exhibit them here simply as observations.

Proposition 4.4. *Let triples of relations Q, R, S or A, B, C be given, and assume
that the constructs are well-formed. Then they will always satisfy the*

    Schröder rule $A ; B \subseteq C \iff A^{\mathsf{T}} ; \overline{C} \subseteq \overline{B} \iff \overline{C} ; B^{\mathsf{T}} \subseteq \overline{A}$

and the

Dedekind rule $\qquad R \,{}_{,}\, S \cap Q \subseteq (R \cap Q \,{}_{,}\, S^{\mathsf{T}}) \,{}_{,}\, (S \cap R^{\mathsf{T}} \,{}_{,}\, Q).$ $\qquad\qquad\qquad$ $\square$

Some other rules one might have expected follow from these, not least the well-known $(R \,{}_{,}\, S)^{\mathsf{T}} = S^{\mathsf{T}} \,{}_{,}\, R^{\mathsf{T}}$ for transposing products.

We convince ourselves that the first of the Schröder equivalences is correct considering pairs of points of the relation – which will later be strictly avoided.

$A \,{}_{,}\, B \subseteq C$

$\qquad\qquad\qquad\qquad\qquad\qquad\qquad\qquad\qquad\qquad$ interpreting point-wise

$\Longleftrightarrow \quad \forall x, y : \; (A \,{}_{,}\, B)_{xy} \to C_{xy}$

$\qquad\qquad\qquad\qquad\qquad\qquad\qquad\qquad$ definition of relation composition

$\Longleftrightarrow \quad \forall x, y : \; \big[\exists z : \; A_{xz} \wedge B_{zy}\big] \to C_{xy}$

$\qquad\qquad\qquad\qquad\qquad\qquad\qquad\qquad\qquad\qquad a \to b = \neg a \vee b$

$\Longleftrightarrow \quad \forall x, y : \; \overline{\exists z : \; A_{xz} \wedge B_{zy}} \vee C_{xy}$

$\qquad\qquad\qquad\qquad\qquad\qquad\qquad\qquad \neg\big(\exists x : p(x)\big) = \forall x : \neg p(x)$

$\Longleftrightarrow \quad \forall x, y : \; \big[\forall z : \overline{A}_{xz} \vee \overline{B}_{zy}\big] \vee C_{xy}$

$\qquad\qquad\qquad\qquad\qquad\qquad\qquad \big\{\forall x : q(x)\big\} \vee a = \forall x : \big(q(x) \vee a\big)$

$\Longleftrightarrow \quad \forall x, y : \; \forall z : \overline{A}_{xz} \vee \overline{B}_{zy} \vee C_{xy}$

$\qquad\qquad\qquad\qquad\qquad\qquad\qquad\qquad$ transposing and rearranging

$\Longleftrightarrow \quad \forall y, z : \forall x : \; \overline{A}^{\mathsf{T}}_{zx} \vee C_{xy} \vee \overline{B}_{zy}$

$\qquad\qquad\qquad\qquad\qquad\qquad \forall x : \big(q(x) \vee a\big) = \big\{\forall x : q(x)\big\} \vee a$

$\Longleftrightarrow \quad \forall y, z : \; \big[\forall x : \overline{A}^{\mathsf{T}}_{zx} \vee C_{xy}\big] \vee \overline{B}_{zy}$

$\qquad\qquad\qquad\qquad\qquad\qquad\qquad \forall x : \neg p(x) = \neg\big(\exists x : p(x)\big)$

$\Longleftrightarrow \quad \forall y, z : \; \overline{\exists x : A^{\mathsf{T}}_{zx} \wedge \overline{C}_{xy}} \vee \overline{B}_{zy}$

$\qquad\qquad\qquad\qquad\qquad\qquad\qquad\qquad$ definition of composition

$\Longleftrightarrow \quad \forall y, z : \; \overline{(A^{\mathsf{T}} \,{}_{,}\, \overline{C})_{zy}} \vee \overline{B}_{zy}$

$\qquad\qquad\qquad\qquad\qquad\qquad\qquad\qquad\qquad \neg a \vee b = a \to b$

$\Longleftrightarrow \quad \forall y, z : \; (A^{\mathsf{T}} \,{}_{,}\, \overline{C})_{zy} \to \overline{B}_{zy}$

$\qquad\qquad\qquad\qquad\qquad\qquad\qquad\qquad$ proceeding to point-free form

$\Longleftrightarrow \quad A^{\mathsf{T}} \,{}_{,}\, \overline{C} \subseteq \overline{B}$

In a similar way – but presented differently – the Dedekind rule may be traced back to predicate-logic form:

$(R \,{}_{,}\, S \cap Q)_{xy}$

$\quad = (R \,{}_{,}\, S)_{xy} \wedge Q_{xy} \qquad\qquad\qquad\qquad\qquad$ definition of intersection

$\quad = \big[\exists z : \; R_{xz} \wedge S_{zy}\big] \wedge Q_{xy} \qquad\qquad\qquad$ definition of composition

$\quad = \exists z : \; R_{xz} \wedge S_{zy} \wedge Q_{xy} \qquad\qquad\qquad$ associative, distributive

$\quad = \exists z : \; R_{xz} \wedge Q_{xy} \wedge S^{\mathsf{T}}_{yz} \wedge S_{zy} \wedge R^{\mathsf{T}}_{zx} \wedge Q_{xy} \qquad$ doubling, transposing

$\quad \to \exists z : \; R_{xz} \wedge (\exists u : \; Q_{xu} \wedge S^{\mathsf{T}}_{uz}) \wedge S_{zy} \wedge (\exists v : \; R^{\mathsf{T}}_{zv} \wedge Q_{vy})$

$\qquad\qquad\qquad\qquad\qquad\qquad\qquad\qquad$ new existential quantifiers

$$= \exists z: \; R_{xz} \wedge (Q \,;\, S^\mathsf{T})_{xz} \wedge S_{zy} \wedge (R^\mathsf{T} \,;\, Q)_{zy} \qquad \text{definition of composition}$$
$$= \exists z: \; (R \cap Q \,;\, S^\mathsf{T})_{xz} \wedge (S \cap R^\mathsf{T} \,;\, Q)_{zy} \qquad \text{definition of intersection}$$
$$= \left[(R \cap Q \,;\, S^\mathsf{T}) \,;\, (S \cap R^\mathsf{T} \,;\, Q) \right]_{xy} \qquad \text{definition of composition}$$

Exercises

4.1 Convince yourself that $A \subseteq A \,;\, A^\mathsf{T} \,;\, A$ for an arbitrary relation A.

4.2 Prove that $A^\mathsf{T} \,;\, C \subseteq D$ implies $A \,;\, B \cap C \subseteq A \,;\, (B \cap D)$.

4.4 Composite operations

Several other operations are defined building on those mentioned so far. It is often more comfortable to use such composite operations[7] instead of aggregates built from the original ones. The following is often used just as an abbreviation.

Definition 4.5. For R an arbitrary relation, $R^\mathsf{d} := \overline{R}^\mathsf{T}$ is called its **dual**. □

This definition draws its entitlement mainly from the fact that many people dislike negation, in particular when working in an unrestricted universe of discourse as often occurs in an environment which is not carefully typed. Consequently, methods to circumvent negation, for which the dual serves, have been sought.

Given a multiplication operation, one will ask whether there exists the quotient of one relation with respect to another in the same way as, for example, $12 \cdot n = 84$ results in $n = 7$. To cope with this question, the concept of residuals will now be introduced; we need a *left* and a *right* residual because composition is not commutative.

Quotients do not always exist as we know from division by 0, or from trying to invert singular matrices. Nevertheless will we find for relations some constructs that behave sufficiently similarly to quotients.

Definition 4.6. Given two possibly heterogeneous relations R, S with coinciding source, we define their

left residual $\qquad R\backslash S := \overline{R^\mathsf{T} \,;\, \overline{S}}.$

---

[7] With respect to their broad applicability, we have presented residuals. However, $R \dagger S := \overline{\overline{R} \,;\, \overline{S}}$, another operation, has often been used in classical texts. It offers long lists of formulae in symmetry to others already obtained for "$\,;\,$".

Given two possibly heterogeneous relations P, Q with coinciding target, we define their

right residual $\qquad Q/P := \overline{\overline{Q} \,;\, P^{\mathsf{T}}}.$ $\qquad\qquad\qquad\qquad\qquad$ □

The left residual $R\backslash S$ is the greatest of all relations X with $R \,;\, X \subseteq S$ – frequently satisfying $R\,;\,(R\backslash S) = S$ in addition. This may be proved applying the Schröder rule

$$R \,;\, X \subseteq S \quad\Longleftrightarrow\quad R^{\mathsf{T}}\,;\,\overline{S} \subseteq \overline{X} \quad\Longleftrightarrow\quad X \subseteq \overline{R^{\mathsf{T}}\,;\,\overline{S}} = R\backslash S.$$

The symbol "\" has been chosen to symbolize that R is divided from S on the left side. As one will easily see in Fig. 4.8, the residual $R\backslash S$ always sets into relation a column of R precisely with those columns of S containing it. The column of the *Queen* in R is, for example, contained in the columns of *Jan, Feb, Dec*. The columns 6,9,10, and A of R are empty, and thus contained in every column of S.

```
                 Jan Feb Mar Apr May Jun Jul Aug Sep Oct Nov Dec
          US   / 0   0   0   1   0   1   1   1   0   1   0   0 \
      French  |  1   0   0   1   0   0   1   0   0   1   0   0  |
S = German    |  1   1   0   0   1   1   0   1   0   0   0   1  |
     British  |  1   1   0   0   0   0   1   0   1   0   1   1  |
     Spanish   \ 0   0   0   1   0   1   1   1   0   0   0   0 /

                 A  K  Q  J  10  9  8  7  6  5  4  3  2
          US   / 0  0  0  0  0   0  0  0  0  0  0  0  1 \
      French  |  0  1  0  0  0   0  1  0  0  0  0  0  0  |
R = German    |  0  0  1  0  0   0  1  1  0  1  0  1  0  |
     British  |  0  1  1  0  0   0  0  1  0  0  0  0  1  |
     Spanish   \ 0  0  0  1  0   0  1  0  0  1  1  0  1 /

                 Jan Feb Mar Apr May Jun Jul Aug Sep Oct Nov Dec
          A    / 1   1   1   1   1   1   1   1   1   1   1   1 \
          K   |  1   0   0   0   0   0   1   0   0   0   0   0  |
          Q   |  1   1   0   0   0   0   0   0   0   0   0   1  |
          J   |  0   0   0   1   0   1   1   1   0   0   0   0  |
          10  |  1   1   1   1   1   1   1   1   1   1   1   1  |
R\S =     9   |  1   1   1   1   1   1   1   1   1   1   1   1  |
          8   |  0   0   0   0   0   0   1   0   0   0   0   0  |
          7   |  1   0   0   0   0   0   0   0   0   0   0   0  |
          6   |  1   1   1   1   1   1   1   1   1   1   1   1  |
          5   |  0   0   0   0   0   1   0   1   0   0   0   0  |
          4   |  0   0   0   1   0   1   1   1   0   0   0   0  |
          3   |  1   1   0   0   1   1   0   1   0   0   0   1  |
          2    \ 0   0   0   0   0   0   1   0   0   0   0   0 /
```

Fig. 4.8 Left residuals show how columns of the relation R below the fraction backslash are contained in columns of the relation S above

We derive the relational form indicated in Fig. 4.8 from the predicate-logic version, starting from column i of R being contained in column k of S:

$$\forall n : \big[(n,i) \in R \;\rightarrow\; (n,k) \in S\big]$$

$\qquad\qquad\qquad\qquad\qquad\qquad$ $a \rightarrow b = \neg a \vee b$, transposition

$$\Longleftrightarrow \quad \forall n : \big[(i,n) \notin R^{\mathsf{T}} \vee (n,k) \in S\big]$$

$\qquad\qquad\qquad\qquad\qquad\qquad$ negation over a quantifier

$$\Longleftrightarrow \quad \neg\big\{\exists n : [(i,n) \in R^{\mathsf{T}} \wedge (n,k) \notin S]\big\}$$

$\qquad\qquad\qquad\qquad\qquad\qquad$ definition of composition

$$\Longleftrightarrow \quad \overline{R^{\mathsf{T}}\,;\,\overline{S}}_{i,k}$$

$\qquad\qquad\qquad\qquad\qquad\qquad$ definition of left residual

$$\Longleftrightarrow \quad (R\backslash S)_{i,k}$$

Correspondingly, we divide P from Q on the right in the right residual which denotes the greatest of all solutions Y satisfying

$$Y \,;\, P \subseteq Q \quad\Longleftrightarrow\quad \overline{Q}\,;\,P^{\mathsf{T}} \subseteq \overline{Y} \quad\Longleftrightarrow\quad Y \subseteq \overline{\overline{Q}\,;\,P^{\mathsf{T}}} = Q/P.$$

As one may see in Fig. 4.9, the residual Q/P sets into relation a row of Q with those rows of P it contains. The row of the *King* in Q, for example, contains in the rows belonging to *Mar, Sep, Nov*. The row of *Mar* is empty, and thus contained in every row of Q, leading to a column with all 1s in Q/P.

$$Q =$$

	US	French	German	British	Spanish
A	0	0	0	0	0
K	0	1	0	1	0
Q	0	0	1	1	0
J	0	0	0	0	1
10	0	0	0	0	0
9	0	0	0	0	0
8	0	0	1	0	1
7	0	1	1	1	0
6	0	0	0	0	0
5	0	0	1	0	1
4	0	0	0	0	1
3	0	0	1	0	0
2	1	0	0	1	1

$$P =$$

	US	French	German	British	Spanish
Jan	0	1	1	1	0
Feb	0	0	1	1	0
Mar	0	0	0	0	0
Apr	1	1	0	0	1
May	0	0	1	0	0
Jun	1	0	1	0	1
Jul	1	1	0	1	1
Aug	1	0	1	0	1
Sep	0	0	0	1	0
Oct	1	1	0	0	0
Nov	0	0	0	1	0
Dec	0	0	1	1	0

$$= Q/P$$

	Jan	Feb	Mar	Apr	May	Jun	Jul	Aug	Sep	Oct	Nov	Dec
A	0	0	1	0	0	0	0	0	0	0	0	0
K	0	0	1	0	0	0	0	0	1	0	1	0
Q	0	1	1	0	1	0	0	0	1	0	1	1
J	0	0	1	0	0	0	0	0	0	0	0	0
10	0	0	1	0	0	0	0	0	0	0	0	0
9	0	0	1	0	0	0	0	0	0	0	0	0
8	0	0	1	0	1	0	0	0	0	0	0	0
7	1	1	1	0	1	0	0	0	1	0	1	1
6	0	0	1	0	0	0	0	0	0	0	0	0
5	0	0	1	0	1	0	0	0	0	0	0	0
4	0	0	1	0	0	0	0	0	0	0	0	0
3	0	0	1	0	1	0	0	0	0	0	0	0
2	0	0	1	0	0	0	0	0	1	0	1	0

Fig. 4.9 The right residual describes how rows of the relation Q above the fraction slash contain rows of the second relation P below

We will use the following identities that may be seen as a first test of the quotient properties.

Proposition 4.7. *The following identities hold for arbitrary relations R, X*
$$R_{;}\left(R\backslash(R_{;}X)\right) = R_{;}X \qquad ((X_{;}R)/R)_{;}R = X_{;}R. \qquad \square$$

These results are intuitive. Given a product $R_{;}X$ – observe that the numerator has a rather specific structure – we may divide R from the left and re-multiply it obtaining again the numerator we started with. Instead of a proof, we recall that this means in fact
$$R_{;}\overline{R^{\mathsf{T}}{}_{;}\overline{R_{;}X}} = R_{;}X \qquad \overline{\overline{X_{;}R}_{;}R^{\mathsf{T}}{}_{;}R} = X_{;}R$$
and subsumes to a very general principle that will be explained later in Section 8.5.

Once one has the possibility of comparing, in the way just described, whether a column is contained in another column, one will also try to look for equality of columns, or rows, respectively. The following concept of a symmetric quotient applies a left residual to A as well as to $\overline{A}$ and intersects the results. This leads to the following point-free formulation.

Definition 4.8. Given two possibly heterogeneous relations A, B with coinciding source, we define their **symmetric quotient**[8] as

---

[8] Observe that the symmetric quotient is *not* a symmetric relation, nor need it even be homogeneous. The name describes that it is defined in some symmetric way.

$$\mathsf{syq}\,(A,B) := \overline{A^{\mathsf{T}};\overline{B}} \cap \overline{\overline{A}^{\mathsf{T}};B}. \hspace{3cm} \square$$

The construct syq[9] thus obtained has been used very successfully in various application fields.

The symmetric quotient can be illustrated as follows. For the two relations $R : X \longrightarrow Y$ and $S : X \longrightarrow Z$, it relates an element $y \in Y$ to an element $z \in Z$ precisely when y and z have the same set of 'inverse images' with respect to R or S respectively. Thus

$$(y,z) \in \mathsf{syq}\,(R,S) \quad \Longleftrightarrow \quad \forall x : [(x,y) \in R \leftrightarrow (x,z) \in S].$$

In terms of matrices, the above condition for y and z means that the corresponding columns of A and B are equal; therefore, the symmetric quotient often serves the purpose of set comprehension and set comparison. Finding equal columns i, k of relations R, S is here defined in predicate logic and then transferred to relational form. We consider columns i of R and k of S over all rows n and formulate that column i equals column k:

$$\forall n : \big[(n,i) \in R \leftrightarrow (n,k) \in S\big]$$

$$\hspace{4cm} a \leftrightarrow b = \big[(a \to b) \wedge (b \to a)\big]$$

$$\Longleftrightarrow \quad \forall n : \big[(n,i) \in R \to (n,k) \in S\big] \wedge \big[(n,k) \in S \to (n,i) \in R\big]$$

$$\hspace{3cm} \forall n : \big(a(n) \wedge b(n)\big) = \{\forall n : a(n)\} \wedge \{\forall n : b(n)\}$$

$$\Longleftrightarrow \quad \big\{\forall n : (n,i) \in R \to (n,k) \in S\big\} \wedge \big\{\forall n : (n,k) \in S \to (n,i) \in R\big\}$$

$$\hspace{6cm} a \to b = \neg a \vee b$$

$$\Longleftrightarrow \quad \big\{\forall n : (n,i) \notin R \vee (n,k) \in S\big\} \wedge \big\{\forall n : (n,i) \in R \vee (n,k) \notin S\big\}$$

$$\hspace{5cm} \forall x : \neg p(x) = \overline{\exists x : p(x)}$$

$$\Longleftrightarrow \quad \overline{\exists n : (n,i) \in R \wedge (n,k) \notin S} \ \wedge \ \overline{\exists n : (n,i) \notin R \wedge (n,k) \in S}$$

$$\hspace{6cm} \text{transposing}$$

$$\Longleftrightarrow \quad \overline{\exists n : (i,n) \in R^{\mathsf{T}} \wedge (n,k) \notin S} \ \wedge \ \overline{\exists n : (i,n) \notin R^{\mathsf{T}} \wedge (n,k) \in S}$$

$$\hspace{4cm} \text{definition of composition}$$

$$\Longleftrightarrow \quad (i,k) \in \overline{R^{\mathsf{T}};\overline{S}} \ \wedge \ (i,k) \in \overline{\overline{R}^{\mathsf{T}};S}$$

---

[9] The heterogeneous construct syq seems to have first been published in [19]. It dates back, however, at least to the report [122] of March 1985. The earliest handwritten concepts have been found dating from November 13, 1982. They emerged during the cooperation of the present author with Hans Zierer, aiming at his diploma thesis on relational semantics, finalized May 1983. Many pages of proofs dated July to November 1982 work with the corresponding term, but do not yet employ the algebraic properties of a symmetric quotient, that had first been named syD in a paper of November 3, 1982. Algebraically handled forerunners of syq were the constructs of a univalent part and a multivalent part of a relation in July 1981 [121]. They are described in detail in Section 4.2 of [123, 124]. Even earlier was the *noyau* of Jacques Riguet [105], a special homogeneous and unary case of the symmetric quotient that already anticipated many of its properties.

intersecting relations

$$\Longleftrightarrow \quad (i,k) \in \overline{R^{\mathsf{T}} ; \overline{S}} \cap \overline{\overline{R}^{\mathsf{T}} ; S}$$

definition of symmetric quotient

$$\Longleftrightarrow \quad (i,k) \in \mathsf{syq}(R,S)$$

Obviously, $\mathsf{syq}(A,B) = A \backslash B \cap \overline{A} \backslash \overline{B}$. The symmetric quotient leads to many formulae resembling cancellation. These will be presented in Section 8.5.

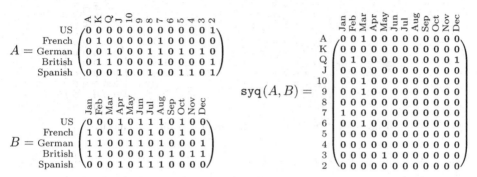

Fig. 4.10 The symmetric quotient shows which columns of the left are equal to columns of the right relation of syq

A closer examination of $\mathsf{syq}(A,B)$ shows that its matrix can be divided into constant boxes after suitably rearranging its rows and columns. One just has to arrange equal columns side by side. Figure 4.11 shows with a different example how this is done. We shall later formulate this property algebraically and discuss some consequences of it.

$$R = \begin{array}{l} \\ \text{US} \\ \text{French} \\ \text{German} \\ \text{British} \\ \text{Spanish} \end{array} \begin{pmatrix} 1&1&0&1&0&1&0&0&1&1&1 \\ 0&0&1&1&0&1&0&1&1&1&0&1 \\ 1&0&1&0&1&0&1&0&0&1&1&0 \\ 0&1&1&1&0&1&0&1&1&1&0&1&1 \\ 0&0&1&1&0&1&0&1&1&1&0&0&1 \end{pmatrix}$$

$$S = \begin{array}{l} \\ \text{US} \\ \text{French} \\ \text{German} \\ \text{British} \\ \text{Spanish} \end{array} \begin{pmatrix} 1&1&1&0&0&1&1&1&1&1&1 \\ 1&0&0&1&1&1&0&0&1&1&0&1 \\ 0&0&1&0&0&0&1&0&0&0&0&0 \\ 1&1&0&1&1&1&0&1&1&1&1&1 \\ 1&0&0&1&1&1&0&0&0&1&0&0 \end{pmatrix}$$

$$\mathsf{syq}(R,S) = \begin{array}{l} 3 \\ K \\ J \\ 6 \\ 9 \\ 7 \\ 10 \\ 4 \\ 8 \\ A \\ 2 \\ 5 \\ Q \end{array} \begin{pmatrix} 1&1&1&0&0&0&0&0&0&0&0&0 \\ 1&1&1&0&0&0&0&0&0&0&0&0 \\ 0&0&0&1&1&0&0&0&0&0&0&0 \\ 0&0&0&1&1&0&0&0&0&0&0&0 \\ 0&0&0&1&1&0&0&0&0&0&0&0 \\ 0&0&0&1&1&0&0&0&0&0&0&0 \\ 0&0&0&0&0&1&1&0&0&0&0&0 \\ 0&0&0&0&0&1&1&0&0&0&0&0 \\ 0&0&0&0&0&1&1&0&0&0&0&0 \\ 0&0&0&0&0&1&1&0&0&0&0&0 \\ 0&0&0&0&0&0&0&1&1&1&0&0 \\ 0&0&0&0&0&0&0&0&0&0&0&0 \\ 0&0&0&0&0&0&0&0&0&0&0&0 \end{pmatrix}$$

Fig. 4.11 Symmetric quotients may always be rearranged to block-diagonal shape

Figure 4.12 shows the details of the intersecting parts of a symmetric quotient.

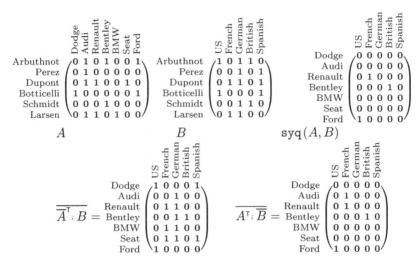

Fig. 4.12 The intersecting parts of a symmetric quotient

More about the algebra underlying this possibility of rearranging will be given in Section 10.4.

Exercises

4.3 Prove the rules relating residuals with their transposed versions:

$$(Q/P)^{\mathsf{T}} = P^{\mathsf{T}} \backslash Q^{\mathsf{T}} \qquad (R/S)^{\mathsf{T}} = S^{\mathsf{T}} \backslash R^{\mathsf{T}}.$$

5

Order and Function: The Standard View

Many relations enjoy specific properties. The most important properties are related to orderings and to functions. That *at most one* or *precisely one* value is assigned to an argument is characteristic of functions or mappings, respectively. We have also developed a multitude of concepts to reason about something that is *better than* or is *more attractive than* something else or *similar to* something else. Concepts related to functions or to orderings lead to a huge pile of formulae and interdependencies. In this chapter, we recall the classics; first in the heterogeneous setting of functions and mappings and afterwards in a homogeneous setting of orderings and equivalences. At the end of this chapter, we will develop our first methods of structural comparison. Although they are necessary in various situations, the names homomorphism, simulation, and congruence often seem to frighten people, giving them the idea that all this may be too theoretical – which it is not.

5.1 Functions

A notational distinction between a (partially defined) function and a mapping (i.e., a totally defined function) is regrettably not commonly made. Many people use both words as synonyms and usually mean a totally defined function. Traditionally, we speak of functions

$$f(x) = \tfrac{1}{x-1} \quad \text{or} \quad g(y) = \sqrt{y - 3}$$

and then discuss separately their peculiar behavior at $x = 1$ or the fact that there exist no, one, or two results for $y < 3, y = 3$, and $y > 3$ – if not assuming it to be the positive branch by convention. Relations typically assign arbitrary sets of results. We introduce a definition to concentrate on the more general case.

Definition 5.1. Let a relation $R : V \longrightarrow W$ be given. We say that the relation

$$R \text{ is } \textbf{univalent} \quad :\Longleftrightarrow \quad \forall x \in V : \forall y, z \in W : [\,(x,y) \in R \wedge (x,z) \in R\,] \rightarrow y = z$$
$$\Longleftrightarrow \quad R^{\mathsf{T}} {\,}_{\scriptscriptstyle\fullmoon}\, R \subseteq \mathbb{I}.$$

Univalent relations are often called (partially defined) **functions**.[1] A relation R is **injective** if its converse is univalent.   □

A univalent relation R associates to every $x \in V$ *at most* one $y \in W$. In slightly more formal terms: assuming that y as well as z should be attached as values to some x, it follows that $y = z$. This is a rather long formulation, which may be shortened to the point-free version below, i.e., with quantifiers and individual variables hidden behind the algebraic formulation. We give a sketch of how the first version of the definition above can successively be made point-free – and thus be transformed into a TITUREL predicate. These transition steps shall only give an idea to the reader how the second shorthand version[2] is invented; it should be stressed that they do not constitute some sort of a 'calculus' we aim at.

$$\forall y, z \in W : \forall x \in V : [(x,y) \in R \wedge (x,z) \in R] \to y = z$$

$$a \to b = \neg a \vee b$$

$$\Longleftrightarrow \quad \forall y, z \in W : \forall x \in V : \overline{(x,y) \in R \wedge (x,z) \in R} \vee y = z$$

$$\forall x : \big(p(x) \vee c\big) = \big\{\forall x : p(x)\big\} \vee c$$

$$\Longleftrightarrow \quad \forall y, z \in W : \Big\{\forall x \in V : \overline{(x,y) \in R \wedge (x,z) \in R}\Big\} \vee y = z$$

$$\forall x : \neg p(x) = \neg \big[\exists x : p(x)\big]$$

$$\Longleftrightarrow \quad \forall y, z \in W : \neg\Big\{\exists x \in V : (x,y) \in R \wedge (x,z) \in R\Big\} \vee y = z$$

$$\text{definition of transposition}$$

$$\Longleftrightarrow \quad \forall y, z \in W : \neg\Big\{\exists x \in V : (y,x) \in R^{\mathsf{T}} \wedge (x,z) \in R\Big\} \vee y = z$$

$$\text{definition of composition}$$

$$\Longleftrightarrow \quad \forall y, z \in W : \neg\big\{(y,z) \in R^{\mathsf{T}}\,{}_{^\vdots} R\big\} \vee y = z$$

$$\neg a \vee b = a \to b \text{ and definition of identity}$$

$$\Longleftrightarrow \quad \forall y, z \in W : (y,z) \in R^{\mathsf{T}}\,{}_{^\vdots} R \to (y,z) \in \mathbb{I}$$

$$\text{transition to point-free version}$$

$$\Longleftrightarrow \quad R^{\mathsf{T}}\,{}_{^\vdots} R \subseteq \mathbb{I}$$

$$\text{transferred to TITUREL}$$

$$\Longleftrightarrow \quad \texttt{(Convs r) :***: r :<==: Ident (tgt r)}$$

One may consider the univalency condition from the point of view of triangles, as in Fig. 5.1. When going back from an image point to one of its arguments and proceeding forward to the image side again, one will always arrive at the point one

---

[1] The property of being univalent or a (partially defined) function is met very frequently and most often radically simplifies the setting. Postulating a (totally defined) mapping brings additional power in reasoning, but not too much; therefore functions deserve to be studied separately.

[2] From [38], we cite a remark of Alfred North Whitehead: ... *by relieving the brain of all unnecessary work, a good notation sets it free to concentrate on more advanced problems, and, in effect, increases the mental power* .... The book then continues: *Any statement in formal logic such as ... can, in principle, be expanded back into primitive atomic form. In practice, this cannot be carried out, because the symbol strings quickly become so long that errors in reading and processing become unavoidable.*

started from – if the relation is assumed to be univalent. The thin bent arrow with a double-arrow pointer will help to indicate which direction of reasoning is intended.

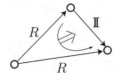

Fig. 5.1 Triangle to define the essence of being univalent

When considering a univalent relation R as a function f_R, it is typically denoted in prefix notation:

$$f_R(x) = y \quad \text{respectively} \quad f_R(x) = \text{undefined}$$

instead of expressing it as

$$(x, y) \in R \quad \text{respectively} \quad \forall y \in W : (x, y) \notin R.$$

	0	1	2	3	4
0					
1	×				
2		×			
3			×		
4				×	

$$\begin{pmatrix} & 0 & 1 & 2 & 3 & 4 \\ 0 & 0 & 0 & 0 & 0 & 0 \\ 1 & 1 & 0 & 0 & 0 & 0 \\ 2 & 0 & 1 & 0 & 0 & 0 \\ 3 & 0 & 0 & 1 & 0 & 0 \\ 4 & 0 & 0 & 0 & 1 & 0 \end{pmatrix}$$

0	undefined
1	0
2	1
3	2
4	3

Fig. 5.2 Table, matrix and table of values for a (partially defined) function

Already from school it is known that composing two mappings results in a mapping again. Now we learn that two functions need not be defined everywhere; composition of (partially defined) functions will always result in (partially defined) functions.

Proposition 5.2 (Univalency).

(i) $Q \,\fatsemi\, R$ is univalent, whenever Q and R are.
(ii) R univalent $\iff$ $R \,\fatsemi\, \overline{\mathbb{I}} \subseteq \overline{R}$.
(iii) $R \subseteq Q$, Q univalent, $R \,\fatsemi\, \mathbb{T} \supseteq Q \,\fatsemi\, \mathbb{T}$ $\implies$ $R = Q$.

Proof: See B.1. □

To interpret (ii), we simply read the formula with the help of Fig. 5.3. Whenever we follow some transition offered by a univalent relation R, and then proceed to a *different* element on the image side, we can be sure that this element cannot be reached directly via R from our starting point. For an intuitive visualization the so-called **dashed arrow convention** has been used: a dashed arrow is conceived to be a *forbidden arrow*; it belongs, thus, to the *negated* relation.

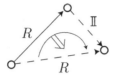

Fig. 5.3 Diagram characterizing the essence of being univalent according to Prop. 5.2.ii

Also (iii) admits such an explanation. If a relation R is contained in a (partially defined) function Q, but is defined for at least as many arguments, then the two will coincide. We recall that relations $R : X \longrightarrow Y$ are here studied with types, i.e., with source $X = \mathtt{src}\,R$ and target $Y = \mathtt{tgt}\,R$. The domain $R_{;}\mathbb{T}_Y \subseteq \mathbb{T}_X$ of a relation must be distinguished from its source, which is often not self-evident when studying relations in a homogeneous environment.

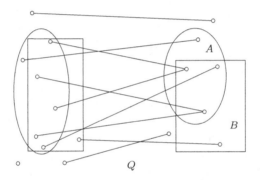

Fig. 5.4 Inverse image of an intersection for a univalent relation

We now study images and inverse images of intersections with respect to univalent relations. For a (partially defined) function, i.e., for a univalent relation, the inverse image of an intersection of images equals the intersection of the inverse images as in Fig. 5.4. This is what the following proposition expresses. Precisely, this distributivity[3] simplifies many sequences of reasoning.

Proposition 5.3. Q *univalent* $\implies$ $Q_{;}(A \cap B) = Q_{;}A \cap Q_{;}B.$

Proof: See B.2. □

---

[3] To what extent such point-free formulae are unknown even to specialists may be estimated from the fact that they were simply wrong as stated in the broadly known formula collection Bronstein–Semendjajew [30] – at least in the author's 24th edition of 1979.

Proposition 5.3 is a rather clear statement while Fig. 5.4, when considered point-wise, is confusing.

A set of persons among which the women may have given birth to other persons will give further intuition. If every woman in the group considered has had at most one child, the following is clear:

'women with blond son' = 'women with a son' ∩ 'women with blond child'

In the case that there exist women with more than one child, it is possible that "⊆" but "≠" holds instead. Indeed, there may exist a woman with a dark-haired son and in addition a blond daughter.

We now give this example a slightly different flavor and consider a set of persons among which may be pregnant women. The set

'Gives birth to a person that is at the same time a son and a daughter'

will be empty – hermaphrodites are assumed to be excluded. In contrast, the set of women obtained by the intersection

'Gives birth to a son' ∩ 'Gives birth to a daughter'

need not be empty in the case of certain twin pregnancies; i.e., if Q is not univalent.

Proposition 5.3 was about *inverse images* of an intersection. Images of an intersection are a bit more difficult to handle. Only when one of the intersecting constituents is 'cylindric over its image', do we find a nice formula to relate both sides. Then the image of the intersection equals the image of the other constituent intersected with the basis of the cylinder.

Proposition 5.4. Q *univalent* $\implies$ $A \cap B_;Q = (A_;Q^\mathsf{T} \cap B)_;Q.$

Proof: See B.3. $\qquad\qquad\qquad\qquad\qquad\qquad\qquad\qquad\qquad\qquad\qquad\qquad\quad$ □

Figure 5.5 presents an example before moving on to the more extended Example 5.5.

Example 5.5. To illustrate the idea behind Prop. 5.4, we provide an example which is easier to grasp in Fig. 5.6. Let A, B resemble sets, however, in different representations. Let Q denote the relation assigning to persons their nationalities – assuming that everybody has at most one. Then assume A as a column-constant relation derived from the subset of European nations and B as a diagonal-shaped relation derived from the subset of women among the persons considered. So $A_;Q^\mathsf{T}$ describes persons with European nationality and $B_;Q$ are the nations assigned to the women in the set of persons considered. The result will now give the same relation for the following two operations:

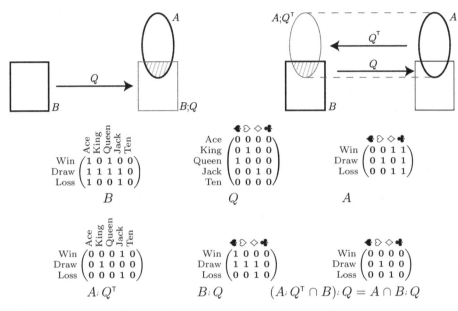

Fig. 5.5 Image of a 'cylindric' intersection

— 'determine European persons that are female and relate them to their nationalities',
— 'determine the nationalities for women and extract those that are European'.

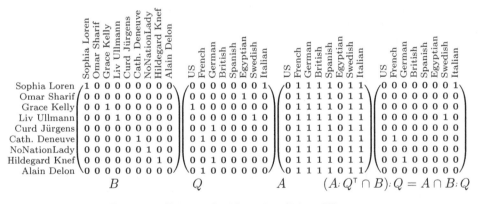

Fig. 5.6 Prop. 5.4 illustrated with nationalities of European women

Assuming in a purely theoretical way Cathérine Deneuve to have French *and also* US citizenship, thus making Q a non-univalent relation, the first set would be bigger than the second.   □

It is obviously difficult to make the intention of Prop. 5.4 clear in spoken language. In a sense, the algebraic form is the only way to make such things precise – with the additional advantage that they then may be formally manipulated.

Now we study the special way in which a univalent relation behaves when composed in combination with negation. In Prop. 5.6, we compose a univalent relation Q from the left-hand side with the complement of A. By the univalent relation Q elements are assigned *at most one* image; we consider the case of an existing and a non-existing image separately. If an element of the source has no image, the row of the composed matrix is empty, and when complementing, this results in a row full of 1s. If an image exists, the row in the product is precisely the row of the second factor determined by this image; the result stays the same if it is complemented either before or after composition. Already now, we expect this rule to simplify when it is restricted to totally defined functions, i.e., mappings; see Prop. 5.13.

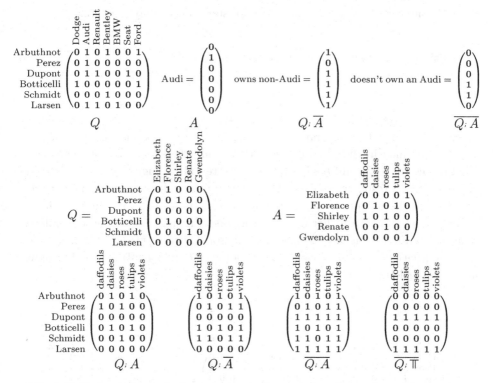

Fig. 5.7 Non-univalent relation versus univalent relation Q together with negation

Proposition 5.6.

(i) Q *univalent* $\implies$ $Q{;}\overline{A} = Q{;}\mathbb{T} \cap \overline{Q{;}A}$.

(ii) Q *univalent* $\implies$ $\overline{Q\,;A} = Q\,;\overline{A} \cup \overline{Q\,;\mathbb{T}}.$

Proof: See B.4.   □

Assume a set of persons owning at most one car each – described by a univalent relation Q. Now consider those who own a car that is not an Audi resembling $Q\,;\overline{A}$. They certainly own a car, $Q\,;\mathbb{T}$, and it is not the case that they own an Audi as expressed by $\overline{Q\,;A}$. Should there, as in Fig. 5.7, exist persons owning more than one car, the equation may not hold. A person may own an Audi as well as a BMW; in this case, $Q\,;\overline{A}$ may be true although $\overline{Q\,;A}$ is not.

When we consider the univalent relation Q, we easily see that one may toggle between $Q\,;\overline{A}$ and $\overline{Q\,;A}$ by simply adding $\overline{Q\,;\mathbb{T}}$ or intersecting with $Q\,;\mathbb{T}$.

There is one further concept that will often show up, namely matchings.

Definition 5.7. A relation λ will be called a **matching** if it is at the same time univalent and injective.   □

Later, matchings are mainly presented as Q-matchings when a (possibly) heterogeneous relation Q of, for example, *sympathy* is given and one looks for a matching $\lambda \subseteq Q$ contained in the given relation; this would then be a possible marriage. A local match-maker would obviously be interested in arranging as many marriages as possible in a respective village. The number of matched pairs is, however, in no way part of the definition of a matching. So the empty relation contained in Q is always a matching – a rather uninteresting one. Figure 5.8 shows an example of a relation Q, a matching $\lambda \subseteq Q$, and this relation Q arranged so as to visualize the one-to-one situation in an obvious form as a (often only partial) diagonal beginning at the upper left.

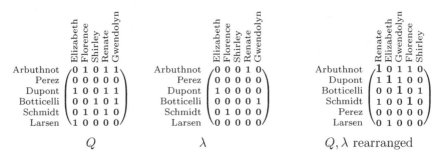

Fig. 5.8 Sympathy Q that leads to marriages λ; in original and in rearranged form

5.2 Mappings and points

From now on, we no longer accept that the functions are just partially defined; we demand instead that they be totally defined. Then some more specialized formulae hold.

Definition 5.8. Let R be a possibly heterogeneous relation.[4]

  (i) R **total** $:\Longleftrightarrow$ $\forall x \in X, \exists y \in Y : (x,y) \in R$ $\Longleftrightarrow$ $\mathbb{T} \subseteq R \,\mathbb{;}\, \mathbb{T}$.

  (ii) R **surjective** $:\Longleftrightarrow$ R^{T} total.

  (iii) A **mapping** is a total and univalent relation.

  (iv) A relation is **bijective** if it is a surjective and injective relation.   □

Total relations and mappings may also be characterized by some other properties; the proofs are left to the reader, who may also consult Section 4.2 of [123, 124].

Proposition 5.9. *Let a possibly heterogeneous relation R be given.*

  (i) R **total** $\Longleftrightarrow$ $\mathbb{T} = R \,\mathbb{;}\, \mathbb{T}$ $\Longleftrightarrow$ $\mathbb{I} \subseteq R \,\mathbb{;}\, R^{\mathsf{T}}$ $\Longleftrightarrow$ $\overline{R} \subseteq R \,\mathbb{;}\, \overline{\mathbb{I}}$

                        $\Longleftrightarrow$ *For all relations S, from $S \,\mathbb{;}\, R = \mathbb{\bot}$ follows $S = \mathbb{\bot}$.*

  (ii) R **surjective** $\Longleftrightarrow$ $\mathbb{T} = \mathbb{T} \,\mathbb{;}\, R$ $\Longleftrightarrow$ $\mathbb{I} \subseteq R^{\mathsf{T}} \,\mathbb{;}\, R$ $\Longleftrightarrow$ $\overline{R} \subseteq \overline{\mathbb{I}} \,\mathbb{;}\, R$

                        $\Longleftrightarrow$ *For all relations S, from $R \,\mathbb{;}\, S = \mathbb{\bot}$ follows $S = \mathbb{\bot}$.*

  (iii) R **mapping** $\Longleftrightarrow$ $R \,\mathbb{;}\, \overline{\mathbb{I}} = \overline{R}$.   □

Another result is immediate: composition of two mappings, thus, results in a mapping again as is now stated formally.

Proposition 5.10. *$Q \,\mathbb{;}\, R$ is a mapping, provided Q and R are mappings.*   □

The concepts of surjectivity and injectivity now being available, we may proceed to define using algebraical methods what intuitively corresponds to an element. This is why we have chosen to denote it with a lower-case letter.

Definition 5.11. Let a relation $x : V \longrightarrow W$ be given. We call

---

[4] There is a point that should be mentioned here, although it is of merely theoretical interest. While in the predicate-logic form of (i) the target Y is explicitly mentioned, the question arises of whether this should or must also be the case in the relation-algebraic version. In more detail: should this read $\mathbb{T}_{X,Y} = R \,\mathbb{;}\, \mathbb{T}_{Y,Y}$, or $\forall Z : \mathbb{T}_{X,Z} = R \,\mathbb{;}\, \mathbb{T}_{Y,Z}$? While these seem to mean the same, they do not. There exist relation algebras in which not necessarily $\mathbb{T}_{A,B} \,\mathbb{;}\, \mathbb{T}_{B,C} = \mathbb{T}_{A,C}$!

(i) v a **vector** $:\iff$ v is row-constant, i.e., $v = v; \mathbb{T}$,

(ii) x a **point** $:\iff$ x is row-constant, injective, and surjective. □

One will find out that a point is something like a transposed mapping of the 1-element set $\mathbb{1}$ into V. Once one has decided to reason point-free, working with elements or points becomes more intricate than one would expect. Observe that the definition given here is slightly more restricted compared to the one chosen in our general reference texts [123, 124].[5] The traditional definition of a point given there reads as follows: x is a point if it is row-constant ($x = x; \mathbb{T}$), injective ($x; x^\mathsf{T} \subseteq \mathbb{I}$), and non-zero ($\mathbb{\perp} \neq x$).[6]

	Win	Draw	Loss
Monday	0	0	0
Tuesday	1	1	1
Wednesday	0	0	0
Thursday	1	1	1
Friday	1	1	1
Saturday	0	0	0
Sunday	0	0	0

	Win	Draw	Loss
Monday	0	0	0
Tuesday	1	1	1
Wednesday	0	0	0
Thursday	0	0	0
Friday	0	0	0
Saturday	0	0	0
Sunday	0	0	0

	January	February	March	April	May	June	July	August	September	October	November	December
Monday	0	0	0	0	0	0	0	0	0	0	0	0
Tuesday	1	1	1	1	1	1	1	1	1	1	1	1
Wednesday	0	0	0	0	0	0	0	0	0	0	0	0
Thursday	0	0	0	0	0	0	0	0	0	0	0	0
Friday	0	0	0	0	0	0	0	0	0	0	0	0
Saturday	0	0	0	0	0	0	0	0	0	0	0	0
Sunday	0	0	0	0	0	0	0	0	0	0	0	0

Fig. 5.9 A vector and two row-constant relations that may be considered the same point

On the face if it, the following is a trivial result. However, the proof contains more mathematical delicacy than one would expect at first sight. One is required to employ the equivalence of the two characterizations of totality of f, namely $\mathbb{I} \subseteq f; f^\mathsf{T}$ and $\mathbb{T} \subseteq f; \mathbb{T}$.

Proposition 5.12 (Shunting). *Let R, S be relations for which the following constructs in connection with x, y and f exist.*

(i) *If f is a mapping,* $R \subseteq S; f^\mathsf{T} \iff R; f \subseteq S$.

(ii) *If x is a point,* $R \subseteq S; x \iff R; x^\mathsf{T} \subseteq S$.

(iii) *If x, y are points,* $y \subseteq S; x \iff x \subseteq S^\mathsf{T}; y$.

Proof: See B.5 □

[5] The new form does not make any difference for the practical applications we aim at. It makes a difference for certain models of relation algebra. There exist relation algebras that follow all the rules explained here, but which are non-standard. This means not least that one can prove that they contain elements that cannot be conceived as a relation as we know it – much in the same way as there exist non-Euclidian geometries.

[6] The form chosen here avoids what is known as the Tarski rule.

The interpretation of this result (which is often referred to as 'shunting' the mapping f) using Fig. 5.10 runs as follows. The relations given are R, f, S indicating the cars owned, the country where the respective car is produced, and the preference of the owner as to the country his car should come from. We see that everybody has bought cars following his preference, $R \, ; f \subseteq S$. Should an owner intend to buy yet another car, he might wish to consult $S \, ; f^\mathsf{T}$ which indicates the car types restricted by his preferences.

$$R = \begin{matrix} \text{Arbuthnot} \\ \text{Perez} \\ \text{Dupont} \\ \text{Botticelli} \\ \text{Schmidt} \\ \text{Larsen} \end{matrix} \begin{pmatrix} 0 & 1 & 0 & 1 & 0 & 0 & 1 \\ 0 & 1 & 0 & 0 & 0 & 0 & 0 \\ 0 & 1 & 1 & 0 & 0 & 1 & 0 \\ 1 & 0 & 0 & 0 & 0 & 0 & 1 \\ 0 & 0 & 0 & 1 & 0 & 0 & 0 \\ 0 & 1 & 1 & 0 & 1 & 0 & 0 \end{pmatrix}$$

$$f = \begin{matrix} \text{Dodge} \\ \text{Audi} \\ \text{Renault} \\ \text{Bentley} \\ \text{BMW} \\ \text{Fiat} \\ \text{Ford} \end{matrix} \begin{pmatrix} 1 & 0 & 0 & 0 & 0 \\ 0 & 0 & 1 & 0 & 0 \\ 0 & 1 & 0 & 0 & 0 \\ 0 & 0 & 0 & 1 & 0 \\ 0 & 0 & 1 & 0 & 0 \\ 0 & 0 & 0 & 0 & 1 \\ 1 & 0 & 0 & 0 & 0 \end{pmatrix}$$

$$S = \begin{matrix} \text{Arbuthnot} \\ \text{Perez} \\ \text{Dupont} \\ \text{Botticelli} \\ \text{Schmidt} \\ \text{Larsen} \end{matrix} \begin{pmatrix} 1 & 0 & 1 & 1 & 0 \\ 0 & 0 & 1 & 0 & 1 \\ 0 & 1 & 1 & 0 & 1 \\ 1 & 0 & 0 & 0 & 1 \\ 0 & 0 & 1 & 1 & 0 \\ 0 & 1 & 1 & 0 & 0 \end{pmatrix}$$

$$R \, ; f = \begin{matrix} \text{Arbuthnot} \\ \text{Perez} \\ \text{Dupont} \\ \text{Botticelli} \\ \text{Schmidt} \\ \text{Larsen} \end{matrix} \begin{pmatrix} 1 & 0 & 1 & 1 & 0 \\ 0 & 0 & 1 & 0 & 0 \\ 0 & 1 & 1 & 0 & 1 \\ 1 & 0 & 0 & 0 & 0 \\ 0 & 0 & 0 & 1 & 0 \\ 0 & 1 & 1 & 0 & 0 \end{pmatrix}$$

$$S \, ; f^\mathsf{T} = \begin{matrix} \text{Arbuthnot} \\ \text{Perez} \\ \text{Dupont} \\ \text{Botticelli} \\ \text{Schmidt} \\ \text{Larsen} \end{matrix} \begin{pmatrix} 1 & 1 & 0 & 1 & 1 & 0 & 1 \\ 0 & 1 & 0 & 0 & 1 & 1 & 0 \\ 0 & 1 & 1 & 0 & 1 & 1 & 0 \\ 1 & 0 & 0 & 0 & 0 & 1 & 1 \\ 0 & 1 & 0 & 1 & 1 & 0 & 0 \\ 0 & 1 & 1 & 0 & 1 & 0 & 0 \end{pmatrix} = \overline{\overline{S} \, ; f^\mathsf{T}}$$

Fig. 5.10 Rolling a mapping to the other side of a containment

Anticipating the following proposition, we mention in passing that $S \, ; f^\mathsf{T} = \overline{\overline{S} \, ; f^\mathsf{T}}$ for a mapping f. The slightly more complicated construct would allow us to handle also the case of cars not stemming from just one country, i.e., f not necessarily a mapping.

We add one more fact on mappings, specializing Prop. 5.6: for a mapping, the inverse image of the complement is equal to the complement of the inverse image. In more primitive words: a mapping may slip below a negation bar when multiplied from the left – and a transposed mapping when multiplied from the right.

Proposition 5.13. *Let A, R be possibly heterogeneous relations:*

(i) f *mapping* $\implies$ $f \, ; \overline{A} = \overline{f \, ; A}$,

(ii) x *is a point* $\implies$ $\overline{R \, ; x} = \overline{R} \, ; x$. $\square$

In view of Prop. 5.6, this result is so evident that it does not need a written proof. We visualize this result with the two situations of Fig. 5.11. On the left, a mapping f is shown while on the right the fat arrows show that Q is not a mapping since it is not univalent.

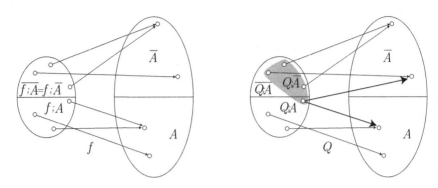

Fig. 5.11 Inverse of the complement equals the complement of the inverse for a mapping

With these concepts we will now handle a classic riddle (see [35]). One man looks at a second man and says:

'Brothers and sisters I have none, but that man's father is my father's son.'

We need, of course, relations such as *child-of, son-of*, or here, f meaning *has-as-father*, that follow the intended scheme, meaning that they are irreflexive, for example. They are assumed to hold on a set of individuals, from which a, b are two different elements. Brothers or sisters of a person we get with $f \,; f^\mathsf{T}$. Since a has neither brothers nor sisters we have $f \,; f^\mathsf{T} \cap a \,; \mathbb{T} \subseteq \mathbb{I}$, meaning that there is just one when looking for those with the same father as a. The question is about the relationship between a and b specified as $f \,; f^\mathsf{T} \,; f^\mathsf{T} \cap a \,; \mathbb{T}$. Starting from b, we have to proceed with f to his father and then again with f to a father (because he is qualified as a son) and finally with f^T to a because it is supposed to be a's father. This product is, however, transposed in order to go from a to b:

$$f \,; f^\mathsf{T} \,; f^\mathsf{T} \cap a \,; \mathbb{T} \subseteq (f \,; f^\mathsf{T} \cap a \,; \mathbb{T} \,; f) \,; (f^\mathsf{T} \cap f \,; f^\mathsf{T} \,; a \,; \mathbb{T}) \subseteq \mathbb{I} \,; f^\mathsf{T} = f^\mathsf{T}.$$

Applying the Dedekind rule, we get that the other person is the son of a.

Exercises

5.1 Prove that a relation R is a mapping precisely when there exists a relation S satisfying $S \,; R \subseteq \mathbb{I}$ and $\mathbb{I} \subseteq R \,; S$.

5.2 Prove that a finite homogeneous relation R satisfying $R \,; S = \mathbb{I}$ for some S is necessarily a permutation with $R^\mathsf{T} = S$. Provide an example showing that finiteness is necessary.

5.3 Order and strictorder

Until the 1930s, orderings were usually studied along the linearly ordered real axis $\mathbb{R}$, but soon diverse other concepts such as weakorder, semiorder, intervalorder, etc. were also handled. Here the basic properties of orderings and equivalences are recalled in predicate form as well as in a point-free manner. The transition between the two levels is more or less immediate. We have also included several variants of the algebraic forms of the definitions, which are so obviously equivalent that they do not justify mention in separate propositions. Later, in Chapter 12, unifying but more advanced concepts will be presented.

Reflexivity

Identity on a set is a homogeneous relation; source and target are of the same type. For homogeneous relations some properties in connection with the identity are standard.

Definition 5.14 (Reflexivity properties). Given an arbitrary homogeneous relation $R : V \longrightarrow V$, we call

$$R \text{ \textbf{reflexive}} \quad :\Longleftrightarrow \quad \forall x \in V : (x,x) \in R \quad :\Longleftrightarrow \quad \mathbb{I} \subseteq R,$$

$$R \text{ \textbf{irreflexive}}[7] \quad :\Longleftrightarrow \quad \forall x \in V : (x,x) \notin R \quad :\Longleftrightarrow \quad R \subseteq \overline{\mathbb{I}}. \qquad \square$$

It should be immediately clear in which way the identity or diagonal relation models that all the pairs (x,x) are in relation R – or are not. A relation on the 1-set is either reflexive or irreflexive, and there is no third possibility as in the example of Fig. 5.12.

$$\begin{pmatrix} 1 & 1 & 1 & 1 \\ 0 & 1 & 0 & 0 \\ 1 & 1 & 1 & 1 \\ 0 & 1 & 0 & 1 \end{pmatrix} \qquad \begin{pmatrix} 1 & 0 & 0 & 1 \\ 0 & 0 & 0 & 0 \\ 1 & 0 & 0 & 1 \\ 0 & 0 & 0 & 1 \end{pmatrix} \qquad \begin{pmatrix} 0 & 0 & 0 & 1 \\ 1 & 0 & 1 & 1 \\ 1 & 1 & 0 & 1 \\ 0 & 0 & 0 & 0 \end{pmatrix}$$

Fig. 5.12 Relations that are reflexive, neither/nor, and irreflexive

Let R be an arbitrary relation. Then the residual R/R is trivially reflexive. Because $R/R = \overline{\overline{R} \,; R^\mathsf{T}}$, this means after negation that $\overline{R} \,; R^\mathsf{T} \subseteq \overline{\mathbb{I}}$; this, however, follows directly from Schröder's equivalences since $\mathbb{I} \,; R \subseteq R$. We have $(r_1, r_2) \in R/R$ precisely when row r_1 is equal to or contains row r_2. Similarly, $(c_1, c_2) \in R\backslash R$ if and only if column c_1 is equal to or is contained in column c_2. The residuals thus clearly contain the full diagonal.

---

[7] In French sometimes *antiréflexif*; see [94].

$$
\begin{array}{c}
\begin{array}{ccccc} \scriptstyle 1 & \scriptstyle 2 & \scriptstyle 3 & \scriptstyle 4 & \scriptstyle 5 \end{array} \\
\begin{array}{c} 1 \\ 2 \\ 3 \\ 4 \\ 5 \end{array}
\left(\begin{array}{ccccc}
1 & 1 & 1 & 1 & 1 \\
0 & 1 & 0 & 0 & 0 \\
0 & 1 & 1 & 0 & 0 \\
0 & 1 & 0 & 1 & 1 \\
0 & 1 & 0 & 1 & 1
\end{array}\right)
\end{array}
\quad
\begin{array}{c}
\begin{array}{ccccc} \scriptstyle 1 & \scriptstyle 2 & \scriptstyle 3 & \scriptstyle 4 & \scriptstyle 5 \end{array} \\
\begin{array}{c} 1 \\ 2 \\ 3 \\ 4 \\ 5 \end{array}
\left(\begin{array}{ccccc}
1 & 0 & 0 & 1 & 0 \\
0 & 0 & 0 & 0 & 0 \\
1 & 0 & 0 & 0 & 0 \\
0 & 0 & 0 & 1 & 0 \\
0 & 0 & 0 & 1 & 0
\end{array}\right)
\end{array}
\quad
\begin{array}{c}
\begin{array}{ccccc} \scriptstyle 1 & \scriptstyle 2 & \scriptstyle 3 & \scriptstyle 4 & \scriptstyle 5 \end{array} \\
\begin{array}{c} 1 \\ 2 \\ 3 \\ 4 \\ 5 \end{array}
\left(\begin{array}{ccccc}
1 & 0 & 0 & 0 & 0 \\
1 & 1 & 1 & 1 & 1 \\
1 & 1 & 1 & 1 & 1 \\
0 & 0 & 0 & 1 & 0 \\
1 & 1 & 1 & 1 & 1
\end{array}\right)
\end{array}
$$

Fig. 5.13 Example relations $R/R, \quad R, \quad R\backslash R$

One will also see that $(R/R)\,{}_\vartriangleright\,R \subseteq R$ and $R\,{}_\vartriangleright\,(R\backslash R) \subseteq R$. The relation R/R is the greatest among the relations X with $X\,{}_\vartriangleright\,R \subseteq R$. Also $R\backslash R$ is the greatest among the relations Y with $R\,{}_\vartriangleright\,Y \subseteq R$. So indeed, some sort of calculation with fractions is possible when interpreting "/" and "\" as division symbols and comparing with $\frac{2}{3}$. Relations do not multiply commutatively, so that we have to indicate on which side division has taken place. One will find similarity of notation with $\frac{2}{3} \times 3 = 2$ and $3 \times \frac{2}{3} = 2$.

It is worth noting that by simultaneous permutation of rows and columns a block-staircase form may be obtained for the left and the right residuals. The original R underwent different permutations for rows and columns. We will later find out that this is not just incidental.

$$
\begin{array}{c}
\begin{array}{ccccc} \scriptstyle 1 & \scriptstyle 3 & \scriptstyle 4 & \scriptstyle 5 & \scriptstyle 2 \end{array} \\
\begin{array}{c} 1 \\ 3 \\ 4 \\ 5 \\ 2 \end{array}
\left(\begin{array}{c|c|cc|c}
1 & 1 & 1 & 1 & 1 \\
\hline
0 & 1 & 0 & 0 & 1 \\
\hline
0 & 0 & 1 & 1 & 1 \\
0 & 0 & 1 & 1 & 1 \\
\hline
0 & 0 & 0 & 0 & 1
\end{array}\right)
\end{array}
\quad
\begin{array}{c}
\begin{array}{ccccc} \scriptstyle 2 & \scriptstyle 3 & \scriptstyle 5 & \scriptstyle 1 & \scriptstyle 4 \end{array} \\
\begin{array}{c} 1 \\ 3 \\ 4 \\ 5 \\ 2 \end{array}
\left(\begin{array}{ccc|c|c}
0 & 0 & 0 & 1 & 1 \\
\hline
0 & 0 & 0 & 1 & 0 \\
\hline
0 & 0 & 0 & 0 & 1 \\
0 & 0 & 0 & 0 & 1 \\
\hline
0 & 0 & 0 & 0 & 0
\end{array}\right)
\end{array}
\quad
\begin{array}{c}
\begin{array}{ccccc} \scriptstyle 2 & \scriptstyle 3 & \scriptstyle 5 & \scriptstyle 1 & \scriptstyle 4 \end{array} \\
\begin{array}{c} 2 \\ 3 \\ 5 \\ 1 \\ 4 \end{array}
\left(\begin{array}{ccc|c|c}
1 & 1 & 1 & 1 & 1 \\
1 & 1 & 1 & 1 & 1 \\
1 & 1 & 1 & 1 & 1 \\
\hline
0 & 0 & 0 & 1 & 0 \\
\hline
0 & 0 & 0 & 0 & 1
\end{array}\right)
\end{array}
$$

Fig. 5.14 Permuted forms of $R/R, \quad R, \quad R\backslash R$

The following property asserts whether a relation together with its converse fills the whole matrix. This property also requires homogeneous relations since we demand that R and R^{T} are of the same type, or 'may be united'.

Definition 5.15 (Connexity properties). Let a homogeneous relation R be given.

R **semi-connex**	$:\Longleftrightarrow$	$\forall x, y : x \neq y \rightarrow \big[(x,y) \in R \lor (y,x) \in R\big]$
	$\Longleftrightarrow$	$\overline{\mathbb{I}} \subseteq R \cup R^{\mathsf{T}},$
R **connex**	$:\Longleftrightarrow$	$\forall x, y : (x,y) \notin R \rightarrow (y,x) \in R$
	$\Longleftrightarrow$	$\mathbb{T} \subseteq R \cup R^{\mathsf{T}}.$

In the literature, semi-connex will often be found denoted as *complete* and connex[8] as *strongly complete*. A relation R is connex precisely when $\overline{R} \subseteq R^{\mathsf{T}}$. $\qquad\square$

---

[8] In French, one may also find *relation totale* or in view of tournaments, a *match*; see [94].

The formal definition does not say anything about the diagonal of a semi-connex relation; nor does it require just one or both of (x, y) and (y, x) to be contained in the relation R.

$$\begin{pmatrix} 0 & 0 & 0 & 1 \\ 1 & 1 & 1 & 1 \\ 1 & 0 & 0 & 1 \\ 0 & 1 & 0 & 0 \end{pmatrix} \qquad \begin{pmatrix} 1 & 1 & 1 & 1 \\ 0 & 1 & 0 & 0 \\ 1 & 1 & 1 & 1 \\ 0 & 1 & 0 & 1 \end{pmatrix}$$

Fig. 5.15 A semi-connex and a connex relation

Symmetry

Relations are also traditionally classified according to their symmetry properties. Again, we have from the definition that the relation and its converse are of the same type, which makes them homogeneous relations.

Definition 5.16 (Symmetry properties). Given again a homogeneous relation $R : V \longrightarrow V$, we call

R **symmetric**	$:\Longleftrightarrow$	$\forall x, y : (x, y) \in R \to (y, x) \in R$
	$\Longleftrightarrow$	$R^{\mathsf{T}} \subseteq R \quad \Longleftrightarrow \quad R^{\mathsf{T}} = R,$
R **asymmetric**	$:\Longleftrightarrow$	$\forall x, y : (x, y) \in R \to (y, x) \notin R$
	$\Longleftrightarrow$	$R \cap R^{\mathsf{T}} \subseteq \mathbb{1} \quad \Longleftrightarrow \quad R^{\mathsf{T}} \subseteq \overline{R},$
R **antisymmetric**	$:\Longleftrightarrow$	$\forall x, y : x \neq y \to \{(x, y) \notin R \vee (y, x) \notin R\}$
	$\Longleftrightarrow$	$R \cap R^{\mathsf{T}} \subseteq \mathbb{I} \quad \Longleftrightarrow \quad R^{\mathsf{T}} \subseteq \overline{R} \cup \mathbb{I}.$ □

A first minor consideration relates reflexivity with symmetry as follows. An asymmetric relation R is necessarily irreflexive. This results easily from the predicate-logic version of the definition of asymmetry when specializing the two variables to $x = y$. Then $\forall x : (x, x) \in R \to (x, x) \notin R$ which implies that $(x, x) \in R$ cannot be satisfied for any x.

$$\begin{pmatrix} 0 & 1 & 1 & 0 \\ 0 & 0 & 0 & 0 \\ 0 & 1 & 0 & 1 \\ 0 & 1 & 0 & 0 \end{pmatrix} \qquad \begin{pmatrix} 0 & 0 & 0 & 1 \\ 1 & 1 & 1 & 0 \\ 0 & 0 & 0 & 0 \\ 0 & 1 & 0 & 1 \end{pmatrix}$$

Fig. 5.16 An asymmetric and an antisymmetric relation

The following observation will often be used, not least when working with orderings and in preference modelling.

Proposition 5.17. *Every homogeneous relation R may be decomposed into an asymmetric part $A := R \cap \overline{R}^{\mathsf{T}}$ and a symmetric part $S := R \cap R^{\mathsf{T}}$, such that $R = A \cup S$ and $\mathbb{1} = A \cap S$.*

Proof: Obviously, $R = R \cap \mathbb{T} = R \cap (\overline{R}^{\mathsf{T}} \cup R^{\mathsf{T}}) = (R \cap \overline{R}^{\mathsf{T}}) \cup (R \cap R^{\mathsf{T}}) =: A \cup S$. $\square$

Symmetry concepts enable us to define

R **tournament** $\quad :\Longleftrightarrow \quad R$ is asymmetric and semi-connex $\quad \Longleftrightarrow \quad R \cup R^{\mathsf{T}} = \overline{\mathbb{I}}.$

It is indeed typical in sports tournaments that a team cannot play against itself. Just one round is assumed to take place and the draw result is assumed to be impossible. Then a tournament describes a possible outcome of win or loss.

Transitivity

Transitivity is used in many application areas; it is central for defining an ordering and an equivalence. We recall the definition more formally and give its algebraic form.

Definition 5.18. Let R be a homogeneous relation. We call

R **transitive** $\quad :\Longleftrightarrow \quad \forall x, y, z \in V : \{(x,y) \in R \wedge (y,z) \in R\} \rightarrow (x,z) \in R$
$\qquad\qquad\qquad \Longleftrightarrow \quad R \,\mathbin{;}\, R \subseteq R.$ $\square$

Intuitively, a relation R is transitive if whenever x is related to y and y is related to z, then also x will be related to z.

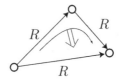

Fig. 5.17 Triangular situation for transitivity

We observe the triangle structure in Fig. 5.17 and show how the two rather different looking variants of the definition are formally related. On the right we roughly indicate the well-known rules we have applied for the respective transition.

$$\forall x, y, z \in V : \Big[(x,y) \in R \wedge (y,z) \in R\Big] \rightarrow (x,z) \in R$$

$$a \rightarrow b = \neg a \vee b, \text{ arranging quantification}$$

$$\Longleftrightarrow \quad \forall x, z \in V : \forall y \in V : \Big\{\overline{(x,y) \in R \wedge (y,z) \in R} \;\vee\; (x,z) \in R\Big\}$$

$$\forall v : \big[p(v) \vee c\big] = \big[\forall v : p(v)\big] \vee c$$

$$\Longleftrightarrow \quad \forall x, z \in V : \Big\{\forall y \in V : \overline{(x,y) \in R \wedge (y,z) \in R}\Big\} \vee (x,z) \in R$$

$$\forall v : \neg p(v) = \neg \big[\exists v : p(v)\big]$$

$$\Longleftrightarrow \quad \forall x, z \in V : \neg \Big\{\exists y \in V : (x,y) \in R \wedge (y,z) \in R\Big\} \vee (x,z) \in R$$

definition of composition

$$\Longleftrightarrow \quad \forall x, z \in V : \overline{(x,z) \in R\, ; R} \vee (x,z) \in R$$

$$\neg a \vee b = a \to b$$

$$\Longleftrightarrow \quad \forall x, z \in V : (x,z) \in R\, ; R \;\to\; (x,z) \in R$$

point-free formulation

$$\Longleftrightarrow \quad R\, ; R \subseteq R$$

We remember that an asymmetric relation turned out to be irreflexive. Now we obtain as a first easy exercise a result in the opposite direction.

Proposition 5.19. *A transitive relation is irreflexive precisely when it is asymmetric.* □

This proposition will be recalled with a proof in a more general context as Prop. 12.5. It is also interesting to study transitivity in combination with reflexivity, for which purpose we provide the following often used definition.

Definition 5.20. Let a homogeneous relation R be given. We call

R a **preorder** $\quad:\Longleftrightarrow\quad$ R reflexive and transitive

$$\Longleftrightarrow \quad \mathbb{I} \subseteq R, \;\; R^2 \subseteq R \quad\Longleftrightarrow\quad \mathbb{I} \subseteq R, \;\; R^2 = R.$$

A preorder is often also called a *partial preorder* or a *quasiorder*. □

Another property of an arbitrary relation R is that its residuals R/R as well as $R\backslash R$ are trivially transitive. In the first case this may be shown by remembering

$$(R/R)\, ; R \subseteq R$$

which immediately implies

$$(R/R)\, ; (R/R)\, ; R \subseteq (R/R)\, ; R \subseteq R$$

so that $(R/R)\, ; (R/R)$ is one of the relations that composed with R are contained in R. Therefore, it is contained in the greatest such relation, namely (R/R).

Since $(r_1, r_2) \in R/R$ indicates that row r_1 is equal to or contains row r_2, transitivity is not surprising at all. What needs to be stressed is that such properties may be dealt with in algebraic precision.

Orderings

Using these elementary constituents, we now build well-known composite definitions. We have chosen to list several equivalent versions leaving the proofs of equivalence to the reader.

Definition 5.21. Let homogeneous relations $E, C : V \longrightarrow V$ be given. We call[9]

E **order** $:\Longleftrightarrow$ E transitive, antisymmetric, reflexive

  $\Longleftrightarrow$ $E\,\!;\! E \subseteq E, \quad E \cap E^{\mathsf{T}} \subseteq \mathbb{I}, \quad \mathbb{I} \subseteq E,$

C **strictorder** $:\Longleftrightarrow$ C transitive and asymmetric

  $\Longleftrightarrow$ $C\,\!;\! C \subseteq C, \quad C \cap C^{\mathsf{T}} \subseteq \mathbb{\bar{I}}$

  $\Longleftrightarrow$ $C\,\!;\! C \subseteq C, \quad C^{\mathsf{T}} \subseteq \overline{C}$

  $\Longleftrightarrow$ C transitive and irreflexive

  $\Longleftrightarrow$ $C\,\!;\! C \subseteq C, \quad C \subseteq \overline{\mathbb{I}}.$ □

As an example we show the divisibility ordering on the set $\{1,\ldots,12\}$. From
Fig. 5.18, we may in an obvious way deduce the reflexive order as well as the
strictorder.[10]

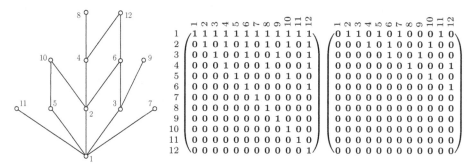

Fig. 5.18 Division ordering on numbers $1\ldots12$: Hasse diagram, ordering, Hasse relation

Orders and strictorders *always* appear together as "$\leq$" and "$<$" do; however, their
algebraic properties must not be mixed up. Orders and strictorders occur so often
in various contexts that it is helpful to have at least a hint in the notation. We have
chosen to use E for orders and C for strictorders as E is closer to "$\leq$" and C is
closer to "$<$".

---

[9] An order is often also called *partial order*.

[10] There is one point to mention when we talk about orderings. People often speak of a 'strict
ordering'. This is not what they intend it to mean, as it is not an *ordering* with the added property
of being *strict* or *asymmetric*. By definition, this cannot be. So we have chosen to use a German
style compound word *strictorder*. A strictorder is not an order! Later, in the same way, a *preorder*
need not be an order.

Definition 5.22. Order and strictorder are related as follows.

(i) Given an ordering E, we call $C_E := \overline{\mathbb{I}} \cap E$ its **associated strictorder**.

(ii) Given a strictorder C, we call $E_C := \mathbb{I} \cup C$ its **associated ordering**.

(iii) Given an ordering E or a strictorder C, we call $H := C_E \cap \overline{C_E^2}$, respectively $H := C \cap \overline{C^2}$, the **corresponding Hasse relation**, usually called the **Hasse diagram** when presented graphically. $\square$

Indeed, it is easily shown that E_C is an ordering and C_E is a strictorder. Obviously, we have

$$E_{C_E} = E \quad \text{and} \quad C_{E_C} = C.$$

When representing an ordering as a graph, one will obtain something like the left figure of Fig. 5.19. It is, however, not a good idea to draw all these arrows on a blackboard when giving a lesson. So one will usually omit arrow heads, replacing them with the convention that arrows always lead upwards. What one cannot reinvent in a unique manner are loops. For an ordering every vertex carries a loop, whereas for a strictorder none of the vertices carries a loop. But as these two always occur in pairs, this is not a big problem, and the loops are usually omitted as in the middle and the right diagrams of Fig. 5.19.

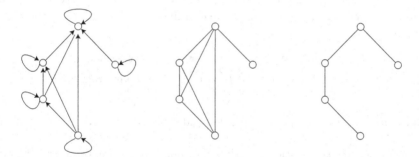

Fig. 5.19 Order with associated strictorder and Hasse diagram

There is one further traditional way of reducing effort when representing orderings. It is economical to represent the ordering by its Hasse diagram, i.e., omitting all those arrows that may be represented by following a sequence of arrows. In general, however, the relationship of an ordering with its Hasse diagram H is less obvious than expected. For the finite case, we will give a definitive answer once the transitive closure is introduced and prove that $C = H^+$ and $E = H^*$. In a finite strictorder, one may ask whether there are immediate successors or predecessors. Figure 5.19 shows an ordering, its associated strictordering and its corresponding Hasse diagram.[11]

---

[11] In the case that immediate successors and predecessors exist, i.e., for discrete orderings in the sense of [16], the ordering can be *generated* by its Hasse diagram.

Linear orderings

In earlier times it was not at all clear that orderings need not be *linear* orderings. But since the development of lattice theory in the 1930s it became more and more evident that most of our reasoning with orderings was also possible when they failed to be linear. So it is simply a matter of economy to present orders first and then specialize to linear orderings.[12]

Definition 5.23. Given again homogeneous relations $E, C : V \longrightarrow V$, we call

$$
\begin{aligned}
E \text{ \textbf{linear order}} \quad &:\Longleftrightarrow \quad E \text{ is an ordering which is also connex,} \\
&\Longleftrightarrow \quad E \,\fatsemi\, E \subseteq E, \quad E \cap E^{\mathsf{T}} \subseteq \mathbb{I}, \quad \mathbb{T} = E \cup E^{\mathsf{T}}, \\
C \text{ \textbf{linear strictorder}} \quad &:\Longleftrightarrow \quad C \text{ is a strictorder which is also semi-connex,} \\
&\Longleftrightarrow \quad C \,\fatsemi\, C \subseteq C, \quad C \cap C^{\mathsf{T}} \subseteq \mathbb{l}, \quad \overline{\mathbb{I}} = C \cup C^{\mathsf{T}}. \quad\square
\end{aligned}
$$

Sometimes a linear order is called a *total order* or a *chain* and a linear strictorder is called a *strict total order, complete order,* or *strict complete order*.

When nicely arranged, the matrix of a linear order has the upper right triangle and the diagonal filled with **1**s, leaving the lower left triangle full of **0**s. It is rather obvious, see Prop. 5.24, that the negative of a linear order is a linear strictorder in reverse direction.

$$
\begin{array}{c}
\diamondsuit \\ \heartsuit \\ \spadesuit \\ \clubsuit
\end{array}
\begin{pmatrix}
1 & 1 & 1 & 1 \\
0 & 1 & 1 & 1 \\
0 & 0 & 1 & 1 \\
0 & 0 & 0 & 1
\end{pmatrix}
\qquad
\begin{array}{c}
\diamondsuit \\ \heartsuit \\ \spadesuit \\ \clubsuit
\end{array}
\begin{pmatrix}
0 & 1 & 1 & 1 \\
0 & 0 & 1 & 1 \\
0 & 0 & 0 & 1 \\
0 & 0 & 0 & 0
\end{pmatrix}
\qquad
\begin{array}{c}
\diamondsuit \\ \heartsuit \\ \spadesuit \\ \clubsuit
\end{array}
\begin{pmatrix}
0 & 1 & 0 & 0 \\
0 & 0 & 1 & 0 \\
0 & 0 & 0 & 1 \\
0 & 0 & 0 & 0
\end{pmatrix}
$$

Fig. 5.20 Linear order, linear strictorder, and Hasse relation for Skat suit ranking

We now exhibit characteristic properties of a linear order. As prominent terms we often find $\overline{E}^{\mathsf{T}}$ for the linear order E as well as $\overline{C}^{\mathsf{T}}$ for its associated strictorder C. These are the first indications of the utility of a **dual** $R^{\mathrm{d}} := \overline{R}^{\mathsf{T}}$ of some relation R.

Proposition 5.24. *The following hold around linear order and strictorder.*

(i) *A linear order E and its associated strictorder C satisfy $\overline{E}^{\mathsf{T}} = C$.*

(ii) *A linear order E satisfies $E \,\fatsemi\, \overline{E}^{\mathsf{T}} \,\fatsemi\, E = C \subseteq E$.*

(iii) *A linear strictorder C satisfies $C \,\fatsemi\, C \,\fatsemi\, \overline{C}^{\mathsf{T}} = C \,\fatsemi\, \overline{C}^{\mathsf{T}} \,\fatsemi\, C = C^2 \subseteq C$.*

(iv) *E is a linear order precisely when $E^{\mathrm{d}} = \overline{E}^{\mathsf{T}}$ is a linear strictorder.*

(v) *C is a linear strictorder precisely when $C^{\mathrm{d}} = \overline{C}^{\mathsf{T}}$ is a linear order.*

---

[12] Today a tendency may be observed that makes the even less restricted preorder the basic structure.

Proof: See B.6. □

Of course, $\overline{C}^{\mathsf{T}} = E$ is an equivalent form of (i). With (ii,iii), we gave the result (i) a more complicated form, thus anticipating a property of Ferrers relations, see Section 10.5. The equation $E \mathbin{;} \overline{E}^{\mathsf{T}} \mathbin{;} E = E \mathbin{;} \overline{E}^{\mathsf{T}} = \overline{E}^{\mathsf{T}}$ holds for arbitrary orderings E.

We now anticipate a result that seems rather obvious, at least for the finite case. Nevertheless, it needs a proof which we will give later; see Prop. 12.14. In computer science, one routinely speaks of *topological sorting* of an ordering. This may be expressed colloquially by demanding that there exists a permutation such that the resulting matrix resides in the upper right triangle, in which case the Szpilrajn extension is completely obvious.

Proposition 5.25 (Topological sorting). *For every order E there exists a linear order E_1, with $E \subseteq E_1$, a so-called Szpilrajn extension.*[13] □

The idea of a proof (which will be presented in full detail along with Prop. 12.14) is easy to communicate. Assume E is not yet linear, i.e., $E \cup E^{\mathsf{T}} \neq \mathbb{T}$. Then there will exist two elements x, y with $x \mathbin{;} y^{\mathsf{T}} \subseteq \overline{E \cup E^{\mathsf{T}}}$ that might be called incomparable elements. With these, we define an ordering relation as $E_1 := E \cup E \mathbin{;} x \mathbin{;} y^{\mathsf{T}} \mathbin{;} E$. In the finite case, this argument may be iterated, and E_n will eventually become linear.

$$
E = \begin{array}{r}
\text{Alfred} \\ \text{Barbara} \\ \text{Christian} \\ \text{Donald} \\ \text{Eugene} \\ \text{Frederick} \\ \text{George}
\end{array}
\begin{pmatrix}
1 & 1 & 0 & 0 & 0 & 1 & 0 \\
0 & 1 & 0 & 0 & 0 & 0 & 0 \\
1 & 1 & 1 & 1 & 1 & 1 & 0 \\
0 & 0 & 0 & 1 & 0 & 1 & 0 \\
0 & 0 & 0 & 0 & 1 & 1 & 0 \\
0 & 0 & 0 & 0 & 0 & 1 & 0 \\
1 & 1 & 0 & 1 & 1 & 1 & 1
\end{pmatrix}
$$

columns (E): Alfred, Barbara, Christian, Donald, Eugene, Frederick, George

$$
E_1 = \begin{pmatrix}
1 & 1 & 0 & 1 & 1 & 1 & 0 \\
0 & 1 & 0 & 1 & 1 & 1 & 0 \\
1 & 1 & 1 & 1 & 1 & 1 & 1 \\
0 & 0 & 0 & 1 & 1 & 1 & 0 \\
0 & 0 & 0 & 0 & 1 & 1 & 0 \\
0 & 0 & 0 & 0 & 0 & 1 & 0 \\
1 & 1 & 0 & 1 & 1 & 1 & 1
\end{pmatrix}
$$

columns (E_1): Christian, George, Alfred, Barbara, Donald, Eugene, Frederick

$$
E\ \text{rearranged} = \begin{array}{r}
\text{Christian} \\ \text{George} \\ \text{Alfred} \\ \text{Barbara} \\ \text{Donald} \\ \text{Eugene} \\ \text{Frederick}
\end{array}
\begin{pmatrix}
1 & 0 & 1 & 1 & 1 & 1 & 1 \\
0 & 1 & 1 & 1 & 1 & 1 & 1 \\
0 & 0 & 1 & 1 & 0 & 0 & 1 \\
0 & 0 & 0 & 1 & 0 & 0 & 0 \\
0 & 0 & 0 & 0 & 1 & 0 & 1 \\
0 & 0 & 0 & 0 & 0 & 1 & 1 \\
0 & 0 & 0 & 0 & 0 & 0 & 1
\end{pmatrix}
$$

columns (E rearranged): Christian, George, Alfred, Barbara, Donald, Eugene, Frederick

Fig. 5.21 An order rearranged inside the upper triangle of its Szpilrajn extension

We study the technique of a Szpilrajn extension considering Fig. 5.21. Choosing Barbara, Donald, for example, one will add in the first step connections Barbara $\mapsto$ Donald and Alfred $\mapsto$ Donald.

---

[13] Edward Marczewski (1907–1976) was a Polish mathematician who used the surname Szpilrajn until 1940. He was a member of the Warsaw School of Mathematics. His life and work after the Second World War were connected with Wrocław, where he was among the creators of the Polish scientific center (Wikipedia).

5.3 Prove equivalence of the variants occurring in Def. 5.23.

5.4 Show that every relation R contained in $\mathbb{I}$ is symmetric.

5.5 Prove that $E^{\mathsf{T}};\overline{E;X} = \overline{E;X}$ for an ordering E and an arbitrary relation X.

5.6 A relation is called **idempotent** provided $R = R^2$. Prove $R \cap \mathbb{I} \neq \mathbb{L}$ for any finite idempotent relation $R \neq \mathbb{L}$. Provide an example to show that the finiteness condition is necessary.

5.4 Equivalence and quotient

As a first application of both symmetry, reflexivity, and transitivity together with the function concept, we study equivalences and their corresponding natural projections. Much of this will be used in Section 7.4 when constructing quotient domains. Concerning natural projections, we recall here only that $x \mapsto [x]_\Xi$ means proceeding from an element x to the class according to the equivalence Ξ. Here we prepare the algebraic formalism, but do not yet mention the algebraic rules which a natural projection obeys.

Definition 5.26. Let $\Xi : V \longrightarrow V$ be a homogeneous relation. We call

Ξ **equivalence** $\quad :\Longleftrightarrow\quad$ Ξ is reflexive, transitive, and symmetric,

$\qquad\qquad\qquad\Longleftrightarrow\quad \mathbb{I} \subseteq \Xi, \quad \Xi;\Xi \subseteq \Xi, \quad \Xi^{\mathsf{T}} \subseteq \Xi.$ $\qquad\qquad$ □

We now prove some rules which are useful for calculations involving equivalence relations; they deal with the effect of multiplication by an equivalence relation with regard to intersection and negation.

Proposition 5.27. *Let Ξ be an equivalence and let A, B, R be arbitrary relations.*

(i) $\Xi;(\Xi;A \cap B) = \Xi;A \cap \Xi;B = \Xi;(A \cap \Xi;B).$
(ii) $\Xi;\overline{\Xi;R} = \overline{\Xi;R}.$

Proof: See B.7. $\qquad\qquad\qquad\qquad\qquad\qquad\qquad\qquad\qquad\qquad\qquad\qquad$ □

Trying to interpret these results, let us say that $\Xi;A$ is the relation A *saturated* with the equivalence Ξ on the source side. Should one element of a class be related to some element, then all are related to that element. In such a case, the saturation of the intersection of a saturated relation with another one is the same as the intersection of the two saturated relations. This is close to a distributivity law for composition with an equivalence. The complement of a source-saturated relation is again source-saturated as shown in Fig. 5.22.

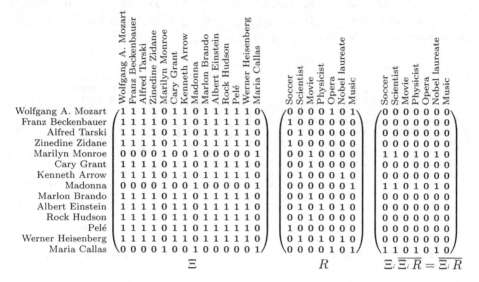

Fig. 5.22 Celebrities partitioned as to their gender, qualified, and class-negated

Very often, we will be concerned with relations in which several rows, respectively columns, look identical. We provide an algebraic mechanism to handle this case. As presented in Def. 5.28, this allows an intuitively clear interpretation.

Definition 5.28. For a (possibly heterogeneous) relation R, we define its corresponding[14]

'row-contains' preorder $\mathcal{R}(R) := \overline{\overline{R};R^{\mathsf{T}}} = R/R$

'column-is-contained' preorder $\mathcal{C}(R) := \overline{R^{\mathsf{T}};\overline{R}} = R\backslash R$

together with

row equivalence $\Xi(R) := \mathsf{syq}\,(R^{\mathsf{T}}, R^{\mathsf{T}}) = \mathcal{R}(R) \cap \mathcal{R}(\overline{R})$

column equivalence $\Psi(R) := \mathsf{syq}\,(R, R) = \mathcal{C}(R) \cap \mathcal{C}(\overline{R})$

and in the case of a homogeneous relation also the

'section' preorder $\mathcal{S}(R) := \mathcal{R}(R) \cap \mathcal{C}(R)$ □

One may wonder why we have chosen different containment directions for rows and

[14] In French: préordre finissant, préordre commençant, respectively préordre des sections.

columns, respectively. As defined here, some other concepts to be defined only later have a simple form. That the letters Ξ, Ψ are chosen so as to represent rows, and columns respectively, is evident.

We visualize row and column equivalence first and concentrate on containment afterwards. Let somebody who plays bridge every day remember his main winning cards over the week; see the relation R of Fig. 5.23.

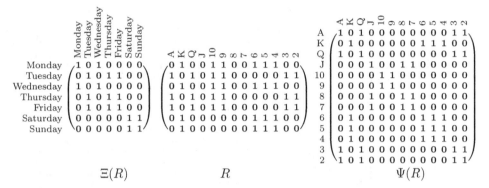

Fig. 5.23 A relation R with its row and column equivalences $\Xi(R), \Psi(R)$

The situation will become more clear when we arrange the matrix of the relation R so as to have equal rows, and columns respectively, side by side as in Fig. 5.24. Also, it is obvious how to condense the representation of such information.

	Monday	Wednesday	Tuesday	Thursday	Friday	Saturday	Sunday
Monday	1	1	0	0	0	0	0
Wednesday	1	1	0	0	0	0	0
Tuesday	0	0	1	1	1	0	0
Thursday	0	0	1	1	1	0	0
Friday	0	0	1	1	1	0	0
Saturday	0	0	0	0	0	1	1
Sunday	0	0	0	0	0	1	1

	A	Q	3	2	K	6	5	4	J	8	7	10	9
	0	0	0	0	1	1	1	1	0	0	0	1	1
	0	0	0	0	1	1	1	1	0	0	0	1	1
	1	1	1	1	0	0	0	0	0	0	0	1	1
	1	1	1	1	0	0	0	0	0	0	0	1	1
	1	1	1	1	0	0	0	0	0	0	0	1	1
	0	0	0	0	1	1	1	1	0	0	0	0	0
	0	0	0	0	1	1	1	1	0	0	0	0	0

	[A]	[K]	[J]	[10]
[Monday]	0	1	0	1
[Tuesday]	1	0	0	1
[Saturday]	0	1	0	0

	A	Q	3	2	K	6	5	4	J	8	7	10	9
A	1	1	1	1	0	0	0	0	0	0	0	0	0
Q	1	1	1	1	0	0	0	0	0	0	0	0	0
3	1	1	1	1	0	0	0	0	0	0	0	0	0
2	1	1	1	1	0	0	0	0	0	0	0	0	0
K	0	0	0	0	1	1	1	1	0	0	0	0	0
6	0	0	0	0	1	1	1	1	0	0	0	0	0
5	0	0	0	0	1	1	1	1	0	0	0	0	0
4	0	0	0	0	1	1	1	1	0	0	0	0	0
J	0	0	0	0	0	0	0	0	1	1	1	0	0
8	0	0	0	0	0	0	0	0	1	1	1	0	0
7	0	0	0	0	0	0	0	0	1	1	1	0	0
10	0	0	0	0	0	0	0	0	0	0	0	1	1
9	0	0	0	0	0	0	0	0	0	0	0	1	1

Fig. 5.24 Relation of Fig. 5.23 rearranged to block form

In view of the rearranged representation, the following results are almost immediate.

Proposition 5.29. *For an arbitrary relation R and its row and column preorder and equivalence, always*

(i) $\Xi(R); R = R = R; \Psi(R)$

(ii) $\mathcal{R}(\overline{R}) = \big(\mathcal{R}(R)\big)^{\mathsf{T}} \qquad \Xi(\overline{R}) = \Xi(R) \qquad \mathcal{R}(R^{\mathsf{T}}) = \mathcal{C}(\overline{R})$

(iii) $\Xi(R) = \Xi(\mathcal{R}(R))$ *or, equivalently,* $\mathtt{syq}(R^{\mathsf{T}}, R^{\mathsf{T}}) = \mathtt{syq}(\overline{\overline{R}; R^{\mathsf{T}}}^{\mathsf{T}}, \overline{\overline{R}; R^{\mathsf{T}}}^{\mathsf{T}})$.

Proof: See B.8. □

Part (ii) of this proposition is quite an intuitive result: negation reverses row containment, but does not change row equivalence. Transposition converts a row consideration into a column consideration; in addition, we have chosen different ordering directions for rows and columns 'row-contains' versus 'column-is-contained'.

Figure 5.25 gives an example of a relation on the left together with its section preorder.

$$R = \begin{array}{c|ccccccccccccc} & 1 & 2 & 3 & 4 & 5 & 6 & 7 & 8 & 9 & 10 & 11 & 12 & 13 \\ \hline 1 & 0&0&1&1&1&0&1&1&1&1&1&1&1 \\ 2 & 0&0&1&1&1&0&1&1&1&1&1&1&1 \\ 3 & 0&0&0&0&1&0&0&0&1&0&1&1&1 \\ 4 & 0&0&0&0&0&0&1&0&1&1&1&1&1 \\ 5 & 0&0&0&0&0&0&0&0&1&0&1&0&1 \\ 6 & 0&0&0&0&0&0&0&0&0&0&0&0&0 \\ 7 & 0&0&0&0&0&0&0&0&0&0&0&0&0 \\ 8 & 0&0&0&0&0&0&0&0&0&1&0&1&1 \\ 9 & 0&0&0&0&0&0&0&0&0&0&1&0&1 \\ 10 & 0&0&0&0&0&0&0&0&0&0&0&1&1 \\ 11 & 0&0&0&0&0&0&0&0&0&0&0&0&1 \\ 12 & 0&0&0&0&0&0&0&0&0&0&0&0&1 \\ 13 & 0&0&0&0&0&0&0&0&0&0&0&0&0 \end{array} \qquad \begin{array}{c|ccccccccccccc} & 1 & 2 & 3 & 4 & 5 & 6 & 7 & 8 & 9 & 10 & 11 & 12 & 13 \\ \hline 1 & 1&1&1&1&1&1&1&1&1&1&1&1&1 \\ 2 & 1&1&1&1&1&1&1&1&1&1&1&1&1 \\ 3 & 0&0&1&0&1&0&1&0&1&1&1&1&1 \\ 4 & 0&0&0&1&1&0&1&1&1&1&1&1&1 \\ 5 & 0&0&0&0&1&0&0&0&1&0&1&1&1 \\ 6 & 0&0&0&0&0&1&1&0&0&0&0&0&1 \\ 7 & 0&0&0&0&0&0&1&0&0&0&0&0&1 \\ 8 & 0&0&0&0&0&0&1&1&0&1&1&1&1 \\ 9 & 0&0&0&0&0&0&0&0&1&0&1&0&1 \\ 10 & 0&0&0&0&0&0&0&0&0&1&0&1&1 \\ 11 & 0&0&0&0&0&0&0&0&0&0&1&0&1 \\ 12 & 0&0&0&0&0&0&0&0&0&0&0&1&1 \\ 13 & 0&0&0&0&0&0&0&0&0&0&0&0&1 \end{array} = \mathcal{S}(R)$$

Fig. 5.25 Section preorder of a relation (which is a block-transitive strictorder)

Difunctional relations

The concept of a difunctional relation is concerned with 'block decomposition'. It generalizes such concepts as being univalent and at the same time being injective to heterogeneous block-versions thereof. Also, the concept of an equivalence relation is generalized in as far as source and target need no longer be identical and the relation proper need not necessarily be total or surjective. The astonishing fact is that several well-known formulae stay the same or are only slightly modified to catch up with the new situation as indicated on the right of Fig. 5.26.

$$
\begin{array}{c}
\; j \qquad\qquad m \\
\begin{array}{r}
1\,2\,3\,4\,5\,6\,7\,8\,9\,10\,11\,12\,13\,14 \\
\begin{array}{r}
1 \\ 2 \\ 3 \\ k\;\; 4 \\ 5 \\ 6 \\ 7 \\ 8 \\ 9 \\ i\;\;10 \\ 11 \\ 12
\end{array}
\left(\begin{array}{cccccccccccccc}
1&0&1&0&1&0&0&1&1&0&1&1&0&1\\
0&1&0&1&0&0&1&0&0&1&0&0&0&0\\
0&0&0&0&0&1&0&0&0&0&0&0&1&0\\
1&0&1&0&1&0&0&1&1&0&1&1&0&1\\
0&1&0&1&0&0&1&0&0&1&0&0&0&0\\
1&0&1&0&1&0&0&1&1&0&1&1&0&1\\
0&1&0&1&0&0&1&0&0&1&0&0&0&0\\
0&0&0&0&0&0&0&0&0&0&0&0&0&0\\
0&0&0&0&0&1&0&0&0&0&0&0&1&0\\
1&0&1&0&1&0&0&1&1&0&1&1&0&1\\
0&0&0&0&0&1&0&0&0&0&0&0&1&0\\
1&0&1&0&1&0&0&1&1&0&1&1&0&1
\end{array}\right)
\end{array}
\end{array}
$$

Right matrix (columns: 1 3 5 8 9 11 12 14 2 4 7 10 6 13):

$$
\begin{array}{r}
\begin{array}{r}
1 \\ k\;\;4 \\ 6 \\ i\;\;10 \\ 12 \\ 2 \\ 5 \\ 7 \\ 3 \\ 9 \\ 11 \\ 8
\end{array}
\left(\begin{array}{cccccccccccccc}
1&1&1&1&1&1&1&1&0&0&0&0&0&0\\
1&1&1&1&1&1&1&1&0&0&0&0&0&0\\
1&1&1&1&1&1&1&1&0&0&0&0&0&0\\
1&1&1&1&1&1&1&1&0&0&0&0&0&0\\
1&1&1&1&1&1&1&1&0&0&0&0&0&0\\
0&0&0&0&0&0&0&0&1&1&1&1&0&0\\
0&0&0&0&0&0&0&0&1&1&1&1&0&0\\
0&0&0&0&0&0&0&0&1&1&1&1&0&0\\
0&0&0&0&0&0&0&0&0&0&0&0&1&1\\
0&0&0&0&0&0&0&0&0&0&0&0&1&1\\
0&0&0&0&0&0&0&0&0&0&0&0&1&1\\
0&0&0&0&0&0&0&0&0&0&0&0&0&0
\end{array}\right)
\end{array}
$$

Fig. 5.26 Predicate-logic interpretation for a relation to be difunctional

That R will assume block-diagonal form when suitably rearranging rows and columns independently cannot immediately be seen from Def. 5.30. We expand the relational formula $R_{,}R^{\mathsf{T}}{}_{,}R \subseteq R$ of the definition to

$$
\forall\, i,m: \quad \big[\exists j,k : (i,j) \in R \,\wedge\, (j,k) \in R^{\mathsf{T}} \wedge (k,m) \in R\big] \quad \to \quad (i,m) \in R
$$

in order to get a more detailed view of what the difunctionality property actually means. In Fig. 5.26 one may indeed see some aspects of it.

Definition 5.30. Let Q be a possibly heterogeneous relation.

Q **difunctional** $\quad :\Longleftrightarrow\quad Q_{,}Q^{\mathsf{T}}{}_{,}Q \subseteq Q \quad \Longleftrightarrow \quad Q_{,}Q^{\mathsf{T}}{}_{,}Q = Q$

$\qquad\qquad\qquad \Longleftrightarrow \quad Q_{,}\overline{Q}^{\mathsf{T}}{}_{,}Q \subseteq \overline{Q}$

$\qquad\qquad\qquad \Longleftrightarrow \quad Q$ has block-diagonal form when suitably rearranging rows and columns independently. $\qquad\square$

With Q, the relation Q^{T} is obviously difunctional also. The equivalence of the definition variants is straightforward. The concept of being *difunctional* is called a *matching relation* or simply a *match* in [46].

Proposition 5.31. *If Q is a difunctional relation, the following holds for arbitrary relations A, B:*

$$
Q_{,}(A \cap Q^{\mathsf{T}}{}_{,}B) = Q_{,}A \cap Q_{,}Q^{\mathsf{T}}{}_{,}B.
$$

Proof: See B.9. $\qquad\qquad\qquad\qquad\qquad\qquad\qquad\qquad\qquad\qquad\qquad\square$

This result obviously unifies what has already been proved in Prop. 5.3, 5.4, and Prop. 5.27.i, following tradition. Figure 5.27 visualizes the effect of Prop. 5.31.

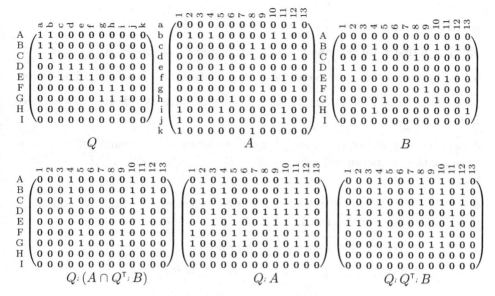

Fig. 5.27 Visualizing Prop. 5.31

For the difunctional case handled here, we have more symmetry than in Prop. 5.3 and 5.4. They may be converted from one to another so that $(A_i Q^\mathsf{T} \cap B)_i Q = A_i Q^\mathsf{T}_i Q \cap B_i Q$. Difunctional means – accepting the block aspect – univalent as well as injective.

The formulae of Prop. 5.6 and Prop. 5.27.ii concerning negation also have an analogous form.

Proposition 5.32. *If Q is a difunctional relation, the following holds for every relation A:*

(i) $Q_i \overline{Q^\mathsf{T}_i A} = Q_i \mathbb{T} \cap \overline{Q_i Q^\mathsf{T}_i A}$ *and* $\overline{Q_i Q^\mathsf{T}_i A} = Q_i \overline{Q^\mathsf{T}_i A} \cup \overline{Q_i \mathbb{T}}$

(ii) $Q_i \overline{Q^\mathsf{T}_i A} = \overline{Q_i Q^\mathsf{T}_i A}$ *in the case that in addition Q is total.*

Proof: See B.10.   □

Exercises

5.7 Prove the following as a generalization of Prop. 5.12. Given arbitrary relations R, S and an equivalence Ω satisfying $S_i \Omega = S$, then a total f such that $f^\mathsf{T}_i f \subseteq \Omega$ satisfies

$$R \subseteq S; f^{\mathsf{T}} \qquad \Longleftrightarrow \qquad R; f \subseteq S.$$

5.8 Prove that for an arbitrary order E always $\mathcal{R}(E) = E = \mathcal{C}(E)$.

5.9 Prove that total relations F contained in an equivalence Θ satisfy $F; \Theta = \Theta$.

5.10 Prove that if F is univalent and Θ is an equivalence, then $F; \Theta$ is difunctional.

5.11 Prove that R is an equivalence if and only if it is reflexive and satisfies $R; R^{\mathsf{T}} \subseteq R$.

5.12 Prove the following statement: for every relation Q contained in an equivalence relation Ξ and for any other relation R the equation $Q; (R \cap \Xi) = Q; R \cap \Xi$ holds. (Compare this result to the modular law of lattice theory.)

5.13 Prove the following statement: for every reflexive and transitive relation A there exists a relation R such that $A = \overline{R^{\mathsf{T}}; \overline{R}}$.

5.14 Show that for R total the row-equivalence is contained in $R; R^{\mathsf{T}}$.

5.15 Show that R is an equivalence precisely when $R = \mathsf{syq}(R, R)$.

5.16 Show that for connex R always $\mathcal{C}(R \cap \overline{R}^{\mathsf{T}}) = \mathcal{R}(R)$.

5.5 Transitive closure

It is folklore that every homogeneous relation R has a *transitive closure* R^{+} which is the least transitive relation it is contained in. Alternatively, one might say that it is the result of 'making R increasingly more transitive'.

Definition 5.33. Given any homogeneous relation R, we define its **transitive closure**
$$R^{+} := \inf \{X \mid R \subseteq X, \ X; X \subseteq X\} = \sup_{i \geq 1} R^{i}$$
in two forms and give also the two definitions of its **reflexive-transitive closure**
$$R^{*} := \inf \{X \mid \mathbb{I} \cup R \subseteq X, \ X; X \subseteq X\} = \sup_{i \geq 0} R^{i}. \qquad \Box$$

We do not recall the proof justifying that one may indeed use two versions, i.e., that the relations thus defined will always coincide; see, for example, [123, 124]. We mention, however, the way in which the reflexive-transitive and the transitive closure are related:
$$R^{+} = R; R^{*} \qquad R^{*} = \mathbb{I} \cup R^{+}.$$

These interrelationships are by now well known. Typically, the infimum definition (the 'descriptive version') *will be made use of in proofs* while the supremum version *is computed* (the 'operational version'). To use the descriptive part to compute the closure would be highly complex because the intersection of all transitive relations above R is formed, which can only be done for toy examples. A better idea for computing R^+ is to approximate it from below as $R \subseteq R \cup R^2 \subseteq R \cup R^2 \cup R^3 \ldots$ in the operational version. Even better are the traditional algorithms such as the Warshall algorithm.

Proposition 5.34. *The following holds for an arbitrary finite homogeneous relation R on a set of n elements:*

(i) $R^n \subseteq (\mathbb{I} \cup R)^{n-1}$,

(ii) $R^* = \sup_{0 \leq i < n} R^i$,

(iii) $R^+ = \sup_{0 < i \leq n} R^i$,

(iv) $(\mathbb{I} \cup R) ; (\mathbb{I} \cup R^2) ; (\mathbb{I} \cup R^4) ; (\mathbb{I} \cup R^8) ; \ldots ; (\mathbb{I} \cup R^{2^{\lfloor \log n \rfloor}}) = R^*$.

Proof: See B.11. □

The construct R^* is obviously a preorder. Two equivalences are traditionally given with any relation.

Definition 5.35. For a (possibly heterogeneous) relation R we define

$\Omega := (R ; R^\mathsf{T})^*$ the **left equivalence** and

$\Omega' := (R^\mathsf{T} ; R)^*$ the **right equivalence**. □

One may also call Ω an equivalence generated by common results and Ω' an equivalence generated by common arguments. Some results follow immediately.

Proposition 5.36. *Let some (possibly heterogeneous) relation R be given and consider Ω, Ω', its left and right equivalences.*

(i) *Ω and Ω' are equivalences.*

(ii) *$\Omega ; R = R ; \Omega'$.*

(iii) *$\mathbb{I} \cup R^\mathsf{T} ; \Omega ; R = \Omega'$.*

(iv) *$\mathbb{I} \cup R ; \Omega' ; R^\mathsf{T} = \Omega$.*

Proof: See B.12. □

From the point of view of Section 5.6, Ω, Ω' constitute an R-congruence; see the forthcoming Example 5.40, where this is also visualized in more detail.

Strongly connected components

Closely related to transitive closure is the determination of strongly connected components. Let R be a homogeneous relation and consider the preorder R^* generated by R. Then $R^* \cap R^{*^\mathsf{T}}$ is the equivalence generated by R which provides a partition of rows as well as columns.

$$R =$$

```
        1 2 3 4 5 6 7 8 9 10 11 12 13
    1 / 1 0 0 0 0 0 0 0 0 0  0  0  1 \
    2 | 0 0 0 0 0 0 0 0 0 0  0  1  1 |
    3 | 0 0 1 0 0 1 0 0 0 1  0  0  0 |
    4 | 0 0 0 0 0 0 0 0 0 0  0  0  0 |
    5 | 0 0 0 0 0 0 0 0 0 1  0  0  0 |
    6 | 0 0 0 0 0 1 0 0 0 0  0  1  0 |
    7 | 0 0 0 0 0 0 0 0 1 0  0  0  0 |
    8 | 0 1 0 0 0 0 0 0 0 0  0  0  0 |
    9 | 0 0 0 0 1 0 0 0 1 0  0  0  0 |
   10 | 0 0 0 0 0 0 0 0 1 0  0  0  0 |
   11 | 0 0 0 0 0 0 0 0 1 0  0  0  0 |
   12 | 1 0 1 0 0 0 0 0 0 0  0  0  0 |
   13 \ 0 0 1 0 0 0 0 0 0 1  0  0  0 /
```

$$= R^*$$

```
         8 2 1 3 6 12 13 4 7 10 5 9 11
    8 / 1|1|1 1 1 1 1|0|0|1|1 1 1 \
    2 | 0|1|1 1 1 1 1|0|0|1|1 1 1 |
    1 | 0|0|1 1 1 1 1|0|0|1|1 1 1 |
    3 | 0|0|1 1 1 1 1|0|0|1|1 1 1 |
    6 | 0|0|1 1 1 1 1|0|0|1|1 1 1 |
   12 | 0|0|1 1 1 1 1|0|0|1|1 1 1 |
   13 | 0|0|1 1 1 1 1|0|0|1|1 1 1 |
    4 | 0|0|0 0 0 0 0|1|0|0|0 0 0 |
    7 | 0|0|0 0 0 0 0|0|1|0|1 1 1 |
   10 | 0|0|0 0 0 0 0|0|0|1|1 1 1 |
    5 | 0|0|0 0 0 0 0|0|0|0|1 1 1 |
    9 | 0|0|0 0 0 0 0|0|0|0|1 1 1 |
   11 \ 0|0|0 0 0 0 0|0|0|0|1 1 1 /
```

Fig. 5.28 A relation with transitive closure arranged by simultaneous permutation

The arrangement of the transitive closure as an upper right block-triangle allows transfer of the arrangement to the original relation itself. So one will obtain a rearrangement as in Fig. 5.29. The subdivisions are justified by the transitive closure. Without these, one would not easily capture the underlying structure. This is, therefore, considered a basic technique of knowledge acquisition.

```
         8 2 1 3 6 12 13 4 7 10 5 9 11
    8 / 0|1|0 0 0 0 0|0|0|0|0 0 0 \
    2 | 0|0|0 0 1 1 0|0|0|0|0 0 0 |
    1 | 0|0|1 0 0 0 1|0|0|0|0 0 0 |
    3 | 0|0|0 1 1 0 0|0|1|0|0 0 0 |
    6 | 0|0|0 0 1 1 0|0|0|0|0 0 0 |
   12 | 0|0|1 1 0 0 0|0|0|0|0 0 0 |
   13 | 0|0|0 1 0 0 0|0|0|0|0 0 1 |
    4 | 0|0|0 0 0 0 0|0|0|0|0 0 0 |
    7 | 0|0|0 0 0 0 0|0|0|0|0 1 0 |
   10 | 0|0|0 0 0 0 0|0|0|0|0 1 0 |
    5 | 0|0|0 0 0 0 0|0|0|0|0 0 1 |
    9 | 0|0|0 0 0 0 0|0|0|0|1 1 0 |
   11 \ 0|0|0 0 0 0 0|0|0|0|0 1 0 /
```

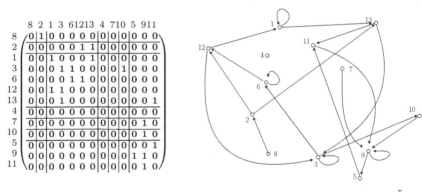

Fig. 5.29 The original relation of Fig. 5.28 grouped according to $R^* \cap R^{*^\mathsf{T}}$

The following observation mainly expresses that on looking for the quotient after having determined the strongly connected components, one will obtain an ordering that may be topologically sorted by Szpilrajn extension, which in turn results in the block-diagonal form mentioned.

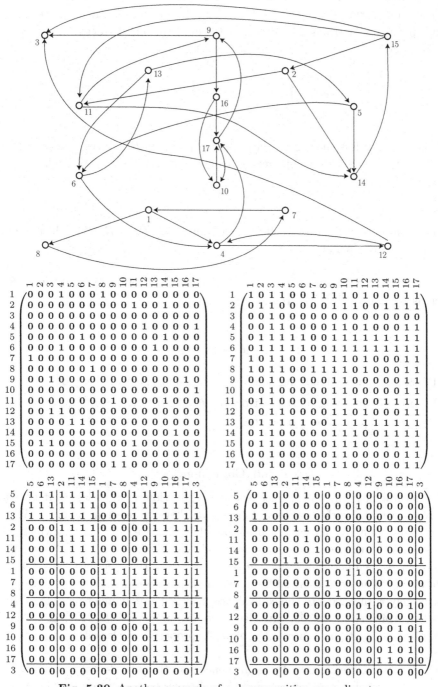

Fig. 5.30 Another example of a decomposition according to strongly connected components

Observation 5.37. By simultaneously permuting rows and columns, any given homogeneous relation R can be transformed into a matrix of matrices in the following form: it has upper triangular pattern with square diagonal blocks

$$\begin{pmatrix} \square & * & * & * \\ \bot\!\!\!\bot & \square & * & * \\ \bot\!\!\!\bot & \bot\!\!\!\bot & \square & * \\ \bot\!\!\!\bot & \bot\!\!\!\bot & \bot\!\!\!\bot & \square \end{pmatrix}$$

where $* = \bot\!\!\!\bot$ unless the generated permuted preorder R^* allows entries $\neq \bot\!\!\!\bot$. The reflexive-transitive closure of every diagonal block is the universal relation $\top\!\!\!\top$ for that block. $\qquad\qquad\qquad\qquad\qquad\qquad\qquad\qquad\qquad\qquad\qquad\qquad\quad\square$

When considering R^*, one should bear in mind that $R^* = \mathcal{R}(R^*)$, which is trivial to prove when expanded.

Figure 5.30 shows yet another example, giving the graph first, and then R, R^* as well as R^* and finally R rearranged.

Exercises

5.17 Show that the following hold for every homogeneous relation R

(i) $\inf \{ H \mid R \cup R \mathbin{;} H \subseteq H \} = \inf \{ H \mid R \cup R \mathbin{;} H = H \} = R^+,$

(ii) $\inf \{ H \mid S \cup R \mathbin{;} H \subseteq H \} = R^* \mathbin{;} S.$

5.6 Congruences

In a natural way, any equivalence leads to a partitioning into classes. It is usually more efficient to work with sets of classes as they are fewer in number. We have learned to compute 'modulo' a prime number; i.e., we only care to which class, determined by the remainder, a number belongs. However, this is only possible in cases where the operations aimed at behave nicely with respect to the subdivision into classes. The operations addition and multiplication do in this sense cooperate adequately with regard to the equivalence 'have_same_remainder_modulo_5'. If one tried an arbitrary subdivision into classes, for example, $\{1, 5, 99, 213\}, \{2, 3, 4, 6\} \ldots$, addition and multiplication would not work 'properly' on classes.

We ask what it means that the equivalence and the intended operations 'behave nicely' and develop very general algebraic characterizations. In such cases, we are accustomed to call them *congruences*. While the study of *mappings* that respect a certain equivalence is a classical topic, only recently have *relations* respecting equivalences also been studied in *simulations* and *bisimulations*. This is well known for algebraic structures, i.e., those defined by mappings on or between sets. In the non-algebraic case, relations are possibly neither univalent nor total. While the basic idea is known from many application fields, the following general concepts may provide a new abstraction.

Definition 5.38. Let B be a relation and Ξ, Θ equivalences. The pair (Ξ, Θ) is called a

$$B\text{-}\mathbf{congruence} \quad :\Longleftrightarrow \quad \Xi;B \subseteq B;\Theta. \qquad \square$$

We are going to show how this containment formula describes what we mean when saying that a 'structure' B between sets X und Y is respected somehow by equivalences Ξ on X and Θ on Y. To this end, consider an element x having an equivalent element x' which is in relation B with y. In all these cases, there shall exist for x an element y' to which it is in relation B, and which is in addition equivalent to y. This may also be written down in predicate-logic form as

$$\forall x \in X : \forall y \in Y :$$
$$\big[\exists x' \in X : (x, x') \in \Xi \wedge (x', y) \in B\big] \quad \rightarrow \quad \big[\exists y' \in Y : (x, y') \in B \wedge (y', y) \in \Theta\big].$$

We will not show here how the relational formula and the predicate-logic expression correspond to one another, but they do. Some examples will illustrate what a congruence is meant to capture.

Example 5.39. Consider a pairset together with the corresponding projection to the second component, $\rho : X \times Y \longrightarrow Y$ and define the equivalence Ξ as 'Have common second component'. Take for the other equivalence the identity on Y. Then $(\Xi, \mathbb{I})$ constitute a ρ-congruence as shown in Fig. 5.31. $\square$

$$\Xi = \begin{array}{c}(a,1)\\(a,2)\\(a,3)\\(a,4)\\(b,1)\\(b,2)\\(b,3)\\(b,4)\\(c,1)\\(c,2)\\(c,3)\\(c,4)\end{array}\begin{pmatrix}
1&0&0&0&1&0&0&0&1&0&0&0\\
0&1&0&0&0&1&0&0&0&1&0&0\\
0&0&1&0&0&0&1&0&0&0&1&0\\
0&0&0&1&0&0&0&1&0&0&0&1\\
1&0&0&0&1&0&0&0&1&0&0&0\\
0&1&0&0&0&1&0&0&0&1&0&0\\
0&0&1&0&0&0&1&0&0&0&1&0\\
0&0&0&1&0&0&0&1&0&0&0&1\\
1&0&0&0&1&0&0&0&1&0&0&0\\
0&1&0&0&0&1&0&0&0&1&0&0\\
0&0&1&0&0&0&1&0&0&0&1&0\\
0&0&0&1&0&0&0&1&0&0&0&1
\end{pmatrix}
\qquad
\rho = \begin{pmatrix}
1&0&0&0\\
0&1&0&0\\
0&0&1&0\\
0&0&0&1\\
1&0&0&0\\
0&1&0&0\\
0&0&1&0\\
0&0&0&1\\
1&0&0&0\\
0&1&0&0\\
0&0&1&0\\
0&0&0&1
\end{pmatrix}
\qquad
\mathbb{I} = \begin{pmatrix}
1&0&0&0\\
0&1&0&0\\
0&0&1&0\\
0&0&0&1
\end{pmatrix}$$

Fig. 5.31 $\Xi, \mathbb{I}$ is a congruence with respect to ρ

If B were a binary operation on a given set and we had $\Xi = \Theta$, we would say that

B 'has the substitution property with regard to Ξ'. Figure 5.32 shows schematically that when the two arguments of a binary mapping are varied inside their equivalence classes then the image may vary also, but is confined to its congruence class.

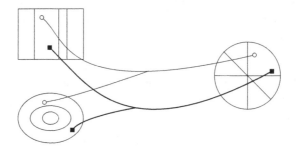

Fig. 5.32 Binary mapping satisfying the substitution property

Example 5.40. The example of Fig. 5.33 is taken from the field of knowledge acquisition. Assume an arbitrary relation R resulting out of some investigation or polling procedure. If it is a rather sparse matrix, the following idea becomes interesting. If it is not sparse, the method is still correct but it will deliver less interesting results. One asks which row entries are related to other row entries simply by the fact that they have a column entry associated with R in common; we consider, thus, $R \,;\, R^\mathsf{T}$. In a symmetric way, we are interested in $R^\mathsf{T} \,;\, R$, i.e., in column entries combined by the property that they have a row entry in common assigned by R.

$$
\begin{array}{c}
\ \\
1\\2\\3\\4\\5\\6\\7\\8\\9\\10\\11\\12\\13\\14
\end{array}
\begin{array}{c}
\scriptstyle 1\ 2\ 3\ 4\ 5\ 6\ 7\ 8\ 9\ 10\ 11\ 12\ 13\ 14\\
\left(\begin{array}{cccccccccccccc}
1&0&0&0&0&1&0&0&0&0&0&0&0&0\\
0&1&0&0&0&0&0&0&0&1&0&0&0&0\\
0&0&0&0&0&0&0&0&0&0&0&0&0&0\\
0&0&0&1&1&0&0&0&0&0&0&0&0&0\\
0&0&0&1&1&0&0&1&0&0&0&0&0&0\\
1&0&0&0&0&1&0&0&0&0&0&0&1&0\\
0&0&0&0&0&0&0&0&0&0&0&0&0&0\\
0&0&0&0&1&0&0&1&0&0&0&1&0&0\\
0&0&0&0&0&0&0&0&0&0&0&0&0&0\\
0&0&0&0&0&0&0&0&0&1&1&0&0&1\\
0&1&0&0&0&0&0&0&0&1&1&0&0&0\\
0&0&0&0&0&0&1&0&0&0&1&0&0&0\\
0&0&0&0&0&1&0&0&0&0&0&0&1&0\\
0&0&0&0&0&0&0&0&1&0&0&0&0&1
\end{array}\right)
\end{array}
\quad
\begin{array}{c}
\ \\
1\\2\\3\\4\\5\\6\\7\\8\\9\\10\\11\\12\\13\\14
\end{array}
\begin{array}{c}
\scriptstyle a\ b\ c\ d\ e\ f\ g\ h\ i\ j\ k\\
\left(\begin{array}{ccccccccccc}
0&1&0&0&0&0&0&0&1&0&0\\
0&0&0&0&0&0&0&0&0&1&0\\
0&0&0&0&0&0&0&0&0&0&0\\
0&0&0&0&0&0&0&1&0&0&0\\
1&0&0&0&0&0&0&1&0&0&0\\
0&0&0&0&0&0&1&0&1&0&0\\
0&0&0&0&0&0&0&0&0&0&0\\
1&0&0&0&1&0&0&0&0&0&0\\
0&0&0&0&0&0&0&0&0&0&0\\
0&0&1&0&0&1&0&0&0&0&0\\
0&0&1&0&0&0&0&0&0&1&0\\
0&0&0&0&1&0&0&0&0&0&0\\
0&0&0&1&0&0&1&0&0&0&0\\
0&0&0&0&0&1&0&0&0&0&0
\end{array}\right)
\end{array}
\quad
\begin{array}{c}
\ \\
a\\b\\c\\d\\e\\f\\g\\h\\i\\j\\k
\end{array}
\begin{array}{c}
\scriptstyle a\ b\ c\ d\ e\ f\ g\ h\ i\ j\ k\\
\left(\begin{array}{ccccccccccc}
1&0&0&0&1&0&0&1&0&0&0\\
0&1&0&0&0&0&0&0&1&0&0\\
0&0&1&0&0&1&0&0&0&1&0\\
0&0&0&1&0&0&1&0&0&0&0\\
1&0&0&0&1&0&0&0&0&0&0\\
0&0&1&0&0&1&0&0&0&0&0\\
0&0&0&1&0&0&1&0&1&0&0\\
1&0&0&0&0&0&0&1&0&0&0\\
0&1&0&0&0&0&1&0&1&0&0\\
0&0&1&0&0&0&0&0&0&1&0\\
0&0&0&0&0&0&0&0&0&0&0
\end{array}\right)
\end{array}
$$

Fig. 5.33 The construct $R \,;\, R^\mathsf{T}$, for arbitrary R, and $R^\mathsf{T} \,;\, R$

Rearranging rows and columns simultaneously, a procedure that does not change the relation, the equivalences are better visualized. Then one can check easily that the left and the right equivalence

$$\Omega := (R; R^{\mathsf{T}})^*, \qquad \Omega' := (R^{\mathsf{T}}; R)^*,$$

as defined in Def. 5.35, satisfy

$$\Omega; R = R; \Omega',$$

i.e., that Ω, Ω' constitute an R-congruence; see Prop. 5.36.ii. The accumulated knowledge means that one is now in a position to say that the first group of elements on the left is in relation only to the first group of elements on the right, etc. □

	1	6	13	2	10	11	14	4	5	8	12	3	7	9
1	1	1	1	0	0	0	0	0	0	0	0	0	0	0
6	1	1	1	0	0	0	0	0	0	0	0	0	0	0
13	1	1	1	0	0	0	0	0	0	0	0	0	0	0
2	0	0	0	1	1	1	1	0	0	0	0	0	0	0
10	0	0	0	1	1	1	1	0	0	0	0	0	0	0
11	0	0	0	1	1	1	1	0	0	0	0	0	0	0
14	0	0	0	1	1	1	1	0	0	0	0	0	0	0
4	0	0	0	0	0	0	0	1	1	1	1	0	0	0
5	0	0	0	0	0	0	0	1	1	1	1	0	0	0
8	0	0	0	0	0	0	0	1	1	1	1	0	0	0
12	0	0	0	0	0	0	0	1	1	1	1	0	0	0
3	0	0	0	0	0	0	0	0	0	0	0	1	0	0
7	0	0	0	0	0	0	0	0	0	0	0	0	1	0
9	0	0	0	0	0	0	0	0	0	0	0	0	0	1

	b	d	g	i	c	f	j	a	e	h	k
1	1	0	0	1	0	0	0	0	0	0	0
6	0	0	1	1	0	0	0	0	0	0	0
13	0	1	1	0	0	0	0	0	0	0	0
2	0	0	0	0	0	0	1	0	0	0	0
10	0	0	0	0	1	1	0	0	0	0	0
11	0	0	0	0	1	0	1	0	0	0	0
14	0	0	0	0	0	1	0	0	0	0	0
4	0	0	0	0	0	0	0	0	0	1	0
5	0	0	0	0	0	0	0	1	0	1	0
8	0	0	0	0	0	0	0	1	1	0	0
12	0	0	0	0	0	0	0	0	1	0	0
3	0	0	0	0	0	0	0	0	0	0	0
7	0	0	0	0	0	0	0	0	0	0	0
9	0	0	0	0	0	0	0	0	0	0	0

	b	d	g	i	c	f	j	a	e	h	k
b	1	1	1	1	0	0	0	0	0	0	0
d	1	1	1	1	0	0	0	0	0	0	0
g	1	1	1	1	0	0	0	0	0	0	0
i	1	1	1	1	0	0	0	0	0	0	0
c	0	0	0	0	1	1	1	0	0	0	0
f	0	0	0	0	1	1	1	0	0	0	0
j	0	0	0	0	1	1	1	0	0	0	0
a	0	0	0	0	0	0	0	1	1	1	0
e	0	0	0	0	0	0	0	1	1	1	0
h	0	0	0	0	0	0	0	1	1	1	0
k	0	0	0	0	0	0	0	0	0	0	1

Left equivalence $\Omega := (R; R^{\mathsf{T}})^*$, the given R, and right equivalence $\Omega' := (R^{\mathsf{T}}; R)^*$

$\eta_\Omega =$

1	1	0	0	0	0	0
6	1	0	0	0	0	0
13	1	0	0	0	0	0
2	0	1	0	0	0	0
10	0	1	0	0	0	0
11	0	1	0	0	0	0
14	0	1	0	0	0	0
4	0	0	1	0	0	0
5	0	0	1	0	0	0
8	0	0	1	0	0	0
12	0	0	1	0	0	0
3	0	0	0	1	0	0
7	0	0	0	0	1	0
9	0	0	0	0	0	1

$S =$

	[b,d,g,i]	[c,f,j]	[a,e,h]	[k]
[1, 6, 13]	1	0	0	0
[2,10,11,14]	0	1	0	0
[4,5,8,12]	0	0	1	0
[3]	0	0	0	0
[7]	0	0	0	0
[9]	0	0	0	0

$\eta_{\Omega'} =$

b	1	0	0	0
d	1	0	0	0
g	1	0	0	0
i	1	0	0	0
c	0	1	0	0
f	0	1	0	0
j	0	1	0	0
a	0	0	1	0
e	0	0	1	0
h	0	0	1	0
k	0	0	0	1

Fig. 5.34 Factorizing a relation by the congruences of its difunctional closure

Remark 5.41. At this point, we insert this remark referring forward to Fig. 10.23. There, the row equivalence $\Xi(h_{\text{difu}}(R))$ and the column equivalence $\Psi(h_{\text{difu}}(R))$ of the difunctional closure $h_{\text{difu}}(R)$ of R are also considered, not just the left equivalence $\Omega := (R; R^{\mathsf{T}})^*$ and right equivalence $\Omega' := (R^{\mathsf{T}}; R)^*$. These also form a congruence. While $\Psi(h_{\text{difu}}(R))$ and Ω' coincide in the present special case, $\Xi(h_{\text{difu}}(R))$ differs from Ω. The lower right 3×3-matrix of $\Xi(h_{\text{difu}}(R))$ would be a 3×3-block of **1**s. The relation S of Fig. 5.34 formed correspondingly, would then have only one empty last line. □

For an account of applications of congruences in software engineering, we refer to [50].

5.7 Homomorphisms

To present the following material so soon may seem irritating. It is, however, necessary for many phenomenological considerations to which we have confined ourselves in this chapter. The term *homomorphism* should not be considered as referring to a deeper mathematical concept. We will here have a look at it from a basic relational point of view.

Let any two 'structures' be given; here for simplicity we assume a relation R_1 between sets X_1, Y_1 and a relation R_2 between sets X_2, Y_2. Such structures may be conceived as addition or multiplication in a group, or as an ordering, an equivalence, or simply a graph on a set, or may describe multiplication of a vector with a scalar in a vector space. As a preparation, we recall the isotonicity of orderings.

Definition 5.42. Let any two ordered sets $X_1, \leq_1$ and $X_2, \leq_2$, respectively $E_i : X_i \longrightarrow X_i, i = 1, 2$, be given as well as a mapping $\varphi : X_1 \longrightarrow X_2$. Then we have the following equivalent possibilities to qualify the mapping as **isotonic** (also **monotonic**).

(i) For all $x, y \in X_1$ satisfying $x \leq_1 y$ we have $\varphi(x) \leq_2 \varphi(y)$, or, writing it slightly more formally, as
$$\forall x \in X_1 : \forall y \in X_1 : (x, y) \in E_1 \rightarrow (\varphi(x), \varphi(y)) \in E_2.$$
(ii) $E_1 {}^{\cdot} \varphi \subseteq \varphi {}^{\cdot} E_2.$ □

It seems obvious that version (ii) is much shorter, less error-prone, and may be supported more easily by a computer algebra system.

Example 5.43. In this example, the structures E_1, E_2 are different orderings on 4-element sets as shown in Fig. 5.35. We consider φ which describes a homomorphism corresponding to horizontal transition from one graph to the other.

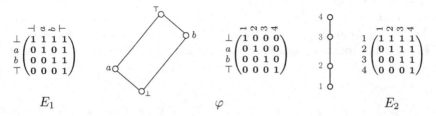

Fig. 5.35 Isotone mapping from a diamond ordering to a linear ordering

Careful observation shows that homomorphy holds only for the ordering, but not for the least upper or greatest lower bounds `lub`, `glb` formed therein. As an example take the two medium elements in Fig. 5.35.

Whenever two elements x, y on the left are in the order relation $(x, y) \in E_1$, their images on the right are also in an order relation, $(\varphi(x), \varphi(y)) \in E_2$. The elements a, b are unrelated on the left-hand side, their images $2, 3$ on the right, however, are related as $(2, 3) \in E_2$. This is definitely allowed in an order homomorphism. But it indicates already that this will not be an isomorphism. For the reverse direction, $(2, 3)$ are related, but their inverse images are not. □

When we strive to compare any two given such structures conceived as a plexus of relations, we must be in a position to relate them somehow. This means, typically, that two mappings $\varphi : X_1 \longrightarrow X_2$ and $\psi : Y_1 \longrightarrow Y_2$ are provided 'mapping the first structure into the second'.

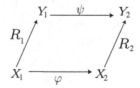

Fig. 5.36 Basic concept of a homomorphism

Once such mappings φ, ψ are given, they are said to form a homomorphism of the first into the second structure if the following holds. Whenever any two elements x, y are related by the first relation R_1, their images $\varphi(x), \psi(y)$ are related by the second relation R_2. This is captured by the lengthy predicate-logic formulation

$$\forall x \in X_1 : \forall y \in Y_1 : (x, y) \in R_1 \to (\varphi(x), \psi(y)) \in R_2$$

which will now be converted to a shorter relational form. A point to mention is that we assume the structure relations R_1, R_2 and the mappings φ, ψ to reside in one relation algebra.

Definition 5.44. Let two 'structures' be given, a relation R_1 between the sets X_1, Y_1 and a relation R_2 between the sets X_2, Y_2. Mappings $\varphi : X_1 \longrightarrow X_2$ and $\psi : Y_1 \longrightarrow Y_2$ from the structure on the left side to the structure on the right are called a **homomorphism** (φ, ψ) of the first into the second structure if the following holds:

$R_1 \,\raise.1ex\hbox{$\scriptstyle\,$}\, \psi \subseteq \varphi \,\raise.1ex\hbox{$\scriptstyle\,$}\, R_2.$

With $\varphi^{\mathsf{T}} \,\raise.1ex\hbox{$\scriptstyle\,$}\, \varphi \subseteq \mathbb{I}, \quad \mathbb{I} \subseteq \varphi \,\raise.1ex\hbox{$\scriptstyle\,$}\, \varphi^{\mathsf{T}}, \quad \psi^{\mathsf{T}} \,\raise.1ex\hbox{$\scriptstyle\,$}\, \psi \subseteq \mathbb{I}, \quad \mathbb{I} \subseteq \psi \,\raise.1ex\hbox{$\scriptstyle\,$}\, \psi^{\mathsf{T}}$, one might also describe the fact that φ, ψ are mappings in a relational form. $\qquad \square$

Often one has structures 'on a set', for which the mappings φ, ψ coincide. Since it is not appealing to talk of the homomorphism (φ, φ), one simply denotes the homomorphism as φ.

This concept of homomorphism – solely based on inclusion $R_1 \,\raise.1ex\hbox{$\scriptstyle\,$}\, \psi \subseteq \varphi \,\raise.1ex\hbox{$\scriptstyle\,$}\, R_2$ – is a very general one; it is extremely broadly applicable. As we have done already on several occasions, we show how the lengthy predicate-logic form translates to the shorter relational form:

$$\forall x_1 \in X_1 : \forall y_1 \in Y_1 : \forall x_2 \in X_2 : \forall y_2 \in Y_2 :$$
$$(x_1, y_1) \in R_1 \wedge (x_1, x_2) \in \varphi \wedge (y_1, y_2) \in \psi \rightarrow (x_2, y_2) \in R_2$$
$$u \vee v = v \vee u, \ a \rightarrow b = \neg a \vee b$$

$$\Longleftrightarrow \quad \forall x_1 \in X_1 : \forall y_1 \in Y_1 : \forall y_2 \in Y_2 : \forall x_2 \in X_2 :$$
$$(x_1, y_1) \notin R_1 \vee (y_1, y_2) \notin \psi \vee (x_1, x_2) \notin \varphi \vee (x_2, y_2) \in R_2$$
$$a \vee [\forall x : p(x)] = \forall x : [a \vee p(x)]$$

$$\Longleftrightarrow \quad \forall x_1 \in X_1 : \forall y_1 \in Y_1 : \forall y_2 \in Y_2 : (x_1, y_1) \notin R_1 \vee (y_1, y_2) \notin \psi \vee$$
$$\big(\forall x_2 \in X_2 : (x_1, x_2) \notin \varphi \vee (x_2, y_2) \in R_2\big)$$
$$\text{by rearrangement, } \forall a : p(a) = \neg[\exists a : \neg p(a)]$$

$$\Longleftrightarrow \quad \forall x_1 \in X_1 : \forall y_2 \in Y_2 : \forall y_1 \in Y_1 : (x_1, y_1) \notin R_1 \vee (y_1, y_2) \notin \psi \vee$$
$$\neg\big(\exists x_2 \in X_2 : (x_1, x_2) \in \varphi \wedge (x_2, y_2) \notin R_2\big)$$
$$\text{definition of composition}$$

$$\Longleftrightarrow \quad \forall x_1 \in X_1 : \forall y_2 \in Y_2 :$$
$$\neg\big(\exists y_1 \in Y_1 : (x_1, y_1) \in R_1 \wedge (y_1, y_2) \in \psi\big) \vee \neg\big((x_1, y_2) \in \varphi \,\raise.1ex\hbox{$\scriptstyle\,$}\, \overline{R_2}\big)$$
$$f \,\raise.1ex\hbox{$\scriptstyle\,$}\, \overline{R} = \overline{f \,\raise.1ex\hbox{$\scriptstyle\,$}\, R} \text{ for a mapping } f$$

$$\Longleftrightarrow \quad \forall x_1 \in X_1 : \forall y_2 \in Y_2 : \neg\big((x_1, y_2) \in R_1 \,\raise.1ex\hbox{$\scriptstyle\,$}\, \psi\big) \vee \big((x_1, y_2) \in \varphi \,\raise.1ex\hbox{$\scriptstyle\,$}\, R_2\big)$$
$$a \rightarrow b = \neg a \vee b$$

$$\Longleftrightarrow \quad \forall x_1 \in X_1 : \forall y_2 \in Y_2 : \big((x_1, y_2) \in R_1 \,\raise.1ex\hbox{$\scriptstyle\,$}\, \psi\big) \rightarrow \big((x_1, y_2) \in \varphi \,\raise.1ex\hbox{$\scriptstyle\,$}\, R_2\big)$$
$$\text{transition to point-free form}$$

$$\Longleftrightarrow \quad R_1 \,\raise.1ex\hbox{$\scriptstyle\,$}\, \psi \subseteq \varphi \,\raise.1ex\hbox{$\scriptstyle\,$}\, R_2$$

The result looks quite similar to a commutativity rule; it is, however, not an equality but containment. When one follows the structural relation R_1 and then proceeds with ψ to the set Y_2, it is always possible to go to the set X_2 first and then follow the structural relation R_2 to reach the same element.

A nice comparison is possible with a congruence: should the equivalences Ξ, Ω constitute a congruence for the mapping f, i.e., $\Xi \mathbin{;} f \subseteq f \mathbin{;} \Omega$, this is precisely the condition for f to be a homomorphism from Ξ to Ω.

The homomorphism condition has four variants which may be used interchangeably.

Proposition 5.45. *If φ, ψ are mappings, then*

$$R_1 \mathbin{;} \psi \subseteq \varphi \mathbin{;} R_2 \iff R_1 \subseteq \varphi \mathbin{;} R_2 \mathbin{;} \psi^\mathsf{T} \iff \varphi^\mathsf{T} \mathbin{;} R_1 \subseteq R_2 \mathbin{;} \psi^\mathsf{T} \iff \varphi^\mathsf{T} \mathbin{;} R_1 \mathbin{;} \psi \subseteq R_2. \square$$

The proof is immediate in view of the mapping properties $\varphi^\mathsf{T} \mathbin{;} \varphi \subseteq \mathbb{I}$ etc.

One may wonder why the condition for homomorphy is just a containment and not an equation. In group theory, for example, $\varphi(x +_1 y) = \varphi(x) +_2 \varphi(y)$ would be demanded. The reason is that we aim at algebraic as well as relational structures. The condition for homomorphy, $R_1 \mathbin{;} \psi \subseteq \varphi \mathbin{;} R_2$, is suited to both. Whenever R_1 as well as R_2 are mappings, making the structures algebraic ones, the homomorphism will automatically satisfy an equation. We fix this as a separate proposition.

Proposition 5.46. *If (φ, ψ) is a homomorphism from the structure R_1 into the structure R_2, and if R_1, R_2 are mappings, then*

$$R_1 \mathbin{;} \psi = \varphi \mathbin{;} R_2.$$

Proof: $R_1 \mathbin{;} \psi \subseteq \varphi \mathbin{;} R_2$ holds by assumption; "$\supseteq$" follows from Prop. 5.2.iii. $\qquad\square$

As usual, isomorphisms are also introduced.

Definition 5.47. We call (φ, ψ) an **isomorphism** between the two relations R_1, R_2, if it is a homomorphism from R_1 to R_2 and if in addition $(\varphi^\mathsf{T}, \psi^\mathsf{T})$ is a homomorphism in the reverse direction from R_2 to R_1. $\qquad\square$

An isomorphism of structures is thus defined to be a 'homomorphism in both directions'. In this case, φ, ψ will be bijective mappings.

The following simple lemma will sometimes help in identifying an isomorphism.

Lemma 5.48. *Let relations R_1, R_2 be given together with a homomorphism (φ, ψ) from R_1 to R_2 such that*

φ, ψ *are bijective mappings and* $R_1 \mathbin{;} \psi = \varphi \mathbin{;} R_2$.

Then φ, ψ is an isomorphism.

Proof: $R_2 \mathbin{;} \psi^{\mathsf{T}} = \varphi^{\mathsf{T}} \mathbin{;} \varphi \mathbin{;} R_2 \mathbin{;} \psi^{\mathsf{T}} = \varphi^{\mathsf{T}} \mathbin{;} R_1 \mathbin{;} \psi \mathbin{;} \psi^{\mathsf{T}} = \varphi^{\mathsf{T}} \mathbin{;} R_1$. □

We provide several illustrations of the concept of a homomorphism.

Example 5.49. We study a structure-preserving mapping of a graph into another graph.

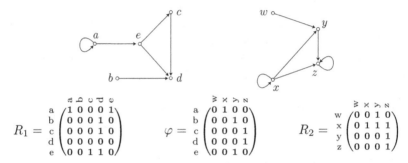

Fig. 5.37 A graph homomorphism

Indeed, going from a to e and then to y, it is also possible to map first from a to x and then proceed to y. Of course, not only this case has to be considered; it has to be checked for all such configurations. We see, for example, that not every point on the right is an image point. We see also that the arrow $c \to d$ is mapped to a loop in z. Starting in a, there was no arrow to c; it exists, however, between the images x and z. Of course φ is not an isomorphism. □

Proposition 5.50. *Let any relation R between sets V and W be given and assume that Ξ, Ω is an R-congruence. Denoting the natural projections as η_Ξ, η_Ω, respectively, we form the quotient sets and consider the relation $S := \eta_\Xi^{\mathsf{T}} \mathbin{;} R \mathbin{;} \eta_\Omega$ between V_Ξ and W_Ω. Then η_Ξ, η_Ω is a homomorphism of the structure R into the structure S that satisfies $R \mathbin{;} \eta_\Omega = \eta_\Xi \mathbin{;} S$.*

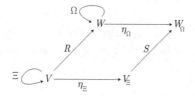

Fig. 5.38 Natural projections as homomorphisms

Proof: See B.13. □

Of course, $\eta_{\Xi}, \eta_{\Omega}$ need not constitute an isomorphism; they are normally not bijective.

Example 5.51. The company of Fig. 5.39 provides an example with *two* structural relations, namely the material flow between its 12 locations and the internal reporting structure. Two consultant firms of this company have been asked to propose models of further subdivision. Their results are shown as the two smaller graphs below.

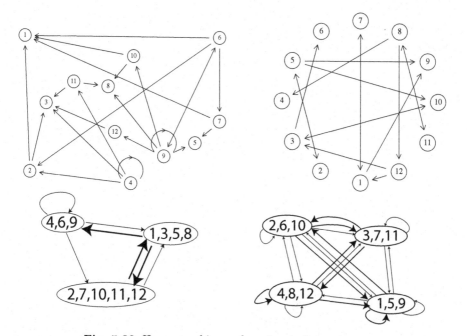

Fig. 5.39 Homomorphisms when structuring a company

It is indicated how both these 12-vertex graphs are mapped onto the left as well as onto the right smaller graphs. Images from the left side are thin lines while those from the right are fat. If it were intended to switch to a more efficient structure of the company, one would probably decide for the subdivision on the left.   □

We have so far been studying homomorphisms. *Mappings* φ, ψ will not exist in every case when one is about to compare structures; sometimes one just has *relations*. For this less favorable situation there exist concepts allowing the comparison of structures, albeit only with minor precision. One basic aspect we will lose is that of being able to 'roll' the conditions so as to have four different but equivalent forms according to Prop. 5.45.

6
Relations and Vectors

When working with real-valued matrices, it is normal to try to determine their eigenvalues and eigenvectors. In the relational case one cannot hope for interesting properties of eigenvalues, as these would be restricted to being $\mathbf{0}, \mathbf{1}$. Nevertheless, there is an important field of study of the equivalent of eigenvectors satisfying $A; x \subseteq x, A; \overline{x} \subseteq x$, or $A; x \subseteq \overline{x}$. This topic can be included in an advanced study of general Galois mechanisms, which is rather involved and, thus, postponed to Chapter 16.

There are, however, some very simple concepts related to a diversity of applications that can be introduced here on a phenomenological basis, in order to obtain ways of denoting several effects coherently. In every case, we will consider one or two vectors together with the relation in question. Quite frequently, one can see that the definitions are chosen adequately as some visualizations immediately reveal intuitive effects.

6.1 Domain and codomain

Relations are used mainly because they allow us to express more than is possible with functions or mappings. They may assign one or more values to an argument or none. It is of course interesting to determine which arguments are assigned values and which are not. In the same way, one will ask which elements on the target side occur as images and which do not. The following rather simplistic definition provides an algebraic formulation for the zones just indicated.

Definition 6.1. Given a (possibly heterogeneous) relation R, we always have the vectors

$$\mathsf{dom}(R) := R; \mathbb{T} \qquad \mathsf{cod}(R) := R^{\mathsf{T}}; \mathbb{T}$$

called the **domain** and the **codomain** of R, respectively. □

With regard to typing, this was a bit sloppy. A fully typed definition would start from a relation $R : X \longrightarrow Y$ with source and target mentioned explicitly, and would qualify the universal subsets as in

$$\texttt{dom}(R) := R;\mathbb{T}_Y \qquad \texttt{cod}(R) := R^{\mathsf{T}};\mathbb{T}_X.$$

Then we obviously have

$$\texttt{dom}(R^{\mathsf{T}}) = \texttt{cod}(R).$$

Next, it is important to be able to denote where the relations behave as functions, i.e., assign no more than one value.

Definition 6.2. Given a (possibly heterogeneous) relation R, we always have its

univalent part $\qquad \texttt{upa}(R) := R \cap \overline{R;\overline{\mathbb{I}}}$

multivalent part $\qquad \texttt{mup}(R) := R \cap R;\overline{\mathbb{I}}.$

Using these two, we define the vectors

$$\texttt{univalentZone}(R) := \texttt{dom}(\texttt{upa}(R)) := \big(R \cap \overline{R;\overline{\mathbb{I}}}\big);\mathbb{T}$$

$$\texttt{multivalentZone}(R) := \texttt{dom}(\texttt{mup}(R)) := \big(R \cap R;\overline{\mathbb{I}}\big);\mathbb{T}. \qquad \Box$$

The univalent part of R collects those assignments via R in which *at most* one image point is assigned, which means that it is *not* the case that there is assigned an image point *non-identical* with the assigned one. On the right of Fig. 6.1, we see that $\texttt{upa}(R)$ can be formed by cutting out all the **1**s belonging to the third horizontal zone. Therefore, the univalent zone is made up of rows $\{b, f, g, h\}$ and the multivalent zone of $\{a, c, d, e, k\}$. United, the latter two result in the domain $\texttt{dom}R$.

	1	2	3	4	5	6	7	8	9	10	11	12	13	14	15	16	17	18	19
a	0	0	0	0	0	0	0	0	0	0	0	0	0	1	1	0	0	1	0
b	0	0	0	0	0	0	0	0	0	0	0	0	0	0	0	1	0	0	0
c	0	0	1	1	0	0	1	0	0	0	0	0	0	0	0	1	0	0	0
d	0	0	0	0	0	0	0	0	0	1	0	0	0	0	0	0	1	0	0
e	0	0	0	0	0	0	0	1	0	1	0	0	0	1	0	0	0	0	0
f	0	0	0	0	0	1	0	0	0	0	0	0	0	0	0	0	0	0	0
g	0	0	0	0	0	0	0	0	0	0	0	1	0	0	0	0	0	0	0
h	0	0	0	0	0	1	0	0	0	0	0	0	0	0	0	0	0	0	0
i	0	0	0	0	0	0	0	0	0	0	0	0	0	0	0	0	0	0	0
j	0	0	0	0	0	0	0	0	0	0	0	0	0	0	0	0	0	0	0
k	0	0	0	0	0	0	1	0	1	0	0	0	0	0	0	0	0	1	0

	1	2	5	11	13	19	3	4	8	9	12	15	17	6	7	10	14	16	18
i	0	0	0	0	0	0	0	0	0	0	0	0	0	0	0	0	0	0	0
j	0	0	0	0	0	0	0	0	0	0	0	0	0	0	0	0	0	0	0
b	0	0	0	0	0	0	0	0	0	0	0	0	0	0	0	0	0	1	0
f	0	0	0	0	0	0	0	0	0	0	0	0	0	1	0	0	0	0	0
g	0	0	0	0	0	0	0	0	0	0	1	0	0	0	0	0	0	0	0
h	0	0	0	0	0	0	0	0	0	0	0	0	0	1	0	0	0	0	0
a	0	0	0	0	0	0	0	0	0	0	0	1	0	0	0	0	1	0	1
c	0	0	0	0	0	0	1	1	0	0	0	0	0	0	1	0	0	1	0
d	0	0	0	0	0	0	0	0	0	0	0	0	1	0	0	1	0	0	0
e	0	0	0	0	0	0	0	0	1	0	0	0	0	0	0	1	1	0	0
k	0	0	0	0	0	0	0	0	0	1	0	0	0	0	1	0	0	0	1

Fig. 6.1 An arbitrary relation subdivided according to definedness, univalence, etc.

Considering just where the relation assigns values, where it behaves univalently, and where it does not, already gives a certain overview of its structure, and may lead to reductions and simplifications or additional preconditions in theorems; see for instance Fig. 6.1. The first zone of rows, respectively columns, contains the empty lines, the second the univalent rows, respectively injective columns. The third zone always has more than one entry per row, respectively per column. Be aware, however, that row d while containing 2 entries, does not have both of these in the lower right sub-rectangle. Also, column 6 considered only in the lower right rectangle is empty.

Circuits and cycles

Circuits and cycles are so omnipresent in graph theory that fully formal definitions are often assumed to be commonly known – and are not always repeated with the necessary care. In particular, when treating all this relation-algebraically, we will have to introduce everything from scratch. We will assume cycles to be unoriented and circuits to have an orientation.

Definition 6.3. Given a homogeneous relation $R : V \longrightarrow V$, we consider a relation $\mathcal{C} \subseteq R$ and call it a **simple circuit**, provided

(i) $\mathcal{C}^\mathsf{T}{}_;\mathcal{C} \subseteq \mathbb{I}, \qquad \mathcal{C}{}_;\mathcal{C}^\mathsf{T} \subseteq \mathbb{I},$

(ii) $\mathcal{C}^+ = \mathcal{C}{}_;\mathbb{T}{}_;\mathcal{C}^\mathsf{T},$

(iii) $\mathcal{C}^2 \subseteq \overline{\mathbb{I}}.$

More specifically, the relation R may be the adjacency $\Gamma : V \longrightarrow V$ of a simple graph and, thus, symmetric. Given the simple circuit $\mathcal{C}$, we then call the relation $\mathcal{C} \cup \mathcal{C}^\mathsf{T}$ the corresponding **simple cycle**. □

Due to (i), $\mathcal{C}$ is univalent and injective. Condition (ii) guarantees that every vertex of the circuit can be reached from any other vertex. Anticipating the domain construction of extrusion of Section 7.5, one may also say that $\mathcal{C}$ with its vertices extruded will turn out to be strongly connected. Condition (iii) serves to exclude trivialities from the discussion, namely circuits of length 2, i.e., consisting of following an arc and then going back along the transposed arc. Figure 6.2 shows how this definition is meant. The graph for adjacency Γ will be found in Fig. 12.36, where the circuit is Feb, Mar, Jul, Oct, Jun.

This definition of a circuit is slightly different compared with the one traditionally used in graph theory, which is given by introducing a sequence $a_0, a_1, \ldots, a_n = a_0$ of vertices and postulating that there is no back or forward branching, etc. Our definition lends itself more easily to relational treatment but models the same idea.

Fig. 6.2 An adjacency Γ with a simple circuit $\mathcal{C}$ and its simple cycle $\mathcal{C} \cup \mathcal{C}^{\mathsf{T}}$

First matrix:

	Jan	Feb	Mar	Apr	May	Jun	Jul	Aug	Sep	Oct	Nov	Dec
Jan	0	1	0	0	1	0	0	0	0	0	0	0
Feb	1	0	1	0	1	1	0	0	0	0	0	0
Mar	0	1	0	1	0	0	1	1	0	0	0	0
Apr	0	0	1	0	0	0	0	1	0	0	0	0
May	1	1	0	0	0	1	0	0	1	1	0	0
Jun	0	1	0	0	1	0	1	0	0	1	0	0
Jul	0	0	1	0	0	1	0	1	0	1	1	0
Aug	0	0	1	1	0	0	1	0	0	0	1	1
Sep	0	0	0	0	1	0	0	0	0	1	0	0
Oct	0	0	0	0	1	1	1	0	1	0	1	0
Nov	0	0	0	0	0	0	1	1	0	1	0	1
Dec	0	0	0	0	0	0	1	0	0	1	0	0

Second matrix:

	Jan	Feb	Mar	Apr	May	Jun	Jul	Aug	Sep	Oct	Nov	Dec
Jan	0	0	0	0	0	0	0	0	0	0	0	0
Feb	0	0	1	0	0	0	0	0	0	0	0	0
Mar	0	0	0	0	0	0	1	0	0	0	0	0
Apr	0	0	0	0	0	0	0	0	0	0	0	0
May	0	0	0	0	0	0	0	0	0	0	0	0
Jun	0	1	0	0	0	0	0	0	0	0	0	0
Jul	0	0	0	0	0	0	0	0	1	0	0	0
Aug	0	0	0	0	0	0	0	0	0	0	0	0
Sep	0	0	0	0	0	0	0	0	0	0	0	0
Oct	0	0	0	0	0	1	0	0	0	0	0	0
Nov	0	0	0	0	0	0	0	0	0	0	0	0
Dec	0	0	0	0	0	0	0	0	0	0	0	0

Third matrix:

	Jan	Feb	Mar	Apr	May	Jun	Jul	Aug	Sep	Oct	Nov	Dec
Jan	0	0	0	0	0	0	0	0	0	0	0	0
Feb	0	0	1	0	0	1	0	0	0	0	0	0
Mar	0	1	0	0	0	0	1	0	0	0	0	0
Apr	0	0	0	0	0	0	0	0	0	0	0	0
May	0	0	0	0	0	0	0	0	0	0	0	0
Jun	0	1	0	0	0	0	0	0	0	1	0	0
Jul	0	0	1	0	0	0	0	0	0	1	0	0
Aug	0	0	0	0	0	0	0	0	0	0	0	0
Sep	0	0	0	0	0	0	0	0	0	0	0	0
Oct	0	0	0	0	0	1	1	0	0	0	0	0
Nov	0	0	0	0	0	0	0	0	0	0	0	0
Dec	0	0	0	0	0	0	0	0	0	0	0	0

Using an example, we will demonstrate where circuits (sometimes called 'picycles') occur in modelling preference and indifference. In Fig. 6.3, we have a relation R and a circuit in it shown first. The right side shows what will later be recognized as a Ferrers relation that is obtained as $R \cap \overline{R}^{\mathsf{T}} =: P$ and presented with rows and columns permuted independently. One will see that the circuit takes transitions $(5, 4), (7, 3)$ from P and transitions $(4, 7), (3, 1), (1, 5)$ from $R \cap \overline{P}$. If one had intended to model preference $P =$ *is better than* and indifference $R \cap \overline{P} =$ *indifferent about*, one would probably dislike such circuits.

Left matrix:

	1	2	3	4	5	6	7
1	1	0	1	1	1	1	1
2	1	1	1	1	1	1	1
3	1	0	1	0	0	1	0
4	1	0	1	1	0	1	1
5	1	1	1	1	1	1	1
6	0	0	0	0	0	1	0
7	1	0	1	1	1	1	1

Middle matrix:

	1	2	3	4	5	6	7
1	0	0	0	0	1	0	0
2	0	0	0	0	0	0	0
3	1	0	0	0	0	0	0
4	0	0	0	0	0	0	1
5	0	0	0	1	0	0	0
6	0	0	0	0	0	0	0
7	0	0	1	0	0	0	0

Right matrix:

	2	5	1	7	4	3	6
2	0	0	1	1	1	1	1
5	0	0	0	0	1	1	1
4	0	0	0	0	0	1	1
7	0	0	0	0	0	1	1
1	0	0	0	0	0	0	1
3	0	0	0	0	0	0	1
6	0	0	0	0	0	0	0

Fig. 6.3 A circuit through preference and indifference

6.2 Rectangular zones

For an order, we easily observe that every element of the set u of elements smaller than some element e is related to every element of the set v of elements greater than e; see Fig. 6.4, for example, with numbers dividing 6 (i.e., $1, 2, 3, 6$) and divided by 6 (i.e., $6, 12$). Also for equivalences and preorders, square zones in the block-diagonal representation have been shown to be important, accompanied possibly by rectangular zones off diagonal.

$$
\begin{array}{c}
\begin{array}{cccccccccccc}
1&2&3&4&5&6&7&8&9&10&11&12
\end{array}\\
\begin{array}{c}
1\\2\\3\\4\\5\\6\\7\\8\\9\\10\\11\\12
\end{array}
\left(\begin{array}{cccccccccccc}
1&1&1&1&1&1&1&1&1&1&1&1\\
0&1&0&1&0&1&0&1&0&1&0&1\\
0&0&1&0&0&1&0&0&1&0&0&1\\
0&0&0&1&0&0&0&1&0&0&0&1\\
0&0&0&0&1&0&0&0&0&1&0&0\\
0&0&0&0&0&1&0&0&0&0&0&1\\
0&0&0&0&0&0&1&0&0&0&0&0\\
0&0&0&0&0&0&0&1&0&0&0&0\\
0&0&0&0&0&0&0&0&1&0&0&0\\
0&0&0&0&0&0&0&0&0&1&0&0\\
0&0&0&0&0&0&0&0&0&0&1&0\\
0&0&0&0&0&0&0&0&0&0&0&1
\end{array}\right)
\end{array}
\qquad
\begin{array}{c}
(0\;0\;0\;0\;0\;1\;0\;0\;0\;0\;0\;1)\\[2pt]
\left(\begin{array}{c}1\\1\\1\\0\\0\\1\\0\\0\\0\\0\\0\\0\end{array}\right)
\left(\begin{array}{cccccccccccc}
1&1&1&1&1&1&1&1&1&1&1&1\\
0&1&0&1&0&1&0&1&0&1&0&1\\
0&0&1&0&0&1&0&0&1&0&0&1\\
0&0&0&1&0&0&0&1&0&0&0&1\\
0&0&0&0&1&0&0&0&0&1&0&0\\
0&0&0&0&0&1&0&0&0&0&0&1\\
0&0&0&0&0&0&1&0&0&0&0&0\\
0&0&0&0&0&0&0&1&0&0&0&0\\
0&0&0&0&0&0&0&0&1&0&0&0\\
0&0&0&0&0&0&0&0&0&1&0&0\\
0&0&0&0&0&0&0&0&0&0&1&0\\
0&0&0&0&0&0&0&0&0&0&0&1
\end{array}\right)
\end{array}
$$

Fig. 6.4 Rectangle $\{1,2,3,6\} \times \{6,12\}$ in the divisibility order on numbers $1 \ldots 12$

In Fig. 6.4, the column vector indicates divisors of 6 while the row vector shows multiples of 6.

Definition 6.4. Given two vectors $u \subseteq X$ and $v \subseteq Y$, together with (possibly heterogeneous) universal relations $\mathbb{T}$, we call the relation

$$R := u_; v^{\mathsf{T}} = u_; \mathbb{T} \cap (v_; \mathbb{T})^{\mathsf{T}}$$

a **rectangular** relation or, simply, a **rectangle**.[1] Given this setting, we call u the **source vector** of the rectangle and v its **target vector**.

For a homogeneous relation, we will call the rectangle $R := u_; u^{\mathsf{T}}$ a **square**. □

This may be introduced in a slightly more general form with $u : X \longrightarrow Z$ and $v : Y \longrightarrow Z$ assumed as row-constant relations, i.e., $u = u_; \mathbb{T}_{Z,Z}$ and $v = v_; \mathbb{T}_{Z,Z}$, to obtain

$$u_; v^{\mathsf{T}} \quad \text{or} \quad u_; \mathbb{T}_{Z,Y} \cap (v_; \mathbb{T}_{Z,X})^{\mathsf{T}}.$$

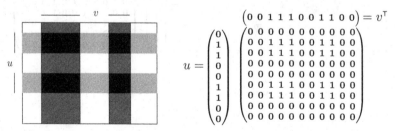

Fig. 6.5 Rectangle from intersection of a row-constant and a column-constant relation

In Def. 6.4, we had two definitional variants, and we should convince ourselves that

---

[1] This notation seems to stem from Jacques Riguet who spoke of a *rélation rectangle* (in French). If R is also symmetric, he called it a **square-shaped** relation or *rélation carré* in French. There are variant notations.

they mean the same. While "$\subseteq$" is clear, the other direction is involved and is postponed to Appendix B.14. The next proposition provides even more characterizations, partly stemming from [97].

Proposition 6.5. *For a relation R the following are equivalent:*

(i) R *is a rectangle,*

(ii) $R_; \mathbb{T}_; R \subseteq R$,

(iii) $R_; \mathbb{T}_; R = R$,

(iv) $R_; \overline{R}^\mathsf{T}_; R = \mathbb{1}$,

(v) *for any fitting pair A, B, the Dedekind rule becomes an equality:*
$$A_; B \cap R = (A \cap R_; B^\mathsf{T})_; (B \cap A^\mathsf{T}_; R).$$

Proof: See B.15. □

An immediate consequence of Prop. 6.5 is that – finite or infinite – intersections of rectangles inside a relation again constitute a rectangle in that relation.

As we will see, rectangles are often studied in combination with some given relation. They may reside inside that relation, may be outside, or contain or frame that relation.

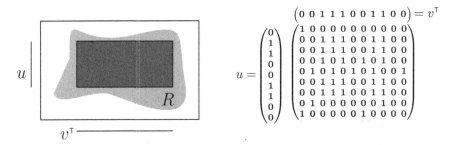

Fig. 6.6 Rectangles inside a relation

Definition 6.6. Given a relation R, we say that the vectors u, v, define a

(i) **rectangle inside R** $:\Longleftrightarrow u_; v^\mathsf{T} \subseteq R \Longleftrightarrow \overline{R}_; v \subseteq \overline{u} \Longleftrightarrow \overline{R}^\mathsf{T}_; u \subseteq \overline{v}$,

(ii) **rectangle around R** $:\Longleftrightarrow R \subseteq u_; v^\mathsf{T}$,

(iii) **rectangle outside R** $:\Longleftrightarrow u, v$ is a rectangle inside $\overline{R}$.

In the context of bipartitioned graphs, a rectangle inside a relation is often called a **block**; see, for example [65]. Reference [77] speaks of cross vectors. □

One will observe that there are three variants given for relations *inside*, but only one for relations containing R. This is not just written without care – so many essentially different variants do not exist. Among the rectangles containing a relation, the smallest is particularly important.

Definition 6.7. The **rectangular closure** of a relation R is defined by

$h_{\text{rect}}(R) := \inf \{ H \mid R \subseteq H, H \text{ is rectangular} \}.$ □

Besides this 'descriptive' definition, we provide an 'operational' form to obtain the rectangular closure.

Proposition 6.8. *Given a relation R, the subsets $u := \text{dom}(R), v := \text{cod}(R)$ together constitute the smallest rectangle containing R, i.e.,*

$h_{\text{rect}}(R) = u_i v^\mathsf{T} = R_i \mathbb{T}_i R = R_i \mathbb{T} \cap \mathbb{T}_i R.$

Proof: See B.16. □

As a frequently used variant we mention what happens in the case of symmetry. A clique in graph theory,[2] i.e., a square inside the adjacency relation $\Gamma := \overline{\mathbb{I}} \cap B$ of the graph, is a set of vertices all pairs of which are linked.

Definition 6.9. Given a reflexive and symmetric relation B, we call the subset

$u \text{ clique of } B \quad :\Longleftrightarrow \quad u_i u^\mathsf{T} \subseteq B \quad \Longleftrightarrow \quad \overline{B}_i u \subseteq \overline{u} \quad \Longleftrightarrow \quad \overline{B}^\mathsf{T}_i u \subseteq \overline{u}.$ □

Cliques are particularly interesting when they cannot be enlarged without losing the clique property. This will be studied extensively in Chapter 10, where also an impressive algebraic characterization will be derived.

$$
\begin{array}{c}
\begin{array}{cccccc} \scriptstyle 1 & \scriptstyle 2 & \scriptstyle 3 & \scriptstyle 4 & \scriptstyle 5 & \scriptstyle 6 \end{array} \\
\begin{array}{c} 1 \\ 2 \\ 3 \\ 4 \\ 5 \\ 6 \end{array}
\begin{pmatrix}
1 & 0 & 1 & 1 & 1 & 1 \\
0 & 1 & 0 & 0 & 0 & 1 \\
1 & 0 & 1 & 1 & 1 & 1 \\
1 & 0 & 1 & 1 & 0 & 0 \\
1 & 0 & 1 & 0 & 1 & 1 \\
1 & 1 & 1 & 0 & 1 & 1
\end{pmatrix}
\end{array}
\quad
\begin{pmatrix} 0 \\ 0 \\ 1 \\ 0 \\ 0 \\ 1 \end{pmatrix}
\quad
\begin{pmatrix} 1 \\ 0 \\ 1 \\ 1 \\ 0 \\ 0 \end{pmatrix}
\quad
\begin{pmatrix} 1 \\ 0 \\ 1 \\ 0 \\ 1 \\ 1 \end{pmatrix}
$$

Fig. 6.7 Three cliques; the last two do not admit bigger ones

---

[2] One will observe that we do not postulate an irreflexive relation as traditionally in graph theory. This is in order to avoid writing $u_i u^\mathsf{T} \subseteq R \cup \mathbb{I}$ all the time, i.e., with the diagonal added.

Exercises

6.1 Determine all the cliques of the following relations:

$$
\begin{array}{c c}
\begin{array}{c}
\quad\; \rotatebox{90}{Mon}\;\rotatebox{90}{Tue}\;\rotatebox{90}{Wed}\;\rotatebox{90}{Thu}\;\rotatebox{90}{Fri}\;\rotatebox{90}{Sat} \\
\begin{array}{c}
\text{Mon} \\ \text{Tue} \\ \text{Wed} \\ \text{Thu} \\ \text{Fri} \\ \text{Sat}
\end{array}
\left(
\begin{array}{cccccc}
1 & 0 & 0 & 0 & 0 & 0 \\
0 & 1 & 0 & 1 & 1 & 0 \\
0 & 0 & 1 & 1 & 1 & 1 \\
0 & 1 & 1 & 1 & 1 & 0 \\
0 & 1 & 1 & 1 & 1 & 0 \\
0 & 0 & 1 & 0 & 0 & 1
\end{array}
\right)
\end{array}
&
\begin{array}{c}
\begin{array}{c}
\text{US} \\ \text{French} \\ \text{German} \\ \text{British} \\ \text{Spanish} \\ \text{Japanese} \\ \text{Italian} \\ \text{Czech}
\end{array}
\left(
\begin{array}{cccccccc}
1 & 0 & 0 & 1 & 1 & 1 & 0 & 1 \\
0 & 1 & 1 & 0 & 0 & 0 & 1 & 1 \\
0 & 1 & 1 & 1 & 1 & 1 & 0 & 0 \\
1 & 0 & 1 & 1 & 0 & 0 & 0 & 0 \\
1 & 0 & 1 & 0 & 1 & 0 & 0 & 0 \\
1 & 0 & 1 & 0 & 0 & 1 & 1 & 1 \\
0 & 1 & 0 & 0 & 0 & 1 & 1 & 1 \\
1 & 1 & 0 & 0 & 0 & 1 & 1 & 1
\end{array}
\right)
\end{array}
\end{array}
$$

6.3 Independent pair of sets and covering pair of sets

Traditionally, the basic concept is studied in many variations. We will investigate two forms it takes in graph theory, namely independent pairs of sets and covering pairs of sets. Regarding Fig. 6.8, we first forbid elements of u to be in relation with elements from v. In the other variant, we forbid the elements of $\overline{s}$ to be in relation with elements from $\overline{t}$, which is the same, but formulated for the complements.

Definition 6.10. Let a relation A be given and consider pairs (u,v) or (s,t) of subsets with s, u taken from the source and t, v from the target side.

(i) (u,v) **independent pair of sets** $:\Longleftrightarrow$ $A_{;}v \subseteq \overline{u}$

$\Longleftrightarrow$ u, v is a rectangle outside A.

(ii) (s,t) **covering pair of sets** $:\Longleftrightarrow$ $A_{;}\overline{t} \subseteq s$

$\Longleftrightarrow$ $\overline{s}, \overline{t}$ is a rectangle outside A.

A definition variant calls (u,v) an independent pair of sets if $A \subseteq \overline{u_{;}\mathbb{T}} \cup \overline{v_{;}\mathbb{T}}^{\mathsf{T}}$. Correspondingly, (s,t) is called a covering pair of sets if $A \subseteq s_{;}\mathbb{T} \cup \mathbb{T}_{;}t^{\mathsf{T}}$.   □

In Fig. 6.8, rows and columns are permuted so that the property is directly visible. Rows and columns have obviously been permuted independently.

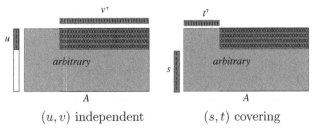

| (u,v) independent | (s,t) covering |

Fig. 6.8 The complement of an independent pair of sets is a covering pair of sets

The idea for the following statement is immediate.

Proposition 6.11. *For a given relation A together with a pair (s,t) we have*

(s,t) *covering pair of sets* $\iff$ $(\overline{s}, \overline{t})$ *independent pair of sets.* $\qquad\qquad$ $\square$

On the right side of Fig. 6.8, (s,t) is indeed a covering pair of sets, since the columns of t together with the rows of s cover all the 1s of the relation A. The covering property $A_i \overline{t} \subseteq s$ follows directly from the algebraic condition: when one follows relation A and finds oneself ending outside t, then the starting point is covered by s. The algebraic form $A \subseteq s_i \mathbb{T} \cup \mathbb{T}_i t^\mathsf{T}$ expresses directly that rows according to s and columns according to t cover all of A.

In the same way, one will find no relation between elements of rows of u with elements of columns of v on the left side. We can indeed read this directly from the condition $A_i v \subseteq \overline{u}$: When following the relation A and ending in v, it turns out that one has started from outside u. With Schröder's rule we immediately arrive at $u_i v^\mathsf{T} \subseteq \overline{A}$; this also expresses that from u to v, there is no relationship according to A.

It is a trivial fact that with (u,v) an independent pair of sets and $u' \subseteq u$, $v' \subseteq v$, then the smaller (u',v') will also be an independent pair of sets. In the same way, if (s,t) is a covering pair of sets and $s' \supseteq s$, $t' \supseteq t$, then (s',t') will also be a covering pair of sets. For independent pairs of sets, one is therefore interested in (cardinality-)maximum sets and for covering pairs one is interested in (cardinality-)minimum sets. For algebraic conditions see Chapter 10. However, there is no simple minimality (respectively maximality) criterion as one may see in Fig. 10.6.

6.4 Reducing vectors

While we have treated possibly heterogeneous relations so far, we now switch to the homogeneous case. Anyone about to solve a system of n linear equations with n variables is usually happy when this system turns out to have a special structure allowing solution of a system of m linear equations with m variables, $m < n$, first and then after resubstitution the solution of an $n - m$ system. It is precisely this that the concept of reducibility captures. When an arrow according to A ends in the set r, then it must already have started in r. Thus, arrows from $\overline{r}$ to r, as symbolized with the dashed arrow convention in Fig. 6.9, are forbidden. The index set $r = \{4,5\}$ reduces the matrix as non-zero connections from $\{1,2,3\}$ to $\{4,5\}$ do not exist.

$$\begin{pmatrix} 2 & -1 & 3 & 0 & 0 \\ 4 & 6 & -2 & 0 & 0 \\ -3 & 0 & -1 & 0 & 0 \\ -3 & 1 & 0 & 2 & -5 \\ 0 & 4 & -2 & 3 & -2 \end{pmatrix} \begin{pmatrix} x \\ y \\ z \\ u \\ v \end{pmatrix} = \begin{pmatrix} 17 \\ 26 \\ -14 \\ -13 \\ 5 \end{pmatrix}$$

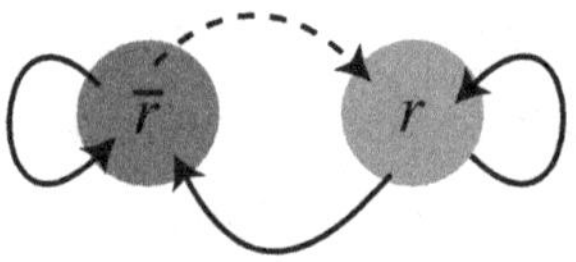

Fig. 6.9 Schema of a reducing vector, also shown with the dashed arrow convention

What we have defined here is a slight variation of the more general topic of covering pairs of sets or independent pairs of sets. In Fig. 6.8, we had permuted rows and columns *independently* so as to obtain a contiguous rectangle of **0**s. The additional restriction we are going to obey now so as to obtain a simpler – inherently homogeneous – case is that rows and columns be permuted *simultaneously*. The aim is then the same, namely arriving at a contiguous rectangle of **0**s.

We discover an algebraic flavor in this property starting from the predicate-logic version

$$\forall x, y : (x, y) \in A \wedge y \in r \rightarrow x \in r$$

$ a \rightarrow b = \neg a \vee b, \text{ arranging quantifiers}$

$$\Longleftrightarrow \quad \forall x : \forall y : (x, y) \notin A \vee y \notin r \vee x \in r$$

$ \forall y : \big(p(y) \vee c\big) = \{\forall y : p(y)\} \vee c$

$$\Longleftrightarrow \quad \forall x : \{\forall y : (x, y) \notin A \vee y \notin r\} \vee x \in r$$

$ \forall y : p(y) = \overline{\exists y : \neg p(y)}, \neg a \vee b = a \rightarrow b$

$$\Longleftrightarrow \quad \forall x : \big(\exists y : (x, y) \in A \wedge y \in r\big) \rightarrow x \in r$$

$ \text{definition of composition}$

$$\Longleftrightarrow \quad \forall x : x \in A_; r \rightarrow x \in r$$

$ \text{transition to point-free form}$

$$\Longleftrightarrow \quad A_; r \subseteq r$$

and derive from this the following two point-free definitions.

Definition 6.12. Let a homogeneous relation A and vectors r, q be given. We say that

(i) r **reduces** A :$\Longleftrightarrow$ $A_; r \subseteq r$ $\Longleftrightarrow$ $A^\mathsf{T}_; \overline{r} \subseteq \overline{r}$ $\Longleftrightarrow$ $A \subseteq \overline{\overline{r}_; r^\mathsf{T}}$,

(ii) q **is contracted by** A :$\Longleftrightarrow$ $A^\mathsf{T}_; q \subseteq q$ $\Longleftrightarrow$ $A_; \overline{q} \subseteq \overline{q}$ $\Longleftrightarrow$ $A \subseteq \overline{q_; \overline{q}^\mathsf{T}}$. □

Because $A_; r \subseteq r \Longleftrightarrow A^\mathsf{T}_; \overline{r} \subseteq \overline{r}$, a relation A is reduced by a set r precisely when its transpose A^T contracts its complement $\overline{r}$. A vector q is, thus, contracted by A precisely when its complement reduces A. The condition $A_; r \subseteq r$ is trivially satisfied for vectors $\bot, \top$, so that interest concentrates mainly on non-trivial reducing vectors, i.e., satisfying $\bot \neq r \neq \top$.

Figure 6.9 indicates that arrows of the graph according to A ending in the subset r will always start in r. It is easy to see that the reducing vectors r (this time, of course, including the extremal ones, $\mathbb{L}, \mathbb{T}$) form a lattice. The essence of the reducibility condition is much more visible after determining a permutation P that sends the 1-entries of r to the end. Applying this *simultaneously*[3] on rows and columns, we obtain the shape $P_i A_i P^{\mathsf{T}} = \begin{pmatrix} A_{11} & \mathbb{L} \\ A_{21} & A_{22} \end{pmatrix}$ as well as $P_i r = \begin{pmatrix} \mathbb{L} \\ \mathbb{T} \end{pmatrix}$. The reduction condition in the form $A \subseteq \overline{r}_i r^{\mathsf{T}}$ indicates directly that it is not the case that from outside r to r there exists a connection according to A.

When visualizing contraction, we decide to show the **0**-field in the lower left part. This means that it resides in the upper right of B^{T}. The result may be obtained by algebraic visualization as mentioned in Appendix C.

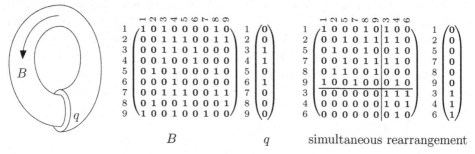

$$\begin{array}{c} B \end{array} \qquad \begin{array}{c} q \end{array} \qquad \text{simultaneous rearrangement}$$

Fig. 6.10 Contraction $B^{\mathsf{T}}_i q \subseteq q$ visualized as a self-filling funnel and via rearrangement

In a similar form, the essence of the contraction condition is now made visible. We determine a permutation P that sends the 1-entries of q to the end as shown on the right side of Fig. 6.10. Applying this *simultaneously* on rows and columns we obtain a shape for the matrix representing B with a lower left area of **0**s. This indicates immediately that $B^{\mathsf{T}}_i q$ can never have an entry outside the zone of 1s of the vector q.

6.5 Progressively infinite subsets

The condition for a subset to reduce a relation is a nice and highly desirable property, and has, thus, important applications. Now, we modify the condition $A_i x \subseteq x$ only slightly, demanding that $y \subseteq A_i y$ holds. One will immediately observe that there are not as many variants of this definition as there are, for example, for reducibility according to $A_i x \subseteq x \iff A^{\mathsf{T}}_i \overline{x} \subseteq \overline{x}$. In general, a product on the greater side of some containment is much more difficult to handle. Nevertheless there is an easily comprehended interpretation of this condition in terms of graph theory.

---

[3] Thus making it a *cogredient* permutation in the sense of [54].

One often looks for loops in a graph or for infinite paths. The task arises of characterizing the point set of starting points of an infinite path of a graph in an algebraic fashion. We assume a homogeneous relation $A : V \longrightarrow V$, and a subset $y \subseteq V$ to be given.

— $A{\,;\,}y \subseteq y$ expresses that all predecessors of y also belong to y. This is easily seen in the corresponding predicate-logic form

$$\forall p : [\exists q : (p,q) \in A \land q \in y] \to p \in y.$$

— $y \subseteq A{\,;\,}y$ expresses that every point of y precedes a point of y. This may again be seen in the predicate-logic form

$$\forall p : p \in y \to [\exists q : (p,q) \in A \land q \in y].$$

If y is a set of starting vertices of infinite paths, $y \subseteq A{\,;\,}y$ is obviously satisfied; for q, one may choose the next vertex of one of the infinite paths starting from p that are guaranteed to exist. This gives rise to an eigenvector consideration of $A{\,;\,}y = y$, characterizing y by analogy with an eigenvector x satisfying $Ax = \lambda x$ in matrix analysis. We concentrate also on the complements. Looking for non-infinite sequences in the execution of programs, for example, means being interested in the terminating sequences. This will be studied later in more detail in Section 16.2. We give the following definition.

Definition 6.13. Let a homogeneous relation A be given and the subsets v, y. We say that

(i) y is **progressively infinite** with respect to A

$:\Longleftrightarrow \quad \forall p : p \in y \to \{\exists q : (p,q) \in A \land q \in y\} \quad \Longleftrightarrow \quad y \subseteq A{\,;\,}y,$

(ii) v is **complement-expanded by** A

$:\Longleftrightarrow \quad \forall p : p \notin v \to \{\exists q : (p,q) \in A \land q \notin v\} \quad \Longleftrightarrow \quad \overline{v} \subseteq A{\,;\,}\overline{v}.$ ☐

A related question is if one starts in a set x with strict progress in the graph, whether one will then unavoidably enter the set y. In two respects this is a question of theoretical importance. Firstly, one may be interested in avoiding some error states in system dynamics. On the other hand, one may wish to be sure of reaching some set eventually. This is also related to correctness investigations of programs.

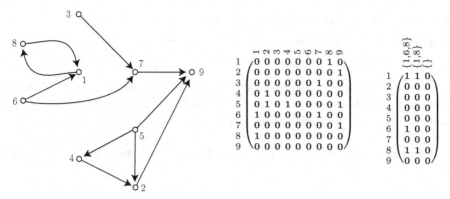

Fig. 6.11 All progressively infinite subsets y of A, i.e., those satisfying $y \subseteq A; y$

In the small example of Fig. 6.11, one will easily observe that all three progressively infinite subsets – which are here combined to form a 3-column matrix – constitute a set of sets closed under forming the union. Therefore, the greatest subset – or the supremum when non-finite – is of particular interest. Looking at the set $\{1, 6, 8\}$ of Fig. 6.11, we see that it characterizes precisely those vertices of the underlying graph from which one may run into an infinite path.

$$
\begin{array}{c c}
 & \begin{matrix} 5 & 4 & 2 & 3 & 7 & 9 & 1 & 6 & 8 \end{matrix} \\
\begin{matrix} 5 \\ 4 \\ 2 \\ 3 \\ 7 \\ 9 \\ 1 \\ 6 \\ 8 \end{matrix} &
\left(\begin{array}{cccccc|ccc}
0 & 1 & 1 & 0 & 0 & 1 & 0 & 0 & 0 \\
0 & 0 & 1 & 0 & 0 & 0 & 0 & 0 & 0 \\
0 & 0 & 0 & 0 & 0 & 1 & 0 & 0 & 0 \\
0 & 0 & 0 & 0 & 1 & 0 & 0 & 0 & 0 \\
0 & 0 & 0 & 0 & 0 & 1 & 0 & 0 & 0 \\
0 & 0 & 0 & 0 & 0 & 0 & 0 & 0 & 0 \\
\hline
0 & 0 & 0 & 0 & 0 & 0 & 0 & 0 & 1 \\
0 & 0 & 0 & 0 & 1 & 0 & 1 & 0 & 0 \\
0 & 0 & 0 & 0 & 0 & 0 & 1 & 0 & 0
\end{array}\right)
\end{array}
$$

Fig. 6.12 The same rearranged with the progressively infinite part at the end

In Fig. 6.12, one will easily observe that the upper right part is an empty relation and that the lower left is arbitrary. The diagonal blocks are characterized as follows. The upper left square allows only finite progress because it is located in its strict upper right *triangle*. The lower right diagonal block is a total relation, and will thus allow infinite progress.

6.6 Stable and absorbant sets

Yet another form of distinguished sets are stable and absorbant sets. They are defined in the homogeneous context, and when visualized will be presented via simultaneous permutation.

Definition 6.14. Given any homogeneous relation B, or equivalently any 1-graph with associated relation B, together with a set x, we call

$$x \text{ stable} \quad :\Longleftrightarrow \quad \forall p : \left\{ \exists q : (p,q) \in B \wedge q \in x \right\} \to p \notin x \quad \Longleftrightarrow \quad B{:}x \subseteq \bar{x}. \qquad \square$$

Instead of a *stable* set, this is often called an *internally stable* set; it is characterized by the fact that the pair x, x is a rectangle (in fact a square) outside B. One can also say that no arrow is allowed to exist from a vertex of x to a vertex of x.

The following condition looks quite similar, but is in fact structurally rather different.

Definition 6.15. Given any homogeneous relation B, or equivalently any 1-graph with associated relation B, together with a set x, we call

$$x \text{ absorbant} \quad :\Longleftrightarrow \quad \forall p : p \notin x \to \left\{ \exists q : (p,q) \in B \wedge q \in x \right\} \quad \Longleftrightarrow \quad \bar{x} \subseteq B{:}x. \qquad \square$$

Instead of an *absorbant* set, one speaks also of an *externally stable* set; it has no easy characterization in terms of rectangles. The characteristic property in graph-theoretical terms is: from every vertex outside x there exists an arrow leading into x.

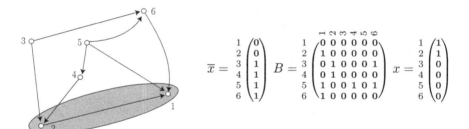

Fig. 6.13 An absorbant set

The following concept of a kernel has been studied extensively in our general reference texts [123, 124], where a lot of game examples, not least in chess, are also presented. So, here we summarize briefly the basic definitions.

Definition 6.16. Given any homogeneous relation B, or equivalently any 1-graph with associated relation B, together with a set x, we call the vector

$$x \text{ a kernel} \quad :\Longleftrightarrow \quad \forall p : p \notin x \leftrightarrow \left\{ \exists q : (p,q) \in B \wedge q \in x \right\} \quad \Longleftrightarrow \quad \bar{x} = B{:}x. \qquad \square$$

A kernel is at the same time a stable and an absorbant set. Some further information is contained in Section 16.4. The determination of kernels, even the question of whether a kernel exists, is very difficult in general.

Some of this may also be presented under the heading of a covering point set or edge set.

Definition 6.17. Given a homogeneous relation $B : X \longrightarrow X$, we call a vector

$$v \text{ a } \textbf{point-covering} \quad :\Longleftrightarrow \quad \forall p : \{\exists q : (p,q) \in B \land q \notin v\} \to p \in v$$
$$\Longleftrightarrow \quad B; \overline{v} \subseteq v. \qquad \qquad \square$$

The vector v is, thus, a point-covering of B precisely when v, v is a covering pair of sets according to Def. 6.10.ii. In particular, one is interested in minimal covering point sets; see Fig. 6.14.

```
        a b c d e f g h i j k              {b,c,d,e,f,h}
a  / 0 0 1 1 0 0 0 1 0 0 0 \        a  / 0 0 0 1 0 1 1 0 1 \
b  | 0 0 0 0 1 0 1 1 1 1 0 |        b  | 1 1 0 1 1 1 1 1 0 |
c  | 1 0 0 0 0 0 0 0 0 0 1 |        c  | 1 1 1 0 1 0 0 1 0 |
d  | 1 0 0 0 0 0 0 0 0 1 1 |        d  | 1 1 1 1 1 0 0 1 0 |
e  | 0 1 0 0 0 0 0 0 0 0 1 |        e  | 1 1 1 0 0 0 0 0 1 |
f  | 0 0 0 0 0 0 1 0 1 0     |      f  | 1 0 0 1 1 1 0 0 0 |
g  | 0 1 0 0 0 0 0 0 0 0 0 |        g  | 0 0 1 0 0 0 0 0 1 |
h  | 1 1 0 0 0 1 0 0 0 0 1 |        h  | 1 1 1 0 1 0 1 1 1 |
i  | 0 1 0 0 0 0 0 0 0 0 0 |        i  | 0 0 1 0 0 0 0 0 1 |
j  | 0 1 0 1 0 1 0 0 0 0 0 |        j  | 0 1 1 0 0 1 1 1 1 |
k  \ 0 0 1 1 1 0 0 1 0 0 0 /        k  \ 0 0 0 1 1 1 1 1 1 /
```

Column headers (right matrix): {b,c,d,e,f,h}, {b,c,d,e,h,j}, {c,d,e,g,h,i,j}, {a,b,d,f,k}, {b,c,d,f,h,k}, {a,b,f,j,k}, {a,b,h,j,k}, {b,c,d,h,j,k}, {a,e,g,h,i,j,k}

Fig. 6.14 A relation and all its minimal covering point sets

7

Domain Construction

It has been shown in Chapters 2 and 3 how moderately sized sets (termed basesets when we want to stress that they are linearly ordered, non-empty, and finite), elements, vectors, and relations can be represented. There is a tendency to try to extend these techniques indiscriminately to all finite situations. We do not follow this trend. Instead, sets, elements, vectors, or relations – beyond those related to ground sets – will carefully be constructed, in particular if they are 'larger'. Only a few generic techniques are necessary. They are presented here in detail as appropriate.

These techniques are far from being new. We have routinely applied them in an informal way since our school days. What is new in the approach chosen here is that we begin to take these techniques seriously: pair forming, *if–then–else–fi*-handling of variants, quotient forming, etc. For pairs, we routinely look for the first and second components; when a set is considered modulo an equivalence, we work with the corresponding equivalence classes and obey carefully the rule that our results should not depend on the specific representative chosen, etc.

What has been indicated here, however, requires a more detailed language to be expressed. This in turn means that a distinction between language and interpretation is suddenly important, which one would like to abstract from when handling relations 'directly'. It turns out that only one or two generically defined relations are necessary for each construction step with quite simple and intuitive algebraic properties. It is important that *the same* generically defined relations will serve to define new elements, new vectors, and new relations – assuming that we know how this definition works on the constituents.

The point to stress is that once a new category object has been constructed, all elements, vectors, and relations touching this domain will be defined using the generic tools while others *simply cannot be formulated.*

7.1 Domains of ground type

All our constructions are supposed to start with **ground types**, which we assume to be given explicitly as shown earlier in Section 2.1. Demanding that they are non-empty is not really a restriction. Case distinctions as to being empty or not have to be made anyway, and we postulate that these distinctions are made *before* entering into further constructions. Relations on ground types are typically given explicitly, i.e., with one of the methods mentioned in Chapter 3. This is where we usually start from. Also elements and vectors are given explicitly and not constructed; this follows the lines of Chapter 2.

Between ground types, however, specific relations may already be constructed. Such a relation will start from vectors $v \subseteq X$ and $w \subseteq Y$ and result in the rectangular relation $v ; w^\mathsf{T} : X \longrightarrow Y$. In Fig. 7.1, we thus define the relation between all the red Bridge suits and all the picture-carrying ones among the Skat card levels Ass, König, Dame, Bube, 10, 9, 8, 7. It is a *constructed* relation although between ground sets – in contrast to R, a relation given by marking arbitrarily.

$$e = \begin{array}{c} \spadesuit \\ \heartsuit \\ \diamondsuit \\ \clubsuit \end{array} \begin{pmatrix} 0 \\ 0 \\ 1 \\ 0 \end{pmatrix} \quad v = \begin{pmatrix} 0 \\ 1 \\ 1 \\ 0 \end{pmatrix} \quad w = \begin{array}{c} \mathrm{A} \\ \mathrm{K} \\ \mathrm{D} \\ \mathrm{B} \\ 10 \\ 9 \\ 8 \\ 7 \end{array} \begin{pmatrix} 0 \\ 1 \\ 1 \\ 1 \\ 0 \\ 0 \\ 0 \\ 0 \end{pmatrix} \quad v ; w^\mathsf{T} = \begin{array}{c} \\ \spadesuit \\ \heartsuit \\ \diamondsuit \\ \clubsuit \end{array} \begin{array}{c} \begin{array}{cccccccc} \mathrm{A} & \mathrm{K} & \mathrm{D} & \mathrm{B} & 10 & 9 & 8 & 7 \end{array} \\ \begin{pmatrix} 0&0&0&0&0&0&0&0 \\ 0&1&1&1&0&0&0&0 \\ 0&1&1&1&0&0&0&0 \\ 0&0&0&0&0&0&0&0 \end{pmatrix} \end{array} \quad R = \begin{array}{c} \begin{array}{cccccccc} \mathrm{A} & \mathrm{K} & \mathrm{D} & \mathrm{B} & 10 & 9 & 8 & 7 \end{array} \\ \begin{pmatrix} 0&0&0&0&0&1&0&0 \\ 0&0&1&1&0&0&1&0 \\ 0&1&0&1&0&0&0&1 \\ 0&0&0&1&0&0&0&0 \end{pmatrix} \end{array}$$

Fig. 7.1 Element, vectors, constructed rectangle; explicit relation R of ground type

7.2 Direct product

Pairs, triples, and n-tuples are omnipresent in everyday situations. All pairs with elements of sets X, Y as components make up a Cartesian product $X \times Y$ of these sets. The notation $(x, y) \in X \times Y$ for pairs, with parentheses and separating comma in between, is very common and need not be explained any further.

We have, however, introduced the concept of types `t1, t2` which may – possibly only later – be interpreted by sets. So we have to show in a generic fashion how such pair-forming can be achieved. Assuming a type `t1` and a type `t2` to be given, we constructively generate their direct product. As an example we provide different interpretations for these types, namely for `t1`

BridgeHonorValues = {A,K,Q,J,10}   or   SkatCardLevels = {A,K,D,B,10,9,8,7}

and in addition for `t2`

BridgeSuits = {♠,♡,♢,♣}   or   SkatSuits = {♣,♠,♡,♢}.

Constructively generating the direct product of `t1`, `t2` means obtaining, depending on the interpretation, either one of the following Bridge honor cards

$$
\begin{aligned}
\{&(\text{Ace},\spadesuit),\ (\text{King},\spadesuit),\ (\text{Queen},\spadesuit),\ (\text{Jack},\spadesuit),\ (10,\spadesuit),\\
 &(\text{Ace},\heartsuit),\ (\text{King},\heartsuit),\ (\text{Queen},\heartsuit),\ (\text{Jack},\heartsuit),\ (10,\heartsuit),\\
 &(\text{Ace},\diamondsuit),\ (\text{King},\diamondsuit),\ (\text{Queen},\diamondsuit),\ (\text{Jack},\diamondsuit),\ (10,\diamondsuit),\\
 &(\text{Ace},\clubsuit),\ (\text{King},\clubsuit),\ (\text{Queen},\clubsuit),\ (\text{Jack},\clubsuit),\ (10,\clubsuit)\},
\end{aligned}
$$

or Skat cards

$$
\begin{aligned}
\{&(\text{A},\clubsuit),\ (\text{K},\clubsuit),\ (\text{D},\clubsuit),\ (\text{B},\clubsuit),\ (10,\clubsuit),\ (9,\clubsuit),\ (8,\clubsuit),\ (7,\clubsuit),\\
 &(\text{A},\spadesuit),\ (\text{K},\spadesuit),\ (\text{D},\spadesuit),\ (\text{B},\spadesuit),\ (10,\spadesuit),\ (9,\spadesuit),\ (8,\spadesuit),\ (7,\spadesuit),\\
 &(\text{A},\heartsuit),\ (\text{K},\heartsuit),\ (\text{D},\heartsuit),\ (\text{B},\heartsuit),\ (10,\heartsuit),\ (9,\heartsuit),\ (8,\heartsuit),\ (7,\heartsuit),\\
 &(\text{A},\diamondsuit),\ (\text{K},\diamondsuit),\ (\text{D},\diamondsuit),\ (\text{B},\diamondsuit),\ (10,\diamondsuit),\ (9,\diamondsuit),\ (8,\diamondsuit),\ (7,\diamondsuit)\}.
\end{aligned}
$$

Projections

What we actually need are the projections from a pair to its components. In both cases, we can easily observe the projection on the first, respectively second, component of the direct product which we denote as π, ρ. This is, however, a bit sloppy in the same sense as the notation $\mathbb{I}$ for the identity was: we have not indicated the product we are working with as in the full form

$$\pi : X \times Y \longrightarrow X \qquad \text{and} \qquad \rho : X \times Y \longrightarrow Y,$$

which seems overly precise when X, Y and $X \times Y$ are already known from the context.

It is obviously necessary to regulate this in a fully schematic way. In the language TITUREL we are offered the possibility of denoting in full, namely

`DirPro t1 t2,`   the **product domain**, corresponding to $X \times Y$

`Pi t1 t2` and `Rho t1 t2,`   the **left** and the **right projection**.

Baseorder of a direct product

One may have wondered concerning the order of elements in the direct product of Fig. 7.2 and the left- respectively right-hanging 'mandrels'. In both cases π has **1**s in lines of $+45$ degrees beginning with length 1, length 2, increased to length of X and going down to length 1 again. Correspondingly, both ρ follow this scheme along -45 degrees. We will now discuss this effect in a detailed example.

Assume workers play some game during the noon break and start from two sets

`workDays` = {Mon, Tue, Wed, Thu, Fri}

and

`bsGameQuali` = {Win,Draw,Loss}.

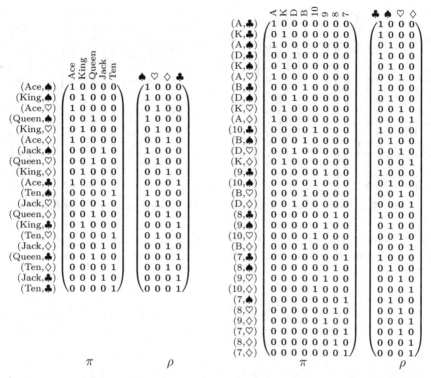

Fig. 7.2 Generic projection relations π and ρ interpreted differently

For the present text, we have chosen to present pairsets as equipped with a baseorder as follows:

workDays × bsGameQuali =

{(Mon,Win),(Tue,Win),(Mon,Draw),(Wed,Win),(Tue,Draw),(Mon,Loss),

(Thu,Win),(Wed,Draw),(Tue,Loss),(Fri,Win),

(Thu,Draw),(Wed,Loss),(Fri,Draw),(Thu,Loss),(Fri,Loss)}.

The projections shown on the right of Fig. 7.3 are compared with another possible arrangement that might seem even more evident. However, it is less symmetric, as one will see from the projections on the right.

	Mon	Tue	Wed	Thu	Fri		Win	Draw	Loss
(Mon,Win)	1	0	0	0	0		1	0	0
(Tue,Win)	0	1	0	0	0		1	0	0
(Mon,Draw)	1	0	0	0	0		0	1	0
(Wed,Win)	0	0	1	0	0		1	0	0
(Tue,Draw)	0	1	0	0	0		0	1	0
(Mon,Loss)	1	0	0	0	0		0	0	1
(Thu,Win)	0	0	0	1	0		1	0	0
(Wed,Draw)	0	0	1	0	0		0	1	0
(Tue,Loss)	0	1	0	0	0		0	0	1
(Fri,Win)	0	0	0	0	1		1	0	0
(Thu,Draw)	0	0	0	1	0		0	1	0
(Wed,Loss)	0	0	1	0	0		0	0	1
(Fri,Draw)	0	0	0	0	1		0	1	0
(Thu,Loss)	0	0	0	1	0		0	0	1
(Fri,Loss)	0	0	0	0	1		0	0	1

	Mon	Tue	Wed	Thu	Fri		Win	Draw	Loss
(Mon,Win)	1	0	0	0	0		1	0	0
(Mon,Draw)	1	0	0	0	0		0	1	0
(Mon,Loss)	1	0	0	0	0		0	0	1
(Tue,Win)	0	1	0	0	0		1	0	0
(Tue,Draw)	0	1	0	0	0		0	1	0
(Tue,Loss)	0	1	0	0	0		0	0	1
(Wed,Win)	0	0	1	0	0		1	0	0
(Wed,Draw)	0	0	1	0	0		0	1	0
(Wed,Loss)	0	0	1	0	0		0	0	1
(Thu,Win)	0	0	0	1	0		1	0	0
(Thu,Draw)	0	0	0	1	0		0	1	0
(Thu,Loss)	0	0	0	1	0		0	0	1
(Fri,Win)	0	0	0	0	1		1	0	0
(Fri,Draw)	0	0	0	0	1		0	1	0
(Fri,Loss)	0	0	0	0	1		0	0	1

Fig. 7.3 Projections π, ρ of direct product shown in two arrangements

Of course, having the same first, respectively second, component is an equivalence relation as one may see from Fig. 7.4 for the left variant.

$\pi;\pi^{\mathsf{T}}$

	(Mon,Win)	(Tue,Win)	(Mon,Draw)	(Wed,Win)	(Tue,Draw)	(Mon,Loss)	(Thu,Win)	(Wed,Draw)	(Tue,Loss)	(Fri,Win)	(Thu,Draw)	(Wed,Loss)	(Fri,Draw)	(Thu,Loss)	(Fri,Loss)
(Mon,Win)	1	0	1	0	0	1	0	0	0	0	0	0	0	0	0
(Tue,Win)	0	1	0	0	1	0	0	0	1	0	0	0	0	0	0
(Mon,Draw)	1	0	1	0	0	1	0	0	0	0	0	0	0	0	0
(Wed,Win)	0	0	0	1	0	0	0	1	0	0	0	1	0	0	0
(Tue,Draw)	0	1	0	0	1	0	0	0	1	0	0	0	0	0	0
(Mon,Loss)	1	0	1	0	0	1	0	0	0	0	0	0	0	0	0
(Thu,Win)	0	0	0	0	0	0	1	0	0	0	1	0	0	1	0
(Wed,Draw)	0	0	0	1	0	0	0	1	0	0	0	1	0	0	0
(Tue,Loss)	0	1	0	0	1	0	0	0	1	0	0	0	0	0	0
(Fri,Win)	0	0	0	0	0	0	0	0	0	1	0	0	1	0	1
(Thu,Draw)	0	0	0	0	0	0	1	0	0	0	1	0	0	1	0
(Wed,Loss)	0	0	0	1	0	0	0	1	0	0	0	1	0	0	0
(Fri,Draw)	0	0	0	0	0	0	0	0	0	1	0	0	1	0	1
(Thu,Loss)	0	0	0	0	0	0	1	0	0	0	1	0	0	1	0
(Fri,Loss)	0	0	0	0	0	0	0	0	0	1	0	0	1	0	1

$\rho;\rho^{\mathsf{T}}$

	(Mon,Win)	(Tue,Win)	(Mon,Draw)	(Wed,Win)	(Tue,Draw)	(Mon,Loss)	(Thu,Win)	(Wed,Draw)	(Tue,Loss)	(Fri,Win)	(Thu,Draw)	(Wed,Loss)	(Fri,Draw)	(Thu,Loss)	(Fri,Loss)
(Mon,Win)	1	1	0	1	0	0	1	0	0	1	0	0	0	0	0
(Tue,Win)	1	1	0	1	0	0	1	0	0	1	0	0	0	0	0
(Mon,Draw)	0	0	1	0	1	0	0	1	0	0	1	0	1	0	0
(Wed,Win)	1	1	0	1	0	0	1	0	0	1	0	0	0	0	0
(Tue,Draw)	0	0	1	0	1	0	0	1	0	0	1	0	1	0	0
(Mon,Loss)	0	0	0	0	0	1	0	0	1	0	0	1	0	1	1
(Thu,Win)	1	1	0	1	0	0	1	0	0	1	0	0	0	0	0
(Wed,Draw)	0	0	1	0	1	0	0	1	0	0	1	0	1	0	0
(Tue,Loss)	0	0	0	0	0	1	0	0	1	0	0	1	0	1	1
(Fri,Win)	1	1	0	1	0	0	1	0	0	1	0	0	0	0	0
(Thu,Draw)	0	0	1	0	1	0	0	1	0	0	1	0	1	0	0
(Wed,Loss)	0	0	0	0	0	1	0	0	1	0	0	1	0	1	1
(Fri,Draw)	0	0	1	0	1	0	0	1	0	0	1	0	1	0	0
(Thu,Loss)	0	0	0	0	0	1	0	0	1	0	0	1	0	1	1
(Fri,Loss)	0	0	0	0	0	1	0	0	1	0	0	1	0	1	1

Fig. 7.4 Relations of having the same first, respectively second, component

The equivalence aspect of $\pi;\pi^{\mathsf{T}}$ can be seen far more easily in the case when the diagonal squares are arranged diagonally as in Fig. 7.5, as a consequence of choosing the second variant for the projections in Fig. 7.3 – at the cost of $\rho;\rho^{\mathsf{T}}$ not looking so neat.

Since we have chosen to conceive basesets as *ordered* entities, we had also to decide on an ordering of the direct product of two basesets. The most naïve way to present projections would have been the right version of Fig. 7.3. But because this is not sufficiently symmetric with regard to the components, we have decided for the left

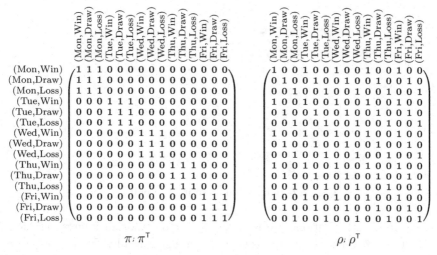

$$\pi;\pi^{\mathsf{T}} \qquad\qquad \rho;\rho^{\mathsf{T}}$$

Fig. 7.5 Relations for equal first, respectively second, component of Fig. 7.3 rearranged

side as a result of Cantor's diagonal enumeration. This diagonal-oriented sequence seems rather difficult to handle; but to a certain extent it allows us to visualize the pairset even when non-finite sets are involved.

The sequence of presenting its elements in π is depicted in the diagonal schema of Fig. 7.6.

	Win	Draw	Loss
Mon	1	3	6
Tue	2	5	9
Wed	4	8	12
Thu	7	11	14
Fri	10	13	15

Fig. 7.6 Enumeration scheme for a direct product

The projection relations look different now. One will not recognize so easily how they are built and that they are built in a very regular fashion. Their algebraic properties, however, prevail.

Whatever we do with a direct product shall henceforth be expressed via the projections mentioned. In particular, we are allowed to define elements just by projections in combination with elements in the constituent sets. The same holds true for the definition of vectors.

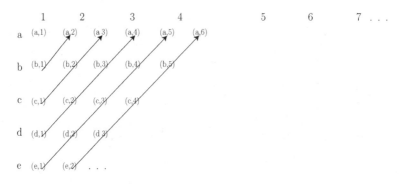

Fig. 7.7 Exhaustion of an infinite pair set

Since we work with relations, we have the opportunity of an interesting exercise to relate the two versions and to visualize them algebraically. The transition from the presentation we have chosen for this book to the presentation as in Fig. 7.3 is easily achieved. Assume B_1, B_2 to be the baseorders in the first, respectively second, factor. Then the lexicographic order of the product is

$$L := \pi_i C_1{}_i \pi^{\mathsf{T}} \cup (\pi_i \pi^{\mathsf{T}} \cap \rho_i B_2{}_i \rho^{\mathsf{T}}),$$

where $C_1 := B_1 \cap \overline{\mathbb{I}}$ is the strictorder corresponding to B_1. In this form, a pair is less than or equal to another pair either if its first component is strictly inferior to the first component of the other or if the first components coincide and its second component is less than or equal to the second component of the other. The relation L is easily shown to be a linear ordering to which the permutation to the upper triangle (see Appendix C) must be determined, which finally gives the transition between the two types of presentation. (Be aware that the two relations have different column type.)

Normally, we abstract from these details of the representation. But whenever we have to present such details, we have to decide on some form. Also, when programs to handle relations are conceived, one has to decide, as these require representation inside the memory of the computer. Only now have personal computers become fast enough to handle relations and thus to provide support in algebraic considerations, and, thus, only now do these decisions *have* to be made. Once they are made, one has additional options, for example to compute the permutations necessary and to try to be more independent of the representation again.

Elements in a direct product

The question is now how to introduce elements of a direct product type. We assume that we know how elements are defined for the component types. We need both e_X

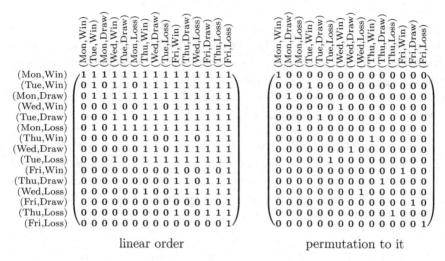

linear order   permutation to it

Fig. 7.8 The linear ordering L and the permutation to it

and e_Y denoting elements in X and in Y, respectively. Then (e_X, e_Y) is notational standard for an element in the direct product $X \times Y$ obtained algebraically as $\pi_; e_X \cap \rho_; e_Y$.

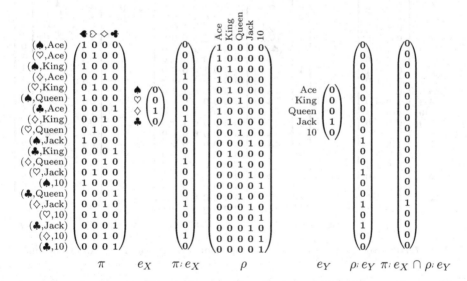

Fig. 7.9 Two elements, projection relations and pair as element in direct product

With these techniques, every element in the product may be denoted.

Vectors in a direct product

In a similar way, we now strive to denote a vector in the direct product. From Part I we know that it is easy to express a vector for ground sets: we use the explicit enumeration or the marking techniques of Section 2.3. The question becomes more involved when the type is a direct product. Basic forms of vectors may then be defined in the following ways. Of course, other forms may also be obtained with intersections, unions, complements, etc.

If v_X and v_Y stand for expressible vectors defining subsets in X or Y, respectively, then in the first place only $\pi \,\dot{}\, v_X$ and $\rho \,\dot{}\, v_Y$ are expressible subsets in the direct product $X \times Y$. From these, further vectors may be obtained by Boolean operations. In the following Skat example, we first project to red cardsuits, then to value cards and finally obtain red value cards via intersection.

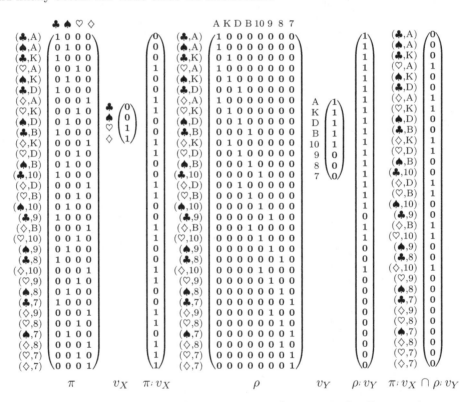

Fig. 7.10 Projections to vectors in components and vectors in the direct product

There is another important consequence of the restriction to notation derived with projections. When we construct vector denotations from those in the finite enumerated basesets, we cannot formulate arbitrarily complex sets in the product, for example. Rather, we are restricted to the constructions that are offered.

Relations starting or ending in a direct product

Let two relations $R : A \longrightarrow B$ and $S : A \longrightarrow Y$, as on the left of Fig. 7.12 below, be given, i.e., with the same source. Then a typical task requires us to construct the relations $R_; \pi^\mathsf{T} : A \longrightarrow B \times Y$ and $S_; \rho^\mathsf{T} : A \longrightarrow B \times Y$ ending in the direct product of the targets. From these, others such as $R_; \pi^\mathsf{T} \cup \overline{S_; \rho^\mathsf{T}}$, for example, may be built using Boolean operators. A particularly interesting construction is the **strict fork operator**[1] with type

$$(R \otimes S) : A \longrightarrow B \times Y \qquad \text{defined as} \qquad (R \otimes S) := R_; \pi^\mathsf{T} \cap S_; \rho^\mathsf{T}.$$

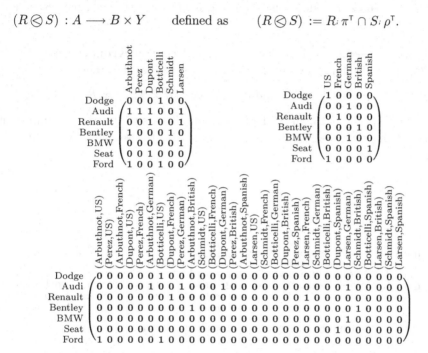

Fig. 7.11 Strict fork operator $(R \otimes S)$ applied to relations with common source

In a symmetric way, we construct relations with the same target starting in a direct product. Let two relations $R : B \longrightarrow A$ and $S : Y \longrightarrow A$ be given. Then it is normal to construct the relations $\pi_; R : B \times Y \longrightarrow A$ and $\rho_; S : B \times Y \longrightarrow A$ ending in the common target. From these further relations may be built using Boolean operators, among which the converses of the strict fork are important, namely the **strict join operator** with type

$$(R \otimes S) : B \times Y \longrightarrow A \qquad \text{defined as} \qquad (R \otimes S) := \pi_; R \cap \rho_; S.$$

---

[1] This should not be confused with the *non-strict* fork operator. The difference is best seen in the requirement that $(R \mathrel{<} S)_; \pi = R$ regardless of whether S is total or not.

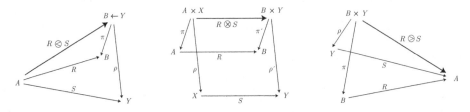

Fig. 7.12 Typing of product-related operations $(R \otimes S)$, $(R \otimes S)$, and $(R \otimes S)$

In addition, there exists what we have decided to denote the **Kronecker product**.[2] Given $R : A \longrightarrow B$ and $S : X \longrightarrow Y$, the Kronecker product is typed

$$(R \otimes S) : A \times X \longrightarrow B \times Y \quad \text{and defined as} \quad (R \otimes S) := \pi; R; {\pi'}^{\mathsf{T}} \cap \rho; S; {\rho'}^{\mathsf{T}}.$$

In the example of Fig. 7.13, one will observe that all rows with c as first component are rows of **0**s, thus destroying any information on the second component in a Kronecker product.

	red	green	blue	orange
a	1	0	0	1
b	0	1	1	1
c	0	0	0	0
d	1	0	0	1

	1	2	3	4
male	1	1	1	1
female	1	0	0	1

Columns: (red,1) (green,1) (red,2) (blue,1) (green,2) (red,3) (blue,2) (green,3) (red,4) (orange,2) (blue,3) (green,4) (orange,3) (blue,4) (orange,4)

	(red,1)	(green,1)	(red,2)	(blue,1)	(green,2)	(red,3)	(blue,2)	(green,3)	(red,4)	(orange,2)	(blue,3)	(green,4)	(orange,3)	(blue,4)	(orange,4)
(a,male)	1 0 1 0 0 1 1 0 0 1 1 0 0 1 0 1													
(b,male)	0 1 0 1 1 0 1 1 1 0 1 1 1 1 1 1													
(a,female)	1 0 0 0 0 1 0 0 1 0 0 1 0 0 0 1													
(c,male)	0 0 0 0 0 0 0 0 0 0 0 0 0 0 0 0													
(b,female)	0 1 0 1 0 0 1 0 0 0 0 0 1 0 1 1													
(d,male)	1 0 1 0 0 1 1 0 0 1 1 0 0 1 0 1													
(c,female)	0 0 0 0 0 0 0 0 0 0 0 0 0 0 0 0													
(d,female)	1 0 0 0 0 1 0 0 1 0 0 1 0 0 0 1													

Fig. 7.13 Kronecker product $(R \otimes S)$ of two relations without any typing interrelation

When one is about to define binary mappings, one may also sometimes employ the direct product. We anticipate here as an example the well-known non-modular lattice of Fig. 7.14, given via its ordering relation $E : X \longrightarrow X$. The task is to obtain the least upper, respectively greatest lower, bound (i.e., join and meet) as binary mappings $\mathcal{J} : X \times X \longrightarrow X$ and $\mathcal{M} : X \times X \longrightarrow X$. One introduces the direct product $\mathtt{DirPro\ x\ x}$ corresponding to $X \times X$, together with the two projections π, ρ and simply forms

$$\mathcal{J} := \mathtt{lubR}_E(\pi \cup \rho) \qquad \text{and} \qquad \mathcal{M} := \mathtt{glbR}_E(\pi \cup \rho).$$

---

[2] Also here, we must be careful with regard to denotation, for which *tensor product, parallel composition* (as in [74]), and *tupeling* have also been used. It should be recognized that we mean the strict operation as opposed to the non-strict operation for which parallel composition is acceptable.

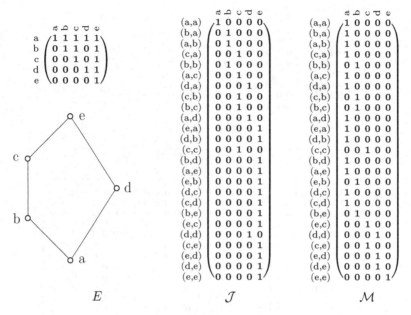

Fig. 7.14 Determining the join and meet operators out of the ordering E of the lattice

Algebraic properties of the generic projection relations

Product definitions have long been investigated by relation algebraists and computer scientists. A considerable part of [140] is devoted to various aspects of this subject. The following definition of a direct product produces a Cartesian product of sets that is characterized in an essentially unique way.[3]

Mathematicians observed that whatever variant of projection they decided for, certain algebraic rules had been satisfied. This was then turned around and it was asked whether this was characteristic for projections – which it is.

Definition 7.1. If any two heterogeneous relations π, ρ with common source are given, they are said to form a **direct product**[4] if

$$\pi^{\mathsf{T}}{;}\pi = \mathbb{I}, \quad \rho^{\mathsf{T}}{;}\rho = \mathbb{I}, \quad \pi{;}\pi^{\mathsf{T}} \cap \rho{;}\rho^{\mathsf{T}} = \mathbb{I}, \quad \pi^{\mathsf{T}}{;}\rho = \mathbb{T}.$$

In particular, π, ρ are mappings, usually called **projections.** □

---

[3] We avoid here speaking of 'up to isomorphism', which would be the mathematically correct form. In this chapter, we try to convince the reader and we give visual help, not yet fully formal proofs.

[4] Such pairs of relations were studied and axiomatized in a not yet satisfactory form by Alfred Tarski in his note [139].

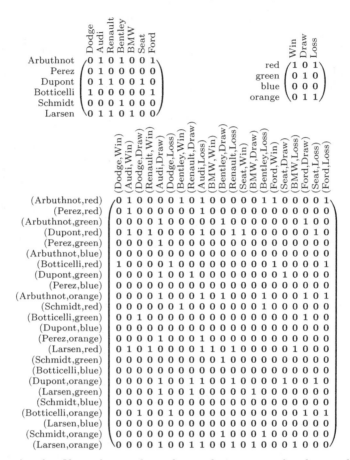

Fig. 7.15 Another Kronecker product of two relations – unrelated as to their types

The first two conditions require π, ρ to be univalent and surjective.[5] Interpreting the condition $\pi \,;\pi^{\mathsf{T}} \cap \rho\,;\rho^{\mathsf{T}} \subseteq \mathbb{I}$ in the case of two sets A, B and their Cartesian product $A \times B$, it ensures that there is *at most* one pair with given images in A and B. In addition, "$= \mathbb{I}$" means that π, ρ are total, i.e., that there are no 'unprojected pairs'. Finally, the condition $\pi^{\mathsf{T}}\!;\rho = \mathbb{T}$ implies that for *every* element in A and *every* element in B there exists a pair in $A \times B$.

Assume that π, ρ as well as some others, for example, π', ρ' of Fig. 7.16, presented by a student, satisfy these formulae. The targets of π, π' coincide indeed, as well

---

[5] One or other category theorist will complain concerning surjectivity being demanded. In mathematics indeed, the direct product of $A \times B$ with A the empty set and B non-empty will not have a surjective projection ρ on B. When we demand surjectivity here, this means that the question of being empty or not – that must be handled anyway – has to be handled *before* entering into the direct product construction.

as those of ρ, ρ'. Then we can explicitly construct a bijection that realizes π' as a permuted version of π and ρ' of ρ, namely

$$P := \pi; \pi'^{\mathsf{T}} \cap \rho; \rho'^{\mathsf{T}}.$$

For the example of Fig. 7.3, this is shown in Fig. 7.16.

	Mon	Tue	Wed	Thu	Fri	Win	Draw	Loss	pair	A	B	C	D	E	F	G	H	I	J	K	L	M	N	O
A	0	1	0	0	0	0	1	0	(Mon,Win)	0	0	0	0	0	0	0	0	0	0	0	1	0	0	0
B	0	0	0	0	1	0	1	0	(Tue,Win)	0	0	0	0	0	1	0	0	0	0	0	0	0	0	0
C	0	1	0	0	0	0	0	1	(Mon,Draw)	0	0	0	1	0	0	0	0	0	0	0	0	0	0	0
D	1	0	0	0	0	0	1	0	(Wed,Win)	0	0	0	0	0	0	0	0	0	0	0	0	0	1	0
E	0	0	0	0	1	0	0	1	(Tue,Draw)	1	0	0	0	0	0	0	0	0	0	0	0	0	0	0
F	0	1	0	0	0	1	0	0	(Mon,Loss)	0	0	0	0	0	0	0	1	0	0	0	0	0	0	0
G	0	0	1	0	0	0	1	0	(Thu,Win)	0	0	0	0	0	0	0	0	0	0	0	0	0	0	1
H	1	0	0	0	0	0	0	1	(Wed,Draw)	0	0	0	0	0	0	1	0	0	0	0	0	0	0	0
I	0	0	0	0	1	1	0	0	(Tue,Loss)	0	0	1	0	0	0	0	0	0	0	0	0	0	0	0
J	0	0	0	1	0	0	1	0	(Fri,Win)	0	0	0	0	0	0	0	0	1	0	0	0	0	0	0
K	0	0	1	0	0	0	0	1	(Thu,Draw)	0	0	0	0	0	0	0	0	0	1	0	0	0	0	0
L	1	0	0	0	0	1	0	0	(Wed,Loss)	0	0	0	0	0	0	0	0	0	0	1	0	0	0	0
M	0	0	0	1	0	0	0	1	(Fri,Draw)	0	1	0	0	0	0	0	0	0	0	0	0	0	0	0
N	0	0	1	0	0	1	0	0	(Thu,Loss)	0	0	0	0	0	0	0	0	0	0	0	0	1	0	0
O	0	0	0	1	0	1	0	0	(Fri,Loss)	0	0	0	0	1	0	0	0	0	0	0	0	0	0	0

Fig. 7.16 Other projections π', ρ' for Fig. 7.3,
bijection $\pi; \pi'^{\mathsf{T}} \cap \rho; \rho'^{\mathsf{T}}$ to relate to these

What we have achieved when constructively generating the direct product `DirPro t1 t2` is essentially unique, which suffices for practical work. In the TITUREL system, for example, one version is implemented. If somebody else implemented another version independently, this would probably be different. Given both, however, we are in a position to construct the bijection that relates them.

An advanced view of the direct product

We recommend that this subsection be totally skipped at first reading; so there is no longer reason to hide the proofs and postpone them to the appendix. The following touches a difficult issue concerning direct products and is explained together with vectorization.

First, we provide several formulae in connection with the direct product, Kronecker product, and projections. They are explained here and visualized. Even their delicate proofs are given – in contrast to the policy followed elsewhere in this chapter.

Proposition 7.2. *Let relations $R : A \longrightarrow B$ and $S : X \longrightarrow Y$ be given and consider the direct products on the source and on the target side of these relations, so that*

$$\pi : A \times X \longrightarrow A \quad \rho : A \times X \longrightarrow X \quad \text{and} \quad \pi' : B \times Y \longrightarrow B \quad \rho' : B \times Y \longrightarrow Y$$

are the respective projections. Then the following hold:

(i) $(R \otimes S) \pi' = \pi; R \cap \rho; S; \mathbb{T}_{Y,B} \subseteq \pi; R$

$\quad (R \otimes S); \rho' = \rho; S \cap \pi; R; \mathbb{T}_{B,Y} \subseteq \rho; S,$

(ii) $(R \otimes S); (P \otimes Q) \subseteq (R; P \otimes S; Q).$

Proof: (i) may be proved using mainly Prop. 5.4:

$$\begin{aligned}
(R \otimes S) \pi' &= (\pi; R; \pi'^{\mathsf{T}} \cap \rho; S; \rho'^{\mathsf{T}}); \pi' && \text{by definition} \\
&= \pi; R \cap \rho; S; \rho'^{\mathsf{T}}; \pi' && \text{Prop. 5.4} \\
&= \pi; R \cap \rho; S; \mathbb{T} && \text{following Def. 7.1}
\end{aligned}$$

(ii) $\begin{aligned}[t]
(R \otimes S); (P \otimes Q) &= (\pi; R; \pi'^{\mathsf{T}} \cap \rho; S; \rho'^{\mathsf{T}}); (\pi'; P; \pi''^{\mathsf{T}} \cap \rho'; Q; \rho''^{\mathsf{T}}) && \text{expanded} \\
&\subseteq \pi; R; \pi'^{\mathsf{T}}; \pi'; P; \pi''^{\mathsf{T}} \cap \rho; S; \rho'^{\mathsf{T}}; \rho'; Q; \rho''^{\mathsf{T}} && \text{monotony} \\
&= \pi; R; P; \pi''^{\mathsf{T}} \cap \rho; S; Q; \rho''^{\mathsf{T}} && \text{using Def. 7.1 for the projections } \pi', \rho' \\
&= (R; P \otimes S; Q) && \text{by definition} \qquad \square
\end{aligned}$

In a similar way, it is even easier to prove the following.

Corollary 7.3.

(i) *Consider the strict fork relation* $(R \ominus S) : A \longrightarrow B \times Y$ *built from the two relations* $R : A \longrightarrow B$ *and* $S : A \longrightarrow Y$. *Then*

$$(R \ominus S); \pi = R \cap S; \mathbb{T} \qquad \text{and} \qquad (R \ominus S); \rho = S \cap R; \mathbb{T}.$$

(ii) *Consider the strict join relation* $(R \ominus S) : B \times Y \longrightarrow A$ *built from the two relations* $R : B \longrightarrow A$ *and* $S : Y \longrightarrow A$. *Then*

$$\pi^{\mathsf{T}}; (R \ominus S) = R \cap \mathbb{T}; S \qquad \text{and} \qquad \rho^{\mathsf{T}}; (R \ominus S) = S \cap \mathbb{T}; R. \qquad \square$$

To Prop. 7.2 and Cor. 7.3 belongs a rather evident corollary.

Corollary 7.4. *Assuming the typing of Prop. 7.2 and Cor. 7.3, the following hold:*

(i) $(R \otimes S); \pi' = \pi; R$ *provided S is total*

$\quad (R \otimes S); \rho' = \rho; S$ *provided R is total*

(ii) $(R \ominus S); \pi = R$ *provided S is total*

$\quad (R \ominus S); \rho = S$ *provided R is total*

(iii) $\pi^{\mathsf{T}}; (R \ominus S) = R$ *provided S is surjective*

$\quad \rho^{\mathsf{T}}; (R \ominus S) = S$ *provided R is surjective* $\square$

This is a situation where we should pause a little and have a look around. In view of the c-related rows of Fig. 7.13, one will probably accept that in Prop. 7.2.i equality does not hold in general. Later we will find additional conditions that guarantee equality. But what about Prop. 7.2.ii? With Fig. 7.17, we get a hint that here

$$R = \begin{array}{c} \text{male} \\ \text{female} \end{array} \begin{array}{cccc} a & b & c & d \\ \begin{pmatrix} 0 & 1 & 0 & 1 \\ 1 & 0 & 0 & 1 \end{pmatrix} \end{array}$$

$$P = \begin{array}{c} a \\ b \\ c \\ d \end{array} \begin{pmatrix} 1 & 0 & 0 & 1 \\ 0 & 1 & 1 & 1 \\ 0 & 0 & 0 & 0 \\ 1 & 0 & 0 & 1 \end{pmatrix} \quad \text{(red, green, blue, orange)}$$

$$S = \begin{array}{c} 1 \\ 2 \\ 3 \\ 4 \end{array} \begin{pmatrix} 0 & 1 & 0 \\ 0 & 0 & 0 \\ 0 & 1 & 0 \\ 1 & 0 & 0 \end{pmatrix} \quad \text{(Win, Draw, Loss)}$$

$$Q = \begin{array}{c} \text{Win} \\ \text{Draw} \\ \text{Loss} \end{array} \begin{pmatrix} 0 & 1 & 1 & 0 & 1 \\ 0 & 1 & 1 & 1 & 0 \\ 1 & 0 & 0 & 1 & 0 \end{pmatrix} \quad \text{(Mon, Tue, Wed, Thu, Fri)}$$

$$R;P = \begin{array}{c} \text{male} \\ \text{female} \end{array} \begin{pmatrix} 1 & 1 & 1 & 1 \\ 1 & 0 & 0 & 1 \end{pmatrix} \quad \text{(red, green, blue, orange)}$$

$$S;Q = \begin{array}{c} 1 \\ 2 \\ 3 \\ 4 \end{array} \begin{pmatrix} 0 & 1 & 1 & 1 & 0 \\ 0 & 0 & 0 & 0 & 0 \\ 0 & 1 & 1 & 1 & 0 \\ 0 & 1 & 1 & 0 & 1 \end{pmatrix} \quad \text{(Mon, Tue, Wed, Thu, Fri)}$$

$(R \otimes S) =$ rows $(\text{male},1),(\text{female},1),(\text{male},2),(\text{female},2),(\text{male},3),(\text{female},3),(\text{male},4),(\text{female},4)$; columns $(a,\text{Win}),(b,\text{Win}),(a,\text{Draw}),(c,\text{Win}),(b,\text{Draw}),(a,\text{Loss}),(d,\text{Win}),(c,\text{Draw}),(b,\text{Loss}),(d,\text{Draw}),(c,\text{Loss}),(d,\text{Loss})$

$$\begin{pmatrix}
0 & 0 & 0 & 0 & 1 & 0 & 0 & 0 & 0 & 1 & 0 & 0 \\
0 & 0 & 1 & 0 & 0 & 0 & 0 & 0 & 0 & 1 & 0 & 0 \\
0 & 0 & 0 & 0 & 0 & 0 & 0 & 0 & 0 & 0 & 0 & 0 \\
0 & 0 & 0 & 0 & 0 & 0 & 0 & 0 & 0 & 0 & 0 & 0 \\
0 & 0 & 0 & 0 & 1 & 0 & 0 & 0 & 0 & 1 & 0 & 0 \\
0 & 0 & 1 & 0 & 0 & 0 & 0 & 0 & 0 & 1 & 0 & 0 \\
0 & 1 & 0 & 0 & 0 & 0 & 1 & 0 & 0 & 0 & 0 & 0 \\
1 & 0 & 0 & 0 & 0 & 0 & 1 & 0 & 0 & 0 & 0 & 0
\end{pmatrix}$$

$(P \otimes Q) =$ rows $(a,\text{Win}),(b,\text{Win}),(a,\text{Draw}),(c,\text{Win}),(b,\text{Draw}),(a,\text{Loss}),(d,\text{Win}),(c,\text{Draw}),(b,\text{Loss}),(d,\text{Draw}),(c,\text{Loss}),(d,\text{Loss})$; columns $(\text{red,Mon}),(\text{green,Mon}),(\text{red,Tue}),(\text{blue,Mon}),(\text{green,Tue}),(\text{red,Wed}),(\text{orange,Mon}),(\text{blue,Tue}),(\text{green,Wed}),(\text{red,Thu}),(\text{orange,Tue}),(\text{blue,Wed}),(\text{green,Thu}),(\text{red,Fri}),(\text{orange,Wed}),(\text{blue,Thu}),(\text{green,Fri}),(\text{orange,Thu}),(\text{blue,Fri}),(\text{orange,Fri})$

$$\begin{pmatrix}
0&0&1&0&0&1&0&0&0&0&1&0&0&1&1&0&0&0&0&1 \\
0&0&0&0&1&0&0&1&1&0&1&1&0&0&1&0&1&0&1&1 \\
0&0&1&0&0&1&0&0&0&1&1&0&0&0&1&0&0&1&0&0 \\
0&0&0&0&0&0&0&0&0&0&0&0&0&0&0&0&0&0&0&0 \\
0&0&0&0&1&0&0&1&1&0&1&1&1&0&1&1&0&1&0&0 \\
1&0&0&0&0&0&1&0&0&1&0&0&0&0&0&0&0&1&0&0 \\
0&0&1&0&0&1&0&0&0&1&0&0&1&0&1&1&0&0&0&1 \\
0&0&0&0&0&0&0&0&0&0&0&0&0&0&0&0&0&0&0&0 \\
0&1&0&1&0&0&1&0&0&0&0&0&1&0&0&1&0&1&0&0 \\
0&0&1&0&0&1&0&0&0&1&1&0&0&0&1&0&0&1&0&0 \\
0&0&0&0&0&0&0&0&0&0&0&0&0&0&0&0&0&0&0&0 \\
1&0&0&0&0&0&1&0&0&1&0&0&0&0&0&0&0&1&0&0
\end{pmatrix}$$

$(R;P \otimes S;Q) = (R \otimes S);(P \otimes Q) =$ rows $(\text{male},1),(\text{female},1),(\text{male},2),(\text{female},2),(\text{male},3),(\text{female},3),(\text{male},4),(\text{female},4)$; same 20 columns as above

$$\begin{pmatrix}
0&0&1&0&1&1&0&1&1&1&1&1&1&0&1&1&0&1&0&0 \\
0&0&1&0&0&1&0&0&0&1&1&0&0&0&1&0&0&1&0&0 \\
0&0&0&0&0&0&0&0&0&0&0&0&0&0&0&0&0&0&0&0 \\
0&0&0&0&0&0&0&0&0&0&0&0&0&0&0&0&0&0&0&0 \\
0&0&1&0&1&1&0&1&1&1&1&1&1&0&1&1&0&1&0&0 \\
0&0&1&0&0&1&0&0&0&1&1&0&0&0&1&0&0&1&0&0 \\
0&0&1&0&1&1&0&1&1&0&1&1&0&1&1&0&1&0&1&1 \\
0&0&1&0&0&1&0&0&0&1&0&0&1&0&0&1&1&0&0&1
\end{pmatrix}$$

Fig. 7.17 An example of equality in Prop. 7.2.ii even in the case of most general typing

equality may hold. Any other such example will underpin this conjecture. So one may be tempted to try a proof. A proof using the methods of predicate logic is easily established. Assume any point

$$((x,y),(u,v)) \in (R;P \otimes S;Q),$$

which means by definition

$$((x,y),(u,v)) \in \pi;R;P;\pi'' \cap \rho;S;Q;\rho''^{\mathsf{T}}.$$

Then obviously

$$\big((x,y),(u,v)\big) \in \pi_{;}R_{;}P_{;}\pi''^{\mathsf{T}} \qquad \text{and} \qquad \big((x,y),(u,v)\big) \in \rho_{;}S_{;}Q_{;}\rho''^{\mathsf{T}},$$

meaning

$$(x,u) \in R_{;}P \qquad \text{and} \qquad (y,v) \in S_{;}Q,$$

so that

$$\exists a : (x,a) \in R \wedge (a,u) \in P \qquad \text{and} \qquad \exists b : (y,b) \in S \wedge (b,v) \in Q.$$

From a,b thus obtained, one can recombine an existing pair (a,b) in between to arrive at

$$\big((x,y),(a,b)\big) \in \pi_{;}R_{;}\pi'^{\mathsf{T}} \cap \rho_{;}S_{;}\rho'^{\mathsf{T}} \qquad \text{and} \qquad \big((a,b),(u,v)\big) \in \pi'_{;}P_{;}\pi''^{\mathsf{T}} \cap \rho'_{;}Q_{;}\rho''^{\mathsf{T}}$$

which results in

$$\big((x,y),(u,v)\big) \in \big(\pi_{;}R_{;}\pi'^{\mathsf{T}} \cap \rho_{;}S_{;}\rho'^{\mathsf{T}}\big)_{;}\big(\pi'_{;}P_{;}\pi''^{\mathsf{T}} \cap \rho'_{;}Q_{;}\rho''^{\mathsf{T}}\big)$$

and, thus, finally in

$$\big((x,y),(u,v)\big) \in (R \otimes S)_{;}(P \otimes Q).$$

Today, we know that a proof of Prop. 7.2.ii with equality in the point-free relation-algebraic style maintained so far **does not exist**.[6] What does this mean? One response is to abandon all the algebraic axiomatization developed in the present book as being inadequate and to return to predicate-logic reasoning. We will – of course – not react in this way. Rather, we remember that in geometry, over two thousand years, we have not been able to prove that there is precisely one parallel to a given line through a given point outside of that line. This seemed very obviously to be satisfied but could not be proved in the Euclidian axiomatization. Only in the first half of the nineteenth century, after earlier work by Gauß, János Bolyai and Nikolai Ivanovich Lobachevsky found independently that there exist non-Euclidian geometries where this may indeed not hold. There existed, thus, an intricate borderline between facts that could be proved and other facts that seemed obvious, but could not be proved. The study of such variant geometries later enhanced the power of mathematical modelling in the course of introducing relativity theory.

We may be in a similar position here. Do there exist non-standard models where

---

[6] In [14], the non-provability of Prop. 7.2.ii is called the *unsharpness problem*. It was Rodrigo Cardoso on November 26, 1982, who during his diploma thesis with the present author declared himself definitely unable to prove, in a point-free manner, what had at that time been considered merely a class exercise. This was the start of a deeper study. As late as October 20–27, 1991, during the 'Semester on Algebraic Logic' at the Stefan Banach Mathematical Center in Warsaw, Roger Maddux when consulted concerning this problem, immediately proposed a point-free proof of the equation. This was correct but was considered insufficient in a letter of the present author to him of March 8, 1992 for not having adhered to most general typing. Roger Maddux felt triggered to construct a small finite example in a non-representable relation algebra that indeed violates equality. Already in 1992, he communicated on this and later explained it in detail to the present author during a common stay in Rio de Janeiro in August 1994. It may be found, for example, in [91] and in computer-checkable form in Section 3.2 of [73].

Prop. 7.2.ii holds in the 'weak' form, precisely as proved here, that provide us with additional modelling power? Part of this question has been answered positively, but cannot be elaborated here.

In the case that one of the relations involved in Prop. 7.2.ii is an identity $\mathbb{I}$, the situation changes considerably, which we explain in the next proposition. The typing for the respective first variants of the formulae is then heavily specialized as anticipated in Fig. 7.18.

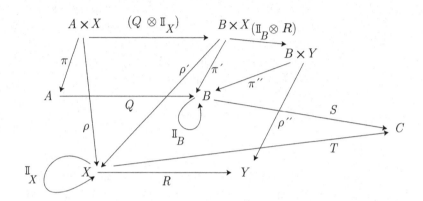

Fig. 7.18 Indicating the specialization of typing for Prop. 7.5

Proposition 7.5. *Assume the typing specialized as indicated in Fig. 7.18 for the left formulae:*

 (i) $(Q \otimes \mathbb{I}_X) \,;\pi' = \pi \,;Q$ $(\mathbb{I} \otimes M) \,;\rho' = \rho \,;M,$

 (ii) $(Q \otimes \mathbb{I}_X) \,;(S \oslash T) = ((Q \,;S) \oslash T)$ $(\mathbb{I} \otimes M) \,;(K \oslash L) = (K \oslash (M \,;L)),$

 (iii) $(Q \otimes \mathbb{I}_X) \,;(\mathbb{I}_B \otimes R) = (Q \otimes R)$ $(\mathbb{I} \otimes M) \,;(K \otimes \mathbb{I}) = (M \otimes K).$

Proof: (i) Follows from Cor. 7.4, since $\mathbb{I}$ is total.

(ii) Direction "$\subseteq$" uses mainly monotony

$(Q \otimes \mathbb{I}) \,;(S \oslash T) = (\pi \,;Q \,;\pi'^{\mathsf{T}} \cap \rho \,;\mathbb{I} \,;\rho'^{\mathsf{T}}) \,;(\pi' \,;S \cap \rho' \,;T)$ expanding

$\subseteq \pi \,;Q \,;\pi'^{\mathsf{T}} \pi' \,;S \cap \rho \,;\rho'^{\mathsf{T}} \,;\rho' \,;T$ monotony

$= \pi \,;Q \,;S \cap \rho \,;T$ since π', ρ' form a direct product

$= ((Q \,;S) \oslash T)$ by definition

To prove direction "$\supseteq$" is a bit more involved:

$((Q \,;S) \oslash T) = \pi \,;Q \,;S \cap \rho \,;T$ expanding

$= [\pi \,;Q \,;\pi'^{\mathsf{T}} \cap \rho \,;\mathbb{I} \,;\rho'^{\mathsf{T}}] \,;\pi' \,;S \cap \rho \,;T$ Prop. 7.4.i since $\mathbb{I}$ is total

$$\subseteq \left(\left[\pi_\cdot Q_\cdot \pi'^{\mathsf{T}} \cap \rho_\cdot \mathbb{I}_\cdot \rho'^{\mathsf{T}}\right] \cap \rho_\cdot T_\cdot S^{\mathsf{T}}_\cdot \pi'^{\mathsf{T}}\right)_\cdot \left(\pi'_\cdot S \cap \left[\pi_\cdot Q_\cdot \pi'^{\mathsf{T}} \cap \rho_\cdot \mathbb{I}_\cdot \rho'^{\mathsf{T}}\right]^{\mathsf{T}}_\cdot \rho_\cdot T\right)$$
$$\text{Dedekind}$$
$$\subseteq \left(\pi_\cdot Q_\cdot \pi'^{\mathsf{T}} \cap \rho_\cdot \mathbb{I}_\cdot \rho'^{\mathsf{T}}\right)_\cdot \left(\pi'_\cdot S \cap \rho'_\cdot \rho^{\mathsf{T}}_\cdot \rho_\cdot T\right) \qquad \text{monotony}$$
$$= \left(\pi_\cdot Q_\cdot \pi'^{\mathsf{T}} \cap \rho_\cdot \mathbb{I}_\cdot \rho'^{\mathsf{T}}\right)_\cdot \left(\pi'_\cdot S \cap \rho'_\cdot T\right) \qquad \text{since } \rho \text{ is univalent and surjective}$$
$$= (Q \otimes \mathbb{I})_\cdot (S \oslash T) \qquad \text{by definition}$$

(iii) Aiming at a use of (ii), we specialize $C := B \times Y$, as well as $S := \mathbb{I}_\cdot \pi''^{\mathsf{T}}$ and $T := R_\cdot \rho''^{\mathsf{T}}$ in order to get

$$
\begin{aligned}
(Q \otimes \mathbb{I})_\cdot (\mathbb{I} \otimes R) &= (Q \otimes \mathbb{I})_\cdot \left(\pi'_\cdot \mathbb{I}_\cdot \pi''^{\mathsf{T}} \cap \rho'_\cdot R_\cdot \rho''^{\mathsf{T}}\right) \qquad \text{expanding} \\
&= (Q \otimes \mathbb{I})_\cdot \left(\pi'_\cdot S \cap \rho'_\cdot T\right) \qquad \text{abbreviated} \\
&= (Q \otimes \mathbb{I})_\cdot (S \oslash T) \qquad \text{by definition} \\
&= ((Q_\cdot S) \oslash T) \qquad \text{due to (ii)} \\
&= \pi_\cdot Q_\cdot S \cap \rho_\cdot T \qquad \text{expanded again} \\
&= \pi_\cdot Q_\cdot \pi''^{\mathsf{T}} \cap \rho_\cdot R_\cdot \rho''^{\mathsf{T}} \qquad \text{abbreviations cancelled} \\
&= (Q \otimes R) \qquad \text{by definition} \qquad \square
\end{aligned}
$$

It will easily be seen from Prop. 7.5.iii, that this construct should not be called parallel composition although it resembles part of the idea. Let us consider (iii) expanded by the right variant

$$(Q \otimes \mathbb{I}_X)_\cdot (\mathbb{I}_B \otimes R) = (Q \otimes R) = (\mathbb{I}_A \otimes R)_\cdot (Q \otimes \mathbb{I}_Y).$$

This does express correctly that Q and R may *with one execution thread* be executed *in either order*. However, *no two execution threads* are provided to execute in parallel. Modelling truly parallel computation needs ideas that are explained only later in Section 19.1. It touches deep concepts of informatics that even today are not commonly understood. Recall the predicate-logic proof of the unsharpness result and assume the two 'parallel' processes $R_\cdot P$ and $S_\cdot Q$ to have taken place in different houses or even on different continents. Then 'observability', i.e., the technical problem of fixing a, b, would have been really difficult. Even the problem of relative speed of the two processes and the speed compared with the speed of communicating their results becomes important.

Vectorization

The concept of vectorization has already shown up in Fig. 3.1 where a relation was also represented as a vector. In several areas of algebra, one speaks of vectorization when a matrix is converted into a column vector by a linear transformation. It may also be applied fruitfully in the present context of a direct product, not least because it matches nicely with the Kronecker product.

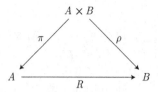

Fig. 7.19 Typing for vectorization in general

Definition 7.6. Let any relation $R : A \longrightarrow B$ be given and form the direct product $A \times B$ with projections denoted as π, ρ. Then the vector resulting out of the operation

$$\mathsf{vec}(R) := (\pi; R \cap \rho); \mathbb{T}_B = (R \otimes \mathbb{I}); \mathbb{T}_B$$

is called the **vectorization** of R. □

The notational variant is obvious in view of the definition of the strict join. One will identify the operation $\mathsf{rel}(v) := \pi^{\mathsf{T}}; (v; \mathbb{T} \cap \rho)$ as the inverse operation. This is elaborated in detail in [123, 124].

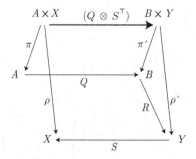

Fig. 7.20 Typing for the vectorization formula of Prop. 7.7

We are now in a position to prove a technical result (see [74]) on vectorization of relations in combination with the Kronecker product.

Proposition 7.7. *Let relations* $Q : A \longrightarrow B$, $S : Y \longrightarrow X$ *and* $R : B \longrightarrow Y$ *be given and consider the direct products* $A \times X$ *and* $B \times Y$ *with*

$$\pi : A \times X \longrightarrow A \quad \rho : A \times X \longrightarrow X \quad and \quad \pi' : B \times Y \longrightarrow B \quad \rho' : B \times Y \longrightarrow Y$$

the respective projections. Then the following holds:

$$\mathsf{vec}(Q; R; S) = (Q \otimes S^{\mathsf{T}}); \mathsf{vec}(R)$$

Proof: Below, containments are in cyclical order so that equality holds everywhere in between:

$$(Q \otimes S^\mathsf{T}) ; \mathsf{vec}(R) = (Q \otimes \mathbb{I}) ; (\mathbb{I} \otimes S^\mathsf{T}) ; (R \oslash \mathbb{I}) ; \mathbb{T} \qquad \text{Prop. 7.5.iii; definition}$$
$$= (Q \otimes \mathbb{I}) ; (R \oslash S^\mathsf{T}) ; \mathbb{T} \qquad \text{Prop. 7.5.ii}$$
$$= ((Q ; R) \oslash S^\mathsf{T}) ; \mathbb{T} \qquad \text{Prop. 7.5.ii}$$
$$= (\rho ; S^\mathsf{T} \cap \pi ; Q ; R) ; \mathbb{T} \qquad \text{definition with intersection terms exchanged } (*)$$
$$\subseteq (\rho \cap \pi ; Q ; R ; S) ; (S^\mathsf{T} \cap \rho^\mathsf{T} ; \pi ; Q ; R) ; \mathbb{T} \qquad \text{Dedekind rule}$$
$$\subseteq (\pi ; Q ; R ; S \cap \rho) ; \mathbb{T} = \mathsf{vec}(Q ; R ; S) \qquad \text{monotony gives what had to be proved}$$
$$\subseteq (\pi ; Q ; R \cap \rho ; S^\mathsf{T}) ; (S \cap R^\mathsf{T} ; Q^\mathsf{T} ; \pi^\mathsf{T} ; \rho) ; \mathbb{T} \qquad \text{Dedekind rule}$$
$$\subseteq (\pi ; Q ; R \cap \rho ; S^\mathsf{T}) ; \mathbb{T} \qquad \text{monotony closing the cycle to } (*) \qquad \square$$

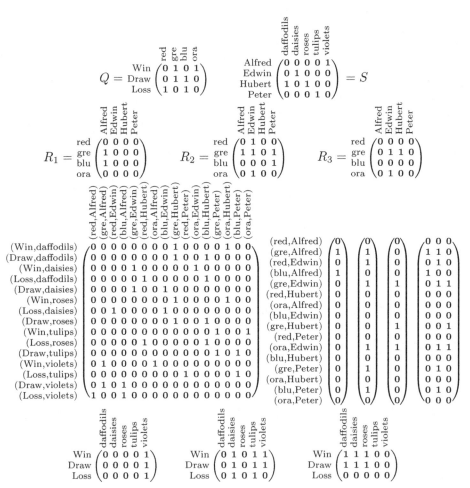

Fig. 7.21 Vectorization of relations R_1, R_2, R_3 in order to handle them simultaneously

We try to visualize the idea of vectorization with Fig. 7.21. Following the notation of Prop. 7.7, the two uppermost relations are Q, S. They are applied to the relations R_1, R_2, R_3 on the second line. In order to handle these simultaneously, a Kronecker product is formed and the relations are vectorized as shown in line three. The last 3-column-relation $=: M$ has been obtained simply by putting the three vectors side by side to be handled later simultaneously, i.e., like just one matrix. The results of the treatment $(Q \otimes S^\mathsf{T}) \,\mathbf{;}\, \mathsf{vec}(M)$ are shown – already singled out again – and transformed back to the relations $Q \,\mathbf{;}\, R_i \,\mathbf{;}\, S$ of the lowest line.

Exercises

7.1 Prove that the Kronecker product is "$\cup$"-distributive in both of its components. Show that if R, S have the generalized inverses G, H, then $(G \otimes H)$ is the generalized inverse of $(R \otimes S)$.

7.2 Prove for the highly specialized Kronecker product of $Q : A \longrightarrow B$ and $R : B \longrightarrow A$ that

$$(Q \,\mathbf{;}\, R \otimes \mathbb{I}) \,\mathbf{;}\, \mathbb{T}_{A \times A, A} = (Q \otimes R^\mathsf{T}) \,\mathbf{;}\, \mathbb{T}_{B \times B, A} = \pi \,\mathbf{;}\, Q \,\mathbf{;}\, R \,\mathbf{;}\, \mathbb{T}_{A,A}.$$

7.3 Direct sum

While product, or pairset, forming is quite well known even to non-specialists, variant handling is known far less.[7] We encounter it as *if–then–else–fi* or in case distinctions.[8] Often it is met as a *disjoint union*.

Again, we will handle this constructively and in a generic form. With the term **direct sum** we mean the type `DirSum t1 t2` which may be interpreted as, for example,

    nationalities + colors

or else as

    germanSoccer + internationalSoccer.

While the notation is rather obvious in the first case,

    nationalities + colors =
        {US,French,German,British,Spanish,red,green,blue,orange},

we have to take care of overlapping element notations in the second:

    germanSoccer + internationalSoccer =

---

[7] Even in the design of programming languages (such as Algol, Pascal, Modula, Ada, for example) variants were largely neglected or not handled in the necessary pure form as they are nowadays in HASKELL.

[8] It should be made clear, however, that we aim at 'structural' distinctions and not at such wilful distinctions as, for example, *income* $>$ £ 123.45 in tax tables.

[Bayern München<,	>Arsenal London,
Borussia Dortmund<,	>FC Chelsea,
	>Juventus Turin,
Werder Bremen<,	>Manchester United,
Schalke 04<,	>Bayern München,
	>Borussia Dortmund,
	>FC Liverpool,
	>Ajax Amsterdam,
VfB Stuttgart<,	>Real Madrid,
	>Olympique Lyon] .

Borussia Dortmund was formerly considered a German soccer team as well as an international soccer team. It is wise to keep these two concepts separate. One therefore introduces some piece of notation to make clear from which side the elements come. Here, we have chosen angle marks. Whenever the two sets are disjoint, we will not use these marks.

There is one further point to observe. In some examples sets will be infinite. In such cases it would be boring to see elements of just the first set infinitely often before any element of the second shows up. The procedure of showing elements running over the computer screen has to be interrupted anyway. In order to see elements from both variants and so to get an increasing impression of the variant set, we have chosen to show elements of the two variants alternately.

Injections

The constructive generation of the direct sum out of types `t1, t2` is achieved in TituRel with the two generic constructs

`DirSum t1 t2`,	the **variant domain**, corresponding to $X + Y$
`Iota t1 t2, Kappa t1 t2`,	the **left** respectively **right injection**.

In a mathematical text, we will use the sloppier form ι, κ, or sometimes with typing $\iota : X \longrightarrow X+Y$ and $\kappa : Y \longrightarrow X+Y$. Figure 7.22 shows an example interpretation of injections.

$$\iota = \begin{array}{c} \\ \spadesuit \\ \heartsuit \\ \diamondsuit \\ \clubsuit \end{array} \begin{array}{ccccccccccc} \spadesuit & \text{Mon} & \heartsuit & \text{Tue} & \diamondsuit & \text{Wed} & \clubsuit & \text{Thu} & \text{Fri} & \text{Sat} & \text{Sun} \\ 1&0&0&0&0&0&0&0&0&0&0 \\ 0&0&1&0&0&0&0&0&0&0&0 \\ 0&0&0&0&1&0&0&0&0&0&0 \\ 0&0&0&0&0&0&1&0&0&0&0 \end{array} \qquad \kappa = \begin{array}{c} \\ \text{Mon} \\ \text{Tue} \\ \text{Wed} \\ \text{Thu} \\ \text{Fri} \\ \text{Sat} \\ \text{Sun} \end{array} \begin{array}{ccccccccccc} \spadesuit & \text{Mon} & \heartsuit & \text{Tue} & \diamondsuit & \text{Wed} & \clubsuit & \text{Thu} & \text{Fri} & \text{Sat} & \text{Sun} \\ 0&1&0&0&0&0&0&0&0&0&0 \\ 0&0&0&1&0&0&0&0&0&0&0 \\ 0&0&0&0&0&1&0&0&0&0&0 \\ 0&0&0&0&0&0&0&1&0&0&0 \\ 0&0&0&0&0&0&0&0&1&0&0 \\ 0&0&0&0&0&0&0&0&0&1&0 \\ 0&0&0&0&0&0&0&0&0&0&1 \end{array}$$

Fig. 7.22 Injection relations $\iota \approx$ `Iota t1 t2` and $\kappa \approx$ `Kappa t1 t2`

Elements in a direct sum

We need either e_X denoting an element in X or e_Y denoting an element in Y. Then $\iota(e_X)$ or $\kappa(e_Y)$, respectively, denote an element in the direct sum $X + Y$ when denoting traditionally with injection mappings. As an example we consider the direct sum of Bridge suits and Bridge honor cards, thereby switching to relational notation

$\iota^{\mathsf{T}};e_X$ and $\kappa^{\mathsf{T}};e_Y$.

$$e_X = \begin{array}{c} \spadesuit \\ \heartsuit \\ \diamondsuit \\ \clubsuit \end{array}\begin{pmatrix} 0 \\ 0 \\ 1 \\ 0 \end{pmatrix} \quad \iota = \begin{array}{c} \\ \\ \\ \\ \end{array}\begin{array}{ccccccccc} \spadesuit & \text{Ace} & \heartsuit & \text{King} & \diamondsuit & \text{Queen} & \clubsuit & \text{Jack} & \text{Ten} \end{array} \begin{pmatrix} 1&0&0&0&0&0&0&0&0 \\ 0&0&1&0&0&0&0&0&0 \\ 0&0&0&0&1&0&0&0&0 \\ 0&0&0&0&0&0&1&0&0 \end{pmatrix}$$

$$e_Y = \begin{array}{r} \text{Ace} \\ \text{King} \\ \text{Queen} \\ \text{Jack} \\ \text{Ten} \end{array}\begin{pmatrix} 0 \\ 0 \\ 0 \\ 1 \\ 0 \end{pmatrix} \quad \kappa = \begin{array}{ccccccccc} \spadesuit & \text{Ace} & \heartsuit & \text{King} & \diamondsuit & \text{Queen} & \clubsuit & \text{Jack} & \text{Ten} \end{array} \begin{pmatrix} 0&1&0&0&0&0&0&0&0 \\ 0&0&0&1&0&0&0&0&0 \\ 0&0&0&0&0&1&0&0&0 \\ 0&0&0&0&0&0&0&1&0 \\ 0&0&0&0&0&0&0&0&1 \end{pmatrix}$$

$$(0\;0\;0\;0\;1\;0\;0\;0\;0) = e_X^{\mathsf{T}};\iota \qquad e_Y^{\mathsf{T}};\kappa = (0\;0\;0\;0\;0\;0\;0\;1\;0)$$

Fig. 7.23 An element, the injection relations and the injected elements as row vectors

The elements $\iota^{\mathsf{T}};e_X$ and $\kappa^{\mathsf{T}};e_Y$ in the direct sum have been transposed to horizontal presentation. This is mainly for reasons of space, but also since one may then better recognize the idea of marking.

Vectors in a direct sum

If v_X and v_Y stand for an expressible vector in X or Y, respectively, then in the first place only $\iota(v_X)$ or $\kappa(v_Y)$ are expressible vectors in the direct sum $X + Y$. This is based on injections conceived as mappings and their usual notation. We may, however, also denote this as $\iota^{\mathsf{T}};v_X$ and $\kappa^{\mathsf{T}};v_Y$ in a relational environment. In the example of Fig. 7.24, we have first the set of red suits among all four suits in the game of Skat and then the set of value-carrying cards among all the cards. The rightmost column shows a union of such vectors.

$$v_X = \begin{array}{c} \clubsuit \\ \spadesuit \\ \heartsuit \\ \diamondsuit \end{array}\begin{pmatrix} 0 \\ 0 \\ 1 \\ 1 \end{pmatrix} \quad \iota = \begin{array}{cccccccccccc} \clubsuit & A & \spadesuit & K & \heartsuit & D & \diamondsuit & B & 10 & 9 & 8 & 7 \end{array} \begin{pmatrix} 1&0&0&0&0&0&0&0&0&0&0&0 \\ 0&0&1&0&0&0&0&0&0&0&0&0 \\ 0&0&0&0&1&0&0&0&0&0&0&0 \\ 0&0&0&0&0&0&1&0&0&0&0&0 \end{pmatrix} \quad \iota^{\mathsf{T}};v_X = \begin{array}{c} \clubsuit \\ A \\ \spadesuit \\ K \\ \heartsuit \\ D \\ \diamondsuit \\ B \\ 10 \\ 9 \\ 8 \\ 7 \end{array}\begin{pmatrix} 0\\0\\0\\0\\1\\0\\1\\0\\0\\0\\0\\0 \end{pmatrix}$$

$$v_Y = \begin{array}{c} A \\ K \\ D \\ B \\ 10 \\ 9 \\ 8 \\ 7 \end{array}\begin{pmatrix} 1\\1\\1\\1\\1\\0\\0\\0 \end{pmatrix} \quad \kappa = \begin{array}{cccccccccccc} \clubsuit & A & \spadesuit & K & \heartsuit & D & \diamondsuit & B & 10 & 9 & 8 & 7 \end{array} \begin{pmatrix} 0&1&0&0&0&0&0&0&0&0&0&0 \\ 0&0&0&1&0&0&0&0&0&0&0&0 \\ 0&0&0&0&0&1&0&0&0&0&0&0 \\ 0&0&0&0&0&0&0&1&0&0&0&0 \\ 0&0&0&0&0&0&0&0&1&0&0&0 \\ 0&0&0&0&0&0&0&0&0&1&0&0 \\ 0&0&0&0&0&0&0&0&0&0&1&0 \\ 0&0&0&0&0&0&0&0&0&0&0&1 \end{pmatrix}$$

$$\kappa^{\mathsf{T}};v_Y = \begin{array}{c} \clubsuit \\ A \\ \spadesuit \\ K \\ \heartsuit \\ D \\ \diamondsuit \\ B \\ 10 \\ 9 \\ 8 \\ 7 \end{array}\begin{pmatrix} 0\\1\\0\\1\\0\\1\\0\\1\\1\\0\\0\\0 \end{pmatrix} \quad \text{union} = \begin{pmatrix} 0\\1\\0\\1\\1\\1\\1\\1\\1\\0\\0\\0 \end{pmatrix}$$

Fig. 7.24 Injections of left and right variant, vector in direct sum, and union of such

Relations starting or ending in a direct sum

In connection with a direct sum, one will construct relations as follows. Given two relations $R : U \longrightarrow W$ and $S : V \longrightarrow W$, i.e., with the same target, it is normal to construct the relations $\iota^{\mathsf{T}}\!; R : U + V \longrightarrow W$ and $\kappa^{\mathsf{T}}\!; S : U + V \longrightarrow W$ starting from the direct sum. From these, others such as $\iota^{\mathsf{T}}\!; R \cup \kappa^{\mathsf{T}}\!; S$ may then be built using Boolean operators.

In a rather similar way we also build relations with the same source ending in a direct sum. Let two relations $R : U \longrightarrow V$ and $S : U \longrightarrow W$ be given, i.e., with the same source. Then it is normal to construct the relations $R\!; \iota : U \longrightarrow V + W$ and $S\!; \kappa : U \longrightarrow V + W$ ending in the direct sum. From these, further relations such as $R\!; \iota \cup \overline{S\!; \kappa}$ may be built using Boolean operators.

There is a comfortable way of representing relations from a direct sum to a direct sum, namely as matrices of relations. We will make use of this on several occasions, for example, on pages 141, 208 and 232. Assume natural injections $\iota : A \longrightarrow A + B$ and $\kappa : B \longrightarrow A + B$ as well as $\iota' : X \longrightarrow X + Y$ and $\kappa' : Y \longrightarrow X + Y$ together with several relations $P : A \longrightarrow X$, $Q : A \longrightarrow Y$, $R : B \longrightarrow X$, and $S : B \longrightarrow Y$. Then the relations $C, C' : A + B \longrightarrow X + Y$ defined as

$$C := \iota^{\mathsf{T}}\!; P\!; \iota' \ \cup \ \iota^{\mathsf{T}}\!; Q\!; \kappa' \ \cup \ \kappa^{\mathsf{T}}\!; R\!; \iota' \ \cup \ \kappa^{\mathsf{T}}\!; S\!; \kappa'$$
$$C' := \iota'^{\mathsf{T}}\!; K\!; \iota'' \ \cup \ \iota'^{\mathsf{T}}\!; L\!; \kappa'' \ \cup \ \kappa'^{\mathsf{T}}\!; M\!; \iota'' \ \cup \ \kappa'^{\mathsf{T}}\!; N\!; \kappa''$$

may be described as

$$C = \begin{pmatrix} P & Q \\ R & S \end{pmatrix}, \qquad C' = \begin{pmatrix} K & L \\ M & N \end{pmatrix}$$

and multiplied to

$$C\!; C' = \begin{pmatrix} P\!; K \cup Q\!; M & P\!; L \cup Q\!; N \\ R\!; K \cup S\!; M & R\!; L \cup S\!; N \end{pmatrix}$$

which is much better suited for human perception. A considerable advantage is that nearly everybody is trained in multiplication of matrices. Another advantage is that composition of the matrices of relations directly resembles the direct sum construct: elaborating the product and using all the rules for direct sum injections (to be introduced below in Def. 7.8) would show precisely this result. Matrix computations are always built from a sum-like operation extended over the result of many product-like operations:

— "$+$" over "$*$" in the case of real- or complex-valued matrices,
— "$\vee$" over "$\wedge$" in the case of Boolean matrices,
— "$\cup$" over "$;$" in the case of matrices of relations.

Not least, one observes that composition of matrices of relations strictly preserves our typing rules.

Algebraic properties of the generic injection relations

As already mentioned, work with the direct sum resembles the *if–then–else–fi* and other case distinctions. The direct sum is often called a coproduct. Regardless of the respective example, the injections satisfy what is demanded in the formal definition.

Definition 7.8. Any two heterogeneous relations ι, κ with common target are said to form the left, respectively right, **injection** of a **direct sum** if

$$\iota_{;}\iota^{\mathsf{T}} = \mathbb{I}, \quad \kappa_{;}\kappa^{\mathsf{T}} = \mathbb{I}, \quad \iota^{\mathsf{T}}{}_{;}\iota \cup \kappa^{\mathsf{T}}{}_{;}\kappa = \mathbb{I}, \quad \iota_{;}\kappa^{\mathsf{T}} = \mathbb{L}. \qquad \square$$

Thus, ι, κ have to be injective mappings with disjoint value sets in the sum as visualized in Fig. 7.23, for example. Given their sources, ι and κ are essentially uniquely defined. This is an important point. For the TITUREL interpretation we have decided on a specific form. Without a fully formal proof we give a hint here that for any other pair ι', κ' satisfying the laws, we are in a position to construct a bijection that relates the injections, namely

$$P := \iota^{\mathsf{T}}{}_{;}\iota' \cup \kappa^{\mathsf{T}}{}_{;}\kappa'.$$

In Fig. 7.25, we show this idea. With the postulates of Def. 7.8, we have obviously $\iota_{;}P = \iota'$ and $\kappa_{;}P = \kappa'$.

$$
\spadesuit\;\begin{pmatrix} 0 & 0 & 0 & 1 & 0 & 0 & 0 & 0 & 0 \\ 1 & 0 & 0 & 0 & 0 & 0 & 0 & 0 & 0 \\ 0 & 0 & 0 & 0 & 0 & 0 & 1 & 0 & 0 \\ 0 & 0 & 1 & 0 & 0 & 0 & 0 & 0 & 0 \end{pmatrix}
$$

$$
\begin{matrix} \text{Ace} \\ \text{King} \\ \text{Queen} \\ \text{Jack} \\ \text{Ten} \end{matrix}\begin{pmatrix} 0 & 0 & 0 & 0 & 1 & 0 & 0 & 0 & 0 \\ 0 & 1 & 0 & 0 & 0 & 0 & 0 & 0 & 0 \\ 0 & 0 & 0 & 0 & 0 & 0 & 1 & 0 & 0 \\ 0 & 0 & 0 & 0 & 0 & 0 & 0 & 1 & 0 \\ 0 & 0 & 1 & 0 & 0 & 0 & 0 & 0 & 0 \end{pmatrix}
$$

$$
\begin{matrix} \text{Ace} \\ \heartsuit \\ \text{King} \\ \diamondsuit \\ \text{Queen} \\ \clubsuit \\ \text{Jack} \\ \text{Ten} \end{matrix}\begin{pmatrix} 0 & 0 & 0 & 1 & 0 & 0 & 0 & 0 & 0 \\ 0 & 0 & 0 & 0 & 0 & 1 & 0 & 0 & 0 \\ 1 & 0 & 0 & 0 & 0 & 0 & 0 & 0 & 0 \\ 0 & 1 & 0 & 0 & 0 & 0 & 0 & 0 & 0 \\ 0 & 0 & 0 & 0 & 0 & 0 & 1 & 0 & 0 \\ 0 & 0 & 0 & 0 & 0 & 0 & 0 & 1 & 0 \\ 0 & 0 & 1 & 0 & 0 & 0 & 0 & 0 & 0 \\ 0 & 0 & 0 & 0 & 0 & 0 & 0 & 0 & 1 \\ 0 & 0 & 1 & 0 & 0 & 0 & 0 & 0 & 0 \end{pmatrix}
$$

Fig. 7.25 Other injections ι', κ' than in Fig, 7.22; bijection $P := \iota^{\mathsf{T}}{}_{;}\iota' \cup \kappa^{\mathsf{T}}{}_{;}\kappa'$ to relate to these

7.4 Quotient domain

Equivalence relations are omnipresent in all our thinking and reasoning. We are accustomed to consider quotient sets modulo an equivalence since learning in school how to add or multiply natural numbers modulo 5, for example. This quotient set does not exist so independently as other sets. On the other hand, we are interested in using the quotient set for further constructions in the same way as the sets introduced earlier. The introduction of the quotient set will employ the natural projection relation, in fact a mapping.

When forming the quotient set of `politicians` modulo `nationality`, for example, one will as usual put a representative in square brackets and thus get its corresponding class:

$$\{\,[\texttt{Bush}]\,,[\texttt{Chirac}]\,,[\texttt{Schmidt}]\,,[\texttt{Thatcher}]\,,[\texttt{Zapatero}]\,\}.$$

The notation should be a bit more precise here as in $[\texttt{Thatcher}]_{\text{SameNationality}}$, mentioning the equivalence relation used. But usually the relation is known from the context so we do not mention it explicitly every time. The mapping $\eta : x \mapsto [x]$ is called the natural projection. This natural projection, however, raises minor problems as one obtains more notation than classes, since, for example,

$$[\texttt{Bush}] = [\texttt{Clinton}].$$

So $\eta : \texttt{Bush} \mapsto [\texttt{Bush}]$ as well as $\eta : \texttt{Bush} \mapsto [\texttt{Clinton}]$ is correct. When defining functions or mappings with an element of a quotient set as an argument, mathematicians are accustomed to show that their results are 'independent of the choice of the representative'.

When trying to generate quotients and natural projections constructively, one will need a type `t`, later interpreted by some set X, for example, and a relation `xi` on `t`, later interpreted by a relation Ξ on X *that must be an equivalence*. As all our relations carry their typing directly with them, the language TITUREL allows us to express this simply as

`QuotMod xi`,   the **quotient domain**, corresponding to X/Ξ, and

`Project xi`,   the **natural projection**, corresponding to $\eta : X \longrightarrow X/\Xi$.

Using this generic construction, the source of `Project xi` will be `t` and the target will be `QuotMod xi`. Figure 7.26 visualizes the intended meaning. Elements, vectors, and relations will then be formulated only via this natural projection.

$$\Xi =$$

	Blair	Bush	Chirac	Clinton	Kohl	Major	Mitterand	Schmidt	Schröder	Thatcher	Zapatero
Blair	1	0	0	0	0	1	0	0	0	1	0
Bush	0	1	0	1	0	0	0	0	0	0	0
Chirac	0	0	1	0	0	0	1	0	0	0	0
Clinton	0	1	0	1	0	0	0	0	0	0	0
Kohl	0	0	0	0	1	0	0	1	1	0	0
Major	1	0	0	0	0	1	0	0	0	1	0
Mitterand	0	0	1	0	0	0	1	0	0	0	0
Schmidt	0	0	0	0	1	0	0	1	1	0	0
Schröder	0	0	0	0	1	0	0	1	1	0	0
Thatcher	1	0	0	0	0	1	0	0	0	1	0
Zapatero	0	0	0	0	0	0	0	0	0	0	1

$$\eta =$$

	[Blair]	[Chirac]	[Schröder]	[Zapatero]	[Bush]
Blair	1	0	0	0	0
Bush	0	0	0	0	1
Chirac	0	1	0	0	0
Clinton	0	0	0	0	1
Kohl	0	0	1	0	0
Major	1	0	0	0	0
Mitterand	0	1	0	0	0
Schmidt	0	0	1	0	0
Schröder	0	0	1	0	0
Thatcher	1	0	0	0	0
Zapatero	0	0	0	1	0

Fig. 7.26 Quotient set and natural projection $\eta \approx$ `Project xi`

When working practically with a quotient domain, one will be obliged to prove that it was indeed an *equivalence* one has divided out. As long as only a relational term is presented, this may not yet be possible, so that it is merely given as an assertion. Immediately before interpretation, however, the proof obligation has to be invoked.

Elements in a quotient domain modulo Ξ

If e_X stands for expressible elements in X, then in the first place classes $[e_X]_\Xi$ are expressible elements in the quotient domain X_Ξ. As an example consider the element $e_X = \mathtt{Chirac}$ of politicians and the corresponding element among the classes $[e_X]_\Xi = [\mathtt{Chirac}]$ modulo the nationalities equivalence.

The transition from an element in a set to the corresponding element in the quotient set is not made very often, so we need not provide an elegant method of denotation. We explicitly apply the natural projection to the column vector representing the element $\eta^\mathsf{T}{;}\, e$ obtaining a vector, and convert back to an element (Fig. 7.27).

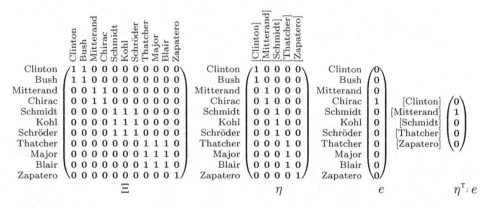

Fig. 7.27 Equivalence, natural projection, element, class with different representative

That $\eta^\mathsf{T}{;}\, e$ for Chirac results in [Mitterand] is irritating only at first sight. But the choice of the representatives when denoting the class may have been executed much earlier and, thus, completely independently. We have to develop the appropriate technology to handle this situation, not least in never trying to write something down using methods outside the language agreed upon.

Vectors in a quotient domain modulo Ξ

If v_X stands for an expressible subset in X, then only classes $\eta^\mathsf{T}{;}\, v_X$ corresponding to $\{[e_X] \mid e_X \in v_X\}$ are expressible subsets in the quotient X_Ξ.

Relations starting from or ending in a quotient domain

All relation terms starting from a quotient set shall begin with η^T. All relational terms ending in a quotient set shall correspondingly terminate with $\eta = \texttt{Project xi}$. Of course, this will often be hidden inside more complex nested constructions. Examples may be found in Fig. 10.24 and Fig. 11.1, for example.

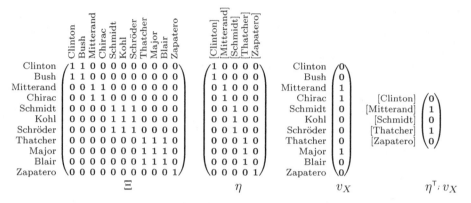

Fig. 7.28 Equivalence, natural projection, subset and subset of classes

Algebraic properties of the generic natural projection relation

It was a novel situation that we had to define a set from an already available set *together with an equivalence relation*. In earlier cases, constructions started from one or more sets. We have to introduce the requirements that must hold for the quotient set and the natural projection.

Definition 7.9. Given an arbitrary equivalence Ξ, a relation η will be called the **natural projection** onto the quotient domain modulo Ξ, provided

$$\Xi = \eta\, \eta^\mathsf{T}, \qquad \eta^\mathsf{T}\, \eta = \mathbb{I}.$$ $\qquad\qquad\Box$

One need not give the set on which the equivalence is defined to hold since every relation carries its source and target information with it. In a very natural way, the question arises to what extent η is uniquely defined. (For a proof that this is an essentially unique definition, see Prop. B.17 in Appendix B.) If the two natural projections η, ξ are presented as in Fig. 7.29, then one will immediately have the bijection $\xi^\mathsf{T}\, \eta$ between their targets.

Ξ

	Jan	Feb	Mar	Apr	May	Jun	Jul	Aug	Sep	Oct	Nov	Dec
Jan	1	1	1	0	0	0	0	0	0	0	0	1
Feb	1	1	1	0	0	0	0	0	0	0	0	1
Mar	1	1	1	0	0	0	0	0	0	0	0	1
Apr	0	0	0	1	1	0	0	0	0	0	0	0
May	0	0	0	1	1	0	0	0	0	0	0	0
Jun	0	0	0	0	0	1	1	1	0	0	0	0
Jul	0	0	0	0	0	1	1	1	0	0	0	0
Aug	0	0	0	0	0	1	1	1	0	0	0	0
Sep	0	0	0	0	0	0	0	0	1	1	1	0
Oct	0	0	0	0	0	0	0	0	1	1	1	0
Nov	0	0	0	0	0	0	0	0	1	1	1	0
Dec	1	1	1	0	0	0	0	0	0	0	0	1

η

	[Jan]	[Apr]	[Jun]	[Sep]
Jan	1	0	0	0
Feb	1	0	0	0
Mar	1	0	0	0
Apr	0	1	0	0
May	0	1	0	0
Jun	0	0	1	0
Jul	0	0	1	0
Aug	0	0	1	0
Sep	0	0	0	1
Oct	0	0	0	1
Nov	0	0	0	1
Dec	1	0	0	0

ξ

	Summer	Autumn	Winter	Spring
Jan	0	0	1	0
Feb	0	0	1	0
Mar	0	0	1	0
Apr	0	0	0	1
May	0	0	0	1
Jun	1	0	0	0
Jul	1	0	0	0
Aug	1	0	0	0
Sep	0	1	0	0
Oct	0	1	0	0
Nov	0	1	0	0
Dec	0	0	1	0

$\xi^{\mathsf{T}};\eta$

	[Jan]	[Apr]	[Jun]	[Sep]
Summer	0	0	1	0
Autumn	0	0	0	1
Winter	1	0	0	0
Spring	0	1	0	0

Fig. 7.29 The quotient is defined in an essentially unique way: Ξ, η, ξ and $\xi^{\mathsf{T}};\eta$

We report here yet another important rule regulating work with natural projections. At first sight, it looks similar to Prop. 5.3, but differs considerably because η is now transposed.

Proposition 7.10. *Let an equivalence Ξ be given and consider its natural projection η. If any two relations A, B are presented, one of which satisfies $\Xi; A = A$, the following holds:*

$$\eta^{\mathsf{T}};(A \cap B) = \eta^{\mathsf{T}};A \cap \eta^{\mathsf{T}};B.$$

Proof: See B.18. □

We refer forward to Prop. 8.18, where further useful formulae concerning the quotient domain and its natural projection will be proved that are here still out of reach.

7.5 Subset extrusion

We have stressed the distinction between a set and a subset of a set; a subset shall *only* exist *relative to a set*. With a bit of formalism, however, a subset can be converted so that it is also a set in its own right which we have decided to call an 'extruded subset'. To this end, we observe how the subset {Bush,Chirac,Kohl,Blair} can be injected into its corresponding baseset {Clinton,Bush,Mitterand,Chirac, Schmidt,Kohl,Schröder,Thatcher,Major,Blair,Zapatero}; see Fig. 7.30. We assume subsets that are extruded to be non-empty. When they are the result of some computation, the decision empty/non-empty is always handled prior to extruding. Because one cannot know the outcome of such a computation in advance, a proof obligation must be propagated as long as interpretation is not yet possible.

When trying to constructively generate extruded subsets and natural injections, one will need a type t, later interpreted by some set X, for example, and a vector u of type t, later interpreted by some non-empty subset $U \subseteq X$. Since all our relations carry their typing directly with them, the language TITUREL allows us to formulate

Extrude u,   the **extruded subset**, corresponding to a new baseset D_U, and

Inject   u,   the **natural injection**, corresponding to $\iota_U : D_U \longrightarrow X$.

Using this generic construction, the source of Inject u will be Extrude u and the target will be t. Figure 7.30 visualizes the intended meaning. Elements, vectors, and relations will then be formulated only via this natural injection. In order to demonstrate that the subset is now a set of equal right, we apply the powerset construction to a smaller example.

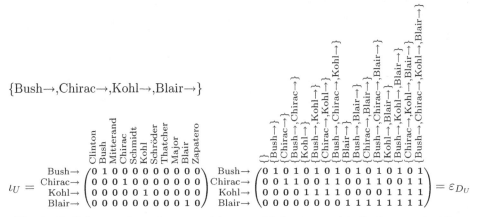

Fig. 7.30 Subset as dependent set with natural injection, and in further construction

One should observe the difference between, for example, Kohl $\in$ t and Kohl$\rightarrow$ $\in D_U$. On looking for a possibility to make the difference visible, we have decided to use the appended arrow.

Elements in an extruded subset

If e stands for expressible elements in X, then only injected elements $e{\rightarrow}$ are expressible in the injected subset $U \subseteq X$, provided $e \in U$. In standard mathematical notation, we have $e{\rightarrow} := \iota_U^{-1}(e)$; denoting algebraically, this becomes $e{\rightarrow} := \iota_U; e$. As an example consider the politician Major, first as one among our set of politicians and then as one of the extruded subset of British politicians.

$$
e = \begin{array}{r}
\text{Clinton}\\ \text{Bush}\\ \text{Mitterand}\\ \text{Chirac}\\ \text{Schmidt}\\ \text{Kohl}\\ \text{Schröder}\\ \text{Thatcher}\\ \text{Major}\\ \text{Blair}\\ \text{Zapatero}
\end{array}
\begin{pmatrix}0\\0\\0\\0\\0\\0\\0\\0\\1\\0\\0\end{pmatrix}
\qquad
v_U = \begin{pmatrix}0\\0\\0\\0\\0\\0\\0\\1\\1\\1\\0\end{pmatrix}
\qquad
\iota_U = \begin{array}{r} \\ \text{Thatcher}\rightarrow\\ \text{Major}\rightarrow\\ \text{Blair}\rightarrow \end{array}
\begin{array}{c}
\begin{array}{ccccccccccc} \rotatebox{90}{Clinton} & \rotatebox{90}{Bush} & \rotatebox{90}{Mitterand} & \rotatebox{90}{Chirac} & \rotatebox{90}{Schmidt} & \rotatebox{90}{Kohl} & \rotatebox{90}{Schröder} & \rotatebox{90}{Thatcher} & \rotatebox{90}{Major} & \rotatebox{90}{Blair} & \rotatebox{90}{Zapatero} \end{array}\\
\begin{pmatrix}
0&0&0&0&0&0&0&1&0&0&0\\
0&0&0&0&0&0&0&0&1&0&0\\
0&0&0&0&0&0&0&0&0&1&0
\end{pmatrix}
\end{array}
\qquad
e_{\rightarrow} = \begin{pmatrix}0\\1\\0\end{pmatrix}
$$

Fig. 7.31 Element e, vector $U \subseteq X$, injection ι_U, and injected element $e_{\rightarrow}$

Vectors in an extruded subset

If v stands for expressible subsets of elements in X, then only injected subsets $v_{\rightarrow}$ with $v \subseteq U$ are expressible in the injected subset $U \subseteq X$ in the first place. Of course, further iterated Boolean constructions based on this as a start may be formulated. As long as we stay in the area of finite basesets, no problem will arise. These constructions become more difficult when infinite sets are considered.

Relations starting or ending in an extruded subset

All relations starting from an extruded subset begin with the natural injection ι_U. All relations ending in an extruded set correspondingly terminate with ι_U^{T}. Of course, this will often not be directly visible inside more complex constructions.

Algebraic properties of the generic natural injection relation

Subset extrusion has hardly ever been considered a domain construction and has stayed in the area of free-hand mathematics. Nevertheless, we will collect the algebraic properties and show that the concept of an extruded subset is defined in an essentially unique form.

Definition 7.11. Let any subset $\perp\!\!\!\perp \neq U \subseteq V$ of some baseset V be given. Whenever a relation $\iota_U : D_U \longrightarrow V$, satisfies the properties

$$\iota_U ; \iota_U^{\mathsf{T}} = \mathbb{I}_{D_U}, \quad \iota_U^{\mathsf{T}} ; \iota_U = \mathbb{I}_V \cap U ; \mathbb{T}_{V,V},$$

it will be called a **natural injection** of the newly introduced domain D_U. □

The new domain D_U, is defined in an essentially unique form. This is best understood assuming two students are given the task of extruding some given $U \subseteq V$.

The first works in the style proposed here, the other in his own style. They return after a while with their solutions ι_U, χ as in Fig. 7.32.

$$
\iota_U = \begin{array}{c}
\quad\; \text{A K D B 10 9 8 7}\\
\begin{array}{c} \text{A}\to\\ 10\to\\ 9\to\\ 8\to\\ 7\to \end{array}
\left(\begin{array}{cccccccc}
1&0&0&0&0&0&0&0\\
0&0&0&1&0&0&0&0\\
0&0&0&0&0&1&0&0\\
0&0&0&0&0&0&1&0\\
0&0&0&0&0&0&0&1
\end{array}\right)
\end{array}
$$

$$
\chi = \begin{array}{c}
\quad\; \text{A K D B 10 9 8 7}\\
\begin{array}{c} \text{V}\\ \text{W}\\ \text{X}\\ \text{Y}\\ \text{Z} \end{array}
\left(\begin{array}{cccccccc}
0&0&0&0&0&0&1&0\\
1&0&0&0&0&0&0&0\\
0&0&0&0&0&0&0&1\\
0&0&0&0&1&0&0&0\\
0&0&0&0&0&1&0&0
\end{array}\right)
\end{array}
$$

$$
P = \begin{array}{c}
\quad\; \text{A 10 9 8 7}\\
\begin{array}{c} \text{V}\\ \text{W}\\ \text{X}\\ \text{Y}\\ \text{Z} \end{array}
\left(\begin{array}{ccccc}
0&0&0&1&0\\
1&0&0&0&0\\
0&0&0&0&1\\
0&1&0&0&0\\
0&0&1&0&0
\end{array}\right)
\end{array}
$$

Fig. 7.32 Two different extrusions ι_U, χ of a subset $U \subseteq V$ and the bijection $P := \iota_U ; \chi^\mathsf{T}$.

Both students demonstrate that the algebraic conditions are met by ι_U, respectively χ, and thus claim that their respective solution is the correct one. The professor then takes these solutions and constructs the bijection P based solely on the material they offered him as $P := \iota_U ; \chi^\mathsf{T}$. Using the postulates of Def. 7.11, then obviously $P ; \chi = \iota_U$ (see Appendix B.19).

A point to mention is that subset extrusion allows us to switch from a set-theoretic consideration to an algebraic one. When using a computer and a formula manipulation system or a theorem prover, this means a considerable restriction in expressivity which is compensated for by much better precision, and even efficiency.

We mention here a result which captures the essence of extrusion in an intuitive way.

Proposition 7.12 (Framing by extrusion). *Let any relation $R : X \longrightarrow Y$ be given with $R \neq \mathbb{L}$ and consider the extrusion $S := \iota ; R ; \iota'^\mathsf{T}$ according to its domain* $\mathrm{dom}(R) = R ; \mathbb{T}$ *and codomain* $\mathrm{cod}(R) = R^\mathsf{T} ; \mathbb{T}$, *i.e., the relations*

$$\iota := \mathtt{Inject}\,(R ; \mathbb{T}) \quad \text{and} \quad \iota' := \mathtt{Inject}\,(R^\mathsf{T} ; \mathbb{T}).$$

Then the following hold:

(i) $\iota^\mathsf{T} ; \mathbb{T} = R ; \mathbb{T} \qquad \iota'^\mathsf{T} ; \mathbb{T} = R^\mathsf{T} ; \mathbb{T}$,
(ii) S *is total and surjective,*
(iii) $\iota^\mathsf{T} ; \iota ; R = R = R ; \iota'^\mathsf{T} ; \iota'$,
(iv) $\iota^\mathsf{T} ; S ; \iota' = R$,
(v) $\iota^\mathsf{T} ; \iota = \mathbb{I} \cap R ; \mathbb{T}$.

Proof: See B.20. □

These small results will be applied surprisingly often. Figure 7.33 tries to visualize the effects. In the second line, the relation R is shown, from which everything starts. The injections ι, ι' extrude domain and range, respectively, and allow S to be constructed. For reasons of space, S and ι are shown in transposed form. With a permutation of rows and columns, one will achieve an even nicer representation that makes S appear as the lower right subrectangle of R.

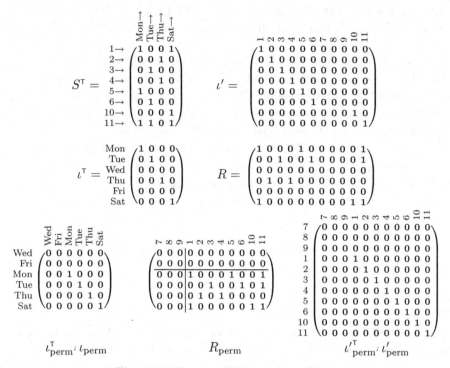

Fig. 7.33 Framing with domain and range

Exercises

7.3 An important application of extrusion is the concept of **tabulation**; see [53, 72], for example. An arbitrary relation $R : X \longrightarrow Y$ with $R \neq \bot\!\!\!\bot$ is said to be tabulated by relations (as a result of the following characterization, they turn out to be mappings) P, Q if

$$P^{\mathsf{T}}\!{;}Q = R \quad P^{\mathsf{T}}\!{;}P = \mathbb{I}_X \cap R\!{;}\mathbb{T}_{YX} \quad Q^{\mathsf{T}}\!{;}Q = \mathbb{I}_Y \cap R^{\mathsf{T}}\!{;}\mathbb{T}_{XY} \quad P\!{;}P^{\mathsf{T}} \cap Q\!{;}Q^{\mathsf{T}} = \mathbb{I}_{X \times Y}.$$

Show that tabulations are unique up to isomorphism, and provide a construction extruding the subset of related pairs.

7.4 Given a symmetric idempotent relation $Q \neq \bot\!\!\!\bot$, one calls a relation R a **splitting** of Q, provided $Q = R^{\mathsf{T}}\!{;}R$ and $R\!{;}R^{\mathsf{T}} = \mathbb{I}$. Prove that splittings are defined uniquely up to isomorphism and provide a construction.

7.6 Direct power

The direct power construction is what we employ when forming the powerset of some set. As this means going from n elements to 2^n, people are usually frightened, so that this is often avoided to the extent that handling this in an algebraic form is not a common ability. The powerset $\mathbf{2}^X$ or $\mathcal{P}(X)$ for the 5-element set $X :=$ {US,French,German,British,Spanish}, for example, looks like

> {{},{US},{French},{US,French},{German},{US,German}, {French,German},
> {US,French,German},{British},{US,British},{French,British},{US,French,British},
> {German,British},{US,German,British},{French,German,British},
> {US,French,German,British},{Spanish},{US,Spanish},{French,Spanish},
> {US,French,Spanish},{German,Spanish},{US,German,Spanish},
> {French,German,Spanish},{US,French,German,Spanish},{British,Spanish},
> {US,British,Spanish},{French,British,Spanish},{US,French,British,Spanish},
> {German,British,Spanish},{US,German,British,Spanish},
> {French,German,British,Spanish},{US,French,German,British,Spanish}}.

Generic membership relations

Membership $e \in U$ of elements in subsets is part of everyday mathematics. The relation "$\in$" is made point-free with the membership relation

$\varepsilon_X : X \longrightarrow \mathbf{2}^X.$

It can easily be generated generically as we will visualize below. To this end we generate membership relations $\varepsilon : X \longrightarrow \mathbf{2}^X$ together with $\Omega : \mathbf{2}^X \longrightarrow \mathbf{2}^X$, the corresponding powerset ordering, in a fractal style. The following relations resemble the 'is_element_of' relation and the powerset ordering using the 3-element set of game qualifications.

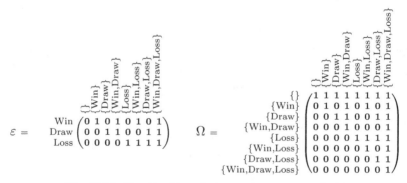

Fig. 7.34 Membership relation and powerset ordering

We immediately observe their method of construction in a fractal fashion that may be characterized by $\varepsilon_0 = ()$, i.e., the 'rowless matrix with one column' or, easier to start with,

$$\varepsilon_1 = (0\ 1) \qquad \varepsilon_2 = \begin{pmatrix} 0 & 1 & 0 & 1 \\ 0 & 0 & 1 & 1 \end{pmatrix} \qquad \varepsilon_{n+1} = \begin{pmatrix} \varepsilon_n & & \varepsilon_n & \\ 0 & \dots & 0 & 1 & \dots & 1 \end{pmatrix}.$$

We can also show how the corresponding powerset order is generated recursively – and thereby determine an easily realizable baseorder for the powerset:

$$\Omega_0 = (1) \qquad \Omega_1 = \begin{pmatrix} 1 & 1 \\ 0 & 1 \end{pmatrix} \qquad \Omega_2 = \begin{pmatrix} 1 & 1 & 1 & 1 \\ 0 & 1 & 0 & 1 \\ 0 & 0 & 1 & 1 \\ 0 & 0 & 0 & 1 \end{pmatrix} \qquad \Omega_{n+1} = \begin{pmatrix} \Omega_n & \Omega_n \\ \mathbb{\perp} & \Omega_n \end{pmatrix}$$

On the set $\{\spadesuit, \heartsuit, \diamondsuit, \clubsuit\}$ of Bridge suits, we form the membership relation and the powerset ordering as shown in Fig. 7.35.

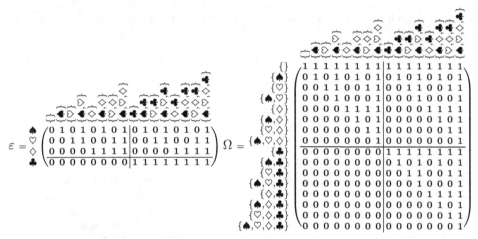

Fig. 7.35 Another membership relation with powerset ordering

Although quite different sets have been used in this example and the one before, one can immediately follow the construction principle explained above. We proceed even further to a 5-element set in Fig. 7.36.

Recognize again the fractal style of this presentation. This gives a basis for the intended generic construction. When trying to generate a direct power and the corresponding membership relation constructively, one will need a type t, later interpreted by some set X, for example. The language TITUREL then allows the formulation of

DirPow t,   the **direct power**, corresponding to the powerset 2^X, and

Member t,   the **membership relation**, corresponding to $\varepsilon : X \longrightarrow 2^X$.

By this generic construction, the source of Member t will be t and the target will be DirPow t. Figure 7.36 visualizes the intended meaning. Elements, vectors, and

142   *Domain construction*

relations will then be formulated only via this membership relation and domain construction.

$$\varepsilon = \begin{array}{c} \text{US} \\ \text{French} \\ \text{German} \\ \text{British} \\ \text{Spanish} \end{array} \left(\begin{array}{c} 0\,1\,0\,1\,0\,1\,0\,1\,0\,1\,0\,1\,0\,1\,0\,1\,0\,1\,0\,1\,0\,1\,0\,1\,0\,1\,0\,1\,0\,1\,0\,1 \\ 0\,0\,1\,1\,0\,0\,1\,1\,0\,0\,1\,1\,0\,0\,1\,1\,0\,0\,1\,1\,0\,0\,1\,1\,0\,0\,1\,1\,0\,0\,1\,1 \\ 0\,0\,0\,0\,1\,1\,1\,1\,0\,0\,0\,0\,1\,1\,1\,1\,0\,0\,0\,0\,1\,1\,1\,1\,0\,0\,0\,0\,1\,1\,1\,1 \\ 0\,0\,0\,0\,0\,0\,0\,0\,1\,1\,1\,1\,1\,1\,1\,1\,0\,0\,0\,0\,0\,0\,0\,0\,1\,1\,1\,1\,1\,1\,1\,1 \\ 0\,0\,0\,0\,0\,0\,0\,0\,0\,0\,0\,0\,0\,0\,0\,0\,1\,1\,1\,1\,1\,1\,1\,1\,1\,1\,1\,1\,1\,1\,1\,1 \end{array}\right) \subseteq X \times \mathbf{2}^X$$

Fig. 7.36 Membership relation

Elements in a direct power

To get an element in the direct power $\mathbf{2}^X$, we need the denotation v of a subset of elements in X. Then a transition is possible to an element in the direct power `DirPow x`, in traditional mathematics simply in the powerset $\mathcal{P}(X)$ or $\mathbf{2}^X$.

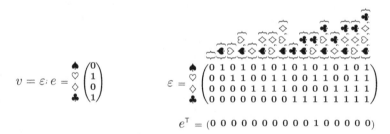

$$v = \varepsilon_{\,;}\, e = \begin{array}{c} \spadesuit \\ \heartsuit \\ \diamondsuit \\ \clubsuit \end{array}\!\begin{pmatrix} 0 \\ 1 \\ 0 \\ 1 \end{pmatrix} \qquad\qquad \varepsilon = \begin{array}{c} \spadesuit \\ \heartsuit \\ \diamondsuit \\ \clubsuit \end{array}\!\left(\begin{array}{c} 0\,1\,0\,1\,0\,1\,0\,1\,0\,1\,0\,1\,0\,1\,0\,1 \\ 0\,0\,1\,1\,0\,0\,1\,1\,0\,0\,1\,1\,0\,0\,1\,1 \\ 0\,0\,0\,0\,1\,1\,1\,1\,0\,0\,0\,0\,1\,1\,1\,1 \\ 0\,0\,0\,0\,0\,0\,0\,0\,1\,1\,1\,1\,1\,1\,1\,1 \end{array}\right)$$

$$e^{\mathsf{T}} = (0\ 0\ 0\ 0\ 0\ 0\ 0\ 0\ 0\ 0\ 1\ 0\ 0\ 0\ 0\ 0)$$

Fig. 7.37 A vector, membership relation, powerset element (transposed to a row vector)

In the visualization of Fig. 7.37, it is immediately clear how to proceed: take the vector v, move it horizontally over the relation ε, and mark the column that equals v. It is more intricate to establish this as an algebraic method. We remember, however, that 'column comparison' has already been introduced as an algebraic operation in Section 4.4. Using this, it simply reads

$$e = \mathsf{syq}\,(\varepsilon, v).$$

Frequently, however, we will also have to go from an element e in the powerset `DirPow t` to its corresponding subset. This is simply composition $v := \varepsilon_{\,;}\, e$. We stress that a subset or vector has two forms of existence, e and v. While mathematicians traditionally abstract over the two – and in many cases have reason to do so – *relational* mathematics cannot and has to make the transitions explicit.

Vectors in a direct power

Expressible subsets of the direct power of X stem from (different!) finite sets $(v_i)_{i \in \mathcal{I}}$ of vectors in X, which we have already defined somehow. The example in Fig. 7.38 shows the red, black, and the extreme-valued suits in the game of Skat, these then comprehended in one matrix, as well as in the corresponding vector in the direct power, which is obtained as

$$v := \sup \big\{ \, \mathsf{syq}(\varepsilon, v_i) \mid i \in \mathcal{I} \big\}.$$

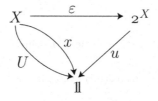

$$\varepsilon = \begin{array}{c} \clubsuit \\ \spadesuit \\ \heartsuit \\ \diamondsuit \end{array} \begin{pmatrix} 0\,1\,0\,1\,0\,1\,0\,1\,0\,1\,0\,1\,0\,1\,0\,1 \\ 0\,0\,1\,1\,0\,0\,1\,1\,0\,0\,1\,1\,0\,0\,1\,1 \\ 0\,0\,0\,0\,1\,1\,1\,1\,0\,0\,0\,0\,1\,1\,1\,1 \\ 0\,0\,0\,0\,0\,0\,0\,0\,1\,1\,1\,1\,1\,1\,1\,1 \end{pmatrix}$$

$$v^{\mathsf{T}} = (0\;0\;0\;1\;0\;0\;0\;0\;0\;1\;0\;0\;1\;0\;0\;0)$$

$$\begin{array}{c}\clubsuit\\\spadesuit\\\heartsuit\\\diamondsuit\end{array}\begin{pmatrix}0\\0\\1\\1\end{pmatrix}\begin{pmatrix}1\\1\\0\\0\end{pmatrix}\begin{pmatrix}1\\0\\0\\1\end{pmatrix}\quad R = \begin{pmatrix}1\,1\,0\\1\,0\,0\\0\,0\,1\\0\,1\,1\end{pmatrix}\quad \text{Inject } v = \begin{array}{c}\{\clubsuit,\spadesuit\}\to\\\{\clubsuit,\diamondsuit\}\to\\\{\heartsuit,\diamondsuit\}\to\end{array}\begin{pmatrix}0\,0\,0\,1\,0\,0\,0\,0\,0\,0\,0\,0\,0\,0\,0\,0\\0\,0\,0\,0\,0\,0\,0\,0\,0\,1\,0\,0\,0\,0\,0\,0\\0\,0\,0\,0\,0\,0\,0\,0\,0\,0\,0\,0\,1\,0\,0\,0\end{pmatrix}$$

Fig. 7.38 Set of vectors determining a single vector along the direct power

Of course, one will lose information on the sequence in which the vectors were originally given. The way back is easier, to a relation that is made up of the vectors v_i, putting them side by side $R := \varepsilon_i (\texttt{Inject } v)^{\mathsf{T}}$. From this matrix, the original vectors may be obtained by selecting a column element e as $R_i e$.

It may have become clear that we are not ready for such operations and that a routine denotation has not yet been agreed upon. Many would try to write a free-style program while there exist algebraically sound methods. The problem is not least that we are accustomed to abstract over the two natures $(v_{i \in \mathcal{I}})_{\mathcal{I}}, v$ of a subset.

$$\begin{array}{ccc} X & \xrightarrow{\;\;\varepsilon\;\;} & 2^X \\[2pt] & {}^{x}\!\!\searrow \;\; \downarrow\;\; \swarrow^{u} & \\ U & \;\mathbb{1}\; & \end{array}$$

Fig. 7.39 Element contained in subset

The transition from some predicate-logic formula to an appropriate relational version is not really difficult – but sometimes it is hard to communicate. There is no traditional notation at hand to make the difference visible between a subset $U \subseteq X$ and the corresponding element $u \in \mathbf{2}^X$ in the powerset. So, in the present environment we have to face the problem of too many overly detailed notations for situations the mathematician traditionally abstracts over. We remember

$$u = \mathsf{syq}(\varepsilon, U) \qquad \text{and} \qquad U = \varepsilon \mathbin{;} u$$

and recall:

$x \in U$,   traditional form, element x contained in a (sub)set U

ε_{xu},   points x and u for U in membership relation, the latter conceived as a matrix; here only used when deducing a relational form from a predicate-logic version

$x \mathbin{;} u^{\mathsf{T}} \subseteq \varepsilon$,   the relational version thereof

$x \subseteq \varepsilon \mathbin{;} u$,   variant of the relational version obtained by shunting via Prop. 5.12.ii.

Relations starting or ending in a direct power

We demand that relations ending in a direct power always be formed using the membership relation $\varepsilon := \varepsilon_X$, in programs denoted as `Member x`. This often takes the more composite form

$$\mathsf{syq}(\varepsilon, R) = \overline{\overline{\varepsilon^{\mathsf{T}} \mathbin{;} R}} \cap \overline{\varepsilon^{\mathsf{T}} \mathbin{;} \overline{R}}$$

for some relation R and the membership ε for its source.

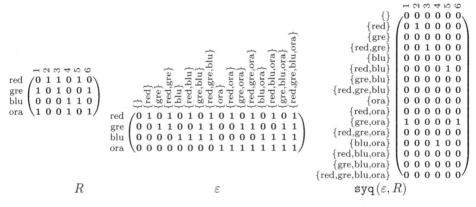

Fig. 7.40 Symmetric quotient showing $R = \varepsilon \mathbin{;} \mathsf{syq}(\varepsilon, R)$

Typically, we also use four composite relations, namely the containment ordering on the powerset

$$\Omega := \overline{\varepsilon^{\mathsf{T}} \mathbin{;} \overline{\varepsilon}} = \varepsilon \backslash \varepsilon,$$

the complement transition in the powerset

$$N := \mathsf{syq}(\varepsilon, \overline{\varepsilon}),$$

the singleton injection into the powerset

$$\sigma := \mathsf{syq}(\mathbb{I}, \varepsilon),$$

and the characterization of atoms of the powerset along the diagonal

$$a := \sigma^{\mathsf{T}} {}_{\displaystyle ;} \sigma.$$

For these relations we mention some of their properties:

$$\sigma_{\displaystyle ;}\Omega = \varepsilon_{\displaystyle ;}\Omega = \varepsilon \qquad \varepsilon_{\displaystyle ;}N = \overline{\varepsilon}.$$

The latter formula in particular provides an important transition: negation as a relation N versus negation toggling matrix coefficients between **0** and **1**. It is intuitively clear that the injected singleton can be enlarged to any set containing the element, etc.

More advanced relations ending and/or starting in a direct power are the power transpose, the power relator, and the existential image taken of a given relation, which we discuss only later in Section 19.1.

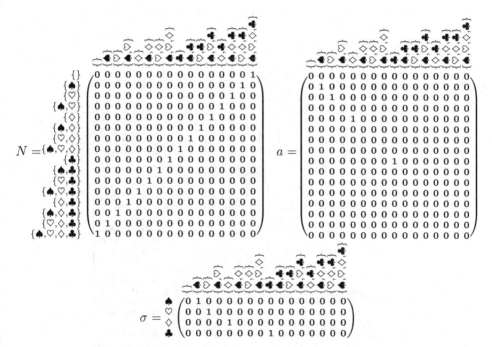

Fig. 7.41 Negation N, atoms a, and singleton injection σ

One will observe the position numbers $2, 3, 5, 9$ in a, corresponding to $2^i + 1$ for $i \geq 0$.

Algebraic properties of the generic membership relation

The domain construction of the direct power is designed to give a relational analog to the situation between a set A and its powerset $\mathbf{2}^A$. In particular, the 'is element of' membership relation, ε or `Member` A, between A and $\mathbf{2}^A$ is specified in the following definition, which has already been given in [13, 19, 20] as follows.

Definition 7.13. A relation ε is called a **membership relation** and its target is called the **direct power** if it satisfies the following properties:

(i) $\mathsf{syq}(\varepsilon,\varepsilon) \subseteq \mathbb{I}$,   (in fact $\mathsf{syq}(\varepsilon,\varepsilon) = \mathbb{I}$, due to (ii)),

(ii) $\mathsf{syq}(\varepsilon, X)$ is surjective for every relation X. $\square$

Instead of (ii), one may also say that $\mathsf{syq}(X, \varepsilon)$ shall be a mapping, i.e., univalent and total. With Cor. 8.17, we will later be able to convince ourselves also that $\mathsf{syq}(R_{;}\varepsilon, R_{;}X)$ is surjective for arbitrary relations R.

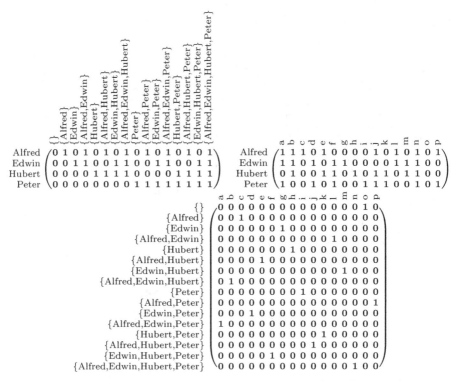

Fig. 7.42 Direct power uniquely defined up to isomorphism: $\varepsilon, \varepsilon'$, related by $P := \mathsf{syq}(\varepsilon, \varepsilon')$

The direct power is also defined in an essentially unique fashion. We demonstrate this with Fig. 7.42. Assume that somebody presents a second relation ε', shown on the upper right, claiming that his relation is the correct membership relation. Then we first check the properties mentioned in Def. 7.13. Since they are satisfied, we can easily construct from the two, $\varepsilon, \varepsilon'$, the bijection $P := \mathsf{syq}(\varepsilon, \varepsilon')$ relating the two proposed versions of the direct power. Using Def. 7.13, then obviously $\varepsilon_i P = \varepsilon'$. The proof is postponed as always in this early phase of our presentation. It requires formulae derived only later. Thus, ε is sufficient for a standard representation.

For better reference, we collect here several useful algebraic properties of the membership relation ε.

Proposition 7.14. *Let a membership relation $\varepsilon : A \longrightarrow 2^A$ be given. Then*

(i) $X = \varepsilon_i \mathsf{syq}(\varepsilon, X) = \overline{\varepsilon_i \overline{\varepsilon^{\mathsf{T}}_i X}}$,

(ii) $\overline{X} = \overline{\varepsilon}_i \mathsf{syq}(\varepsilon, X) = \overline{\varepsilon}_i \overline{\overline{\varepsilon}^{\mathsf{T}}_i X}$.

Proof: See B.21.   □

Matrix 1 — Hasse relation of powerset order

	{}	{r}	{g}	{r,g}	{b}	{r,b}	{g,b}	{r,g,b}	{o}	{r,o}	{g,o}	{r,g,o}	{b,o}	{r,b,o}	{g,b,o}	{r,g,b,o}
{}	0	1	1	0	1	0	0	0	1	0	0	0	0	0	0	0
{r}	0	0	0	1	0	1	0	0	0	1	0	0	0	0	0	0
{g}	0	0	0	1	0	0	1	0	0	0	1	0	0	0	0	0
{r,g}	0	0	0	0	0	0	0	1	0	0	0	1	0	0	0	0
{b}	0	0	0	0	0	1	1	0	0	0	0	0	1	0	0	0
{r,b}	0	0	0	0	0	0	0	1	0	0	0	0	0	1	0	0
{g,b}	0	0	0	0	0	0	0	1	0	0	0	0	0	0	1	0
{r,g,b}	0	0	0	0	0	0	0	0	0	0	0	0	0	0	0	1
{o}	0	0	0	0	0	0	0	0	0	1	1	0	1	0	0	0
{r,o}	0	0	0	0	0	0	0	0	0	0	0	1	0	1	0	0
{g,o}	0	0	0	0	0	0	0	0	0	0	0	1	0	0	1	0
{r,g,o}	0	0	0	0	0	0	0	0	0	0	0	0	0	0	0	1
{b,o}	0	0	0	0	0	0	0	0	0	0	0	0	0	1	1	0
{r,b,o}	0	0	0	0	0	0	0	0	0	0	0	0	0	0	0	1
{g,b,o}	0	0	0	0	0	0	0	0	0	0	0	0	0	0	0	1
{r,g,b,o}	0	0	0	0	0	0	0	0	0	0	0	0	0	0	0	0

Matrix 2 — cardinality preorder

	{}	{r}	{g}	{r,g}	{b}	{r,b}	{g,b}	{r,g,b}	{o}	{r,o}	{g,o}	{r,g,o}	{b,o}	{r,b,o}	{g,b,o}	{r,g,b,o}
{}	1	1	1	1	1	1	1	1	1	1	1	1	1	1	1	1
{r}	0	1	1	1	1	1	1	1	1	1	1	1	1	1	1	1
{g}	0	1	1	1	1	1	1	1	1	1	1	1	1	1	1	1
{r,g}	0	0	0	1	0	1	1	1	0	1	1	1	1	1	1	1
{b}	0	1	1	1	1	1	1	1	1	1	1	1	1	1	1	1
{r,b}	0	0	0	1	0	1	1	1	0	1	1	1	1	1	1	1
{g,b}	0	0	0	1	0	1	1	1	0	1	1	1	1	1	1	1
{r,g,b}	0	0	0	0	0	0	0	1	0	0	1	0	0	1	0	1
{o}	0	1	1	1	1	1	1	1	1	1	1	1	1	1	1	1
{r,o}	0	0	0	1	0	1	1	1	0	1	1	1	1	1	1	1
{g,o}	0	0	0	1	0	1	1	1	0	1	1	1	1	1	1	1
{r,g,o}	0	0	0	0	0	0	0	1	0	0	1	0	0	1	0	1
{b,o}	0	0	0	1	0	1	1	1	0	1	1	1	1	1	1	1
{r,b,o}	0	0	0	0	0	0	0	1	0	0	1	0	0	1	0	1
{g,b,o}	0	0	0	0	0	0	0	1	0	0	1	0	0	1	0	1
{r,g,b,o}	0	0	0	0	0	0	0	0	0	0	0	0	0	0	0	1

Matrix 3 — mapping to quotient

	{}	{r}	{r,g}	{r,g,b}	{r,g,b,o}
{}	1	0	0	0	0
{r}	0	1	0	0	0
{g}	0	1	0	0	0
{r,g}	0	0	1	0	0
{b}	0	1	0	0	0
{r,b}	0	0	1	0	0
{g,b}	0	0	1	0	0
{r,g,b}	0	0	0	1	0
{o}	0	1	0	0	0
{r,o}	0	0	1	0	0
{g,o}	0	0	1	0	0
{r,g,o}	0	0	0	1	0
{b,o}	0	0	1	0	0
{r,b,o}	0	0	0	1	0
{g,b,o}	0	0	0	1	0
{r,g,b,o}	0	0	0	0	1

Fig. 7.43 Hasse relation of powerset order, cardinality preorder, mapping to quotient

In addition to the powerset ordering Ω, we have the preorder $O_{||}$ by cardinality. It is easily shown that it is closely related to the powerset ordering as follows. Define the powerset strictorder $C_\Omega := \overline{\mathbb{I}} \cap \Omega$ and form the Hasse relation $H := C_\Omega \cap \overline{C_\Omega^2}$ for it. Then

$$O_{||} = (H^{\mathsf{T}}_i H)^*_i \Omega.$$

With $H^\mathsf{T}\mathbin{;}H$, an arbitrary element of the subset in question may be exchanged by a, possibly, different element. Figure 7.43 shows the Hasse relation of the powerset order and cardinality preorder on the set of colors {red, green, blue, orange}. (In the non-finite case, the Hasse relation can be $\mathbb{\perp\!\!\!\perp}$ and this construction will not work.)

Figure 7.43 shows the powerset strictordering as well as the preorder by cardinality, together with the mapping onto the classes of 0-, 1-, 2-, 3-, 4-element sets. The projection is obtained as $\eta := \texttt{Project}\ (O_{||} \cap O_{||}^\mathsf{T})$, where the latter turns out to be the equivalence of having the same cardinality.

Exercises

7.5 Prove that, given a direct power $\varepsilon : V \longrightarrow 2^V$, there exists a one-to-one correspondence between the set of relations $R : V \longrightarrow W$ and the set of mappings $f : W \longrightarrow 2^V$.

7.6 A multirelation $R : X \longrightarrow 2^Y$ has been defined in [103, 104] as a relation satisfying $R\mathbin{;}\Omega = R$ with powerset ordering $\Omega : 2^Y \longrightarrow 2^Y$. For multirelations, a special form of composition is declared

$$R \circ S := R\mathbin{;}\overline{\varepsilon^\mathsf{T}\mathbin{;}\overline{S}}.$$

Prove that this composition is an associative operation.

7.7 Domain permutation

A set in mathematics is *not* equipped with an ordering of its elements. In nearly all conceivable circumstances, this is an adequate abstraction, from which we will here deviate a little bit. When presenting a finite set on a sheet of paper or on a blackboard, one will always do it in some order. Since this is more or less unavoidable, we are going to make it explicit. Once this decision is taken, one will try to represent a set or a relation so as to make perception more easy – the idea underlying the technique of *algebraic visualization* frequently used in the present book. This means that the sequence chosen may depend on the context in which the set is to be presented. In such a case, the necessary permutation will be deduced from that context and applied immediately before presentation.

We are now going to introduce the necessary tools. Because all our relations carry their type with them, we only need a bijective mapping **p** ending in some target in order to define a rearranged target. It is a strict proof obligation that **p** must turn out to be a bijective mapping. We derive the rearranged presentation of the target in the most convincing way from **p**. To this end, we demand that the relation **p** will appear after rearrangement as a matrix of diagonal shape. Elements, vectors, and relations over a permuted target will, of course, stay the same as far as their algebraic status is concerned, but they will appear differently when presented via a marking vector or as a matrix, respectively. For this purpose, the language TITUREL allows us to formulate

PermTgt p,   the **permuted target**, and

ReArrTo p,   the **permutation relation** achieving the permutation intended.

Using this generic construction, the source of the relation ReArrTo p will be the original target of the relation p, while its target is the newly introduced PermTgt p. Figure 7.44 visualizes the intended meaning. Elements, vectors, and relations will then be formulated only via these generic constructs. These two simple additions will be the basic building blocks of the algebraic visualization applied throughout this book; see Appendix C.

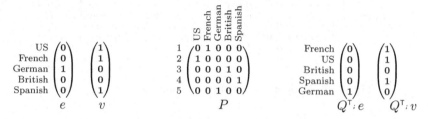

$$
\begin{array}{cc}
& e \\
\begin{array}{c} \text{US} \\ \text{French} \\ \text{German} \\ \text{British} \\ \text{Spanish} \end{array}
\begin{pmatrix} 0 \\ 0 \\ 1 \\ 0 \\ 0 \end{pmatrix}
\end{array}
\quad
\begin{array}{c} v \\ \begin{pmatrix} 1 \\ 1 \\ 0 \\ 0 \\ 1 \end{pmatrix} \end{array}
$$

Fig. 7.44 Element, vector, and bijective mapping P into a baseset as well as permuted versions thereof with $Q := $ ReArrTo P of Fig. 7.45 below

In the example of Figs. 7.44 and 7.45, the original target of p (or P) is the baseset

{US,French,German,British,Spanish}

which is then also the source of $Q := $ ReArrTo p, while the target of the latter is

{French,US,British,Spanish,German}.

One will observe that the source of p may be quite arbitrary in this process. What matters is that somehow a bijective mapping has occurred according to which we intend to rearrange the target. The new sequence is obtained scanning P row-wise from top to bottom, thereby looking for the occurrence of **1**s; here indicating French first, and then US etc.

$\leftarrow$ permuted target

original target $\rightarrow$

Fig. 7.45 ReArrTo p starting in *unpermuted*, ending in the *permuted* target and $P \,; Q$

The relation $Q := $ ReArrTo p, considered as a matrix only, coincides with $\mathbf{p}^\mathsf{T}$, but no longer has as target the target of p but `PermTgt` p. Its source is its original target. It serves as the generic means to generate elements in the permuted source, vectors ranging over the permuted source and relations starting or ending there. An element $e_X \in X$ gives rise to the element $Q^\mathsf{T} {}_; e_X$. The vector v_X leads to the vector $Q^\mathsf{T} {}_; v_X$. A relation $R : X \longrightarrow Y$ will become a relation $Q^\mathsf{T} {}_; R$ with source `PermTgt` p. A relation $S : Z \longrightarrow X$ will become a relation $S {}_; Q$ with target `PermTgt` p.

With Fig. 7.46, we show how a relation is permuted on its target side using Q.

	US	French	German	British	Spanish			French	US	British	Spanish	German
Clinton	1	0	0	0	0		Clinton	0	1	0	0	0
Bush	1	0	0	0	0		Bush	0	1	0	0	0
Mitterand	0	1	0	0	0		Mitterand	1	0	0	0	0
Chirac	0	1	0	0	0		Chirac	1	0	0	0	0
Schmidt	0	0	1	0	0		Schmidt	0	0	0	0	1
Kohl	0	0	1	0	0		Kohl	0	0	0	0	1
Schröder	0	0	1	0	0		Schröder	0	0	0	0	1
Thatcher	0	0	0	1	0		Thatcher	0	0	1	0	0
Major	0	0	0	1	0		Major	0	0	1	0	0
Blair	0	0	0	1	0		Blair	0	0	1	0	0
Zapatero	0	0	0	0	1		Zapatero	0	0	0	1	0

original: permuted:

Fig. 7.46 Nationality of politicians unpermuted and permuted

Algebraic properties of domain permutation

The algebraic characterization we routinely present is rather simple. The relation $P : X \longrightarrow Y$, we have been starting from, must satisfy

$$P {}_; P^\mathsf{T} = \mathbb{I}_X \qquad P^\mathsf{T} {}_; P = \mathbb{I}_Y$$

in order to indicate that it is a bijective mapping.[9]

During the other domain constructions we have always been careful to show that their result is determined in an essentially unique way (i.e., up to isomorphism). This holds true trivially in this case, owing to the above algebraic characterization. As long as this is uninterpreted, one will have to keep in mind the obligation to prove that P is a bijective mapping.[10]

7.8 Remarks on further constructions

Just for completeness with respect to some theoretical issues, we introduce the 1-element unit set `UnitOb`. This can also be characterized among the non-empty sets

---

[9] Observe that `ReArrTo` p is then no longer homogeneous in the strong typing sense of TITUREL – although the matrix is square.

[10] There is a point to mention, though. It may be the case that the relations P, P' are finally interpreted in the same way, but are given in two syntactically different forms. Prior to interpretation, one may not yet be able to decide that they are equal, and thus has to handle two, potentially different, permuted targets.

in an essentially unique form. The characterization is simply $\mathbb{T}_{\texttt{UnitOb},\texttt{UnitOb}} = \mathbb{I}_{\texttt{UnitOb}}$, i.e., the universal relation equals the identity relation on such domains. The single element in such a set is usually called $\texttt{Unit-1}$; it may be found at several occasions in the present book.

More involved is the recursive construction of sets with which we may obtain infinite sets for the first time. Then stacks can be specified, or lists, for example, which is outside the scope of the present book.

Universal versus equational characterizations

At several occasions, we have presented a 'universal characterization', saying that the direct sum, the direct product or the direct power structure is *uniquely characterized up to isomorphism*. Normally in mathematical folklore, such a universal characterization ranges over all sets C carrying the structure in question and all mappings R, S. Some sort of a preordering of (C, R, S) via the possibility of factorizing is introduced and the definition asserts that some sort of a supremum will be obtained; see Fig. 7.47.

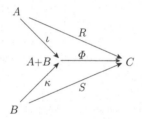

Fig. 7.47 Universal characterization of the direct sum

This method is, thus, purely descriptive. Even if a sum candidate is presented, it cannot be tested against this definition: quantification runs *over all sets* carrying the structure and *over all mappings* leading to these; the characterization is not even first-order.

So it is important that, when working with heterogeneous relations, one is able to give an *equational* definition and is able to provide candidates constructively. Comparison with other candidates presented can simply be executed computationally. Over a long period of time, relation algebraists have been accustomed to working homogeneously; see not least [140]. This made concepts difficult, since the well-established typing mechanisms a computer scientist applies routinely had to be replaced developing ad hoc mathematics.

It seems that homomorphisms of heterogeneous structures (for example, graphs, programs) were first formalized in relational form during the winter term 1974/75 in lectures on *Graphentheorie* by the present author, the notes [119] of which have been printed as an internal report of the *Institut für Informatik* at *Technische Universität München*. This was then used in [108, 109, 110, 111].

Once homomorphisms had been formalized, the characterizing formulae for direct sums, direct products, and direct powers were rather straightforward and could be handled in diploma theses at Technische Universität München. Initiated by the present author together with Rudolf Berghammer, these were studied by Rodrigo Cardoso (*Untersuchung von parallelen Programmen mit relationenalgebraischen Methoden*), finished December 1982, and Hans Zierer (*Relationale Semantik*), finished May 1983.

The first publication of the equational characterizations seems to have been presented with the series of publications [19, 20, 116, 123, 124, 141], not least [21, 150], which followed the diploma theses mentioned.

7.9 Equivalent representations

It should be mentioned that we have introduced constructions in finer detail than usual and we have abandoned some abstractions. In particular, we have several possibilities of representing 'the same' item classically abstracted over. So this approach will probably suffer from obstruction of those people who traditionally use abstractions thrown away here.

We give an example using the domain constructions we have just introduced to show that concepts may be represented in quite different ways. While from a certain point of view it is desirable to abstract from details of representation, one often loses intuition when abstraction is taken too far. In particular, one will not immediately see in which cases a situation is 'linear', and thus offers itself to being handled relationally.

In Fig. 7.48, we present representations of a relation as a matrix, as a vector along the product space and finally as an element in the powerset of the product space. The latter concept, however, will hardly ever be used. We also indicate how the transition between the versions can be managed with the generic projection relations π, ρ and the membership relation ε.

(Diagram: nodes $\mathbb{1}$, $2^{X\times Y}$, $X\times Y$, X, Y with arrows labelled e, t, ε, π, ρ, R.)

$$
\begin{array}{c}
\begin{array}{ccc} \text{Win} & \text{Draw} & \text{Loss} \end{array}\\
\begin{array}{c}\text{male}\\ \text{female}\end{array}
\begin{pmatrix} 0 & 1 & 0 \\ 1 & 0 & 1 \end{pmatrix} = R
\end{array}
$$

$$
\begin{array}{c}
(\text{male,Win})\\ (\text{female,Win})\\ (\text{male,Draw})\\ (\text{female,Draw})\\ (\text{male,Loss})\\ (\text{female,Loss})
\end{array}
\begin{pmatrix} 0 \\ 1 \\ 1 \\ 0 \\ 0 \\ 1 \end{pmatrix} = t
$$

$$R = \pi^{\mathsf{T}}; (\mathbb{I} \cap t; \mathbb{T}); \rho$$

$$t = (\pi; R \cap \rho); \mathbb{T}$$

$$e = \mathsf{syq}(\varepsilon, t)$$

$$t = \varepsilon; e$$

{}	0
{(male,Win)}	0
{(female,Win)}	0
{(male,Win),(female,Win)}	0
{(male,Draw)}	0
{(male,Win),(male,Draw)}	0
{(female,Win),(male,Draw)}	0
{(male,Win),(female,Win),(male,Draw)}	0
{(female,Draw)}	0
{(male,Win),(female,Draw)}	0
{(female,Win),(female,Draw)}	0
{(male,Win),(female,Win),(female,Draw)}	0
{(male,Draw),(female,Draw)}	0
{(male,Win),(male,Draw),(female,Draw)}	0
{(female,Win),(male,Draw),(female,Draw)}	0
{(male,Win),(female,Win),(male,Draw),(female,Draw)}	0
{(male,Loss)}	0
{(male,Win),(male,Loss)}	0
{(female,Win),(male,Loss)}	0
{(male,Win),(female,Win),(male,Loss)}	0
{(male,Draw),(male,Loss)}	0
{(male,Win),(male,Draw),(male,Loss)}	0
{(female,Win),(male,Draw),(male,Loss)}	0
{(male,Win),(female,Win),(male,Draw),(male,Loss)}	0
{(female,Draw),(male,Loss)}	0
{(male,Win),(female,Draw),(male,Loss)}	0
{(female,Win),(female,Draw),(male,Loss)}	0
{(male,Win),(female,Win),(female,Draw),(male,Loss)}	0
{(male,Draw),(female,Draw),(male,Loss)}	0
{(male,Win),(male,Draw),(female,Draw),(male,Loss)}	0
{(female,Win),(male,Draw),(female,Draw),(male,Loss)}	0
{(male,Win),(female,Win),(male,Draw),(female,Draw),(male,Loss)}	0
{(female,Loss)}	0
{(male,Win),(female,Loss)}	0
{(female,Win),(female,Loss)}	0
{(male,Win),(female,Win),(female,Loss)}	0
{(male,Draw),(female,Loss)}	0
{(male,Win),(male,Draw),(female,Loss)}	0
{(female,Win),(male,Draw),(female,Loss)}	1
{(male,Win),(female,Win),(male,Draw),(female,Loss)}	0
{(female,Draw),(female,Loss)}	0
{(male,Win),(female,Draw),(female,Loss)}	0
{(female,Win),(female,Draw),(female,Loss)}	0
{(male,Win),(female,Win),(female,Draw),(female,Loss)}	0
{(male,Draw),(female,Draw),(female,Loss)}	0
{(male,Win),(male,Draw),(female,Draw),(female,Loss)}	0
{(female,Win),(male,Draw),(female,Draw),(female,Loss)}	0
{(male,Win),(female,Win),(male,Draw),(female,Draw),(female,Loss)}	0
{(male,Loss),(female,Loss)}	0
{(male,Win),(male,Loss),(female,Loss)}	0
{(female,Win),(male,Loss),(female,Loss)}	0
{(male,Win),(female,Win),(male,Loss),(female,Loss)}	0
{(male,Draw),(male,Loss),(female,Loss)}	0
{(male,Win),(male,Draw),(male,Loss),(female,Loss)}	0
{(female,Win),(male,Draw),(male,Loss),(female,Loss)}	0
{(male,Win),(female,Win),(male,Draw),(male,Loss),(female,Loss)}	0
{(female,Draw),(male,Loss),(female,Loss)}	0
{(male,Win),(female,Draw),(male,Loss),(female,Loss)}	0
{(female,Win),(female,Draw),(male,Loss),(female,Loss)}	0
{(male,Win),(female,Win),(female,Draw),(male,Loss),(female,Loss)}	0
{(male,Draw),(female,Draw),(male,Loss),(female,Loss)}	0
{(male,Win),(male,Draw),(female,Draw),(male,Loss),(female,Loss)}	0
{(female,Win),(male,Draw),(female,Draw),(male,Loss),(female,Loss)}	0
{(male,Win),(female,Win),(male,Draw),(female,Draw),(male,Loss),(female,Loss)}	0

$= e$

Fig. 7.48 Relating representations of a relation as a matrix R, a vector t, and as an element e

PART III
ALGEBRA

In this part of the present text, we reach a third level of abstraction. Recall that in Part I relations were observed as they occur in real life situations. We then made a step forward using point-free algebraic formulation in Part II; however, we did not introduce the respective algebraic proofs immediately. Instead, we visualized the effects and tried to construct with relations. In a sense, this corresponds to what one always finds in a book treating eigenvectors and eigenvalues of real- or complex-valued matrices, or their invariant subspaces: usually this is heavily supported with visualizing matrix situations. We did this with full mathematical rigor but have not yet convinced the reader concerning this fact. Proofs, although rather trivial in that beginning phase, have been postponed so as to establish an easy line of understanding first.

As gradually more advanced topics are handled, we will now switch to a fully formal style with proofs immediately appended. However, the reader will be in a position to refer to the first two parts and to see there the effects.

Formulating only in algebraic terms – or what comes close to that, formulating in the relational language TITUREL – means that we are far more restricted in expressivity. On the other hand this will improve the precision considerably. Restricting ourselves to the relational language will later allow computer-aided transformations and proofs.

A point should be stressed for theoreticians: from now on, we develop relational mathematics as an axiomatic theory with, respectively without, the Point Axiom.

8

Relation Algebra

Concerning syntax and notation, everything is now available to work with. We take this opportunity to have a closer look at the algebraic laws of relation algebra. In particular, we will be interested in how they can be traced back to a small subset of rules which can serve as axioms. We present them now and discuss them immediately afterwards.

We should stress that we work with heterogeneous relations. This contrasts greatly with the traditional work of the relation algebra community which is almost completely restricted to a homogeneous environment – possibly enhanced by cylindric algebra considerations. Some of the constructs which follow simply do not exist in a homogeneous context, for example the direct power and the membership relation. At first glance, it seems simpler to study homogeneous as opposed to heterogeneous relations. But attempting domain constructions in the homogeneous setting immediately leads necessarily to non-finite models. Also deeper problems, such as the fact that $\mathbb{T}_{A,B}; \mathbb{T}_{B,C} = \mathbb{T}_{A,C}$ does not necessarily hold, have only recently come to attention; this applies also to unsharpness.

8.1 Laws of relation algebra

The set of axioms for an abstract (possibly heterogeneous) relation algebra is nowadays generally agreed upon, and it is rather short. When we use the concept of a category, this does not mean that we are introducing a higher concept. Rather, it is used here as a mathematically acceptable way to prevent multiplying a 7×5-matrix with a 4×6-matrix.

A category may, in the naïve sense to which we restrict ourselves here, be conceived as based on a complete 1-graph. Every vertex of its (mostly finite) set of vertices means a type, and this type can be interpreted with a baseset. Every arrow (p, q) means, thus, a (directed) pair of types, and can be interpreted with the set of relations from the baseset interpreting p to the baseset interpreting q, called the morphism set $\mathcal{M}_{pq}$. Only relations belonging to consecutive arrows $(p, q), (q, r)$ can be composed with "$;$" so as to obtain a relation belonging to the shortcutting arrow (p,r) in the complete 1-graph. The loops in the 1-graph, i.e., the arrows (p,p), contain in particular the identity relation $\mathbb{I}_p$ on the baseset interpreting the type p. The only two assumptions for a category are as follows: composition with "$;$", restricted as just explained, is associative and the identities act as such.

Definition 8.1. A **heterogeneous relation algebra**[1] is defined as a structure that

— is a category with respect to composition "$;$" and identities $\mathbb{I}$,
— has complete atomic Boolean lattices with $\cup, \cap, {}^{-}, \bot\!\!\!\bot, \top\!\!\!\top, \subseteq$ as morphism sets,
— obeys rules for transposition in connection with the latter two that may be stated in either one of the following two ways:

Dedekind[2] $R\,;S \cap Q \subseteq (R \cap Q\,;S^\mathsf{T})\,;(S \cap R^\mathsf{T}\,;Q)$ or

Schröder $A\,;B \subseteq C \iff A^\mathsf{T}\,;\overline{C} \subseteq \overline{B} \iff \overline{C}\,;B^\mathsf{T} \subseteq \overline{A}.$ □

Composition, union, etc., are thus only partial operations and one has to be careful not to violate the composition rules. In order to avoid clumsy presentation, we shall adhere to the following policy in a heterogeneous algebra with its only partially defined operations: whenever we say 'For every R ...', we mean 'For every R for which the construct in question is defined ...' We are often a bit sloppy, writing $\top\!\!\!\top$ when $\top\!\!\!\top_{X,Y}$, for example, would be more precise.

The Dedekind rule as well as the Schröder rule are widely unknown. They are as important as the rule of associativity, the rule demanding distributivity, or the De Morgan rule, and therefore deserve to be known much better. One may ask why they are so little known. Is it because they are so complicated? Consider, for example,

$a * (b + c) = a * b + a * c$ distributivity,
$a + (b + c) = (a + b) + c$ associativity,
$\overline{a \cup b} = \overline{a} \cap \overline{b}$ De Morgan law.

When comparing the Dedekind and the Schröder rules with these commonly known mathematical rules, they are indeed a little bit longer as text. But already the effort of checking associativity[3] is tremendous compared with the pictorial examples below, making them seem simpler.

One should also mention that the Schröder rule – still without the name which

[1] See, for example, [117] and the short presentation of theoretical background material in [71].
[2] Also the shortened version $R\,;S \cap Q \subseteq (R \cap Q\,;S^\mathsf{T})\,;S$, known as the modular law, would suffice.
[3] Who has ever checked associativity for a group table? We simply take it for granted, or given by construction.

was only attributed to it later – has its origin *in the same paper* as the famous De Morgan[4] law, namely in [40]. Most people probably have not read far enough in De Morgan's paper. Rather, they seem to have confined themselves to using slightly simplified and restricted forms of these rules. These versions describe functions, orderings, and equivalences. School education favors functions. Even today, when teaching orderings at university level it is not clear from the beginning that orderings need not be linear. Thinking in relations is mainly avoided because people prefer the assignment of *just one* as traditionally in personal relations.

Exercises

8.1 Prove using relation-algebraic techniques that
$$Q ; \overline{R} ; S \subseteq \overline{Q ; R ; S} \quad \text{and} \quad \overline{Q ; \overline{R} ; S} \subseteq \overline{Q ; R ; S}.$$

8.2 Visualizing the algebraic laws

For an intuitive description we recall that as a result of the dashed arrow convention a dotted or dashed arrow indicates the *negated* relation. The thin bent arrow with a double-arrow pointer shows the direction of reasoning intended.

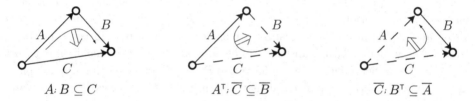

$$A ; B \subseteq C \qquad\qquad A^{\mathsf{T}} ; \overline{C} \subseteq \overline{B} \qquad\qquad \overline{C} ; B^{\mathsf{T}} \subseteq \overline{A}$$

Fig. 8.1 The dashed arrow convention in the case of Schröder equivalences

Visualizing the Schröder equivalences

It is now easy to memorize and understand the Schröder rule.

- The containment $A ; B \subseteq C$ means that for consecutive arrows of A and B, there will always exist a 'shortcutting' arrow in C.
- But assuming this to hold, we have the following situation: when following A in reverse direction and then a non-existent arrow of C, there can be no shortcutting arrow in B. Of course not; it would be an arrow consecutive to one of A with the consequence of an arrow in C which is impossible.

---

[4] He was Scottish (of Huguenot origin), not French, born in Madura (India) (according to others, Madras).

- Let us consider a non-existent arrow from C (i.e., an arrow from $\overline{C}$), followed by an arrow from B in reverse direction. Then it cannot happen that this can be shortcut by an arrow of A because then there would be consecutive arrows of A and B without a shortcut in C.

On thinking about this for a while, one will indeed understand Schröder's rule. What one cannot assess at this point is that Schröder's rule (in combination with the others mentioned in Def. 8.1) is strong enough to span all our standard methods of thinking and reasoning. The next three examples will at least illustrate this claim.

If we are about to work with a function, we mean that it must never assign *two different* images to an argument; therefore, a function F satisfies $F^{\mathsf{T}} \mathbin{;} F \subseteq \mathbb{I}$. So, when going back from an image to its argument and then going forward again to an image, it will turn out that one will arrive at the same element. But one may also follow the function to an image and then proceed to *another* element on the image side, which then *cannot* be the image of the starting point: $F \mathbin{;} \overline{\mathbb{I}} \subseteq \overline{F}$. This transition, which is difficult to formulate in spoken language, is again simple using Schröder's rule.

Consider the equivalence $\Xi := $ *'plays in the same team as'* defined over the set of players in some national sports league. When a and b play in the same team, and also b and c play in the same team, we reason that a will play in the same team with c. Altogether we have transitivity of an equivalence relation, meaning not least $\Xi \mathbin{;} \Xi \subseteq \Xi$. Using Schröder's rule, we have also $\Xi^{\mathsf{T}} \mathbin{;} \overline{\Xi} \subseteq \overline{\Xi}$ meaning that when b plays in the same team with a, but a does not play in the same team with c, then b and c cannot play in the same team.

Yet another example assumes C as a comparison of persons according to their height. If a is taller than b and b in turn is taller than c, it is clear for us that a is taller than c, i.e. $C \mathbin{;} C \subseteq C$. We routinely use 'is smaller than' to denote the converse. We will probably also be able to reason that $C^{\mathsf{T}} \mathbin{;} \overline{C} \subseteq \overline{C}$; namely when a is smaller than b and b is not taller than c, then a is not taller than c.

Nearly all of human thinking is based on such triangle situations; mankind does not seem capable of achieving more. If everyday situations are concerned, we handle this routinely, most often using dedicated denotation such as 'is taller than', 'is godfather of', 'has as grandson', etc. The mathematical aspect is often obscured when operations are to a certain extent integrated in our language as

- concerning negation in
— 'is married' versus 'is bachelor' as a unary relation, i.e., a vector

— 'is linked to' versus 'are unlinked'
- concerning converses in
— 'is taller than' versus 'is smaller than'
— 'sells to' versus 'buys from'
- concerning composition in
— 'is father of' ⨾ 'is parent of' = 'is grandfather of'
— 'is brother of' ⨾ 'is parent of' = 'is uncle of'.

This brings additional problems[5] because natural languages often handle such situations in quite different ways. Think of the word 'brother' as used in Arabic or 'godfather' in the Mafia context. When new relations come into play for which no standard notation has been agreed, we often fail to handle this properly. Big problems usually show up when more than one relation is envisaged such as 'plays in the same team as', 'is taller than', or 'is mapped to' and different relations are involved as in the following examples.

Example 8.2. Let some gentlemen $\mathcal{G}$, ladies $\mathcal{L}$, and flowers $\mathcal{F}$ be given, as well as the three relations that are typically neither univalent nor total or surjective:

- has sympathy for $S : \mathcal{G} \longrightarrow \mathcal{L}$,
- likes flower $L : \mathcal{L} \longrightarrow \mathcal{F}$,
- has bought $B : \mathcal{G} \longrightarrow \mathcal{F}$.

Then the following will hold

$$S \mathbin{;} L \subseteq B \quad\Longleftrightarrow\quad S^{\mathsf{T}} \mathbin{;} \overline{B} \subseteq \overline{L} \quad\Longleftrightarrow\quad \overline{B} \mathbin{;} L^{\mathsf{T}} \subseteq \overline{S}.$$

None of these containments need be satisfied. In any case, however, either all three or none of them will be true. The following three propositions are, thus, logically equivalent.

- All gentlemen g buy at least those flowers f that are liked by some of the ladies sympathetic to them.
- Whenever a gentleman feels sympathy for a lady l, and does not buy flower f, lady l is not fond of flower f.
- Whenever a gentleman g does not buy a flower that lady l likes, then g has no sympathy for l.

Since we will only in rare cases have that $S \mathbin{;} L \subseteq B$, this may seem rather artificial – but it is not. Assume, for example, a situation where a gentleman does not buy a flower liked by one of the ladies sympathetic to him. Then the other two statements will also not be satisfied. For instance: *Hubert* feels sympathy for *Elizabeth*, but does

---

[5] Complaints of this kind have a long history; see Appendix D.4.

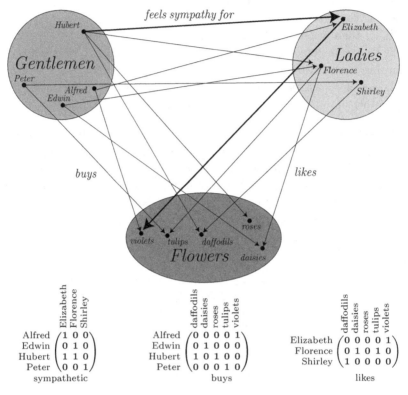

	Elizabeth	Florence	Shirley
Alfred	1	0	0
Edwin	0	1	0
Hubert	1	1	0
Peter	0	0	1

sympathetic

	daffodils	daisies	roses	tulips	violets
Alfred	0	0	0	0	1
Edwin	0	1	0	0	0
Hubert	1	0	1	0	0
Peter	0	0	0	1	0

buys

	daffodils	daisies	roses	tulips	violets
Elizabeth	0	0	0	0	1
Florence	0	1	0	1	0
Shirley	1	0	0	0	0

likes

Fig. 8.2 Visualization of the Schröder rule

not buy *violets*; i.e., $S_\cdot L \subseteq B$ does not hold. But then, obviously, neither $S^\mathsf{T}_\cdot \overline{B} \subseteq \overline{L}$ holds, nor $\overline{B}_\cdot L^\mathsf{T} \subseteq \overline{S}$. □

To memorize the Schröder rule, we concentrate on *triangles*. As indicated in Fig. 8.2, a triangle may be formed by a cycle *is sympathetic–likes–buys*. Once we have found that *every* arrow belongs to some triangle, it is straightforward to state that over every line there exists an element in the third set such that a triangle can be formed – regardless of which line has been chosen to start with.

Example 8.3. Look at the situation with three relations on one set of human beings, namely

- B 'is brother of'
- P 'is parent of'
- G 'is godfather of'.

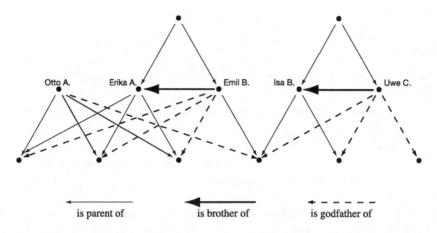

Fig. 8.3 Another visualization of the Schröder equivalences

Now we have according to the Schröder rule

$$B_i\, P \subseteq G \quad \Longleftrightarrow \quad B^{\mathsf{T}}_i\, \overline{G} \subseteq \overline{P} \quad \Longleftrightarrow \quad \overline{G}_i\, P^{\mathsf{T}} \subseteq \overline{B}.$$

This is applicable to every group of human beings, stating that the following three propositions are logically equivalent. So either all three of them are true or none.

- Uncleship implies godfathership.

- A person with a brother who is not godfather of a child, will never be parent of that child.

- If someone is not the godfather of a child of somebody, he will not be that person's brother.

Now assume the leftmost godfathership arrow starting at **Emil B.** is not indicated. Then obviously $B_i\, P \subseteq G$ is *not* satisfied. But then also $\overline{G}_i\, P^{\mathsf{T}} \subseteq \overline{B}$ is not satisfied, namely: take the missing arrow and follow parenthood back to **Erika A.** To this person a brother arrow exists, so that $\overline{G}_i\, P^{\mathsf{T}} \subseteq \overline{B}$ cannot be true. □

Visualizing the Dedekind formula

Observe that the setting is a little bit different now. The Dedekind formula is satisfied for any (well-defined) three-relation configuration whatsoever. We recall Fig. 8.2 as an example. From the beginning, we admit in this case a situation where $S_i\, L \subseteq B$ *may not* be satisfied.

In an arbitrary such configuration the inclusion

$$S_; L \cap B \subseteq (S \cap B_; L^{\mathsf{T}})_; (L \cap S^{\mathsf{T}}_; B)$$

is satisfied:

— when a gentleman g buys a flower f, which *some* lady sympathetic
  to him likes,
  — there will exist some lady l sympathetic to him, liking *some*
    flower he bought,
  and
  — this lady l will like flower f, that *some* gentleman with
    sympathy for her has bought.

Observe how fuzzily this is formulated concerning existence of 'some ...'. One
need not learn this by heart. Just consider the triangles of three arrows. Then at
least the transpositions follow immediately. Written in predicate-logic form, this
looks intimidatingly complex:

$$\forall g : \forall f : \left[\left\{ \exists l' : (g, l') \in S \wedge (l', f) \in L \right\} \wedge (g, f) \in B \right] \longrightarrow$$
$$\left\{ \exists l : \left[(g, l) \in S \wedge \left(\exists f' : (g, f') \in B \wedge (l, f') \in L \right) \right] \wedge \right.$$
$$\left. \left[(l, f) \in L \wedge \left(\exists g' : (g', l) \in S \wedge (g', f) \in B \right) \right] \right\}.$$

At this point, one will begin to appreciate the algebraic shorthand; one is less
exposed to the likelihood of committing errors. It is particularly difficult to express
such relations in spoken language, not least because it depends heavily on the
language chosen – or available.

The Dedekind rule, interpreted for godfather, brother and uncle as

$$B_; P \cap G \subseteq (B \cap G_; P^{\mathsf{T}})_; (P \cap B^{\mathsf{T}}_; G),$$

says that the following holds *for every configuration of human beings*:

— an uncle u of a child c, who is at the same time a godfather of c,
  — is a brother of *some* person p, who is parent of *one* of his
    godchildren,
  and
  — p is at the same time parent of child c and *has a* brother,
    who is godfather of c.

One should again emphasize that at three positions there is an ambivalence. The
person p is parent of *one of his* godchildren. There *exists* a brother of p who is
godfather of c. The third is hidden with our language in uncle $B_; P$. These three are

not identified more than these descriptions allow. The very general quantifications, however, make this sufficiently precise. Another illustration is added as Fig. 8.4.

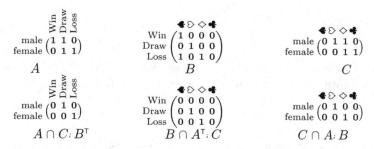

Fig. 8.4 Matrix visualization of the Dedekind formula

In the matrix visualization of Fig. 8.4, we observe how several entries are cut out of the original relations A, B, C.

8.3 Elementary properties of relations

The following are elementary properties of operations on relations. We mention them here without proofs which can be found in our standard reference [123, 124].

Proposition 8.4 (Standard rules).

(i) $\mathbb{L}_{;}R = R_{;}\mathbb{L} = \mathbb{L}$

(ii) $R \subseteq S \implies Q_{;}R \subseteq Q_{;}S, \quad R_{;}Q \subseteq S_{;}Q$ *(monotonicity)*

(iii) $Q_{;}(R \cap S) \subseteq Q_{;}R \cap Q_{;}S, \quad (R \cap S)_{;}Q \subseteq R_{;}Q \cap S_{;}Q$ *(∩-subdistributivity)*

    $Q_{;}(R \cup S) = Q_{;}R \cup Q_{;}S, \quad (R \cup S)_{;}Q = R_{;}Q \cup S_{;}Q$ *(∪-distributivity)*

(iv) $(R_{;}S)^{\mathsf{T}} = S^{\mathsf{T}}_{;}R^{\mathsf{T}}$ □

While (i) is trivial, (iv) is already known for matrices in linear algebra. In (iii) it is stated that composition distributes over union, but only 'subdistributivity' holds for composition over intersection.

$$\left[\begin{pmatrix} 1 & 0 \\ 0 & 1 \end{pmatrix} \cap \begin{pmatrix} 0 & 1 \\ 1 & 0 \end{pmatrix} \right]_{;} \begin{pmatrix} 1 & 1 \\ 1 & 1 \end{pmatrix} = \begin{pmatrix} 0 & 0 \\ 0 & 0 \end{pmatrix} \quad \subseteq \quad \begin{pmatrix} 1 & 1 \\ 1 & 1 \end{pmatrix} = \begin{pmatrix} 1 & 0 \\ 0 & 1 \end{pmatrix}_{;} \begin{pmatrix} 1 & 1 \\ 1 & 1 \end{pmatrix} \cap \begin{pmatrix} 0 & 1 \\ 1 & 0 \end{pmatrix}_{;} \begin{pmatrix} 1 & 1 \\ 1 & 1 \end{pmatrix}$$

Fig. 8.5 Example of subdistributivity

In Prop. 8.5, special care is necessary with regard to typing as one can see from the indices of the universal relations.

Proposition 8.5 (Row and column masks). *The following formulae hold for arbitrary relations* $P : V \longrightarrow X, Q : U \longrightarrow V, R : U \longrightarrow X, S : V \longrightarrow W$:

(i) $(Q \cap R \mathbin{;} \mathbb{T}_{X,V}) \mathbin{;} S = Q \mathbin{;} S \cap R \mathbin{;} \mathbb{T}_{X,W}$,

(ii) $(Q \cap (P \mathbin{;} \mathbb{T}_{X,U})^{\mathsf{T}}) \mathbin{;} S = Q \mathbin{;} (S \cap P \mathbin{;} \mathbb{T}_{X,W})$.   □

An interpretation of these masking formulae with $P \mathbin{;} \mathbb{T}$ meaning 'is red-haired', Q meaning 'is brother of', S meaning 'is parent of', and $R \mathbin{;} \mathbb{T}$ meaning 'is bald' reads as follows.

(i) The result is the same when looking for *bald brothers* b of parents of some child c or when first looking for *nephewships* (b, c) and afterwards selecting those with b bald.

(ii) This is hard to reproduce in colloquial form:

   (a brother of a red-haired person) who is parent of child c

   is a brother of (a red-haired parent of child c).

Another example is shown in Fig. 8.6.

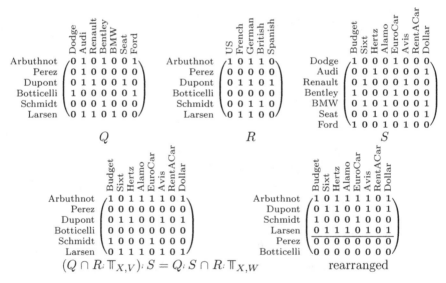

$$(Q \cap R \mathbin{;} \mathbb{T}_{X,V}) \mathbin{;} S = Q \mathbin{;} S \cap R \mathbin{;} \mathbb{T}_{X,W} \qquad \text{rearranged}$$

Fig. 8.6 Composition and row mask, the case of Prop. 8.5.i; also with rows permuted

Now we visualize (ii), starting from the relations Q, P, and S in Fig. 8.7. The interesting factor relations are then shown with columns, respectively rows, permuted.

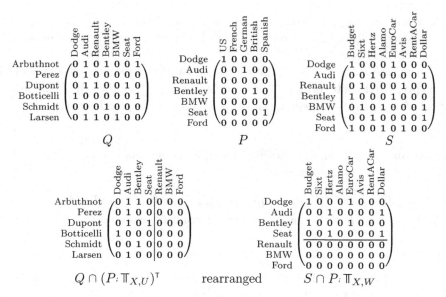

Fig. 8.7 Composition and row mask, the case of Prop. 8.5.ii, visualized with permutation

Exercises

8.2 In [88], R. Duncan Luce claimed the existence of some X with the properties
$$X \subseteq S, \quad Q \subseteq \top_i X, \quad \text{and} \quad Q_i X^\top \subseteq R$$
to be sufficient for $Q \subseteq R_i S$. Prove this claim using the Dedekind formula.

8.3 Prove that $\quad X \subseteq \overline{P_i Y_i Q^\top} \quad \Longleftrightarrow \quad Y \subseteq \overline{P^\top_i X_i Q}$,
i.e., that the functionals $X \mapsto \overline{P^\top_i X_i Q}$ and $Y \mapsto \overline{P_i Y_i Q^\top}$ constitute a Galois correspondence.

8.4 Show the following identities for an arbitrary vector $s : X \longrightarrow Y$:
$$\overline{s_i s^\top} \cap \mathbb{I}_{X,X} = \overline{s_i \top_{Y,X}} \cap \mathbb{I}_{X,X} = \overline{s}_i \top_{Y,X} \cap \mathbb{I}_{X,X}$$
$$\overline{\overline{s}_i \overline{s}^\top} \cap \mathbb{I}_{X,X} = \overline{\overline{s}_i \top_{Y,X}} \cap \mathbb{I}_{X,X} = s_i \top_{Y,X} \cap \mathbb{I}_{X,X}.$$

8.4 Cancellation properties of residuals and cones

After having introduced relations in a formal way in Chapter 5, we immediately presented the main rules and formulae concerning orders and functions to cover what is more or less commonly known. In principle, they should have been included at the present location. Our aim was to work with that and to give sufficient intuition, while we have postponed proofs to the appendix.

In the present chapter we have taken yet another step towards relation algebra, building on axioms and giving fully formal proofs. Here, it seems appropriate to present some less commonly known rules and formulae, most of which are novel.

The formulae we have in mind are really general, but have been studied only in more specialized contexts so far. Our aim is, therefore, to prove them and also to get rid of any additional assumptions that are unnecessary and just tradition.

Definition 8.6. Given any (possibly heterogeneous) relation $R : X \longrightarrow Y$, we define for it two functionals, namely

(i) $\mathtt{ubd}_R(U) := \overline{\overline{R^\mathsf{T}}\, ; U}$,   the **upper bound cone functional** and
(ii) $\mathtt{lbd}_R(V) := \overline{\overline{R}\, ; V}$,   the **lower bound cone functional**.   □

The reader will later easily find them identical with the upper, respectively lower, bound construct of Def. 9.2, however, without assuming the relation R to be an ordering. Some very important properties can already be shown without the ordering requirement. For an intuitive interpretation, we also refer forward to Section 9.2.

For the explanations that follow, let us assume R to be an arbitrary relation between objects and their properties. All our spoken discourse seems to follow a typical quantifying scheme. Whenever somebody talks about a subset U of objects, one automatically asks the question

'Which of the properties do all of objects U have in common?'

Analogously in the other direction: given some subset W of properties, one will routinely ask

'Which objects enjoy all the properties of W?'

The following statements are then immediate.

- If the set of objects is increased to $U' \supseteq U$, there may be equally many or fewer, but definitely not more, properties that they commonly enjoy.
- If one takes more properties $W' \supseteq W$, there may be equally many or fewer objects, but definitely not more, that share all these properties.

This will now be studied more formally. We start with a set of objects U and determine the set of all the properties W_U common to them:

$$W_U := \{w \mid \forall u \in U : (u, w) \in R\}$$
$$= \{w \mid \forall u \in X : u \in U \rightarrow (u, w) \in R\} \quad \text{restricting } U \subseteq X \text{ in the source}$$
$$= \{w \mid \forall u \in X : u \notin U \lor (u, w) \in R\} \quad a \rightarrow b \Longleftrightarrow \neg a \lor b$$

$$= \{w \mid \neg\neg[\forall u \in X : (w,u) \in R^\mathsf{T} \vee u \notin U]\} \qquad \text{transposition, rearranging}$$
$$= \{w \mid \neg[\exists u \in X : (w,u) \notin R^\mathsf{T} \wedge u \in U]\} \qquad \neg\forall \ldots = \exists\neg \ldots$$
$$= \{w \mid \neg(w \in \overline{R}^\mathsf{T}\,;U)\} \qquad \text{definition of composition}$$
$$= \{w \mid w \in \overline{\overline{R}^\mathsf{T}\,;U}\} \qquad \text{lifting negation to point-free form}$$
$$= \mathtt{ubd}_R(U) \qquad \text{by definition}$$

Figure 8.8 shows first a relation R of objects – this time celebrities – and some of the properties attributed to them. The second relation P presents four sets of persons, grouped into the four columns of a matrix. The fifth relation presents $\mathtt{ubd}_R(P)$, thus indicating the set of properties common to all members of the respective set of celebrities. Person sets 2 and 4 are obviously too diverse to allow the members to share common properties. In much the same way, we can start from the fourth matrix Q, and consider it as representing three sets of properties arranged into one matrix. Then the third relation is $\mathtt{lbd}_R(Q)$, the sets of people enjoying all the respective properties of the set. Of course, there does not exist a person who is at the same time related with Soccer, Movie, and Music, as the property set 2 demands.

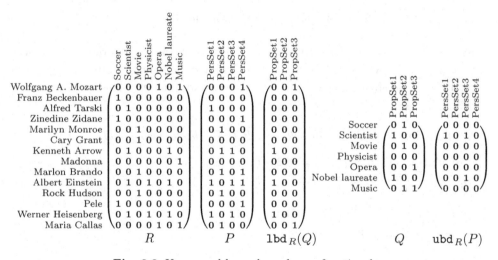

Fig. 8.8 Upper and lower bound cone functional

Very often, upper bounds and lower bounds are formed one after the other. Then important formulae hold that we now exhibit.

Definition 8.7. Given any relations R, U, we may form

$$\bigwedge\nolimits_R(U) := \mathtt{lbd}_R(\mathtt{ubd}_R(U)),$$

the R-**contact closure** of U.   □

One will immediately see that $U \subseteq \bigwedge_R(U)$; this follows from applying the Schröder equivalences. If R happens to be an ordering, we can see the closure property more easily. Then first the set of all upper bounds is considered and for these the minorants are taken which make up a lower cone including U. The graphic part of Fig. 8.9 may help us to understand how a contact should be understood although R is not an order.

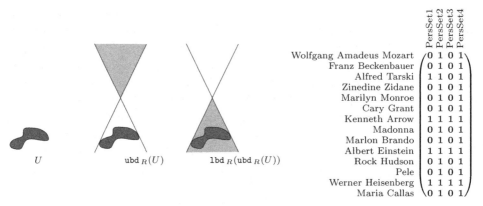

	PersSet1	PersSet2	PersSet3	PersSet4
Wolfgang Amadeus Mozart	0	1	0	1
Franz Beckenbauer	0	1	0	1
Alfred Tarski	1	1	0	1
Zinedine Zidane	0	1	0	1
Marilyn Monroe	0	1	0	1
Cary Grant	0	1	0	1
Kenneth Arrow	1	1	1	1
Madonna	0	1	0	1
Marlon Brando	0	1	0	1
Albert Einstein	1	1	1	1
Rock Hudson	0	1	0	1
Pele	0	1	0	1
Werner Heisenberg	1	1	1	1
Maria Callas	0	1	0	1

Fig. 8.9 The R-contact closure for P of Fig. 8.8

Figure 8.9 shows that contact holds for all in person sets 2 and 4 owing to the fact that the respective set of persons did not enjoy any common property. In person set 1, all scientists are collected while set 3 contains all Nobel laureates.

Proposition 8.8. *Given any relations* R, U, V, *the following hold:*

(i) $\operatorname{ubd}_R(\operatorname{lbd}_R(\operatorname{ubd}_R(U))) = \operatorname{ubd}_R(U)$

(ii) $\overline{\overline{R}^{\mathsf{T}}{}_{\mathsf{;}} R_{\mathsf{;}} \overline{\overline{R}^{\mathsf{T}}{}_{\mathsf{;}} U}} = \overline{\overline{R}^{\mathsf{T}}{}_{\mathsf{;}} U}$

(iii) $\operatorname{lbd}_R(\operatorname{ubd}_R(\operatorname{lbd}_R(V))) = \operatorname{lbd}_R(V)$

(iv) $\overline{\overline{R}{}_{\mathsf{;}} \overline{R}^{\mathsf{T}}{}_{\mathsf{;}} \overline{R}{}_{\mathsf{;}} V} = \overline{R{}_{\mathsf{;}} V}$

(v) $\operatorname{lbd}_\varepsilon(\operatorname{ubd}_\varepsilon(U)) = U$

(vi) $\overline{\varepsilon{}_{\mathsf{;}} \overline{\varepsilon^{\mathsf{T}}{}_{\mathsf{;}} U}} = U$

Proof: Obviously, (ii,iv) are nothing other than expansions of (i,iii). We decide, for example, to prove (ii) and remove the outer negations

$$\overline{\overline{R}^{\mathsf{T}}{}_{\mathsf{;}} R_{\mathsf{;}} \overline{\overline{R}^{\mathsf{T}}{}_{\mathsf{;}} U}} = \overline{R}^{\mathsf{T}}{}_{\mathsf{;}} U.$$

Now, "$\subseteq$" needs just one application of the Schröder rule. "$\supseteq$" follows with monotony from

$$\overline{\overline{R} \, ; \overline{\overline{R}^{\mathsf{T}}} \, ; U} \supseteq U = \overline{\overline{U}}$$

which is true; this can be seen again with the Schröder equivalence after removing the outermost negations.

(v,vi) are obtained as special cases applying Prop. 7.14. □

We also give another interpretation of essentially the same result, writing it down with residuals and, thus, in a form more reminiscent of semigroup theory.

Proposition 8.9. *Given any relations* R, X, Y*, the following hold:*

(i) $\mathtt{ubd}_R(U) = R^{\mathsf{T}}/U^{\mathsf{T}}$

(ii) $\Big(R/(U\backslash R)\Big)\backslash R = U\backslash R$

(iii) $\mathtt{lbd}_R(V) = R/V^{\mathsf{T}}$

(iv) $R/\Big((R/V)\backslash R\Big) = R/V$

Proof: (i,iii) are simply expansions of Def. 4.6, while (ii,iv) reformulate Prop. 8.8 correspondingly. □

The results (ii,iv) may best be interpreted conceiving the cone functionals as quotients. They remind us of what holds for natural numbers:

$$r \; : \; \frac{r}{r\,/\,x} = r \cdot \frac{r\,/\,x}{r} = \frac{r}{x}.$$

The results in the present generalized form are relatively new and are now increasingly known. They seem useful for abbreviating proofs at several occasions. We will learn more on orderings in Chapter 9, but will soon see that we then study in traditional form a specialized case. Whatever is formulated concerning least upper bounds and cones, does not require the relation studied to be an ordering. Therefore, we have decided to place this result here since it is a very basic one.

8.5 Cancellation properties of the symmetric quotient

Algebraic properties of the symmetric quotient are very important, but far from being broadly known. Therefore, they are recalled here – some newly invented – proved, and also visualized.

	Greenpeace	Red Cross	Amnesty International	World Wildlife Fund	Attac	Doctors Without Borders	Red Crescent
Arbuthnot	1	0	0	1	0	0	1
Perez	0	0	0	0	0	0	0
Dupont	0	1	1	0	1	0	0
Botticelli	1	0	0	1	0	0	1
Schmidt	0	1	1	0	1	0	0
Larsen	0	0	0	0	0	0	0

	General Motors	Siemens	Vivendi	Lloyds	British Airways	Apple	Vodafone	Microsoft
Arbuthnot	1	0	0	0	1	0	1	0
Perez	0	0	0	0	0	0	0	0
Dupont	1	0	1	1	0	1	0	0
Botticelli	1	0	0	1	1	0	1	0
Schmidt	0	0	1	0	0	1	0	0
Larsen	0	0	0	0	0	0	0	0

	General Motors	Siemens	Vivendi	Lloyds	British Airways	Apple	Vodafone	Microsoft
Greenpeace	0	0	0	0	1	0	1	0
Red Cross	0	0	1	0	0	1	0	0
Amnesty International	0	0	1	0	0	1	0	0
World Wildlife Fund	0	0	0	0	1	0	1	0
Attac	0	0	1	0	0	1	0	0
Doctors Without Borders	0	1	0	0	0	0	0	1
Red Crescent	0	0	0	0	1	0	1	0

Fig. 8.10 The symmetric quotient: A, B and $\mathtt{syq}(A, B)$

In Fig. 8.10, we assume a group of individuals and two relations in which these are involved and that are typically neither functions nor mappings: the non-governmental organizations they give donations to and the stocks they own. It is immediately clear – and does not need a written proof – that (i,ii) of Prop. 8.10 hold. Since the symmetric quotient of Def. 4.8 compares columns as to equality, it is immaterial whether it compares the columns or their complements. Also, when comparing the columns of A with those of B, one will obtain the converse when comparing the columns of B with those of A.

Proposition 8.10. *Let A, B be arbitrary relations with the same source, so that the symmetric quotient may be formed.*

(i) $\mathtt{syq}(\overline{A}, \overline{B}) = \mathtt{syq}(A, B)$

(ii) $\mathtt{syq}(B, A) = \left[\mathtt{syq}(A, B)\right]^{\mathsf{T}}$ □

For truly heterogeneous relations, in general $\mathtt{syq}(A^{\mathsf{T}}, B^{\mathsf{T}})$ cannot be built for typing reasons. At least, A and B need the same source. In the case that $A = B$, however, even more can be proved, since then also $\mathtt{syq}(A^{\mathsf{T}}, A^{\mathsf{T}})$ is defined; see Fig. 8.11. We demonstrate this with the not necessarily symmetric relation 'has sent a letter to' on a set of individuals.

	Arbuthnot	Perez	Dupont	Botticelli	Schmidt	Larsen
Arbuthnot	1	0	1	1	0	1
Perez	0	1	1	0	1	1
Dupont	0	1	1	0	1	1
Botticelli	1	0	1	1	0	1
Schmidt	0	0	1	0	0	1
Larsen	0	1	1	0	1	1

A

	Arbuthnot	Perez	Dupont	Botticelli	Schmidt	Larsen
	1	0	0	1	0	0
	0	1	1	0	0	1
	0	1	1	0	0	1
	1	0	0	1	0	0
	0	0	0	0	1	0
	0	1	1	0	0	1

$\mathsf{syq}(A^{\mathsf{T}}, A^{\mathsf{T}})$

	Arbuthnot	Perez	Dupont	Botticelli	Schmidt	Larsen
	1	0	0	1	0	0
	0	1	0	0	1	0
	0	0	1	0	0	1
	1	0	0	1	0	0
	0	1	0	0	1	0
	0	0	1	0	0	1

$\mathsf{syq}(A, A)$

Fig. 8.11 Symmetric quotients of a homogeneous relation give their row respectively column equivalence

Rearranging the relations will make clear, that $\mathsf{syq}(A^{\mathsf{T}}, A^{\mathsf{T}})$ as well as $\mathsf{syq}(A, A)$ are equivalences; we identify the row respectively column congruence of A according to Def. 5.28. The name A of the relation has not been changed, because the relation has not changed[6] – its matrix presentation obviously has.

$A =$

	Arbuthnot	Botticelli	Perez	Schmidt	Dupont	Larsen
Arbuthnot	1	1	0	0	1	1
Botticelli	1	1	0	0	1	1
Perez	0	0	1	1	1	1
Dupont	0	0	1	1	1	1
Larsen	0	0	1	1	1	1
Schmidt	0	0	0	0	1	1

	Arbuthnot	Botticelli	Perez	Dupont	Larsen	Schmidt
	1	1	0	0	0	0
	1	1	0	0	0	0
	0	0	1	1	1	0
	0	0	1	1	1	0
	0	0	1	1	1	0
	0	0	0	0	0	1

$= \mathsf{syq}(A^{\mathsf{T}}, A^{\mathsf{T}})$

$\neq$

	Arbuthnot	Botticelli	Perez	Schmidt	Dupont	Larsen
Arbuthnot	1	1	0	0	0	0
Botticelli	1	1	0	0	0	0
Perez	0	0	1	1	0	0
Schmidt	0	0	1	1	0	0
Dupont	0	0	0	0	1	1
Larsen	0	0	0	0	1	1

$= \mathsf{syq}(A, A)$

Fig. 8.12 Symmetric quotients of a homogeneous relation rearranged

Observe that the sequence of names has changed in the right relation. From the form given in Fig. 8.12 one can easily convince oneself that (i,ii) of Prop. 8.11 are indeed satisfied.

Proposition 8.11. *Let A be an arbitrary (possibly heterogeneous) relation.*

(i) $A_{;}\, \mathsf{syq}(A, A) = A$

(ii) $\mathbb{I}_{\mathsf{tgt}A} \subseteq \mathsf{syq}(A, A)$

Proof: The first part "⊆" of (i) follows from $A_{;}\,\mathsf{syq}(A, A) \subseteq A_{;}\, \overline{A^{\mathsf{T}}_{;}\, \overline{A}}$ and $\ldots \subseteq A$ using the Schröder equivalence. The second part "⊇", together with (ii), holds since $\mathbb{I} \subseteq \overline{A^{\mathsf{T}}_{;}\, \overline{A}}$ for all relations A again using the Schröder equivalences. □

---

[6] This is not true for the language TITUREL where the permuted version is considered to have a different source.

The next propositions will show that a symmetric quotient indeed behaves to a certain extent as a quotient usually does. The first results show how dividing and then multiplying again leads back to the origin.

Proposition 8.12. *Assuming arbitrary relations A, B, C, always*

(i) $A \, \syq(A, B) = B \cap \mathbb{T} \, \syq(A, B)$,

(ii) $A \, \syq(A, B) \subseteq B$,

(iii) $\syq(A, B)$ *surjective* $\implies$ $A \, \syq(A, B) = B$.

Proof: (i) We have
$$B \cap \mathbb{T} \, \syq(A, B) = \big(B \cap A \, \syq(A, B)\big) \cup \big(B \cap \overline{A} \, \syq(A, B)\big) = A \, \syq(A, B)$$
since very obviously
$$A \, \syq(A, B) \subseteq A \, \overline{A^{\mathsf{T}} \, \overline{B}} \subseteq B \quad \text{and}$$
$$\overline{A} \, \syq(A, B) = \overline{A} \, \syq(\overline{A}, \overline{B}) \subseteq \overline{A} \, \overline{\overline{A}^{\mathsf{T}} \, B} \subseteq \overline{B}.$$

(ii) is trivial. (iii) follows directly from (i) using the definition of surjectivity. $\square$

We have, thus, analyzed that $A \, \syq(A, B)$ can differ from B only in one way: a column of $A \, \syq(A, B)$ is either equal to the column of B or is zero.

So far, we have shown cancelling of the type $p \times \frac{q}{p} = q$ known for integers; now we try to find also something like $\frac{q}{p} \times \frac{p}{r} = \frac{q}{r}$ for symmetric quotients.

Proposition 8.13. *For arbitrary relations A, B, C we have*

(i) $\syq(A, B) \, \syq(B, C) = \syq(A, C) \cap \syq(A, B) \, \mathbb{T}$
$$\qquad\qquad\qquad\qquad\quad = \syq(A, C) \cap \mathbb{T} \, \syq(B, C) \subseteq \syq(A, C),$$

(ii) $\syq(A, B) \, \syq(B, C) = \syq(A, C)$ *if $\syq(A, B)$ is total, or*
$$\qquad\qquad\qquad\qquad\qquad\qquad\qquad\quad \text{if } \syq(B, C) \text{ is surjective.}$$

Proof: (i) Without loss of generality we concentrate on the first equality sign. Direction "$\subseteq$" follows from the Schröder equivalence via
$$(A^{\mathsf{T}} \, \overline{C} \cup \overline{A}^{\mathsf{T}} \, C) \, [\syq(B, C)]^{\mathsf{T}} = A^{\mathsf{T}} \, \overline{C} \, \syq(\overline{C}, \overline{B}) \cup \overline{A}^{\mathsf{T}} \, C \, \syq(C, B) \subseteq A^{\mathsf{T}} \, \overline{B} \cup \overline{A}^{\mathsf{T}} \, B$$
using Prop. 8.12. Direction "$\supseteq$" can be obtained using Prop. 8.11, the Dedekind rule and the result just proved:
$$\syq(A, B) \, \mathbb{T} \cap \syq(A, C)$$
$$\subseteq \big(\syq(A, B) \cap \syq(A, C) \, \mathbb{T}^{\mathsf{T}}\big) \, \big(\mathbb{T} \cap [\syq(A, B)]^{\mathsf{T}} \, \syq(A, C)\big)$$

$\subseteq \mathsf{syq}\,(A,B)\,{}_\circ\mathsf{syq}\,(B,A)\,{}_\circ\mathsf{syq}\,(A,C) \subseteq \mathsf{syq}\,(A,B)\,{}_\circ\mathsf{syq}\,(B,C)$

(ii) follows immediately from (i). □

Thus, we have again found some sort of a quotient behavior; if it is not directly satisfied, it is at least replaced by sub-cancellability.

We list some special cases.

Proposition 8.14.

(i) $\mathsf{syq}\,(A,A)\,{}_\circ\mathsf{syq}\,(A,B) = \mathsf{syq}\,(A,B)$
(ii) $\mathsf{syq}\,(A,A)$ *is always an equivalence relation.*
(iii) $\mathsf{syq}\,(A,B)$ *is always a difunctional relation.*

Proof: (i) follows from Prop. 8.13.ii with Prop. 8.11.ii.

(ii) $\mathsf{syq}\,(A,A)$ is reflexive due to Prop. 8.11.ii, symmetric by construction, and transitive according to (i).

(iii) $\mathsf{syq}\,(A,B)\,{}_\circ[\mathsf{syq}\,(A,B)]^\mathsf{T}\,{}_\circ\mathsf{syq}\,(A,B) = \mathsf{syq}\,(A,B)\,{}_\circ\mathsf{syq}\,(B,A)\,{}_\circ\mathsf{syq}\,(A,B)$ from where we may proceed using the cancellation rule Prop. 8.13.i two times. □

According to (iii) above, a symmetric quotient can always be rearranged similar to Fig. 5.26, i.e., as a (possibly only partial) block-diagonal relation. We apply these results to the complex of row and column equivalences and containments.

Proposition 8.15.

(i) $\mathsf{syq}\,(\overline{R_{\,{}_\circ}R^\mathsf{T}},\overline{R_{\,{}_\circ}R^\mathsf{T}}) = \Xi(R)$ $\mathsf{syq}\,(\overline{R^\mathsf{T}{}_\circ\overline{R}},\overline{R^\mathsf{T}{}_\circ\overline{R}}) = \Psi(R)$
(ii) $\mathcal{R}(R)\,{}_\circ\Xi(R) = \mathcal{R}(R)$ $\Xi(R)\,{}_\circ\mathcal{R}(\overline{R}) = \mathcal{R}(\overline{R})$
         $\mathcal{C}(R)\,{}_\circ\Psi(R) = \mathcal{C}(R)$ $\Psi(R)\,{}_\circ\mathcal{C}(\overline{R}) = \mathcal{C}(\overline{R})$

Proof: (i) According to Def. 5.28, we have $\Xi(R) = \mathsf{syq}\,(R^\mathsf{T},R^\mathsf{T}) = \overline{R_{\,{}_\circ}R^\mathsf{T}} \cap \overline{R_{\,{}_\circ}\overline{R}^\mathsf{T}}$. Now we replace $\overline{R}$ two times by the equal, but more complicated, version according to Prop. 8.8.ii in transposed form with $U := \mathbb{I}$

$$\overline{R} = \overline{R_{\,{}_\circ}\overline{R}^\mathsf{T}{}_\circ R}$$

so as to obtain the result (up to trivial manipulations).

(ii) We prove only the very first equality and use $\Xi(R)$ according to (i). With Prop. 8.11, it is now trivial because

$$\overline{R_; R^\mathsf{T}}; \mathsf{syq}\,(\overline{\overline{R_; R^\mathsf{T}}}, \overline{\overline{R_; R^\mathsf{T}}}) = \overline{R_; R^\mathsf{T}}.$$ $\square$

When one is asked the question whether the row equivalence of the row-is-contained preorder of R is equal to the row equivalence of R, one will need some time for an answer. This answer is given completely formally with (i).

In proofs or computations it is often useful to know how composition with a relation leads to an effect on the symmetric quotient; similarly for composition with a mapping. This is captured in the following proposition.

Proposition 8.16.

(i) $\quad\mathsf{syq}\,(A, B) \subseteq \mathsf{syq}\,(C_; A, C_; B)$ *for every C*

$\qquad\mathsf{syq}\,(A, B) = \mathsf{syq}\,(F_; A, F_; B)$ *for a surjective mapping F*

(ii) $\quad F_; \mathsf{syq}\,(A, B) = \mathsf{syq}\,(A_; F^\mathsf{T}, B)$ *for every mapping F*

(iii) $\quad\mathsf{syq}\,(A, B)_; F^\mathsf{T} = \mathsf{syq}\,(A, B_; F^\mathsf{T})$ *for every mapping F*

Proof: (i) We show, for example,

$$C_; B \subseteq C_; B \iff C^\mathsf{T}_; \overline{C_; B} \subseteq \overline{B} \implies (C_; A)^\mathsf{T}_; \overline{C_; B} \subseteq A^\mathsf{T}_; \overline{B}.$$

In the case of a surjective mapping F, we have $(F_; A)^\mathsf{T}_; \overline{F_; B} = A^\mathsf{T}_; F^\mathsf{T}_; \overline{F_; B} = A^\mathsf{T}_; \overline{B}$, since the mapping F may slip out of the negation and $F^\mathsf{T}_; F = \mathbb{I}$ holds for a surjective mapping.

(ii) Following Prop. 5.13, we have $F_; \overline{S} = \overline{F_; S}$ for a mapping F and arbitrary S, so that with Prop. 5.3

$$F_; \mathsf{syq}\,(A, B)= F_; \overline{A^\mathsf{T}_; \overline{B}} \cap F_; \overline{\overline{A}^\mathsf{T}_; B} = \overline{F_; A^\mathsf{T}_; \overline{B}} \cap \overline{F_; \overline{A}^\mathsf{T}_; B}$$

$$= \overline{(A_; F^\mathsf{T})^\mathsf{T}_; \overline{B}} \cap \overline{\overline{A_; F^\mathsf{T}}^\mathsf{T}_; B} = \mathsf{syq}\,(A_; F^\mathsf{T}, B).$$

(iii) is the transposed version of (ii). $\square$

The following corollary is concerned with the membership relation and refers back to the definition of the latter in Def. 7.13.

Corollary 8.17. *For arbitrary relations R, X and the membership relation ε of R on its target side, the construct $\mathsf{syq}\,(R_; \varepsilon, R_; X)$ is surjective.*

Proof: Using Prop. 8.16.i, we have $\mathsf{syq}\,(R_; \varepsilon, R_; X) \supseteq \mathsf{syq}\,(\varepsilon, X)$, which is surjective by definition of ε. $\square$

We mention further identities of a symmetric quotient in combination with a natural projection that are sometimes very helpful. They belong inherently with the introduction of the quotient domain in Def. 7.9, but could not be proved at that early time.

Proposition 8.18. *Consider an arbitrary equivalence together with its natural projection, $\Xi = \eta \,;\, \eta^{\mathsf{T}}$, and some relation A satisfying $A = A \,;\, \Xi$. Then the following identities hold:*

(i) $\eta^{\mathsf{T}} \,;\, \mathsf{syq}(A, B) = \mathsf{syq}(A \,;\, \eta, B)$

(ii) $\overline{\overline{A \,;\, \eta}} = A \,;\, \eta$

Proof: (i) $\eta^{\mathsf{T}} \,;\, \mathsf{syq}(A, B) = \eta^{\mathsf{T}} \,;\, \left[\overline{\overline{A^{\mathsf{T}}} \,;\, B} \cap \overline{A^{\mathsf{T}} \,;\, \overline{B}} \right] = \eta^{\mathsf{T}} \,;\, \left[\overline{\overline{\Xi \,;\, A^{\mathsf{T}}} \,;\, B} \cap \overline{\Xi \,;\, A^{\mathsf{T}} \,;\, \overline{B}} \right]$

$= \eta^{\mathsf{T}} \,;\, \left[\overline{\overline{\eta \,;\, \eta^{\mathsf{T}} \,;\, A^{\mathsf{T}}} \,;\, B} \cap \overline{\eta \,;\, \eta^{\mathsf{T}} \,;\, A^{\mathsf{T}} \,;\, \overline{B}} \right]$

$= \eta^{\mathsf{T}} \,;\, \left[\eta \,;\, \overline{\overline{\eta^{\mathsf{T}} \,;\, A^{\mathsf{T}}} \,;\, B} \cap \eta \,;\, \overline{\eta^{\mathsf{T}} \,;\, A^{\mathsf{T}} \,;\, \overline{B}} \right]$

$= \eta^{\mathsf{T}} \,;\, \eta \,;\, \left[\overline{\overline{\eta^{\mathsf{T}} \,;\, A^{\mathsf{T}}} \,;\, B} \cap \overline{\eta^{\mathsf{T}} \,;\, A^{\mathsf{T}} \,;\, \overline{B}} \right] = \mathbb{I} \,;\, \mathsf{syq}(A \,;\, \eta, B)$

(ii) $\overline{\overline{A \,;\, \eta}} = \overline{\overline{A \,;\, \Xi \,;\, \eta}} = \overline{\overline{A \,;\, \eta \,;\, \eta^{\mathsf{T}} \,;\, \eta}} = \overline{\overline{A \,;\, \eta}} \,;\, \eta^{\mathsf{T}} \,;\, \eta = \overline{\overline{A \,;\, \eta}} \,;\, \mathbb{I} = A \,;\, \eta$ □

One will find out that these formulae amend what is already known for mappings, extending it to their transpose in the case of a natural projection: Prop. 8.16.ii and Prop. 5.13.i.

The next formulae resemble what might be called *shift-inverting a cone*. When trying to understand this intuitively, one should have a look at Fig. 8.9, which visualizes the two-fold application $R \mapsto \mathsf{lbd}_R(\mathsf{ubd}_R(Y))$. In (ii), one can shift this functional as visualized in Fig. 8.13 and replace it with $\mathsf{ubd}_R(\mathsf{lbd}_R(X))$.

Proposition 8.19. *We assume three arbitrary relations R, X, Y, requiring only that $R \,;\, X$ and $R^{\mathsf{T}} \,;\, Y$ can be formed.*

(i) $\mathsf{syq}(\mathsf{lbd}_R(X), \mathsf{lbd}_R(\mathsf{ubd}_R(Y))) = \mathsf{syq}(\mathsf{ubd}_R(\mathsf{lbd}_R(X)), \mathsf{ubd}_R(Y))$

(ii) $\mathsf{syq}(\overline{\overline{R \,;\, X}}, \overline{R \,;\, \overline{R^{\mathsf{T}}} \,;\, Y}) = \mathsf{syq}(\overline{R^{\mathsf{T}}} \,;\, \overline{R \,;\, X}, \overline{R^{\mathsf{T}}} \,;\, Y)$

(iii) $\mathsf{syq}(R/X^{\mathsf{T}}, R/(Y\backslash R)) = \mathsf{syq}(((R/X^{\mathsf{T}})\backslash R)^{\mathsf{T}}, (Y\backslash R)^{\mathsf{T}})$

Proof: (i) We expand both sides

$$\mathsf{syq}(\mathsf{lbd}_R(X), \mathsf{lbd}_R(\mathsf{ubd}_R(Y))) = \overline{\overline{X^{\mathsf{T}} \,;\, \overline{R^{\mathsf{T}}} \,;\, \overline{R \,;\, \overline{R^{\mathsf{T}}} \,;\, Y}}} \cap \overline{X^{\mathsf{T}} \,;\, \overline{R^{\mathsf{T}}} \,;\, \overline{R \,;\, \overline{R^{\mathsf{T}}} \,;\, Y}}$$

$$\mathtt{syq}\,(\mathtt{ubd}\,_R(\mathtt{lbd}\,_R(X)),\mathtt{ubd}\,_R(Y)) = \overline{\overline{X^\mathsf{T}{}_;\overline{R}^\mathsf{T}{}_;\overline{R}{}_;\overline{R}^\mathsf{T}{}_;Y}} \cap \overline{\overline{X^\mathsf{T}{}_;\overline{R}^\mathsf{T}{}_;\overline{R}{}_;\overline{R}^\mathsf{T}{}_;Y}}.$$

Now, the first term in the first case equals the second term in the second case. The other terms can be transformed into one another, applying twice the trivially satisfied equation of the following type

$$\overline{\overline{R}^\mathsf{T}{}_;\overline{\overline{R}{}_;\overline{R}^\mathsf{T}{}_;Y}} = \overline{\overline{R}^\mathsf{T}{}_;Y}.$$

(ii) and (iii) are the same as (i), but written differently.   □

When considering Fig. 8.14, one will observe that the pairs of relations

| $\mathtt{lbd}\,_R(X)$ | versus | $\mathtt{lbd}\,_R(\mathtt{ubd}\,_R(Y))$ |
| $\mathtt{ubd}\,_R(\mathtt{lbd}\,_R(X))$ | versus | $\mathtt{ubd}\,_R(Y)$ |

that are compared via the symmetric quotient differ even in their source types. Nonetheless the relation marked $\mathtt{syq}$ in Fig. 8.14 relates *either one* of these pairs.

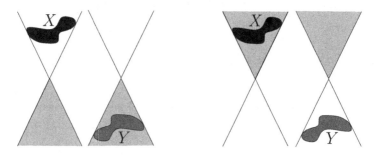

Fig. 8.13 Shift-inverting the cone

There are several other formulae concerning the symmetric quotient. They may also be understood as some sort of cancelling. Again, we provide an analogy: so far, we have had the first two ways of cancelling, and now we aim at the third

$$p \times \tfrac{q}{p} = q \qquad\qquad \tfrac{q}{p} \times \tfrac{p}{r} = \tfrac{q}{r} \qquad\qquad \tfrac{r}{p} : \tfrac{r}{q} = \tfrac{q}{p}.$$

In total, symmetric quotients offer more or less the same division formulae, provided one has certain totality and/or surjectivity properties. If not, estimates are possible so as to establish 'sub-cancellability'.

Proposition 8.20. *Let relations X, Y, Z be given for which the constructs can be formed.*

(i) $\mathtt{syq}\,(X,Y) \subseteq \mathtt{syq}\,(Z,X) \setminus \mathtt{syq}\,(Z,Y)$

(ii) $\mathtt{syq}\,(\mathtt{syq}\,(X,Y),\mathtt{syq}\,(X,Z)) \supseteq \mathtt{syq}\,(Y,Z)$

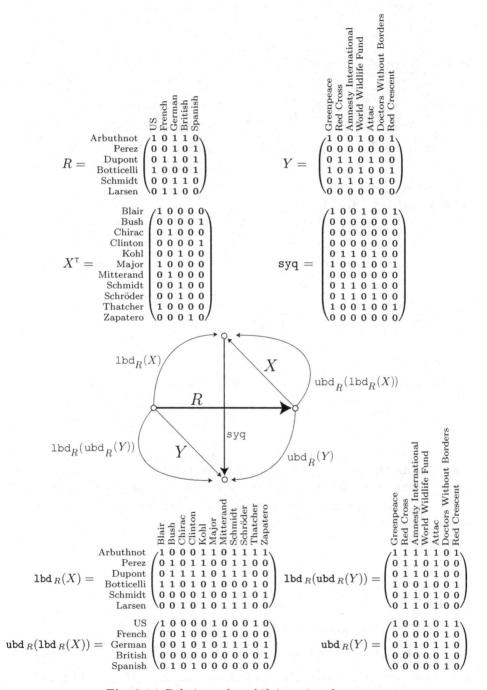

Fig. 8.14 Relations when shift-inverting the cone

(iii) $\operatorname{syq}(\operatorname{syq}(X,Y),\operatorname{syq}(X,Z)) = \operatorname{syq}(Y,Z)$ *if* $\operatorname{syq}(X,Y)$ *and* $\operatorname{syq}(X,Z)$
                                                                are surjective

Proof: (i) $\operatorname{syq}(Z,X) \backslash \operatorname{syq}(Z,Y)$

$= \left[\operatorname{syq}(Z,X)\right]^{\mathsf{T}} ; \overline{\operatorname{syq}(Z,Y)}$ expanded

$= \operatorname{syq}(X,Z) ; \overline{\left[\overline{Z}^{\mathsf{T}} ; Y \cup Z^{\mathsf{T}} ; \overline{Y}\right]}$ transposed, expanded, negated

$= \operatorname{syq}(X,Z) ; \overline{\overline{Z}^{\mathsf{T}} ; Y} \cap \overline{\operatorname{syq}(X,Z) ; Z^{\mathsf{T}} ; \overline{Y}}$ distributive, De Morgan

$= \operatorname{syq}(\overline{X},\overline{Z}) ; \overline{\overline{Z}^{\mathsf{T}} ; Y} \cap \overline{\operatorname{syq}(X,Z) ; Z^{\mathsf{T}} ; \overline{Y}}$ since always $\operatorname{syq}(A,B) = \operatorname{syq}(\overline{A},\overline{B})$

$\supseteq \overline{\overline{X}^{\mathsf{T}} ; Y} \cap \overline{X^{\mathsf{T}} ; \overline{Y}}$ two times Prop. 8.12.ii

$= \operatorname{syq}(X,Y)$ by definition

(ii) $\operatorname{syq}(\operatorname{syq}(X,Y),\operatorname{syq}(X,Z))$

$= \overline{\overline{\operatorname{syq}(X,Y)}^{\mathsf{T}} ; \operatorname{syq}(X,Z)} \cap \overline{\operatorname{syq}(X,Y)^{\mathsf{T}} ; \overline{\operatorname{syq}(X,Z)}}$ partly expanded

$= \overline{\operatorname{syq}(Y,X) ; \operatorname{syq}(X,Z)} \cap \overline{\operatorname{syq}(Y,X) ; \overline{\operatorname{syq}(X,Z)}}$ transposed

$= \overline{\left[\overline{Y}^{\mathsf{T}} ; X \cup Y^{\mathsf{T}} ; \overline{X}\right] ; \operatorname{syq}(X,Z)} \cap \overline{\operatorname{syq}(Y,X) ; \left[\overline{X}^{\mathsf{T}} ; Z \cup X^{\mathsf{T}} ; \overline{Z}\right]}$ further expanded

$= \overline{\overline{Y}^{\mathsf{T}} ; X ; \operatorname{syq}(X,Z)} \cap \overline{Y^{\mathsf{T}} ; \overline{X} ; \operatorname{syq}(X,Z)} \cap$

         $\overline{\operatorname{syq}(Y,X) ; \overline{X}^{\mathsf{T}} ; Z} \cap \overline{\operatorname{syq}(Y,X) ; X^{\mathsf{T}} ; \overline{Z}}$ distributivity, De Morgan

$\supseteq \overline{\overline{Y}^{\mathsf{T}} ; Z} \cap \overline{Y^{\mathsf{T}} ; \overline{X} ; \operatorname{syq}(\overline{X},\overline{Z})} \cap \overline{\operatorname{syq}(\overline{Y},\overline{X}) ; \overline{X}^{\mathsf{T}} ; Z} \cap \overline{Y^{\mathsf{T}} ; \overline{Z}}$ cancelled

$\supseteq \overline{\overline{Y}^{\mathsf{T}} ; Z} \cap \overline{Y^{\mathsf{T}} ; \overline{Z}}$ cancelled again

$= \operatorname{syq}(Y,Z)$

(iii) In the case of surjectivity, equality will hold according to Prop. 8.12.iii.   □

This was quite a lot of mostly new formulae so that one should ask whether it was worth proving all these. The author is absolutely confident that the answer is 'yes!'. In view of the many publications of computer scientists on programming semantics, power transposes, and existential images that are often just postulational, we here favor Leibniz' style of saying 'Calculemus!'.

Exercises

8.5 Prove that $\operatorname{syq}(R,R) \subseteq R^{\mathsf{T}} ; R$ whenever R is surjective.

8.6 Prove that for injective A, B additional properties hold:

$A^{\mathsf{T}} ; B \subseteq \operatorname{syq}(A,B)$ $A ; \operatorname{syq}(A,B) = A ; \mathbb{T} \cap B$ $B \subseteq \operatorname{syq}(A^{\mathsf{T}}, \operatorname{syq}(A,B))$.

8.7 Prove that for arbitrary relations R, X the constructs $U := \bigwedge_R(X)$ and $V := \operatorname{ubd}_R(X)$ satisfy the requirements $\overline{R} ; V = \overline{U}$ and $\overline{R}^{\mathsf{T}} ; U = \overline{V}$ for a non-enlargeable rectangle.

8.6 Tarski rule and Point Axiom

In the preceding sections, a stock of rules for working with relations has been put together and many examples have shown how these rules cooperate. Unexpectedly, however, we are now about to arrive at an important theoretical border, so this section should be skipped at first reading. The presentation may best be started with the following at first sight trivial result.

Proposition 8.21 (Tarski rule). $\quad R \neq \mathbb{L} \implies \mathbb{T}_; R_; \mathbb{T} = \mathbb{T}.$ $\hfill \square$

For any relation given as a matrix, it will obviously hold; see Fig. 8.15.

$$R = \begin{pmatrix} 0 & 0 & 0 & 0 \\ 0 & 0 & 1 & 0 \\ 0 & 0 & 0 & 0 \end{pmatrix} \qquad R_; \mathbb{T} = \begin{pmatrix} 0 & 0 & 0 & 0 \\ 1 & 1 & 1 & 1 \\ 0 & 0 & 0 & 0 \end{pmatrix} \qquad \mathbb{T}_; R = \begin{pmatrix} 0 & 0 & 1 & 0 \\ 0 & 0 & 1 & 0 \\ 0 & 0 & 1 & 0 \end{pmatrix}$$

Fig. 8.15 Illustrating the Tarski rule

Another (theoretically equivalent) version can be given.

Proposition 8.22 (Variant of the Tarski rule). *Always*
$$R_; \mathbb{T} = \mathbb{T} \quad or \quad \mathbb{T}_; \overline{R} = \mathbb{T}.$$
$\hfill \square$

This says what sounds completely trivial when given as a Boolean matrix, namely that a relation R is total or its negation $\overline{R}$ is surjective, i.e., has an entry $\mathbf{1}$ in every column. Of course, in our standard interpretation, every row that assigns no value is a row full of $\mathbf{0}$s and will provide for a row of $\mathbf{1}$s in the complement.

Any attempt, however, to prove Prop. 8.21 or Prop. 8.22 – from the axioms adopted with Def. 8.1 – will fail. Here, we cannot convince the reader in this regard. We can, however, report that small finite examples of relation algebras exist (to be found in, for example, [73]) in which these propositions do not hold. In these relation algebras the elements *cannot* be conceived as sets of pairs.

Another issue of this kind is concerned with *representable* versus *non-representable* relation algebras and may be explained as follows. In the proofs that occur later in this book, we hardly ever have to resort to the relation as a set of pairs, and thus, to the following so-called Point Axiom.

Proposition 8.23 (Point Axiom). *For every relation R the following holds:*
$$R \neq \mathbb{L} \iff x_; y^\mathsf{T} \subseteq R \text{ for certain points } x, y.$$
$\hfill \square$

Around all these facts, a remarkable amount of literature has appeared. One can
work with a set-up of relation algebras which does, or which does not, in addition
to Def. 8.1 contain Prop. 8.21, Prop. 8.22, and/or Prop. 8.23 as an axiom. The
interdependency forms a highly interrelated complex, and we do not elaborate on it
further. Only a close examination shows the important consequences[7] of postulating
or not postulating the Tarski rule.

A first consequence is also given with the following proposition.

Proposition 8.24 (Intermediate Point Theorem). *Let any relations $R : X \longrightarrow Y$
and $S : Y \longrightarrow Z$ be given together with a point $x \subseteq X$ and a point $z \subseteq Z$. Assuming
the Point Axiom to hold, the following are equivalent:*

- $x \subseteq R \, ; S \, ; z$,
- $x \subseteq R \, ; w$ *and* $w \subseteq S \, ; z$ *for some point w.*

Proof: Since the step back from the second to the first statement is trivial, we
concentrate on the other direction. The vector $S \, ; z \cap \overline{R^{\mathsf{T}} \, ; x}$ will turn out to be
non-empty and, thus, contain a point w according to the Point Axiom. To prove
this claim, assume $S \, ; z \cap \overline{R^{\mathsf{T}} \, ; x} = \bot\!\!\!\bot$, which via shunting a point is equivalent with
$R \, ; S \, ; z \subseteq \overline{x}$ contradicting the first statement with a point x. To establish the second
statement is now easy:

$$w \subseteq S \, ; z \cap \overline{R^{\mathsf{T}} \, ; x} \subseteq S \, ; z$$

follows with monotony. Furthermore, $w \subseteq S \, ; z \cap \overline{R^{\mathsf{T}} \, ; x} \subseteq \overline{R^{\mathsf{T}} \, ; x}$ so that $x \subseteq R \, ; w$
using the Schröder rule and shunting a point.   □

At a further advanced level, we can find an even more sophisticated result in
Prop. 19.12 that also seems interesting.

---

[7] Alfred Tarski observed that the rule has to do with the simplicity of the algebra.

9

Orders and Lattices

Lattices penetrate all our life. Early in school we learn about divisibility and look for the greatest common divisor as well as for the least common multiple. Later we learn about Boolean lattices and use union and intersection "$\cup, \cap$" for sets as well as disjunction and conjunction "$\vee, \wedge$" for predicates. Concept lattices give us orientation in all our techniques of discourse in everyday life – usually without coming to our attention. Nevertheless, lattices completely dominate our imagination. Mincut lattices originate from flow optimization, assignment problems, and further optimization tasks. It is well known that an ordering can always be embedded into a complete lattice. Several embedding constructions are conceivable: examples are cut completion and ideal completion.

We introduce all this step by step, concentrating first on order-theoretic functionals. Then we give several examples determining the maximal, minimal, greatest, and least elements of subsets. Thus we learn to work with orderings and compute with them.

9.1 Maxima and minima

Whenever an ordering is presented, one will be interested in finding maximal and minimal elements of subsets. Of course, we do not think of linear orderings only. It is a pity that many people – even educated people – colloquially identify the *maximal* element with the *greatest*. What makes this even worse is that the two concepts often indeed coincide. However, the definitions and meanings become different in nature as soon as orderings are not just finite and linear. We suffer here from historical development: in former times, people hardly ever considered orderings other than linear or powerset orderings.

- An element is *a maximal element* of a subset when there does not exist any strictly greater element in this subset.
- An element is *the greatest element* of a subset if it is greater than (or equal to) any other element of this subset.

Having announced this basic difference, we concentrate on maximal (and minimal) elements first.

Definition 9.1. Let a set V be given with a strictorder "$<$", as well as an arbitrary subset $U \subseteq V$.

(i) The element $m \in U$ is called **a maximal element** of U, if no element of U is strictly greater than m; in predicate-logic form:
$$m \in U \,\wedge\, \forall u \in U : m \not< u.$$

(ii) Correspondingly, the element $m \in U$ is called **a minimal element** of U, if no element of U is strictly less than m; in predicate-logic form:
$$m \in U \,\wedge\, \forall u \in U : m \not> u.$$

(iii) If the strictorder is given as a relation C, the set of maximal, respectively minimal, elements of the set U is defined as
$$\max{}_C(U) := U \cap \overline{C \,;\, U}, \qquad \text{respectively} \qquad \min{}_C(U) := U \cap \overline{C^{\mathsf{T}} \,;\, U}. \qquad \square$$

When looking for maximal elements, one should be prepared to find as a result *a set of elements* which may thus be empty, a 1-element set, or a multi-element set. The algebraic definition uses the strictorder C instead of "$<$". Then we move gradually to the point-free form:

$$m \in U \,\wedge\, \forall u \in U : m \not< u$$

          proceeding to quantification over the whole domain

$$\Longleftrightarrow \quad m \in U \,\wedge\, \forall u : u \in U \to m \not< u$$

          $a \to b = \neg a \vee b$

$$\Longleftrightarrow \quad m \in U \,\wedge\, \forall u : u \notin U \vee m \not< u$$

          $\forall x : p(x) = \neg\big(\exists x : \neg p(x)\big)$

$$\Longleftrightarrow \quad m \in U \,\wedge\, \neg\big(\exists u : u \in U \wedge m < u\big)$$

          $a \wedge b = b \wedge a$

$$\Longleftrightarrow \quad m \in U \,\wedge\, \neg\big(\exists u : m < u \wedge u \in U\big)$$

          corresponding matrix and vector form

$$\Longleftrightarrow \quad U_m \cap \neg\big(\exists u : C_{mu} \wedge U_u\big)$$

          composition of relation and vector

$$\Longleftrightarrow \quad U_m \cap \overline{C \,;\, U}_m$$

So we have justified the point-free definition
$$\max{}_C(U) := U \cap \overline{C \,;\, U}$$
in (iii) and analogously,
$$\min{}_C(U) := U \cap \overline{C^{\mathsf{T}} \,;\, U}.$$

Usually, people are heavily oriented towards predicate-logic formulations. However, one can also read the present relation-algebraic form directly: maximal elements are elements from U for which it is not the case that one can go a step up according to C and reach an element from U.

$$\begin{array}{c} \quad\;\; \scriptstyle 1\;2\;3\;4\;5 \\ \begin{matrix}1\\2\\3\\4\\5\end{matrix} \begin{pmatrix} 1&1&1&1&1\\0&1&0&1&1\\0&0&1&0&1\\0&0&0&1&0\\0&0&0&0&1 \end{pmatrix} \end{array} \qquad \begin{matrix}1\\2\\3\\4\\5\end{matrix}\begin{pmatrix}1\\1\\0\\0\\0\end{pmatrix}\mapsto\begin{pmatrix}0\\1\\0\\0\\0\end{pmatrix} \qquad \begin{matrix}1\\2\\3\\4\\5\end{matrix}\begin{pmatrix}0\\1\\0\\1\\1\end{pmatrix}\mapsto\begin{pmatrix}0\\0\\0\\1\\1\end{pmatrix} \qquad \begin{matrix}1\\2\\3\\4\\5\end{matrix}\begin{pmatrix}1\\1\\1\\0\\0\end{pmatrix}\mapsto\begin{pmatrix}0\\1\\1\\0\\0\end{pmatrix}$$

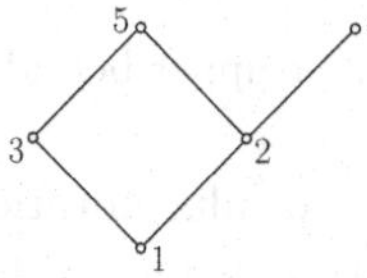

Fig. 9.1 An ordering with three sets and their sets of maxima

9.2 Bounds and cones

An upper bound of a subset $U \subseteq V$ is some element s that is greater than or equal to all elements of U, regardless of whether the element s itself belongs to U. A lower bound is defined accordingly. We first recall this definition in a slightly more formal predicate-logic way.

Definition 9.2. Let an ordering "$\leq$" be given on a set V.

  (i) The element $s \in V$ is an **upper bound** (also **majorant**) of the set $U \subseteq V$ if
$$\forall u \in U : u \leq s.$$
 (ii) The element $s \in V$ is a **lower bound** (also **minorant**) of the set $U \subseteq V$ if
$$\forall u \in U : s \leq u. \qquad \square$$

Often, we are not interested in just one upper bound, but have in view the set of all upper bounds. It is more than evident that an element above an upper bound will also be an upper bound. This motivates us to define in a point-free relational manner the concept of an upper cone as a set, where given an element, all its superiors belong to the set.

Definition 9.3. For an ordering E on a set V, the set of upper, respectively lower, bounds of the set U is
$$\mathrm{ubd}_E(U) = \overline{\overline{E^\mathsf{T}}\,{;}\,U}, \quad \text{respectively} \quad \mathrm{lbd}_E(U) = \overline{\overline{E}\,{;}\,U}$$
for which we will use $\mathrm{ubd}(U), \mathrm{lbd}(U)$, when E is already agreed upon. $\qquad \square$

$$
\begin{array}{c}
\;1\;2\;3\;4\;5 \\
\begin{array}{c}1\\2\\3\\4\\5\end{array}
\begin{pmatrix}
1 & 1 & 1 & 1 & 1 \\
0 & 1 & 0 & 1 & 1 \\
0 & 0 & 1 & 0 & 1 \\
0 & 0 & 0 & 1 & 0 \\
0 & 0 & 0 & 0 & 1
\end{pmatrix}
\end{array}
\qquad
\begin{array}{c}1\\2\\3\\4\\5\end{array}
\begin{pmatrix}1\\1\\0\\0\\0\end{pmatrix}\mapsto\begin{pmatrix}0\\1\\0\\1\\1\end{pmatrix}
\qquad
\begin{array}{c}1\\2\\3\\4\\5\end{array}
\begin{pmatrix}0\\0\\0\\1\\1\end{pmatrix}\mapsto\begin{pmatrix}0\\0\\0\\0\\0\end{pmatrix}
\qquad
\begin{array}{c}1\\2\\3\\4\\5\end{array}
\begin{pmatrix}0\\1\\1\\0\\1\end{pmatrix}\mapsto\begin{pmatrix}0\\0\\0\\0\\1\end{pmatrix}
$$

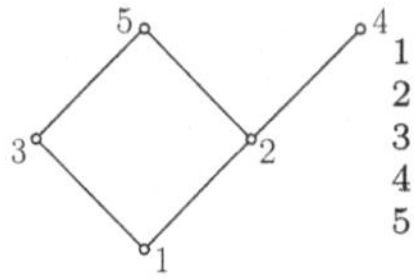

Fig. 9.2 An ordering with three sets and their upper bound sets

It is, however, no longer convenient to work with the infix notation "$\leq$" for the ordering in question. Algebraic considerations make us switch to the letter $E_\leq$, or simply E, to denote the ordering relation. Let $\mathsf{ubd}_E(U)$ be the set of all upper bounds of $U \subseteq V$, i.e.,

$$\mathsf{ubd}_E(U) := \{s \in V \mid \forall x \in U : x \leq s\}.$$

When transforming the right-hand side of this upper bound definition, we arrive at an algebraic condition:

$s \in \mathsf{ubd}_E(U)$

                                                           see above

$\Longleftrightarrow \quad \forall x \in U : x \leq s$

                                $\forall a \in A : p(a) = \forall a : a \in A \to p(a)$

$\Longleftrightarrow \quad \forall x : x \in U \to x \leq s$

                                      $a \to b = \neg a \vee b$

$\Longleftrightarrow \quad \forall x : x \notin U \vee x \leq s$

                              $\forall x : p(x) = \neg(\exists x : \neg p(x))$

$\Longleftrightarrow \quad \neg(\exists x : x \in U \wedge x \not\leq s)$

$\Longleftrightarrow \quad \neg(\exists x : s \not\geq x \wedge x \in U)$

                                         $a \wedge b = b \wedge a$

$\Longleftrightarrow \quad \neg(s \in \overline{E}^{\mathsf{T}}{}_{;}U)$

                                  definition of composition

$\Longleftrightarrow \quad s \in \overline{\overline{E}^{\mathsf{T}}{}_{;}U}$

                                  shifting negation

$\Longleftrightarrow \quad \left[\overline{\overline{E}^{\mathsf{T}}{}_{;}U}\right]_s$

                              transfer to point-free notation

This justifies the preceding Def. 9.3. With this form, we are in a position easily to compute sets of upper bounds in the form of an operation on matrices and vectors known from matrix analysis.

Definition 9.4. Let an ordering E be given on a set V. A subset $U \subseteq V$ is said to satisfy the **upper cone property** if $U = E^{\mathsf{T}}{}_{;}U$. In analogy, U has the **lower cone property** if $U = E_{;}U$. □

One can interpret $U = E^\mathsf{T} {}_; U$ by saying that when one has stepped down according to E ending in a point of the set U, then one necessarily started in U. That this holds true when $U := \mathrm{ubd}_E(X)$ is an upper bound set can be proved algebraically, recalling that always $E {}_; \overline{E}^\mathsf{T} = \overline{E}^\mathsf{T}$ and $E {}_; \overline{\overline{E}^\mathsf{T} {}_; X} = \overline{\overline{E}^\mathsf{T} {}_; X}$ for an order E:

$$E^\mathsf{T} {}_; U = E^\mathsf{T} {}_; \mathrm{ubd}_E(X) = E^\mathsf{T} {}_; \overline{\overline{E}^\mathsf{T} {}_; X} = \overline{\overline{E}^\mathsf{T} {}_; X} = \mathrm{ubd}_E(X) = U$$

Exercises

9.1 Prove that for points x, y, the condition $\mathrm{lbd}_E(x) = \mathrm{lbd}_E(y)$ implies $x = y$.

9.3 Least and greatest elements

As we have already noticed, it is necessary to distinguish the maximal elements from the greatest element of a set. If the latter exists, it will turn out to be the only maximal element. The predicate-logic form of a definition of a greatest element is as follows.

Definition 9.5. Let a set V be given that is ordered with the relation "$\leq$", and an arbitrary subset $U \subseteq V$. The element $g \in U$ is called the **greatest element** of U if for all elements $e \in U$ we have $e \leq g$. The element $l \in U$ is called the **least element** of U if for all elements $e \in U$ we have $e \geq l$.   □

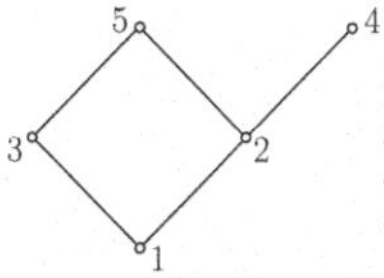

Fig. 9.3 An ordering with three subsets and their greatest element sets

It is important to note that g, l must be elements of the set U in question; otherwise, they cannot be either the greatest or the least element of U. Such elements may exist or may not; so it is wise to talk of the *set of* greatest elements, for example. This set can always be computed and may turn out to be empty or not. In any case, we will get a result when we define it in a point-free relational form as follows.

Definition 9.6. Given an ordering E, we define for a subset U

$$\mathrm{gre}_E(U) := U \cap \mathrm{ubd}_E(U) \quad \text{and} \quad \mathrm{lea}_E(U) := U \cap \mathrm{lbd}_E(U).$$ □

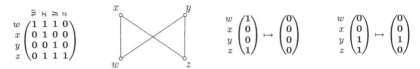

Fig. 9.4 Order, Hasse diagram, subsets, and non-existing or existing greatest elements

Greatest elements may also be formulated with the symmetric quotient.

Proposition 9.7. *Given an order E, greatest elements of a set v may also be determined using the symmetric quotient as*

$$\operatorname{gre}_E(v) = \operatorname{syq}(E, E{:}v) \qquad \operatorname{lea}_E(v) = \operatorname{syq}(E^{\mathsf{T}}, E^{\mathsf{T}}{:}v).$$

Proof: Recall that $\overline{E}^{\mathsf{T}}{:}E = \overline{E}^{\mathsf{T}}$ and $E^{\mathsf{T}}{:}\overline{E{:}X} = \overline{E{:}X}$ for an ordering E and any X.

$$
\begin{aligned}
\operatorname{syq}(E, E{:}v) &= \overline{\overline{E}^{\mathsf{T}}{:}E{:}v} \cap \overline{E^{\mathsf{T}}{:}\overline{E{:}v}} \\
&= \overline{\overline{E}^{\mathsf{T}}{:}v} \cap \overline{\overline{E{:}v}} \\
&= \overline{\overline{E}^{\mathsf{T}}{:}v} \cap E{:}v \\
&= \overline{\overline{E}^{\mathsf{T}}{:}v} \cap \left(E \cap (E^{\mathsf{T}} \cup \overline{E}^{\mathsf{T}})\right){:}v \\
&= \overline{\overline{E}^{\mathsf{T}}{:}v} \cap \left((E \cap E^{\mathsf{T}}) \cup (E \cap \overline{E}^{\mathsf{T}})\right){:}v \\
&= \overline{\overline{E}^{\mathsf{T}}{:}v} \cap \left((E \cap E^{\mathsf{T}}){:}v \cup (E \cap \overline{E}^{\mathsf{T}}){:}v\right) \\
&= \overline{\overline{E}^{\mathsf{T}}{:}v} \cap \left(\mathbb{I}{:}v \cup (E \cap \overline{E}^{\mathsf{T}}){:}v\right) \qquad \text{antisymmetry} \\
&= \overline{\overline{E}^{\mathsf{T}}{:}v} \cap v \qquad \text{last term vanishes by intersection} \\
&= \operatorname{ubd}_E(v) \cap v = \operatorname{gre}_E(v) \qquad\qquad\qquad\qquad\qquad \square
\end{aligned}
$$

9.4 Greatest lower and least upper bounds

Among the set of upper bounds of some set, there may exist a least element in the same way as there is a least element among all multiples of a finite set of natural numbers – known as a least common multiple. Starting from here, traditional functionals can be obtained, namely the least upper bound of u (also *supremum*), i.e., the at most 1-element set of least elements among the set of all upper bounds of u. In contrast to our expectation that a least upper bound may exist or not, it will *always* exist as a vector; it may, however be the null vector, resembling non-existence, or else a 1-element vector.

Definition 9.8. (i) Let an ordered set $V, \leq$ be given and a subset $U \subseteq V$. An element l is called the **least upper bound** of U if it is the least element in the set of all upper bounds of U. An element g is called the **greatest lower bound** of U if it is the greatest element in the set of all lower bounds of U.

(ii) The **least upper** and **greatest lower bounds** may be defined in a point-free relational form as

$$\texttt{lub}_E(U) := \texttt{ubd}_E(U) \cap \texttt{lbd}_E(\texttt{ubd}_E(U)) := \overline{\overline{E}^\top;U} \cap \overline{\overline{E};\overline{\overline{E}^\top;U}},$$

$$\texttt{glb}_E(U) := \texttt{lbd}_E(U) \cap \texttt{ubd}_E(\texttt{lbd}_E(U)) := \overline{\overline{E};U} \cap \overline{\overline{E}^\top;\overline{\overline{E};U}}. \qquad \square$$

As mentioned previously, these functionals are always defined; the results may, however, be null vectors. It is an easy task to prove that $\texttt{lub}, \texttt{glb}$ are always injective, resembling the fact that such bounds are uniquely defined if they exist, see Chapter 3 of [123, 124]. As an example we compute the least upper bound of the relation E itself, employing the well-known facts $\overline{E}^\top;E = \overline{E}^\top$ and $\overline{E};E^\top = \overline{E}$ as well as antisymmetry of E:

$$\texttt{lub}_E(E) = \overline{\overline{E}^\top;E} \cap \overline{\overline{E};\overline{\overline{E}^\top;E}} = E^\top \cap \overline{\overline{E};E^\top} = E^\top \cap E = \mathbb{I}$$

Considering E as a set of column vectors, every column represents the lower cone hanging below the respective element, which is then, of course, the least upper bound thereof.

$$\begin{array}{c|ccccc} & 1 & 2 & 3 & 4 & 5 \\ \hline 1 & 1 & 1 & 1 & 1 & 1 \\ 2 & 0 & 1 & 0 & 1 & 1 \\ 3 & 0 & 0 & 1 & 0 & 1 \\ 4 & 0 & 0 & 0 & 1 & 0 \\ 5 & 0 & 0 & 0 & 0 & 1 \end{array}$$

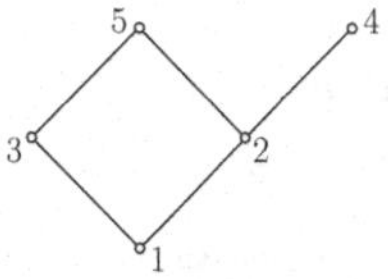

$$\begin{pmatrix}1\\1\\0\\0\\0\end{pmatrix} \mapsto \begin{pmatrix}0\\1\\0\\0\\0\end{pmatrix}, \quad \begin{pmatrix}0\\1\\1\\0\\0\end{pmatrix} \mapsto \begin{pmatrix}0\\0\\0\\0\\1\end{pmatrix}, \quad \begin{pmatrix}0\\1\\1\\1\\0\end{pmatrix} \mapsto \begin{pmatrix}0\\0\\0\\0\\0\end{pmatrix}$$

Fig. 9.5 An ordering with three subsets and their least upper bound sets

Traditionally, a vector is often a column vector. In many cases, however, a row vector would be more convenient. We decided to introduce a variant notation for order-theoretic functionals working on row vectors:

$$\texttt{lubR}_E(X) := [\texttt{lub}_E(X^\top)]^\top, \text{ etc.}$$

For convenience we introduce notation for least and greatest elements as

$$0_E = \texttt{glb}_E(\mathbb{T}) = \texttt{lea}_E(\mathbb{T}), \quad 1_E = \texttt{lub}_E(\mathbb{T}) = \texttt{gre}_E(\mathbb{T}).$$

When using $0_E, 1_E$ it is understood that the underlying vectors are not null vectors.

Proposition 9.9 (Connecting $\texttt{syq}$ and $\texttt{lub}$). *Given an order E, the least upper bound, respectively the greatest lower bound, of a set v may also be determined using the symmetric quotient as*

$$\operatorname{lub}_E(v) = \operatorname{syq}(E^{\mathsf{T}}, \overline{\overline{E}^{\mathsf{T}}{}_;v}) \qquad \operatorname{glb}_E(v) = \operatorname{syq}(E, \overline{\overline{E_;v}}).$$

Proof: The proof is based on Prop. 9.7. By definition,

$$\operatorname{lub}_E(v) = \operatorname{lea}_E(\operatorname{ubd}_E(v)) = \operatorname{syq}(E^{\mathsf{T}}, E^{\mathsf{T}}{}_;\operatorname{ubd}_E(v)) = \operatorname{syq}(E^{\mathsf{T}}, \operatorname{ubd}_E(v)),$$

where the last step uses $E^{\mathsf{T}}{}_;\overline{\overline{E}^{\mathsf{T}}{}_;v} = \overline{\overline{E}^{\mathsf{T}}{}_;v}$; similarly for glb. $\qquad\square$

This result is quite intuitive when we recall that the symmetric quotient acts as column comparison. Then we have, in the first case, formed the upper bound set and compared it with the cones above an element; this element is then, of course, the least in this cone.

In the case when the ordering is a powerset ordering, the symmetric quotient and least upper bound are more intimately related.

Proposition 9.10 (Connecting syq and lub in a powerset). *In a powerset lattice we have for all relations X with the product $\varepsilon_; X$ defined that*

$$\operatorname{lub}_\Omega(X) = \operatorname{syq}(\varepsilon, \varepsilon_; X) \qquad\qquad \operatorname{glb}_\Omega(X) = \operatorname{syq}(\overline{\varepsilon}, \overline{\varepsilon}_; X),$$
$$\varepsilon_;\operatorname{lub}_\Omega(X) = \varepsilon_;\operatorname{syq}(\varepsilon, \varepsilon_; X) = \varepsilon_; X \qquad \overline{\varepsilon}_;\operatorname{glb}_\Omega(X) = \overline{\varepsilon}_;\operatorname{syq}(\overline{\varepsilon}, \overline{\varepsilon}_; X) = \overline{\varepsilon}_; X.$$

Proof:

$$
\begin{aligned}
\operatorname{syq}(\varepsilon, \varepsilon_; X) &= \overline{\overline{\varepsilon}^{\mathsf{T}}{}_;\varepsilon_; X} \cap \overline{\varepsilon^{\mathsf{T}}{}_;\overline{\varepsilon_; X}} && \text{expanding } \operatorname{syq} \\
&= \overline{\overline{\varepsilon}^{\mathsf{T}}{}_;\varepsilon_; X} \cap \overline{\varepsilon^{\mathsf{T}}{}_;\overline{\varepsilon}_;\overline{\varepsilon}^{\mathsf{T}}{}_;\varepsilon_; X} && \text{Prop. 7.14} \\
&= \overline{\overline{\Omega}^{\mathsf{T}}{}_; X} \cap \overline{\Omega_;\overline{\Omega}^{\mathsf{T}}{}_; X} && \text{definition of powerset ordering} \\
&= \operatorname{lub}_\Omega(X) && \text{Def. 9.8} \qquad\square
\end{aligned}
$$

9.5 Lattices

We have seen that least upper or greatest lower bounds are important. The case when they exist for every 2-element subset – or, more generally, for every subset – deserves separate consideration. The standard definition is as follows.

Definition 9.11. An order E is called a **lattice**[1] if for every 2-element subset there exists a greatest lower as well as a least upper bound. The ordering E is called a **complete lattice** when every subset has a least upper bound. $\qquad\square$

---

[1] In German *Verband* and in French *treillis*.

For the lattice, we refer back to Fig. 7.14, where we defined the join and meet based on a given ordering E, and employing the direct product, as

$$\mathcal{J} := \text{lubR}_E(\pi \cup \rho) \quad \text{and} \quad \mathcal{M} := \text{glbR}_E(\pi \cup \rho).$$

The requirement for the ordering E to be a lattice is that $\mathcal{J}, \mathcal{M}$ be mappings. Using the membership ε, we may also define E to be a complete lattice in a point-free relational form, namely as follows.

Definition 9.12. E is called a **complete lattice** if $\text{lub}_E(\varepsilon)$ is surjective. $\qquad\square$

We may also say that E is a complete lattice if

$$f := \left[\text{lub}_E(\varepsilon) \right]^{\mathsf{T}} = \text{syq}(\overline{E^{\mathsf{T}}; \varepsilon}, E^{\mathsf{T}})$$

is always a mapping. Figure 9.6 does not show a lattice since the two elements w, z have two upper bounds but among these not a least one. This may easily be seen in the figure: the broad relation shows all occurring upper bound sets; the right relation shows the transposed ordering, thus visualizing all upper cones above an element. The relation f, which compares columns between the two, obviously cannot be a mapping.

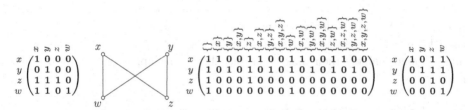

Fig. 9.6 An ordering which is not a lattice: E, Hasse diagram, $\overline{E^{\mathsf{T}}; \varepsilon}$, and E^{T}

Proposition 9.13. *Every powerset ordering* $\Omega = \overline{\varepsilon^{\mathsf{T}}; \overline{\varepsilon}}$ *gives rise to a complete lattice.*

Proof: For the membership relation ε' along the powerset we have

$$
\begin{aligned}
\text{lub}_\Omega(\varepsilon') &= \text{syq}(\Omega^{\mathsf{T}}, \overline{\Omega^{\mathsf{T}}; \varepsilon'}) && \text{Prop. 9.9} \\
&= \text{syq}(\overline{\varepsilon^{\mathsf{T}}; \varepsilon}, \overline{\overline{\varepsilon^{\mathsf{T}}; \varepsilon}; \varepsilon'}) && \text{expanded} \\
&= \text{syq}(\overline{\varepsilon^{\mathsf{T}}; \varepsilon}, \overline{\varepsilon^{\mathsf{T}}; \varepsilon; \varepsilon'}) && \text{negated} \\
&\supseteq \text{syq}(\varepsilon, \varepsilon; \varepsilon') && \text{cancelling a common left factor, Prop. 8.16.i}
\end{aligned}
$$

so as to arrive at a relation which is surjective by definition. $\qquad\square$

Concerning complete lattices, there is some slightly tricky folklore. Every subset –

this quantification includes the empty subset! – is required to have a least upper bound. Every element is an upper bound of the empty set, which can easily be seen from the definition of the upper bound set $\mathbf{ubd}_E(\mathbb{1}) = \overline{\overline{E^\mathsf{T}}\,;\mathbb{1}} = \mathbb{T}$. This upper bound set is in turn required to have a least element, namely the least upper bound of the empty set, meaning that there exists the least element of the whole set. This, however, guarantees that every lower bound set is non-empty, which allows the greatest lower bounds to be formed as the least upper bound of the, thus, non-empty set of lower bounds. Taking all this together, it was not a mistake in Def. 9.12 to demand only that least upper bounds exist for a complete lattice; the greatest lower bounds will then automatically exist.

Staying completely in the relational setting, this might be communicated more easily as

$$\mathbf{glb}_E(\varepsilon) = \overline{\overline{E\,;\varepsilon}\cap\overline{E^\mathsf{T}}\,;\overline{\overline{E\,;\varepsilon}}} \qquad \text{Prop. 9.8}$$
$$= \overline{\overline{E\,;\overline{E^\mathsf{T}\,;\overline{\overline{E\,;\varepsilon}}}}\cap\overline{E^\mathsf{T}}\,;\overline{\overline{E\,;\varepsilon}}} \qquad \text{Prop. 8.8: } \mathbf{lbd}\,(X) = \mathbf{lbd}\,(\mathbf{ubd}\,(\mathbf{lbd}\,(X)))$$
$$= \mathbf{lub}_E(\overline{\overline{E\,;\varepsilon}}) \qquad \text{Prop. 9.8 again}$$

Thus $\mathbf{glb}_E(\varepsilon)$ will be surjective because $\mathbf{lub}_E(\overline{\overline{E\,;\varepsilon}})$ is surjective for every complete lattice according to Def. 9.11.

In the finite case, the two concepts of a *lattice* and of a *complete lattice* will turn out to coincide. When considering the natural numbers $\mathbb{N}$ ordered by "$\leq$", one will see that this is not the case for non-finite ordered sets: every two numbers have a least upper bound, namely their maximum, but there is no upper bound of *all* natural numbers.

The following examples and remarks will convince the reader that lattices indeed occur unexpectedly at many places.

The concept of a (complete) lattice has already been formulated in a point-free fashion. We will now generalize this slightly.

Proposition 9.14. *Let a relation $E : X \longrightarrow X$ on a set X be given together with the membership relation $\varepsilon : X \longrightarrow \mathbf{2}^X$ between X and its powerset. Then E is a complete lattice precisely when $\left[\mathbf{lub}_E(R)\right]^\mathsf{T}$ is a mapping for all relations for which $E\,;R$ may be formed.*

Proof: We recall how we have expressed that all subsets have a least upper bound using ε, because ε 'has all subsets as its columns'. Now we assume any relation R for which $E\,;R$ may be formed. Obviously

$$\mathbf{lub}_E(R) = \mathbf{syq}(E^\mathsf{T}, \overline{\overline{E^\mathsf{T}}\,;R}) \qquad \text{Prop. 9.9}$$

$$= \mathsf{syq}\,(E^\mathsf{T}, \overline{\overline{E}^\mathsf{T} \,_{;}\varepsilon\,_{;} \mathsf{syq}\,(\varepsilon, R)}) \qquad \text{Prop. 7.14}$$

$$= \mathsf{syq}\,(E^\mathsf{T}, \overline{\overline{E}^\mathsf{T} \,_{;}\varepsilon\,_{;} f^\mathsf{T}}) \qquad \text{abbreviating } f := \big[\, \mathsf{syq}\,(\varepsilon, R)\,\big]^\mathsf{T}$$

$$= \mathsf{syq}\,(E^\mathsf{T}, \overline{\overline{E}^\mathsf{T} \,_{;}\varepsilon}\,)_{;} f^\mathsf{T} \qquad \text{mapping } f \text{ slips out of negation, Prop. 8.16.iii}$$

$$= \mathrm{lub}\,_E(\varepsilon)_{;} f^\mathsf{T} \qquad \text{Prop. 9.9}$$

Composed of two transposed mappings, this is itself a transposed mapping. □

We add here a concept of structural comparison for lattices that is helpful and important in many cases.

Definition 9.15. Given orderings E, E' and a mapping f, we call

f **(lattice-)continuous** $:\Longleftrightarrow$ $f^\mathsf{T}{}_{;}\mathrm{lub}\,_E(X) = \mathrm{lub}\,_{E'}(f^\mathsf{T}{}_{;}X)$ for every X. □

In order to prove that continuity of f implies isotony, we estimate

$$f^\mathsf{T} = f^\mathsf{T}{}_{;}\mathbb{I} = f^\mathsf{T}{}_{;}\mathrm{lub}\,_E(E) = \mathrm{lub}\,_{E'}(f^\mathsf{T}{}_{;}E) \subseteq \mathrm{ubd}\,_{E'}(f^\mathsf{T}{}_{;}E) = \overline{\overline{E'}^\mathsf{T}{}_{;}f^\mathsf{T}{}_{;}E}$$

and apply the Schröder equivalence and Prop. 5.13 to obtain $E_{;} f \subseteq f_{;} E'$.

Corollary 9.16. *Every surjective image of a continuous mapping applied to a powerset ordering results in a complete lattice.*

Proof: Let the continuous and surjective mapping be $f : 2^X \longrightarrow Y$ and the orderings $\Omega : 2^X \longrightarrow 2^X$ and $E : Y \longrightarrow Y$. We consider $\varepsilon : Y \longrightarrow 2^Y$.

$$\mathrm{lub}\,_E(\varepsilon) = \mathrm{lub}\,_E(f^\mathsf{T}{}_{;} f_{;} \varepsilon) \qquad \text{since } f \text{ is a surjective mapping}$$

$$= f^\mathsf{T}{}_{;}\mathrm{lub}\,_\Omega(f_{;} \varepsilon) \qquad \text{employing continuity}$$

The result follows because $\mathrm{lub}\,_\Omega(f_{;} \varepsilon)$ is always surjective by definition. □

The idea of continuity is that applying f and forming the least upper bound commute. Illustrating examples of continuity will be found in relational integration in Section 14.3.

Mincut lattice

A lattice often met in applications is the so-called mincut lattice. Consider any directed graph whose arrows are ℝ-labelled with positive *capacities*; think of pipe widths, for example. Then a typical task in operations research asks, given any two different vertices s, t considered as source and sink, what is the maximal amount of flow that can be assigned to the arrows indicating the maximum amount of oil or any other commodity that may be pumped from the source to the sink.

Of course, one does not allow commodities to emerge other than in s and to vanish other than in t. The capacities assigned to the arrows delimit the amount that may be pumped from s to t. In a complicated way, this total limit has its origin in added limits of arcs traversed. Consider the example shown in Fig. 9.7. The source is f and the target is g.

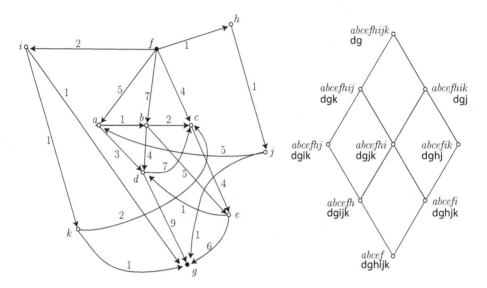

Fig. 9.7 A flow maximization task and mincuts encountered forming a lattice

It is impossible to pump more than 17 units of commodity per time unit from f to g. One can identify 9 pairs of complementary subsets, the first containing f, the second containing g, so that indeed the amount of 17 units can be pumped from the first set to the second:

$\{a, b, c, e, f\}$	$\{d, g, h, i, j, k\}$
$\{a, b, c, e, f, i\}$	$\{d, g, h, j, k\}$
$\{a, b, c, e, f, i, k\}$	$\{d, g, h, j\}$
$\{a, b, c, e, f, h\}$	$\{d, g, i, j, k\}$
$\{a, b, c, e, f, h, j\}$	$\{d, g, i, k\}$
$\{a, b, c, e, f, h, i\}$	$\{d, g, j, k\}$
$\{a, b, c, e, f, h, i, k\}$	$\{d, g, j\}$
$\{a, b, c, e, f, h, i, j\}$	$\{d, g, k\}$
$\{a, b, c, e, f, h, i, j, k\}$	$\{d, g\}$

All these pairs are called **mincuts** around the target g. The smallest contains just d, g. The lattice structure of all these mincuts can be recognized in Fig. 9.8: the nine fat gray borderlines indicate precisely the nine mincuts listed above. Whenever one follows such a gray line from left to right and sums up the capacities of crossing arrows that lead upwards, one will always end up with 17 in total.

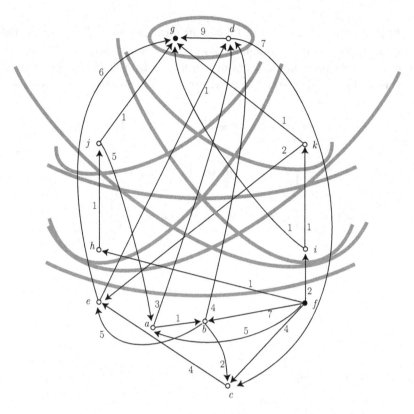

Fig. 9.8 A flow maximization rearranged according to mincut lattice

Antichain lattice

Given an order, we have so far concentrated on subsets and their greatest elements, maximal elements, etc. We are now interested in subsets of elements that form a linear order or, contrarily, in subsets of elements that are completely unrelated to each other. In the first case one speaks of a chain; in the second of an antichain. We provide a formal definition.

Definition 9.17. We assume an order E. Let a non-empty subset v be given with its corresponding natural injection $\iota := \mathtt{Inject}\, v$, so that $v = \iota^\mathsf{T}; \mathbb{T}$, and consider the correspondingly restricted order $E_v := \iota; E; \iota^\mathsf{T}$. We call

(i) v a **chain** $:\Longleftrightarrow$ E_v is a linear order,
(ii) v an **antichain** $:\Longleftrightarrow$ E_v is a trivial order, i.e., $E_v = \mathbb{I}$.   □

A justification of this definition in first-order form can easily be given for the first case, for example. To qualify a non-empty subset v as a chain, consider an element $x \in v$ in comparison to all elements y; if y is also in v, it has to be greater than or equal to or less than or equal to x. We transform this stepwise into a relational form:

$$\forall x : x \in v \rightarrow \left\{ \forall y : y \in v \rightarrow \Big[(x,y) \in E \vee (y,x) \in E \Big] \right\}$$

$$\text{transposition}$$

$$\Longleftrightarrow \quad \forall x : x \in v \rightarrow \left\{ \forall y : y \in v \rightarrow \Big[(x,y) \in (E \cup E^{\mathsf{T}}) \Big] \right\}$$

$$a \rightarrow b = \neg a \vee b$$

$$\Longleftrightarrow \quad \forall x : x \in v \rightarrow \left\{ \forall y : y \notin v \vee \Big[(x,y) \in (E \cup E^{\mathsf{T}}) \Big] \right\}$$

$$\forall z : p(z) = \neg \big\{ \exists z : \neg p(z) \big\}$$

$$\Longleftrightarrow \quad \forall x : x \in v \rightarrow \neg \left\{ \exists y : y \in v \wedge \Big[(x,y) \in \overline{E \cup E^{\mathsf{T}}} \Big] \right\}$$

$$a \wedge b = b \wedge a, \text{ definition of composition}$$

$$\Longleftrightarrow \quad \forall x : x \in v \rightarrow \neg \Big[x \in \overline{E \cup E^{\mathsf{T}}} \, \mathbin{;} v \Big]$$

$$a \rightarrow \neg b = b \rightarrow \neg a$$

$$\Longleftrightarrow \quad \forall x : \Big[x \in \overline{E \cup E^{\mathsf{T}}} \, \mathbin{;} v \Big] \rightarrow x \notin v$$

$$\text{transfer to point-free notation}$$

$$\Longleftrightarrow \quad \overline{E \cup E^{\mathsf{T}}} \, \mathbin{;} v \subseteq \overline{v}$$

$$\text{Schröder rule}$$

$$\Longleftrightarrow \quad v \mathbin{;} v^{\mathsf{T}} \subseteq E \cup E^{\mathsf{T}}$$

$$\text{extruding } v$$

$$\Longleftrightarrow \quad \iota^{\mathsf{T}} \mathbin{;} \mathbb{T} \mathbin{;} (\iota^{\mathsf{T}} \mathbin{;} \mathbb{T})^{\mathsf{T}} \subseteq E \cup E^{\mathsf{T}}$$

$$\text{transposing}$$

$$\Longleftrightarrow \quad \iota^{\mathsf{T}} \mathbin{;} \mathbb{T} \mathbin{;} \iota \subseteq E \cup E^{\mathsf{T}}$$

$$\iota \text{ is a mapping}$$

$$\Longleftrightarrow \quad \mathbb{T} \subseteq \iota \mathbin{;} (E \cup E^{\mathsf{T}}) \mathbin{;} \iota^{\mathsf{T}} = E_v \cup E_v^{\mathsf{T}}$$

To derive a simple relational formula describing all chains of an ordering, we start from $\overline{E \cup E^{\mathsf{T}}} \, \mathbin{;} v \subseteq \overline{v}$, so that applying this condition to all subsets simultaneously, we obtain the vector

$$chains(E) := \Big[\overline{\overline{E \cup E^{\mathsf{T}}} \, \mathbin{;} \varepsilon \cap \varepsilon} \Big]^{\mathsf{T}} \mathbin{;} \mathbb{T}$$

among which in Section 15.3 one will wish to concentrate on the inclusion-maximal chains. Those obtained as cardinality-maximal chains disregard too much of the ordering as one may see in Fig. 9.12; shorter ones are also needed to cover all vertices according to the Dilworth Chain Decomposition Theorem, Prop. 9.19.

In a similar way, we proceed for antichains. The set of all antichains of an ordering E (with corresponding strictorder C) can also be described in a simple formula using the membership relation ε. The condition that a subset v is an antichain reads $C\, {:}\, v \subseteq \bar{v}$. This expresses that when one proceeds to another point along C and arrives in v one cannot have started in v. Using the membership relation ε, we apply this condition to all subsets simultaneously with $C\, {:}\, \varepsilon \subseteq \bar{\varepsilon}$ and characterize the set of antichains with the vector

$$antichains(E) := \overline{\left[C\, {:}\, \varepsilon \cap \varepsilon\right]^{\mathsf{T}}\, {:}\, \mathbb{T}}$$

from which afterwards cardinality-maximum antichains can be extracted using the cardinality preorder $O_{\|}$ on the powerset. This time, those obtained as inclusion-maximal antichains are uninteresting as one may see in Fig. 9.12.

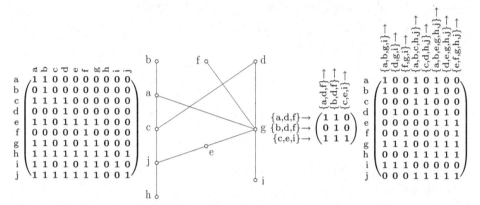

Fig. 9.9 An ordering with its antichain lattice and all inclusion-maximal chains

Figure 9.9 shows an example where we find out that the cardinality-maximum antichains obtained form a lattice.

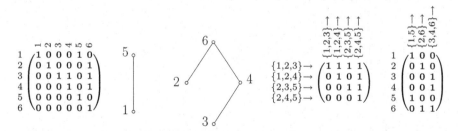

Fig. 9.10 An ordering with its antichain lattice and all inclusion-maximal chains

Also in Fig. 9.10, we observe a lattice structure on the set of cardinality-maximum antichains, which we now trace back to the element level. The set of all cardinality-maximum antichains is an interesting example of a lattice.

Let $A := (v_i)_{i \in J}$ be the (finite) set of cardinality-maximum antichains for E. The cardinalities of these sets are, of course, equal. For such a v it is a characteristic property that

$$v = v \cap \overline{C_i\, v} = (v \cup C_i\, v) \cap \overline{C_i\, v} = E_i\, v \cap \overline{C_i\, E_i\, v} = \mathtt{max}\,(E_i\, v).$$

Proposition 9.18. *Given any finite order* $E = \mathbb{I} \cup C$, *the set of cardinality-maximum antichains forms a lattice with order, infima, and suprema as*

$$v_1 \preceq v_2 \quad :\Longleftrightarrow \quad v_1 \subseteq E_i\, v_2$$

$$v_1 \smile v_2 := \mathtt{max}\,(E_i\, v_1 \cup E_i\, v_2),$$

$$v_1 \frown v_2 := \mathtt{max}\,(E_i\, v_1 \cap E_i\, v_2).$$

Proof: "$\preceq$" is obviously reflexive and transitive. Antisymmetry follows from

$$v_1 \subseteq E_i\, v_2,\ v_2 \subseteq E_i\, v_1 \quad \Longrightarrow \quad E_i\, v_1 = E_i\, v_2 \quad \Longrightarrow$$
$$v_1 = \mathtt{max}\,(E_i\, v_1) = \mathtt{max}\,(E_i\, v_2) = v_2.$$

To prove that every supremum is again an antichain, we first evaluate

$$C_i\,(E_i\, v_1 \cup E_i\, v_2) = C_i\, E_i\, v_1 \cup C_i\, E_i\, v_2 = C_i\, v_1 \cup C_i\, v_2$$
$$\mathtt{max}\,(E_i\, v_1 \cup E_i\, v_2) = (E_i\, v_1 \cup E_i\, v_2) \cap \overline{C_i\,(E_i\, v_1 \cup E_i\, v_2)}$$
$$= \big[v_1 \cup C_i\, v_1 \cup v_2 \cup C_i\, v_2\big] \cap \overline{C_i\, v_1} \cap \overline{C_i\, v_2} = (v_1 \cup v_2) \cap \overline{C_i\, v_1} \cap \overline{C_i\, v_2}.$$

For the latter expression, the condition $C_i\, v \subseteq \bar{v}$ characterizing an antichain is obviously satisfied.

It requires some additional effort to prove that the supremum so defined is indeed a cardinality-maximum antichain.   □

The tiny example of Fig. 9.11 visualizes this result.

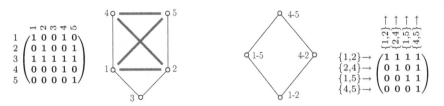

Fig. 9.11 An ordering with the lattice of its cardinality-maximum antichains

The closely related Kuratowski Theorem states that every chain in an ordered set is included in a maximal chain. This is not really interesting in the finite case. The general proof requires the Axiom of Choice, or some similarly powerful argument, but will not be elaborated on here.

Proposition 9.19 (Dilworth Chain Decomposition Theorem). *If v is a cardinality-maximum antichain in an order E, it is possible to find a chain through each of its points such that their union is sufficient to cover all vertices.* □

We do not prove this here; rather we consider Fig. 9.9 and observe that one may indeed start from such antichains and find unions of chains that cover all vertices:

$$\{a,d,f\} \to \qquad \mapsto \qquad \{i,g,\underline{a},b,\} \to, \quad \{h,j,c,\underline{d}\} \to, \quad \{h,j,e,g,\underline{f}\} \to$$
$$\{b,d,f\} \to \qquad \mapsto \qquad \{i,g,a,\underline{b}\} \to, \quad \{h,j,c,\underline{d}\} \to, \quad \{h,j,e,g,\underline{f}\} \to$$
$$\{c,e,i\} \to \qquad \mapsto \qquad \{h,j,\underline{c},a,b\} \to, \quad \{h,j,\underline{e},g,f\} \to, \quad \{\underline{i},g,d\} \to$$

There exist many more examples of lattices occurring in practice, and we had originally intended to show more of them. In the meantime, however, the brilliant new book [11] by Rudolf Berghammer has appeared that treats lattices thoroughly and also with relational means, so we can restrict our presentation here.

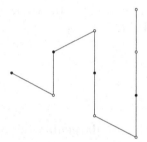

Fig. 9.12 Cardinality-maximum antichains versus inclusion-maximal chains

Exercises

9.2 Determine the antichain lattice as well as all inclusion-maximal chains for the following two orderings:

	1	2	3	4	5	6	7
1	1	0	1	1	1	0	1
2	0	1	0	0	0	0	0
3	0	0	1	0	0	0	0
4	0	0	0	1	0	0	0
5	0	0	1	0	1	0	0
6	0	1	0	1	0	1	0
7	0	0	0	1	0	0	1

	a	b	c	d	e	f	g	h	i
a	1	0	1	0	0	1	0	0	1
b	0	1	0	0	0	0	0	1	0
c	0	0	1	0	0	0	0	0	0
d	0	1	1	1	1	0	0	1	0
e	0	0	1	0	1	0	1	0	0
f	0	0	0	0	0	1	0	0	0
g	0	1	0	0	0	0	1	1	0
h	0	0	0	0	0	0	0	1	0
i	0	0	1	0	0	1	0	0	1

10

Rectangles, Fringes, Inverses

Although not many scientists seem to be aware of this fact, a significant amount of our reasoning is concerned with 'rectangles' in/of a relation. Rectangles are handled at various places from a theoretical point of view as well as from a practical viewpoint. Among the application areas are equivalences, preorders, concept lattices, clustering methods, and measuring, to mention just a few seemingly unrelated topics. In most cases, rectangles are treated in the respective application environment, i.e., together with certain additional properties. So it is not clear which results stem from their status as rectangles as such and which employ these additional properties. Here we try to formulate the properties of rectangles before going into the various applications and hope, thus, to present several concepts only once, and to reduce the overall amount of work.

10.1 Non-enlargeable rectangles

In Def. 6.6, we have already introduced rectangles inside a relation (sometimes called blocks) on a rather phenomenological basis. We are going to investigate them here in more detail and start by recalling that u, v form a rectangle inside R if either one, and thus all, of the following equivalent containments are satisfied:

$$u_{;}v^{\mathsf{T}} \subseteq R \quad \Longleftrightarrow \quad \overline{R}_{;}v \subseteq \overline{u} \quad \Longleftrightarrow \quad \overline{R}^{\mathsf{T}}_{;}u \subseteq \overline{v}.$$

The main idea is to look for the, in some sense, maximal rectangles inside a relation.

Definition 10.1. Let a relation R be given together with subsets u, v that form a rectangle inside R. The rectangle u, v is said to be **non-enlargeable** inside R if there does not exist a *different* rectangle u', v' inside R such that $u \subseteq u'$ and $v \subseteq v'$. For brevity, non-enlargeable rectangles will also be called **dicliques**.[1] □

---

[1] In [65], this denotation was originally used in the case that R was a homogeneous relation – preferably with $u \neq \bot\!\bot$ as well as $v \neq \bot\!\bot$ to exclude trivialities.

Non-enlargeable rectangles are maximal, but they need not be the greatest rectangles. The property of constituting a non-enlargeable rectangle has an elegant algebraic characterization.

Proposition 10.2. (i) *Let u, v define a rectangle inside the relation R. Precisely when both*

$$\overline{R \,;\, v} \supseteq \overline{u} \quad and \quad \overline{R^{\mathsf{T}} \,;\, u} \supseteq \overline{v}$$

are satisfied in addition, there will exist no strictly greater rectangle u', v' inside R.
(ii) *Vectors u, v constitute a non-enlargeable rectangle inside R if and only if*

$$\overline{R \,;\, v} = \overline{u} \quad and \quad \overline{R^{\mathsf{T}} \,;\, u} = \overline{v}.$$

Proof: (i) Let us assume a rectangle u, v inside R, i.e., satisfying $\overline{R \,;\, v} \subseteq \overline{u}$ and, equivalently, $\overline{R^{\mathsf{T}} \,;\, u} \subseteq \overline{v}$, that does not satisfy, for example, the first inclusion mentioned in the proposition so that $\overline{u} \not\supseteq \overline{R \,;\, v}$. Then $u' := \overline{R \,;\, v} \not\supseteq u$ and $v' := v$ obviously constitute a strictly greater rectangle:

$$\overline{R \,;\, v'} = \overline{R \,;\, v} = \overline{u'} \quad and \quad \overline{R^{\mathsf{T}} \,;\, u'} = \overline{R^{\mathsf{T}} \,;\, \overline{R \,;\, v}} \subseteq \overline{v} = \overline{v'}.$$

Consider for the opposite direction a rectangle u, v inside R satisfying the two inclusions together with another rectangle u', v' inside R such that $u \subseteq u'$ and $v \subseteq v'$. Then we may conclude with monotony and an application of the Schröder rule that $\overline{v'} \supseteq \overline{R^{\mathsf{T}} \,;\, u'} \supseteq \overline{R^{\mathsf{T}} \,;\, u} \supseteq \overline{v}$. This results in $v' = v$. In a similar way it is shown that $u = u'$. To sum up, u', v' cannot be strictly greater than u, v.

(ii) means amalgamating (i) with the condition for a rectangle inside R. □

Note that both of the equations $\overline{R \,;\, v} = \overline{u}$ and $\overline{R^{\mathsf{T}} \,;\, u} = \overline{v}$ are used in this proof; via Schröder's rule they are *not* equivalent with one another as their "$\subseteq$"-versions are. We may express (ii) using residuals and obtain

$$u = R/v^{\mathsf{T}} \quad and \quad v^{\mathsf{T}} = u \backslash R.$$

Translation to predicate logic may help in intuitively understanding this result. The condition $\overline{u} \subseteq \overline{R \,;\, v}$ then reads

$$\forall x : \;\; \bigl(x \notin u \;\;\rightarrow\;\; \exists y : (x, y) \notin R \wedge y \in v\bigr),$$

so that any attempt to enlarge u by adding x, which is not yet contained in u, will fail as such a y is guaranteed to exist.

Consider a pair (x, y) of elements related by some relation R. Due to Prop. 5.12, this may be expressed a little more algebraically as $x \,;\, y^{\mathsf{T}} \subseteq R$ or, equivalently, as $x \subseteq R \,;\, y$. It is immediately clear that y may or may not be the only point related with x. With $R^{\mathsf{T}} \,;\, x$ we have the set of all elements of the target side related with x. Because we have been starting with $(x, y) \in R$, it is non-empty, i.e., $\mathbb{\perp} \neq y \subseteq R^{\mathsf{T}} \,;\, x$.

For reasons we shall accept shortly, it is advisable to use the identity $R^{\mathsf{T}} {\,}_{!} x = \overline{\overline{R}^{\mathsf{T}}} {\,}_{!} x$ which holds due to Prop. 5.13 if x is a point. We then see that a whole rectangle – which may be 1-element only – is contained in R, namely the one given by rows

$$u_x := \overline{\overline{R}{\,}_{!} \overline{R}^{\mathsf{T}}} {\,}_{!} x = \overline{\overline{R}{\,}_{!} R^{\mathsf{T}}} {\,}_{!} x \qquad \text{on the source side together with columns}$$
$$v_x := \overline{\overline{R}^{\mathsf{T}}} {\,}_{!} x = R^{\mathsf{T}} {\,}_{!} x \qquad \text{on the target side.}$$

The right variants are obtained since x, y had been assumed to be points. One application of the Schröder equivalence shows that indeed $u_x {\,}_{!} v_x^{\mathsf{T}} \subseteq R$. Some preference has been given here to x, so that we expect something similar to hold when starting from y. This is indeed the case; for the rectangle defined by

$$u_y := \overline{\overline{R}{\,}_{!} y} = R {\,}_{!} y \qquad \text{on the source side}$$
$$v_y := \overline{\overline{R}^{\mathsf{T}} {\,}_{!} \overline{R}{\,}_{!} y} = \overline{R}^{\mathsf{T}} {\,}_{!} R {\,}_{!} y \qquad \text{on the target side}$$

we have analogously $u_y {\,}_{!} v_y^{\mathsf{T}} \subseteq R$. Figure 10.1 indicates how these rectangles turn out to be non-enlargeable.

For easier reference, we collect these ideas in a proposition.

Proposition 10.3. *A point $x {\,}_{!} y^{\mathsf{T}} \subseteq R$ in a (possibly heterogeneous) relation R gives rise to*

(i) *a non-enlargeable rectangle inside R* **started horizontally**
$$u_x := \overline{\overline{R}{\,}_{!} \overline{R}^{\mathsf{T}}} {\,}_{!} x = \overline{\overline{R}{\,}_{!} R^{\mathsf{T}}} {\,}_{!} x \supseteq x, \qquad v_x := \overline{\overline{R}^{\mathsf{T}}} {\,}_{!} x = R^{\mathsf{T}} {\,}_{!} x \supseteq y,$$
(ii) *a non-enlargeable rectangle inside R* **started vertically**
$$u_y := \overline{\overline{R}{\,}_{!} y} = R {\,}_{!} y \supseteq x, \qquad v_y := \overline{\overline{R}^{\mathsf{T}} {\,}_{!} \overline{R}{\,}_{!} y} = \overline{R}^{\mathsf{T}} {\,}_{!} R {\,}_{!} y \supseteq y.$$

Proof: In both cases, we have non-enlargeable rectangles inside R according to Prop. 10.2.ii, since indeed

$$\overline{R}{\,}_{!} v_x = \overline{u_x} \quad \text{and} \quad \overline{R}^{\mathsf{T}} {\,}_{!} u_x = \overline{v_x}$$

as well as

$$\overline{R}{\,}_{!} v_y = \overline{u_y} \quad \text{and} \quad \overline{R}^{\mathsf{T}} {\,}_{!} u_x = \overline{v_y}.$$

This may easily be seen directly, respectively via the very general identities such as $\overline{R{\,}_{!} \overline{R}^{\mathsf{T}} {\,}_{!} \overline{R}{\,}_{!} X} = \overline{R}{\,}_{!} X$. $\qquad\qquad\qquad \square$

The rectangles started horizontally and started vertically may coincide – a case to be handled soon. Regarding Def. 9.3, one will find out that, although R has not been defined as an ordering, the construct is similar to those defining upper bound sets and lower bound sets of upper bound sets.

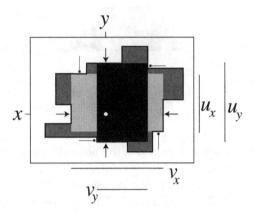

Fig. 10.1 Non-enlargeable rectangles

In Fig. 10.1, let the relation R in question be the 'non-white' area, inside which we consider an arbitrary pair (x, y) of elements related by R. To illustrate the pair (u_x, v_x), let the point (x, y) first slide inside R horizontally over the maximum distance v_x, limited as indicated by $\rightarrow \leftarrow$. Then move the full subset v_x as far as possible inside R vertically, obtaining u_x, and thus, the light-shaded rectangle. Symbols like ⬛ indicate where the light rectangle cannot be enlarged in the vertical direction.

In much the same way, let the point on column y slide as far as possible inside R, obtaining u_y, limited by $\downarrow$ and $\uparrow$. This vertical interval is then moved horizontally inside R as far as possible resulting in v_y and in the black rectangle, confined by the symbol ▪—.

Observe that the non-enlargeable rectangles need not be coherent in the general case; nor need there be just two. Figure 10.2, where the relation considered is assumed to be precisely the union of all rectangles, shows a point contained in five non-enlargeable rectangles. What will also become clear is that with those rectangles obtained by looking for the maximum horizontal or vertical extensions first, one finds extreme cases.

Many important concepts concerning relations depend heavily on such non-enlargeable rectangles. Not least, a decomposition into a set of non-enlargeable rectangles, or dicliques, offers an efficient way of storing information on a system; see, for example, [65].

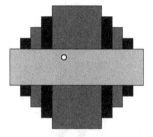

Fig. 10.2 A point contained in five non-enlargeable rectangles

Figure 10.3 will be used to illustrate our exposition several times. It shows all non-enlargeable rectangles of the relation on the left. The vertical vectors as well as the horizontal vectors of the non-enlargeable rectangles are extruded as one may see from the row and column annotations. The bijection shows how they belong to one another.

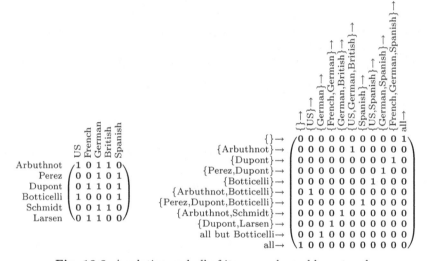

Left relation (columns: US, French, German, British, Spanish):

	US	French	German	British	Spanish
Arbuthnot	1	0	1	1	0
Perez	0	0	1	0	1
Dupont	0	1	1	0	1
Botticelli	1	0	0	0	1
Schmidt	0	0	1	1	0
Larsen	0	1	1	0	0

Right matrix (columns: {}→, {US}→, {German}→, {French,German}→, {German,British}→, {US,German,British}→, {Spanish}→, {US,Spanish}→, {German,Spanish}→, {French,German,Spanish}→, all→):

	{}	{US}	{German}	{French,German}	{German,British}	{US,German,British}	{Spanish}	{US,Spanish}	{German,Spanish}	{French,German,Spanish}	all
{}→	0	0	0	0	0	0	0	0	0	0	1
{Arbuthnot}→	0	0	0	0	0	1	0	0	0	0	0
{Dupont}→	0	0	0	0	0	0	0	0	0	1	0
{Perez,Dupont}→	0	0	0	0	0	0	0	0	1	0	0
{Botticelli}→	0	0	0	0	0	0	0	1	0	0	0
{Arbuthnot,Botticelli}→	0	1	0	0	0	0	0	0	0	0	0
{Perez,Dupont,Botticelli}→	0	0	0	0	0	0	1	0	0	0	0
{Arbuthnot,Schmidt}→	0	0	0	0	1	0	0	0	0	0	0
{Dupont,Larsen}→	0	0	0	1	0	0	0	0	0	0	0
all but Botticelli→	0	0	1	0	0	0	0	0	0	0	0
all→	1	0	0	0	0	0	0	0	0	0	0

Fig. 10.3 A relation and all of its non-enlargeable rectangles

We note in passing that the relation on the right is not a homogeneous one, but has a square matrix.

The symmetric case: maxcliques

We now investigate what happens in the case when R is symmetric. In the first place, dicliques u, v may be discussed as before. It is more interesting to concentrate on cliques as defined in Def. 6.9 by one, and thus all, of the conditions

$$u_\cdot u^\mathsf{T} \subseteq R \quad \Longleftrightarrow \quad \overline{R}_\cdot u \subseteq \overline{u} \quad \Longleftrightarrow \quad \overline{R}^{\mathsf{T}}_\cdot u \subseteq \overline{u},$$

i.e., imposing the requirement $u = v$.

Definition 10.4. Given a symmetric relation R and a clique u in R, we call it a **maxclique**[2] provided that for every clique $v \supseteq u$ necessarily $u = v$. □

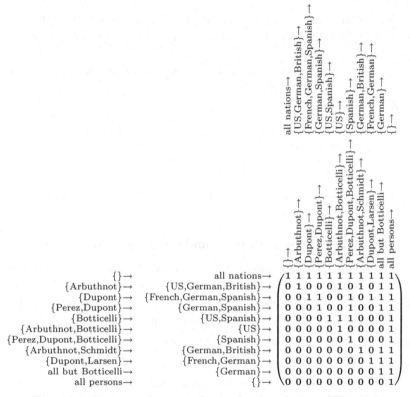

Fig. 10.4 Rearranged concept lattice for the relation of Fig. 10.3

Examples of non-enlargeable cliques of symmetric and reflexive relations, i.e., max-cliques, may be found in Fig. 6.7. Also at this point, we will find a nice algebraic characterization. Cliques have traditionally been investigated in graph theory and not so often in relation algebra. Graph theorists usually assume an irreflexive adjacency Γ. For us, it would be inconvenient always to cut out the diagonal, so we work with the reflexive $R := \mathbb{I} \cup \Gamma$.

---

[2] Observe that a diclique is a maximal block by definition. Because a clique in graph theory need not be maximal, we have coined the word 'maxclique'.

Proposition 10.5. *Let any symmetric and reflexive relation R be given and consider a clique u inside R. If we assume the Point Axiom to hold, then*

$$u \text{ maxclique} \quad \Longleftrightarrow \quad \overline{R} ; u = \overline{u}.$$

Proof: "$\Longrightarrow$": If u is a maxclique, then it is a clique, i.e., $\overline{R} ; u \subseteq \overline{u}$. Assume the equation is not satisfied, i.e., $\overline{R} ; u \subsetneq \overline{u}$. Using the Point Axiom, there exists a point $x \subseteq \overline{\overline{R} ; u} \cap \overline{u}$. With x, we find the strictly greater $v := u \cup x$, which turns out to be a clique because

$$\overline{R} ; v = \overline{R} ; (u \cup x) = \overline{R} ; u \cup \overline{R} ; x \subseteq \overline{u} \cap \overline{x} = \overline{v}.$$

The necessary four containments may be proved using symmetry, reflexivity, and the Schröder rule. Altogether this contradicts the maximality of u.

"$\Longleftarrow$": A u satisfying $\overline{R} ; u = \overline{u}$ is a clique. In addition, assume a clique v with $u \subseteq v$. Then

$$u = \overline{\overline{R} ; u} \supseteq \overline{\overline{R} ; v} \supseteq v,$$

because of the assumption, applying $u \subseteq v$, and using that v is a clique. Thus, v cannot be strictly greater. □

Beginning with Def. 11.7, we will investigate maxcliques in considerable depth.

10.2 Independent pairs and covering pairs of sets

We resume the study of non-enlargeable rectangles from a slightly different point of view. Recall from Def. 6.10, that a relation A is given and pairs of subsets are considered with the first taken from the source side and the second from the target side. This may be met in two complementary forms as an

independent pair u, v of sets $\quad \Longleftrightarrow \quad A ; v \subseteq \overline{u} \quad \Longleftrightarrow \quad A^{\mathsf{T}} ; u \subseteq \overline{v} \quad$ or as a

covering pair s, t of sets $\quad \Longleftrightarrow \quad A ; \overline{t} \subseteq s \quad \Longleftrightarrow \quad A^{\mathsf{T}} ; \overline{s} \subseteq t.$

Both will here have a more detailed treatment asking for the extremal cases. An independent pair of sets is a rectangle outside A. When concentrating on complements of the sets, we get a covering pair of sets $u := \overline{s}$ and $v := \overline{t}$.

On the right side of Fig. 6.8 or of Fig. 10.5, (s, t) is indeed a covering pair of sets, because the columns of t together with the rows of s cover all the 1s of the relation A. The covering property $A ; \overline{t} \subseteq s$ follows directly from the algebraic condition: when one follows relation A and finds oneself ending outside t, then the starting point is covered by s. Algebraic transformation shows that $A \subseteq s ; \mathbb{T} \cup \mathbb{T} ; t^{\mathsf{T}}$ is an equivalent form, expressing that rows according to s and columns according to t cover all of A.

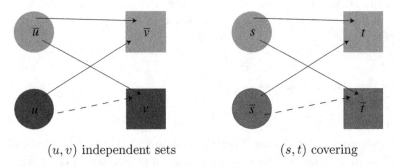

(u, v) independent sets (s, t) covering

Fig. 10.5 Visualizing independent sets and covering sets

Both concepts allow for enlarging the pair, or reducing it, so as to arrive at an equation.

Proposition 10.6. (i) *Consider a pair s, t of sets together with the relation A. Precisely when both*

$$A{:}\overline{t} = s \qquad and \qquad A^{\mathsf{T}}{:}\overline{s} = t$$

are satisfied, s, t will constitute a minimal covering pair of sets of A; i.e., there will exist no covering pair x, y of sets of A satisfying $x \subseteq s$, $y \subseteq t$ and $(x, y) \neq (s, t)$.

(ii) *Consider a pair s, t of sets together with the relation A. Precisely when both*

$$A{:}t = \overline{s} \qquad and \qquad A^{\mathsf{T}}{:}s = \overline{t}$$

are satisfied, s, t will be a maximal independent pair of sets of the relation A; i.e., there will exist no independent pair x, y of sets satisfying $x \supseteq s$, $y \supseteq t$ and $(x, y) \neq (s, t)$.

Proof: This is nothing more than a re-formulation of Prop. 10.2 in the present context. □

A relation may be decomposed along the independent or covering pairs of sets as already indicated in Fig. 6.8.

The diversity of independent pairs of sets shown in Fig. 10.6 suggests looking for the following line-covering possibility. For the moment, call rows and columns, respectively, lines. Then together with the $|x| \times |y|$ zone of $\mathbf{0}$s, we are able to cover all entries $\mathbf{1}$ by $|\overline{x}|$ horizontal plus $|\overline{y}|$ vertical lines. It is standard to try to minimize the number of lines required to cover all $\mathbf{1}$s of the relation.

Definition 10.7. Given a relation A, the **term rank** is defined as the minimum number of lines necessary to cover all entries $\mathbb{1}$ in A, i.e.,

$$\min\{|s| + |t| \mid A\overline{\,;\,t} \subseteq s\}. \qquad \square$$

Consider this schematically as

$$\begin{pmatrix} A_{11} & \mathbb{\bot\bot} \\ A_{21} & A_{22} \end{pmatrix} ; \begin{pmatrix} \mathbb{\bot\bot} \\ \mathbb{T} \end{pmatrix} = \begin{pmatrix} \overline{\mathbb{T}} \\ \overline{\mathbb{\bot\bot}} \end{pmatrix}.$$

Hoping to arrive at fewer lines than the columns of A_{11} and the rows of A_{22} to cover all of A, one might start a first naïve attempt and try to cover with s and t but with row i, for example, omitted. If (s,t) is already minimal, there will be an entry in row i of A_{22} containing a $\mathbb{1}$. Therefore, A_{22} is a total relation. In the same way, A_{11} turns out to be surjective.

But we may also try to get rid of a (not just singleton-)set $x \subseteq s$ of rows and accept that a set of columns be added instead. It follows from minimality that regardless of how we choose $x \subseteq s$, there will be at least as many columns necessary to cover what has been left out. This leads to the following famous definition.

Definition 10.8. Let a relation $Q : X \longrightarrow Y$ with a point set $x \subseteq X$ on the source side be given.

x satisfies the **Hall condition**

$:\Longleftrightarrow \quad |z| \leq |Q^{\mathsf{T}}; z|$ for every subset $z \subseteq x$.

$\Longleftrightarrow \quad$ for every subset $z \subseteq x$ there exists a matching ρ

$\qquad$ with $\rho ; \mathbb{T} = z$, $\rho^{\mathsf{T}} ; \mathbb{T} \subseteq Q^{\mathsf{T}} ; z$.

x can be **saturated**

$:\Longleftrightarrow \quad$ there exists a matching $\lambda \subseteq Q$ with $\lambda ; \mathbb{T} = x$. $\qquad \square$

Since we have nearly completely refrained from using natural numbers in all this text, it looks funny to see an argument 'strictly more' which is seemingly based on natural numbers. However, this is not really the case, since it can be formulated via existence, or non-existence, of a matching relation.

Summarizing, if we have a line-covering with $|s| + |t|$ minimal, then A_{11}^{T} as well as A_{22} will satisfy the Hall condition. We will later learn how to find minimum line-coverings and maximum independent sets without just checking them all exhaustively. Then a better visualization will also be possible; see Fig. 16.13. Additional structure will be extracted employing assignment mechanisms.

In Fig. 10.6, we observe that apart from the upper-right rectangle of $\mathbf{0}$s, in every case the lower left of these rectangles touches a diagonal. This diagonal in turn indicates how a matching will always exist. For now, it is just an observation that we can arrange rows and columns so as to present these diagonals; we will need additional theoretical concepts in order to obtain results around Fig. 16.13.

(Figure 10.6: a grid of 20 binary matrices, arranged in 4 columns × 5 rows. The first matrix is the base relation; the remaining are its cardinality‑minimum coverings, each shown with a partition bar. Best‑effort transcription below.)

Matrix 1 (base relation) — columns 1 2 3 4 5 6 7

```
    1 2 3 4 5 6 7
a   0 0 0 0 1 0 0
b   0 0 0 0 0 0 0
c   0 0 0 1 0 0 0
d   0 0 0 1 0 0 0
e   0 1 0 0 0 0 0
f   0 0 0 1 0 0 0
g   0 1 1 0 1 1 0
h   0 0 0 0 0 0 1
i   1 0 0 0 0 0 0
j   0 1 0 0 0 0 0
k   0 0 0 0 0 1 0
```

Column-1 coverings (rows 2–5)

```
    4 7 1 6|5 3 2
b   0 0 0 0|0 0 0
d   1 0 0 0|0 0 0
e   0 0 1 0|0 0 0
f   1 0 0 0|0 0 0
c   1 0 0 0|0 0 0
h   0 1 0 0|0 0 0
j   0 0 1 0|0 0 0
k   0 0 0 1|0 0 0
a   0 0 0 0|1 0 0
g   0 0 1 1|1 1 0
i   0 0 0 0|0 0 1
```
```
    5 4 7 2 6|3 1
b   0 0 0 0 0|0 0
d   0 1 0 0 0|0 0
e   0 0 0 1 0|0 0
f   0 1 0 0 0|0 0
a   1 0 0 0 0|0 0
c   0 1 0 0 0|0 0
h   0 0 1 0 0|0 0
j   0 0 0 1 0|0 0
k   0 0 0 0 1|0 0
g   1 0 0 1 1|1 0
i   0 0 0 0 0|0 1
```
```
    4 7 1 2|5 3 6
b   0 0 0 0|0 0 0
d   1 0 0 0|0 0 0
e   0 0 0 1|0 0 0
f   1 0 0 0|0 0 0
c   1 0 0 0|0 0 0
h   0 1 0 0|0 0 0
i   0 0 1 0|0 0 0
j   0 0 0 1|0 0 0
a   0 0 0 0|1 0 0
g   0 0 1 1|1 1 1
k   0 0 0 0|0 0 1
```
```
    5 4 7 1 2|3 6
b   0 0 0 0 0|0 0
d   0 1 0 0 0|0 0
e   0 0 0 1 0|0 0
f   0 1 0 0 0|0 0
a   1 0 0 0 0|0 0
c   0 1 0 0 0|0 0
h   0 0 1 0 0|0 0
i   0 0 0 1 0|0 0
j   0 0 0 0 1|0 0
g   1 0 0 1 1|1 1
k   0 0 0 0 0|0 1
```

Column-2 coverings (rows 1–5)

```
    4 2|5 3 7 1 6
b   0 0|0 0 0 0 0
c   1 0|0 0 0 0 0
d   1 0|0 0 0 0 0
e   0 1|0 0 0 0 0
f   1 0|0 0 0 0 0
j   0 1|0 0 0 0 0
a   0 0|1 0 0 0 0
g   0 1|1 1 0 0 1
h   0 0|0 0 1 0 0
i   0 0|0 0 0 1 0
k   0 0|0 0 0 0 1
```
```
    5 4 2|3 7 1 6
b   0 0 0|0 0 0 0
c   0 0 1|0 0 0 0
d   0 0 1|0 0 0 0
j   0 1 0|0 0 0 0
a   1 0 0|0 0 0 0
e   0 1 0|0 0 0 0
f   0 0 1|0 0 0 0
g   1 1 0|1 0 0 1
h   0 0 0|0 1 0 0
i   0 0 0|0 0 1 0
k   0 0 0|0 0 0 1
```
```
    5 4 3 2 6|7 1
b   0 0 0 0 0|0 0
d   0 1 0 0 0|0 0
e   0 0 0 1 0|0 0
f   0 1 0 0 0|0 0
a   1 0 0 0 0|0 0
c   0 1 0 0 0|0 0
h   0 0 1 0 0|0 0
j   0 0 0 1 0|0 0
k   0 0 0 0 1|0 0
g   1 0 1 1 1|0 0
i   0 0 0 0 0|0 1
```
```
    4 1 2 6|5 3 7
b   0 0 0 0|0 0 0
d   1 0 0 0|0 0 0
e   0 0 1 0|0 0 0
f   1 0 0 0|0 0 0
c   1 0 0 0|0 0 0
i   0 1 0 0|0 0 0
j   0 0 1 0|0 0 0
k   0 0 0 1|0 0 0
a   0 0 0 0|1 0 0
g   0 0 1 1|1 0 1
h   0 0 0 0|0 1 0
```
```
    5 4 1 2 6|3 7
b   0 0 0 0 0|0 0
d   0 1 0 0 0|0 0
e   0 0 0 1 0|0 0
f   0 1 0 0 0|0 0
a   1 0 0 0 0|0 0
c   0 1 0 0 0|0 0
i   0 0 1 0 0|0 0
j   0 0 0 1 0|0 0
k   0 0 0 0 1|0 0
g   1 0 0 1 1|1 0
h   0 0 0 0 0|0 1
```

Column-3 coverings (rows 1–5)

```
    2 4 7|5 3 1 6
b   0 0 0|0 0 0 0
c   0 1 0|0 0 0 0
d   0 1 0|0 0 0 0
j   1 0 0|0 0 0 0
e   1 0 0|0 0 0 0
f   0 1 0|0 0 0 0
h   0 0 1|0 0 0 0
a   0 0 0|1 0 0 0
g   1 0 0|1 1 0 1
i   0 0 0|0 0 1 0
k   0 0 0|0 0 0 1
```
```
    5 2 4 7|3 1 6
b   0 0 0 0|0 0 0
c   0 0 1 0|0 0 0
d   0 0 1 0|0 0 0
j   0 1 0 0|0 0 0
a   1 0 0 0|0 0 0
e   0 1 0 0|0 0 0
f   0 0 1 0|0 0 0
h   0 0 0 1|0 0 0
g   1 1 0 0|1 0 1
i   0 0 0 0|0 1 0
k   0 0 0 0|0 0 1
```
```
    5 4 3 7 2|6 1
b   0 0 0 0 0|0 0
d   0 1 0 0 0|0 0
e   0 0 0 1 0|0 0
f   0 1 0 0 0|0 0
a   1 0 0 0 0|0 0
c   0 1 0 0 0|0 0
g   1 0 1 0 1|1 0
h   0 0 0 1 0|0 0
j   0 0 0 0 1|0 0
k   0 0 0 0 0|1 0
i   0 0 0 0 0|0 1
```
```
    4 7 1 2|6 5 3
b   0 0 0 0|0 0 0
d   1 0 0 0|0 0 0
e   0 0 0 1|0 0 0
f   1 0 0 0|0 0 0
c   1 0 0 0|0 0 0
h   0 1 0 0|0 0 0
i   0 0 1 0|0 0 0
j   0 0 0 1|0 0 0
k   0 0 0 0|1 0 0
a   0 0 0 0|0 1 0
g   0 0 1 1|1 1 1
```
```
    5 4 7 1 2|6 3
b   0 0 0 0 0|0 0
d   1 0 0 0 0|0 0
e   0 0 0 1 0|0 0
f   1 0 0 0 0|0 0
a   1 0 0 0 0|0 0
c   0 1 0 0 0|0 0
h   0 0 1 0 0|0 0
i   0 0 0 1 0|0 0
j   0 0 0 0 1|0 0
k   0 0 0 0 0|1 0
g   1 0 0 1 1|1 1
```

Column-4 coverings (rows 1–5)

```
    4 2 6|5 3 7 1
b   0 0 0|0 0 0 0
d   1 0 0|0 0 0 0
e   0 1 0|0 0 0 0
f   1 0 0|0 0 0 0
c   1 0 0|0 0 0 0
j   0 1 0|0 0 0 0
k   0 0 1|0 0 0 0
a   0 0 0|1 0 0 0
g   0 1 1|1 1 0 0
h   0 0 0|0 0 1 0
i   0 0 0|0 0 0 1
```
```
    5 2 4 7|6 3 1
b   0 0 0 0|0 0 0
d   1 0 0 0|0 0 0
e   0 0 1 0|0 0 0
f   1 0 0 0|0 0 0
c   1 0 0 0|0 0 0
i   0 1 0 0|0 0 0
j   0 0 1 0|0 0 0
a   0 0 0 1|0 0 0
g   0 0 1 1|1 0 1
h   0 0 0 0|0 1 0
k   0 0 0 0|0 0 1
```
```
    5 4 7 1|2 6 3
b   0 0 0 0|0 0 0
d   1 0 0 0|0 0 0
e   0 0 1 0|0 0 0
f   1 0 0 0|0 0 0
c   1 0 0 0|0 0 0
i   0 1 0 0|0 0 0
j   0 0 1 0|0 0 0
a   0 0 0 1|0 0 0
g   0 0 1 1|1 0 1
h   0 0 0 0|0 1 0
k   0 0 0 0|0 0 1
```
```
    5 2 4 7|1 6 3
b   0 0 0 0|0 0 0
c   0 0 1 0|0 0 0
d   0 0 1 0|0 0 0
j   0 1 0 0|0 0 0
a   1 0 0 0|0 0 0
e   0 1 0 0|0 0 0
f   0 0 1 0|0 0 0
i   0 0 0 1|0 0 0
g   1 1 0 0|1 0 1
h   0 0 0 0|0 1 0
k   0 0 0 0|0 0 1
```
```
    5 4 3 1 2 6|7
b   0 0 0 0 0 0|0
d   0 1 0 0 0 0|0
e   0 0 0 0 1 0|0
f   0 1 0 0 0 0|0
a   1 0 0 0 0 0|0
c   0 1 0 0 0 0|0
g   1 0 1 0 1 1|0
i   0 0 0 1 0 0|0
k   0 0 0 0 0 1|0
h   0 0 0 0 0 0|1
```

Fig. 10.6 One relation of term rank 7 together with all its cardinality-minimum coverings

10.3 Fringes

We now specialize our study of rectangles and define the fringe of an arbitrary relation to be the set of those entries that belong to *just one* non-enlargeable rectangle.
This means that the rectangles started horizontally and those started vertically
will coincide. As a result several properties will hold. This will then lead to some
semigroup related considerations, such as generalized inverses and Moore–Penrose
inverses, for example, in Section 10.7.

As already announced, we now study the circumstances under which a point (x, y)
is contained in precisely one non-enlargeable rectangle.

Proposition 10.9. *For a pair (x, y) related by a relation R the following are equivalent:*

(i) *(x, y) is contained in precisely one non-enlargeable rectangle inside R,*
(ii) $x_i y^{\mathsf{T}} \subseteq R \cap \overline{R_i \overline{R^{\mathsf{T}}_i} R}$.

Proof: We recall Prop. 10.3 and start with an easy to prove equivalence

$$\overline{R_i \overline{R^{\mathsf{T}}_i} x} \supseteq \overline{R_i y}$$

$$ \quad \text{negated}$$

$$\Longleftrightarrow \quad R_i \overline{R^{\mathsf{T}}_i} x \subseteq \overline{R_i y}$$

$$ \quad \text{Schröder rule}$$

$$\Longleftrightarrow \quad \overline{R^{\mathsf{T}}_i \overline{R_i y}} \subseteq \overline{R^{\mathsf{T}}_i x}$$

$$ \quad \text{since } x, y \text{ are points; see Prop. 5.13.ii}$$

$$\Longleftrightarrow \quad \overline{R^{\mathsf{T}}_i R_i y} \subseteq \overline{R^{\mathsf{T}}_i x}$$

$$ \quad \text{Schröder rule}$$

$$\Longleftrightarrow \quad R^{\mathsf{T}}_i \overline{R_i R^{\mathsf{T}}_i x} \subseteq \overline{y}$$

$$ \quad \text{Schröder rule}$$

$$\Longleftrightarrow \quad y_i x^{\mathsf{T}} \subseteq \overline{R^{\mathsf{T}}_i \overline{R_i} R^{\mathsf{T}}}$$

$$ \quad \text{transposed}$$

$$\Longleftrightarrow \quad x_i y^{\mathsf{T}} \subseteq \overline{R_i \overline{R^{\mathsf{T}}_i} R}$$

(i) $\Longrightarrow$ (ii): If there is just one non-enlargeable rectangle for $x_i y^{\mathsf{T}} \subseteq R$, the extremal
rectangles according to Prop. 10.3 will coincide, but we use only that

$$u_x = \overline{R_i \overline{R^{\mathsf{T}}_i} x} \supseteq \overline{R_i y} = u_y,$$

so that we can apply the equivalence mentioned above.

(ii) $\Longrightarrow$ (i): The assumption splits into $x_i y^{\mathsf{T}} \subseteq R$ and $x_i y^{\mathsf{T}} \subseteq \overline{R_i \overline{R^{\mathsf{T}}_i} R}$, from which
the first shows that x, y is inside R and

$$x_i y^\mathsf{T} \subseteq R \iff y \subseteq R^\mathsf{T}_i x \implies u_x = \overline{R_i \overline{R^\mathsf{T}_i}}_i x = \overline{\overline{R_i} R^\mathsf{T}_i} x \subseteq \overline{R_i} y = u_y.$$

With this equivalence, the second condition means $u_x \supseteq u_y$, altogether resulting in equality. □

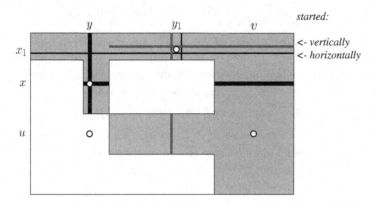

Fig. 10.7 Point (x_1, y_1) inside two and point (x, y) inside just one non-enlargeable rectangle

Figure 10.7 shows the point (x_1, y_1) with two and the point (x, y) with just one non-enlargeable rectangle around it. The points admitting just one non-enlargeable rectangle inside a relation R play an important rôle, so we introduce a notation for them.

Definition 10.10. For a (possibly heterogeneous) relation R we define its

$$\mathtt{fringe}(R) := R \cap \overline{R_i \overline{R^\mathsf{T}_i} R}.$$ □

A quick inspection shows that $\mathtt{fringe}(R^\mathsf{T}) = [\mathtt{fringe}(R)]^\mathsf{T}$. We will quite frequently abbreviate $\mathtt{fringe}(R)$ to ∇_R or even ∇ when there is no doubt concerning R. This symbol nicely reflects the situation of Fig. 10.7. There, and in other figures to come, the relation is rearranged by algebraic visualization and shows the fringe as block diagonal with other parts arranged to the upper right triangle delimited by the fringe.

Figure 10.8 shows a fringe consisting of just two entries in the matrix. One will find out that {Dupont, Larsen} × {French, German} is indeed the only non-enlargeable rectangle containing the entry (Larsen, French); correspondingly for {Arbuthnot, Schmidt} × {German, British} around (Schmidt, British).

	US	French	German	British	Spanish
Arbuthnot	1	0	1	1	0
Perez	0	0	1	0	1
Dupont	0	1	1	0	1
Botticelli	1	0	0	0	1
Schmidt	0	0	1	1	0
Larsen	0	1	1	0	0

	US	French	German	British	Spanish
Arbuthnot	0	0	0	0	0
Perez	0	0	0	0	0
Dupont	0	0	0	0	0
Botticelli	0	0	0	0	0
Schmidt	0	0	0	1	0
Larsen	0	1	0	0	0

Fig. 10.8 Relation of Fig. 10.3 together with its fringe

The concept of a fringe has unexpectedly many applications. As a first example we mention that the fringe of an ordering is the identity, since

$$\texttt{fringe}(E) = E \cap \overline{E; \overline{E^{\mathsf{T}}}; E} = E \cap \overline{\overline{E^{\mathsf{T}}}; E} = E \cap \overline{\overline{E}}^{\mathsf{T}} = E \cap E^{\mathsf{T}} = \mathbb{I}.$$

This applies not least to the ordering "$\leq$" on $\mathbb{R}$. But fringes are usually not that simple, which can be seen already from the linear strictorder "$<$" on $\mathbb{R}$. This strictorder is, of course, transitive $C; C \subseteq C$, but satisfies also $C \subseteq C; C$, meaning that whatever element relationship one chooses, for example $3.7 < 3.8$, one will find an element in between, $3.7 < 3.75 < 3.8$. Let us for a moment call C a **dense** relation if it satisfies $C \subseteq C; C$. The following proposition resembles a result of Michael Winter [147]. To be a dense relation means not least that the Hasse diagram vanishes.

This result together with the block-diagonal part of Fig. 10.7, where obviously the points with just one non-enlargeable rectangle reside, shows that a fringe has indeed something in common with a border zone. Given a relation R and its fringe ∇, we have the following visualization. If $(x, y) \in \nabla$, and for some point (x_1, v) also $(x, v) \in R$ and $(x_1, y) \in R$, then always $(x_1, v) \in R$. On the contrary, consider $(x_1, y_1) \in R$, however $\notin \nabla$, and (u, y). Obviously, $(x_1, y) \in R$ and $(u, y_1) \in R$, but $(u, y) \notin R$.

Proposition 10.11. *A dense linear strictordering has an empty fringe.*

Proof: Using the standard property Prop. 5.24 of a linear strictordering $\overline{C}^{\mathsf{T}} = E = \mathbb{I} \cup C$, we have

$$C; \overline{C}^{\mathsf{T}}; C = C; (\mathbb{I} \cup C); C = C; C \cup C; C; C = C,$$

so that $\texttt{fringe}(C) = C \cap \overline{C; \overline{C}^{\mathsf{T}}; C} = C \cap \overline{C} = \bot\!\!\!\bot.$ $\qquad \square$

The existence of a non-empty fringe thus heavily depends on finiteness or at least discreteness, giving this some sort of topological flavor. Fringe considerations are central for difunctional, Ferrers, and block-transitive relations.

We now have a look at the nicely arranged

- identity relation $\mathbb{I}$ which is simply reproduced

$$\mathbb{I} = \begin{matrix} & \begin{smallmatrix}1&2&3&4&5&6&7\end{smallmatrix} \\ \begin{matrix}1\\2\\3\\4\\5\\6\\7\end{matrix} & \begin{pmatrix} 1&0&0&0&0&0&0 \\ 0&1&0&0&0&0&0 \\ 0&0&1&0&0&0&0 \\ 0&0&0&1&0&0&0 \\ 0&0&0&0&1&0&0 \\ 0&0&0&0&0&1&0 \\ 0&0&0&0&0&0&1 \end{pmatrix} \end{matrix} \qquad \mathtt{fringe}(\mathbb{I}) = \begin{matrix} & \begin{smallmatrix}1&2&3&4&5&6&7\end{smallmatrix} \\ \begin{matrix}1\\2\\3\\4\\5\\6\\7\end{matrix} & \begin{pmatrix} 1&0&0&0&0&0&0 \\ 0&1&0&0&0&0&0 \\ 0&0&1&0&0&0&0 \\ 0&0&0&1&0&0&0 \\ 0&0&0&0&1&0&0 \\ 0&0&0&0&0&1&0 \\ 0&0&0&0&0&0&1 \end{pmatrix} \end{matrix}$$

Fig. 10.9 Fringe of an identity

- partial identity relation which is reproduced again

$$R = \begin{matrix} & \begin{smallmatrix}1&2&3&4&5&6&7\end{smallmatrix} \\ \begin{matrix}1\\2\\3\\4\\5\\6\\7\end{matrix} & \begin{pmatrix} 1&0&0&0&0&0&0 \\ 0&1&0&0&0&0&0 \\ 0&0&0&0&0&0&0 \\ 0&0&0&0&0&0&0 \\ 0&0&0&0&1&0&0 \\ 0&0&0&0&0&1&0 \\ 0&0&0&0&0&0&1 \end{pmatrix} \end{matrix} \qquad \mathtt{fringe}(R) = \begin{matrix} & \begin{smallmatrix}1&2&3&4&5&6&7\end{smallmatrix} \\ \begin{matrix}1\\2\\3\\4\\5\\6\\7\end{matrix} & \begin{pmatrix} 1&0&0&0&0&0&0 \\ 0&1&0&0&0&0&0 \\ 0&0&0&0&0&0&0 \\ 0&0&0&0&0&0&0 \\ 0&0&0&0&1&0&0 \\ 0&0&0&0&0&1&0 \\ 0&0&0&0&0&0&1 \end{pmatrix} \end{matrix}$$

Fig. 10.10 Fringe of a partial identity

- full block diagonal which is also reproduced

$$R = \begin{matrix} & \begin{smallmatrix}1&2&3&4&5&6&7\end{smallmatrix} \\ \begin{matrix}1\\2\\3\\4\\5\\6\\7\end{matrix} & \begin{pmatrix} 1&1&1&0&0&0&0 \\ 1&1&1&0&0&0&0 \\ 1&1&1&0&0&0&0 \\ 0&0&0&1&1&0&0 \\ 0&0&0&1&1&0&0 \\ 0&0&0&0&0&1&1 \\ 0&0&0&0&0&1&1 \end{pmatrix} \end{matrix} \qquad \mathtt{fringe}(R) = \begin{matrix} & \begin{smallmatrix}1&2&3&4&5&6&7\end{smallmatrix} \\ \begin{matrix}1\\2\\3\\4\\5\\6\\7\end{matrix} & \begin{pmatrix} 1&1&1&0&0&0&0 \\ 1&1&1&0&0&0&0 \\ 1&1&1&0&0&0&0 \\ 0&0&0&1&1&0&0 \\ 0&0&0&1&1&0&0 \\ 0&0&0&0&0&1&1 \\ 0&0&0&0&0&1&1 \end{pmatrix} \end{matrix}$$

Fig. 10.11 Fringe of a block-diagonal relation

- full heterogeneous block diagonal: difunctional which is reproduced

$$R = \begin{matrix} & \begin{smallmatrix}1&2&3&4&5&6&7&8&9&10&11&12&13\end{smallmatrix} \\ \begin{matrix}1\\2\\3\\4\\5\\6\\7\end{matrix} & \begin{pmatrix} 1&1&1&1&1&0&0&0&0&0&0&0&0 \\ 1&1&1&1&1&0&0&0&0&0&0&0&0 \\ 1&1&1&1&1&0&0&0&0&0&0&0&0 \\ 0&0&0&0&0&1&0&0&0&0&0&0&0 \\ 0&0&0&0&0&1&0&0&0&0&0&0&0 \\ 0&0&0&0&0&0&1&1&1&1&1&1 \\ 0&0&0&0&0&0&1&1&1&1&1&1 \end{pmatrix} \end{matrix}$$

$$\mathtt{fringe}(R) = \begin{matrix} & \begin{smallmatrix}1&2&3&4&5&6&7&8&9&10&11&12&13\end{smallmatrix} \\ \begin{matrix}1\\2\\3\\4\\5\\6\\7\end{matrix} & \begin{pmatrix} 1&1&1&1&1&0&0&0&0&0&0&0&0 \\ 1&1&1&1&1&0&0&0&0&0&0&0&0 \\ 1&1&1&1&1&0&0&0&0&0&0&0&0 \\ 0&0&0&0&0&1&0&0&0&0&0&0&0 \\ 0&0&0&0&0&1&0&0&0&0&0&0&0 \\ 0&0&0&0&0&0&1&1&1&1&1&1 \\ 0&0&0&0&0&0&1&1&1&1&1&1 \end{pmatrix} \end{matrix}$$

Fig. 10.12 Fringe of a heterogeneous difunctional relation

- partial heterogeneous block diagonal which is reproduced

$$R = \begin{array}{c} \\ 1 \\ 2 \\ 3 \\ 4 \\ 5 \\ 6 \\ 7 \end{array}\begin{array}{c}\scriptstyle 1\,2\,3\,4\,5\,6\,7\,8\,9\,10\,11\,12\,13\\ \left(\begin{array}{ccccccccccccc} 1&1&1&0&0&0&0&0&0&0&0&0&0 \\ 1&1&1&0&0&0&0&0&0&0&0&0&0 \\ 1&1&1&0&0&0&0&0&0&0&0&0&0 \\ 0&0&0&0&0&1&0&0&0&0&0&0&0 \\ 0&0&0&0&0&1&0&0&0&0&0&0&0 \\ 0&0&0&0&0&0&1&1&1&1&1&1&1 \\ 0&0&0&0&0&0&1&1&1&1&1&1&1 \end{array}\right)\end{array}
\qquad
\texttt{fringe}(R) = \begin{array}{c} \\ 1 \\ 2 \\ 3 \\ 4 \\ 5 \\ 6 \\ 7 \end{array}\begin{array}{c}\scriptstyle 1\,2\,3\,4\,5\,6\,7\,8\,9\,10\,11\,12\,13\\ \left(\begin{array}{ccccccccccccc} 1&1&1&0&0&0&0&0&0&0&0&0&0 \\ 1&1&1&0&0&0&0&0&0&0&0&0&0 \\ 1&1&1&0&0&0&0&0&0&0&0&0&0 \\ 0&0&0&0&0&1&0&0&0&0&0&0&0 \\ 0&0&0&0&0&1&0&0&0&0&0&0&0 \\ 0&0&0&0&0&0&1&1&1&1&1&1&1 \\ 0&0&0&0&0&0&1&1&1&1&1&1&1 \end{array}\right)\end{array}$$

Fig. 10.13 Fringe of a 'partially' difunctional relation

- upper right triangle: linear order from which the diagonal survives

$$R = \begin{array}{c} \\ 1 \\ 2 \\ 3 \\ 4 \\ 5 \\ 6 \\ 7 \end{array}\begin{array}{c}\scriptstyle 1\,2\,3\,4\,5\,6\,7\\ \left(\begin{array}{ccccccc} 1&1&1&1&1&1&1 \\ 0&1&1&1&1&1&1 \\ 0&0&1&1&1&1&1 \\ 0&0&0&1&1&1&1 \\ 0&0&0&0&1&1&1 \\ 0&0&0&0&0&1&1 \\ 0&0&0&0&0&0&1 \end{array}\right)\end{array}
\qquad
\texttt{fringe}(R) = \begin{array}{c} \\ 1 \\ 2 \\ 3 \\ 4 \\ 5 \\ 6 \\ 7 \end{array}\begin{array}{c}\scriptstyle 1\,2\,3\,4\,5\,6\,7\\ \left(\begin{array}{ccccccc} 1&0&0&0&0&0&0 \\ 0&1&0&0&0&0&0 \\ 0&0&1&0&0&0&0 \\ 0&0&0&1&0&0&0 \\ 0&0&0&0&1&0&0 \\ 0&0&0&0&0&1&0 \\ 0&0&0&0&0&0&1 \end{array}\right)\end{array}$$

Fig. 10.14 Fringe of a linear order

- strict upper right triangle: linear strictorder where only the side diagonal survives

$$R = \begin{array}{c} \\ 1 \\ 2 \\ 3 \\ 4 \\ 5 \\ 6 \\ 7 \end{array}\begin{array}{c}\scriptstyle 1\,2\,3\,4\,5\,6\,7\\ \left(\begin{array}{ccccccc} 0&1&1&1&1&1&1 \\ 0&0&1&1&1&1&1 \\ 0&0&0&1&1&1&1 \\ 0&0&0&0&1&1&1 \\ 0&0&0&0&0&1&1 \\ 0&0&0&0&0&0&1 \\ 0&0&0&0&0&0&0 \end{array}\right)\end{array}
\qquad
\texttt{fringe}(R) = \begin{array}{c} \\ 1 \\ 2 \\ 3 \\ 4 \\ 5 \\ 6 \\ 7 \end{array}\begin{array}{c}\scriptstyle 1\,2\,3\,4\,5\,6\,7\\ \left(\begin{array}{ccccccc} 0&1&0&0&0&0&0 \\ 0&0&1&0&0&0&0 \\ 0&0&0&1&0&0&0 \\ 0&0&0&0&1&0&0 \\ 0&0&0&0&0&1&0 \\ 0&0&0&0&0&0&1 \\ 0&0&0&0&0&0&0 \end{array}\right)\end{array}$$

Fig. 10.15 Fringe of a linear strictorder

- upper right block triangle from which the 'block-fringe' remains

$$R = \begin{array}{c} \\ 1 \\ 2 \\ 3 \\ 4 \\ 5 \\ 6 \\ 7 \end{array}\begin{array}{c}\scriptstyle 1\,2\,3\,4\,5\,6\,7\\ \left(\begin{array}{ccccccc} 1&1&1&1&1&1&1 \\ 1&1&1&1&1&1&1 \\ 1&1&1&1&1&1&1 \\ 0&0&0&1&1&1&1 \\ 0&0&0&1&1&1&1 \\ 0&0&0&0&0&1&1 \\ 0&0&0&0&0&1&1 \end{array}\right)\end{array}
\qquad
\texttt{fringe}(R) = \begin{array}{c} \\ 1 \\ 2 \\ 3 \\ 4 \\ 5 \\ 6 \\ 7 \end{array}\begin{array}{c}\scriptstyle 1\,2\,3\,4\,5\,6\,7\\ \left(\begin{array}{ccccccc} 1&1&1&0&0&0&0 \\ 1&1&1&0&0&0&0 \\ 1&1&1&0&0&0&0 \\ 0&0&0&1&1&0&0 \\ 0&0&0&1&1&0&0 \\ 0&0&0&0&0&1&1 \\ 0&0&0&0&0&1&1 \end{array}\right)\end{array}$$

Fig. 10.16 Fringe of an upper block triangle

- irreflexive upper right block triangle from which again the 'block-fringe' remains

$$R = \begin{array}{c} \\ 1 \\ 2 \\ 3 \\ 4 \\ 5 \\ 6 \\ 7 \end{array}\begin{array}{c}\scriptstyle 1\,2\,3\,4\,5\,6\,7\\ \left(\begin{array}{ccccccc} 0&0&0&1&1&1&1 \\ 0&0&0&1&1&1&1 \\ 0&0&0&1&1&1&1 \\ 0&0&0&0&0&1&1 \\ 0&0&0&0&0&1&1 \\ 0&0&0&0&0&0&0 \\ 0&0&0&0&0&0&0 \end{array}\right)\end{array}
\qquad
\texttt{fringe}(R) = \begin{array}{c} \\ 1 \\ 2 \\ 3 \\ 4 \\ 5 \\ 6 \\ 7 \end{array}\begin{array}{c}\scriptstyle 1\,2\,3\,4\,5\,6\,7\\ \left(\begin{array}{ccccccc} 0&0&0&1&1&0&0 \\ 0&0&0&1&1&0&0 \\ 0&0&0&1&1&0&0 \\ 0&0&0&0&0&1&1 \\ 0&0&0&0&0&1&1 \\ 0&0&0&0&0&0&0 \\ 0&0&0&0&0&0&0 \end{array}\right)\end{array}$$

Fig. 10.17 Fringe of an irreflexive upper block triangle

• heterogeneous upper right block triangle: Ferrers

$R =$

	1	2	3	4	5	6	7	8	9	10	11	12	13
1	0	0	1	1	1	1	1	1	1	1	1	1	1
2	0	0	1	1	1	1	1	1	1	1	1	1	1
3	0	0	1	1	1	1	1	1	1	1	1	1	1
4	0	0	0	0	1	1	1	1	1	1	1	1	1
5	0	0	0	0	1	1	1	1	1	1	1	1	1
6	0	0	0	0	0	0	0	0	1	1	1	1	1
7	0	0	0	0	0	0	0	0	0	0	0	0	0

$\texttt{fringe}(R) =$

	1	2	3	4	5	6	7	8	9	10	11	12	13
1	0	0	1	1	0	0	0	0	0	0	0	0	0
2	0	0	1	1	0	0	0	0	0	0	0	0	0
3	0	0	1	1	0	0	0	0	0	0	0	0	0
4	0	0	0	0	1	1	1	1	0	0	0	0	0
5	0	0	0	0	1	1	1	1	0	0	0	0	0
6	0	0	0	0	0	0	0	0	1	1	1	1	1
7	0	0	0	0	0	0	0	0	0	0	0	0	0

Fig. 10.18 Fringe of a non-total Ferrers relation

• an 'arbitrary' heterogeneous relation

R

	Spanish	German	US	Czech	French	British	Japanese	Italian
Mon	0	0	0	1	0	0	1	0
Tue	1	0	0	0	0	1	0	0
Wed	0	1	0	1	1	0	1	1
Thu	1	0	0	0	0	1	0	0
Fri	0	1	0	1	1	0	1	1
Sat	0	0	0	0	0	0	1	1
Sun	0	0	0	1	0	0	0	1

$\texttt{fringe}(R)$

	Spanish	German	US	Czech	French	British	Japanese	Italian
Mon	0	0	0	0	0	0	0	0
Tue	1	0	0	0	0	1	0	0
Wed	0	1	0	0	1	0	0	0
Thu	1	0	0	0	0	1	0	0
Fri	0	1	0	0	1	0	0	0
Sat	0	0	0	0	0	0	0	0
Sun	0	0	0	0	0	0	0	0

Fig. 10.19 An arbitrary relation and its fringe

With the technique of algebraic visualization, we find for the relation of Fig. 10.19 an arrangement as in Fig. 10.20.

$R_{\texttt{rearr}}$

	US	French	German	British	Spanish	Japanese	Italian	Czech
Wed	0	1	1	0	0	1	1	1
Fri	0	1	1	0	0	1	1	1
Tue	0	0	0	1	1	0	0	0
Thu	0	0	0	1	1	0	0	0
Sat	0	0	0	0	0	0	1	1
Mon	0	0	0	0	0	1	0	1
Sun	0	0	0	0	0	1	1	0

$\texttt{fringe}(R_{\texttt{rearr}})$

	US	French	German	British	Spanish	Japanese	Italian	Czech
Wed	0	1	1	0	0	0	0	0
Fri	0	1	1	0	0	0	0	0
Tue	0	0	0	1	1	0	0	0
Thu	0	0	0	1	1	0	0	0
Sat	0	0	0	0	0	0	0	0
Mon	0	0	0	0	0	0	0	0
Sun	0	0	0	0	0	0	0	0

$\Xi_\nabla(R_{\texttt{rearr}})$

	Wed	Fri	Tue	Thu	Sat	Mon	Sun
Wed	1	1	0	0	0	0	0
Fri	1	1	0	0	0	0	0
Tue	0	0	1	1	0	0	0
Thu	0	0	1	1	0	0	0
Sat	0	0	0	0	0	0	0
Mon	0	0	0	0	0	0	0
Sun	0	0	0	0	0	0	0

$\Xi(R_{\texttt{rearr}})$

	Wed	Fri	Tue	Thu	Sat	Mon	Sun
Wed	1	1	0	0	0	0	0
Fri	1	1	0	0	0	0	0
Tue	0	0	1	1	0	0	0
Thu	0	0	1	1	0	0	0
Sat	0	0	0	0	1	0	0
Mon	0	0	0	0	0	1	0
Sun	0	0	0	0	0	0	1

Fig. 10.20 Same relation as in Fig. 10.19 nicely rearranged, with fringe and row equivalence

After all these examples, it is evident that diagonals, even partial ones, are reproduced. This also holds true when the diagonals are just square or even rectangular partial block diagonals. The second class is made up of 'triangles' – be they partial or block oriented. In these cases when finite, the bounding/delimiting part is extracted while the rest of the triangle vanishes. An upper right triangle including the diagonal is converted to the diagonal. A strict upper right triangle is converted to the upper side-diagonal, *provided the relation in question is finite.* This also holds true when the relation is subdivided consistently into rectangular blocks.

$$R = \begin{matrix} 1 \\ 2 \\ 3 \\ 4 \\ 5 \end{matrix}\begin{pmatrix} 0&1&1&1&1 \\ 1&0&1&1&1 \\ 1&1&0&1&1 \\ 1&1&1&0&1 \\ 1&1&1&1&0 \end{pmatrix} \quad \mathtt{fringe}(R) = \begin{matrix} 1 \\ 2 \\ 3 \\ 4 \\ 5 \end{matrix}\begin{pmatrix} 0&0&0&0&0 \\ 0&0&0&0&0 \\ 0&0&0&0&0 \\ 0&0&0&0&0 \\ 0&0&0&0&0 \end{pmatrix} \quad \text{but} \quad R' = \begin{matrix}1\\2\end{matrix}\begin{pmatrix}0&1\\1&0\end{pmatrix} = \mathtt{fringe}(R')$$

Fig. 10.21 An example of a non-empty relation R with an empty fringe

The fringe may also be obtained with the symmetric quotient from the row-contains-preorder and the relation in question.

Proposition 10.12. *For an arbitrary (possibly heterogeneous) relation R, the fringe and the row-contains-preorder satisfy*

$$\mathtt{fringe}(R) = \mathtt{syq}(\overline{\overline{R}\, R^\mathsf{T}}, R) = \mathtt{syq}(\mathcal{R}(R), R).$$

Proof: We expand `fringe`, `syq`, $\mathcal{R}(R)$, and apply trivial operations to obtain

$$R \cap \overline{R\,\overline{R}^\mathsf{T}\, R} = \overline{R\,\overline{R}^\mathsf{T}\, R} \cap \overline{R\,\overline{R}^\mathsf{T}\, R}.$$

It suffices, thus, to convince ourselves that with Prop. 4.7, here $\overline{R} = \overline{R\,\overline{R}^\mathsf{T}\, R}$, the first term on the left side turns out to be equal to the second on the right. □

The fringe, thus, shows which columns of R are made up of columns of the row-contains-preorder. In addition, we are now allowed to make use of cancellation formulae such as Prop. 8.13 that regulate the behavior of a symmetric quotient.

Proposition 10.13. *When a relation S is with f,g mapped surjectively onto another relation R, so is its fringe; i.e.,*

$$S = f\,R\,g^\mathsf{T} \quad \Longrightarrow \quad \nabla_S = f\,\nabla_R\,g^\mathsf{T}.$$

Proof: by simple evaluation:

$$\begin{aligned}
\nabla_S &= S \cap \overline{S\,\overline{S}^\mathsf{T}\, S} && \text{by definition} \\
&= f\,R\,g^\mathsf{T} \cap \overline{f\,R\,g^\mathsf{T}\,\overline{f\,R\,g^\mathsf{T}}^\mathsf{T}\, f\,R\,g^\mathsf{T}} && \text{since } S = f\,R\,g^\mathsf{T} \\
&= f\,R\,g^\mathsf{T} \cap f\,R\,g^\mathsf{T}\,g\,\overline{R}^\mathsf{T}\, f^\mathsf{T}\, f\,R\,g^\mathsf{T} && \text{mappings slipping out of negation} \\
&= f\,R\,g^\mathsf{T} \cap f\,R\,\overline{R}^\mathsf{T}\, R\,g^\mathsf{T} && \text{mappings } f,g \text{ are surjective} \\
&= f\,(R \cap \overline{R\,\overline{R}^\mathsf{T}\, R})\,g^\mathsf{T} && \text{since } f,g \text{ are univalent} \\
&= f\,\nabla_R\,g^\mathsf{T} && \text{by definition} \quad\square
\end{aligned}$$

Fringe-restricted constructs

Next, we present a plexus of formulae that appear difficult. Anyone not yet familiar with the field would probably not dare to try to prove them directly. Nevertheless, they are heavily interrelated and may even be visualized and then understood. The first rather artificial looking functionals are based on the fringe. We show that to a certain extent the row equivalence $\Xi(R)$ we have studied so far may be substituted by $\Xi_\nabla(R)$; both coincide as long as the fringe is total. They may be different, but only in the restricted way that a square diagonal block of the `fringe`-partial row equivalence is either equal to the other or empty.

The fringe then gives rise to two 'partial equivalences' that closely resemble the row and column equivalences.

Definition 10.14. In addition to the fringe $\nabla := \mathtt{fringe}(R) = R \cap \overline{R \, \overline{R}^\mathsf{T} \, R}$, we define for an arbitrary (possibly heterogeneous) relation R

(i) $\Xi_\nabla(R) := \nabla \, \nabla^\mathsf{T}$, the **fringe-partial row equivalence**,

(ii) $\Psi_\nabla(R) := \nabla^\mathsf{T} \, \nabla$, the **fringe-partial column equivalence**. $\qquad\square$

We recall the characterization of a fringe that appeared in Prop. 10.9, namely that the fringe collects those entries of a relation R that are contained in precisely one non-enlargeable rectangle. Notations $\Xi_\nabla(R)$ and $\Psi_\nabla(R)$ are again chosen so as to resemble visually rows and columns, respectively.

Proposition 10.15. *For an arbitrary (possibly heterogeneous) relation R, the fringe $\nabla := \mathtt{fringe}(R)$ and the fringe-partial row respectively column equivalences satisfy the following:*

(i) $\Xi_\nabla(R) = \Xi(R) \cap \nabla \, \mathbb{T}$

(ii) $\Xi(R) \, \nabla = \Xi_\nabla(R) \, \nabla = \nabla = \nabla \, \Psi_\nabla(R) = \nabla \, \Psi(R)$

(iii) $\nabla^\mathsf{T} \, \Xi(R) \, \nabla \subseteq \Psi(R)$

(iv) $\Xi_\nabla(R) \, R \subseteq R \, \nabla^\mathsf{T} \, R \subseteq R$
     $R \, \Psi_\nabla(R) \subseteq R \, \nabla^\mathsf{T} \, R \subseteq R$

(v) $\nabla \, R^\mathsf{T} \cap R \, \nabla^\mathsf{T} = \nabla \, \nabla^\mathsf{T} = \Xi_\nabla(R) \subseteq \Xi(R)$
     $\nabla^\mathsf{T} \, R \cap R^\mathsf{T} \, \nabla = \nabla^\mathsf{T} \, \nabla = \Psi_\nabla(R) \subseteq \Psi(R)$

(vi) $\nabla \subseteq \nabla \, R^\mathsf{T} \, \nabla$

Proof: (i) $\Xi_\nabla(R) = \nabla \, \nabla^\mathsf{T} = \mathtt{syq}(\overline{R \, R^\mathsf{T}}, R) \, \mathtt{syq}(R, \overline{R \, R^\mathsf{T}})$ $\qquad$ Prop. 10.12
$\qquad\quad = \mathtt{syq}(\overline{R \, R^\mathsf{T}}, \overline{R \, R^\mathsf{T}}) \cap \mathtt{syq}(\overline{R \, R^\mathsf{T}}, R) \, \mathbb{T}$ $\qquad$ following Prop. 8.13.i.

$$= \Xi(R) \cap \nabla ; \mathbb{T} \qquad \text{Prop. 8.15}$$

(ii) We expand $\Xi(R)$ with Prop. 8.15 and ∇ with Prop. 10.12

$$\Xi(R) ; \nabla = \mathsf{syq}\,(\overline{\overline{R ; R^{\mathsf{T}}}}, \overline{R ; R^{\mathsf{T}}}) ; \mathsf{syq}\,(\overline{R ; R^{\mathsf{T}}}, R) \subseteq \mathsf{syq}\,(\overline{\overline{R ; R^{\mathsf{T}}}}, R) = \nabla,$$

and apply the cancellation rule Prop. 8.13.i. From the result obtained, we can proceed with

$$\nabla ; \nabla^{\mathsf{T}} ; \nabla = \Xi_{\nabla}(R) ; \nabla \subseteq \Xi(R) ; \nabla \qquad \text{according to (i)}$$
$$\subseteq \nabla \qquad \text{see above}$$
$$\subseteq \nabla ; \nabla^{\mathsf{T}} ; \nabla = \nabla ; \Psi_{\nabla}(R) \qquad \text{since } A \subseteq A ; A^{\mathsf{T}} ; A \text{ for every relation } A$$

obtaining equality everywhere in between.

(iii) $\nabla^{\mathsf{T}} ; \Xi(R) ; \nabla \subseteq \nabla^{\mathsf{T}} ; \nabla \qquad$ see above
$$\qquad\qquad\qquad = \Psi_{\nabla}(R) \qquad \text{by Def. 10.10}$$
$$\qquad\qquad\qquad \subseteq \Psi(R) \qquad \text{applying (i) to } R^{\mathsf{T}}$$

(iv) $R ; \nabla^{\mathsf{T}} ; R = R ; \left[\mathsf{syq}\,(\overline{\overline{R ; R^{\mathsf{T}}}}, R) \right]^{\mathsf{T}} ; R \qquad \text{Prop. 10.12}$
$$\qquad\qquad = R ; \mathsf{syq}\,(R, \overline{\overline{R ; R^{\mathsf{T}}}}) ; R \qquad \text{transposing a symmetric quotient}$$
$$\qquad\qquad \subseteq \overline{\overline{R ; R^{\mathsf{T}}}} ; R \qquad \text{cancelling with Prop. 8.12}$$
$$\qquad\qquad \subseteq R \qquad \text{which holds for every relation}$$

The rest is then simple because $\Xi_{\nabla}(R) = \nabla ; \nabla^{\mathsf{T}} \subseteq R ; \nabla^{\mathsf{T}}$.

(v) We cancel symmetric quotients several times according to Prop. 8.12:
$$\nabla ; R^{\mathsf{T}} = \mathsf{syq}\,(\overline{\overline{R ; R^{\mathsf{T}}}}, R) ; R^{\mathsf{T}} \qquad \text{expanded}$$
$$\qquad\quad = \left[R ; \mathsf{syq}\,(R, \overline{\overline{R ; R^{\mathsf{T}}}}) \right]^{\mathsf{T}} \qquad \text{transposed}$$
$$\qquad\quad = \left[\overline{\overline{R ; R^{\mathsf{T}}}} \cap \mathbb{T} ; \mathsf{syq}\,(R, \overline{\overline{R ; R^{\mathsf{T}}}}) \right]^{\mathsf{T}} \qquad \text{cancelling according to Prop. 8.12.i}$$
$$\qquad\quad = \overline{R ; \overline{R}^{\mathsf{T}}} \cap \mathsf{syq}\,(\overline{\overline{R ; R^{\mathsf{T}}}}, R) ; \mathbb{T} \qquad \text{transposed}$$
$$\qquad\quad = \overline{R ; \overline{R}^{\mathsf{T}}} \cap \nabla ; \mathbb{T} \qquad \text{according to Prop. 10.12}$$

The second term is handled in mainly the same way:
$$R ; \nabla^{\mathsf{T}} = R ; \left[\mathsf{syq}\,(\overline{\overline{R ; R^{\mathsf{T}}}}, R) \right]^{\mathsf{T}} \qquad \text{according to Prop. 10.12}$$
$$\qquad\quad = R ; \mathsf{syq}\,(R, \overline{\overline{R ; R^{\mathsf{T}}}}) \qquad \text{transposed}$$
$$\qquad\quad = \overline{\overline{R ; R^{\mathsf{T}}}} \cap \mathbb{T} ; \mathsf{syq}\,(R, \overline{R ; R^{\mathsf{T}}}) \qquad \text{cancelling according to Prop. 8.12.i}$$
$$\qquad\quad = \overline{\overline{R ; R^{\mathsf{T}}}} \cap \mathbb{T} ; \nabla^{\mathsf{T}} \qquad \text{according to Prop. 10.12}$$

Now, these are put together:
$$R ; \nabla^{\mathsf{T}} \cap \nabla ; R^{\mathsf{T}} = \overline{\overline{R ; R^{\mathsf{T}}}} \cap \mathbb{T} ; \nabla^{\mathsf{T}} \cap \overline{R ; \overline{R}^{\mathsf{T}}} \cap \nabla ; \mathbb{T} \qquad \text{see above}$$
$$\qquad\qquad = \mathsf{syq}\,(R^{\mathsf{T}}, R^{\mathsf{T}}) \cap \mathbb{T} ; \nabla^{\mathsf{T}} \cap \nabla ; \mathbb{T} \qquad \text{definition of symmetric quotient}$$
$$\qquad\qquad = \Xi(R) \cap \mathbb{T} ; \nabla^{\mathsf{T}} \cap \nabla ; \mathbb{T} \qquad \text{Def. 5.28 of row equivalence}$$
$$\qquad\qquad = \Xi_{\nabla}(R) \cap \mathbb{T} ; \nabla^{\mathsf{T}} \qquad \text{due to (i)}$$
$$\qquad\qquad = \nabla ; \nabla^{\mathsf{T}} \cap \mathbb{T} ; \nabla^{\mathsf{T}} \qquad \text{Def. 10.14 of fringe-partial row equivalence}$$
$$\qquad\qquad = \nabla ; \nabla^{\mathsf{T}}$$

(vi) is trivial in view of (ii). $\qquad\square$

Anticipating Def. 10.42, we may say that ∇^{T} is always a sub-inverse of R. The formulae presented here go far beyond those semigroup-related formulae known around inverses, for example.

$$R = \begin{array}{c} \\ \text{Win} \\ \text{Draw} \\ \text{Loss} \end{array}\begin{array}{c} \text{\small red gre blu ora} \\ \begin{pmatrix} 0 & 1 & 0 & 1 \\ 0 & 1 & 1 & 0 \\ 1 & 0 & 1 & 0 \end{pmatrix}\end{array} \qquad \begin{array}{c} \\ \text{Win} \\ \text{Draw} \\ \text{Loss} \end{array}\begin{array}{c} \text{\small Win Draw Loss} \\ \begin{pmatrix} 1 & 0 & 0 \\ 0 & 0 & 0 \\ 0 & 0 & 1 \end{pmatrix}\end{array} = \Xi_{\nabla}(R) \subseteq \Xi(R) = \begin{array}{c} \text{\small Win Draw Loss} \\ \begin{pmatrix} 1 & 0 & 0 \\ 0 & 1 & 0 \\ 0 & 0 & 1 \end{pmatrix}\end{array}$$

$$\mathtt{fringe}(R) = \begin{array}{c} \\ \text{Win} \\ \text{Draw} \\ \text{Loss} \end{array}\begin{array}{c} \\ \begin{pmatrix} 0 & 0 & 0 & 1 \\ 0 & 0 & 0 & 0 \\ 1 & 0 & 0 & 0 \end{pmatrix}\end{array} \quad \begin{array}{c} \\ \text{red} \\ \text{gre} \\ \text{blu} \\ \text{ora} \end{array}\begin{array}{c} \text{\small red gre blu ora} \\ \begin{pmatrix} 1 & 0 & 0 & 0 \\ 0 & 0 & 0 & 0 \\ 0 & 0 & 0 & 0 \\ 0 & 0 & 0 & 1 \end{pmatrix}\end{array} = \Psi_{\nabla}(R) \subseteq \Psi(R) = \begin{array}{c} \text{\small red gre blu ora} \\ \begin{pmatrix} 1 & 0 & 0 & 0 \\ 0 & 1 & 0 & 0 \\ 0 & 0 & 1 & 0 \\ 0 & 0 & 0 & 1 \end{pmatrix}\end{array}$$

Fig. 10.22 Row equivalence and fringe-partial row equivalence

The following proposition relates the fringe of the row-contains-preorder with the row equivalence.

Proposition 10.16. *We have for every (possibly heterogeneous) relation R, that*

(i) $\mathtt{fringe}(\mathcal{R}(R)) = \mathtt{fringe}(\overline{R\,;R^{\mathsf{T}}}) = \mathrm{syq}(R^{\mathsf{T}}, R^{\mathsf{T}}) = \Xi(R)$,
(ii) $\mathtt{fringe}(\mathcal{C}(R)) = \mathtt{fringe}(\overline{R^{\mathsf{T}}\,;R}) = \mathrm{syq}(R, R) = \Psi(R)$.

Proof: In both cases, only the equality in the middle is important because the rest is just expansion of definitions. Thus reduced, the first identity, for example, requires us to prove that

$$\overline{R\,;R^{\mathsf{T}}} \cap \overline{\overline{R\,;R^{\mathsf{T}}}\,;R\,;R^{\mathsf{T}}}\,;\overline{R\,;R^{\mathsf{T}}} = \overline{R\,;R^{\mathsf{T}}} \cap \overline{R\,;R^{\mathsf{T}}}.$$

The first term on the left equals the first on the right. In addition, the second terms are equal, which is also trivial according to Prop. 4.7. $\qquad\square$

The next corollary shows that the fringe may indeed be important since it relates the fringe with difunctionality.

Corollary 10.17. *We assume the present context.*

(i) *For arbitrary R, the construct $\mathtt{fringe}(R)$ is difunctional.*
(ii) *A relation R is difunctional precisely when $R = \mathtt{fringe}(R)$.*
(iii) *Forming the fringe is an idempotent operation, i.e.,*

$$\texttt{fringe}(\texttt{fringe}(R)) = \texttt{fringe}(R).$$

Proof: (i) This is simply a reformulation of Prop. 10.15.ii.

(ii) If R equals its fringe $\nabla := \texttt{fringe}(R)$, then R must also be difunctional according to (i). In the reverse direction, we use the third variant of Def. 5.30 of being difunctional:

$$\nabla = R \cap \overline{R\, ; \overline{R}^{\mathsf{T}}\, ; R} \supseteq R \cap \overline{\overline{R}} = R.$$

(iii) This is shown in a way similar to the proof of (ii), thereby using (i):

$$\texttt{fringe}(\nabla) = \nabla \cap \overline{\nabla\, ; \overline{\nabla}^{\mathsf{T}}\, ; \nabla} \supseteq \nabla \cap \overline{\overline{\nabla}} = \nabla. \qquad \square$$

Exercises

10.1 Prove

$$\Xi_\nabla(R)\, ; \Xi(R) = \Xi_\nabla(R) = \Xi(R)\, ; \Xi_\nabla(R),$$
$$\Psi_\nabla(R)\, ; \Psi(R) = \Psi_\nabla(R) = \Psi(R)\, ; \Psi_\nabla(R).$$

10.2 Prove that for an arbitrary (possibly heterogeneous) relation R, always

$$\Xi(R) \subseteq \Xi(\texttt{fringe}(R)).$$

10.4 Difunctional relations

Relations that are identical with their fringe have turned out to be important. They are called difunctional relations. The concept of a difunctional relation has already been formulated for heterogeneous relations in a phenomenological form in Def. 5.30. It generalizes the concept of a (possibly partial) equivalence relation in as far as source and target need no longer be identical. It also generalizes the concept of a matching to a block form. We recall that a (possibly heterogeneous) relation R is called **difunctional** if $R\, ; R^{\mathsf{T}}\, ; R \subseteq R$ (which is the essential requirement, meaning in fact $R\, ; R^{\mathsf{T}}\, ; R = R$ since $R\, ; R^{\mathsf{T}}\, ; R \supseteq R$ is true for every relation), or equivalently, if $R\, ; \overline{R}^{\mathsf{T}}\, ; R \subseteq \overline{R}$.

The property of being difunctional is not often met when arbitrary relations are considered. Nevertheless, this concept is really important as we will see later that every relation possesses a difunctional closure with respect to which the relation can be decomposed. These decompositions in turn give rise to all the applications. They are widely used in data analysis as well to optimize data bases, for knowledge discovery in data bases, unsupervised learning, data warehousing, etc. We give a first idea with the rearrangement of Fig. 10.25. There we also indicate how one may factorize the relation – so far without proof.

There exist many equivalent ways of expressing the property of being difunctional.

Proposition 10.18. *For any relation $R : X \longrightarrow Y$ the following are equivalent:*

  (i) *R is difunctional.*
 (ii) *$R = \texttt{fringe}(R)$.*
(iii) *Every pair $(x,y) \in R$ is contained in precisely one non-enlargeable rectangle of R.*
(iv) *R is the union of rectangles with row and column intervals pairwise non-overlapping.*

Proof: (i) $\Longleftrightarrow$ (ii) is already known from Cor. 10.17.ii.

(ii) $\Longrightarrow$ (iii) is trivial according to Prop. 10.9. Therefore, every point in R is contained in just one non-enlargeable rectangle.

(iii) $\Longrightarrow$ (iv): Every relation is in a trivial way the union of non-enlargeable rectangles, at least by one-point rectangles. Assume that two different non-enlargeable rectangles intersect in a non-trivial way. Then a point in the intersection is obviously contained in more than one non-enlargeable rectangle – contradicting assumption (iii). Even if the rectangles do not intersect and only their source vectors, for example, intersect, this would give rise to a point on the source side with horizontal extension touching both rectangles.

(iv) $\Longrightarrow$ (i): It is trivial to convince oneself by matrix computation that $R\,R^\mathsf{T}\,R = R$ whenever R is the union of axis-parallel rectangles with non-overlapping source and target sides. $\qquad\square$

The block-diagonal structure is investigated in the following proposition remembering row and column equivalence of Def. 5.28 as well as the left and the right equivalence of Def. 5.35. Most of the important properties of difunctional relations rest on the close similarity of the row equivalence and the left equivalence (and similarly of the column equivalence and the right equivalence) that we are now going to investigate.

Proposition 10.19. *Given any difunctional relation R, we have*

  (i) $(R\,R^\mathsf{T})^{+} = R\,R^\mathsf{T} = \Xi(R) \cap R\,\mathbb{T} = \Xi(R) \cap \mathbb{T}\,R^\mathsf{T} \subseteq \Xi(R).$

(ii) $R; R^\mathsf{T} \cup \overline{R; \mathbb{T}}; \overline{R; \mathbb{T}}^\mathsf{T} = \Xi(R)$.

(iii) $(R; R^\mathsf{T})^* \subseteq \Xi(R)$

$\quad (R; R^\mathsf{T})^* = \Xi(R) \qquad$ *if R is total.*

Proof: (i) The first equality is trivial in the case when R is difunctional. We concentrate on proving, for example

$$R; R^\mathsf{T} \subseteq \Xi(R) = \mathsf{syq}(R^\mathsf{T}, R^\mathsf{T}) = \overline{\overline{R; R^\mathsf{T}}} \cap \overline{R; \overline{R}^\mathsf{T}},$$

with $\Xi(R)$ already expanded by definition. But this is trivial, because, for example

$$R; R^\mathsf{T} \subseteq \overline{\overline{R; R^\mathsf{T}}} \iff \overline{R; R^\mathsf{T}}; R \subseteq \overline{R} \iff R; R^\mathsf{T}; R \subseteq R \iff R \text{ difunctional.}$$

Containment $R; R^\mathsf{T} \supseteq \Xi(R) \cap \mathbb{T}; R^\mathsf{T}$, for example, follows from

$$\mathbb{T} = \mathbb{T}; R^\mathsf{T} \cup \overline{\mathbb{T}; R^\mathsf{T}} = (\overline{R} \cup R); R^\mathsf{T} \cup \overline{\mathbb{T}; R^\mathsf{T}} = \overline{R}; R^\mathsf{T} \cup R; R^\mathsf{T} \cup \overline{\mathbb{T}; R^\mathsf{T}}$$
$$\iff \overline{\overline{R; R^\mathsf{T}}} \cap \mathbb{T}; R^\mathsf{T} \subseteq R; R^\mathsf{T}.$$

(ii) "$\subseteq$" follows partly from (i) and partly from

$$\overline{R; \mathbb{T}}; \overline{R; \mathbb{T}}^\mathsf{T} \subseteq \Xi(R) = \overline{\overline{R; R^\mathsf{T}}} \cap \overline{R; \overline{R}^\mathsf{T}}$$
$$\iff (\overline{R; R^\mathsf{T}} \cup R; \overline{R}^\mathsf{T}); \overline{R; \mathbb{T}} \subseteq R; \mathbb{T}.$$

Now, inclusion for the first term holds because $R^\mathsf{T}; \overline{R; \mathbb{T}} \subseteq \bot\!\!\!\bot$. The second is trivially contained in $R; \mathbb{T}$.

"$\supseteq$" holds for arbitrary relations:

$$R; R^\mathsf{T} \cup \overline{R; \mathbb{T}}; \overline{R; \mathbb{T}}^\mathsf{T} \supseteq \overline{\overline{R; R^\mathsf{T}}} \cap \overline{R; \overline{R}^\mathsf{T}}$$
$$\iff \mathbb{T} = R; R^\mathsf{T} \cup \overline{R; \mathbb{T}}; \overline{R; \mathbb{T}}^\mathsf{T} \cup \overline{R; R^\mathsf{T}} \cup R; \overline{R}^\mathsf{T}$$
$$\iff \mathbb{T} = \mathbb{T}; R^\mathsf{T} \cup R; \mathbb{T} \cup \overline{R; \mathbb{T}}; \overline{R; \mathbb{T}}^\mathsf{T}$$
$$\iff \mathbb{T} = (R; \mathbb{T})^\mathsf{T} \cup R; \mathbb{T} \cup \overline{R; \mathbb{T}}; \overline{R; \mathbb{T}}^\mathsf{T}$$
$$\iff \overline{R; \mathbb{T}}^\mathsf{T} \cap \overline{R; \mathbb{T}} \subseteq \overline{R; \mathbb{T}}; \overline{R; \mathbb{T}}^\mathsf{T}.$$

Using $\overline{R; \mathbb{T}}; \mathbb{T} = \overline{R; \mathbb{T}}$,[3] this may be proved as in the proof B.14 in the appendix.

(iii) Reflexivity together with (i) suffices. In the case of totality, $R; \mathbb{T} = \mathbb{T}$.   □

Of course, all this holds correspondingly for the column side. We leave the respective reformulation as an exercise.

It is sometimes tricky to handle the differences resulting from totality or non-totality of R.

Both the row and column equivalence $\Xi(R), \Psi(R)$ as well as the left and the right

---

[3] This is again a point where we need that $\mathbb{T}; \mathbb{T} = \mathbb{T}$, which is not satisfied in an arbitrary relation algebra.

equivalence $(R\,R^\mathsf{T})^*, (R\,R^\mathsf{T})^*$, constitute congruences. For arbitrary relations, they are rather different, while they are close to identical for difunctional relations. The classes belonging to non-empty rows coincide. While the former collects all empty rows, if any, into one class, the latter keeps them separated in 1-element classes.

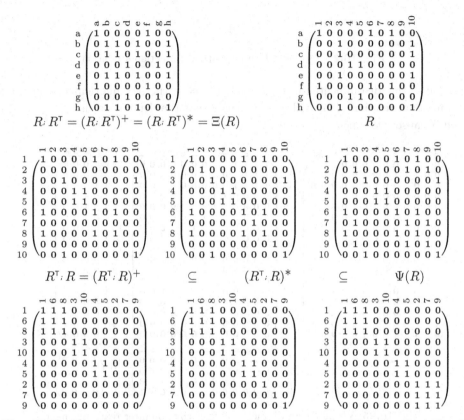

$$R\,R^\mathsf{T} = (R\,R^\mathsf{T})^+ = (R\,R^\mathsf{T})^* = \Xi(R) \qquad\qquad R$$

$$R^\mathsf{T}\,R = (R^\mathsf{T}\,R)^+ \qquad \subseteq \qquad (R^\mathsf{T}\,R)^* \qquad \subseteq \qquad \Psi(R)$$

Fig. 10.23 Coinciding and non-coinciding equivalences around a difunctional relation R

In Fig. 10.23, the possible situations are shown. The left or source side is trivial in as far as everything coincides; in contrast, on the target or right side: there are three relations

$$R^\mathsf{T}\,R = (R^\mathsf{T}\,R)^+ \subseteq (R^\mathsf{T}\,R)^* \subseteq \Psi(R)$$

that are indeed different. The rearranged relations give better intuition concerning the lower right area described by $\overline{R^\mathsf{T}\,\mathbb{T}}\,,\overline{R^\mathsf{T}\,\mathbb{T}}^\mathsf{T}$.

These situations shall now be considered in more detail forming quotients according to Fig. 10.24.

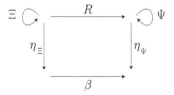

Fig. 10.24 Quotient forming with a difunctional relation

We follow the idea indicated in Fig. 10.24, and construct the quotient relation of a difunctional relation. This will give evidence that the denotation has been chosen well because two functions are involved.

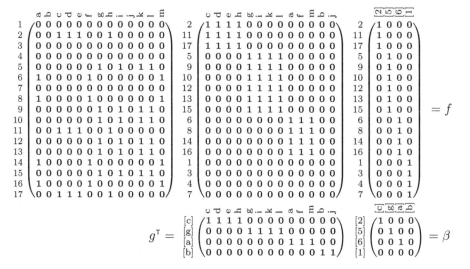

Fig. 10.25 Visualizing a difunctional factorization after rearrangement

Proposition 10.20. *Let some difunctional relation R be given together with its row equivalence and its column equivalence (see Def. 5.28),*

$$\Xi := \Xi(R) := \mathsf{syq}(R^{\mathsf{T}}, R^{\mathsf{T}}) \qquad \Psi := \Psi(R) := \mathsf{syq}(R, R),$$

as well as with the corresponding natural projection mappings η_{Ξ}, η_{Ψ}. Then the quotient

$$\beta := \eta_{\Xi}^{\mathsf{T}}{}_{;} R {}_{;} \eta_{\Psi}$$

of R modulo row and column equivalence is a matching, i.e., a univalent and injective relation.

Proof: The results of Prop. 10.19 extend to

$R^\mathsf{T} ; \Xi ; R = R^\mathsf{T} ; R \subseteq \Psi.$

This in turn allows us to calculate, for example,

$$
\begin{aligned}
\beta^\mathsf{T} ; \beta &= \eta_\Psi^\mathsf{T} ; R^\mathsf{T} ; \eta_\Xi ; \eta_\Xi^\mathsf{T} ; R ; \eta_\Psi \\
&= \eta_\Psi^\mathsf{T} ; R^\mathsf{T} ; \Xi ; R ; \eta_\Psi \\
&\subseteq \eta_\Psi^\mathsf{T} ; \Psi ; \eta_\Psi \\
&= \eta_\Psi^\mathsf{T} ; \eta_\Psi ; \eta_\Psi^\mathsf{T} ; \eta_\Psi \\
&= \mathbb{I} ; \mathbb{I} = \mathbb{I}. \qquad\qquad\qquad\qquad\qquad\qquad \Box
\end{aligned}
$$

We can add several easy consequences. In the case when R is total, β will be a mapping; in the case when R is surjective, β will be a surjective matching; in the case when R is total and surjective, β will be a bijective mapping.

Corollary 10.21.

R *difunctional* $\iff$ *There exist functions* f, g *such that* $R = f ; g^\mathsf{T}.$

Proof: Following Prop. 10.20, take $f := \eta_\Xi ; \beta$ and $g := \eta_\Psi$ to establish "$\Longrightarrow$". The opposite direction "$\Longleftarrow$" is trivial. $\qquad\qquad\qquad\qquad\qquad\qquad\qquad\qquad\qquad \Box$

The functions f, g need not be mappings as in the next corollary which is a variant included in order to match with future analogous decompositions.

Corollary 10.22.

R *difunctional* $\iff$ *There exist surjective mappings* f_0, g *and a matching* β *such that* $R = f_0 ; \beta ; g^\mathsf{T}.$

Proof: We simply choose $f_0 := \eta_\Xi$ in the proof of Prop. 10.20. $\qquad\qquad\qquad \Box$

Difunctional closures

The importance of difunctionality rests mainly on the fact that every relation admits a difunctional closure, i.e., a smallest difunctional relation containing it. Blowing up R with the left equivalence $\Omega := (R ; R^\mathsf{T})^*$ and with the right equivalence $\Omega' = (R^\mathsf{T} ; R)^*$ so as to obtain $\Omega ; R ; \Omega'$, will bring about what is known as the difunctional closure. The definition proper, however, is descriptive in nature.

$$
\begin{array}{c}
\begin{array}{c}\;1\,2\,3\,4\,5\,6\,7\,8\,9\end{array}\\
\begin{array}{c}
a\\b\\c\\d\\e\\f\\g\\h\\i\\j\\k
\end{array}
\left(\begin{array}{ccccccccc}
1&0&1&0&1&0&0&0&0\\
0&0&0&1&0&0&0&0&0\\
0&1&0&0&0&0&0&0&0\\
0&0&0&0&1&0&0&0&0\\
0&0&0&0&0&0&1&0&0\\
0&0&0&0&0&0&0&0&1\\
0&0&0&0&0&0&0&0&1\\
0&0&0&0&0&0&0&0&1\\
0&0&0&1&0&0&0&0&1\\
0&0&0&0&0&0&0&0&0\\
0&1&0&0&0&0&1&0&0
\end{array}\right)\\
R
\end{array}
\quad
\begin{array}{c}
\begin{array}{c}\;a\,b\,c\,d\,e\,f\,g\,h\,i\,j\,k\end{array}\\
\left(\begin{array}{ccccccccccc}
1&0&0&1&0&0&0&0&0&0&0\\
0&1&0&0&0&1&1&1&1&0&0\\
0&0&1&0&1&0&0&0&0&0&1\\
1&0&0&1&0&0&0&0&0&0&0\\
0&0&1&0&1&0&0&0&0&0&1\\
0&1&0&0&0&1&1&1&1&0&0\\
0&1&0&0&0&1&1&1&1&0&0\\
0&1&0&0&0&1&1&1&1&0&0\\
0&1&0&0&0&1&1&1&1&0&0\\
0&0&0&0&0&0&0&0&0&1&0\\
0&0&1&0&1&0&0&0&0&0&1
\end{array}\right)\\
\Xi(h_{\mathrm{difu}}(R))
\end{array}
$$

$$
\begin{array}{c}
\begin{array}{c}\;1\,2\,3\,4\,5\,6\,7\,8\,9\end{array}\\
\left(\begin{array}{ccccccccc}
1&0&1&0&1&0&0&0&0\\
0&0&0&1&0&0&0&0&1\\
0&1&0&0&0&0&1&0&0\\
1&0&1&0&1&0&0&0&0\\
0&1&0&0&0&0&1&0&0\\
0&0&0&1&0&0&0&0&1\\
0&0&0&1&0&0&0&0&1\\
0&0&0&1&0&0&0&0&1\\
0&0&0&1&0&0&0&0&1\\
0&0&0&0&0&0&0&0&0\\
0&1&0&0&0&0&1&0&0
\end{array}\right)\\
h_{\mathrm{difu}}(R)
\end{array}
\quad
\begin{array}{c}
\begin{array}{c}\;1\,2\,3\,4\,5\,6\,7\,8\,9\end{array}\\
\begin{array}{c}1\\2\\3\\4\\5\\6\\7\\8\\9\end{array}
\left(\begin{array}{ccccccccc}
1&0&1&0&1&0&0&0&0\\
0&1&0&0&0&0&1&0&0\\
1&0&1&0&1&0&0&0&0\\
0&0&0&1&0&0&0&0&1\\
1&0&1&0&1&0&0&0&0\\
0&0&0&0&0&1&0&1&0\\
0&1&0&0&0&0&1&0&0\\
0&0&0&0&0&1&0&1&0\\
0&0&0&1&0&0&0&0&1
\end{array}\right)\\
\Psi(h_{\mathrm{difu}}(R))
\end{array}
$$

Fig. 10.26 R with its difunctional closure between row and column equivalence

Here again, we are confronted with the close similarity between $(R\,\raise0.3ex\hbox{;}\,R^{\mathsf T})^*$ and $\Xi(R)$ of Prop. 10.19 and Fig. 10.23. In the rightmost relation $\Psi(h_{\mathrm{difu}}(R))$ of Fig. 10.27, we have just one equivalence class $[6,8]$. Using $\Omega' = (R^{\mathsf T}\,\raise0.3ex\hbox{;}\,R)^*$ for the arrangement into groups, there would have been two classes $[6]$ and $[8]$.

$$
\begin{array}{c}
\begin{array}{c}\;1\,3\,5\,4\,9\,2\,7\,6\,8\end{array}\\
\begin{array}{c}a\\d\\b\\f\\g\\h\\i\\c\\e\\k\\j\end{array}
\left(\begin{array}{ccccccccc}
1&1&1&0&0&0&0&0&0\\
1&1&1&0&0&0&0&0&0\\
0&0&0&1&1&0&0&0&0\\
0&0&0&1&1&0&0&0&0\\
0&0&0&1&1&0&0&0&0\\
0&0&0&1&1&0&0&0&0\\
0&0&0&1&1&0&0&0&0\\
0&0&0&0&0&1&1&0&0\\
0&0&0&0&0&1&1&0&0\\
0&0&0&0&0&1&1&0&0\\
0&0&0&0&0&0&0&0&0
\end{array}\right)
\end{array}
\quad
\begin{array}{c}
\begin{array}{c}\;1\,3\,5\;4\,9\;2\,7\;6\,8\end{array}\\
\begin{array}{c}a\\d\\b\\f\\g\\h\\i\\c\\e\\k\\j\end{array}
\left(\begin{array}{ccc|cc|cc|cc}
1&1&1&0&0&0&0&0&0\\
0&0&1&0&0&0&0&0&0\\
0&0&0&1&0&0&0&0&0\\
0&0&0&0&1&0&0&0&0\\
0&0&0&0&1&0&0&0&0\\
0&0&0&0&1&0&0&0&0\\
0&0&0&1&1&0&0&0&0\\
0&0&0&0&0&1&0&0&0\\
0&0&0&0&0&1&0&0&0\\
0&0&0&0&0&1&1&0&0\\
0&0&0&0&0&0&0&0&0
\end{array}\right)
\end{array}
$$

Fig. 10.27 Difunctional closure of Fig. 10.26 rearranged; original arranged accordingly

Another example of a difunctional block decomposition is shown in Fig. 10.28. In the case when every block has its own identity, one can store the information concerning the relation much more easily than in the normal matrix form with $m \times n$ Boolean coefficients.

$$
A =
\begin{array}{c}
\begin{array}{c}\;1\,2\,3\,4\,5\,6\,7\,8\;9\,10\,11\,12\,13\,14\end{array}\\
\begin{array}{c}1\\2\\3\\4\\5\\6\\7\\8\\9\\10\\11\\12\end{array}
\left(\begin{array}{cccccccccccccc}
1&0&1&0&1&0&0&1&1&0&1&1&0&1\\
0&1&0&1&0&0&1&0&0&1&0&0&0&0\\
0&0&0&0&0&1&0&0&0&0&0&0&1&0\\
1&0&1&0&1&0&0&1&1&0&1&1&0&1\\
0&1&0&1&0&0&1&0&0&1&0&0&0&0\\
1&0&1&0&1&0&0&1&1&0&1&1&0&1\\
0&1&0&1&0&0&1&0&0&1&0&0&0&0\\
0&0&0&0&0&0&0&0&0&0&0&0&0&0\\
0&0&0&0&0&1&0&0&0&0&0&0&1&0\\
1&0&1&0&1&0&0&1&1&0&1&1&0&1\\
0&0&0&0&0&1&0&0&0&0&0&0&1&0\\
1&0&1&0&1&0&0&1&1&0&1&1&0&1
\end{array}\right)
\end{array}
$$

$$
A_{\mathrm{rearranged}} =
\begin{array}{c}
\begin{array}{c}\;1\,3\,5\,8\;9\,11\,12\,14\;2\,4\,7\,10\;6\,13\end{array}\\
\begin{array}{c}1\\4\\6\\10\\12\\2\\5\\7\\3\\9\\11\\8\end{array}
\left(\begin{array}{cccccccc|cccc|cc}
1&1&1&1&1&1&1&1&0&0&0&0&0&0\\
1&1&1&1&1&1&1&1&0&0&0&0&0&0\\
1&1&1&1&1&1&1&1&0&0&0&0&0&0\\
1&1&1&1&1&1&1&1&0&0&0&0&0&0\\
1&1&1&1&1&1&1&1&0&0&0&0&0&0\\
0&0&0&0&0&0&0&0&1&1&1&1&0&0\\
0&0&0&0&0&0&0&0&1&1&1&1&0&0\\
0&0&0&0&0&0&0&0&1&1&1&1&0&0\\
0&0&0&0&0&0&0&0&0&0&0&0&1&1\\
0&0&0&0&0&0&0&0&0&0&0&0&1&1\\
0&0&0&0&0&0&0&0&0&0&0&0&1&1\\
0&0&0&0&0&0&0&0&0&0&0&0&0&0
\end{array}\right)
\end{array}
$$

Fig. 10.28 A difunctional relation with block rearrangement

Based on this idea for storing difunctional relations, one may first proceed to a difunctional closure and then apply the decomposition according to the latter; see yet another example in Fig. 10.29.

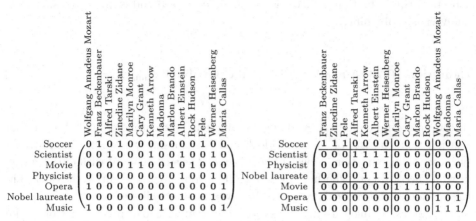

Fig. 10.29 A relation with block rearrangement according to its difunctional closure

Thus prepared, we proceed to characterize the difunctional closure.

Proposition 10.23. *Let any (possibly heterogeneous) relation R be given and consider its so-called* **difunctional closure**

$$h_{\mathrm{difu}}(R) := \inf \{\, H \mid R \subseteq H, H \text{ is difunctional} \,\},$$

i.e., the smallest difunctional relation containing R.

(i) $h_{\mathrm{difu}}(R)$ *is well defined.*

(ii) $h_{\mathrm{difu}}(R) = R\, (R^{\mathsf{T}}\, R)^{+} = (R\, R^{\mathsf{T}})^{+}\, R = (R\, R^{\mathsf{T}})^{+}\, R\, (R^{\mathsf{T}}\, R)^{+}.$

(iii) $h_{\mathrm{difu}}(R) = R\, (R^{\mathsf{T}}\, R)^{*} = (R\, R^{\mathsf{T}})^{*}\, R = (R\, R^{\mathsf{T}})^{*}\, R\, (R^{\mathsf{T}}\, R)^{*}.$

Proof: (i) The selection condition '$R \subseteq H$, H is difunctional' always delivers a non-empty set since $\mathbb{T}$ is certainly difunctional. It is infimum-hereditary since whenever $R \subseteq H_i$ for $i = 1, 2$, we have obviously $R \subseteq H_1 \cap H_2$, and, similarly, for infinite intersections. It is also easily shown that the intersection of difunctional relations will be difunctional again.

(ii) is shown with standard methods of regular algebra. We concentrate on the first variant. Difunctionality of $R\, (R^{\mathsf{T}}\, R)^{+}$ is trivial, so that certainly $D := h_{\mathrm{difu}}(R) \subseteq R\, (R^{\mathsf{T}}\, R)^{+}$. To show containment in the other direction, we start with $R \subseteq D$. But then also $R\, R^{\mathsf{T}}\, R \subseteq D\, D^{\mathsf{T}}\, D \subseteq D$ holds, since D is difunctional. Iteratively applied, we get $R\, (R^{\mathsf{T}}\, R)^{+} \subseteq D$.

(iii) follows from (ii) since only $R\, \mathbb{I}$ is added, for example. $\qquad\square$

Meanwhile, we have some feeling of what a difunctional matrix looks like. We also know the algebraic characterization from the definition. Now we ask for the practical aspects of this definition, which has long been discussed and is known in numerical analysis as chainability.

Definition 10.24. A relation R is called **chainable** if $h_{\text{difu}}(R) = \mathbb{T}$. □

In the preceding definition, we have for simplicity concentrated on a difunctional relation that consists of just one block and has no empty rows or columns. For purposes of visualization, conceive the relation R as a chessboard with dark squares or white according to whether R_{ik} is **1** or **0**. A rook operates on this chessboard in the horizontal and vertical directions; however, it is only allowed to change direction on **1**-squares. Using this interpretation, the definition declares a relation to be chainable if the non-vanishing entries (i.e., the dark squares) can all be reached from one another by a sequence of 'rook moves'.

```
     a b c d e f g h i j k l m n
 1  /0 0 1 0 0 0 0 1 0 0 0 0 0 0\
 2  | 0 1 0 1 0 0 0 0 1 0 0 0 0 0 |
 3  | 0 0 0 0 0 0 0 0 0 0 0 0 1 0 |
 4  | 1 0 0 0 1 0 0 0 1 0 1 0 0 1 |
 5  | 0 0 0 0 0 1 0 0 0 0 0 0 0 0 |
 6  | 1 0 1 0 0 0 0 0 0 0 0 1 0 1 |
 7  | 0 0 0 1 0 0 1 0 0 0 0 0 0 0 |
 8  | 0 0 0 0 0 0 0 0 0 0 0 0 0 0 |
 9  | 0 0 0 0 0 1 0 0 0 0 0 0 1 0 |
10  | 0 0 0 0 0 0 1 0 0 1 0 0 1 0 |
11  | 0 0 0 0 0 0 0 0 0 0 0 0 1 0 |
12  \0 0 0 0 0 0 0 1 0 0 1 0 0 0/
```

```
       a c e h i k l n  b d g j  f m
 1  /0 1 0 1 0 0 0 0 | 0 0 0 0 | 0 0\
 4  | 1 0 1 0 1 1 0 1 | 0 0 0 0 | 0 0 |
 6  | 1 1 0 0 0 0 1 1 | 0 0 0 0 | 0 0 |
10  | 0 0 0 1 0 1 0 0 | 0 0 0 0 | 0 0 |
12  | 0 0 0 0 1 0 1 0 | 0 0 0 0 | 0 0 |
 2  | 0 0 0 0 0 0 0 0 | 1 1 0 1 | 0 0 |
 5  | 0 0 0 0 0 0 0 0 | 0 0 1 0 | 0 0 |
 7  | 0 0 0 0 0 0 0 0 | 0 1 1 0 | 0 0 |
 3  | 0 0 0 0 0 0 0 0 | 0 0 0 0 | 0 1 |
 9  | 0 0 0 0 0 0 0 0 | 0 0 0 0 | 1 1 |
11  | 0 0 0 0 0 0 0 0 | 0 0 0 0 | 0 1 |
 8  \0 0 0 0 0 0 0 0 | 0 0 0 0 | 0 0/
```

Fig. 10.30 A decomposition according to the difunctional closure visualized via permutation and partitioning

Figure 10.30 shows these rook moves in every diagonal block. Obviously, one cannot reach another block in this way. We illustrate this definition further, mentioning a related concept. For the moment, the relation R is conceived as a hypergraph-incidence between hyperedges (rows) and vertices (columns). Then $K := \overline{\mathbb{I}} \cap R\, ; R^{\mathsf{T}}$ is the so-called edge-adjacency, see for example [123, 124]. Since adjacencies are traditionally conceived as being irreflexive, we face minor technical difficulties in managing the transition from $R\, ; R^{\mathsf{T}}$ to K.

Proposition 10.25. *A total and surjective relation R is chainable precisely when its edge-adjacency K is strongly connected.*

Proof: Total is $\mathbb{I} \subseteq R\, ; R^{\mathsf{T}}$, and thus $\mathbb{I} \cup R\, ; R^{\mathsf{T}} = R\, ; R^{\mathsf{T}}$. First we relate R with the edge-adjacency and show $K^* = (R R^\mathsf{T})^*$, using the formula $(A \cup B)^* = (A^* B)^* A^*$, well-known from regular algebra. Then

$$(R; R^\mathsf{T})^* = \left[\mathbb{I} \cup (\overline{\mathbb{I}} \cap R; R^\mathsf{T})\right]^* = \left[\mathbb{I}^*; (\overline{\mathbb{I}} \cap R; R^\mathsf{T})\right]^*; \mathbb{I}^* = (\mathbb{I}; K)^*; \mathbb{I} = K^*.$$

In the first direction, we have from chainability, $h_{\mathrm{difu}}(R) = (R; R^\mathsf{T})^*; R = \mathbb{T}$, immediately that $K^* = (R; R^\mathsf{T})^* \supseteq (R; R^\mathsf{T})^+ = (R; R^\mathsf{T})^*; R; R^\mathsf{T} = \mathbb{T}; R^\mathsf{T} = \mathbb{T}$, since R is total.

The other direction is $h_{\mathrm{difu}}(R) = (R; R^\mathsf{T})^*; R = K^*; R = \mathbb{T}; R = \mathbb{T}$, and we see that R must be surjective, as otherwise there might exist an isolated vertex unrelated to all the edges. □

Application to knowledge acquisition

There are several fields of data mining in which difunctionality is heavily used. One such field is predictive modelling, which means classification and regression. Another is clustering, i.e., grouping similar objects. Finally, summarization should be mentioned which means discovering association rules from given data. The input to a corresponding algorithm is a single flat table – usually of real numbers, but in our case of truth values. So we start from an arbitrary relation that may be heterogeneous. If data from several relations are studied, one can subsume this under the aforementioned case introducing products, for example.

The task is to output a *pattern* valid in the given data. Such a pattern is simply a proposition describing relationships among the facts of the given data. It is useful only when it is simpler than the enumeration of all the given facts.

Many algorithms of this type have been developed for machine learning and for the interpretation of statistics. Machine learning does not mean the intelligent creation of propositions; rather the algorithms are fed with a set of hypotheses – the patterns – that may be valid in the data, and they perform an exhaustive or heuristic search for valid hypotheses. A pattern is like a proposition or a predicate and so splits the dataset according to whether or not it is satisfied.

The idea may best be captured by looking at the example in Fig. 10.31. First, the difunctional closure of some given relation R is determined. The relation we start with is the one on the top line. The difunctional closure thus obtained gives rise to a row as well as a column equivalence, here positioned accordingly in the second line. Once these congruences have been found, they can be turned into rules and predicates that are here only presented with our standard technique of algebraic visualization.

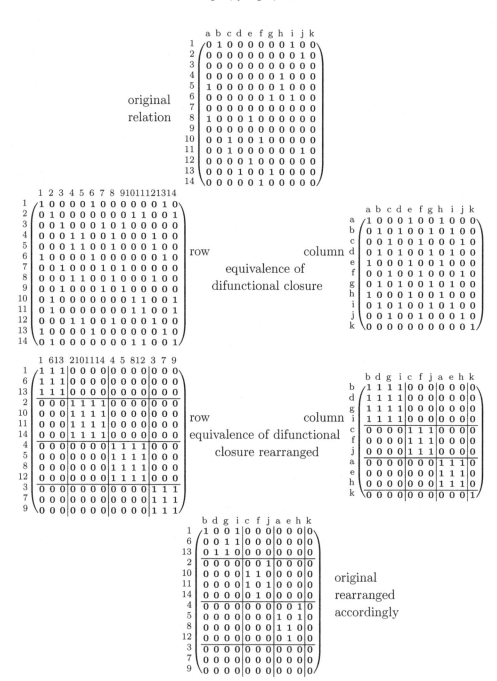

Fig. 10.31 Using difunctional decomposition for knowledge acquisition

In the subdivided form one will recognize subsets and predicates: classes according to the row and the column equivalence first. Propositions may also easily be deduced. For example, the elements of the first subset on the source side are only related to elements of the first subset on the range side, etc. The subsets have been determined during the running of the algorithm and afterwards made visible using the technique of algebraic visualization we have already applied several times. One should bear in mind that all this has been deduced using standard relational techniques from the initial relation.

Difunctionality and independence

The concept of independent sets also has a close connection to difunctionality. The proof will, however, employ the Point Axiom.

Proposition 10.26. *Let a finite (possibly heterogeneous) relation A be given. Then A is either chainable or it admits a pair (s,t) which is non-trivial, i.e., $s \neq \bot\!\bot$ and $t \neq \bot\!\bot$, such that both, s,t as well as $\bar{s}, \bar{t}$, constitute at the same time an independent pair of sets and a covering pair of sets, i.e.*

$$A_{;}t \subseteq \bar{s} \qquad and \qquad A_{;}\bar{t} \subseteq s.$$

Proof: Consider the difunctional closure $G := h_{\mathsf{difu}}(A)$. The dichotomy is as to whether $G \neq \mathbb{T}$ or $G = \mathbb{T}$, in which case A is chainable according to Def. 10.24. Assume the former; then there exist points x, y satisfying $x_{;} y^{\mathsf{T}} \subseteq \overline{G}$, for which we construct the non-enlargeable rectangle outside G started horizontally according to Prop. 10.3.i. Then

$$x \subseteq u_x = \overline{G_{;}\overline{G^{\mathsf{T}}_{;}x}} \qquad and \qquad y \subseteq v_x = \overline{G^{\mathsf{T}}_{;}x}.$$

Setting $s := u_x$ and $t := v_x$, we find that the two formulae are satisfied, for which we use that G is difunctional so that for arbitrary Y always $G^{\mathsf{T}}_{;}G_{;}\overline{G^{\mathsf{T}}_{;}Y} \subseteq \overline{G^{\mathsf{T}}_{;}Y}$.

By construction $\bot\!\bot \neq x \subseteq s$ and $\bot\!\bot \neq y \subseteq t$. □

Difunctionality and covering pairs of sets are related basically in the following way.

Proposition 10.27. *If and only if a relation A admits a pair (x,y) such that (x,y) and $(\bar{x}, \bar{y})$ is a covering pair of sets, its difunctional closure will admit this covering pair of sets.*

Proof: Let $H := h_{\mathsf{difu}}(A)$. It is trivial to conclude from H to A since $A \subseteq H$.

From $A_{;}\bar{y} \subseteq x$ and $A_{;}\overline{\bar{y}} \subseteq \bar{x}$, or equivalently, $A^{\mathsf{T}}_{;}x \subseteq \bar{y}$, we derive $A \subseteq \mathsf{syq}\,(x^{\mathsf{T}}, \bar{y}^{\mathsf{T}})$. Since this symmetric quotient is some difunctional relation above A, it is above H, resulting in $H_{;}\bar{y} \subseteq x$ and $H_{;}y \subseteq \bar{x}$. □

Exercises

10.3 Let R be any difunctional relation. We consider the direct sum of its source and target, and thereon the composite relation

$$R_c := \begin{pmatrix} \mathbb{I} \cup R \, R^\mathsf{T} & R \\ R^\mathsf{T} & \mathbb{I} \cup R^\mathsf{T} \, R \end{pmatrix}.$$

Prove that it is an equivalence.

10.4 Prove that $R^\mathsf{T} \, R \, S \subseteq S$ implies $R \, \overline{S} = R \, \mathbb{T} \cap \overline{R \, S}$ (this should be compared with Prop. 5.6).

10.5 If f and g are univalent and Ξ is an equivalence relation, then $f \, \Xi \, g^\mathsf{T}$ is difunctional.

10.6 Prove or disprove: if A and B are chainable, then $A \, B$ is chainable.

10.5 Ferrers relations

We have seen that a difunctional relation corresponds to a (possibly partial) block-diagonal relation. So the question arose as to whether there was a counterpart of a linear order with rectangular block-shaped matrices. In this context, the Ferrers[4] property of a relation is studied.

Definition 10.28. We say that a (possibly heterogeneous) relation

A is **Ferrers** if $A \, \overline{A}^\mathsf{T} \, A \subseteq A$. □

Equivalent forms of the condition proper, which are easy to prove, are

$$\overline{R} \, R^\mathsf{T} \, \overline{R} \subseteq \overline{R} \quad \text{and} \quad R^\mathsf{T} \, \overline{R} \, R^\mathsf{T} \subseteq R^\mathsf{T};$$

i.e., R is Ferrers, if and only if $\overline{R}$ or R^T are Ferrers.[5]

The meaning of the algebraic condition just presented will now be visualized and interpreted.

[4] Introduced by Jacques Riguet in [106].
[5] From Wikipedia, the free encyclopedia: Norman Macleod Ferrers (1829–1903) was a British mathematician at Cambridge's Gonville and Caius College (vice chancellor of Cambridge University 1884) who now seems to be remembered mainly for pointing out a conjugacy in integer partition diagrams, which are accordingly called Ferrers graphs and are closely related to Young diagrams.
N. M. Ferrers, *An Elementary Treatise on Trilinear Coordinates*, London, 1861.
N. M. Ferrers (ed.), *Mathematical Papers of the Late George Green*, 1871.

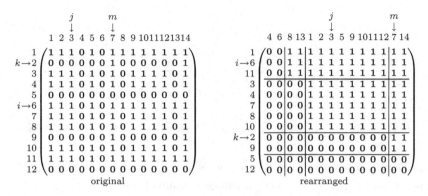

Fig. 10.32 A Ferrers relation

It is at first sight not at all clear that the matrix representing A may, owing to the Ferrers property, be written in staircase (or echelon) block form after suitably rearranging rows and columns independently. The graph interpretation is as follows. Given any two arrows, there exists an arrow from one of the starting points leading to the ending point of the other arrow.

$$\forall i,k \in X, \forall j,m \in Y : \big\{(i,j) \in A \wedge (k,m) \in A \quad \to \quad (i,m) \in A \vee (k,j) \in A\big\}$$

$$\iff \forall i,j,k,m : \big\{(i,j) \in A \wedge (j,k) \notin A^\mathsf{T} \wedge (k,m) \in A \quad \to \quad (i,m) \in A\big\}$$

$$\iff \forall i,m : \big\{\big[\exists k : (\exists j : (i,j) \in A \wedge (j,k) \notin \overline{A}^\mathsf{T}) \wedge (k,m) \in A\big] \to (i,m) \in A\big\}$$
$$\text{definition of composition}$$

$$\iff \forall i,m : \big\{\big[\exists k : (i,k) \in A\,\overline{A}^\mathsf{T} \wedge (k,m) \in A\big] \to (i,m) \in A\big\}$$
$$\text{definition of composition}$$

$$\iff \forall i,m : \big\{(i,m) \in A\,\overline{A}^\mathsf{T}\,A \quad \to \quad (i,m) \in A\big\}$$
$$\text{transition to point-free form}$$

$$\iff A\,\overline{A}^\mathsf{T}\,A \subseteq A$$

In Fig. 10.33 again, we explain this using the dashed arrow convention. The sentence above said: 'Given any two arrows, there exists an arrow from one of the starting points leading to the ending point of the other arrow'. This is modified only slightly. If from one starting point of the two arrows an arrow to the endpoint of the other arrow does not exist – indicated by the dashed arrow – then it will exist from the starting point of the other arrow.

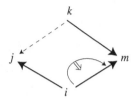

Fig. 10.33 Ferrers property represented with the dashed arrow convention

Being Ferrers has something in common with a linear (strict-)order, however, of the row-contains-preorder or the column-is-contained-preorder.

Proposition 10.29. *Let R be an arbitrary (possibly heterogeneous) relation.*

(i) R *Ferrers* $\implies$ $R_{;}\overline{R}^{\mathsf{T}}$ *as well as* $\overline{R}^{\mathsf{T}}_{;}R$ *are transitive and Ferrers.*

(ii) R *Ferrers* $\iff$ $\mathcal{R}(R)$ *connex* $\iff$ $\mathcal{C}(R)$ *connex.*[6]

(iii) R *Ferrers* $\implies$ $R_{;}\overline{R}^{\mathsf{T}}_{;}R$ *is Ferrers again.*

Proof: (i) Transitivity of the residuals is a trivial consequence of the Ferrers property of R; furthermore

$$R_{;}\overline{R}^{\mathsf{T}}_{;}\overline{R_{;}\overline{R}^{\mathsf{T}}}^{\mathsf{T}}_{;}R_{;}\overline{R}^{\mathsf{T}} \subseteq R_{;}\overline{R}^{\mathsf{T}}_{;}R_{;}\overline{R}^{\mathsf{T}} \subseteq R_{;}\overline{R}^{\mathsf{T}}.$$

(ii) We employ Schröder equivalences to obtain

$$R\overline{R}^{\mathsf{T}}_{;}R \subseteq R \iff \overline{R}R^{\mathsf{T}}_{;}\overline{R} \subseteq \overline{R} \iff R\overline{R}^{\mathsf{T}} \subseteq \overline{\overline{R}_{;}R^{\mathsf{T}}} \iff \overline{R/R}^{\mathsf{T}} \subseteq R/R$$

which means that $\mathcal{R}(R)$ is connex. The second equivalence follows analogously.

(iii) We use the assumption $R_{;}\overline{R}^{\mathsf{T}}_{;}R \subseteq R$ that R is Ferrers at the beginning and continue with easy applications of the Schröder rule to estimate

$$R\overline{R}^{\mathsf{T}}_{;}R_{;}\overline{R_{;}R_{;}\overline{R}^{\mathsf{T}}_{;}R}^{\mathsf{T}}_{;}R_{;}R\overline{R}^{\mathsf{T}}_{;}R \subseteq R\overline{R}^{\mathsf{T}}_{;}R_{;}\overline{R_{;}R_{;}\overline{R}^{\mathsf{T}}_{;}R}^{\mathsf{T}}_{;}R_{;}R \subseteq R\overline{R}^{\mathsf{T}}_{;}R_{;}\overline{R_{;}R^{\mathsf{T}}_{;}\overline{R}}_{;}R \subseteq R_{;}\overline{R}^{\mathsf{T}}_{;}R. \qquad \Box$$

Thus, the row-contains-preorder and the column-is-contained-preorder of the given possibly heterogeneous Ferrers relation turn out to be connex. It is clear that all rows, or columns respectively, are in a linear order with regard to containment – up to the fact that there may exist multiplicities.

---

[6] $R_{;}\overline{R}^{\mathsf{T}}$ and $\overline{R}^{\mathsf{T}}_{;}R$ are in fact weakorders, see Def. 12.4.

Exhausting Ferrers relations

We now prove several properties of Ferrers relations that make them attractive for the purposes of modelling preferences etc. An important contribution to this comes from a detailed study of the behavior of the fringe. Look at Fig. 10.34. It contains a sequence of relations constituted by a Ferrers relation R, followed by the results stemming from repeatedly applying the operator $X \mapsto X_i \overline{X}^{\mathsf{T}}_i X$. It is hard to grasp what really happens on proceeding from one relation to the next. The drawing in the lower right shows in rearranged form how a finite Ferrers relation may gradually be exhausted.

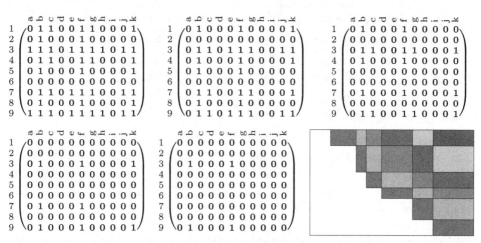

Fig. 10.34 Stepwise fringe exhaustion $X \mapsto X_i \overline{X}^{\mathsf{T}}_i X$ of a Ferrers relation, as a schema

The algebraic apparatus for this exhaustion – which is challenging – is now presented. The main point is that the fringe may be too small, as in the case of $(\mathbb{R}, <)$ where the fringe vanishes completely. Much effort has to be spent on guaranteeing that the fringe is sufficiently big, and thus, that in every round of shelling a non-empty part is removed. The easiest way to do this is to restrict ourselves to the finite case, but we will try to be slightly more general with the following definition introduced by Michael Winter (see [146, 147]).

Definition 10.30. Given a Ferrers relation R with fringe ∇, we call

$$R \text{ strongly Ferrers } :\Longleftrightarrow R = \nabla_i \nabla^{\mathsf{T}}_i R \nabla^{\mathsf{T}}_i \nabla \Longleftrightarrow R = \Xi_\nabla(R)_i R \Psi_\nabla(R). \qquad \square$$

Thus, we will in some sense be independent from always having to determine whether the exhaustion actually peels in a non-trivial way or not. But even this may not be the case as in the strongly Ferrers example of $(\mathbb{R}, \le)$. It does, however, work for $(\mathbb{N}, >)$, although this is non-finite.

Proposition 10.31. *For a finite Ferrers relation R, the following statements hold, in which we abbreviate $\nabla := \texttt{fringe}(R)$.*

(i) *There exists a natural number $k \geq 0$ that gives rise to a strictly increasing exhaustion as*

$$\mathbb{1} = (R\,\overline{R}^{\mathsf{T}})^k \subsetneqq (R\,\overline{R}^{\mathsf{T}})^{k-1} \subsetneqq \ldots \subsetneqq R\,\overline{R}^{\mathsf{T}}\,R\,\overline{R}^{\mathsf{T}} \subsetneqq R\,\overline{R}^{\mathsf{T}}.$$

(ii) *The construct $R\,\overline{R}^{\mathsf{T}}$ is a progressively finite weakorder according to the forthcoming Def. 12.4.*

(iii) $R\,\overline{R}^{\mathsf{T}} = \nabla\,\overline{R}^{\mathsf{T}}$ $\qquad\qquad$ $\overline{R}^{\mathsf{T}}\,R = \overline{R}^{\mathsf{T}}\,\nabla.$

(iv) *R allows a disjoint decomposition as*
$R = \texttt{fringe}(R) \cup \texttt{fringe}(R\,\overline{R}^{\mathsf{T}}\,R) \cup \ldots \cup \texttt{fringe}((R\,\overline{R}^{\mathsf{T}})^k\,R)$ *for some $k \geq 0$.*

(v) *R allows a disjoint decomposition as $R = \nabla \cup \nabla\,\overline{R}^{\mathsf{T}}\,\nabla.$*

Proof: (i) We start the following chain of inclusions from the right applying recursively that R is Ferrers:

$$\mathbb{1} \subseteq (R\,\overline{R}^{\mathsf{T}})^k \subseteq (R\,\overline{R}^{\mathsf{T}})^{k-1} \ldots \subseteq R\,\overline{R}^{\mathsf{T}}\,R\,\overline{R}^{\mathsf{T}} \subseteq R\,\overline{R}^{\mathsf{T}}.$$

Finiteness implies that it will eventually be stationary, i.e., $(R\,\overline{R}^{\mathsf{T}})^{k+1} = (R\,\overline{R}^{\mathsf{T}})^k$. This means in particular that the condition $Y \subseteq (R\,\overline{R}^{\mathsf{T}})\,Y$ holds for $Y := (R\,\overline{R}^{\mathsf{T}})^k$. The construct $R\,\overline{R}^{\mathsf{T}}$ is obviously transitive and irreflexive, so that, in combination with finiteness, it is also progressively bounded.[7] According to Section 6.3 of [123, 124], this means that $Y = (R\,\overline{R}^{\mathsf{T}})^k = \mathbb{1}$.

(ii) Negative transitivity according to the forthcoming Def. 12.3 follows in this way:

$$\overline{R}\,R^{\mathsf{T}} \subseteq \overline{R}\,R^{\mathsf{T}} \quad\Longleftrightarrow\quad \overline{R^{\mathsf{T}}\overline{R}\,R^{\mathsf{T}}} \subseteq \overline{R}^{\mathsf{T}}$$
$$\Longrightarrow\quad R\,\overline{R^{\mathsf{T}}\overline{R}\,R^{\mathsf{T}}} \subseteq R\,\overline{R}^{\mathsf{T}} \quad\Longleftrightarrow\quad \overline{R\,\overline{R}^{\mathsf{T}}}\,\overline{R\,\overline{R}^{\mathsf{T}}} \subseteq R\,\overline{R}^{\mathsf{T}}.$$

(iii) $R\,\overline{R}^{\mathsf{T}} = [(R \cap \overline{R\,\overline{R}^{\mathsf{T}}\,R}) \cup (R \cap R\,\overline{R}^{\mathsf{T}}\,R)]\,\overline{R}^{\mathsf{T}}$
$\qquad = (\nabla \cup R\,\overline{R}^{\mathsf{T}}\,R)\,\overline{R}^{\mathsf{T}}$ $\qquad\qquad$ since R is Ferrers
$\qquad = \nabla\,\overline{R}^{\mathsf{T}} \cup R\,\overline{R}^{\mathsf{T}}\,R\,\overline{R}^{\mathsf{T}}$
$\qquad = \nabla\,\overline{R}^{\mathsf{T}} \cup (\nabla\,\overline{R}^{\mathsf{T}} \cup R\,\overline{R}^{\mathsf{T}}\,R\,\overline{R}^{\mathsf{T}})\,R\,\overline{R}^{\mathsf{T}}$ $\qquad$ applied recursively
$\qquad = \nabla\,\overline{R}^{\mathsf{T}} \cup \nabla\,\overline{R}^{\mathsf{T}}\,R\,\overline{R}^{\mathsf{T}} \cup R\,\overline{R}^{\mathsf{T}}\,R\,\overline{R}^{\mathsf{T}}\,R\,\overline{R}^{\mathsf{T}}$
$\qquad = \nabla\,\overline{R}^{\mathsf{T}} \cup R\,\overline{R}^{\mathsf{T}}\,R\,\overline{R}^{\mathsf{T}}\,R\,\overline{R}^{\mathsf{T}}$ $\qquad\qquad$ since also $\overline{R}^{\mathsf{T}}$ is Ferrers
$\qquad = \ldots = \nabla\,\overline{R}^{\mathsf{T}} \cup \mathbb{1}$ $\qquad\qquad\qquad$ see (ii)

(iv) The disjoint decomposition starts with $R = (R \cap \overline{R\,\overline{R}^{\mathsf{T}}\,R}) \cup (R \cap R\,\overline{R}^{\mathsf{T}}\,R) = \nabla \cup R\,\overline{R}^{\mathsf{T}}\,R$ and can then only be continued finitely many times for the Ferrers relation $R\,\overline{R}^{\mathsf{T}}\,R$.

---

[7] Here, we are in a conflict: yes, being progressively finite is discussed in detail only later in Section 16.2, but it is used already here. We accept this because we aim mainly at the finite situation for the following results. They would, however, be formulated in too narrow a fashion if progressive finiteness were not considered. On the other hand, it would not have been wise to present all the Galois results and only then show this result on Ferrers relations.

(v) We start as in the proof of (iv) and apply (iii). □

It is mainly this effect which enables us to arrive at the results that follow. First, we observe how successively discarding the fringe leaves a decreasing sequence of relations – strictly decreasing when finite or at least not dense. Although they may possibly be heterogeneous, Ferrers relations can in many respects be considered similar to a linear (strict)ordering.

The following proposition is a classic; it may be found not least in [48] and also with a completely different point-free proof in [123, 124]. The proof here is yet another, but a constructive one, which means that one can write the constructs down in the language TITUREL and immediately run this as a program. This is because the constructs are generic and are uniquely characterized up to isomorphism, so that a standard realization for interpretation is possible.

Proposition 10.32. *If $R : X \longrightarrow Y$ is a (possibly heterogeneous) finite relation:*

$$R \text{ Ferrers} \quad \Longleftrightarrow \quad \begin{array}{l} \text{There exist mappings } f, g \text{ and a linear} \\ \text{strictorder } C \text{ such that } R = f \, ; C \, ; g^{\mathsf{T}}. \end{array}$$

The constructive proof of "$\Longrightarrow$" will produce surjective mappings f, g if R is neither total nor surjective.

Proof: "$\Longleftarrow$" follows using several times that mappings may slip below a negation from the left without affecting the result; see Prop. 5.13.

$$
\begin{aligned}
R \, ; \overline{R}^{\mathsf{T}} \, ; R &= f \, ; C \, ; g^{\mathsf{T}} \, ; \overline{f \, ; C \, ; g^{\mathsf{T}}}^{\mathsf{T}} \, ; f \, ; C \, ; g^{\mathsf{T}} && \text{by definition} \\
&= f \, ; C \, ; g^{\mathsf{T}} \, ; g \, ; \overline{C}^{\mathsf{T}} \, ; f^{\mathsf{T}} \, ; f \, ; C \, ; g^{\mathsf{T}} && \text{transposing; } f, g \text{ are mappings} \\
&\subseteq f \, ; C \, ; \overline{C}^{\mathsf{T}} \, ; C \, ; g^{\mathsf{T}} && f, g \text{ are univalent} \\
&\subseteq f \, ; C \, ; g^{\mathsf{T}} && \text{since the linear strictorder } C \text{ is Ferrers} \\
&= R && \text{again by definition}
\end{aligned}
$$

"$\Longrightarrow$" Let $R : X \longrightarrow Y$ be Ferrers. There may exist empty rows or columns in R or not. To take care of this in a general form, we enlarge the source to the direct sum $X + \mathbb{1}$ and the target to the direct sum $\mathbb{1} + Y$ and consider the relation $R' := \iota_X^{\mathsf{T}} \, ; R \, ; \kappa_Y$. In R', there will definitely exist at least one empty row and at least one empty column. It is intuitively clear – and easy to demonstrate – that R' is also Ferrers:

$$
\begin{aligned}
R' \, ; \overline{R'}^{\mathsf{T}} \, ; R' &= \iota_X^{\mathsf{T}} \, ; R \, ; \kappa_Y \, ; \overline{\kappa_Y^{\mathsf{T}} \, ; R^{\mathsf{T}} \, ; \iota_X} \, ; \iota_X^{\mathsf{T}} \, ; R \, ; \kappa_Y && \text{by definition; transposing} \\
&= \iota_X^{\mathsf{T}} \, ; R \, ; \kappa_Y \, ; \overline{\kappa_Y^{\mathsf{T}} \, ; R^{\mathsf{T}}}^{\mathsf{T}} \, ; \iota_X \, ; \iota_X^{\mathsf{T}} \, ; R \, ; \kappa_Y && \text{Prop. 5.13, } \iota_X, \kappa_Y \text{ are mappings} \\
&= \iota_X^{\mathsf{T}} \, ; R \, ; \overline{R}^{\mathsf{T}} \, ; R \, ; \kappa_Y && \iota_X, \kappa_Y \text{ are total and injective} \\
&\subseteq \iota_X^{\mathsf{T}} \, ; R \, ; \kappa_Y && R \text{ is Ferrers} \\
&= R' && \text{again by definition}
\end{aligned}
$$

The relation R' has been constructed so that $\overline{R'}^{\mathsf{T}}$ is total and surjective. Observe that $\overline{R}$ formed inside the upper right sub-rectangle of Fig. 10.35 would *not* have been surjective. Following Prop. 10.31.v, then also $\nabla := \mathtt{fringe}(\overline{R'})$ is total and surjective. Because fringes are always difunctional, ∇ is a block diagonal when nicely arranged, which will – after quotient forming – provide us with the bijective mapping λ.[8]

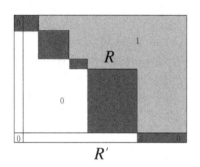
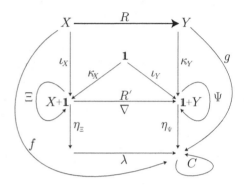

Fig. 10.35 Constructing a Ferrers decomposition: $R : X \longrightarrow Y$ and $R' : X + \mathbb{1} \longrightarrow \mathbb{1} + Y$

We introduce row equivalence $\Xi(R') := \mathtt{syq}(R'^{\mathsf{T}}, R'^{\mathsf{T}}) = \Xi(\overline{R'})$ as well as column equivalence $\Psi(R') := \mathtt{syq}(R', R')$ of R' together with the corresponding natural projections which we call η_Ξ, η_Ψ. Abbreviating $\nabla := \mathtt{fringe}(\overline{R'})$, we define

$$\lambda := \eta_\Xi^{\mathsf{T}} \,;\, \nabla \,;\, \eta_\Psi,$$
$$f := \iota_X \,;\, \eta_\Xi \,;\, \lambda,$$
$$g := \kappa_Y \,;\, \eta_\Psi,$$
$$C := \lambda^{\mathsf{T}} \,;\, \eta_\Xi^{\mathsf{T}} \,;\, R' \,;\, \eta_\Psi.$$

The crucial point has already been proved after Prop. 10.20: as the quotient of a total and surjective difunctional relation, λ is a bijective mapping.

Using all this, we prove that C is a linear strictorder, i.e., that it is transitive, irreflexive, and semi-connex. We start by expressing C more directly with R':

$$\begin{aligned}
C &= \lambda^{\mathsf{T}} \,;\, \eta_\Xi^{\mathsf{T}} \,;\, R' \,;\, \eta_\Psi && \text{by definition of } C \\
&= \eta_\Psi^{\mathsf{T}} \,;\, \nabla^{\mathsf{T}} \,;\, \eta_\Xi \,;\, \eta_\Xi^{\mathsf{T}} \,;\, R' \,;\, \eta_\Psi && \text{by definition of } \lambda \\
&= \eta_\Psi^{\mathsf{T}} \,;\, \nabla^{\mathsf{T}} \,;\, \Xi(R') \,;\, R' \,;\, \eta_\Psi && \text{natural projection} \\
&= \eta_\Psi^{\mathsf{T}} \,;\, \nabla^{\mathsf{T}} \,;\, R' \,;\, \eta_\Psi && \text{since the row equivalence satisfies } \Xi(R') \,;\, R' = R'
\end{aligned}$$

---

[8] By the way, [99] and a whole chapter of [100] are devoted to the algebraic description of the 'holes' or 'hollows' and 'noses' that show up here. We feel that our general generic construction via the extension R' eliminates several case distinctions. Not least, it allows us to formulate the respective terms and then interpret them immediately using TITUREL.

$$= \eta_\Psi^\mathsf{T} ; \overline{R'}^\mathsf{T} ; R' ; \eta_\Psi \qquad \text{according to Prop. 10.31.iii}$$

Up to quotient forming, C is, thus, the column-is-contained weakorder of R', leading to transitivity:

$$
\begin{aligned}
C ; C &= \eta_\Psi^\mathsf{T} ; \overline{R'}^\mathsf{T} ; R' ; \eta_\Psi ; \eta_\Psi^\mathsf{T} ; \overline{R'}^\mathsf{T} ; R' ; \eta_\Psi && \text{see above} \\
&= \eta_\Psi^\mathsf{T} ; \overline{R'}^\mathsf{T} ; R' ; \Psi(R') ; \overline{R'}^\mathsf{T} ; R' ; \eta_\Psi && \text{natural projection} \\
&= \eta_\Psi^\mathsf{T} ; \overline{R'}^\mathsf{T} ; R' ; \overline{R'}^\mathsf{T} ; R' ; \eta_\Psi && \text{column congruence} \\
&\subseteq \eta_\Psi^\mathsf{T} ; \overline{R'}^\mathsf{T} ; R' ; \eta_\Psi && \text{Ferrers} \\
&= C && \text{see above}
\end{aligned}
$$

With $C \cup C^\mathsf{T} = \eta_\Psi^\mathsf{T} ; \overline{\Psi(R')} ; \eta_\Psi = \eta_\Psi^\mathsf{T} ; \overline{\eta_\Psi ; \eta_\Psi^\mathsf{T}} ; \eta_\Psi = \eta_\Psi^\mathsf{T} ; \eta_\Psi ; \overline{\mathbb{I}} ; \eta_\Psi^\mathsf{T} ; \eta_\Psi = \overline{\mathbb{I}}$, it is irreflexive and semi-connex. $\qquad \square$

For a pictorial example assume the reports on soccer teams observed in a newspaper during one week, as given in middle of the upper row of Fig. 10.36. The permutations that allow the Ferrers relation – as well as the linear strictorder obtained via quotient forming – to reside in the upper right triangle are shown at its left and right side. Both have been determined by algebraic means oriented along the techniques of algebraic visualization sketched in Appendix C.

Corollary 10.33. *Assume the same setting as in Prop. 10.32 with R neither total nor surjective, i.e., with f, g surjective. Then*

(i) $\nabla_R := \mathtt{fringe}(R) = f ; H_C ; g^\mathsf{T}$ *with H_C the Hasse relation of C,*
   $\nabla_{\overline{R}} := \mathtt{fringe}(\overline{R}) = f ; g^\mathsf{T}$,
(ii) $\Xi(R) = f ; f^\mathsf{T}, \qquad \Psi(R) = g ; g^\mathsf{T}$.

Proof: (i) We start from ∇_R.

$$
\begin{aligned}
\nabla_R &= f ; C ; g^\mathsf{T} \cap \overline{f ; C ; g^\mathsf{T} ; g ; C^\mathsf{T} ; f^\mathsf{T}} ; f ; C ; g^\mathsf{T} && \text{by definition, transposed} \\
&= f ; \left[C \cap \overline{C ; g^\mathsf{T} ; g ; C^\mathsf{T} ; f^\mathsf{T} ; f ; C} \right] ; g^\mathsf{T} && \text{mappings distributive} \\
&= f ; \left[C \cap C ; g^\mathsf{T} ; g ; \overline{C^\mathsf{T}} ; f^\mathsf{T} ; f ; C \right] ; g^\mathsf{T} && \text{mappings slipping out of negation} \\
&= f ; \left[C \cap C ; \overline{C^\mathsf{T}} ; C \right] ; g^\mathsf{T} && \text{mappings } f, g \text{ are surjective} \\
&= f ; \nabla_C ; g^\mathsf{T} && \text{by definition of } \nabla_C \\
&= f ; H_C ; g^\mathsf{T},
\end{aligned}
$$

since for an arbitrary finite linear strictorder C the fringe ∇_C equals the Hasse relation $H_C = C \cap \overline{C ; C}$

$$
\begin{aligned}
\nabla_{\overline{R}} &= \overline{f ; C ; g^\mathsf{T}} \cap \overline{\overline{f ; C ; g^\mathsf{T}} ; g ; \overline{C^\mathsf{T}} ; f^\mathsf{T} ; \overline{f ; C ; g^\mathsf{T}}} && \text{by definition, transposed} \\
&= f ; \overline{C} ; g^\mathsf{T} \cap \overline{f ; \overline{C} ; g^\mathsf{T} ; g ; C^\mathsf{T} ; f^\mathsf{T} ; f ; \overline{C} ; g^\mathsf{T}} && \text{mappings slipping out of negation} \\
&= f ; \left[\overline{C} \cap \overline{\overline{C} ; g^\mathsf{T} ; g ; C^\mathsf{T} ; f^\mathsf{T} ; f ; \overline{C}} \right] ; g^\mathsf{T} && \text{mappings slipping out of negation} \\
&= f ; \left[\overline{C} \cap \overline{\overline{C} ; C^\mathsf{T} ; \overline{C}} \right] ; g^\mathsf{T} && \text{mappings } f, g \text{ surjective mappings}
\end{aligned}
$$

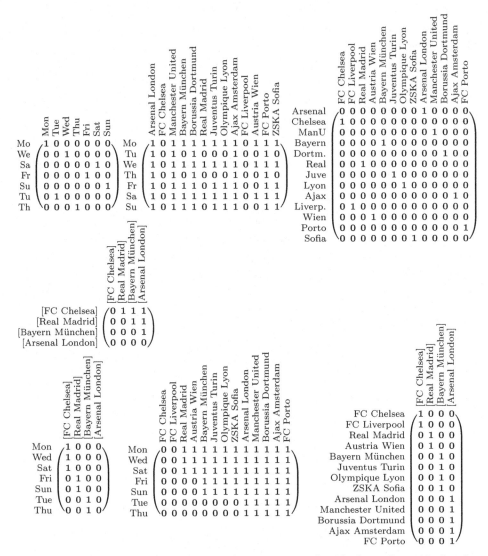

Fig. 10.36 Factorization of a Ferrers relation, presented in the lower line with f and g

$$= f \,\mathbin{;} \nabla_{\overline{C}} \,\mathbin{;} g^{\mathsf{T}} \qquad \text{by definition of } \nabla_{\overline{C}}$$
$$= f \,\mathbin{;} g^{\mathsf{T}} \qquad \text{for a finite linear strictorder } C \text{ the fringe } \nabla_{\overline{C}} \text{ equals the identity}$$

(ii) $\Psi(R) = \mathsf{syq}(R, R)$ by definition
$$= \mathsf{syq}(f \,\mathbin{;} C \,\mathbin{;} g^{\mathsf{T}}, f \,\mathbin{;} C \,\mathbin{;} g^{\mathsf{T}}) \qquad \text{as assumed}$$
$$= \mathsf{syq}(C \,\mathbin{;} g^{\mathsf{T}}, C \,\mathbin{;} g^{\mathsf{T}}) \qquad \text{cancelling a surjective mapping, Prop. 8.16.i}$$

$$= g \mathbin{;} \mathsf{syq}(C,C) \mathbin{;} g^\mathsf{T} \qquad \text{Prop. 8.16.ii,iii}$$
$$= g \mathbin{;} g^\mathsf{T} \qquad \text{since } C \text{ is a finite linear strictorder} \qquad \square$$

The decomposition of Theorem 10.32 is unique when f, g are surjective. Of course, one may choose a much larger target of f, g than necessary.[9]

Proposition 10.34. *Up to isomorphism, there exists just one factorization according to Prop. 10.32 in the case when f, g are required to be surjective.*

Proof: Let any triple f_1, g_1, C_1 be presented with $R = f_1 \mathbin{;} C_1 \mathbin{;} g_1^\mathsf{T}$ and f_1, g_1 surjective. We define
$$\varphi := f^\mathsf{T} \mathbin{;} f_1, \qquad \psi := g^\mathsf{T} \mathbin{;} g_1$$
and convince ourselves that $\varphi = \psi$ and that this provides an isomorphism. From Prop. 10.33.i, we know that $f \mathbin{;} g^\mathsf{T} = \nabla_{\overline{R}} = f_1 \mathbin{;} g_1^\mathsf{T}$. For surjective functions as assumed, this immediately gives
$$f \mathbin{;} \psi = f \mathbin{;} g^\mathsf{T} \mathbin{;} g_1 = \nabla_{\overline{R}} \mathbin{;} g_1 = f_1 \mathbin{;} g_1^\mathsf{T} \mathbin{;} g_1 = f_1,$$
$$f \mathbin{;} \varphi = f \mathbin{;} f^\mathsf{T} \mathbin{;} f_1 = \Xi(R) \mathbin{;} f_1 = f_1 \mathbin{;} f_1^\mathsf{T} \mathbin{;} f_1 = f_1.$$
This results in $f \mathbin{;} \psi = f_1 = f \mathbin{;} \varphi$ for a surjective mapping f, so that $\varphi = \psi$.
$$\varphi \mathbin{;} \varphi^\mathsf{T} = f^\mathsf{T} \mathbin{;} f_1 \mathbin{;} (f^\mathsf{T} \mathbin{;} f_1)^\mathsf{T} = f^\mathsf{T} \mathbin{;} f_1 \mathbin{;} f_1^\mathsf{T} \mathbin{;} f = f^\mathsf{T} \mathbin{;} \Xi(R) \mathbin{;} f = f^\mathsf{T} \mathbin{;} f \mathbin{;} f^\mathsf{T} \mathbin{;} f = \mathbb{I} \mathbin{;} \mathbb{I} = \mathbb{I},$$
$$C \mathbin{;} \varphi = C \mathbin{;} \psi = \varphi \mathbin{;} \varphi^\mathsf{T} \mathbin{;} C \mathbin{;} \psi = \varphi \mathbin{;} f_1^\mathsf{T} \mathbin{;} f \mathbin{;} C \mathbin{;} g^\mathsf{T} \mathbin{;} g_1$$
$$= \varphi \mathbin{;} f_1^\mathsf{T} \mathbin{;} R \mathbin{;} g_1 = \varphi \mathbin{;} f_1^\mathsf{T} \mathbin{;} f_1 \mathbin{;} C_1 \mathbin{;} g_1^\mathsf{T} \mathbin{;} g_1 = \varphi \mathbin{;} C_1. \qquad \square$$

Exercises

10.7 Let R be some Ferrers relation, i.e., $R \mathbin{;} \overline{R}^\mathsf{T} \mathbin{;} R \subseteq R$. Show that
$$\overline{R}^\mathsf{T} \mathbin{;} R \mathbin{;} \overline{R}^\mathsf{T} \subseteq \overline{R}^\mathsf{T} \qquad R \mathbin{;} \overline{R}^\mathsf{T} \cap \overline{R} \mathbin{;} R^\mathsf{T} = \mathbb{\bot\!\bot} \qquad \overline{R}^\mathsf{T} \mathbin{;} R \cap R^\mathsf{T} \mathbin{;} \overline{R} = \mathbb{\bot\!\bot}$$
are equivalent characterizations.

10.8 (i) Prove that every relation R is the union as well as intersection of Ferrers relations.
(ii) Let an arbitrary relation R be given and consider all reflexive and transitive relations Q on the direct sum of $\mathsf{src}(X)$ and $\mathsf{tgt}(Y)$ satisfying $\iota \mathbin{;} Q \mathbin{;} \kappa^\mathsf{T} = R$. Among these,
$$Q_M := \begin{pmatrix} R/R & R \\ \hline R^\mathsf{T} \mathbin{;} \overline{R} \mathbin{;} R^\mathsf{T} & R\backslash R \end{pmatrix}$$
is the greatest.
(iii) Given the setting just studied, R is Ferrers precisely when Q_M is connex.

---
[9] Observe that we did not in general require f, g to be surjective as in [123, 124].

10.6 Block-transitive relations

We are interested in the concepts that we already know for an order or a strictorder, but now study generalized to a heterogeneous environment in which multiple rows or columns may also occur. The starting point is a Ferrers relation. We have seen how in many respects it can be compared with a linear (strict)order. Is it possible to obtain in such a generalized setting similar results for a not necessarily linear (strict)order? This has indeed been found; see [146, 147].

Definition 10.35. A (possibly heterogeneous) relation R is called **block-transitive** if either one of the following equivalent conditions holds, expressed via its fringe $\nabla := \mathtt{fringe}(R)$

(i) $R \subseteq \nabla; \mathbb{T}$ and $R \subseteq \mathbb{T}; \nabla$,

(ii) $R \subseteq \nabla; \mathbb{T}; \nabla$,

(iii) $R = \nabla; \nabla^{\mathsf{T}}; R; \nabla^{\mathsf{T}}; \nabla$.

Proof: (i) $\Longrightarrow$ (ii):

$$R \subseteq \nabla; \mathbb{T} \cap \mathbb{T}; \nabla \subseteq (\nabla \cap \mathbb{T}; \nabla; \nabla^{\mathsf{T}}); (\mathbb{T} \cap \nabla^{\mathsf{T}}; \mathbb{T}; \nabla) \subseteq \nabla; \nabla^{\mathsf{T}}; \mathbb{T}; \nabla \subseteq \nabla; \mathbb{T}; \nabla.$$

(ii) $\Longrightarrow$ (iii): According to Prop. 10.15.iv, "$\supseteq$" holds for arbitrary relations R. In addition

$$R = \nabla; \mathbb{T}; \nabla \cap R \subseteq (\nabla \cap R; \nabla^{\mathsf{T}}; \mathbb{T}); (\mathbb{T}; \nabla \cap \nabla^{\mathsf{T}}; R) \subseteq \nabla; \nabla^{\mathsf{T}}; R.$$

In a similar fashion, we deduce $R \subseteq R; \nabla^{\mathsf{T}}; \nabla$, so that '$\subseteq$' follows.

(iii) $\Longrightarrow$ (i) is trivial. $\square$

We have met this condition already for Ferrers relations, when defining a **strongly Ferrers** relation in Def. 10.30. Being block transitive is mainly a question of how large the fringe is. The fringe must be large enough to 'span with its rectangular closure' the given relation R. The coarsest rectangle containing ∇ must contain R also.

For this concept, Michael Winter had originally (see [147] and earlier) coined the property of being of *order-shape*. We do not use this word here since it may cause misunderstanding: we are always careful to distinguish an order from a strictorder; they have different definitions, that overlap in both being transitive. In the following, we will see that, in a less consistent way, relations may share the property of being block transitive.

Proposition 10.36. *For an arbitrary block-transitive relation R, we again abbreviate* $\nabla := \mathtt{fringe}(R)$ *and obtain*

$$R \,;\, \nabla^{\mathsf{T}} \,;\, R = R.$$

Proof: We start from $R \subseteq \nabla \,;\, \mathbb{T}$.

$$
\begin{aligned}
R = R \cap \nabla \,;\, \mathbb{T} \\
&= \Xi(R) \,;\, R \cap \nabla \,;\, \mathbb{T} && \text{Prop. 5.29} \\
&= (\Xi(R) \cap \nabla \,;\, \mathbb{T}) \,;\, R && \text{masking rule} \\
&= \Xi_\nabla(R) \,;\, R && \text{Prop. 10.15.i} \\
&= \nabla \,;\, \nabla^{\mathsf{T}} \,;\, R && \text{Def. 10.14} \\
&\subseteq R \,;\, \nabla^{\mathsf{T}} \,;\, R && \text{because } \nabla \subseteq R
\end{aligned}
$$

The reverse containment is satisfied for every relation; Prop. 10.15.iv. □

This shows that ∇^{T} is the generalized inverse (Def. 10.42.ii) for a block-transitive relation.

We now give the most specialized examples of a block-transitive relation.

Proposition 10.37. (i) *Any difunctional relation R is block transitive.*

(ii) *Any finite Ferrers relation R is block transitive and, thus, strongly Ferrers.*

Proof: (i) Trivial since $R = \mathtt{fringe}(R)$.

(ii) We abbreviate $\nabla := \mathtt{fringe}(R)$. According to Prop. 10.31.v, we then have that $R = \nabla \cup \nabla \,;\, \overline{R}^{\mathsf{T}} \,;\, \nabla$, so that $R \subseteq \nabla \,;\, \mathbb{T}$ as well as $R \subseteq \mathbb{T} \,;\, \nabla$. □

This is in contrast to $(\mathbb{R}, <)$ which is Ferrers but not block transitive, simply because its fringe has already been shown to be empty.

Block-transitive relations also admit a factorization. In this case, we proceed differently compared with our proof of the Ferrers factorization: instead of looking for existing or non-existing empty rows/columns, we restrict ourselves completely to total and surjective relations.

Proposition 10.38. *Let* $R : X \longrightarrow Y$ *be a total and surjective (possibly heterogeneous) relation.*

$$
R \text{ block transitive} \quad \Longleftrightarrow \quad
\begin{array}{l}
\textit{There exist surjective mappings } f, g \textit{ and} \\
\textit{an order } E \textit{ such that } R = f \,;\, E \,;\, g^{\mathsf{T}}.
\end{array}
$$

Proof: "$\Longleftarrow$" We recall that the fringe ∇_E of an order E is always the identity. With the given condition, fringes of E and R are closely related:

$$
\begin{aligned}
f \,\fatsemi\, g^\mathsf{T} &= f \,\fatsemi\, \nabla_E \,\fatsemi\, g^\mathsf{T} && \text{because } \nabla_E = \mathbb{I} \\
&= f \,\fatsemi\, (E \cap \overline{E \,\fatsemi\, \overline{E}^\mathsf{T} \,\fatsemi\, E}) \,\fatsemi\, g^\mathsf{T} && \text{expanding } \nabla_E \\
&= f \,\fatsemi\, E \,\fatsemi\, g^\mathsf{T} \cap \overline{f \,\fatsemi\, E \,\fatsemi\, \overline{E}^\mathsf{T} \,\fatsemi\, E \,\fatsemi\, g^\mathsf{T}} && \text{since } f, g \text{ are univalent} \\
&= f \,\fatsemi\, E \,\fatsemi\, g^\mathsf{T} \cap \overline{f \,\fatsemi\, E \,\fatsemi\, \overline{E}^\mathsf{T} \,\fatsemi\, E \,\fatsemi\, g^\mathsf{T}} && \text{mappings slipping below negation} \\
&= f \,\fatsemi\, E g^\mathsf{T} \cap \overline{f \,\fatsemi\, E \,\fatsemi\, g^\mathsf{T} \,\fatsemi\, g \,\fatsemi\, \overline{E}^\mathsf{T} f^\mathsf{T} \,\fatsemi\, f \,\fatsemi\, E \,\fatsemi\, g^\mathsf{T}} && \text{since } f, g \text{ are univalent and surjective} \\
&= f \,\fatsemi\, E \,\fatsemi\, g^\mathsf{T} \cap \overline{f \,\fatsemi\, E \,\fatsemi\, g^\mathsf{T} \,\fatsemi\, \overline{f \,\fatsemi\, E \,\fatsemi\, g^\mathsf{T}}^\mathsf{T} \,\fatsemi\, f \,\fatsemi\, E \,\fatsemi\, g^\mathsf{T}} && \text{mappings slipping below negation} \\
&= \mathtt{fringe}(f \,\fatsemi\, E \,\fatsemi\, g^\mathsf{T}) = \nabla_R
\end{aligned}
$$

Using this, we may proceed as follows:

$$
\begin{aligned}
R &= f \,\fatsemi\, E \,\fatsemi\, g^\mathsf{T} && \text{by assumption} \\
&= f \,\fatsemi\, g^\mathsf{T} \,\fatsemi\, g \,\fatsemi\, f^\mathsf{T} \,\fatsemi\, f \,\fatsemi\, E \,\fatsemi\, g^\mathsf{T} \,\fatsemi\, g \,\fatsemi\, f^\mathsf{T} \,\fatsemi\, f \,\fatsemi\, g^\mathsf{T} && \text{since } f \text{ and } g \text{ are univalent and surjective} \\
&= \nabla_R \,\fatsemi\, \nabla_R^\mathsf{T} \,\fatsemi\, R \,\fatsemi\, \nabla_R^\mathsf{T} \,\fatsemi\, \nabla_R
\end{aligned}
$$

"$\Longrightarrow$" Given a total, surjective, and block-transitive R, we have $\nabla \,\fatsemi\, \mathbb{T} = \mathbb{T}$. It is a great help that we may use the following several times, according to Prop. 10.15

$$
(*) \qquad \nabla \,\fatsemi\, \nabla^\mathsf{T} = \Xi_\nabla(R) = \Xi(R) = \eta_\Xi \,\fatsemi\, \eta_\Xi^\mathsf{T} \quad \text{and} \quad \nabla^\mathsf{T} \,\fatsemi\, \nabla = \Psi_\nabla(R) = \Psi(R) = \eta_\Psi \,\fatsemi\, \eta_\Psi^\mathsf{T}
$$

with the corresponding natural projections called η_Ξ, η_Ψ. We define

$$
\begin{aligned}
\lambda &:= \eta_\Xi^\mathsf{T} \,\fatsemi\, \nabla \,\fatsemi\, \eta_\Psi \\
E &:= \lambda^\mathsf{T} \,\fatsemi\, \eta_\Xi^\mathsf{T} \,\fatsemi\, R \,\fatsemi\, \eta_\Psi \\
f &:= \eta_\Xi \,\fatsemi\, \lambda \\
g &:= \eta_\Psi.
\end{aligned}
$$

Now, the conditions have to be proved. As a natural projection, by definition, g is a surjective mapping, and so f will be, once it has been proved that λ is. We show one of the properties; the others follow in a similar form

$$
\begin{aligned}
\lambda^\mathsf{T} \,\fatsemi\, \lambda &= \eta_\Psi^\mathsf{T} \,\fatsemi\, \nabla^\mathsf{T} \,\fatsemi\, \eta_\Xi \,\fatsemi\, \eta_\Xi^\mathsf{T} \,\fatsemi\, \nabla \,\fatsemi\, \eta_\Psi && \text{expanded} \\
&= \eta_\Psi^\mathsf{T} \,\fatsemi\, \nabla^\mathsf{T} \,\fatsemi\, \nabla \,\fatsemi\, \nabla^\mathsf{T} \,\fatsemi\, \nabla \,\fatsemi\, \eta_\Psi && \text{using } (*) \\
&= \eta_\Psi^\mathsf{T} \,\fatsemi\, \eta_\Psi \,\fatsemi\, \eta_\Psi^\mathsf{T} \,\fatsemi\, \eta_\Psi \,\fatsemi\, \eta_\Psi^\mathsf{T} \,\fatsemi\, \eta_\Psi && \text{using } (*) \\
&= \mathbb{I} \,\fatsemi\, \mathbb{I} \,\fatsemi\, \mathbb{I} = \mathbb{I} && \text{a natural projection is surjective and univalent}
\end{aligned}
$$

Transitivity

$$
\begin{aligned}
E \,\fatsemi\, E &= \lambda^\mathsf{T} \,\fatsemi\, \eta_\Xi^\mathsf{T} \,\fatsemi\, R \,\fatsemi\, \eta_\Psi \,\fatsemi\, \lambda^\mathsf{T} \,\fatsemi\, \eta_\Xi^\mathsf{T} \,\fatsemi\, R \,\fatsemi\, \eta_\Psi && \text{expanding } E \\
&= \lambda^\mathsf{T} \,\fatsemi\, \eta_\Xi^\mathsf{T} \,\fatsemi\, R \,\fatsemi\, \eta_\Psi \,\fatsemi\, \eta_\Psi^\mathsf{T} \,\fatsemi\, \nabla^\mathsf{T} \,\fatsemi\, \eta_\Xi \,\fatsemi\, \eta_\Xi^\mathsf{T} \,\fatsemi\, R \,\fatsemi\, \eta_\Psi && \text{expanding } \lambda \\
&= \lambda^\mathsf{T} \,\fatsemi\, \eta_\Xi^\mathsf{T} \,\fatsemi\, R \,\fatsemi\, \nabla^\mathsf{T} \,\fatsemi\, \nabla \,\fatsemi\, \nabla^\mathsf{T} \,\fatsemi\, \nabla \,\fatsemi\, \nabla^\mathsf{T} \,\fatsemi\, R \,\fatsemi\, \eta_\Psi && \text{using } (*) \\
&= \lambda^\mathsf{T} \,\fatsemi\, \eta_\Xi^\mathsf{T} \,\fatsemi\, R \,\fatsemi\, \Psi(R) \,\fatsemi\, \nabla^\mathsf{T} \,\fatsemi\, \Xi(R) \,\fatsemi\, R \,\fatsemi\, \eta_\Psi && \text{using } (*) \text{ again} \\
&= \lambda^\mathsf{T} \,\fatsemi\, \eta_\Xi^\mathsf{T} \,\fatsemi\, R \,\fatsemi\, \nabla^\mathsf{T} \,\fatsemi\, R \,\fatsemi\, \eta_\Psi && \text{Prop. 10.15.iv} \\
&= \lambda^\mathsf{T} \,\fatsemi\, \eta_\Xi^\mathsf{T} \,\fatsemi\, R \,\fatsemi\, \eta_\Psi = E && \text{Prop. 10.36}
\end{aligned}
$$

Reflexivity

$E = \lambda^{\mathsf{T}}{}_{\mathsf{i}}\eta_{\Xi}^{\mathsf{T}}{}_{\mathsf{i}}R{}_{\mathsf{i}}\eta_{\Psi}$ expanding E

$\supseteq \eta_{\Psi}^{\mathsf{T}}{}_{\mathsf{i}}\nabla^{\mathsf{T}}{}_{\mathsf{i}}\eta_{\Xi}{}_{\mathsf{i}}\eta_{\Xi}^{\mathsf{T}}{}_{\mathsf{i}}\nabla{}_{\mathsf{i}}\eta_{\Psi}$ expanding λ, $R \supseteq \nabla$

$= \eta_{\Psi}^{\mathsf{T}}{}_{\mathsf{i}}\nabla^{\mathsf{T}}{}_{\mathsf{i}}\nabla{}_{\mathsf{i}}\nabla^{\mathsf{T}}{}_{\mathsf{i}}\nabla{}_{\mathsf{i}}\eta_{\Psi}$ using (*)

$= \eta_{\Psi}^{\mathsf{T}}{}_{\mathsf{i}}\eta_{\Psi}{}_{\mathsf{i}}\eta_{\Psi}^{\mathsf{T}}{}_{\mathsf{i}}\eta_{\Psi}{}_{\mathsf{i}}\eta_{\Psi}^{\mathsf{T}}{}_{\mathsf{i}}\eta_{\Psi}$ natural projection

$= \mathbb{I}$ three times natural projection

Antisymmetry

$E \cap E^{\mathsf{T}} = \lambda^{\mathsf{T}}{}_{\mathsf{i}}\eta_{\Xi}^{\mathsf{T}}{}_{\mathsf{i}}R{}_{\mathsf{i}}\eta_{\Psi} \cap \eta_{\Psi}^{\mathsf{T}}{}_{\mathsf{i}}R^{\mathsf{T}}{}_{\mathsf{i}}\eta_{\Xi}{}_{\mathsf{i}}\lambda$ expanding E

$= \eta_{\Psi}^{\mathsf{T}}{}_{\mathsf{i}}\nabla^{\mathsf{T}}{}_{\mathsf{i}}\eta_{\Xi}{}_{\mathsf{i}}\eta_{\Xi}^{\mathsf{T}}{}_{\mathsf{i}}R{}_{\mathsf{i}}\eta_{\Psi} \cap \eta_{\Psi}^{\mathsf{T}}{}_{\mathsf{i}}R^{\mathsf{T}}{}_{\mathsf{i}}\eta_{\Xi}{}_{\mathsf{i}}\eta_{\Xi}^{\mathsf{T}}{}_{\mathsf{i}}\nabla{}_{\mathsf{i}}\eta_{\Psi}$ expanding λ

$= \eta_{\Psi}^{\mathsf{T}}{}_{\mathsf{i}}\nabla^{\mathsf{T}}{}_{\mathsf{i}}\nabla{}_{\mathsf{i}}\nabla^{\mathsf{T}}{}_{\mathsf{i}}R{}_{\mathsf{i}}\eta_{\Psi} \cap \eta_{\Psi}^{\mathsf{T}}{}_{\mathsf{i}}R^{\mathsf{T}}{}_{\mathsf{i}}\nabla{}_{\mathsf{i}}\nabla^{\mathsf{T}}{}_{\mathsf{i}}\nabla{}_{\mathsf{i}}\eta_{\Psi}$ using (*)

$= \eta_{\Psi}^{\mathsf{T}}{}_{\mathsf{i}}\nabla^{\mathsf{T}}{}_{\mathsf{i}}R{}_{\mathsf{i}}\eta_{\Psi} \cap \eta_{\Psi}^{\mathsf{T}}{}_{\mathsf{i}}R^{\mathsf{T}}{}_{\mathsf{i}}\nabla{}_{\mathsf{i}}\eta_{\Psi}$ Prop. 10.15.ii

$= (\eta_{\Psi}^{\mathsf{T}}{}_{\mathsf{i}}\nabla^{\mathsf{T}}{}_{\mathsf{i}}R \cap \eta_{\Psi}^{\mathsf{T}}{}_{\mathsf{i}}R^{\mathsf{T}}{}_{\mathsf{i}}\nabla{}_{\mathsf{i}}\eta_{\Psi}{}_{\mathsf{i}}\eta_{\Psi}^{\mathsf{T}}){}_{\mathsf{i}}\eta_{\Psi}$ Prop. 5.4

$= \eta_{\Psi}^{\mathsf{T}}{}_{\mathsf{i}}(\eta_{\Psi}{}_{\mathsf{i}}\eta_{\Psi}^{\mathsf{T}}{}_{\mathsf{i}}\nabla^{\mathsf{T}}{}_{\mathsf{i}}R \cap R^{\mathsf{T}}{}_{\mathsf{i}}\nabla{}_{\mathsf{i}}\eta_{\Psi}{}_{\mathsf{i}}\eta_{\Psi}^{\mathsf{T}}){}_{\mathsf{i}}\eta_{\Psi}$ Prop. 5.4

$= \eta_{\Psi}^{\mathsf{T}}{}_{\mathsf{i}}(\nabla^{\mathsf{T}}{}_{\mathsf{i}}\nabla{}_{\mathsf{i}}\nabla^{\mathsf{T}}{}_{\mathsf{i}}R \cap R^{\mathsf{T}}{}_{\mathsf{i}}\nabla{}_{\mathsf{i}}\nabla^{\mathsf{T}}{}_{\mathsf{i}}\nabla){}_{\mathsf{i}}\eta_{\Psi}$ using (*)

$= \eta_{\Psi}^{\mathsf{T}}{}_{\mathsf{i}}(\nabla^{\mathsf{T}}{}_{\mathsf{i}}R \cap R^{\mathsf{T}}{}_{\mathsf{i}}\nabla){}_{\mathsf{i}}\eta_{\Psi}$ Prop. 10.15.ii again

$= \eta_{\Psi}^{\mathsf{T}}{}_{\mathsf{i}}\nabla^{\mathsf{T}}{}_{\mathsf{i}}\nabla{}_{\mathsf{i}}\eta_{\Psi}$ Prop. 10.15.v

$= \eta_{\Psi}^{\mathsf{T}}{}_{\mathsf{i}}\eta_{\Psi}{}_{\mathsf{i}}\eta_{\Psi}^{\mathsf{T}}{}_{\mathsf{i}}\eta_{\Psi}$ using (*)

$= \mathbb{I}$ natural projection   □

We will now extend the well-known Szpilrajn extension so as to work also for block-transitive relations.

Proposition 10.39 (Szpilrajn-type extension). *Every finite total, surjective, and block-transitive relation R has a* **Ferrers extension**, *i.e., a Ferrers relation $F \supseteq R$ satisfying* $\mathtt{fringe}(F) = \mathtt{fringe}(R)$.

Proof: We use Prop. 10.38, and assume from the beginning that $R = f{}_{\mathsf{i}}E{}_{\mathsf{i}}g^{\mathsf{T}}$ with f, g surjective mappings and E an ordering. Then we apply a Szpilrajn-extension to E, obtaining E_1, from which we define $F := f{}_{\mathsf{i}}E_1{}_{\mathsf{i}}g^{\mathsf{T}}$. Then we may simply compute as follows:

$$\mathtt{fringe}(F) = F \cap \overline{F{}_{\mathsf{i}}\overline{F}^{\mathsf{T}}{}_{\mathsf{i}}F} \quad \text{by definition}$$

$$= f{}_{\mathsf{i}}E_1{}_{\mathsf{i}}g^{\mathsf{T}} \cap \overline{f{}_{\mathsf{i}}E_1{}_{\mathsf{i}}g^{\mathsf{T}}{}_{\mathsf{i}}\overline{f{}_{\mathsf{i}}E_1{}_{\mathsf{i}}g^{\mathsf{T}}}^{\mathsf{T}}{}_{\mathsf{i}}f{}_{\mathsf{i}}E_1{}_{\mathsf{i}}g^{\mathsf{T}}} \quad \text{definition of } F$$

$$= f{}_{\mathsf{i}}E_1{}_{\mathsf{i}}g^{\mathsf{T}} \cap f{}_{\mathsf{i}}\overline{E_1{}_{\mathsf{i}}g^{\mathsf{T}}{}_{\mathsf{i}}g{}_{\mathsf{i}}\overline{E_1}^{\mathsf{T}}{}_{\mathsf{i}}f^{\mathsf{T}}}{}_{\mathsf{i}}f{}_{\mathsf{i}}E_1{}_{\mathsf{i}}g^{\mathsf{T}} \quad \text{maps slip out of negation}$$

$$= f{}_{\mathsf{i}}E_1{}_{\mathsf{i}}g^{\mathsf{T}} \cap f{}_{\mathsf{i}}E_1{}_{\mathsf{i}}\overline{\overline{E_1}^{\mathsf{T}}}{}_{\mathsf{i}}E_1{}_{\mathsf{i}}g^{\mathsf{T}} \quad \text{mappings } f, g \text{ are surjective}$$

$$= f{}_{\mathsf{i}}(E_1 \cap E_1{}_{\mathsf{i}}\overline{\overline{E_1}^{\mathsf{T}}}{}_{\mathsf{i}}E_1){}_{\mathsf{i}}g^{\mathsf{T}} \quad \text{univalency of } f, g$$

$$= f{}_{\mathsf{i}}\mathtt{fringe}(E_1){}_{\mathsf{i}}g^{\mathsf{T}} \quad \text{definition of a fringe}$$

$$= f{}_{\mathsf{i}}g^{\mathsf{T}} \quad \text{fringe of an order is the identity}$$

$$= \mathtt{fringe}(R) \quad \text{see the proof of Prop. 10.38}$$

Now we prove that F is Ferrers, using the fact that the E_1 obtained is a linear order:

$$F \mathbin{;} \overline{F}^{\mathsf{T}} \mathbin{;} F = f \mathbin{;} E_1 \mathbin{;} g^{\mathsf{T}} \mathbin{;} \overline{f \mathbin{;} E_1 \mathbin{;} g^{\mathsf{T}}}^{\mathsf{T}} \mathbin{;} f \mathbin{;} E_1 \mathbin{;} g^{\mathsf{T}} = f \mathbin{;} E_1 \mathbin{;} g^{\mathsf{T}} \mathbin{;} g \mathbin{;} \overline{E_1}^{\mathsf{T}} \mathbin{;} f^{\mathsf{T}} \mathbin{;} f \mathbin{;} E_1 \mathbin{;} g^{\mathsf{T}}$$
$$= f \mathbin{;} E_1 \mathbin{;} \overline{E_1}^{\mathsf{T}} \mathbin{;} E_1 \mathbin{;} g^{\mathsf{T}} \subseteq f \mathbin{;} E_1 \mathbin{;} g^{\mathsf{T}} = F. \qquad \qquad \square$$

While this was an extension of a block-transitive relation, we now take an arbitrary relation and try to find a block-transitive relation therein – which then may be the argument of the Szpilrajn extension.

Definition 10.40. For a (possibly heterogeneous) relation R with fringe $\nabla := \mathtt{fringe}(R)$, we define its **block-transitive kernel** as

$$\mathtt{btk}(R) := R \cap \nabla \mathbin{;} \mathbb{T} \cap \mathbb{T} \mathbin{;} \nabla = \nabla \mathbin{;} \nabla^{\mathsf{T}} \mathbin{;} R \mathbin{;} \nabla^{\mathsf{T}} \mathbin{;} \nabla. \qquad \square$$

We are now going to show that forming the block-transitive kernel of a relation is an idempotent operation.

Proposition 10.41. *For every (possibly heterogeneous) relation R, the fringe $\nabla := \mathtt{fringe}(R)$ does not change when reducing R to its block-transitive kernel:*

$$\nabla = \mathtt{fringe}(R \cap \nabla \mathbin{;} \mathbb{T} \cap \mathbb{T} \mathbin{;} \nabla)$$

Proof: We start with

$$(R \cap \nabla \mathbin{;} \mathbb{T} \cap \mathbb{T} \mathbin{;} \nabla) \mathbin{;} \overline{R \cap \nabla \mathbin{;} \mathbb{T} \cap \mathbb{T} \mathbin{;} \nabla}^{\mathsf{T}} \mathbin{;} (R \cap \nabla \mathbin{;} \mathbb{T} \cap \mathbb{T} \mathbin{;} \nabla)$$
$$= (R \cap \mathbb{T} \mathbin{;} \nabla) \mathbin{;} \left[\overline{R}^{\mathsf{T}} \cup \overline{\mathbb{T} \mathbin{;} \nabla^{\mathsf{T}} \cap \nabla^{\mathsf{T}} \mathbin{;} \mathbb{T}} \right] \mathbin{;} (R \cap \nabla \mathbin{;} \mathbb{T}) \cap \nabla \mathbin{;} \mathbb{T} \cap \mathbb{T} \mathbin{;} \nabla$$
$$\qquad \text{two times masking with Prop. 8.5 and a trivial Boolean step}$$
$$= R \mathbin{;} \left(\left[\overline{R}^{\mathsf{T}} \cup \overline{\mathbb{T} \mathbin{;} \nabla^{\mathsf{T}} \cap \nabla^{\mathsf{T}} \mathbin{;} \mathbb{T}} \right] \cap \nabla^{\mathsf{T}} \mathbin{;} \mathbb{T} \cap \mathbb{T} \mathbin{;} \nabla^{\mathsf{T}} \right) \mathbin{;} R \cap \nabla \mathbin{;} \mathbb{T} \cap \mathbb{T} \mathbin{;} \nabla \qquad \text{masking twice}$$
$$= R \mathbin{;} \left(\overline{R}^{\mathsf{T}} \cap \nabla^{\mathsf{T}} \mathbin{;} \mathbb{T} \cap \mathbb{T} \mathbin{;} \nabla^{\mathsf{T}} \right) \mathbin{;} R \cap \nabla \mathbin{;} \mathbb{T} \cap \mathbb{T} \mathbin{;} \nabla \qquad \text{with purely Boolean argument}$$
$$= (R \cap \mathbb{T} \mathbin{;} \nabla) \mathbin{;} \overline{R}^{\mathsf{T}} \mathbin{;} (R \cap \nabla \mathbin{;} \mathbb{T}) \cap \nabla \mathbin{;} \mathbb{T} \cap \mathbb{T} \mathbin{;} \nabla$$

so that

$$\mathtt{fringe}(R \cap \nabla \mathbin{;} \mathbb{T} \cap \mathbb{T} \mathbin{;} \nabla)$$
$$= R \cap \nabla \mathbin{;} \mathbb{T} \cap \mathbb{T} \mathbin{;} \nabla \cap \overline{(R \cap \mathbb{T} \mathbin{;} \nabla) \mathbin{;} \overline{R}^{\mathsf{T}} \mathbin{;} (R \cap \nabla \mathbin{;} \mathbb{T}) \cap \nabla \mathbin{;} \mathbb{T} \cap \mathbb{T} \mathbin{;} \nabla}$$
$$= R \cap \nabla \mathbin{;} \mathbb{T} \cap \mathbb{T} \mathbin{;} \nabla \cap \overline{\left[(R \cap \mathbb{T} \mathbin{;} \nabla) \mathbin{;} \overline{R}^{\mathsf{T}} \mathbin{;} (R \cap \nabla \mathbin{;} \mathbb{T}) \cup \overline{\nabla \mathbin{;} \mathbb{T} \cap \mathbb{T} \mathbin{;} \nabla} \right]}$$
$$= R \cap \nabla \mathbin{;} \mathbb{T} \cap \mathbb{T} \mathbin{;} \nabla \cap \overline{(R \cap \mathbb{T} \mathbin{;} \nabla) \mathbin{;} \overline{R}^{\mathsf{T}} \mathbin{;} (R \cap \nabla \mathbin{;} \mathbb{T})}$$
$$= R \cap \overline{R \mathbin{;} \overline{R}^{\mathsf{T}} \mathbin{;} R} = \mathtt{fringe}(R)$$

In the last step, "$\supseteq$" is clear. In the other direction, "$\subseteq R$" is also evident. It remains to convince ourselves of "$\subseteq R \mathbin{;} \overline{R}^{\mathsf{T}} \mathbin{;} R$". It then suffices to show that

$$\nabla \mathbin{;} \mathbb{T} \cap \mathbb{T} \mathbin{;} \nabla \cap \overline{(R \cap \mathbb{T} \mathbin{;} \nabla) \mathbin{;} \overline{R}^{\mathsf{T}} \mathbin{;} (R \cap \nabla \mathbin{;} \mathbb{T})} \subseteq R \mathbin{;} \overline{R}^{\mathsf{T}} \mathbin{;} R$$

$$\Longleftrightarrow \quad \nabla_; \mathbb{T} \cap \mathbb{T}_; \nabla \cap R_; \overline{R}^{\mathsf{T}}_; R \subseteq (R \cap \mathbb{T}_; \nabla)_; \overline{R}^{\mathsf{T}}_; (R \cap \nabla_; \mathbb{T}).$$

The latter can indeed be proved via masking. □

This was a really ugly proof which we have nonetheless presented. It employs hardly more than Boolean algebra with the masking formula, but with terms running in opposite directions, so that it would not have been easy for a reader to do this for himself or herself.

10.7 Inverses

The following touches a widely known special topic. Difunctionality is related to the concept of inverses in the context of linear algebra used for numerical problems. This will also provide deeper insight into the structure of a difunctional relation.

Definition 10.42. Let some relation A be given. The relation G is called

(i) a **sub-inverse** of A if
$$A_; G_; A \subseteq A,$$
(ii) a **generalized inverse** of A if
$$A_; G_; A = A,$$
(iii) a **Thierrin–Vagner inverse** of A if the following two conditions hold
$$A_; G_; A = A, \qquad G_; A_; G = G,$$
(iv) a **Moore–Penrose inverse** of A if the following four conditions hold
$$A_; G_; A = A, \qquad G_; A_; G = G, \qquad (A_; G)^{\mathsf{T}} = A_; G, \qquad (G_; A)^{\mathsf{T}} = G_; A.$$

The relation R is called **regular** if it has a generalized inverse. Owing to the symmetric situation in the case of a Thierrin–Vagner inverse G of A, the two relations A, G are also simply called **inverses** of each other. □

In a number of situations, semigroup theory[10] is applicable to relations. Some of the following definitions stem from [77]. A sub-inverse will always exist because $\amalg$ satisfies the requirement. With sub-inverses G, G' also their union $G \cup G'$ or supremum is obviously a sub-inverse so that one will ask which is the largest.

Proposition 10.43. $\overline{R^{\mathsf{T}}_; \overline{R}_; R^{\mathsf{T}}}$ *is the unique largest sub-inverse of R.*

---

[10] In [101], lots of semigroup concepts are studied in the special case of the semigroup of relations, not least all the Green relations along with which the results of semigroup theory may nicely be presented. For relations, however, much reduces to trivialities, while other effects cannot be handled in this way.

Proof: Assuming an arbitrary sub-inverse X of R, it will satisfy $R \mathbin{;} X \mathbin{;} R \subseteq R$ by definition, which is equivalent to

$$\Longleftrightarrow \quad X^{\mathsf{T}} \mathbin{;} R^{\mathsf{T}} \mathbin{;} \overline{R} \subseteq \overline{R} \quad \Longleftrightarrow \quad R \mathbin{;} \overline{R}^{\mathsf{T}} \mathbin{;} R \subseteq \overline{X}^{\mathsf{T}} \quad \Longleftrightarrow \quad X \subseteq \overline{R \mathbin{;} \overline{R}^{\mathsf{T}} \mathbin{;} R}^{\mathsf{T}}. \qquad \square$$

This leads to yet another characterization of a fringe, namely as an intersection of R with the transpose of its largest sub-inverse.

Corollary 10.44. $\mathtt{fringe}(R) = \sup\{Y \mid Y \subseteq R,\ R \mathbin{;} Y^{\mathsf{T}} \mathbin{;} R \subseteq R\}$

Proof: The set of relations over which the supremum is taken is non-empty because it contains $\mathbb{L}$. The selection criterion is obviously "$\cup$"-hereditary, so that the, by now existing, supremum will also satisfy the selection property.

The remaining question is whether the equation holds.

"$\supseteq$": Every Y satisfying the selection criterion satisfies $Y \subseteq R$. From $R \mathbin{;} Y^{\mathsf{T}} \mathbin{;} R \subseteq R$, we deduce as in Prop. 10.43 that $Y \subseteq \overline{R \mathbin{;} \overline{R}^{\mathsf{T}} \mathbin{;} R}$, which extends to the supremum.

"$\subseteq$": Prop. 10.15.iv. $\qquad \square$

A generalized inverse is not uniquely determined. As an example assume a homogeneous $\mathbb{T}$. It obviously has the generalized inverses $\mathbb{I}$ and $\mathbb{T}$. With generalized inverses G_1, G_2 also $G_1 \cup G_2$ is a generalized inverse. Also suprema of inverses are inverses again. There will, thus, exist a greatest inverse, if any. Regular relations, i.e., those with existing generalized inverse, may be characterized precisely by the following containment which is in fact an equation.

Proposition 10.45. R *regular* $\quad \Longleftrightarrow \quad R \subseteq R \mathbin{;} \overline{R \mathbin{;} \overline{R}^{\mathsf{T}} \mathbin{;} R}^{\mathsf{T}} \mathbin{;} R.$

Proof: If R is regular, there exists an X with $R \mathbin{;} X \mathbin{;} R = R$. It is, therefore, a sub-inverse and so $X \subseteq \overline{R \mathbin{;} \overline{R}^{\mathsf{T}} \mathbin{;} R}^{\mathsf{T}}$ according to Prop. 10.43. Then

$$R = R \mathbin{;} X \mathbin{;} R \subseteq R \mathbin{;} \overline{R \mathbin{;} \overline{R}^{\mathsf{T}} \mathbin{;} R}^{\mathsf{T}} \mathbin{;} R.$$

We prove the other direction. From Prop. 10.43, we know that $R \mathbin{;} \overline{R \mathbin{;} \overline{R}^{\mathsf{T}} \mathbin{;} R}^{\mathsf{T}} \mathbin{;} R \subseteq R$ for arbitrary R. The condition is, therefore, in fact an equality, and $X := \overline{R \mathbin{;} \overline{R}^{\mathsf{T}} \mathbin{;} R}^{\mathsf{T}}$ a generalized inverse. $\qquad \square$

Every block-transitive relation according to Def. 10.35 is regular in this sense; see Prop. 10.36.

Proposition 10.46. *If R is a regular relation, its uniquely defined greatest Thierrin–Vagner inverse is*

$$\overline{R \,;\, \overline{R^\mathsf{T}} \,;\, R}^{\mathsf{T}} \,;\, R \,;\, \overline{R \,;\, \overline{R^\mathsf{T}} \,;\, R}^{\mathsf{T}} =: \mathsf{TV}.$$

Proof: Evaluation of $\mathsf{TV} \,;\, R \,;\, \mathsf{TV} = \mathsf{TV}$ and $R \,;\, \mathsf{TV} \,;\, R = R$ using Prop. 10.45 with equality shows that TV is indeed a Thierrin–Vagner inverse. Any Thierrin–Vagner inverse G is in particular a sub-inverse, so that $G \subseteq \overline{R \,;\, \overline{R^\mathsf{T}} \,;\, R}^{\mathsf{T}}$ which implies $G = G \,;\, R \,;\, G \subseteq \mathsf{TV}$. □

We transfer a result on Moore–Penrose inverses, well-known for real-valued matrices, for example, to the relational setting.

Theorem 10.47. *Moore–Penrose inverses are unique provided they exist.*

Proof: Assume two Moore–Penrose inverses G, H of A to be given. Then we may proceed as follows:

$$\begin{aligned}
G &= G \,;\, A \,;\, G = G \,;\, G^\mathsf{T} \,;\, A^\mathsf{T} = G \,;\, G^\mathsf{T} \,;\, A^\mathsf{T} \,;\, H^\mathsf{T} \,;\, A^\mathsf{T} = G \,;\, G^\mathsf{T} \,;\, A^\mathsf{T} \,;\, A \,;\, H = G \,;\, A \,;\, G \,;\, A \,;\, H = G \,;\, A \,;\, H \\
&= G \,;\, A \,;\, H \,;\, A \,;\, H = G \,;\, A \,;\, A^\mathsf{T} \,;\, H^\mathsf{T} \,;\, H = A^\mathsf{T} \,;\, G^\mathsf{T} \,;\, A^\mathsf{T} \,;\, H^\mathsf{T} \,;\, H \\
&= A^\mathsf{T} \,;\, H^\mathsf{T} \,;\, H = H \,;\, A \,;\, H = H.
\end{aligned}$$ □

We now relate these concepts with permutations and difunctionality.

Theorem 10.48. *For a (possibly heterogeneous) relation A, the following are equivalent.*

(i) *A has a Moore–Penrose inverse.*

(ii) *A has A^T as its Moore–Penrose inverse.*

(iii) *A is difunctional.*

(iv) *Any two rows (or columns) of A are either disjoint or identical.*

(v) *There exist permutations P, Q such that $P \,;\, A \,;\, Q$ has block-diagonal form, i.e.,*

$$P \,;\, A \,;\, Q = \begin{pmatrix} B_1 & \mathbb{L} & \mathbb{L} & \mathbb{L} & \mathbb{L} & \mathbb{L} \\ \mathbb{L} & B_2 & \mathbb{L} & \mathbb{L} & \mathbb{L} & \mathbb{L} \\ \mathbb{L} & \mathbb{L} & B_3 & \mathbb{L} & \mathbb{L} & \mathbb{L} \\ \mathbb{L} & \mathbb{L} & \mathbb{L} & B_4 & \mathbb{L} & \mathbb{L} \\ \mathbb{L} & \mathbb{L} & \mathbb{L} & \mathbb{L} & B_5 & \mathbb{L} \\ \mathbb{L} & \mathbb{L} & \mathbb{L} & \mathbb{L} & \mathbb{L} & \mathbb{L} \end{pmatrix}$$

with (not necessarily square) diagonal entries $B_i = \mathbb{T}$.

Proof: For the key step (i)$\Longrightarrow$(ii): $G = G\,\dot{}\,A\,G \subseteq G\,\dot{}\,A\,A^{\mathsf{T}}\,\dot{}\,A\,G = A^{\mathsf{T}}\,\dot{}\,G^{\mathsf{T}}\,\dot{}\,A^{\mathsf{T}}\,\dot{}\,A\,G = (A\,G\,A)^{\mathsf{T}}\,\dot{}\,A\,G = A^{\mathsf{T}}\,\dot{}\,A\,G = A^{\mathsf{T}}\,\dot{}\,G^{\mathsf{T}}\,\dot{}\,A^{\mathsf{T}} = (A\,G\,A)^{\mathsf{T}} = A^{\mathsf{T}}$ and, deduced symmetrically, $A \subseteq G^{\mathsf{T}}$. $\square$

Exercises

10.9 Determine the largest sub-inverse, the greatest Thierrin–Vagner inverse, and the Moore–Penrose inverse, if any, for the following relations:

$$
\begin{array}{c}
a\\b\\c\\d\\e\\f\\g
\end{array}
\begin{pmatrix}
0&0&1&0&0&0&0&0&0&0&0\\
1&0&0&0&0&0&0&0&0&1&0\\
0&0&0&1&0&0&0&0&0&0&0\\
0&0&0&0&0&0&0&0&1&0&0\\
0&0&1&0&0&0&0&0&0&0&0\\
0&0&0&0&0&1&0&0&0&0&1\\
1&0&0&0&0&0&0&0&0&1&1
\end{pmatrix}
\qquad
\begin{array}{c}
a\\b\\c\\d\\e
\end{array}
\begin{pmatrix}
0&1&0&0&0&0&0\\
0&0&0&0&0&0&0\\
0&0&1&0&1&0&0\\
0&0&1&0&0&0&0\\
0&0&0&0&0&0&0
\end{pmatrix}
\qquad
\begin{array}{c}
a\\b\\c\\d\\e\\f\\g
\end{array}
\begin{pmatrix}
0&0&1&1&0&1&0&1&0\\
0&1&0&1&0&0&1&0&1\\
0&1&0&0&0&0&1&0&1\\
0&1&0&0&0&0&0&0&0\\
1&0&1&0&0&0&0&1&1\\
0&1&0&1&1&1&0&1&0\\
1&1&1&0&0&1&0&0&1
\end{pmatrix}
$$

10.10 Prove or disprove: every largest sub-inverse, i.e., every relation $R\,\dot{}\,\overline{R}^{\mathsf{T}}\,\dot{}\,\overline{R}^{\mathsf{T}}$, is regular.

11
Concept Analysis

We are going to denote as *power operations* those operations obtained by employing the membership relation ε between a domain X and its power domain $\mathbf{2}^X$. They are not very well known since there is a tendency to hide them behind traditional abstractions. If there are subsets considered as vectors satisfying $u \subseteq v$, their corresponding elements in the power domain will be in the relation $e_u; e_v^\mathsf{T} \subseteq \Omega$, with Ω the powerset ordering. We concentrate on such transitions from the algebraic side. This makes them more easily accessible to a formal treatment.

Power operations as used, for example, in programming will be studied in Chapter 19. Already here, it will become possible to formulate and prove important factorizations: factorization according to maxcliques, or the diclique factorization, giving rise to the field of concept lattices with all its applications. Also cut completion of an ordering and ideal completion subsume to this setting. To provide easy access, we start with considerations that remind us of vector spaces and their rank. At the end of the chapter, a plexus of interrelationships of relations around any given relation is outlined which we have decided to call the *topography* around that relation.

11.1 Row and column spaces

First, we take a look at something that corresponds to vector spaces over $\mathbb{R}$. Rows of a relation may be joined so as to obtain unions of rows. All finite unions – including the union over the empty set of rows – give the **row union space**; correspondingly the **column union space**. One should not stick too closely to the classical idea of a vector space and its dimension or basis. It can be proved that in the **0-1**-case studied here, there exists just one basis. When given an independent set of vectors, one will not necessarily be able to extend it to a basis, etc. We are thus confronted with an entirely different situation.

It is evident that several combinations of rows may produce the same union. In order to eliminate all multiple rows, we consider the row equivalence $\Xi(\varepsilon^\mathsf{T}; R)$ and determine the natural projection η for it, i.e., we factorize it to $\Xi(\varepsilon^{\mathsf{T}};R) = \eta;\eta^{\mathsf{T}}$. All possible types of row unions are then shown by $\eta^{\mathsf{T}};\varepsilon^{\mathsf{T}};R$. In an analogous form, $R;\varepsilon';\eta'$ shows all possible types of unions of columns. Figure 11.1 gives an example. Some ideas for the following stem from investigations concerning real-valued matrices as presented, for example, in [77].

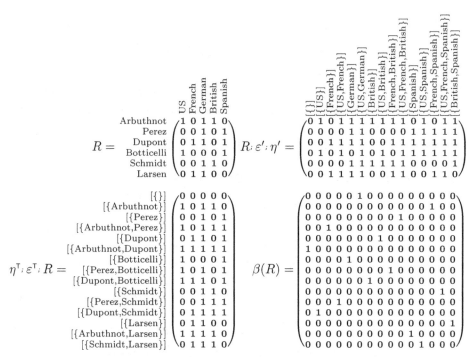

Fig. 11.1 Illustrating row union and column union space of a relation R

Duplicates in Fig. 11.1 are eliminated; i.e., every row and column name is in brackets and stands for the class of all combinations with the same result. A fact that one might not have expected in the first place is that the two spaces obviously have equal cardinality. Even more, Fig. 11.2 shows that row union space and column union space are ordered mainly in the same way, however in counter-running directions. This reminds us somehow of the fact that row rank equals column rank for real-valued matrices. We go even further and establish a bijective mapping between the two, that makes this precise.

The lower right relation of Fig. 11.1 indicates how row types and column types are related: negate the row and move it over the original relation (i.e., upper left); then the indicated column shows the hits. We can formulate this also the other way round: negate any column from the upper right matrix, slide it horizontally over R and mark the hits; this is then the row β indicates in reverse direction.

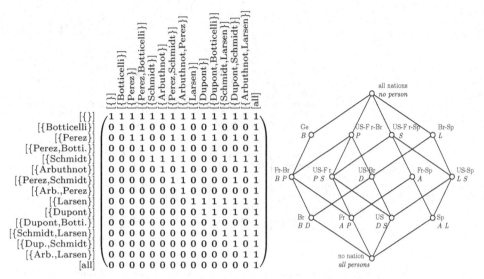

Fig. 11.2 Antitone ordering of row union space and column union space

In the relation of Fig. 11.2, the rows are rearranged to containment order. In the Hasse diagram it is shown that the columns are thereby arranged in reverse direction. The row space is obviously a lattice. In the same way, the column space forms a lattice. The two together have been shown in an example to be anti-isomorphic. A nice result stems from [81]: there are exactly n numbers greater than 2^{n-1} that can serve as the cardinalities of row spaces of Boolean $n \times n$-matrices, namely $2^{n-1} + 2^i$ for $i = 0, \ldots, n-1$.

Figure 11.3 is nearly the same as Fig. 11.1, but now for row, respectively column, intersections. Also in this case, we find a bijection as indicated in the figure. We have applied the same procedure to $\overline{R}$ instead of R and then negated the result, a procedure that more or less closely resembles the De Morgan rule $a \wedge b = \overline{a} \vee \overline{b}$.

The method of comparison is different in this case. Whenever we look at an entry u, v of $\beta(\overline{R})$ in Fig. 11.3, i.e., $uv^{\mathsf{T}} \subseteq \beta(\overline{R})$, we will encounter the same situation that we exemplify with the entry $[\{\text{Schmidt}\}], [\{\text{British}\}]$. The corresponding row intersection in R is $\{\text{German,British}\}$ and the column union is $\{\text{Arbuthnot,Schmidt}\}$. These two make up a non-enlargeable rectangle of R.

We now define constructs that will soon be met quite frequently. The first is a relation established between all sets of rows of a relation R and its columns; a given

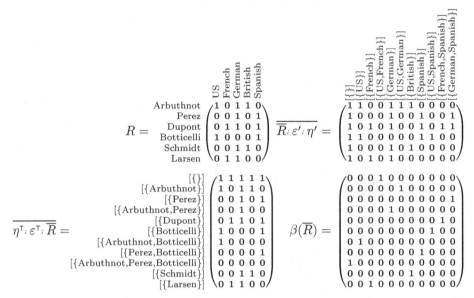

Fig. 11.3 Row intersection space and column intersection space of R related by $\beta(\overline{R})$

set of rows is related to the columns as indicated by its meet or intersection. In a similar way, the second relation is defined between the rows of a given relation and all its sets of columns so that the intersection of the respective columns of a set indicates where the relationship holds.

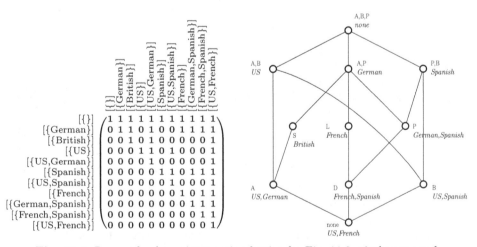

Fig. 11.4 Row and column intersection lattice for Fig. 11.3, nicely arranged

Considering Figs. 11.1 and 11.4, together with the relation that has been treated in both cases, other remarkable facts show up. In Fig. 11.4, dicliques are related as may be seen at {Larsen,Dupont} × {French,German}, for example, which in Fig. 11.4 is abbreviated via representatives to L×French. In contrast, maximal independent pairs of sets in Fig. 11.1 are put into relation, for example, {Perez,Schmidt} × {US,French}.

Definition 11.1. Let a (possibly heterogeneous) relation R be given and consider the membership relations ε on its source side and ε' on its target side. Then we call

   (i) $\Xi^{\vee}(R) := \eta^{\mathsf{T}}{_{;}}\,\varepsilon^{\mathsf{T}}{_{;}}\,R$ the **row set union relation** for R,

   (ii) $\Psi^{\vee}(R) := R{_{;}}\,\varepsilon'{_{;}}\,\eta'$ the **column set union relation** for R,

 (iii) $\Xi^{\wedge}(R) := \overline{\eta^{\mathsf{T}}{_{;}}\,\varepsilon^{\mathsf{T}}{_{;}}\,\overline{R}} = \eta^{\mathsf{T}}{_{;}}\,\varepsilon^{\mathsf{T}}{_{;}}\,\overline{R}$ the **row set intersection relation** for R,

 (iv) $\Psi^{\wedge}(R) := \overline{\overline{R}{_{;}}\,\varepsilon'{_{;}}\,\eta'} = \overline{\overline{R}{_{;}}\,\varepsilon'}{_{;}}\,\eta'$ the **column set intersection relation** for R.

In all these cases, η is the respective projection according to row equivalence of $\Xi(\varepsilon^{\mathsf{T}}{_{;}}R) = \Xi(\varepsilon^{\mathsf{T}}{_{;}}\overline{R})$; in the same way, η' means quotient forming with respect to the column equivalence.   □

The equivalent versions in (iii) and (iv) are legitimate according to Prop. 8.18.ii. We will henceforth give preference to the version with more negation bars! It will turn out to be a good idea to consider the minorants of majorants construct, i.e., the R-contact closure $\bigwedge_R(X)$ of Def. 8.7, to be basic and to make use of the fact that always $\mathsf{ubd}_R(\mathsf{lbd}_R(\mathsf{ubd}_R(X))) = \mathsf{ubd}_R(X)$. One has already a certain idea of $X \mapsto \mathsf{ubd}_R(\mathsf{lbd}_R(X))$ as a closure operation. This gives safer guidance in many ways; not least, one will be prevented from thinking in terms of composition. It is vital here to think in terms of residuation.

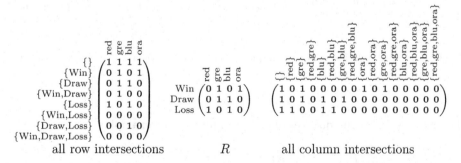

Fig. 11.5 A relation R with all its row intersections and its column intersections

In Fig. 11.5, we provide a simple example without quotient forming by η, η'. The row marked {Win,Draw} is obviously the intersection of the rows marked Win and Draw of R. In a similar way, the column {red,blu} is the intersection of columns red and blu of R. For rows as well as for columns, all possible intersections are shown, starting with the borderline cases of intersections applied to the empty sets {}. It is, however, much nicer to identify all the zero columns on the right and the two zero rows on the left. In addition column {red} and column {red,blu} should be put in one equivalence class, and also {ora} with column {gre,ora}. One will then arrive at 7 different rows and also 7 different columns, between which a one-to-one interrelationship may then indeed be established; see Fig. 11.6.

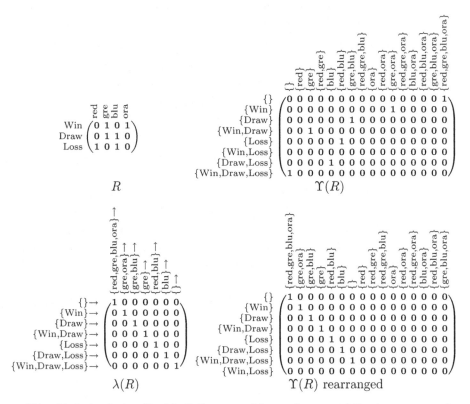

Fig. 11.6 A relation R with diclique matching and concept bijection rearranged

There are several consequences of these definitions, for example

$$\left[\Psi^\wedge(R)\right]^\mathsf{T} = \overline{\Xi^\vee(\overline{R^\mathsf{T}})} \qquad \Psi^\wedge(R) = \left[\Xi^\wedge(R^\mathsf{T})\right]^\mathsf{T}.$$

Yet another result is that $R/\Xi^\wedge(R)$ is an Aumann contact relation according to Def. 11.18.

11.2 Factorizations

Several theorems well-known for real- or complex-valued matrices have their largely unknown counterparts for relations. These are developed in this section.

Diclique factorization

While the visualizations in Figs. 11.1 and 11.4 have been quite convincing, we need a deeper concept for how to relate row and column spaces, be it for unions or for intersections, and a proof that there is indeed a bijection between the two. In the explanations, it will have become clear that dicliques (i.e., non-enlargeable rectangles), or else non-enlargeable independent pairs of sets are always important, so we remind ourselves of the corresponding situations in Chapter 10. A diclique u, v, i.e., a non-enlargeable rectangle inside R, is characterized by two formulae:

$$\overline{R_i v} = \overline{u} \quad \text{and} \quad \overline{R^\mathsf{T}_i u} = \overline{v}.$$

Every diclique u, v is composed of a set u on the source side and a set v on the target side. Relating sets belonging to a diclique means that they satisfy at the same time

$$u = \overline{\overline{R_i v}} \quad \text{and} \quad v = \overline{\overline{R^\mathsf{T}_i u}};$$

see Prop. 10.2. In the following definition, these operations are computed with the symmetric quotient applied to $\varepsilon, \varepsilon'$, i.e., to every subset simultaneously. Intersecting the two results gives precisely the non-enlargeable ones related.

Definition 11.2. Given any (possibly heterogeneous) relation R together with ε, the membership relation starting on the source side, and ε', the corresponding relation starting from the target side, we introduce

(i) $\Upsilon := \Upsilon(R) := \mathsf{syq}(\varepsilon, \overline{\overline{R_i \varepsilon'}}) \cap \mathsf{syq}(\overline{\overline{R^\mathsf{T}_i \varepsilon}}, \varepsilon')$, the **diclique matching**,
(ii) $\lambda := \lambda(R) := \iota_i \Upsilon_i \iota'^\mathsf{T}$, the **concept bijection**.

For (ii), we have extruded those relationships that belong to the non-enlargeable rectangles as

$\iota := \mathtt{Inject}\,(\Upsilon_i \mathbb{T}) \quad$ for the vector $\Upsilon_i \mathbb{T}$ and
$\iota' := \mathtt{Inject}\,(\Upsilon^\mathsf{T}_i \mathbb{T}) \quad$ for the vector $\Upsilon^\mathsf{T}_i \mathbb{T}$. □

The product $\iota_i \Upsilon$ turns out to be total by construction; see Prop. 7.12. We recall that the construct Υ determines the set of all non-enlargeable rectangles – including the trivial ones with one side empty and the other side full. It relates the row sets to the column sets of the non-enlargeable rectangles, i.e., the dicliques. A variant form of the definition would have been

$$\Upsilon := \mathsf{syq}\,(\varepsilon, \mathtt{lbd}\,_R(\varepsilon')) \cap \mathsf{syq}\,(\mathtt{ubd}\,_R(\varepsilon), \varepsilon').$$

The denotation "Υ" has again been chosen so as to be easily memorizable: it is intended to point to an *element* of the powerset on the left and to an *element* of the powerset on the right side.

It is relatively easy to convince ourselves that Υ is indeed a matching and λ is a bijection, thus justifying the denotations chosen.

Proposition 11.3. *Assume the setting of Def. 11.2 in a context with the Point Axiom, then*

(i) Υ *is a matching,*

(ii) λ *is a bijective mapping,*

(iii) *the original relation may be obtained as* $R = \varepsilon_{;} \Upsilon_{;} \varepsilon'^{\mathsf{T}}$,

(iv) $\Upsilon = \iota^{\mathsf{T}}{}_{;} \lambda_{;} \iota'$,

(v) $\overline{R}_{;} \varepsilon' \cap \mathbb{T}_{;} \Upsilon = \overline{\varepsilon}_{;} \Upsilon$ $\overline{R}^{\mathsf{T}}{}_{;} \varepsilon \cap \mathbb{T}_{;} \Upsilon^{\mathsf{T}} = \overline{\varepsilon'}_{;} \Upsilon^{\mathsf{T}}$.

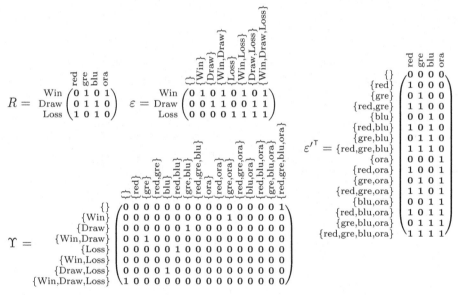

Fig. 11.7 Arbitrary relation R factorized as $R = \varepsilon_{;} \Upsilon_{;} \varepsilon'^{\mathsf{T}}$ according to Prop. 11.3.iii

Proof: (i) We have to convince ourselves that Υ satisfies $\Upsilon^{\mathsf{T}}{}_{;} \Upsilon \subseteq \mathbb{I}$ and $\Upsilon_{;} \Upsilon^{\mathsf{T}} \subseteq \mathbb{I}$. We show one of the cases; the other is proved in a similar way.

$$\Upsilon^{\mathsf{T}}{}_{;} \Upsilon \subseteq \mathsf{syq}(\varepsilon', \overline{\overline{R}^{\mathsf{T}}{}_{;} \varepsilon})_{;} \mathsf{syq}(\overline{\overline{R}^{\mathsf{T}}{}_{;} \varepsilon}, \varepsilon') \qquad \text{definition, transposition, and monotony}$$

$$\subseteq \mathsf{syq}(\varepsilon', \varepsilon') \qquad \text{cancelling symmetric quotients}$$

$$= \mathbb{I} \qquad \text{property of the membership relation}$$

(ii) is more or less immediate: given the matching property of Υ, what has been done in addition was to extrude the area where Υ and Υ^{T} are defined. Totality and surjectivity follow from Prop. 7.12.

(iii) Using Prop. 7.14, it is immediate that

$$\varepsilon_{\,;}\,\Upsilon_{\,;}\,\varepsilon'^{\mathsf{T}} \subseteq \varepsilon_{\,;}\,\mathsf{syq}\,(\varepsilon,\overline{\overline{R_{\,;}\,\varepsilon'}})_{\,;}\,\varepsilon'^{\mathsf{T}} = \overline{\overline{R_{\,;}\,\varepsilon'}}_{\,;}\,\varepsilon'^{\mathsf{T}} = R.$$

It remains to prove the reverse containment, which seems to require the Point Axiom. Assume that there were points x,y with $x_{\,;}\,y^{\mathsf{T}} \subseteq R$ but $x_{\,;}\,y^{\mathsf{T}} \not\subseteq \varepsilon_{\,;}\,\Upsilon_{\,;}\,\varepsilon'^{\mathsf{T}}$. With these, we construct the non-enlargeable rectangle around x,y started horizontally as in Prop. 10.3, i.e.,

$$u := \overline{\overline{R_{\,;}\,v}}, \quad \text{where} \quad v := \overline{R^{\mathsf{T}}}_{\,;}\,x.$$

To these vectors correspond points e_u, e_v in the respective powersets, satisfying

$$e_u := \mathsf{syq}\,(\varepsilon,u) \quad u = \varepsilon_{\,;}\,e_u, \qquad e_v := \mathsf{syq}\,(\varepsilon',v) \quad v = \varepsilon'_{\,;}\,e_v.$$

The crucial idea is to prove that $e_{u\,;}\,e_v^{\mathsf{T}} \subseteq \Upsilon$. Once this has been shown, it immediately leads to $x_{\,;}\,y^{\mathsf{T}} \subseteq u_{\,;}\,v^{\mathsf{T}} = \varepsilon_{\,;}\,e_{u\,;}\,e_v^{\mathsf{T}}{}_{;}\,\varepsilon'^{\mathsf{T}} \subseteq \varepsilon_{\,;}\,\Upsilon_{\,;}\,\varepsilon'^{\mathsf{T}}$, contradicting the assumption that $x_{\,;}\,y^{\mathsf{T}} \not\subseteq \varepsilon_{\,;}\,\Upsilon_{\,;}\,\varepsilon'^{\mathsf{T}}$.

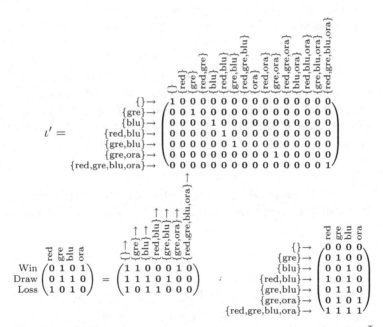

Fig. 11.8 Diclique factorizing an arbitrary relation R as $R = U_{\,;}\,V^{\mathsf{T}}$ according to Prop. 11.4.i

We restrict ourselves to showing containment in one of the two terms of Υ:

$$\begin{aligned} e_u &= \mathsf{syq}\,(\varepsilon,u) \\ &= \mathsf{syq}\,(\varepsilon,\overline{\overline{R_{\,;}\,v}}) \quad \text{see above} \end{aligned}$$

$$= \mathsf{syq}(\varepsilon, \overline{\overline{R \,;\, \varepsilon'\,;\, e_v}}) \qquad \text{because } v = \varepsilon'\,;\, e_v$$
$$= \mathsf{syq}(\varepsilon, \overline{R \,;\, \varepsilon'}\,;\, e_v) \qquad \text{Prop. 5.13, point } e_v \text{ slipping out of negation}$$
$$= \mathsf{syq}(\varepsilon, \overline{R \,;\, \varepsilon'})\,;\, e_v \qquad \text{Prop. 8.16.iii}$$
$$\Longleftrightarrow \quad e_u \,;\, e_v^{\mathsf{T}} \subseteq \mathsf{syq}(\varepsilon, \overline{R \,;\, \varepsilon'}) \qquad \text{shunting with Prop. 5.12.ii}$$

(iv) This follows from Prop. 7.12.iv.

(v) "$\supseteq$" is relatively simple and consists of just one essential estimation:
$$\overline{\varepsilon}\,;\, \Upsilon \subseteq \overline{\varepsilon}\,;\, \mathsf{syq}(\varepsilon, \overline{R \,;\, \varepsilon'}) = \overline{\varepsilon}\,;\, \mathsf{syq}(\overline{\varepsilon}, \overline{R \,;\, \varepsilon'}) \subseteq \overline{R \,;\, \varepsilon'}.$$
For "$\subseteq$", we start with
$$\varepsilon\,;\, \Upsilon \subseteq \varepsilon\,;\, \mathsf{syq}(\varepsilon, \overline{R \,;\, \varepsilon'}) = \overline{R \,;\, \varepsilon'}$$
so that recalling Prop. 5.6
$$\overline{R \,;\, \varepsilon'} \subseteq \overline{\overline{\varepsilon}\,;\, \Upsilon} = \overline{\varepsilon}\,;\, \Upsilon \cup \overline{\mathbb{T}}\,;\, \Upsilon. \qquad \qquad \square$$

Figure 11.7 shows such a factorization of an arbitrary relation R. One immediately feels that one should get rid of the many empty columns and rows in Υ – a topic we have already taken care of with the extrusion to λ in Def. 11.2.

The bijective mapping λ, respectively the matching Υ, thus obtained, may not immediately have a nice appearance. Applying our domain construction for target permutation to the bijective mapping λ, we would be able to arrange it in a nicer form. The natural injection ι' is shown as the upper relation in Fig. 11.8. The rest of that figure shows a factorization into just two factors which is in a somehow similar form also known for real-valued matrices. This will now be treated formally.

The conditions using symmetric quotients in the following proposition mean in an algebraically usable form nothing other than 'columns of U, respectively V, are pairwise different'.

Proposition 11.4 (Diclique factorization). *Every (possibly heterogeneous) relation $R : X \longrightarrow Y$ may be factorized $R = U \,;\, V^{\mathsf{T}}$ into two factors satisfying*
$$\overline{R}\,;\, V = \overline{U}, \qquad \overline{R}^{\mathsf{T}}\,;\, U = \overline{V}, \qquad \mathsf{syq}(U, U) = \mathbb{I}, \qquad \mathsf{syq}(V, V) = \mathbb{I}.$$
This factorization is uniquely determined up to isomorphism.

Proof: In view of Fig. 11.9, this factorization may be defined explicitly as
$$U := \varepsilon\,;\, \iota^{\mathsf{T}}\,;\, \lambda \qquad \text{and} \qquad V := \varepsilon'\,;\, \iota'^{\mathsf{T}}.$$
With Prop. 11.3.iv, one will see that

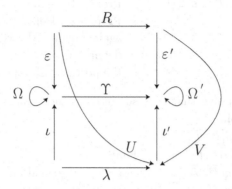

Fig. 11.9 The typing around diclique matching Υ and concept bijection λ

$$U_i V^\mathsf{T} = \varepsilon_i \, \iota^\mathsf{T}_i \, \lambda_i \, \iota'_i \, {\varepsilon'}^\mathsf{T} = \varepsilon_i \, \Upsilon_i \, {\varepsilon'}^\mathsf{T} = R$$

and has in addition to prove the requirements on U, V:

$$
\begin{aligned}
\overline{R}_i V &= \overline{R}_i \, {\varepsilon'}_i \, {\iota'}^\mathsf{T} && \text{expanding} \\
&= (\overline{R}_i \, \varepsilon' \cap \mathbb{T}_i \, \Upsilon)_i \, {\iota'}^\mathsf{T} \cup (\overline{R}_i \, \varepsilon' \cap \overline{\mathbb{T}_i \Upsilon})_i \, {\iota'}^\mathsf{T} && \text{subdividing} \\
&= (\overline{R}_i \, \varepsilon' \cap \mathbb{T}_i \, \Upsilon)_i \, {\iota'}^\mathsf{T} \cup \mathbb{L} && \iota' \text{ extrudes } \Upsilon^\mathsf{T}_i \, \mathbb{T} \\
&= \overline{\varepsilon}_i \, \Upsilon_i \, {\iota'}^\mathsf{T} && \text{Prop. 11.3.v} \\
&= \overline{\varepsilon}_i \, \iota^\mathsf{T}_i \, \lambda_i \, \iota'_i \, {\iota'}^\mathsf{T} && \text{Def. 11.2.ii} \\
&= \overline{\varepsilon}_i \, \iota^\mathsf{T}_i \, \lambda && \text{since } \iota' \text{ is injective and total} \\
&= \overline{\varepsilon_i \, \iota^\mathsf{T}_i \, \lambda} && \iota^\mathsf{T}_i \, \lambda \text{ is a transposed map} \\
&= \overline{U} && \text{by definition} \\
\mathsf{syq}(U, U) &= \mathsf{syq}(\varepsilon_i \, \iota^\mathsf{T}_i \, \lambda, \varepsilon_i \, \iota^\mathsf{T}_i \, \lambda) && \text{by definition} \\
&= \lambda^\mathsf{T}_i \, \iota_i \, \mathsf{syq}(\varepsilon, \varepsilon)_i \, \iota^\mathsf{T}_i \, \lambda && \text{Prop. 8.16.ii} \\
&= \lambda^\mathsf{T}_i \, \iota_i \, \mathbb{I}_i \, \iota^\mathsf{T}_i \, \lambda && \text{Prop. 7.13} \\
&= \lambda^\mathsf{T}_i \, \lambda && \iota \text{ is total and injective} \\
&= \mathbb{I} && \text{Prop. 5.13, } \lambda \text{ is a bijection}
\end{aligned}
$$

Concerning uniqueness, assume a second factorization to be given, $R = U_{1_i} V_1^\mathsf{T}$, with the required properties. The proof must now be executed very carefully, because we do not pretend from the beginning to be able to construct an isomorphism directly between any two arbitrarily given factorizations, but only from the constructively generated factorization to another arbitrary factorization. In particular, we must not use symmetry arguments.

We define $\varphi := \mathsf{syq}(U, U_1)$. With the cancellation rules for the symmetric quotient it is immediate that φ is univalent and injective. It is also total because

$$\varphi_i \, \varphi^\mathsf{T} = \mathsf{syq}(U, U_1)_i \, \mathsf{syq}(U_1, U) \qquad \text{definition and transposition}$$

$$= \mathsf{syq}\,(\varepsilon\,{}_{\mathit{i}}\,\iota^{\mathsf{T}}\,{}_{\mathit{i}}\,\lambda, U_1)\,{}_{\mathit{i}}\,\mathsf{syq}\,(U_1, U) \quad \text{definition of } U$$
$$= \lambda^{\mathsf{T}}\,{}_{\mathit{i}}\,\iota\,{}_{\mathit{i}}\,\mathsf{syq}\,(\varepsilon, U_1)\,{}_{\mathit{i}}\,\mathsf{syq}\,(U_1, U) \quad \text{Prop. 8.16, since } \lambda^{\mathsf{T}}\,{}_{\mathit{i}}\,\iota \text{ is a mapping}$$
$$= \lambda^{\mathsf{T}}\,{}_{\mathit{i}}\,\iota\,{}_{\mathit{i}}\,\mathsf{syq}\,(\varepsilon, U) \quad \text{Prop. 8.12, Def. 7.13}$$
$$= \mathsf{syq}\,(\varepsilon\,{}_{\mathit{i}}\,\iota^{\mathsf{T}}\,{}_{\mathit{i}}\,\lambda, U) \quad \text{again because } \lambda^{\mathsf{T}}\,{}_{\mathit{i}}\,\iota \text{ is a mapping}$$
$$= \mathsf{syq}\,(U, U) \quad \text{definition of } U$$
$$= \mathbb{I} \quad \text{the postulated property of } U$$

Finally, we have to prove that φ is surjective, which seems to require the Point Axiom. Assume a point

$$a\,{}_{\mathit{i}}\,b^{\mathsf{T}} \subseteq \overline{\mathbb{T}\,{}_{\mathit{i}}\,\varphi} = \overline{\mathbb{T}\,{}_{\mathit{i}}\,\mathsf{syq}\,(U, U_1)}.$$
$$\Longleftrightarrow \quad a \subseteq \overline{\mathbb{T}\,{}_{\mathit{i}}\,\mathsf{syq}\,(U, U_1)\,{}_{\mathit{i}}\,b} \quad \text{shunting}$$
$$\Longleftrightarrow \quad a \subseteq \overline{\mathbb{T}\,{}_{\mathit{i}}\,\mathsf{syq}\,(U, U_1)}\,{}_{\mathit{i}}\,b \quad \text{slipping below negation}$$
$$\Longleftrightarrow \quad a \subseteq \overline{\mathbb{T}\,{}_{\mathit{i}}\,\mathsf{syq}\,(U, U_1\,{}_{\mathit{i}}\,b)} \quad \text{since a point is a transposed mapping}$$

Now we concentrate on the vector $u := U_1\,{}_{\mathit{i}}\,b$ and define also $v := \overline{R}^{\mathsf{T}}\,{}_{\mathit{i}}\,u$. It is easy to see that u, v constitute a non-enlargeable rectangle inside R because of the properties required for U_1, V_1. It has, thus, also to be present in U, V by construction.

The φ, thus defined, satisfies $U\,{}_{\mathit{i}}\,\varphi = U_1$ because $U\,{}_{\mathit{i}}\,\varphi = U\,{}_{\mathit{i}}\,\mathsf{syq}\,(U, U_1) \subseteq U_1$ with equality because φ is surjective. It also satisfies

$$\overline{\mathsf{syq}\,(V, V_1)} = \overline{V}^{\mathsf{T}}\,{}_{\mathit{i}}\,V_1 \cup V^{\mathsf{T}}\,{}_{\mathit{i}}\,\overline{V_1} \quad \text{expanding}$$
$$= U^{\mathsf{T}}\,{}_{\mathit{i}}\,\overline{R}\,{}_{\mathit{i}}\,V_1 \cup V^{\mathsf{T}}\,{}_{\mathit{i}}\,\overline{R}^{\mathsf{T}}\,{}_{\mathit{i}}\,U_1 \quad \text{using the postulated requirements}$$
$$= U^{\mathsf{T}}\,{}_{\mathit{i}}\,\overline{U_1} \cup \overline{U}^{\mathsf{T}}\,{}_{\mathit{i}}\,U_1 \quad \text{using the postulated requirements again}$$
$$= \overline{\varphi} \quad \text{as defined} \qquad \square$$

For this factorization, there exists a vast amount of applications, mostly centered around the buzzword of concept lattice analysis.

One may also phrase this a little differently: for every relation R there exists a relation U, uniquely determined up to isomorphism, such that

$$R = U\,{}_{\mathit{i}}\,(U\backslash R), \text{ or, equivalently, } R = U\,{}_{\mathit{i}}\,\overline{U^{\mathsf{T}}\,{}_{\mathit{i}}\,\overline{R}}.$$

In particular, one need *not* employ the order of subsets to determine the largest, as has been reported in [12, 13]; the algebraic qualification as in Prop. 10.2 is sufficient. Figure 10.3 visualizes this result, showing a relation as well as the counter-running connection between rows and columns of non-enlargeable rectangles, a so-called *polarity*.

Example 11.5. We insert here an example of a non-finite relation, namely

$$R := \{(x, y) \mid y < x\} \subseteq \mathbb{R}^2$$

representing the normal strictorder on $\mathbb{R}$. In a first attempt one might be satisfied with Fig. 11.10 which indicates rectangles $\{x \mid x > c\} \times \{y \mid y < c\}$ that are actually sufficient to cover the space strictly below the line $y = x$. Taking only rational c, for example, would also suffice in exhausting the relation. However, neither yet satisfies the algebraic conditions.

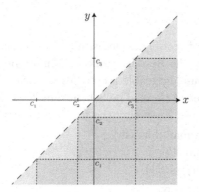

Fig. 11.10 Exhausting the relation ">" on $\mathbb{R}$ with rectangles $\{x \mid x > c\} \times \{y \mid y < c\}$

Taking

$$R = \{(x,y) \mid y < x\} \qquad \text{and} \qquad \overline{R} = \{(x,y) \mid y \geq x\}$$
$$U = \{(x,y) \mid y < x\} \qquad \text{and} \qquad V = \{(x,y) \mid x < y\},$$

we arrive indeed at

$$U \,\raise2pt{;}\, V^{\mathsf{T}} = \{(x,y) \mid \exists c : x > c \land y < c\} = R,$$

but only at

$$\overline{R}\,\raise2pt{;}\, V = \{(x,y) \mid \exists c : x \leq c \land c < y\} \subsetneqq \{(x,y) \mid y \geq x\} = \overline{U}.$$

One has to take $\{x \mid x \geq c\} \times \{y \mid y < c\}$ and $\{x \mid x > c\} \times \{y \mid y \leq c\}$, i.e., inclusive of one closed border.   □

For real-valued matrices, a similar result runs under the name of a **singular-value decomposition**,[1] by which any – not necessarily square – real-valued matrix R may be decomposed as $R = U D V^{\mathsf{T}}$ with D a diagonal matrix of non-negative quasi-diagonal eigenvalues and U, V unitary matrices. An example is as follows:

$$\begin{pmatrix} 1 & 0 & 0 & 0 & 2 \\ 0 & 0 & 5 & 0 & 0 \\ 0 & 0 & 0 & 0 & 0 \\ 0 & 7 & 0 & 0 & 0 \end{pmatrix} = \begin{pmatrix} 0 & 0 & 1 & 0 \\ 0 & 1 & 0 & 0 \\ 0 & 0 & 0 & -1 \\ 1 & 0 & 0 & 0 \end{pmatrix} \cdot \begin{pmatrix} 7 & 0 & 0 & 0 & 0 \\ 0 & 5 & 0 & 0 & 0 \\ 0 & 0 & \sqrt{5} & 0 & 0 \\ 0 & 0 & 0 & 0 & 0 \end{pmatrix} \cdot \begin{pmatrix} 0 & 1 & 0 & 0 & 0 \\ 0 & 0 & 1 & 0 & 0 \\ \sqrt{0.2} & 0 & 0 & 0 & \sqrt{0.8} \\ 0 & 0 & 0 & 1 & 0 \\ -\sqrt{0.8} & 0 & 0 & 0 & \sqrt{0.2} \end{pmatrix}$$

One may find our relational version of this to be even nicer. The singular-value decomposition insists on U, V being square matrices at the cost of D possibly being non-square and admitting only a quasi-diagonal. The relational result is more symmetric. The diagonal need not be mentioned at all as it is the identity relation. The outer factors enjoy very similar properties.

---

[1] The singular-values of R are defined to be the eigenvalues of RR^{T}, respectively $R^{\mathsf{T}}R$.

We mention the following obvious fact without proof.

Proposition 11.6. *Every relation R may be written as a union of non-enlargeable rectangles.* □

The minimum number of rectangles necessary has been named, according to [77], after Boris M. Schein as the **Schein-rank.** It is more or less immediate that the minimum number of non-enlargeable rectangles necessary leads to a so-called matroid structure on the set of these. The Schein-rank for the example of Fig. 10.8 is 5 since in addition to those 2 covering the fringe, 3 others are necessary, namely

({Perez,Dupont}, {German,Spanish}),
({Botticelli}, {US,Spanish}),
({Arbuthnot,Botticelli}, {US}).

Also the following 3 would do:

({Perez,Dupont}, {German,Spanish}),
({Botticelli}, {US,Spanish}),
({Arbuthnot}, {US,German,British}).

Only recently, numerous publications have appeared on computing the Schein-rank, in semigroup context, for non-negative or fuzzy relations, and in image understanding. Relating this to the factorization and to fringes, one may obviously say

number of blocks in the fringe $\leq$ Schein-rank $\leq$ number of columns in U

with equality instead of the first "$\leq$" in the case of a Ferrers relation.

Factorization in the symmetric case

The diclique factorization of Prop. 11.4 specializes considerably in the case when the relation is symmetric and reflexive – and, thus, also homogeneous. We refer to Def. 6.9, the definition of cliques, and also to the definition of maxcliques in Def. 10.4.

Definition 11.7. Given any symmetric and reflexive relation $B : X \longrightarrow X$ together with the membership relation $\varepsilon : X \longrightarrow 2^X$, we introduce

(i) $\Upsilon_{\mathrm{mc}} := \mathsf{syq}(\varepsilon, \overline{\overline{B};\varepsilon}) \cap \mathsf{syq}(\overline{\overline{B};\varepsilon}, \varepsilon) \cap \mathbb{I},$ the **maxclique matching,**
(ii) $\lambda_{\mathrm{mc}} := \iota; \iota^{\mathsf{T}} = \mathbb{I},$ the **maxclique bijection,** now the identity.

For (ii), we have extruded those relationships that belong to the maxcliques as

$\iota := \texttt{Inject}\,(\Upsilon_{\mathrm{mc}};\mathbb{T})$ for the vector $\Upsilon_{\mathrm{mc}};\mathbb{T}.$ □

The λ of Def. 11.2 has a certain meaning: it is a bijection, but between different sets, so that a correspondence is communicated. This is no longer the case – and is also unnecessary – for the symmetric λ_{mc} that fully reduces to an identity.

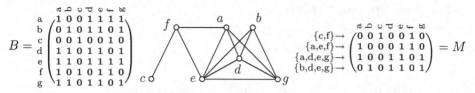

Fig. 11.11 Maxclique factorization $B = M^{\mathsf{T}} \cdot M$ of a symmetric and reflexive relation

One will observe in Fig. 11.11, that there are also dicliques with $u \neq v$, for example, $u = \{a, e\}, v = \{a, d, e, f, g\}$. These are explicitly excluded by the definition of Υ_{mc}. Degeneration is here completely unimportant: the worst case is $B : \mathbb{1} \longrightarrow \mathbb{1}$, reflexive, which is factorized to $M : \mathbb{1} \longrightarrow \mathbb{1}$, the identity.

Proposition 11.8 (Maxclique factorization). *Let any reflexive and symmetric relation B be given. Then*

(i) $\Upsilon^{\mathsf{T}} \subseteq \Upsilon$, *i.e., the Υ of Def. 11.3, is symmetric and Υ_{mc} is its diagonal part,*
(ii) $B = \varepsilon \cdot \Upsilon_{\mathrm{mc}} \cdot \varepsilon^{\mathsf{T}}$,
(iii) $\overline{B} \cdot \varepsilon \cap \mathbb{T} \cdot \Upsilon_{\mathrm{mc}} = \overline{\varepsilon} \cdot \Upsilon_{\mathrm{mc}}$,
(iv) B *may be factorized as $B = M^{\mathsf{T}} \cdot M$ by a maxclique relation, i.e., with M restricted by*
$$M \cdot \overline{B} = \overline{M} \quad \text{and} \quad \overline{M \cdot M^{\mathsf{T}}} = \mathbb{I},$$
and this factorization is uniquely determined up to isomorphism.

Proof: (i) The symmetric B is certainly homogeneous, so that $\varepsilon = \varepsilon'$. Furthermore, $B^{\mathsf{T}} = B$ implies $\overline{B}^{\mathsf{T}} = \overline{B}$. Then, Υ is symmetric because

$$\begin{aligned}
\Upsilon^{\mathsf{T}} &= \big\{ \mathsf{syq}\,(\varepsilon, \overline{B} \cdot \varepsilon') \cap \mathsf{syq}\,(\overline{B}^{\mathsf{T}} \cdot \varepsilon, \varepsilon') \big\}^{\mathsf{T}} && \text{by definition} \\
&= \big\{ \mathsf{syq}\,(\varepsilon, \overline{B} \cdot \varepsilon) \cap \mathsf{syq}\,(\overline{B} \cdot \varepsilon, \varepsilon) \big\}^{\mathsf{T}} && \text{because of symmetry, see above} \\
&= \mathsf{syq}\,(\overline{B} \cdot \varepsilon, \varepsilon) \cap \mathsf{syq}\,(\varepsilon, \overline{B} \cdot \varepsilon) && \text{transposing symmetric quotients} \\
&= \Upsilon && \text{since the latter is obviously symmetric}
\end{aligned}$$

(ii) $\varepsilon \cdot \Upsilon_{\mathrm{mc}} \cdot \varepsilon^{\mathsf{T}} \subseteq \varepsilon \cdot \Upsilon \cdot \varepsilon^{\mathsf{T}} = B$ due to Prop. 11.3.iii.

For the other direction, we distinguish dicliques that are symmetric from the others according to Def. 11.7. Every point $x \cdot y^{\mathsf{T}} \subseteq B$ is part of the clique $x \cup y$:

$$(x \cup y) \cdot (x \cup y)^{\mathsf{T}} = x \cdot x^{\mathsf{T}} \cup x \cdot y^{\mathsf{T}} \cup y \cdot x^{\mathsf{T}} \cup y \cdot y^{\mathsf{T}} \subseteq \mathbb{I} \cup B \cup B^{\mathsf{T}} \cup \mathbb{I} \subseteq B,$$

since B is reflexive and symmetric. Therefore every point of B is also part of a maxclique. These special dicliques necessarily suffice to exhaust B.

(iii) We start with Prop. 5.6 for the injective Υ_{mc}

$$\overline{\varepsilon}\,\mathbf{;}\,\Upsilon_{\mathrm{mc}} = \overline{\varepsilon\,\mathbf{;}\,\Upsilon_{\mathrm{mc}}} \cap \overline{\mathbb{T}\,\mathbf{;}\,\Upsilon_{\mathrm{mc}}}$$

and continue with two statements

$$\overline{\varepsilon}\,\mathbf{;}\,\Upsilon_{\mathrm{mc}} \subseteq \overline{\varepsilon}\,\mathbf{;}\,\mathsf{syq}\,(\varepsilon, \overline{\overline{B}\,\mathbf{;}\,\varepsilon}) = \overline{\varepsilon}\,\mathbf{;}\,\mathsf{syq}\,(\overline{\varepsilon}, \overline{B}\,\mathbf{;}\,\varepsilon) \subseteq \overline{B}\,\mathbf{;}\,\varepsilon,$$

$$\varepsilon\,\mathbf{;}\,\Upsilon_{\mathrm{mc}} \subseteq \varepsilon\,\mathbf{;}\,\mathsf{syq}\,(\varepsilon, \overline{\overline{B}\,\mathbf{;}\,\varepsilon}) \subseteq \overline{B}\,\mathbf{;}\,\varepsilon.$$

(iv) According to Def. 11.7, we distinguish dicliques that are symmetric from the others. Then we define $M := \iota\,\mathbf{;}\,\varepsilon^{\mathsf{T}}$. Therefore

$$
\begin{aligned}
B &= \varepsilon\,\mathbf{;}\,\Upsilon\,\mathbf{;}\,\varepsilon^{\mathsf{T}} && \text{Prop. 11.3.iii} \\
&\supseteq \varepsilon\,\mathbf{;}\,\Upsilon_{\mathrm{mc}}\,\mathbf{;}\,\varepsilon^{\mathsf{T}} && \text{because } \Upsilon_{\mathrm{mc}} \subseteq \Upsilon \\
&= \varepsilon\,\mathbf{;}\,\iota^{\mathsf{T}}\,\mathbf{;}\,\iota\,\mathbf{;}\,\varepsilon^{\mathsf{T}} && \text{because } \iota \text{ extrudes } \Upsilon_{\mathrm{mc}} \subseteq \mathbb{I} \\
&= \varepsilon\,\mathbf{;}\,\iota^{\mathsf{T}}\,\mathbf{;}\,(\varepsilon\,\mathbf{;}\,\iota^{\mathsf{T}})^{\mathsf{T}} && \text{transposition} \\
&= M^{\mathsf{T}}\,\mathbf{;}\,M && \text{by definition}
\end{aligned}
$$

Equality follows because maxcliques suffice to exhaust B which has already been shown in (ii). The conditions of Prop. 11.4 reduce to those given here:

$$
\begin{aligned}
M\,\mathbf{;}\,\overline{B} &= \iota\,\mathbf{;}\,\varepsilon^{\mathsf{T}}\,\mathbf{;}\,\overline{B} && \text{expanded} \\
&= \iota\,\mathbf{;}\,\left[(\varepsilon^{\mathsf{T}}\,\mathbf{;}\,\overline{B} \cap \Upsilon_{\mathrm{mc}}\,\mathbf{;}\,\mathbb{T}) \cup (\varepsilon^{\mathsf{T}}\,\mathbf{;}\,\overline{B} \cap \overline{\Upsilon_{\mathrm{mc}}\,\mathbf{;}\,\mathbb{T}})\right] && \text{subdivided} \\
&= \iota\,\mathbf{;}\,(\varepsilon^{\mathsf{T}}\,\mathbf{;}\,\overline{B} \cap \Upsilon_{\mathrm{mc}}\,\mathbf{;}\,\mathbb{T}) \cup \iota\,\mathbf{;}\,(\varepsilon^{\mathsf{T}}\,\mathbf{;}\,\overline{B} \cap \overline{\Upsilon_{\mathrm{mc}}\,\mathbf{;}\,\mathbb{T}}) && \text{distributivity} \\
&= \iota\,\mathbf{;}\,(\varepsilon^{\mathsf{T}}\,\mathbf{;}\,\overline{B} \cap \Upsilon_{\mathrm{mc}}\,\mathbf{;}\,\mathbb{T}) \cup \mathbb{\perp\!\!\!\perp} && \text{since } \iota \text{ is defined to extrude } \Upsilon_{\mathrm{mc}}\,\mathbf{;}\,\mathbb{T} \\
&= \iota\,\mathbf{;}\,\Upsilon_{\mathrm{mc}}\,\mathbf{;}\,\overline{\varepsilon}^{\mathsf{T}} && \text{according to (iii)} \\
&= \iota\,\mathbf{;}\,\overline{\varepsilon}^{\mathsf{T}} && \text{since } \iota \text{ extrudes } \Upsilon_{\mathrm{mc}}\,\mathbf{;}\,\mathbb{T} \text{ for the symmetric and univalent } \Upsilon_{\mathrm{mc}} \\
&= \overline{\iota\,\mathbf{;}\,\varepsilon^{\mathsf{T}}} && \text{because } \iota \text{ is a mapping} \\
&= \overline{M} && \text{definition of } M
\end{aligned}
$$

The two conditions in Prop. 11.4 with regard to the symmetric quotient reduce to just one, namely

$$\mathsf{syq}\,(M^{\mathsf{T}}, M^{\mathsf{T}}) = \mathbb{I}.$$

But in the present case, this may be simplified even further observing that owing to the maxclique property Prop. 10.5

$$\overline{M}\,\mathbf{;}\,M^{\mathsf{T}} = M\,\mathbf{;}\,\overline{B}\,\mathbf{;}\,M^{\mathsf{T}} = M\,\mathbf{;}\,\overline{M}^{\mathsf{T}}.$$

Therefore,

$$\mathsf{syq}\,(M^{\mathsf{T}}, M^{\mathsf{T}}) = \overline{\overline{M}\,\mathbf{;}\,M^{\mathsf{T}}} \cap \overline{M\,\mathbf{;}\,\overline{M}^{\mathsf{T}}} = \overline{\overline{M}\,\mathbf{;}\,M^{\mathsf{T}}} = \mathbb{I}.$$

The task remains to show that this factorization is essentially unique. Assume another factorization,

$$B = N^{\mathsf{T}}\,\mathbf{;}\,N \quad \text{with} \quad N\,\mathbf{;}\,\overline{B} = \overline{N} \quad \text{and} \quad \overline{N\,\mathbf{;}\,N^{\mathsf{T}}} = \mathbb{I}.$$

The isomorphism may be defined directly as $\varphi := \mathsf{syq}\,(M^{\mathsf{T}}, N^{\mathsf{T}})$. That φ is a bijective mapping satisfying $M = \varphi\,\mathbf{;}\,N$ is shown in the same way as in Prop. 11.4. $\qquad\square$

This result has important applications. Conceiving B as the reflexive version of the adjacency of some simple graph $(X, \overline{\mathbb{I}} \cap B)$, the domain of M is the set of maxcliques (i.e., non-enlargeable cliques). Figure 11.12 shows such a factorization with M already transposed.

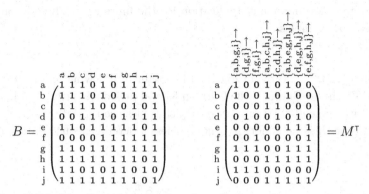

Fig. 11.12 Maxclique factorization of a reflexive and symmetric relation

Remark 11.9. The factorization $B = M^{\mathsf{T}} \, ; M$ reminds us of a result from linear algebra, namely that any positive semi-definite matrix is the so-called Gram matrix or Gramian matrix[2] of a set of vectors. To recall this, let every row, respectively column, correspond to a vector. With the so identified vectors v_i, the symmetric matrix of inner products is formed, whose entries are given by $G_{ij} = v_i^{\mathsf{T}} v_j$. In the case of an orthonormal set of vectors, for example, the Gram matrix is the identity.

Exercises

11.1 Determine the maxclique factorization for the following three relations:

	US	French	German	British	Spanish	Japanese	Italian	Czech
US	1	1	1	1	0	1	0	1
French	1	1	1	1	0	1	0	1
German	1	1	1	0	0	1	0	1
British	1	1	0	1	1	0	1	1
Spanish	0	0	0	1	1	1	1	1
Japanese	1	1	1	0	1	1	1	1
Italian	0	0	0	1	1	1	1	0
Czech	1	1	1	1	1	1	0	1

	Mon	Tue	Wed	Thu	Fri	Sat
Mon	1	0	0	0	0	0
Tue	0	1	0	1	1	0
Wed	0	0	1	1	1	1
Thu	0	1	1	1	1	0
Fri	0	1	1	1	1	0
Sat	0	0	1	0	0	1

	Greenland	Rügen	Cuba	Jamaica	Tasmania	Sumatra	Bali	Guam	Hokkaido	Formosa	Madagaskar	Zanzibar
Greenland	1	0	1	0	0	0	0	0	1	0	1	1
Rügen	0	1	0	0	0	1	1	0	1	0	0	1
Cuba	1	0	1	0	1	1	0	0	1	1	1	0
Jamaica	0	0	0	1	0	0	1	1	0	1	0	0
Tasmania	0	0	1	0	1	1	1	0	1	1	0	1
Sumatra	0	1	1	0	1	1	0	0	0	1	1	1
Bali	0	1	0	1	1	0	1	1	1	1	0	0
Guam	0	0	0	1	0	0	1	1	1	0	0	1
Hokkaido	1	1	1	0	1	0	1	1	1	1	0	1
Formosa	0	0	1	1	1	1	1	0	1	1	0	0
Madagaskar	1	0	1	0	0	1	0	0	0	0	1	1
Zanzibar	1	1	0	0	1	1	0	1	1	0	1	1

[2] According to Wikipedia, Jørgen Pedersen Gram (1850–1916) was a Danish actuary and mathematician.

11.3 Concept lattices

The dicliques or non-enlargeable rectangles, have often been called *concepts* inside a relation. Their investigation is then often called concept analysis. As we will see, it is heavily related to complete lattices. The complete lattice of non-enlargeable rectangles gives an important application for the finite case. These ideas are best introduced via examples.

Example 11.10. The table of Fig. 11.13 contains – following an article of Rudolf Wille and Bernhard Ganter – the German federal presidents prior to Johannes Rau together with some of their properties. (Such tables are typically rectangular; it so happened that this one is not. It is nonetheless a heterogeneous relation.)

	Start $<$ 60	Start $\geq$ 60	One P.	Two P.	CDU	SPD	FDP
Heuß	0	1	0	1	0	0	1
Lübke	0	1	0	1	1	0	0
Heinemann	0	1	1	0	0	1	0
Scheel	1	0	1	0	0	0	1
Carstens	0	1	1	0	1	0	0
Weizsäcker	0	1	0	1	1	0	0
Herzog	0	1	1	0	1	0	0

Fig. 11.13 German federal presidents

The easiest way to work with such a table is simply to ask whether Heinemann, when president, was a member of the SPD, for example. When talking about some topic related to this, one will most certainly after a while also formulate more general propositions that generalize and quantify.

- 'All federal presidents that were members of the CDU have entered office at age 60 or more.'
- 'There existed a federal president who was a member of the FDP and stayed in office for just one period.'
- 'All federal presidents with two periods in office have been members of the CDU.'
- 'All federal presidents who started at less than 60 years of age stayed in office for just one period.'

Such a statement may be satisfied or not. In every case, these quantified observations concerned *sets of* federal presidents and *sets of* their properties – albeit 1-element sets. It was important to determine whether these sets of properties were all satisfied or none of the property set, or whether there existed someone for whom the properties were satisfied.

The interdependency of such propositions follows a certain scheme. This general scheme is completely independent of the fact that federal presidents are concerned or their membership in some political party. The mechanism is quite similar to that with which we have learned to determine majorant and minorant sets of subsets in an ordered set. It is related with the concept of the non-enlargeable rectangle inside the given relation which has already been studied extensively.

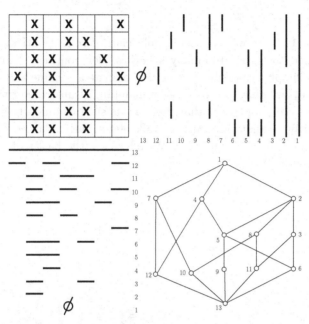

No.	Sets of German federal presidents	Their properties
1	all federal presidents	none
2	all but Scheel	{≥ 60}
3	{Lübke, Carstens, Weizsäcker, Herzog}	{≥ 60, CDU}
4	{Heinemann, Scheel, Carstens, Herzog}	{one period}
5	{Heinemann, Carstens, Herzog}	{≥ 60, one period}
6	{Carstens, Herzog}	{≥ 60, one period, CDU}
7	{Heuß, Scheel}	{FDP}
8	{Heuß, Lübke, Weizsäcker}	{≥ 60, two periods}
9	{Heinemann}	{≥ 60, one period, SPD}
10	{Heuß}	{≥ 60; two periods; FDP}
11	{Lübke, Weizsäcker}	{≥ 60, two periods, CDU}
12	{Scheel}	{< 60, one period, FDP}
13	no federal president	all properties

Fig. 11.14 Concept lattice example of Fig. 11.13 in extended presentation

One will observe that all non-enlargeable rectangles are presented in the Hasse diagram of Fig. 11.14. Proceeding upwards in the Hasse diagram of Fig. 11.14, the sets of objects are ordered increasingly while sets of properties are ordered decreasingly. As occurs often in mathematics, the top and the bottom pairs seem slightly artificial. □

Example 11.11. Next we consider a politician of some political party. We are fully accustomed to the all-too-many speeches they routinely deliver at various occasions. It is simply impossible for them to prepare a genuine and new talk every time. What they are able to achieve is to maintain a substrate of well-formulated speech elements covering this or that topic. From such modules, they arrange their talks. It is tacitly understood that such talks are slightly adapted in order to give a best fit for the respective audience. We offer here the basics of a technique for preparing such talks.

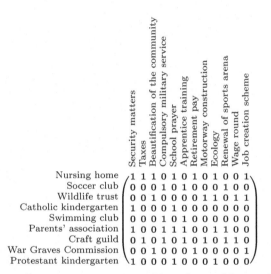

Fig. 11.15 Potential voters versus topics table to be used by a politician

First the expected recipients of all these speeches are subdivided into several groups or subsets. Then one asks which group of persons might be particularly interested in this or that topic, so that a relation as in Fig. 11.15 is obtained. Secondly, the concept lattice is determined as shown in Figs. 11.16 and 11.17. The lattice allows the formation of least upper bounds and greatest lower bounds, which will probably be utilized by the politician: he may well address every relevant topic, but choose supremum in the case that he expects it to raise a positive reaction and infima when expected to sound negative. □

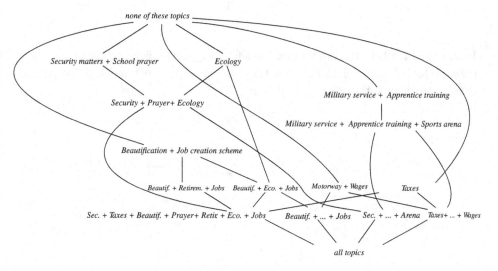

Fig. 11.16 Ordering for concept lattice A

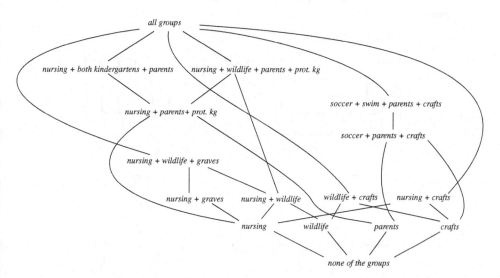

Fig. 11.17 Ordering for concept lattice B

We present yet another rather similar example of a concept lattice analysis.

Example 11.12. In the relation on the left of Fig. 11.18, several properties are attributed to a couple of nations. Then the non-enlargeable rectangles are determined that again show anti-isomorphic lattices.

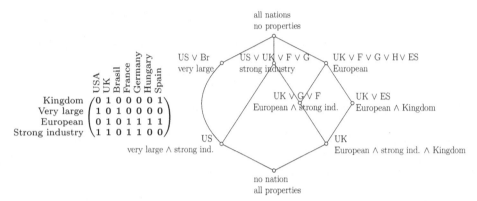

Fig. 11.18 Concept lattice for several nations with some of their important properties

Another example of a lattice of dicliques is given with Fig. 11.19. There is not much interpretation, but it is easy to grasp what the non-enlargeable rectangles look like.

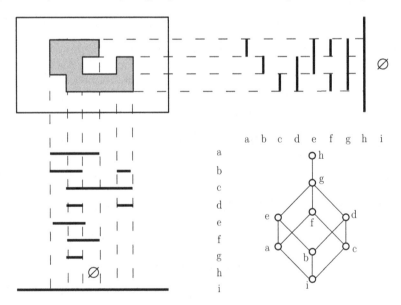

Fig. 11.19 Lattice of dicliques

The rowset-to-columnset difunctional

We have, thus, several times put into relation a subset of the source side with a subset on the target side, selected so that they make up a non-enlargeable rectangle inside R. Here, we resume this task again, and recall Prop. 10.3. Our starting point is a relation $R : X \longrightarrow Y$ together with subsets $u \subseteq X$ and $v \subseteq Y$. This time, an arbitrary subset s on the source side is taken and moved horizontally inside R as far as this is possible so as to determine

$$u_s := \overline{R \,\overline{R^{\mathsf{T}} \,; s}} \supseteq s, \qquad v_s := \overline{R^{\mathsf{T}} \,; s}.$$

Then u_s, v_s form a non-enlargeable rectangle inside R. We might also have taken any set t on the target side and moved it vertically over the maximum distance to obtain

$$u_t := \overline{R \,; t}, \qquad v_t := \overline{R^{\mathsf{T}} \,; \overline{R \,; t}} \supseteq t,$$

again a non-enlargeable rectangle u_t, v_t inside R.

The idea is now to apply these formulae column-wise to the membership relations $\varepsilon, \varepsilon'$ and to compare where the results agree. We establish this together with a bijective mapping between the row union space and the column union space. The symbol "$\bowtie$" – as opposed to "Υ" – has been chosen to symbolize that two vectors, not just points, are matched in both directions.

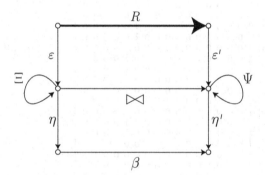

Fig. 11.20 Typing for rowset-to-columnset difunctional and rowspace-to-columnspace bijection

Definition 11.13. For an arbitrary (possibly heterogeneous) relation R, we call

(i) $\bowtie \; := \; \bowtie(R) := \mathsf{syq}\,(\overline{R \,; \overline{R^{\mathsf{T}} \,; \varepsilon}}, \overline{R \,; \varepsilon'})$, **rowset-to-columnset difunctional**,

(ii) $\beta := \beta(R) := \eta^{\mathsf{T}} \,; \bowtie \,; \eta'$, **rowspace-to-columnspace bijection**.

The natural projections according to its row and column equivalence $\Xi(\bowtie)$ and $\Psi(\bowtie)$ are here called η, respectively η'. $\qquad\qquad\qquad\qquad\qquad\square$

Definition 11.13 has given preference to the idea expressed by u_s, v_s above. 'Shift-inverting the cone' according to Prop. 8.19.ii, one may, however, also go the other way round and use the idea expressed via u_t, v_t:

$$\bowtie = \mathsf{syq}(\overline{\overline{R\,;\,\overline{\overline{R}^{\mathsf{T}}\,;\,\varepsilon}}}, \overline{\overline{R\,;\,\varepsilon'}}) = \mathsf{syq}(\overline{\overline{\overline{R}^{\mathsf{T}}\,;\,\varepsilon}}, \overline{\overline{R}^{\mathsf{T}}\,;\,\overline{R\,;\,\varepsilon'}}).$$

Figure 11.21 illustrates these constructs. The difunctional is rearranged while the bijection is not. The bijection holds between classes according to the row, respectively column, equivalence and has, thus, bracketed entry names.

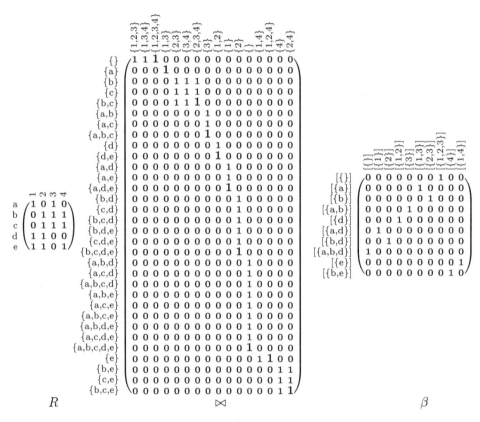

Fig. 11.21 Relation R with difunctional $\bowtie$ rearranged; bijection β in original form

From $\bowtie$, there is a close correspondence to $\Upsilon := \mathsf{syq}(\varepsilon, \overline{R\,;\,\varepsilon'}) \cap \mathsf{syq}(\overline{R}^{\mathsf{T}}\,;\,\varepsilon, \varepsilon')$, the diclique matching that we now exhibit. In comparison with Fig. 11.22, one will see, that Υ marks precisely the prescinded lower right corner points of the rearranged $\bowtie$ of the above Fig. 11.21.

Proposition 11.14. *We consider an arbitrary (possibly heterogeneous) relation R.*

(i) $\bowtie$ *is a total, surjective, and difunctional relation.*

(ii) $\Upsilon \subseteq \bowtie$.

(iii) *The construct* $\bowtie$ *satisfies with respect to* $C := \overline{\overline{R \,\overline{R}^{\mathsf{T}} \,; \varepsilon}}$, *the contact closure:*
$$C \,; \bowtie = \overline{\overline{R \,; \varepsilon'}} \qquad and \qquad C = \overline{\overline{R \,; \varepsilon'}} \,; \bowtie^{\mathsf{T}}.$$

(iv) $\varepsilon \,; \bowtie = \overline{\overline{R \,; \varepsilon'}}$.

(v) $\varepsilon \,; \bowtie \,; \varepsilon'^{\mathsf{T}} = R$.

Proof: (i) The relation $\bowtie$ is difunctional as a symmetric quotient. It is also total and surjective because
$$\bowtie = \mathsf{syq}(\overline{\overline{R}^{\mathsf{T}} \,; \varepsilon}, \overline{R}^{\mathsf{T}} \,; \overline{\overline{R \,; \varepsilon'}}) \qquad \text{by definition}$$
$$= \mathsf{syq}(\overline{R}^{\mathsf{T}} \,; \varepsilon, \overline{R}^{\mathsf{T}} \,; \overline{\overline{R \,; \varepsilon'}}) \qquad \mathsf{syq}(A,B) = \mathsf{syq}(\overline{A},\overline{B}); \text{ Prop. 8.10.i}$$
where the latter is surjective following Cor. 8.17.

(ii)
$$\Upsilon \subseteq \mathsf{syq}(\overline{R}^{\mathsf{T}} \,; \varepsilon, \varepsilon') \qquad \text{by definition and monotony}$$
$$\subseteq \mathsf{syq}(\overline{R \,; \overline{R}^{\mathsf{T}} \,; \varepsilon}, \overline{R \,; \varepsilon'}) \qquad \text{adding a common factor, Prop. 8.16.i}$$
$$= \mathsf{syq}(\overline{\overline{R \,; \overline{R}^{\mathsf{T}} \,; \varepsilon}}, \overline{R \,; \varepsilon'}) \qquad \text{always } \mathsf{syq}(A,B) = \mathsf{syq}(\overline{A},\overline{B}), \text{ Prop. 8.10.i}$$
$$= \bowtie \qquad \text{by definition}$$

(iii) This is in both cases simply cancelling for symmetric quotients according to Prop. 8.12.iii.

(iv)
$$\varepsilon \,; \bowtie = \varepsilon \,; \mathsf{syq}(\overline{\overline{R}^{\mathsf{T}} \,; \varepsilon}, \overline{R}^{\mathsf{T}} \,; \overline{\overline{R \,; \varepsilon'}}) \qquad \text{by definition}$$
$$= \varepsilon \,; \mathsf{syq}(\overline{R}^{\mathsf{T}} \,; \varepsilon, \overline{R}^{\mathsf{T}} \,; \overline{\overline{R \,; \varepsilon'}}) \qquad \text{since always } \mathsf{syq}(A,B) = \mathsf{syq}(\overline{A},\overline{B})$$
$$\supseteq \varepsilon \,; \mathsf{syq}(\varepsilon, \overline{\overline{R \,; \varepsilon'}}) \qquad \text{cancelling a common left factor, Prop. 8.16}$$
$$= \overline{\overline{R \,; \varepsilon'}} \qquad \text{Prop. 7.14}$$

For the direction "$\subseteq$", we recall that $\varepsilon \subseteq C$ and use (iii).

(v) We use (iv) in $\varepsilon \,; \bowtie \,; \varepsilon'^{\mathsf{T}} = \overline{\overline{R \,; \varepsilon'}} \,; \varepsilon'^{\mathsf{T}} = \overline{\overline{R}} = R$, following Prop. 7.14 again. $\qquad\square$

In general, the relation $\bowtie$ is not a bijective mapping. In view of (iii), one may say that it provides a 'block-isomorphism' between $C = \overline{\overline{R \,; \overline{R}^{\mathsf{T}} \,; \varepsilon}}$ and $\overline{\overline{R \,; \varepsilon'}}$.

There is a close interrelationship between Υ, λ on one side and $\bowtie, \beta$ on the other. This interrelationship is now visualized with Fig. 11.22. The matching Υ gives rise to natural injections on either side, while the difunctional relation $\bowtie$ leads to natural projections. The two are different concepts. A detailed discussion is postponed until Prop. 11.25, although the transition between extruded sets and quotient sets is already indicated with the relations δ, δ'.

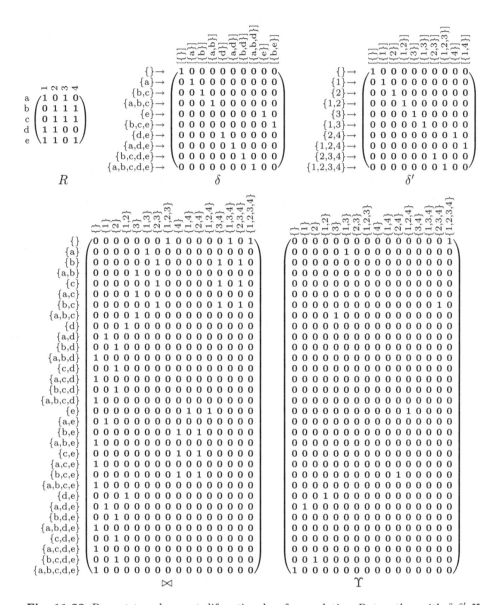

Fig. 11.22 Rowset-to-columnset difunctional ⋈ for a relation R, together with $\delta, \delta', \Upsilon$

11.4 Closure and contact

Closure forming is a very basic technique of everyday life. Looking at a geometric figure, one can immediately imagine what its *convex* closure looks like. Nobody would ever think of its *concave* closure. Even when switching to a more mathematical structure, we have a rather firm feeling what closure means. Given a set of vectors, we know that they span a vector space. A closure is, thus, obviously something that is not smaller than the original, that will stay the same when the closure is formed again, and of course, it will not deliver a smaller result when applied to a larger argument. All this is traditionally captured by the following definition.

Definition 11.15. Let some ordered set $(V, \leq)$ be given. A mapping $\rho : V \longrightarrow V$ is called a **closure operation**, if it is

(i)   expanding $x \leq \rho(x)$,
(ii) isotonic $x \leq y \longrightarrow \rho(x) \leq \rho(y)$,
(iii) idempotent $\rho(\rho(x)) \leq \rho(x)$.   □

Typically, closures are combined with some predicate that is '∩-hereditary' or more precisely 'infimum-hereditary'. By this we mean that applying the predicate to an intersection of two arguments equals the intersected results of the predicate applied to the two arguments; this applies also to infima. (In doing this, a greatest one to apply to is tacitly assumed to exist.) Transitive closures are widely known. But there exist other related closures, some of them central for all our reasoning, and, thus omnipresent. We give several examples:

- $X, Y \subseteq \mathbb{R} \times \mathbb{R}$ convex $\Longrightarrow X \cap Y$ convex   convex closure
- X, Y vector spaces $\Longrightarrow X \cap Y$ vector space   linear closure
- X, Y axis-parallel rectangles $\Longrightarrow X \cap Y$ axis-parallel rectangular closure
- X, Y transitive relations $\Longrightarrow X \cap Y$ transitive   transitive closure
- X, Y difunctional relations $\Longrightarrow X \cap Y$ difunctional   difunctional closure
- X, Y upper cones $\Longrightarrow X \cap Y$ upper cone   upper cone

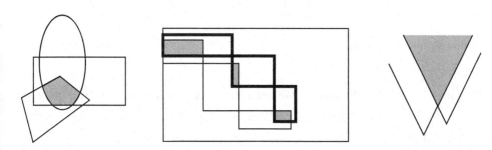

Fig. 11.23 Intersection of convex sets, sets of non-overlapping axis-parallel rectangles, and cones

Starting from such '∩-hereditary' properties, one forms the construct

$$\rho(X) := \inf\{Y \mid X \subseteq Y, Y \text{ satisfies '∩-hereditary' property}\}$$

and thus obtains the convex closure, the spanned vector space, the least encompassing set of axis-parallel rectangles, or the transitive closure, etc.

Proposition 11.16. *Let an arbitrary relation $R : U \longrightarrow V$ be given.*

(i) *Forming R-contact closures*

$$X \mapsto \bigwedge_R(X)$$

for relations $X : U \longrightarrow W$ is a closure operation.

(ii) *The set of R-contact closures*

$$CC_{R,W} := \{Y \mid Y = \bigwedge_R(Z) \quad \text{for some } Z : U \longrightarrow W\}$$

forms a complete lattice, in particular:

— *join over an arbitrary subset $\mathcal{Y} \subseteq CC_{R,W}$ may be given in either form as*

$$\bigwedge_R(\sup\{Y \mid Y \in \mathcal{Y}\}) = \text{lbd}_R\left(\inf\{\text{ubd}_R(Y) \mid Y \in \mathcal{Y}\}\right)$$

— $\bigwedge_R(⊥)$ *is the least element,*

— $⊤$ *is the greatest element.*

Proof: (i) It is trivial to show that $Z \mapsto \bigwedge_R(Z)$ is a closure operation according to Def. 11.15, i.e., expanding, isotonic, and idempotent

$$Z \subseteq \bigwedge_R(Z) \qquad Z \subseteq Z' \implies \bigwedge_R(Z) \subseteq \bigwedge_R(Z') \qquad \bigwedge_R(\bigwedge_R(Z)) = \bigwedge_R(Z).$$

(ii) We show that both definitions are equivalent; the rest is left as an exercise.

$$\begin{aligned}
\text{lbd}_R\left(\inf\{\text{ubd}_R(Y) \mid Y \in \mathcal{Y}\}\right) &= \overline{R\,\overline{}}\,\left(\inf\{\text{ubd}_R(Y) \mid Y \in \mathcal{Y}\}\right) && \text{expanding} \\
&= \overline{R\,\overline{}}\, \sup\{\overline{R^{\mathsf{T}}}\,;Y \mid Y \in \mathcal{Y}\} && \text{because } \inf\{A \mid \ldots\} = \overline{\sup\{\overline{A} \mid \ldots\}} \\
&= \overline{R}\,\overline{R^{\mathsf{T}}}\,; \sup\{Y \mid Y \in \mathcal{Y}\} && \text{composition is distributive} \\
&= \bigwedge_R(\sup\{Y \mid Y \in \mathcal{Y}\}) && \text{by definition} \qquad\qquad \square
\end{aligned}$$

In an obvious way, we now start lifting the closure properties to the relational level.

Proposition 11.17. *We assume the ordering $E : X \longrightarrow X$ and consider a mapping $\rho : X \longrightarrow X$. Then ρ is a closure operator if and only if*

(i) $\rho \subseteq E$,

(ii) $E\,;\rho \subseteq \rho\,;E$,

(iii) $\rho\,;\rho \subseteq \rho$.

Proof: Here again, we cannot give a relation-algebraic proof; (i,ii,iii) have simply to be postulated. What we can do, however, is to convince ourselves that the intentions of Def. 11.15 are met when lifting in this way. To this end, we check (iii), for example, versus Def. 11.15.iii. We do not immediately use that ρ is a mapping, assuming that it is just a relation when starting from

$$\rho(\rho(x)) \leq \rho(x).$$

Adding quantifiers, this reads in the present context

$$\forall x, y, z : \rho_{xy} \wedge \rho_{yz} \rightarrow [\exists w : \rho_{xw} \wedge E_{zw}]$$

transposition, definition of composition

$$\Longleftrightarrow \quad \forall x, y, z : \rho_{xy} \wedge \rho_{yz} \rightarrow (\rho_{.} E^{\mathsf{T}})_{xz}$$

$$a \rightarrow b = \neg a \vee b$$

$$\Longleftrightarrow \quad \forall x, y, z : \neg \rho_{xy} \vee \neg \rho_{yz} \vee (\rho_{.} E^{\mathsf{T}})_{xz}$$

$$\forall x : p(x) = \neg [\exists x : \neg p(x)], \text{ arranging quantifiers}$$

$$\Longleftrightarrow \quad \forall x, z : \neg \big[\exists y : \big(\rho_{xy} \wedge \rho_{yz} \wedge \overline{[\rho_{.} E^{\mathsf{T}}]_{xz}} \big) \big]$$

$$\exists y : [p(y) \wedge c] = [\exists y : p(y)] \wedge c$$

$$\Longleftrightarrow \quad \forall x, z : \neg \big[(\exists y : \rho_{xy} \wedge \rho_{yz}) \wedge \overline{[\rho_{.} E^{\mathsf{T}}]_{xz}} \big]$$

definition of composition

$$\Longleftrightarrow \quad \forall x, z : \neg \big[(\rho_{.} \rho)_{xz} \wedge \overline{[\rho_{.} E^{\mathsf{T}}]_{xz}} \big]$$

$$\neg a \vee b = a \rightarrow b$$

$$\Longleftrightarrow \quad \forall x, z : (\rho_{.} \rho)_{xz} \rightarrow [\rho_{.} E^{\mathsf{T}}]_{xz}$$

transition to point-free form

$$\Longleftrightarrow \quad \rho_{.} \rho \subseteq \rho_{.} E^{\mathsf{T}}$$

This is not yet what we intended to prove. But

$$\rho_{.} E^{\mathsf{T}} \cap \rho_{.} E = \rho_{.} (E^{\mathsf{T}} \cap E) = \rho_{.} \mathbb{I} = \rho$$

since ρ is univalent. Now, $\rho_{.} \rho \subseteq \rho$ is easily obtained using (i) which gives $\rho_{.} \rho \subseteq \rho_{.} E$ and antisymmetry of E. With Prop. 5.2.iii, we even get $\rho_{.} \rho = \rho$. $\qquad\square$

The identity relation $\mathbb{I} : X \longrightarrow X$ is certainly a closure operation.

Contact relations

According to Georg Aumann [4, 5], closures always come with a simpler construct, namely a contact relation, and there is an important correlation between closure forming and contact relations. The idea is rather immediate: forming the convex closure is a mapping of a subset to another subset; what, however, makes this other subset the closure? It cannot be shrunk any further as there are *contact points* preventing this. We will develop this step by step, starting from Fig. 11.24 which shows the basic situation of this idea.

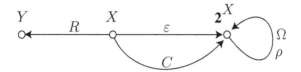

Fig. 11.24 Basic situation between contact and closure

First, we define contact relations as such, i.e., not yet establishing interrelationship with closure forming. This has been generalized in [114].

Definition 11.18. We consider a set related to its powerset, with a membership relation $\varepsilon : X \longrightarrow 2^X$. Then a relation $C : X \longrightarrow 2^X$ is called an **Aumann contact relation**, provided

(i) it contains the membership relation, i.e., $\varepsilon \subseteq C$,
(ii) an element x in contact with a set Y *all of whose elements* are in contact with a set Z, will itself be in contact with Z, the so-called *infectivity* of contact, i.e.,

$$C \,;\, \overline{\varepsilon^{\mathsf{T}} \,;\, \overline{C}} \subseteq C, \quad \text{or equivalently,} \quad C^{\mathsf{T}} \,;\, \overline{C} \subseteq \varepsilon^{\mathsf{T}} \,;\, \overline{C}. \qquad \square$$

The closure relation ρ is of size $2^n \times 2^n$. Slightly more efficient to store than the closure is its contact relation with size $n \times 2^n$.

Denoting the powerset ordering as $\Omega : 2^X \longrightarrow 2^X$, one will easily show that C, considered column-wise, forms an upper cone. From

$$C^{\mathsf{T}} \,;\, \overline{C} \subseteq \varepsilon^{\mathsf{T}} \,;\, \overline{C} \subseteq \varepsilon^{\mathsf{T}} \,;\, \overline{\varepsilon} = \overline{\Omega}$$

one may deduce $C = C \,;\, \Omega$, of which "$\subseteq$" is obvious. Condition (ii) is in fact an equation $C^{\mathsf{T}} \,;\, \overline{C} = \varepsilon^{\mathsf{T}} \,;\, \overline{C}$.[3]

The following result brings Aumann contacts close to previous considerations concerning the R-contact closure $\mathbf{lbd}_R(\mathbf{ubd}_R(\varepsilon))$ of Def. 8.7. Thus, it provides access to a lot of formulae for majorants and minorants.

Proposition 11.19. *Let an arbitrary relation $R : X \longrightarrow Y$ be given. Then the R-contact closure*

---

[3] A slightly more restrictive definition demands $C \subseteq \mathbb{T} \,;\, \varepsilon$, i.e., no element is in contact with the empty set and *distributivity* of contact $C_{x,(Y \cup Z)} \longrightarrow C_{x,Y} \cup C_{x,Z}$, in which case we obtain a topological contact that already leads to every property of a topology.

$$C := \bigwedge_R(\varepsilon) = \overline{\overline{R \mathbin{;} \overline{R}^\mathsf{T} \mathbin{;} \varepsilon}}$$

is always an Aumann contact relation; it satisfies in addition

$$\overline{R}^\mathsf{T} \mathbin{;} C = \overline{R}^\mathsf{T} \mathbin{;} \varepsilon \quad and \quad C = \bigwedge_R(C) = \overline{\overline{R \mathbin{;} \overline{R}^\mathsf{T} \mathbin{;} C}}.$$

Proof: To show this, we prove $\varepsilon \subseteq \overline{\overline{R \mathbin{;} \overline{R}^\mathsf{T} \mathbin{;} \varepsilon}} = C$, which is trivial from the closure property, and

$$C^\mathsf{T} \mathbin{;} \overline{C} = \overline{\overline{R \mathbin{;} \overline{R}^\mathsf{T} \mathbin{;} \varepsilon}}^\mathsf{T} \mathbin{;} \overline{\overline{R \mathbin{;} \overline{R}^\mathsf{T} \mathbin{;} \varepsilon}} \subseteq \varepsilon^\mathsf{T} \mathbin{;} \overline{\overline{R \mathbin{;} \overline{R}^\mathsf{T} \mathbin{;} \varepsilon}} = \varepsilon^\mathsf{T} \mathbin{;} \overline{C}$$

$$\Longleftarrow \quad \overline{\overline{R \mathbin{;} \overline{R}^\mathsf{T} \mathbin{;} \varepsilon}}^\mathsf{T} \mathbin{;} R \subseteq \varepsilon^\mathsf{T} \mathbin{;} \overline{R} \quad \Longleftrightarrow \quad \varepsilon^\mathsf{T} \mathbin{;} \overline{R \mathbin{;} \overline{R}^\mathsf{T}} \subseteq \big[\overline{R \mathbin{;} \overline{R}^\mathsf{T} \mathbin{;} \varepsilon}\big]^\mathsf{T}$$

$$\overline{R}^\mathsf{T} \mathbin{;} C = \overline{R}^\mathsf{T} \mathbin{;} \overline{\overline{R \mathbin{;} \overline{R}^\mathsf{T} \mathbin{;} \varepsilon}} = \overline{R}^\mathsf{T} \mathbin{;} \varepsilon. \qquad \qquad \square$$

The construct $C := \overline{\overline{R \mathbin{;} \overline{R}^\mathsf{T} \mathbin{;} \varepsilon}}$ may be read as follows. It declares those combinations $x \in X$ and $S \subseteq X$ to be in contact C, for which every relationship $(x, y) \notin R$ implies that there exists also an $x' \in S$ in relation $(x', y) \notin R$.

We illustrate contact defined via R assuming a relation between persons and topics as occurs among activists of non-governmental organizations. Let the relation R describe persons p willing to engage for topics t. Then typically activist groups g are spontaneously formed, a process that will now be made relationally precise. A person p is said to be in contact with a group g of persons if, whatever topic t one considers, when everyone in the group g works for t then so does p. This is so delicate to handle in natural language that we again include a predicate-logic justification:

$$(p, g) \in C \quad \Longleftrightarrow \quad (p, g) \in \overline{\overline{R \mathbin{;} \overline{R}^\mathsf{T} \mathbin{;} \varepsilon}}$$

expanding

$$\Longleftrightarrow \quad \neg\Big[\exists t : \overline{R}_{p,t} \wedge \overline{\exists p' : \overline{R}^\mathsf{T}_{t,p'} \wedge \varepsilon_{p',g}}\Big]$$

$\neg[\exists x : q(x)] = \forall x : \neg q(x)$ and switching terms

$$\Longleftrightarrow \quad \forall t : \big[\exists p' : \overline{R}^\mathsf{T}_{t,p'} \wedge \varepsilon_{p',g}\big] \vee R_{p,t}$$

$\neg a \vee b = a \to b$, transposition, switching terms

$$\Longleftrightarrow \quad \forall t : \neg\big[\exists p' : \varepsilon_{p',g} \wedge \overline{R}_{p',t}\big] \to R_{p,t}$$

$\neg[\exists x : q(x)] = \forall x : \neg q(x)$

$$\Longleftrightarrow \quad \forall t : \big[\forall p' : p' \notin g \vee R_{p',t}\big] \to R_{p,t}$$

$\neg a \vee b = a \to b$

$$\Longleftrightarrow \quad \forall t : \big[\forall p' : p' \in g \to R_{p',t}\big] \to R_{p,t}$$

After this illustration, we show how every closure operator gives rise to an Aumann contact relation.

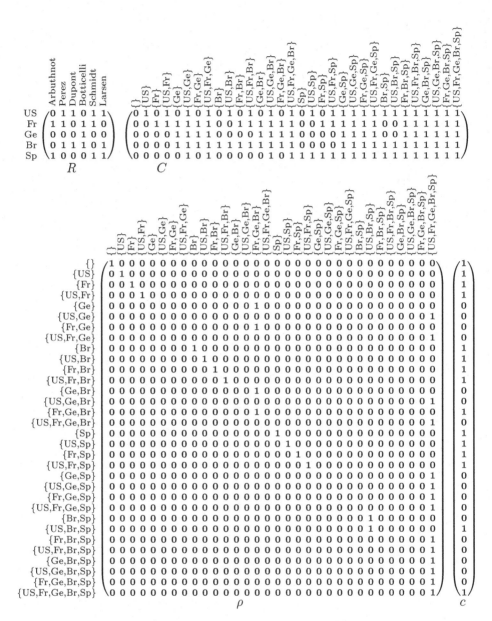

Fig. 11.25 A relation R introducing a contact relation C,
its closure ρ, and its closed sets c

Proposition 11.20. *Given any closure operator* $\rho : 2^X \longrightarrow 2^X$ *on some powerset defined via a membership relation* $\varepsilon : X \longrightarrow 2^X$, *the construct* $C := \varepsilon\,\rho^{\mathsf{T}}$ *turns out to be an Aumann contact relation.*

Proof: $\varepsilon \subseteq C = \varepsilon\,;\rho^{\mathsf{T}}$ follows from $\varepsilon\,;\rho \subseteq \varepsilon\,;\Omega = \varepsilon$ with shunting; see Prop. 5.12.i.

$$
\begin{aligned}
C\,;\overline{\varepsilon^{\mathsf{T}}\,;\overline{C}} &= \varepsilon\,;\rho^{\mathsf{T}}\,;\overline{\varepsilon^{\mathsf{T}}\,;\overline{\varepsilon\,;\rho^{\mathsf{T}}}} && \text{definition} \\
&= \varepsilon\,;\rho^{\mathsf{T}}\,;\overline{\varepsilon^{\mathsf{T}}\,;\overline{\varepsilon}}\,;\rho^{\mathsf{T}} && \text{with mapping } \rho \text{ slipping out of negation} \\
&= \varepsilon\,;\rho^{\mathsf{T}}\,;\Omega\,;\rho^{\mathsf{T}} && \text{definition of } \Omega \\
&\subseteq \varepsilon\,;\Omega\,;\rho^{\mathsf{T}}\,;\rho^{\mathsf{T}} && \text{closure } \rho \text{ considered as an } \Omega\text{-homomorphism, Prop. 5.45} \\
&\subseteq \varepsilon\,;\Omega\,;\rho^{\mathsf{T}} && \text{with the third closure property Prop. 11.17.iii} \\
&= \varepsilon\,;\rho^{\mathsf{T}} && \text{since } \varepsilon\,;\Omega = \varepsilon \\
&= C && \text{definition of } C \qquad\qquad \square
\end{aligned}
$$

However, we may also go in the reverse direction and define a closure operation starting from a contact relation. Particularly interesting is the collection of *closed* subsets as we may reconfigure the closure operation already from these. In Fig. 11.26, we show the powerset ordering restricted to this set.

Proposition 11.21. *Given any Aumann contact relation* $C : X \longrightarrow 2^X$, *forming the construct* $\rho := \mathsf{syq}\,(C,\varepsilon)$ *results in a closure operator.*

Proof: During the proof, we follow the numbering of Prop. 11.17.

(i) $\rho = \mathsf{syq}\,(C,\varepsilon) \subseteq \overline{C^{\mathsf{T}}\,;\overline{\varepsilon}} \subseteq \overline{\varepsilon^{\mathsf{T}}\,;\overline{\varepsilon}} = \Omega$ \qquad Def. 11.18.i

(ii) We prove the second homomorphism condition according to Prop. 5.45:

$$
\begin{aligned}
\Omega &= \overline{\varepsilon^{\mathsf{T}}\,;\overline{\varepsilon}} && \text{by definition} \\
&\subseteq \overline{\varepsilon^{\mathsf{T}}\,;\overline{C}} && \text{Def. 11.18.i} \\
&= \overline{C^{\mathsf{T}}\,;\overline{C}} && \text{Def. 11.18.ii} \\
&= \mathsf{syq}\,(C,\varepsilon)\,;\overline{\varepsilon^{\mathsf{T}}\,;\overline{\varepsilon}}\,;\mathsf{syq}\,(\varepsilon,C) && \text{since always } \varepsilon\,\mathsf{syq}\,(\varepsilon,X) = X, \overline{\varepsilon}\,\mathsf{syq}\,(\varepsilon,Y) = \overline{Y} \\
&= \rho\,;\overline{\varepsilon^{\mathsf{T}}\,;\overline{\varepsilon}}\,;\rho^{\mathsf{T}} && \text{by definition} \\
&= \rho\,;\overline{\varepsilon^{\mathsf{T}}\,;\overline{\varepsilon}}\,;\rho^{\mathsf{T}} && \text{mapping slipping out of negation} \\
&= \rho\,;\Omega\,;\rho^{\mathsf{T}} && \text{by definition}
\end{aligned}
$$

(iii) Two terms are treated separately. With

$$\overline{C}^{\mathsf{T}}\,;C \subseteq \overline{C}^{\mathsf{T}}\,;\varepsilon \qquad \text{following the second property of a contact relation}$$

$$C^{\mathsf{T}}\,;\overline{C} \subseteq C^{\mathsf{T}}\,;\overline{\varepsilon} \qquad \text{since } \varepsilon \subseteq C$$

we prove $\rho \subseteq \mathsf{syq}\,(C,C)$, so that $\rho\,;\rho \subseteq \mathsf{syq}\,(C,C)\,;\mathsf{syq}\,(C,\varepsilon) \subseteq \mathsf{syq}\,(C,\varepsilon) = \rho$. $\quad\square$

In Fig. 11.25, we have marked which subsets stay unchanged when applying ρ, i.e., which are fixed points with respect to ρ. They are identified by forming $c := (\rho \cap \mathbb{I})\,;\mathbb{T}$ and shown as a vector on the right-hand side.

Proposition 11.22. *Assuming the setting developed so far, finite intersections of closed sets are closed again.*

Proof: We show this for any two closed subsets x, y. They obviously correspond to points $e_x := \mathsf{syq}(\varepsilon, x)$ and $e_y := \mathsf{syq}(\varepsilon, y)$ in the powerset. Their intersection $x \cap y$ corresponds to $e_{x \cap y} := \mathsf{syq}(\varepsilon, x \cap y) = \mathsf{syq}(\varepsilon, \varepsilon \mathbin{;} e_x \cap \varepsilon \mathbin{;} e_y)$. We start from $\rho^{\mathsf{T}} \mathbin{;} e_x = e_x$ and $\rho^{\mathsf{T}} \mathbin{;} e_y = e_y$ resembling the fact that x, y are closed and try to prove correspondingly $\rho^{\mathsf{T}} \mathbin{;} e_{x \cap y} = e_{x \cap y}$.

Since the right side is a point and the left side is certainly surjective, it is sufficient to prove

$$\rho^{\mathsf{T}} \mathbin{;} e_{x \cap y} \subseteq e_{x \cap y}$$
$$\Longleftrightarrow \quad \rho \mathbin{;} \overline{e_{x \cap y}} \subseteq \overline{e_{x \cap y}} \qquad \text{Schröder rule}$$
$$\Longleftrightarrow \quad \rho \mathbin{;} \left[\overline{\varepsilon}^{\mathsf{T}} \mathbin{;} (\varepsilon \mathbin{;} e_x \cap \varepsilon \mathbin{;} e_y) \cup \varepsilon^{\mathsf{T}} \mathbin{;} (\overline{\varepsilon \mathbin{;} e_x} \cup \overline{\varepsilon \mathbin{;} e_y}) \right] \subseteq \overline{\varepsilon}^{\mathsf{T}} \mathbin{;} (\varepsilon \mathbin{;} e_x \cap \varepsilon \mathbin{;} e_y) \cup \varepsilon^{\mathsf{T}} \mathbin{;} (\overline{\varepsilon \mathbin{;} e_x} \cup \overline{\varepsilon \mathbin{;} e_y})$$

The last line has been obtained by expanding and applying the De Morgan rule. The first term is now easy to handle remembering that $C = \varepsilon \mathbin{;} \rho^{\mathsf{T}}$:

$$\rho \mathbin{;} \overline{\varepsilon}^{\mathsf{T}} = \overline{\rho \mathbin{;} \varepsilon^{\mathsf{T}}} = \overline{C}^{\mathsf{T}} \subseteq \overline{\varepsilon}^{\mathsf{T}}.$$

As an example for the others we treat e_x:

$$\rho \mathbin{;} \varepsilon^{\mathsf{T}} \mathbin{;} \overline{\varepsilon \mathbin{;} e_x} = \rho \mathbin{;} \varepsilon^{\mathsf{T}} \mathbin{;} \overline{\varepsilon \mathbin{;} \rho^{\mathsf{T}} \mathbin{;} e_x} \qquad \text{closedness of } e_x$$
$$= C^{\mathsf{T}} \mathbin{;} \overline{C \mathbin{;} e_x} \qquad \text{because } C = \varepsilon \mathbin{;} \rho^{\mathsf{T}}$$
$$= C^{\mathsf{T}} \mathbin{;} \overline{C} \mathbin{;} e_x \qquad \text{point slips out of negation}$$
$$= \varepsilon^{\mathsf{T}} \mathbin{;} \overline{C} \mathbin{;} e_x \qquad \text{condition } C^{\mathsf{T}} \mathbin{;} \overline{C} = \varepsilon^{\mathsf{T}} \mathbin{;} \overline{C} \text{ for the contact relation } C$$
$$= \varepsilon^{\mathsf{T}} \mathbin{;} \overline{\varepsilon \mathbin{;} \rho^{\mathsf{T}}} \mathbin{;} e_x \qquad \text{because } C = \varepsilon \mathbin{;} \rho^{\mathsf{T}}$$
$$= \varepsilon^{\mathsf{T}} \mathbin{;} \overline{\varepsilon} \mathbin{;} \rho^{\mathsf{T}} \mathbin{;} e_x \qquad \text{mapping } \rho \text{ slips out of negation}$$
$$= \varepsilon^{\mathsf{T}} \mathbin{;} \overline{\varepsilon} \mathbin{;} e_x \qquad \text{closedness of } e_x$$
$$= \varepsilon^{\mathsf{T}} \mathbin{;} \overline{\varepsilon \mathbin{;} e_x} \qquad \text{point slips below negation} \qquad \square$$

When a powerset ordering is given together with a set c of closed subsets of some closure operation, it is also possible to re-obtain the closure operation ρ from this set c. We do not elaborate this in detail.

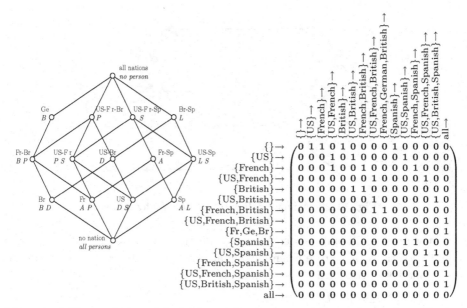

Fig. 11.26 Hasse relation of the lattice of closed subsets with respect to ρ from Fig. 11.25

Looking at the powerset ordering restricted to the sets marked by c, we get a lattice. Astonishingly enough, the lattice obtained is the same as the lattice of the row union space, already shown in the diagram on the right of Fig. 11.2. This may be checked in Fig. 11.26, with the Hasse relation for Fig. 11.25 and the diagram on the right of Fig. 11.2 repeated. The two are defined in completely different ways. The former was obtained by forming the quotient according to a row equivalence, indicated by bracketing in the row and column denotations. The present figure was obtained by extruding the fixed point of some closure forming, indicated with injection arrows in row and column denotations. A question arises immediately: is it possible to establish an isomorphism between the two? This is the basic question in the next subsection.

The topography around an arbitrary relation

So far, we have developed quite a menagerie of relations around any given relation R. We will now arrange the diversity of these into a coherent system. To this end, we recall in Fig. 11.27 the typing information we have gradually developed. The relation initially given is R, for which we consider the by now well-known membership relations on the source and on the target side, in TITUREL denoted as

$$\varepsilon := \texttt{Member(src}(R))\qquad \text{and}\qquad \varepsilon' := \texttt{Member(tgt}(R)).$$

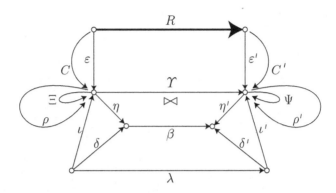

Fig. 11.27 A relation R with its topography around

From now on, we will omit any obviously symmetric counterparts. There were two main observations that arose when looking at non-enlargeable rectangles inside R. On the one hand, there exists the diclique matching

$$\Upsilon := \Upsilon(R) := \mathsf{syq}(\varepsilon, \overline{\overline{R \, ; \, \varepsilon'}}) \cap \mathsf{syq}(\overline{\overline{R^{\mathsf{T}} \, ; \, \varepsilon}}, \varepsilon') = \mathsf{syq}(\varepsilon, \overline{\overline{R \, ; \, C'}}) \cap \mathsf{syq}(\overline{\overline{R^{\mathsf{T}} \, ; \, C}}, \varepsilon')$$

and on the other hand the surjective and total rowset-to-columnset difunctional in which the latter is contained

$$\bowtie := \bowtie(R) := \mathsf{syq}(\overline{\overline{R \, ; \, \overline{\overline{R^{\mathsf{T}} \, ; \, \varepsilon}}}}, \overline{\overline{R \, ; \, \varepsilon'}}).$$

Using the contact closure,

$$C := \bigwedge\nolimits_R(\varepsilon) = \overline{\overline{R \, ; \, \overline{\overline{R^{\mathsf{T}} \, ; \, \varepsilon}}}},$$

which according to Prop. 11.19 is interrelated with the others as $\overline{\overline{R^{\mathsf{T}} \, ; \, C}} = \overline{\overline{R^{\mathsf{T}} \, ; \, \varepsilon}}$, we may also write the rowset-to-columnset difunctional (with Prop. 11.19) as

$$\bowtie = \mathsf{syq}(C, \overline{\overline{R \, ; \, \varepsilon'}}) = \mathsf{syq}(C, \overline{\overline{R \, ; \, C'}}).$$

While this might look as if it is constructed in a non-symmetric way, it is not, because we may shift-invert the cone so as to obtain

$$\bowtie = \mathsf{syq}(\overline{\overline{R^{\mathsf{T}} \, ; \, \varepsilon}}, \overline{\overline{R^{\mathsf{T}} \, ; \, \overline{\overline{R \, ; \, \varepsilon'}}}}) = \mathsf{syq}(\overline{\overline{R^{\mathsf{T}} \, ; \, \varepsilon}}, C') = \mathsf{syq}(\overline{\overline{R^{\mathsf{T}} \, ; \, C}}, C').$$

The following proposition will show how extremely interrelated these constructs are. We will be careful to restrict the number of proofs to a minimum. The key steps must be shown *before* extruding with $\iota := \mathtt{Inject}(\Upsilon \, ; \, \mathbb{T})$ or projecting with $\eta := \mathtt{Project}(\Xi(\bowtie))$. Participation of these latter two relations is fully regulated by their postulated generic properties of Def. 7.8 and Def. 7.11.

One will recognize in (i) that $\overline{\overline{C^{\mathsf{T}} \, ; \, \varepsilon}} \cap \overline{\varepsilon^{\mathsf{T}} \, ; \, \overline{C}}$ looks similar to, but is not, a symmetric quotient. (iii) will show that applying ρ does not change the class according to $\Xi(\bowtie)$.

Proposition 11.23. *We assume the general setting explained so far. Then*

(i) $\Xi(\bowtie) = \bowtie_{;}\bowtie^{\mathsf{T}} = \mathsf{syq}\,(C,C) = \overline{\overline{C}^{\mathsf{T}}_{;}\varepsilon} \cap \overline{\varepsilon^{\mathsf{T}}_{;}\overline{C}} = \Xi(\overline{\varepsilon^{\mathsf{T}}_{;}\overline{R}}) = \rho_{;}\rho^{\mathsf{T}}$,

(ii) $\Xi(\bowtie)_{;}\rho = \rho$ $\quad$ *and* $\quad$ $\rho_{;}\bowtie = \bowtie$,

(iii) $\rho \subseteq \Xi(\bowtie)$,

(iv) $\Xi(\bowtie)_{;}\Omega_{;}\Xi(\bowtie) = \rho_{;}\rho^{\mathsf{T}}_{;}\Omega_{;}\rho_{;}\rho^{\mathsf{T}} = \rho_{;}\Omega_{;}\rho^{\mathsf{T}} = \overline{C^{\mathsf{T}}_{;}\overline{C}} = \overline{\varepsilon^{\mathsf{T}}_{;}\overline{C}}$.

Proof: (i) We will arrive from several starting points at $\mathsf{syq}\,(C,C)$:

$$\Xi(\bowtie) = \mathsf{syq}\,(\bowtie^{\mathsf{T}}, \bowtie^{\mathsf{T}}) \qquad \text{by definition of a row equivalence}$$

$\qquad = \mathsf{syq}\,(\mathsf{syq}\,(\overline{R}_{;}\varepsilon', C), \mathsf{syq}\,(\overline{R}_{;}\varepsilon', C)) \qquad$ expansion of $\bowtie$ transposed

$\qquad = \mathsf{syq}\,(C,C) \qquad$ according to Prop. 8.20.iii, since $\bowtie$ is total and surjective

$\qquad = \overline{\overline{C}^{\mathsf{T}}_{;}C} \cap \overline{C^{\mathsf{T}}_{;}\overline{C}} \qquad$ by definition

$\qquad = \overline{\overline{C}^{\mathsf{T}}_{;}\varepsilon} \cap \overline{\varepsilon^{\mathsf{T}}_{;}\overline{C}} \qquad$ since C is a contact relation

$\bowtie_{;}\bowtie^{\mathsf{T}} = \mathsf{syq}\,(C, \overline{R}_{;}\varepsilon')_{;}\mathsf{syq}\,(\overline{R}_{;}\varepsilon', C) \qquad$ by definition

$\qquad = \mathsf{syq}\,(C,C) \qquad$ cancelling with $\bowtie$ total

$\qquad = \mathsf{syq}\,(\overline{R_{;}\overline{R}^{\mathsf{T}}_{;}\varepsilon}, \overline{R_{;}\overline{R}^{\mathsf{T}}_{;}\varepsilon}) \qquad$ expanding C

$\qquad = \mathsf{syq}\,(\overline{R}^{\mathsf{T}}_{;}R_{;}\overline{\overline{R}^{\mathsf{T}}_{;}\varepsilon}, \overline{R}^{\mathsf{T}}_{;}\varepsilon) \qquad$ shift-inverting the cone

$\qquad = \mathsf{syq}\,(\overline{R}^{\mathsf{T}}_{;}\varepsilon, \overline{R}^{\mathsf{T}}_{;}\varepsilon) \qquad$ since $\mathtt{lbd}_R(\mathtt{ubd}_R(\mathtt{lbd}_R(R))) = \mathtt{lbd}_R(R)$

$\qquad = \Xi(\overline{\varepsilon^{\mathsf{T}}_{;}\overline{R}}) \qquad$ by definition of a row equivalence

$\rho_{;}\rho^{\mathsf{T}} = \mathsf{syq}\,(C,\varepsilon)_{;}\mathsf{syq}\,(\varepsilon, C) \qquad$ by definition and transposition

$\qquad = \mathsf{syq}\,(C,C) \qquad$ cancelling according to Prop. 8.13.ii with ρ total

(ii) $\Xi(\bowtie)_{;}\rho = \mathsf{syq}\,(C,C)_{;}\mathsf{syq}\,(C,\varepsilon) = \mathsf{syq}\,(C,\varepsilon) = \rho \qquad$ cancelling, $\Xi(\bowtie)$ total

$\rho_{;}\bowtie = \rho_{;}\mathsf{syq}\,(\overline{R}^{\mathsf{T}}_{;}\varepsilon, C') = \mathsf{syq}\,(\overline{R}^{\mathsf{T}}_{;}\varepsilon_{;}\rho^{\mathsf{T}}, C') = \mathsf{syq}\,(\overline{R}^{\mathsf{T}}_{;}C, C') = \bowtie$

(iii) We decide to prove $\rho \subseteq \mathsf{syq}\,(C,C)$, i.e.,

$$\overline{\overline{C}^{\mathsf{T}}_{;}\varepsilon} \cap \overline{C^{\mathsf{T}}_{;}\overline{\varepsilon}} \subseteq \overline{\overline{C}^{\mathsf{T}}_{;}C} \cap \overline{C^{\mathsf{T}}_{;}\overline{C}}$$

using the properties of an Aumann contact. The infectivity property

$$C^{\mathsf{T}}_{;}\overline{C} \subseteq \varepsilon^{\mathsf{T}}_{;}\overline{C}$$

implies containment for the first terms. Containment of the second terms is trivial because $\varepsilon \subseteq C$.

(iv) The first equality follows from (i). Now, $\rho_{;}\rho^{\mathsf{T}}_{;}\Omega_{;}\rho_{;}\rho^{\mathsf{T}} = \rho_{;}\Omega_{;}\rho^{\mathsf{T}}$ following (i,iii) and because ρ is isotonic

$$\rho_{;}\Omega_{;}\rho^{\mathsf{T}} = \rho_{;}\overline{\varepsilon^{\mathsf{T}}_{;}\overline{\varepsilon}}_{;}\rho^{\mathsf{T}} = \overline{\rho_{;}\varepsilon^{\mathsf{T}}_{;}\varepsilon_{;}\rho^{\mathsf{T}}} = \overline{C^{\mathsf{T}}_{;}\overline{C}} = \overline{\varepsilon^{\mathsf{T}}_{;}\overline{C}}. \qquad \square$$

There are further relationships around this complex, in particular those concerning Υ; they touch both sides, i.e., employ both ρ', ε'. A result of (ii) is that $\rho \mathbin{;} \Upsilon$ is a mapping, because it is obviously univalent and $\bowtie \mathbin{;} \rho'$ is total.

Proposition 11.24. *We proceed with the notation introduced before.*

(i) $\rho^{\mathsf{T}} \mathbin{;} \Upsilon = \Upsilon$

(ii) $\rho \mathbin{;} \Upsilon = \mathsf{syq}(\overline{\overline{R}^{\mathsf{T}} \mathbin{;} C}, \varepsilon') = \bowtie \mathbin{;} \rho'$ and $\quad \Upsilon \mathbin{;} {\rho'}^{\mathsf{T}} = \mathsf{syq}(\varepsilon, \overline{\overline{R} \mathbin{;} C'}) = \rho^{\mathsf{T}} \mathbin{;} \bowtie$

(iii) $\rho \mathbin{;} \Upsilon \mathbin{;} {\rho'}^{\mathsf{T}} = \bowtie$ and $\quad \Upsilon = \rho^{\mathsf{T}} \mathbin{;} \bowtie \mathbin{;} \rho'$ and $\quad \Upsilon = \rho^{\mathsf{T}} \mathbin{;} \bowtie \cap \bowtie \mathbin{;} \rho'$

(iv) $\bowtie \mathbin{;} \Upsilon^{\mathsf{T}} = \rho$ and $\quad \rho^{\mathsf{T}} \mathbin{;} \rho = \Upsilon \mathbin{;} \Upsilon^{\mathsf{T}}$

(v) $\Upsilon \subseteq \rho \mathbin{;} \Upsilon \subseteq \bowtie$

(vi) $\Xi(\bowtie) \mathbin{;} \Upsilon \mathbin{;} \Psi(\bowtie) = \bowtie$

Proof: (i) We handle $\rho^{\mathsf{T}} \mathbin{;} \Upsilon \subseteq \Upsilon$ first, apply the Schröder rule to obtain $\rho \mathbin{;} \overline{\Upsilon} \subseteq \overline{\Upsilon}$, and expand Υ with distributivity:

$$\rho \mathbin{;} \overline{\overline{\varepsilon}^{\mathsf{T}} \mathbin{;} \overline{R} \mathbin{;} \varepsilon'} \cup \rho \mathbin{;} \varepsilon^{\mathsf{T}} \mathbin{;} \overline{R} \mathbin{;} \varepsilon' \cup \rho \mathbin{;} \overline{\varepsilon^{\mathsf{T}} \mathbin{;} \overline{R} \mathbin{;} \varepsilon'} \cup \rho \mathbin{;} \overline{\varepsilon^{\mathsf{T}} \mathbin{;} \overline{\overline{R}} \mathbin{;} \varepsilon'} \subseteq \overline{\varepsilon}^{\mathsf{T}} \mathbin{;} \overline{R} \mathbin{;} \varepsilon' \cup \varepsilon^{\mathsf{T}} \mathbin{;} \overline{R} \mathbin{;} \varepsilon' \cup \overline{\varepsilon^{\mathsf{T}} \mathbin{;} \overline{R} \mathbin{;} \varepsilon'} \cup \overline{\varepsilon^{\mathsf{T}} \mathbin{;} \overline{\overline{R}} \mathbin{;} \varepsilon'}$$

Letting ρ slip below negation and using $C = \varepsilon \mathbin{;} \rho^{\mathsf{T}}$, $\overline{R}^{\mathsf{T}} \mathbin{;} C = \overline{R}^{\mathsf{T}} \mathbin{;} \varepsilon$, and $\varepsilon \subseteq C$, one will find that this holds indeed.

We have to convince ourselves that equality also holds, for which purpose we need the Point Axiom. Assuming any point $e_u \mathbin{;} e_v^{\mathsf{T}} \subseteq \Upsilon$, we will show $\rho^{\mathsf{T}} \mathbin{;} e_u = e_u$. This is more or less obvious because Υ relates only vertical and horizontal sides u, v with $u = \varepsilon \mathbin{;} e_u, v = \varepsilon \mathbin{;} e_v$ of non-enlargeable rectangles, i.e., satisfying

$$u = \overline{\overline{R} \mathbin{;} \overline{R}^{\mathsf{T}} \mathbin{;} u} \qquad v = \overline{\overline{R}^{\mathsf{T}} \mathbin{;} u}.$$

This serves to find out that

$$\rho^{\mathsf{T}} \mathbin{;} e_u = \mathsf{syq}(\varepsilon, C) \mathbin{;} e_u = \mathsf{syq}(\varepsilon, \overline{\overline{R} \mathbin{;} \overline{R}^{\mathsf{T}} \mathbin{;} \varepsilon}) \mathbin{;} e_u = \mathsf{syq}(\varepsilon, \overline{\overline{R} \mathbin{;} \overline{R}^{\mathsf{T}} \mathbin{;} \varepsilon \mathbin{;} e_u})$$
$$= \mathsf{syq}(\varepsilon, \overline{\overline{R} \mathbin{;} \overline{R}^{\mathsf{T}} \mathbin{;} u}) = \mathsf{syq}(\varepsilon, u) = e_u.$$

(ii) We concentrate on the first formula; the other follows symmetrically:

$$\rho \mathbin{;} \Upsilon = \rho \mathbin{;} \left[\mathsf{syq}(\varepsilon, \overline{\overline{R} \mathbin{;} \varepsilon'}) \cap \mathsf{syq}(\overline{\overline{R}^{\mathsf{T}} \mathbin{;} \varepsilon}, \varepsilon') \right] \qquad \text{by definition}$$
$$= \rho \mathbin{;} \mathsf{syq}(\varepsilon, \overline{\overline{R} \mathbin{;} \varepsilon'}) \cap \rho \mathbin{;} \mathsf{syq}(\overline{\overline{R}^{\mathsf{T}} \mathbin{;} \varepsilon}, \varepsilon') \qquad \text{since } \rho \text{ is univalent}$$
$$= \mathsf{syq}(\varepsilon \mathbin{;} \rho^{\mathsf{T}}, \overline{\overline{R} \mathbin{;} \varepsilon'}) \cap \mathsf{syq}(\overline{\overline{R}^{\mathsf{T}} \mathbin{;} \varepsilon} \mathbin{;} \rho^{\mathsf{T}}, \varepsilon') \qquad \text{Prop. 8.16.ii since } \rho \text{ is a mapping}$$
$$= \mathsf{syq}(\varepsilon \mathbin{;} \rho^{\mathsf{T}}, \overline{\overline{R} \mathbin{;} \varepsilon'}) \cap \mathsf{syq}(\overline{\overline{R}^{\mathsf{T}} \mathbin{;} \varepsilon \mathbin{;} \rho^{\mathsf{T}}}, \varepsilon') \qquad \text{mapping } \rho \text{ slipping below negation}$$
$$= \mathsf{syq}(C, \overline{\overline{R} \mathbin{;} \varepsilon'}) \cap \mathsf{syq}(\overline{\overline{R}^{\mathsf{T}} \mathbin{;} C}, \varepsilon') \qquad \text{because } C = \varepsilon \rho^{\mathsf{T}} \text{ according to Prop. 11.20}$$
$$\subseteq \mathsf{syq}(\overline{\overline{R}^{\mathsf{T}} \mathbin{;} C}, \varepsilon') \qquad \text{monotony}$$
$$= \mathsf{syq}(\overline{\overline{R}^{\mathsf{T}} \mathbin{;} C}, C') \mathbin{;} \mathsf{syq}(C', \varepsilon') \qquad \text{cancelling with first factor } \bowtie \text{ total}$$
$$= \bowtie \mathbin{;} \rho' \qquad \text{by definition}$$

The task remains to show reverse containment at one transition above, for which we provide alternative representations of $\bowtie$:

$$\mathsf{syq}(\overline{R}^{\mathsf{T}}{}_{;}C,\varepsilon') \subseteq \mathsf{syq}(C,\overline{R}{}_{;}\varepsilon') = \bowtie = \mathsf{syq}(\overline{R}^{\mathsf{T}}{}_{;}C,C').$$

Taking the more appropriate version on the right, we expand to

$$\overline{C^{\mathsf{T}}{}_{;}\overline{R}{}_{;}\varepsilon'} \cap \overline{C^{\mathsf{T}}{}_{;}\overline{R}{}_{;}\overline{\varepsilon'}} \subseteq \overline{C^{\mathsf{T}}{}_{;}\overline{R}{}_{;}C'} \cap \overline{C^{\mathsf{T}}{}_{;}\overline{R}{}_{;}\overline{C'}}$$

and can use that $\varepsilon' \subseteq C'$ under the double negations as well as $\overline{R}{}_{;}\varepsilon' = \overline{R}{}_{;}C'$.

(iii) With (ii), $\rho{}_{;}\Upsilon{}_{;}\rho'^{\mathsf{T}} = \mathsf{syq}(\overline{R}^{\mathsf{T}}{}_{;}C,\varepsilon'){}_{;}\rho'^{\mathsf{T}} = \mathsf{syq}(\overline{R}^{\mathsf{T}}{}_{;}C,\varepsilon'{}_{;}\rho'^{\mathsf{T}}) = \mathsf{syq}(\overline{R}^{\mathsf{T}}{}_{;}C,C') = \bowtie$
$\rho^{\mathsf{T}}{}_{;}\bowtie{}_{;}\rho' = \rho^{\mathsf{T}}{}_{;}\rho{}_{;}\Upsilon \subseteq \Upsilon$ using (ii).

But also

$$\rho^{\mathsf{T}}{}_{;}\bowtie{}_{;}\rho' = \rho^{\mathsf{T}}{}_{;}\rho{}_{;}\Upsilon = \rho^{\mathsf{T}}{}_{;}\rho{}_{;}\rho^{\mathsf{T}}{}_{;}\Upsilon \supseteq \rho^{\mathsf{T}}\Upsilon = \Upsilon \text{ using (i,ii) and totality.}$$

Using (ii),

$$\rho^{\mathsf{T}}{}_{;}\bowtie \cap \bowtie{}_{;}\rho = \mathsf{syq}(\varepsilon,\overline{R}{}_{;}C') \cap \mathsf{syq}(\overline{R}^{\mathsf{T}}{}_{;}C,\varepsilon') = \Upsilon.$$

(iv) $\bowtie{}_{;}\Upsilon^{\mathsf{T}} = \mathsf{syq}(C,\overline{R}{}_{;}\varepsilon'){}_{;}\big(\mathsf{syq}(\overline{R}{}_{;}\varepsilon',\varepsilon)\big) \cap \mathsf{syq}(\varepsilon',\overline{R}^{\mathsf{T}}{}_{;}\varepsilon))$
$\subseteq \mathsf{syq}(C,\overline{R}{}_{;}\varepsilon'){}_{;}\mathsf{syq}(\overline{R}{}_{;}\varepsilon',\varepsilon) \cap \ldots = \mathsf{syq}(C,\varepsilon) \cap \ldots \subseteq \rho.$

Again, we need a separate investigation concerning equality

$$\rho^{\mathsf{T}}{}_{;}\rho = \Upsilon{}_{;}\bowtie^{\mathsf{T}}{}_{;}\bowtie{}_{;}\Upsilon^{\mathsf{T}} = \Upsilon{}_{;}\Psi(\bowtie){}_{;}\Upsilon^{\mathsf{T}} = \Upsilon{}_{;}\rho'{}_{;}\rho'^{\mathsf{T}}{}_{;}\Upsilon^{\mathsf{T}} = \Upsilon{}_{;}\Upsilon^{\mathsf{T}}.$$

(v) The first containment follows from (i) with shunting. For the second, we recall that

$$\rho = \mathsf{syq}(C,\varepsilon) \qquad \Upsilon = \mathsf{syq}(\varepsilon,\overline{R}{}_{;}\varepsilon') \cap \mathsf{syq}(\overline{R}^{\mathsf{T}}{}_{;}\varepsilon,\varepsilon') \qquad \bowtie = \mathsf{syq}(C,\overline{R}{}_{;}\varepsilon')$$

and find out that this is just one step of cancelling when concentrating on the first term of Υ.

(vi) The essential part is obtained with the help of (ii,iv) and Prop. 11.23.i

$$\Xi(\bowtie){}_{;}\Upsilon{}_{;}\Psi(\bowtie) = \rho{}_{;}\rho^{\mathsf{T}}{}_{;}\Upsilon{}_{;}\rho'{}_{;}\rho'^{\mathsf{T}} = \rho{}_{;}\Upsilon{}_{;}\rho'{}_{;}\rho'^{\mathsf{T}} = \rho{}_{;}\Upsilon{}_{;}\rho'^{\mathsf{T}} = \bowtie. \qquad \square$$

Proceeding further, we now introduce ι,η. We have thus formed the quotient according to the difunctional relation $\bowtie$ and also extruded according to the matching relation Υ. As a result, we have *with two inherently different constructions* arrived at 'the same' relation: we may either consider the interrelationship of quotients

$$\beta := \beta(R) := \eta^{\mathsf{T}}{}_{;}\bowtie{}_{;}\eta'$$

or we may relate the extruded sets with

$$\lambda := \lambda(R) := \iota{}_{;}\Upsilon{}_{;}\iota'^{\mathsf{T}}.$$

Both have turned out to be bijective mappings. Even more, using the natural injection, respectively projection,

$$\eta := \texttt{Project } \Xi(\bowtie) \qquad \text{and} \qquad \iota := \texttt{Inject } (\Upsilon, \mathbb{T})$$

we find an isomorphism between λ and β.

Proposition 11.25. *Consider $\lambda := \lambda(R)$ together with the injections ι, ι' according to Prop. 11.3 and $\beta := \beta(R)$ together with η, η' as defined in Def. 11.13. Then*

$$\delta := \iota, \eta, \qquad \delta' := \iota', \eta'$$

are bijective mappings that provide an isomorphism between λ and β; i.e.,

$$\lambda, \delta' = \delta, \beta \qquad \text{and} \qquad \beta, \delta'^{\mathsf{T}} = \delta^{\mathsf{T}}, \lambda.$$

Proof: First, we prove the bijective mapping status using $\iota, \rho = \iota$ and $\rho, \eta = \eta$:

$$\delta^{\mathsf{T}}, \delta = \eta^{\mathsf{T}}, \iota^{\mathsf{T}}, \iota, \eta = \eta^{\mathsf{T}}, \Upsilon, \Upsilon^{\mathsf{T}}, \eta = \eta^{\mathsf{T}}, \rho^{\mathsf{T}}, \rho, \eta = \eta^{\mathsf{T}}, \eta = \mathbb{I}$$

$$\delta, \delta^{\mathsf{T}} = \iota, \eta, \eta^{\mathsf{T}}, \iota^{\mathsf{T}} = \iota, \Xi, \iota^{\mathsf{T}} = \iota, \rho, \rho^{\mathsf{T}}, \iota^{\mathsf{T}} = \iota, \iota^{\mathsf{T}} = \mathbb{I}.$$

Then we estimate as follows:

$$
\begin{array}{lll}
\lambda, \delta' = \iota, \Upsilon, \iota'^{\mathsf{T}}, \iota', \eta' & \text{expanded} \\
\subseteq \iota, \Upsilon, \eta' & \text{since } \iota' \text{ is univalent} \\
\subseteq \iota, \bowtie, \eta' & \text{see Prop. 11.14.ii} \\
\subseteq \iota, \eta, \eta^{\mathsf{T}}, \bowtie, \eta' & \text{since } \Xi(\bowtie) = \eta, \eta^{\mathsf{T}} \text{ is the row equivalence of } \bowtie \\
= \delta, \beta & \text{definition of } \delta \text{ and } \beta
\end{array}
$$

Because both δ and δ' are mappings, we have in addition equality. $\square$

We recall the typing with Fig. 11.27. With $\overline{\varepsilon^{\mathsf{T}}, \overline{R}}$, all intersections of rows of R are considered, but in contrast to Def. 11.1.iii, multiplicities are not yet eliminated, so that $\Xi(\overline{\varepsilon^{\mathsf{T}}, \overline{R}})$ in (iii) may be expected to be a non-trivial row equivalence of these intersections.

Corollary 11.26. *The construct β is a bijective mapping enjoying the identities*

$$\overline{R, \left[\Xi^\wedge(R)\right]^{\mathsf{T}}}, \beta = \Psi^\wedge(R) \qquad \text{and} \qquad \overline{R, \left[\Xi^\wedge(R)\right]^{\mathsf{T}}} = \Psi^\wedge(R), \beta^{\mathsf{T}}.$$

Proof: We recall that the quotient $\beta := \eta^{\mathsf{T}}, \bowtie, \eta'$ of $\bowtie$ modulo its row equivalence $\Xi(\bowtie)$ as well its column equivalence turns out to be a bijective mapping due to Prop. 10.20.

$$
\begin{array}{ll}
\overline{R, \left[\Xi^\wedge(R)\right]^{\mathsf{T}}}, \beta = \overline{R, \left[\Xi^\wedge(R)\right]^{\mathsf{T}}}, \beta & \text{because } \beta \text{ is a bijective mapping} \\
= \overline{R, \eta^{\mathsf{T}}, \overline{\varepsilon^{\mathsf{T}}, \overline{R}}^{\mathsf{T}}, \eta^{\mathsf{T}}}, \bowtie, \eta' & \text{expanded} \\
= \overline{R, \overline{R}^{\mathsf{T}}, \varepsilon, \eta}, \eta^{\mathsf{T}}, \bowtie, \eta' & \text{transposed and mapping slipped below negation}
\end{array}
$$

$$= \overline{\overline{R_\varepsilon \overline{R}^\mathsf{T}}_\varepsilon \varepsilon_\varepsilon \eta_\varepsilon \eta^\mathsf{T}_\varepsilon \bowtie_\varepsilon \eta'} \qquad \text{using Prop. 8.18.ii}$$

$$= \overline{\overline{R_\varepsilon \overline{R}^\mathsf{T}}_\varepsilon \varepsilon_\varepsilon \Xi(\bowtie)_\varepsilon \bowtie_\varepsilon \eta'} \qquad \text{by definition of } \eta$$

$$= \overline{\overline{R_\varepsilon \overline{R}^\mathsf{T}}_\varepsilon \varepsilon_\varepsilon \bowtie_\varepsilon \eta'} \qquad \text{property of the row equivalence}$$

$$= \overline{\overline{R_\varepsilon C'}_\varepsilon \eta'} \qquad \text{using Prop. 11.14.iii}$$

$$= \overline{R_\varepsilon \varepsilon'_\varepsilon \eta'} \qquad \qquad \square$$

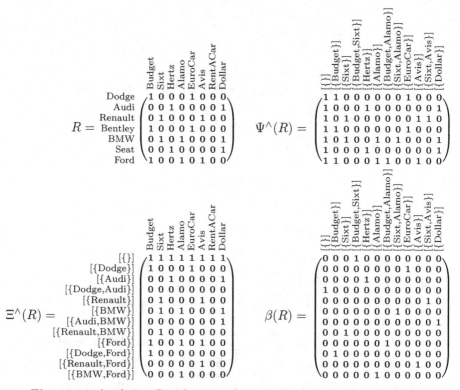

Fig. 11.28 A relation R with some relations of its relational topography

Now we are going to prove formally what seemed evident from the illustrations, namely that complete lattices show up.

Proposition 11.27.

(i) $\Omega_\eta := \eta^\mathsf{T}_\varepsilon \Omega_\varepsilon \eta$ *is an ordering.*

(ii) η *is isotonic.*

(iii) η *is (lattice-)continuous, i.e.,* $\eta^\mathsf{T}_\varepsilon \mathtt{lub}_\Omega(Y) = \mathtt{lub}_{\Omega_\eta}(\eta^\mathsf{T}_\varepsilon Y)$ *for arbitrary* Y.

(iv) Ω_η *is a complete lattice.*

Proof: (i) $\Omega_\eta \supseteq \eta^\mathsf{T}\mathbin{;}\eta = \mathbb{I}$, so that Ω_η is obviously reflexive. To prove transitivity is more involved:

$$
\begin{aligned}
\Omega_\eta\mathbin{;}\Omega_\eta &= \eta^\mathsf{T}\mathbin{;}\Omega\mathbin{;}\eta\mathbin{;}\eta^\mathsf{T}\mathbin{;}\Omega\mathbin{;}\eta && \text{by definition} \\
&= \eta^\mathsf{T}\mathbin{;}\Omega\mathbin{;}\Xi(\bowtie)\mathbin{;}\Omega\mathbin{;}\eta && \text{since } \eta \text{ is the natural projection for } \Xi(\bowtie) \\
&= \eta^\mathsf{T}\mathbin{;}\Omega\mathbin{;}\rho\mathbin{;}\rho^\mathsf{T}\mathbin{;}\Omega\mathbin{;}\eta && \text{Prop. 11.23.i} \\
&\subseteq \eta^\mathsf{T}\mathbin{;}\rho\mathbin{;}\Omega\mathbin{;}\Omega\mathbin{;}\rho^\mathsf{T}\mathbin{;}\eta && \text{because } \rho \text{ as a closure is isotonic} \\
&\subseteq \eta^\mathsf{T}\mathbin{;}\rho\mathbin{;}\Omega\mathbin{;}\rho^\mathsf{T}\mathbin{;}\eta && \text{because } \Omega \text{ is transitive} \\
&\subseteq \eta^\mathsf{T}\mathbin{;}\Omega\mathbin{;}\eta && \text{because } \eta^\mathsf{T}\mathbin{;}\rho \subseteq \eta^\mathsf{T}\mathbin{;}\Xi(\bowtie) = \eta^\mathsf{T}\mathbin{;}\eta\mathbin{;}\eta^\mathsf{T} = \eta^\mathsf{T} \\
&= \Omega_\eta && \text{by definition}
\end{aligned}
$$

Antisymmetry

$$
\begin{aligned}
\Omega_\eta \cap \Omega_\eta^\mathsf{T} &= \eta^\mathsf{T}\mathbin{;}\Omega\mathbin{;}\eta \cap \eta^\mathsf{T}\mathbin{;}\Omega^\mathsf{T}\mathbin{;}\eta && \text{by definition} \\
&= \eta^\mathsf{T}\mathbin{;}\eta\mathbin{;}\eta^\mathsf{T}\mathbin{;}\Omega\mathbin{;}\eta\mathbin{;}\eta^\mathsf{T}\mathbin{;}\eta \cap \eta^\mathsf{T}\mathbin{;}\eta\mathbin{;}\eta^\mathsf{T}\mathbin{;}\Omega^\mathsf{T}\mathbin{;}\eta\mathbin{;}\eta^\mathsf{T}\mathbin{;}\eta && \text{because } \eta^\mathsf{T}\mathbin{;}\eta = \mathbb{I} \\
&= \eta^\mathsf{T}\mathbin{;}(\eta\mathbin{;}\eta^\mathsf{T}\mathbin{;}\Omega\mathbin{;}\eta\mathbin{;}\eta^\mathsf{T} \cap \eta\mathbin{;}\eta^\mathsf{T}\mathbin{;}\Omega^\mathsf{T}\mathbin{;}\eta\mathbin{;}\eta^\mathsf{T})\mathbin{;}\eta && \text{Prop. 7.10} \\
&= \eta^\mathsf{T}\mathbin{;}(\Xi(\bowtie)\mathbin{;}\Omega\mathbin{;}\Xi(\bowtie) \cap \Xi(\bowtie)\mathbin{;}\Omega^\mathsf{T}\mathbin{;}\Xi(\bowtie))\mathbin{;}\eta && \eta \text{ natural projection} \\
&= \eta^\mathsf{T}\mathbin{;}(\overline{C^\mathsf{T}\mathbin{;}\overline{C}} \cap \overline{C^\mathsf{T}\mathbin{;}\overline{C}}^\mathsf{T})\mathbin{;}\eta && \text{Prop. 11.23.iv} \\
&= \eta^\mathsf{T}\mathbin{;}\mathsf{syq}(C,C)\mathbin{;}\eta && \text{by definition} \\
&= \eta^\mathsf{T}\mathbin{;}\Xi(\bowtie)\mathbin{;}\eta && \text{Prop. 11.23.i} \\
&= \eta^\mathsf{T}\mathbin{;}\eta\mathbin{;}\eta^\mathsf{T}\mathbin{;}\eta = \mathbb{I} && \text{natural projection}
\end{aligned}
$$

(ii) $\Omega\mathbin{;}\eta \subseteq \Xi(\bowtie)\mathbin{;}\Omega\mathbin{;}\eta = \eta\mathbin{;}\eta^\mathsf{T}\mathbin{;}\Omega\mathbin{;}\eta = \eta\mathbin{;}\Omega_\eta$

(iii)
$$
\begin{aligned}
\mathsf{lub}_{\Omega_\eta}(\eta^\mathsf{T}\mathbin{;}Y) &= \mathsf{syq}(\Omega_\eta^\mathsf{T}, \overline{\overline{\Omega_\eta}^\mathsf{T}\mathbin{;}\eta^\mathsf{T}\mathbin{;}Y}) && \text{due to Prop. 9.9} \\
&= \mathsf{syq}(\eta^\mathsf{T}\mathbin{;}\Omega^\mathsf{T}\mathbin{;}\eta, \overline{\overline{\eta^\mathsf{T}\mathbin{;}\Omega^\mathsf{T}\mathbin{;}\eta}\mathbin{;}\eta^\mathsf{T}\mathbin{;}Y}) && \text{by definition of } \Omega_\eta \\
&= \mathsf{syq}(\eta^\mathsf{T}\mathbin{;}\Omega^\mathsf{T}\mathbin{;}\eta\mathbin{;}\eta^\mathsf{T}\mathbin{;}\eta, \overline{\overline{\eta^\mathsf{T}\mathbin{;}\Omega^\mathsf{T}\mathbin{;}\eta}\mathbin{;}\eta^\mathsf{T}\mathbin{;}Y}) && \text{since } \eta^\mathsf{T}\mathbin{;}\eta = \mathbb{I} \\
&= \mathsf{syq}(\eta\mathbin{;}\eta^\mathsf{T}\mathbin{;}\Omega^\mathsf{T}\mathbin{;}\eta\mathbin{;}\eta^\mathsf{T}\mathbin{;}\eta, \eta\mathbin{;}\overline{\overline{\eta^\mathsf{T}\mathbin{;}\Omega^\mathsf{T}\mathbin{;}\eta}\mathbin{;}\eta^\mathsf{T}\mathbin{;}Y}) && \text{Prop. 8.16.i} \\
&= \mathsf{syq}(\eta\mathbin{;}\eta^\mathsf{T}\mathbin{;}\Omega^\mathsf{T}\mathbin{;}\eta\mathbin{;}\eta^\mathsf{T}\mathbin{;}\eta, \overline{\eta\mathbin{;}\overline{\eta^\mathsf{T}\mathbin{;}\Omega^\mathsf{T}\mathbin{;}\eta}\mathbin{;}\eta^\mathsf{T}\mathbin{;}Y}) && \text{slipping below negation} \\
&= \mathsf{syq}(\Xi(\bowtie)\mathbin{;}\Omega^\mathsf{T}\mathbin{;}\Xi(\bowtie)\mathbin{;}\eta, \overline{\Xi(\bowtie)\mathbin{;}\Omega^\mathsf{T}\mathbin{;}\Xi(\bowtie)}\mathbin{;}Y) && \text{definition of projection } \eta \\
&= \eta^\mathsf{T}\mathbin{;}\mathsf{syq}(\Xi(\bowtie)\mathbin{;}\Omega^\mathsf{T}\mathbin{;}\Xi(\bowtie), \overline{\Xi(\bowtie)\mathbin{;}\Omega^\mathsf{T}\mathbin{;}\Xi(\bowtie)}\mathbin{;}Y) && \text{following Prop. 8.18.i} \\
&= \eta^\mathsf{T}\mathbin{;}\mathsf{syq}(\overline{\overline{C}^\mathsf{T}\mathbin{;}\varepsilon}, \overline{\overline{C}^\mathsf{T}\mathbin{;}\varepsilon}\mathbin{;}Y) && \text{Prop. 11.23.iv} \\
&= \eta^\mathsf{T}\mathbin{;}\mathsf{syq}(\overline{\rho\mathbin{;}\varepsilon^\mathsf{T}\mathbin{;}\varepsilon}, \overline{\rho\mathbin{;}\varepsilon^\mathsf{T}\mathbin{;}\varepsilon}\mathbin{;}Y) && \text{expanding } C = \varepsilon\mathbin{;}\rho^\mathsf{T} \\
&= \eta^\mathsf{T}\mathbin{;}\mathsf{syq}(\rho\mathbin{;}\overline{\varepsilon^\mathsf{T}\mathbin{;}\varepsilon}, \rho\mathbin{;}\overline{\varepsilon^\mathsf{T}\mathbin{;}\varepsilon}\mathbin{;}Y) && \text{mapping } \rho \text{ slipping out of negation} \\
&\supseteq \eta^\mathsf{T}\mathbin{;}\mathsf{syq}(\overline{\varepsilon^\mathsf{T}\mathbin{;}\varepsilon}, \overline{\varepsilon^\mathsf{T}\mathbin{;}\varepsilon}\mathbin{;}Y) && \text{eliminating common left factor using Prop. 8.16.i} \\
&= \eta^\mathsf{T}\mathbin{;}\mathsf{syq}(\Omega^\mathsf{T}, \overline{\Omega}^\mathsf{T}\mathbin{;}Y) && \text{by definition of } \Omega \\
&= \eta^\mathsf{T}\mathbin{;}\mathsf{lub}_\Omega(Y) && \text{due to Prop. 9.9}
\end{aligned}
$$

In addition equality holds: every lub is injective by construction, and the least upper bound according to the powerset ordering Ω is surjective.

(iv) This shows that the ordering Ω_η is a complete lattice; see Cor. 9.16.   □

Another example of injected subsets related to projected equivalences, both ordered as concept lattices, is given with Fig. 11.29.

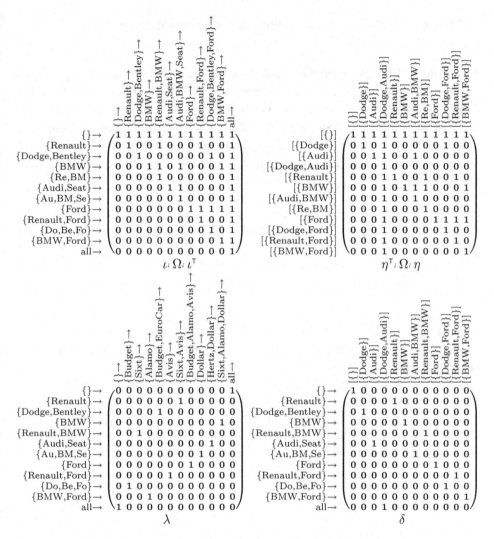

Fig. 11.29 Further items of relational topography for Fig. 11.28

11.5 Completion of an ordering

We have seen that a lattice, and in particular a complete lattice, enjoys pleasing properties. So the question has arisen whether one can embed an ordering into a lattice so as to be able to make use of some of these nice properties. This is called a **cut completion** or of an **ideal completion**. This is a well-established field to which we will only give some hints, restricting ourselves to the finite case which simplifies the situation considerably.

With an example we try to make clear what a cut completion is like.

Example 11.28. The ordered set shall be the 4-element set of Fig. 9.6, shown again in Fig. 11.30. The matrix gives the ordering while the graph presents only the Hasse diagram. Consider the 4-element set $V = \{w, x, y, z\}$, related with ε to its powerset 2^V, and assume it to be ordered by E. Using $\mathtt{ubd}_E(\varepsilon)$, which means forming $\overline{\overline{E}^{\mathsf{T}} \cdot \varepsilon}$, all upper bounds of these sets are determined simultaneously. Following this, for all these upper bound sets, in addition, the lower bounds $\mathtt{lbd}_E(\mathtt{ubd}_E(\varepsilon))$ are formed simultaneously.

A few sets, called cuts, are not changed by this 2-step procedure; they are 'invariant' under forming minorants of majorants. All sets with this characteristic property are marked, then extruded and depicted with their inclusion ordering in the lower left of Fig. 11.30. One will immediately recognize that all the sets unchanged are closed to the downside as, for example, $\{x, w, z\}$. They are, thus, lower cones.

The underlinings in Fig. 11.30 indicate in which way every element of the original ordered set $V = \{w, x, y, z\}$ can be found as the top of some cone directed downwards. However, elements on the middle vertical of the second graph do not carry such markings.

The two diagrams do not look too similar because they are only presented as Hasse diagrams. But indeed, E_{wx} is reflected by the set inclusion $\{\underline{w}\} \subseteq \{\underline{x}, w, z\}$, showing that the ordering E is mapped isotonic/homomorphic into the inclusion ordering of the lower cones obtained. □

What might easily be considered juggling around, follows a general concept. With the technique presented here in the example, one can in principle 'embed' *every* ordered set into a complete lattice. Applying such a technique, one will not least be able to define real numbers in a precise manner. The ancient Greeks already knew that the rational numbers are not sufficiently dense and one has to insert $\sqrt{2}$ or π, for example, in between.

We recall that forming lower bounds of upper bounds, $\bigwedge_E(\varepsilon) = \overline{E \cdot \overline{E}^{\mathsf{T}} \cdot \varepsilon}$, has already been studied intensively. So we have a means to define the cuts just determined and to compare them with down-sets.

Definition 11.29. Whenever some order E is given, we call a subset

(i) c a **cut** $:\Longleftrightarrow$ $\bigwedge_E(c) = c,$
(ii) v a **down-set**[4] $:\Longleftrightarrow$ $E \,{}_;v = v.$ □

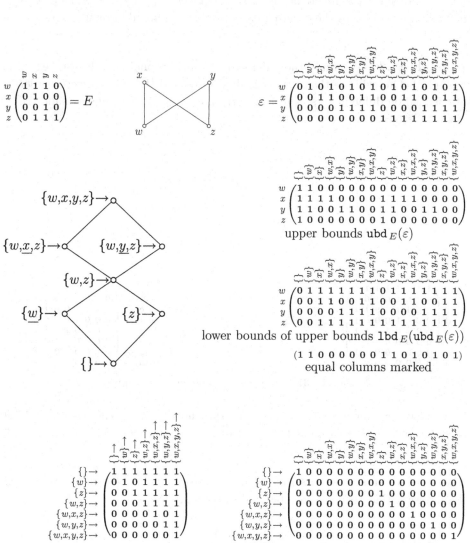

Fig. 11.30 A cut completion

---

[4] Down-sets are also often called *order ideals* or *lower cones*.

It is obvious that every cut is a down-set, but not vice versa. We will speak of a cut completion when cuts are used and of an ideal completion when down-sets (also called order ideals) are used.

First, however, we provide another example, where cut completion is principally unnecessary since a complete lattice is already given. Applying the same procedure nonetheless reproduces the ordering, thereby showing an additional flavor.

Example 11.30. We study the order of the 3-dimensional cube in this regard in Fig. 11.31. One will see that the original ordering and the cut-completed ordering are defined on different sets. The 1 : 1 correspondence of these sets is made visible via bold letters.

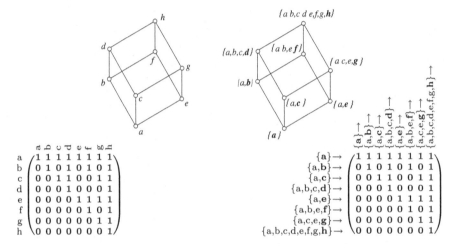

Fig. 11.31 A second cut completion example

In the resulting, obviously isomorphic, ordering, no longer elements are ordered but sets of elements, that turn out to be lower cones, or down-sets.   □

The set of all cuts as well as that of all down-sets can be determined by writing down a term that then can also be used to compute them.

Proposition 11.31. *Given an ordering E, one can determine the cuts and the down-sets by vector characterization as*

$$\Delta_{\mathrm{cc}} := \overline{\left[\bigwedge_E(\varepsilon) \cap \overline{\varepsilon} \right]^{\mathsf{T}} \, ; \mathbb{T}} \qquad \Delta_{\mathrm{ds}} := \overline{\left[E \, ; \varepsilon \cap \overline{\varepsilon} \right]^{\mathsf{T}} \, ; \mathbb{T}}.$$

Once these vectors are determined, we extrude them and define the corresponding natural injections

$$\iota_{cc} := \text{Inject } \Delta_{cc} \qquad \iota_{ds} := \text{Inject } \Delta_{ds}$$

and obtain the complete lattices

$$E_{cc} := \iota_{cc}; \overline{\varepsilon^{\mathsf{T}}; \overline{\varepsilon}}; \iota_{cc}^{\mathsf{T}} \qquad E_{ds} := \iota_{ds}; \overline{\varepsilon^{\mathsf{T}}; \overline{\varepsilon}}; \iota_{ds}^{\mathsf{T}}.$$

Proof: All this is sufficiently similar to the investigations in the preceding sections; we will not repeat the proofs that indeed complete lattices show up. □

The determination of the characterizing vectors has already given a look ahead to Section 15.3, where subsets are computed that are characterized in some way. Here, obviously $\bigwedge_E(\varepsilon) \supseteq \varepsilon$ and also $E; \varepsilon \supseteq \varepsilon$, so that only $\bigwedge_E(\varepsilon) \cap \overline{\varepsilon}$ respectively $E\varepsilon \cap \overline{\varepsilon}$ must be considered. The columns that vanish are made visible by multiplying the converse with $\mathbb{T}$. Negating brings precisely those where equality holds. So one obtains a formula that, when written in the relational language TITUREL, may immediately be evaluated.

It shall not be concealed that things become more complicated in the non-finite case. There, a huge folklore has developed over the last century which is reported in a detailed and knowledgeable form in [11, 37], for example. Here, we only show several examples of cut completion embedded into an ideal completion via down-sets.

Example 11.32. Let the following ordering of Fig. 11.32 be given: the sets $\{1, 2\}$ and $\{1, 2, 4\}$ in Fig. 11.32 are down-sets that do not occur as cuts. In general, cut completion is coarser compared with ideal completion. On the other hand, the latter is more likely to preserve interesting properties of the order. So both have attracted considerable interest. The embedding of the cut completion of Fig. 11.32 into its ideal completion is shown in the lower right. □

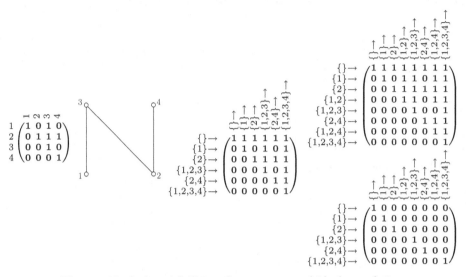

Fig. 11.32 Order with Hasse diagram, cut and ideal completion
as well as the embedding of cut into ideal completion

Example 11.33 is another attempt to compute the cut as well as the ideal completion of an arbitrarily given (finite) ordering.

Example 11.33. Consider, for example, the set $\{1, 2, 3\}$. It is not a cut, but a downset. For the ideal completion, a new element is inserted directly above $1 \approx \{1,3\}$ and $2 \approx \{2,3\}$. The appearance of $\{\} \subseteq \{3\}$ is due to the definition of an order ideal, which is allowed to be an empty set. $\square$

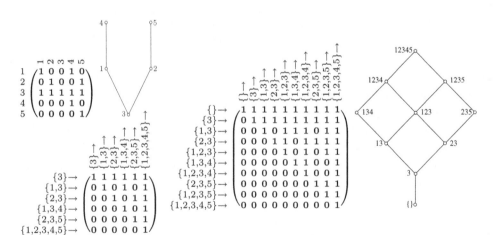

Fig. 11.33 An order with Hasse diagram and its cut and ideal completion

Example 11.34. In the case of Fig. 11.34, the cut completion is a simple one: a top element $\{a, b, c, d, e, f\}$ and a bottom element $\{\}$ are added; see upper part of Fig. 11.35. One should again observe that annotations are now sets of the original vertices. For those cases where this set is a set closed to the down-side, its top element is underlined to make clear in which way the embedding takes place.

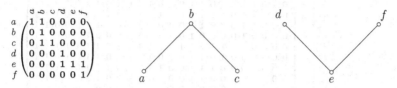

Fig. 11.34 Strictorder on a 6-element set given as a relation and as a Hasse diagram

In this case, the ideal completion needs many more elements. It is shown as a graph in the lower half of Fig. 11.35, and as matrix in Fig. 11.36.

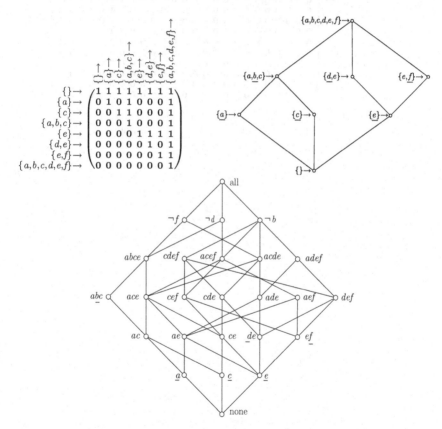

Fig. 11.35 Lattice completion of the order of Fig. 11.34, obtained by cut completion compared to the more detailed lattice obtained by ideal completion

Concept analysis

	{}→	{a}→	{c}→	{a,c}→	{a,b,c}→	{e}→	{a,e}→	{c,e}→	{a,c,e}→	{a,b,c,e}→	{d,e}→	{a,d,e}→	{c,d,e}→	{a,c,d,e}→	{a,b,c,d,e}→	{e,f}→	{a,e,f}→	{c,e,f}→	{a,c,e,f}→	{a,b,c,e,f}→	{d,e,f}→	{a,d,e,f}→	{c,d,e,f}→	{a,c,d,e,f}→	{a,b,c,d,e,f}→
{}→	1	1	1	1	1	1	1	1	1	1	1	1	1	1	1	1	1	1	1	1	1	1	1	1	1
{a}→	0	1	0	1	1	0	1	0	1	1	0	1	0	1	1	0	1	0	1	1	0	1	0	1	1
{c}→	0	0	1	1	1	0	0	1	1	1	0	0	1	1	1	0	0	1	1	1	0	0	1	1	1
{a,c}→	0	0	0	1	1	0	0	0	1	1	0	0	0	1	1	0	0	0	1	1	0	0	0	1	1
{a,b,c}→	0	0	0	0	1	0	0	0	0	1	0	0	0	0	1	0	0	0	0	1	0	0	0	0	1
{e}→	0	0	0	0	0	1	1	1	1	1	1	1	1	1	1	1	1	1	1	1	1	1	1	1	1
{a,e}→	0	0	0	0	0	0	1	0	1	1	0	1	0	1	1	0	1	0	1	1	0	1	0	1	1
{c,e}→	0	0	0	0	0	0	0	1	1	1	0	0	1	1	1	0	0	1	1	1	0	0	1	1	1
{a,c,e}→	0	0	0	0	0	0	0	0	1	1	0	0	0	1	1	0	0	0	1	1	0	0	0	1	1
{a,b,c,e}→	0	0	0	0	0	0	0	0	0	1	0	0	0	0	1	0	0	0	0	1	0	0	0	0	1
{d,e}→	0	0	0	0	0	0	0	0	0	0	1	1	1	1	1	0	0	0	0	0	1	1	1	1	1
{a,d,e}→	0	0	0	0	0	0	0	0	0	0	0	1	0	1	1	0	0	0	0	0	0	1	0	1	1
{c,d,e}→	0	0	0	0	0	0	0	0	0	0	0	0	1	1	1	0	0	0	0	0	0	0	1	1	1
{a,c,d,e}→	0	0	0	0	0	0	0	0	0	0	0	0	0	1	1	0	0	0	0	0	0	0	0	1	1
{a,b,c,d,e}→	0	0	0	0	0	0	0	0	0	0	0	0	0	0	1	0	0	0	0	0	0	0	0	0	1
{e,f}→	0	0	0	0	0	0	0	0	0	0	0	0	0	0	0	1	1	1	1	1	1	1	1	1	1
{a,e,f}→	0	0	0	0	0	0	0	0	0	0	0	0	0	0	0	0	1	0	1	1	0	1	0	1	1
{c,e,f}→	0	0	0	0	0	0	0	0	0	0	0	0	0	0	0	0	0	1	1	1	0	0	1	1	1
{a,c,e,f}→	0	0	0	0	0	0	0	0	0	0	0	0	0	0	0	0	0	0	1	1	0	0	0	1	1
{a,b,c,e,f}→	0	0	0	0	0	0	0	0	0	0	0	0	0	0	0	0	0	0	0	1	0	0	0	0	1
{d,e,f}→	0	0	0	0	0	0	0	0	0	0	0	0	0	0	0	0	0	0	0	0	1	1	1	1	1
{a,d,e,f}→	0	0	0	0	0	0	0	0	0	0	0	0	0	0	0	0	0	0	0	0	0	1	0	1	1
{c,d,e,f}→	0	0	0	0	0	0	0	0	0	0	0	0	0	0	0	0	0	0	0	0	0	0	1	1	1
{a,c,d,e,f}→	0	0	0	0	0	0	0	0	0	0	0	0	0	0	0	0	0	0	0	0	0	0	0	1	1
{a,b,c,d,e,f}→	0	0	0	0	0	0	0	0	0	0	0	0	0	0	0	0	0	0	0	0	0	0	0	0	1

Fig. 11.36 Ideal completion of the strictorder of Fig. 11.34; embedding indicated via underlining

PART IV
APPLICATIONS

So far we have concentrated on the foundations of relational mathematics. Now we switch to applications. A first area of applications concerns all the different variants of orderings as they originated in operations research: weakorders, semiorders, intervalorders, and block-transitive orderings. With the Scott–Suppes Theorem in relational form as well as with the study of the consecutive **1**s property, we here approach research level.

The second area of applications concerns modelling preferences with relations. The hierarchy of orderings is considered investigating indifference and incomparability, often starting from so-called preference structures, i.e., relational outcomes of assessment procedures. A bibliography on early preference considerations is contained in [2].

The area of aggregating preferences with relations, studied as a third field of applications, is relatively new. It presents relational measures and integration in order to treat the trust and belief of the Dempster–Shafer Theory in relational form. In contrast, the fuzzy approach is well known, with the coefficients of matrices stemming from the real interval $[0, 1]$; it comes closer and closer to relational algebra proper. In the present book, a direct attempt is made. Also t-norms and De Morgan triples can be generalized to a relational form.

Then, we study graph theory, where several effects need relational algebra beyond just regular algebra for an adequate description. This leads directly to algorithms for solving many tasks in practice. Finally, we give an account of the broad area of Galois mechanisms. This subject has long been known, but is not often described in relational form. Many seemingly different topics are collected under one common roof: termination, games, matching and assignment, etc.

12
Orderings: An Advanced View

After one's first work with orderings, one will certainly come across a situation in which the concept of an ordering cannot be applied in its initial form, a high jump competition, for example. Here certain heights are given and athletes achieve them or not. Often, more than one athlete will reach the same maximum height, for example a whole class of athletes will jump 2.35 m high. Such a situation can no longer be studied using an order – one will switch to a preorder. We develop the traditional hierarchy of orderings (linear strictorder, weakorder, semiorder, intervalorder) in a coherent and proof-economic way. Intervalorders are treated in more detail since they have attracted much attention in their relationship with interval graphs, transitive orientability, and the consecutive 1s property. Block-transitive strictorders are investigated as a new and algebraically promising concept. Then we study how an ordering of some type can – by slightly extending it – be embedded into some more restrictive type; for example, a semiorder into its weakorder closure.

In a very general way, equivalences are related to preorders, and these in turn are related to measurement theory as used in physics, psychology, economic sciences, and other fields. Scientists have contributed to measurement theory in order to give a firm basis to social sciences or behavioral sciences. The degree to which a science is considered an already developed one depends to a great extent on the ability to measure. To relate utility considerations with qualitative comparisons, factorizations of orderings are presented. Measuring as necessary in social sciences or psychology, for example, is definitely more difficult than measuring in engineering. While the latter may well use real numbers, the former suffer from the necessity of working with qualitative scales.

Then the – by now nearly legendary – differences between the Continental European and the Anglo-Saxon schools is explained. This is mainly the counterplay of direct comparison on the one hand and the utility approach conceived as always coding in $\mathbb{R}$ on the other. With factorization theorems for every field of discourse, we make the transitions explicit and hope, thus, to clarify the difference, but also to provide the joining piece.

12.1 Additional properties of orderings

Sometimes the Ferrers property is also satisfied for a homogeneous relation. It is then nicely embedded in a plexus of formulae that characterize the standard properties of orders and equivalences. When converting the homogeneous Ferrers relation to upper triangular form, this can sometimes be achieved by simultaneous[1] permutation, in other cases it can only be achieved by independently permuting.

A remark is necessary to convince the reader that we should study mainly *irreflexive* relations among the homogeneous Ferrers relations. So far, it has been more or less a matter of taste whether one worked with orders E or with their corresponding strictorders C. Everthing could easily have been reformulated in the other form, using that $E = \mathbb{I} \cup C$ and $C = \overline{\mathbb{I}} \cap E$, i.e., by adding or removing the identity.

Proposition 12.1. *If an order E is Ferrers, then so is its corresponding strictorder $C := \overline{\mathbb{I}} \cap E$. The reverse implication does not hold.*

Proof: An order that is Ferrers is necessarily linear, since we can prove connexity:

$$E \cup E^{\mathsf{T}} \supseteq E \,\mathbin{;} \overline{E}^{\mathsf{T}} \,\mathbin{;} E \cup E^{\mathsf{T}} \supseteq \mathbb{I} \,\mathbin{;} \overline{E}^{\mathsf{T}} \,\mathbin{;} \mathbb{I} \cup E^{\mathsf{T}} = \overline{E}^{\mathsf{T}} \cup E^{\mathsf{T}} = \mathbb{T},$$

applying the Ferrers property and reflexivity. Then we compute

$$
\begin{aligned}
C \,\mathbin{;} \overline{C}^{\mathsf{T}} \,\mathbin{;} C &= C \,\mathbin{;} \overline{\overline{\mathbb{I}} \cap E}^{\mathsf{T}} \,\mathbin{;} C && \text{definition of } C \\
&= C \,\mathbin{;} (\mathbb{I} \cup \overline{E}^{\mathsf{T}}) \,\mathbin{;} C && \text{negation and transposition} \\
&= C \,\mathbin{;} C \cup C \,\mathbin{;} \overline{E}^{\mathsf{T}} \,\mathbin{;} C && \text{distributive composition} \\
&\subseteq C \,\mathbin{;} C \cup C \,\mathbin{;} E \,\mathbin{;} C && E \text{ is connex as a linear order; see above} \\
&= C \,\mathbin{;} C \cup C \,\mathbin{;} (\mathbb{I} \cup C) \,\mathbin{;} C && \text{definition of } C \\
&\subseteq C && \text{transitivity of } C
\end{aligned}
$$

The ordering relation on $\mathbb{B}^2$ in Fig. 12.1 shows that this does not hold in the reverse direction: C is Ferrers but E is not. □

Thus, strictorders with Ferrers property are the more general version compared with orders.

$$E = \begin{pmatrix} 1 & 1 & 1 & 1 \\ 0 & 1 & 0 & 1 \\ 0 & 0 & 1 & 1 \\ 0 & 0 & 0 & 1 \end{pmatrix} = \mathbb{I} \cup C \qquad\qquad C = \begin{pmatrix} 0 & 1 & 1 & 1 \\ 0 & 0 & 0 & 1 \\ 0 & 0 & 0 & 1 \\ 0 & 0 & 0 & 0 \end{pmatrix} = \overline{\mathbb{I}} \cap E$$

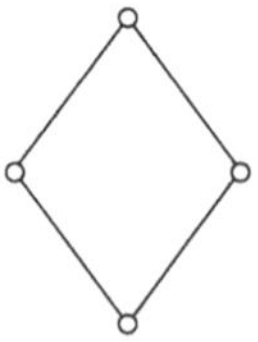

Fig. 12.1 An order that is not Ferrers with the corresponding strictorder that is

---

[1] Frobenius in [54] calls it a cogredient permutation.

We have already given the definition of an order and a strictorder together with a sample of related propositions. For more advanced investigations, we need these folklore concepts combined with the Ferrers property, semi-transitivity, and negative transitivity.

Semi-transitivity

In the graph representing a semi-transitive[2] relation, we have the following property. Given any two consecutive arrows together with an arbitrary vertex w, an arrow will lead from the starting point of the two consecutive arrows to the point w or an arrow will lead from the point w to the end of the two consecutive arrows. Figure 12.2 symbolizes the idea with the dashed arrow convention.

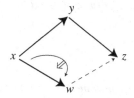

Fig. 12.2 Semi-transitivity expressed with the dashed arrow convention

This idea is captured in the following definition.

Definition 12.2. We call a (necessarily homogeneous) relation

R **semi-transitive** $:\Longleftrightarrow R\mathbin{;}R\mathbin{;}\overline{R}^{\mathsf{T}}\subseteq R \Longleftrightarrow \overline{R}\mathbin{;}R\subseteq \overline{R\mathbin{;}R}$

$\Longleftrightarrow \forall x,y,z,w\in X:$
$$\big\{\big[(x,y)\in R\wedge (y,z)\in R\big]\to \big[(x,w)\in R\vee (w,z)\in R\big]\big\}. \qquad \square$$

The relation-algebraic variants are equivalent via the Schröder equivalence. It is not so easy a task to convince oneself that the algebraic and the predicate-logic versions express the same – although one will appreciate the shorthand form when using it in a proof.

$$\forall x,y,z,w:\big\{\big[(x,y)\in R\wedge (y,z)\in R\big]\to \big[(x,w)\in R\vee (w,z)\in R\big]\big\}$$
$$a\to b=\neg a\vee b$$
$$\Longleftrightarrow \forall x,y,z,w:\big\{(x,y)\notin R\vee (y,z)\notin R\vee (x,w)\in R\vee (w,z)\in R\big\}$$
$$\forall w:c\vee b(w)=c\vee \big[\forall w:b(w)\big]$$
$$\Longleftrightarrow \forall x,y,z:\big\{(x,y)\notin R\vee (y,z)\notin R\vee \big[\forall w:(x,w)\in R\vee (w,z)\in R\big]\big\}$$

---

[2] This notion was introduced by John S. Chipman in [32].

$$\forall w : p(w) = \neg\big[\exists w : \neg p(w)\big]$$

$$\Longleftrightarrow \quad \forall x,y,z : \Big\{(x,y) \notin R \vee (y,z) \notin R \vee \neg\big[\exists w : (x,w) \notin R \wedge (w,z) \notin R\big]\Big\}$$

$$\text{definition of composition}$$

$$\Longleftrightarrow \quad \forall x,y,z : \Big\{(x,y) \notin R \vee (y,z) \notin R \vee \neg\big[(x,z) \in \overline{R}\,\raisebox{0.2ex}{;}\,\overline{R}\big]\Big\}$$

$$\forall y : \big[b(y) \vee a\big] = \big[\forall y : b(y)\big] \vee a$$

$$\Longleftrightarrow \quad \forall x,z : \Big\{\big[\forall y : (x,y) \notin R \vee (y,z) \notin R\big] \vee (x,z) \notin \overline{R}\,\raisebox{0.2ex}{;}\,\overline{R}\Big\}$$

$$\forall y : p(y) = \neg\big[\exists y : \neg p(y)\big]$$

$$\Longleftrightarrow \quad \forall x,z : \Big\{\neg\big[\exists y : (x,y) \in R \wedge (y,z) \in R\big] \vee (x,z) \notin \overline{R}\,\raisebox{0.2ex}{;}\,\overline{R}\Big\}$$

$$\text{definition of composition}$$

$$\Longleftrightarrow \quad \forall x,z : \Big\{\neg\big[(x,z) \in R\,\raisebox{0.2ex}{;}\,R\big] \vee (x,z) \notin \overline{R}\,\raisebox{0.2ex}{;}\,\overline{R}\Big\}$$

$$\text{``}\vee\text{'' is commutative, } a \rightarrow b = \neg a \vee b$$

$$\Longleftrightarrow \quad \forall x,z : \Big\{(x,z) \in \overline{R}\,\raisebox{0.2ex}{;}\,\overline{R} \rightarrow (x,z) \notin R\,\raisebox{0.2ex}{;}\,R\Big\}$$

$$\text{point-free definition of containment}$$

$$\Longleftrightarrow \quad \overline{R}\,\raisebox{0.2ex}{;}\,\overline{R} \subseteq \overline{R\,\raisebox{0.2ex}{;}\,R}$$

Negative transitivity

One often tries to model preferences with relations that do not fully meet the definition of a linear ordering. Linear orderings always draw heavily on their characteristic property $\overline{E} = C^{\mathsf{T}}$ of Prop. 5.24. Because this no longer holds in such cases, we have become interested in the complement, and the property of being *negatively transitive* has also been investigated. A linear order E has a linear strictorder as its complement which is transitive.

Definition 12.3. We call a (necessarily homogeneous) relation

$$R \textbf{ negatively transitive}^3 \quad :\Longleftrightarrow \quad \overline{R}\,\raisebox{0.2ex}{;}\,\overline{R} \subseteq \overline{R}$$

$$\Longleftrightarrow \quad \forall a,b,c : (a,c) \in R \rightarrow \big[(a,b) \in R \vee (b,c) \in R\big]. \qquad \square$$

Figure 12.3 shows the essence of negative transitivity with a dashed arrow diagram.

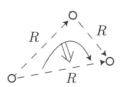

Fig. 12.3 Negative transitivity shown with the dashed arrow convention

---

[3] In the non-point-free version, this has also become known as Chipman's condition, see [1, 32].

The predicate-logic version assumes any two vertices a, c in relation R and considers an arbitrary other point b. Then b will necessarily be 'R-above' a or 'R-below' c. One can deduce the point-free version as follows:

$$\forall a, b, c : \left\{ (a, c) \in R \to \left[(a, b) \in R \vee (b, c) \in R \right] \right\}$$

$$\hspace{4cm} p \to q = \neg p \vee q \text{ and commutations}$$

$$\Longleftrightarrow \quad \forall a, c : \left\{ \forall b : (a, b) \in R \vee (b, c) \in R \vee (a, c) \notin R \right\}$$

$$\hspace{4cm} \left[\forall x : p(x) \vee q \right] = \left[\forall x : p(x) \right] \vee q$$

$$\Longleftrightarrow \quad \forall a, c : \left\{ \left[\forall b : (a, b) \in R \vee (b, c) \in R \right] \vee (a, c) \notin R \right\}$$

$$\hspace{4cm} \forall x : p(x) = \neg \left[\exists x : \neg p(x) \right]$$

$$\Longleftrightarrow \quad \forall a, c : \left\{ \neg \left[\exists b : (a, b) \notin R \wedge (b, c) \notin R \right] \vee (a, c) \notin R \right\}$$

$$\hspace{4cm} \text{definition of composition}$$

$$\Longleftrightarrow \quad \forall a, c : \left\{ \neg \left[(a, c) \in \overline{R} \,\overline{R} \right] \vee (a, c) \notin R \right\}$$

$$\hspace{4cm} p \to q = \neg p \vee q$$

$$\Longleftrightarrow \quad \forall a, c : \left\{ (a, c) \in \overline{R} \,\overline{R} \to (a, c) \in \overline{R} \right\}$$

$$\hspace{4cm} \text{transition to point-free notation}$$

$$\Longleftrightarrow \quad \overline{R} \,\overline{R} \subseteq \overline{R}$$

This concept is, however, mainly interesting for linear orderings. Already the ordering of a Boolean lattice $\mathbb{B}^k$ with $k > 1$ is no longer negatively transitive.

As an illustration of a semi-transitive relation that is not negatively transitive consider Fig. 12.4. The relation is obviously transitive; however, it is not Ferrers, so it does not satisfy the properties of an intervalorder of Def. 12.4. We convince ourselves that it is indeed not Ferrers looking at $(1, 4) \in R, (3, 4) \notin R, (3, 7) \in R$, but $(1, 7) \notin R$. Lack of negative transitivity is also obvious: $(1, 7) \notin R, (7, 6) \notin R$, but $(1, 6) \in R$.

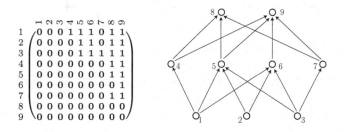

Fig. 12.4 A semi-transitive, but neither Ferrers nor negatively transitive, relation

12.2 The hierarchy of orderings

The following Fig. 12.5 gives the types of orders we are going to deal with in a schematic form. The traditional definitions do not demand so many properties, because some of them result from combinations of others. This applies not least for the first three properties in the table: transitive and asymmetric hold if and only if transitive and irreflexive is satisfied. So, regardless of the version chosen for the definition, all three will hold. It is, however, a good idea to see the chain of specificity showing how researchers began to move from linear strictorder to strictorder.

Whenever we use one of these strictorders in our reasoning, we are allowed to use *all* of the properties marked "•". In a case when we have to convince ourselves that a relation belongs to one of these strictorder classes, we first decide on a convenient set of 'spanning properties', for example those marked "|" or "o", and prove just these.

We have, thus, announced a series of propositions that are intended to prove that the subsets are sufficient to span all the properties of the respective type of strictorder.

Definition 12.4. Bullets "•" in Fig. 12.5 define the concepts of **linear strictorder, weakorder,**[4] **semiorder,**[5] **intervalorder,** and **strictorder** in a redundant way.

	linear strictorder	weakorder	semiorder	interval-order	strictorder
transitive	• | o	•	•	•	• | o
asymmetric	• |	• |	• o	• o	• |
irreflexive	• o	•	• |	• |	• o
Ferrers	•	•	• | o	• | o	—
semi-transitive	•	•	• | o	—	—
negatively transitive	•	• |	—	—	—
semi-connex	• | o	—	—	—	—
Proof with Prop.	12.10	12.9	12.7	12.6	12.5

Fig. 12.5 Types of strictorders with 'spanning' subsets of their properties

Via "|", "o", this diagram also indicates some minimal sets of properties that already suffice for the respective type of strictorder. □

The generalizations have occurred successively from the left to the right. In the

---

[4] In French sometimes *ordre fort.*
[5] In French sometimes also *quasi-ordre* (in order not to be confused with Claude Berge's earlier and different use of the word 'semiorder' in his famous book [10]) or *ordre quasi-fort* in [94].

interest of economy of proof, we proceed from the right to the left. Firstly, we recall as folklore Prop. 5.19 mentioned earlier.

Proposition 12.5. *A transitive relation is irreflexive precisely when it is asymmetric; i.e., in either case it satisfies all items mentioned in Fig. 12.5 for a strictorder.*

Proof: Writing down transitivity and irreflexivity, we obtain $R_{;}R \subseteq R \subseteq \overline{\mathbb{I}}$. Now the Schröder equivalence is helpful because it allows us to obtain directly $R^{\mathsf{T}}_{;}\mathbb{I} \subseteq \overline{R}$, the condition of asymmetry.

For the other direction, we do not even need transitivity,
$$\mathbb{I} = (\mathbb{I} \cap R) \cup (\mathbb{I} \cap \overline{R}) \subseteq (\mathbb{I} \cap R) \cup \overline{R} \subseteq \overline{R}$$
since for the partial diagonal $\mathbb{I} \cap R = \mathbb{I} \cap R^{\mathsf{T}} \subseteq R^{\mathsf{T}} \subseteq \overline{R}$ using asymmetry. □

Slightly more restrictive than strictorders are intervalorders.

Proposition 12.6. (i) *An irreflexive Ferrers relation is transitive and asymmetric, i.e., satisfies all items mentioned in Fig. 12.5 for an intervalorder.*

(ii) *An asymmetric Ferrers relation is transitive and irreflexive, i.e., satisfies all items mentioned in Fig. 12.5 for an intervalorder.*

Proof: (i) We have $R \subseteq \overline{\mathbb{I}}$, or equivalently, $\mathbb{I} \subseteq \overline{R}$. Therefore, the Ferrers property specializes to transitivity:
$$R_{;}R = R_{;}\mathbb{I}_{;}R \subseteq R_{;}\overline{R}^{\mathsf{T}}_{;}R \subseteq R.$$
From Prop. 12.5, we have also that R is asymmetric.

(ii) follows with (i) since *asymmetric* implies *irreflexive*, as shown in the proof of Prop. 12.5. □

The Ferrers property $R_{;}R^{\mathrm{d}}_{;}R \subseteq R$ is in this case studied for homogeneous relations; homogeneous Ferrers relations have often been called **biorders**. Therefore, intervalorders are irreflexive biorders. Intervalorders have been studied thoroughly by Peter C. Fishburn; see, for example [51].

The next step in restricting leads us to semiorders. Semiorders were first introduced in [89] by R. Duncan Luce in 1956. Even if an irreflexive semi-transitive relation is non-Ferrers as in Fig. 12.4, it is necessarily transitive since this may be deduced according to Prop. 12.7.

Proposition 12.7. *We consider a homogeneous relation R.*

(i) *If R is irreflexive and semi-transitive then it is transitive.*
(ii) *An* irreflexive *semi-transitive Ferrers relation is transitive and asymmetric, i.e., satisfies all items mentioned in Fig. 12.5 for a semiorder.*
(iii) *An asymmetric semi-transitive Ferrers relation is transitive and irreflexive, i.e., satisfies all items mentioned in Fig. 12.5 for a semiorder.*

Proof: (i) $R \,; R = R \,; R \,; \mathbb{I} \subseteq R \,; R \,; \overline{R}^\mathsf{T} \subseteq R$

The proof of (ii) now follows immediately from Prop. 12.6. (iii) is a trivial consequence of (ii) since *asymmetric* implies *irreflexive*.   □

The semiorder property of R propagates to its powers R^k as we are going to show now as an aside.

Proposition 12.8. *Let a semiorder R be given together with a natural number $k > 0$. Then R^k is a semiorder as well.*

Proof: The case $k = 1$ is trivial since R is given as a semiorder. The key observation for the general case is that for $k > 1$

$$R \,; \overline{R^k}^\mathsf{T} \subseteq \overline{R^{k-1}}^\mathsf{T}$$

according to the Schröder equivalence. This can now be applied iteratively

$$R^k \,; \overline{R^k}^\mathsf{T} = R^{k-1} \,; (R \,; \overline{R^k}^\mathsf{T}) \subseteq R^{k-1} \,; \overline{R^{k-1}}^\mathsf{T} \subseteq \ldots \subseteq R \,; \overline{R}^\mathsf{T}.$$

The Ferrers condition for R^k can be deduced from the corresponding one for R:

$$R^k \,; \overline{R^k}^\mathsf{T} \,; R^k \subseteq R \,; \overline{R}^\mathsf{T} \,; R^k = R \,; \overline{R}^\mathsf{T} \,; R \,; R^{k-1} \subseteq R \,; R^{k-1} = R^k.$$

The semi-transitivity condition is propagated analogously to powers of R

$$R^k \,; R^k \,; \overline{R^k}^\mathsf{T} \subseteq R^k \,; R \,; \overline{R}^\mathsf{T} = R^{k-1} \,; R \,; R \,; \overline{R}^\mathsf{T} \subseteq R^{k-1} \,; R = R^k.$$ □

A full account of the theory of semiorders is given in [100].

The following result is proved here to demonstrate how a subset of properties suffices to define a weakorder – obtained by another step of restriction.

Proposition 12.9. *An* asymmetric *and negatively transitive relation W is transitive, irreflexive, Ferrers, and semi-transitive, i.e., satisfies all items mentioned in Fig. 12.5 for a weakorder.*

Proof: W is necessarily transitive, because we get $W \,; W \subseteq W$ using the Schröder equivalence and due to asymmetry and negative transitivity from

$$\overline{W} ; W^{\mathsf{T}} \subseteq \overline{W} ; \overline{W} \subseteq \overline{W}.$$

Being negatively transitive may also be written in transposed form as $W ; \overline{W}^{\mathsf{T}} \subseteq W$. To prove that W is semi-transitive and Ferrers, is now easy:

$$W ; W ; \overline{W}^{\mathsf{T}} \subseteq W ; W \subseteq W,$$
$$W ; \overline{W}^{\mathsf{T}} ; W \subseteq W ; W \subseteq W.$$

With Prop. 12.5, we have finally that W is also irreflexive.   □

We note in passing, that 'negatively transitive and irreflexive' does not suffice to establish a weakorder, as the example $W := \overline{\mathbb{I}}$ shows for $n > 1$ rows. The kth power of a finite weakorder need *not* be a weakorder again. Figure 12.6 further illustrates the concept of a weakorder.

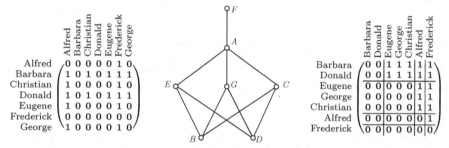

Fig. 12.6 Personal assessment as a weakorder, Hasse diagram and in rearranged form

It remains to convince ourselves concerning the spanning properties of a linear strictorder.

Proposition 12.10. *A semi-connex strictorder is Ferrers, semi-transitive, and negatively transitive, i.e., satisfies all items mentioned in Fig. 12.5 for a linear strictorder.*

Proof: Semi-connex means by definition $\overline{\mathbb{I}} \subseteq C \cup C^{\mathsf{T}}$ or else $\mathbb{T} = \mathbb{I} \cup C \cup C^{\mathsf{T}}$, so that in combination with irreflexivity and asymmetry $\overline{C} = \mathbb{I} \cup C^{\mathsf{T}}$. Negative transitivity follows then from

$$\overline{C} ; \overline{C} = (\mathbb{I} \cup C^{\mathsf{T}}) ; (\mathbb{I} \cup C^{\mathsf{T}}) = \mathbb{I} \cup C^{\mathsf{T}} \cup C^{\mathsf{T}2} = \mathbb{I} \cup C^{\mathsf{T}} = \overline{C},$$

and Prop. 12.9 produces the remaining parts of the proof.   □

Exercises

12.1 Consider a weakorder W together with its dual $P := W^{\mathrm{d}} = \overline{W}^{\mathsf{T}}$ and determine the section preorder T. Prove that $P = T$.

12.2 Let $J := \overline{R} \cap \overline{R}^{\mathsf{T}}$ be the so-called incomparability of an intervalorder R. Then there do not exist any four elements a, b, c, d such that they form a 4-cycle without chord in J.

12.3 Prove the following formulae for an intervalorder:

  (i) $R \mathbin{;} (\overline{R} \cap R^{\mathrm{d}}) \mathbin{;} R \subseteq R$ attributed to Norbert Wiener,
  (ii) $(R \cap R^{\mathrm{d}}) \mathbin{;} (R \cap R^{\mathsf{T}}) \mathbin{;} (R \cap R^{\mathrm{d}}) \subseteq (R \cap R^{\mathrm{d}})$ attributed to R. Duncan Luce,
  (iii) $(R \cap R^{\mathsf{T}}) \mathbin{;} (R \cap R^{\mathrm{d}}) \subseteq \big((R \cap R^{\mathsf{T}}) \mathbin{;} (R \cap R^{\mathrm{d}})\big)^{\mathrm{d}}$.

12.4 Prove that S is a semiorder precisely when it is irreflexive and has a connex section preorder.

12.3 Block-transitive strictorders

The following is an attempt to find an even more general version of a strictorder, that is not necessarily an intervalorder, but nonetheless has appealing algebraic properties.

Proposition 12.11. *We consider an order E together with its strictorder C.*

  (i) *The order E is block-transitive.*
  (ii) *The strictorder C may – but need not – be block-transitive.*
  (iii) *The fringe of the strictorder C is contained in its Hasse relation $H := C \cap \overline{C^2}$.*
  (iv) *The fringe of a linear strictorder C is equal to its Hasse relation $H := C \cap \overline{C^2}$ – and may thus also be empty.*

Proof: (i) The fringe of an order is the identity due to antisymmetry and trivial rules for an ordering, as proved immediately after Def. 10.10. The property of being block-transitive is, thus, trivially satisfied.

(ii) The linear strictorder on three elements, for example, is easily shown to be block transitive. According to Prop. 10.14, the strictorder C describing "$<$" on $\mathbb{R}$ has an empty fringe and, thus, cannot be block transitive.

(iii) We have obviously $\mathtt{fringe}(C) = C \cap \overline{C \mathbin{;} \overline{C}^{\mathsf{T}} \mathbin{;} C} \subseteq C \cap \overline{C^2}$ since by irreflexivity $\mathbb{I} \subseteq \overline{C}^{\mathsf{T}}$.

(iv) In addition to (iii), we have $C \mathbin{;} \overline{C}^{\mathsf{T}} \mathbin{;} C = C \mathbin{;} (\mathbb{I} \cup C) \mathbin{;} C = C^2$. $\qquad\square$

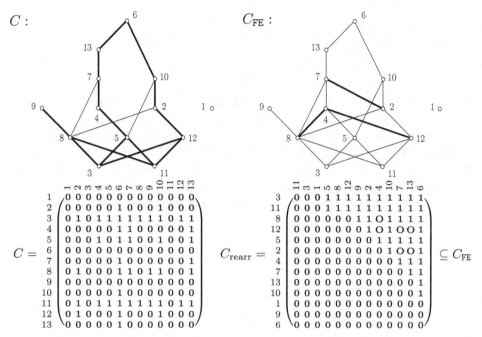

Fig. 12.7 Block-transitive strictorder blown up to a Ferrers relation

On the left side of Fig. 12.7, we see the Hasse diagram of a block-transitive strict-order. The dark lines show its fringe which is strictly contained in the Hasse relation. Underneath one will find the relation underlying the relation presented in two different ways: the original matrix as well as a rearrangement by an independent permutation of rows and columns guided by some Ferrers extension C_{FE} of C. The relation is not an intervalorder because it is not Ferrers – best recognized from its permuted matrix on the right. Additions indicated by dark lines in the right figure convert the relation to a Ferrers extension. In the permuted matrix, one will easily recognize which additions make it a Ferrers relation, i.e., an intervalorder: the dark lines $(8, 4), (12, 4), (2, 7)$ in the Hasse diagram have to be added – and some others resulting simply from transitivity, but hidden in the Hasse diagram.

Remark 12.12. The complement of the relation C on the left of Fig. 12.7 provides another example of a maxclique factorization. It gives already a hint at the deeper investigation of intervalgraphs and transitive orientation in Prop. 12.37. We first consider the obviously reflexive and symmetric relation $B := \overline{C} \cap C^d$ of the block-transitive strictorder on the left of Fig. 12.7. According to Prop. 11.8, it may be factorized $B = M^{\top} M$ by a set of maxcliques which is not shown here. Among these maxcliques are the following four connected with $(4, 7) \in C$ and $(2, 10) \in C$:

$$\{9,7,10,1\}\rightarrow$$

$$\{9,7,2,1\}\rightarrow \qquad\qquad\qquad\qquad \{9,4,10,1\}\rightarrow$$

$$\{9,4,2,1\}\rightarrow$$

Because of this diamond the whole set of maxcliques cannot be *linearly* ordered.

In contrast, consider the interval relation C_{FE} on the upper right of Fig. 12.7, obtained via Szpilrajn extension of the left. Figure 12.8 shows the factorization $B_{\mathrm{FE}} = \overline{C_{\mathrm{FE}}} \cap C_{\mathrm{FE}}^{\mathrm{d}} = M_{\mathrm{FE}}^{\mathsf{T}}\,;M_{\mathrm{FE}}$ into its maxclique relation M_{FE}. It is also shown that the domain of M_{FE} is linearly strictordered with $C_{\Delta} := M_{\mathrm{FE}}^{\mathsf{T}}\,;C_{\mathrm{FE}}\,;M_{\mathrm{FE}}$. The maxclique $\{9,7,2,1\}\rightarrow$ of $B = \overline{C} \cap C^{\mathrm{d}}$ has vanished and no longer prevents this.

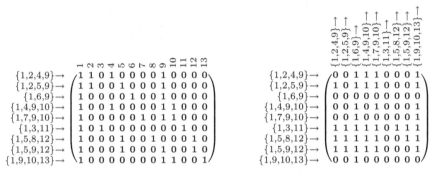

Fig. 12.8 Maxclique factorization M_{FE} and strictorder C_{Δ} on maxcliques introduced by the given intervalorder

One will observe that the greatest element in the strictorder on the right is $\{1,6,9\}\rightarrow$ which is the set of maximal elements. In the same sense, $\{1,3,11\}\rightarrow$ and $\{1,5,8,12\}\rightarrow$ are the lowest and its successor.

It will also become clear that the cardinality-maximum maxcliques, i.e., the 4-element maxcliques, precisely determine the cardinality-maximum antichains. Using Dilworth's Theorem we can reason that the intervalorder C_{FE} – as well as C – can be covered completely by 4 inclusion-maximal chains. In this case, these may be chosen as

$$3 < 8 < 9 \qquad 11 < 5 < 4 < 7 < 13 < 6 \qquad 11 < 12 < 2 < 10 < 6 \qquad 1. \qquad \square$$

Figure 12.9 provides another example, namely a strictorder that is not block transitive. Note that $\nabla\,;\nabla^{\mathsf{T}}\,;C\,;\nabla^{\mathsf{T}}\,;\nabla$ does not have a very close connection to C since it does not contain, for example, $(3,5)$.

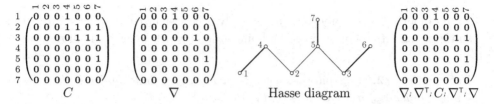

Fig. 12.9 Hasse diagram of a non-block-transitive strictorder
with fringe indicated by dark lines

We know that the fringe of a strictorder is contained in the Hasse relation. This
helps us in giving a characterization of block-transitivity.

Proposition 12.13. *A strictorder is block transitive precisely when every arrow of
H (or, restricted even further, of $H \cap \overline{\nabla}$) has an arrow of its fringe ∇ ending at its
end and an arrow of ∇ beginning at its start.*

Proof: C is block-transitive if and only if $C \subseteq \nabla; \mathbb{T}; \nabla$. Then indeed, every arrow
of C has an arrow of ∇ ending at its end and one arrow of ∇ beginning at its start.
This will then automatically be satisfied also for H.

In the reverse direction, we use that every arrow of C can be replaced by a sequence
$a_1, \ldots, a_p$ of arrows of H. Then those for a_1, a_p provide the arrows demanded to
exist.   □

This is violated in Fig. 12.9: $(2,5)$ is in $H \cap \overline{\nabla}$, but has no fringe arrow ending in
5 and no fringe arrow starting in 2. Block-transitivity of a finite strictorder may,
thus, be decided along its Hasse relation which is usually less effort than deciding
along C itself.

Exercises

12.5 Prove that for any preorder the fringe is precisely its row equivalence.

12.4 Order extensions

The many types of orderings are not unrelated. In this section, we collect how
to obtain the weakorder closure of a semiorder, some semiorder containing a given
intervalorder, and how to find a Szpilrajn extension of a strictorder to a linear order
(topological sorting). With techniques developed earlier, we are also in a position
to embed a block-transitive strictorder into an intervalorder.

Topological sorting

We recall a result with proof that has already been visualized along with Prop. 5.25 and Fig. 5.21. Its proof requires the Point Axiom, which we use only in rare cases.

Proposition 12.14 (Szpilrajn's Theorem). *For every finite[6] order E there exists a linear order E_{Sp} in which it is contained, i.e., $E \subseteq E_{\mathsf{Sp}}$.*

Proof: Assume E is not yet linear, i.e., $E \cup E^\mathsf{T} \neq \mathbb{T}$. Then there exist according to the Point Axiom of Def. 8.23 two points x, y with $x \,; y^\mathsf{T} \subseteq \overline{E \cup E^\mathsf{T}}$.

If we define $E_1 := E \cup E \,; x \,; y^\mathsf{T} \,; E$, it is easy to show that E_1 is an order again: it is reflexive as already E is. For proving transitivity and antisymmetry, we need the following intermediate result:

$$y^\mathsf{T} \,; E^2 \,; x = y^\mathsf{T} \,; E \,; x \subseteq \mathbb{L}.$$

It is equivalent with $y \,; \mathbb{T} \subseteq \overline{E \,; x} = \overline{E} \,; x$ since x is a point, and shunting according to Lemma 5.12.ii with $y \,; x^\mathsf{T} = y \,; \mathbb{T} \,; x^\mathsf{T} \subseteq \overline{E}$ which holds owing to the way x, y have been chosen.

Now transitivity is shown as

$$E_1 \,; E_1 = E^2 \cup E^2 \,; x \,; y^\mathsf{T} \,; E \cup E \,; x \,; y^\mathsf{T} \,; E^2 \cup E \,; x \,; y^\mathsf{T} \,; E^2 \,; x \,; y^\mathsf{T} \,; E = E \cup E \,; x \,; y^\mathsf{T} \,; E = E_1.$$

Antisymmetry may be shown by evaluating the additive parts of $E_1 \cap E_1^\mathsf{T}$ separately. Because E is an order, $E \cap E^\mathsf{T} \subseteq \mathbb{I}$. Furthermore,

$$E \cap E^\mathsf{T} \,; y \,; x^\mathsf{T} \,; E^\mathsf{T} \subseteq E \cap E^\mathsf{T} \,; \overline{E \cup E^\mathsf{T}} \,; E^\mathsf{T} \subseteq E \cap E^\mathsf{T} \,; \overline{E} \,; E^\mathsf{T} = E \cap E^\mathsf{T} \,; \overline{E} = E \cap \overline{E} = \mathbb{L}$$
$$E^\mathsf{T} \,; y \,; x^\mathsf{T} \,; E^\mathsf{T} \cap E \,; x \,; y^\mathsf{T} \,; E \subseteq (E^\mathsf{T} \,; y \cap E \,; x \,; y^\mathsf{T} \,; E \,; E \,; x) \,; (\dots) \subseteq (\dots \cap \mathbb{L}) \,; (\dots) \subseteq \mathbb{L},$$

applying the Dedekind rule in the last line.

In the finite case, this argument may be iterated, and E_n will eventually become linear. □

So, for every finite order E there exists a bijective homomorphism onto a linear order E_{Sp}. Expressed differently, there exists a permutation such that the resulting order resides in the upper right triangle – in which case the Szpilrajn extension is completely obvious.

$$
\begin{array}{c}
\begin{array}{c}{\scriptstyle 1\ 2\ 3\ 4\ 5}\end{array}\\
\begin{array}{c}3\\1\\2\\4\\5\end{array}\!\!\left(\begin{array}{ccccc}0&0&1&0&0\\1&0&0&0&0\\0&1&0&0&0\\0&0&0&1&0\\0&0&0&0&1\end{array}\right)
\end{array}
\ \doteq\
\begin{array}{c}
\begin{array}{c}{\scriptstyle 1\ 2\ 3\ 4\ 5}\end{array}\\
\begin{array}{c}1\\2\\3\\4\\5\end{array}\!\!\left(\begin{array}{ccccc}1&0&0&1&0\\0&1&0&0&1\\1&1&1&1&1\\0&0&0&1&0\\0&0&0&0&1\end{array}\right)
\end{array}
\ \doteq\
\begin{array}{c}
\begin{array}{c}{\scriptstyle 3\ 1\ 2\ 4\ 5}\end{array}\\
\begin{array}{c}1\\2\\3\\4\\5\end{array}\!\!\left(\begin{array}{ccccc}0&1&0&0&0\\0&0&1&0&0\\1&0&0&0&0\\0&0&0&1&0\\0&0&0&0&1\end{array}\right)
\end{array}
\ =\
\begin{array}{c}
\begin{array}{c}{\scriptstyle 3\ 1\ 2\ 4\ 5}\end{array}\\
\begin{array}{c}3\\1\\2\\4\\5\end{array}\!\!\left(\begin{array}{ccccc}1&1&1&1&1\\0&1&0&1&0\\0&0&1&0&1\\0&0&0&1&0\\0&0&0&0&1\end{array}\right)
\end{array}
\ \subseteq E_{\mathsf{Sp}}=\
\begin{array}{c}
\begin{array}{c}{\scriptstyle 3\ 1\ 2\ 4\ 5}\end{array}\\
\begin{array}{c}3\\1\\2\\4\\5\end{array}\!\!\left(\begin{array}{ccccc}1&1&1&1&1\\0&1&1&1&1\\0&0&1&1&1\\0&0&0&1&1\\0&0&0&0&1\end{array}\right)
\end{array}
$$

Fig. 12.10 E presented with permutation as $\pi^\mathsf{T} \,; E \,; \pi$ is contained in a Szpilrajn extension of E

[6] Even more – with Zorn's Lemma, non-finite cases may also be handled.

Already in Prop. 10.39, this result has been generalized considerably: the relation need not be homogeneous and the linear order in which to embed may be taken blockwise.

Intervalorder extension of a block-transitive strictorder

The main part of this task has already been achieved in Prop. 10.39 for the heterogeneous context, when forming a Ferrers extension of a block-transitive relation.

Embedding an intervalorder into a semiorder

The following process of embedding an intervalorder into a semiorder is not uniquely determined; besides the one chosen, also $S' := R \cup \overline{R}^{\mathsf{T}}; R; R$ would do, for example.

Proposition 12.15. *Starting from some intervalorder R, one will always obtain a semiorder when forming $S := R \cup R; R; \overline{R}^{\mathsf{T}}$.*

Proof: In view of Def. 12.4, we decide to show that S is irreflexive, Ferrers, and semi-transitive. First, irreflexivity follows since R is already irreflexive and $R; R; \overline{R}^{\mathsf{T}} \subseteq \overline{\mathbb{I}}$ is via the Schröder rule equivalent with transitivity of R. Now, we prove a transition in advance that is applied more than once. Using monotony and two times the Schröder equivalence, we get

$$R; \overline{R}^{\mathsf{T}}; \overline{\overline{R}; R^{\mathsf{T}}; R^{\mathsf{T}}} \subseteq R; \overline{R^{\mathsf{T}}; R^{\mathsf{T}}} \subseteq \overline{R}^{\mathsf{T}}.$$

To prove that S is Ferrers, means showing

$$(R \cup R; R; \overline{R}^{\mathsf{T}}); (\overline{R}^{\mathsf{T}} \cap \overline{\overline{R}; R^{\mathsf{T}}; R^{\mathsf{T}}}); (R \cup R; R; \overline{R}^{\mathsf{T}}) \subseteq R \cup R; R; \overline{R}^{\mathsf{T}},$$

from which four products are considered separately:

$R; (\overline{R}^{\mathsf{T}} \cap \ldots); R \subseteq R; \overline{R}^{\mathsf{T}}; R \subseteq R$ because R is Ferrers

$R; (\overline{R}^{\mathsf{T}} \cap \ldots); R; R; \overline{R}^{\mathsf{T}} \subseteq R; \overline{R}^{\mathsf{T}}; R; R; \overline{R}^{\mathsf{T}} \subseteq R; R; \overline{R}^{\mathsf{T}}$ same with additional factors

$R; R; \overline{R}^{\mathsf{T}}; (\overline{R}^{\mathsf{T}} \cap \overline{\overline{R}; R^{\mathsf{T}}; R^{\mathsf{T}}}); R \subseteq R; R; \overline{R}^{\mathsf{T}}; \overline{\overline{R}; R^{\mathsf{T}}; R^{\mathsf{T}}}; R \subseteq R; \overline{R}^{\mathsf{T}}; R \subseteq R$ Ferrers

$R; R; \overline{R}^{\mathsf{T}}; (\overline{R}^{\mathsf{T}} \cap \overline{\overline{R}; R^{\mathsf{T}}; R^{\mathsf{T}}}); R; R; \overline{R}^{\mathsf{T}} \subseteq R; R; \overline{R}^{\mathsf{T}}$ same with additional factors

Proving semi-transitivity repeats these steps in a slightly modified fashion to show

$$(R \cup R; R; \overline{R}^{\mathsf{T}}); (R \cup R; R; \overline{R}^{\mathsf{T}}); (\overline{R}^{\mathsf{T}} \cap \overline{\overline{R}; R^{\mathsf{T}}; R^{\mathsf{T}}}) \subseteq R \cup R; R; \overline{R}^{\mathsf{T}},$$

where we again consider four products:

$R; R; (\overline{R}^{\mathsf{T}} \cap \ldots) \subseteq R; R; \overline{R}^{\mathsf{T}}$

$R; R; R; \overline{R}^{\mathsf{T}}; (\overline{R}^{\mathsf{T}} \cap \overline{\overline{R}; R^{\mathsf{T}}; R^{\mathsf{T}}}) \subseteq R; R; R; \overline{R}^{\mathsf{T}}; \overline{\overline{R}; R^{\mathsf{T}}; R^{\mathsf{T}}} \subseteq R; R; \overline{R^{\mathsf{T}}}$ see above

$R; R; \overline{R}^{\mathsf{T}}; R; (\overline{R}^{\mathsf{T}} \cap \ldots) \subseteq R; R; \overline{R}^{\mathsf{T}}$ since R is Ferrers

$$R\,;R\,;\overline{R}^{\mathsf{T}}\,;R\,;R\,;\overline{R}^{\mathsf{T}}\,;(\overline{R}^{\mathsf{T}}\cap\overline{R\,;R^{\mathsf{T}}\,;R^{\mathsf{T}}})\subseteq R\,;R\,;R\,;\overline{R}^{\mathsf{T}}\,;\overline{R\,;R^{\mathsf{T}}\,;R^{\mathsf{T}}}\subseteq R\,;R\,;\overline{R}^{\mathsf{T}}$$

$$\text{see above}\qquad\square$$

The following intervalorder shows the idea behind Prop. 12.15. In Fig. 12.25 it will be studied in detail. Not least, Fig. 12.25 provides a rearranged form with rows and columns permuted independently.

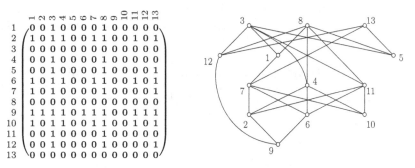

$$
\begin{array}{c|ccccccccccccc}
 & 1 & 2 & 3 & 4 & 5 & 6 & 7 & 8 & 9 & 10 & 11 & 12 & 13 \\
\hline
1 & 0&0&1&0&0&0&0&1&0&0&0&0&0 \\
2 & 1&0&1&1&0&0&1&1&0&0&1&0&1 \\
3 & 0&0&0&0&0&0&0&0&0&0&0&0&0 \\
4 & 0&0&1&0&0&0&0&1&0&0&0&0&0 \\
5 & 0&0&1&0&0&0&0&1&0&0&0&0&1 \\
6 & 1&0&1&1&0&0&1&1&0&0&1&0&1 \\
7 & 1&0&1&0&0&0&0&1&0&0&0&0&1 \\
8 & 0&0&0&0&0&0&0&0&0&0&0&0&0 \\
9 & 1&1&1&1&0&1&1&1&0&0&1&1&1 \\
10 & 1&0&1&1&0&0&1&1&0&0&1&0&1 \\
11 & 0&0&1&0&0&0&0&1&0&0&0&0&0 \\
12 & 0&0&1&0&0&0&0&1&0&0&0&0&1 \\
13 & 0&0&0&0&0&0&0&0&0&0&0&0&0 \\
\end{array}
$$

Fig. 12.11 An intervalorder that is not a semiorder

Figure 12.11 shows the matrix and Hasse diagram of an intervalorder. It cannot be a semiorder as is documented by $2<7<1$, where for 12 we have neither $2<12$ nor $12<1$.

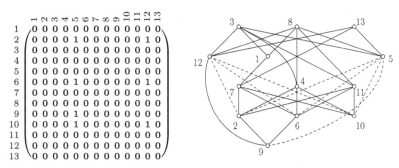

$$
\begin{array}{c|ccccccccccccc}
 & 1 & 2 & 3 & 4 & 5 & 6 & 7 & 8 & 9 & 10 & 11 & 12 & 13 \\
\hline
1 & 0&0&0&0&0&0&0&0&0&0&0&0&0 \\
2 & 0&0&0&1&0&0&0&0&0&0&0&1&0 \\
3 & 0&0&0&0&0&0&0&0&0&0&0&0&0 \\
4 & 0&0&0&0&0&0&0&0&0&0&0&0&0 \\
5 & 0&0&0&0&0&0&0&0&0&0&0&0&0 \\
6 & 0&0&0&1&0&0&0&0&0&0&0&1&0 \\
7 & 0&0&0&0&0&0&0&0&0&0&0&0&0 \\
8 & 0&0&0&0&0&0&0&0&0&0&0&0&0 \\
9 & 0&0&0&1&0&0&0&0&0&0&0&0&0 \\
10 & 0&0&0&1&0&0&0&0&0&0&0&1&0 \\
11 & 0&0&0&0&0&0&0&0&0&0&0&0&0 \\
12 & 0&0&0&0&0&0&0&0&0&0&0&0&0 \\
13 & 0&0&0&0&0&0&0&0&0&0&0&0&0 \\
\end{array}
$$

Fig. 12.12 Semiorder additions to the intervalorder of Fig. 12.11; see also Fig. 12.26

Weakorder closure of a semiorder

Semiorders possess a uniquely defined extension to a weakorder, which is often introduced using the concept of a dual; see Def. 4.5. Duals will later be studied in more detail, starting in Section 13.1.

Definition 12.16. Given a weakorder W, we call a relation T a W-**threshold**, provided

$$W^{\mathrm{d}};T \cup T;W^{\mathrm{d}} = T \subseteq W.$$ □

In other words, T considered row-wise will be an upper cone with respect to the corresponding connex preorder $W^{\mathrm{d}} \supseteq W$, while considered column-wise it will be a lower cone. Since $W^{\mathrm{d}} \supseteq \mathbb{I}$, this means in particular, that also $W^{\mathrm{d}};T = T;W^{\mathrm{d}} = T$. Concerning such thresholds, we prove that they are necessarily Ferrers, transitive, and semi-transitive, and thus, a semiorder:

```
      a b c d e f g h i j k l m            a b c d e f g h i j k l m                  i m a f e b l h g c d j k
  a / 0 1 1 1 0 0 1 1 0 1 1 1 0 \      / 0 1 1 1 1 0 1 1 0 1 1 1 0 \       i / 0 0 1 1 1 1 1 1 1 1 1 1 1 \
  b | 0 0 1 1 0 0 0 0 0 1 1 0 0 |      | 0 0 1 1 0 0 1 1 0 1 1 0 0 |      m | 0 0 1 1 1 1 1 1 1 1 1 1 1 |
  c | 0 0 0 0 0 0 0 0 0 1 1 0 0 |      | 0 0 0 0 0 0 0 0 0 1 1 0 0 |      a | 0 0 0 0 0 1 1 1 1 1 1 1 1 |
  d | 0 0 0 0 0 0 0 0 0 1 1 0 0 |      | 0 0 0 0 0 0 0 0 0 1 1 0 0 |      f | 0 0 0 0 0 1 1 1 1 1 1 1 1 |
  e | 0 0 1 1 0 0 1 0 0 1 1 0 0 |      | 0 1 1 1 0 0 1 1 0 1 1 1 0 |      e | 0 0 0 0 0 0 0 0 1 1 1 1 1 |
  f | 0 1 1 1 0 0 1 1 0 1 1 1 0 |      | 0 1 1 1 1 0 1 1 0 1 1 1 0 |      b | 0 0 0 0 0 0 0 0 1 1 1 1 1 |
  g | 0 0 0 0 0 0 0 0 0 1 1 0 0 |      | 0 0 1 1 0 0 0 0 0 1 1 0 0 |      l | 0 0 0 0 0 0 0 0 1 1 1 1 1 |
  h | 0 0 0 0 0 0 0 0 0 1 1 0 0 |      | 0 0 1 1 0 0 1 0 0 1 1 0 0 |      h | 0 0 0 0 0 0 0 0 0 0 1 1 1 |
  i | 1 1 1 1 1 1 1 1 0 1 1 1 0 |      | 1 1 1 1 1 1 1 1 0 1 1 1 0 |      g | 0 0 0 0 0 0 0 0 0 0 1 1 1 |
  j | 0 0 0 0 0 0 0 0 0 0 0 0 0 |      | 0 0 0 0 0 0 0 0 0 0 0 0 0 |      c | 0 0 0 0 0 0 0 0 0 0 0 1 1 |
  k | 0 0 0 0 0 0 0 0 0 0 0 0 0 |      | 0 0 0 0 0 0 0 0 0 0 0 0 0 |      d | 0 0 0 0 0 0 0 0 0 0 0 1 1 |
  l | 0 0 1 1 0 0 0 0 0 1 1 0 0 |      | 0 0 1 1 0 0 1 1 0 1 1 0 0 |      j | 0 0 0 0 0 0 0 0 0 0 0 0 0 |
  m \ 1 1 1 1 1 1 1 1 0 1 1 1 0 /      \ 1 1 1 1 1 1 1 1 0 1 1 1 0 /      k \ 0 0 0 0 0 0 0 0 0 0 0 0 0 /
```

Fig. 12.13 A threshold T in a weakorder W; rearranged: T does not, W does touch block-diagonal

Proposition 12.17. *A threshold T in a weakorder W is, considered on its own, a semiorder.*

Proof: In view of Prop. 12.4, we decide to prove that T is irreflexive, Ferrers, and semi-transitive. T is irreflexive since $T \subseteq W \subseteq \overline{\mathbb{I}}$. The intermediate result

$$T;\overline{T}^{\top} \subseteq W$$

(obtained using that it is equivalent with $T^{\top};\overline{W} \subseteq T^{\top}$ as well as with $W^{\mathrm{d}};T \subseteq T$, where the latter is satisfied by assumption), will serve in two cases. Firstly, in proving the Ferrers property with

$$T;\overline{T}^{\top};T \subseteq W;T \subseteq W^{\mathrm{d}};T \subseteq T,$$

and secondly in proving semi-transitivity in a rather similar way

$$T;T;\overline{T}^{\top} \subseteq T;W \subseteq T;W^{\mathrm{d}} \subseteq T.$$ □

In fact, every semiorder turns out to be a W-threshold of some weakorder which it uniquely determines. The question is where to enlarge S so as to obtain the weakorder $W \supseteq S$. If S were itself already negatively transitive, it would satisfy $\overline{S};\overline{S} \subseteq \overline{S}$, and we might take $W := S$. If this is not yet satisfied, we have

$$\mathbb{I}\!\mathbb{I} \neq \overline{S};\overline{S} \cap S \subseteq (\overline{S} \cap S;\overline{S}^{\top});(\overline{S} \cap \overline{S}^{\top};S),$$

where on the right side none of the factors will vanish. We add the factors to S in

$$W := S \cup (\overline{S} \cap \overline{S}^\mathsf{T} \mathbin{;} S) \cup (\overline{S} \cap S \mathbin{;} \overline{S}^\mathsf{T}) = S \cup \overline{S}^\mathsf{T} \mathbin{;} S \cup S \mathbin{;} \overline{S}^\mathsf{T} = \overline{S}^\mathsf{T} \mathbin{;} S \cup S \mathbin{;} \overline{S}^\mathsf{T} = S^\mathsf{d} \mathbin{;} S \cup S \mathbin{;} S^\mathsf{d},$$

using that $\mathbb{I} \subseteq \overline{S}$ for an irreflexive S. One obtains what we define as the **weakorder closure** of the semiorder:

Proposition 12.18. (i) *Every semiorder S may be enlarged so as to obtain the weakorder*

$$W := S^\mathsf{d} \mathbin{;} S \cup S \mathbin{;} S^\mathsf{d} \subseteq S^\mathsf{d}$$

in which it is a W-threshold.

(ii) *The weakorder may with the same effect as in (i) also be determined as*

$$W' := I \mathbin{;} S \cup S \mathbin{;} I$$

where $I := S^\mathsf{d} \cap \overline{S}$ will later be called the indifference with respect to S.

Proof: (i) In view of Def. 12.4, we decide to prove that W is asymmetric and negatively transitive. W is asymmetric since with the Ferrers and semi-transitivity property

$$W^\mathsf{T} = S^\mathsf{T} \mathbin{;} (S^\mathsf{d})^\mathsf{T} \cup (S^\mathsf{d})^\mathsf{T} \mathbin{;} S^\mathsf{T} = S^\mathsf{T} \mathbin{;} \overline{S} \cup \overline{S} \mathbin{;} S^\mathsf{T} \subseteq \overline{\overline{S}^\mathsf{T} \mathbin{;} S} \cap \overline{S \mathbin{;} \overline{S}^\mathsf{T}} = \overline{S^\mathsf{d} \mathbin{;} S \cup S \mathbin{;} S^\mathsf{d}} = \overline{W}.$$

W negatively transitive means

$$\overline{W} \mathbin{;} \overline{W} = (\overline{\overline{S}^\mathsf{T} \mathbin{;} S} \cap \overline{S \mathbin{;} \overline{S}^\mathsf{T}}) \mathbin{;} (\overline{\overline{S}^\mathsf{T} \mathbin{;} S} \cap \overline{S \mathbin{;} \overline{S}^\mathsf{T}}) \subseteq \overline{\overline{S}^\mathsf{T} \mathbin{;} S} \cap \overline{S \mathbin{;} \overline{S}^\mathsf{T}} = \overline{W},$$

which holds since $\overline{\overline{S}^\mathsf{T} \mathbin{;} S}$ as well as $\overline{S \mathbin{;} \overline{S}^\mathsf{T}}$ are transitive as row-contains-preorders; see Def. 5.28. It remains to prove that S is indeed a W-threshold

$$W^\mathsf{d} \mathbin{;} S \cup S \mathbin{;} W^\mathsf{d} = S \subseteq W.$$

This is easy because, for example,

$$W^\mathsf{d} \mathbin{;} S = (\overline{S^\mathsf{d} \mathbin{;} S}^\mathsf{T} \cap \overline{S \mathbin{;} S^\mathsf{d}}^\mathsf{T}) \mathbin{;} S = (\dots \cap \overline{S \mathbin{;} S^\mathsf{T}}) \mathbin{;} S \subseteq \overline{\overline{S \mathbin{;} S^\mathsf{T}}} \mathbin{;} S \subseteq S = \mathbb{I} \mathbin{;} S \subseteq W^\mathsf{d} \mathbin{;} S.$$

S is transitive implying $S^\mathsf{T} \mathbin{;} \overline{S} \subseteq \overline{S}$. Therefore, $S^\mathsf{d} \mathbin{;} S = \overline{S}^\mathsf{T} \mathbin{;} S \subseteq \overline{S}^\mathsf{T} = S^\mathsf{d}$, so that $W \subseteq S^\mathsf{d}$.

(ii) By definition, $\mathbb{I} \subseteq I \subseteq S^\mathsf{d}$, so that $W' \subseteq W$. For the other direction, we decompose, for example,

$$S^\mathsf{d} \mathbin{;} S = (S^\mathsf{d} \cap \overline{S}) \mathbin{;} S \cup (S^\mathsf{d} \cap S) \mathbin{;} S \subseteq I \mathbin{;} S \cup S = I \mathbin{;} S \cup S \mathbin{;} \mathbb{I} \subseteq I \mathbin{;} S \cup S \mathbin{;} I = W'. \quad \square$$

Concerning the weakorder closure, [99] speaks of the *underlying weakorder*. There it is mentioned that every upper-diagonal step matrix is a semiorder. Among these semiorders, the weakorders are those for which 'the edge of each step touches the diagonal'; see Fig. 12.13.

Example 12.19. The relation of Fig. 12.14 is a semiorder. It is not a weakorder as the triangle A, F, D, for example, shows that it is not negatively transitive.

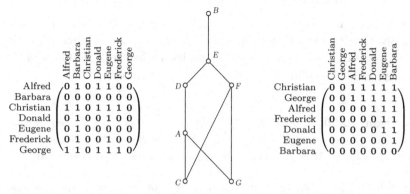

	Alfred	Barbara	Christian	Donald	Eugene	Frederick	George
Alfred	0	1	0	1	1	0	0
Barbara	0	0	0	0	0	0	0
Christian	1	1	0	1	1	1	0
Donald	0	1	0	0	1	0	0
Eugene	0	1	0	0	0	0	0
Frederick	0	1	0	0	1	0	0
George	1	1	0	1	1	1	0

	Christian	George	Alfred	Frederick	Donald	Eugene	Barbara
Christian	0	0	1	1	1	1	1
George	0	0	1	1	1	1	1
Alfred	0	0	0	0	0	1	1
Frederick	0	0	0	0	0	1	1
Donald	0	0	0	0	0	1	1
Eugene	0	0	0	0	0	0	1
Barbara	0	0	0	0	0	0	0

Fig. 12.14 A semiorder, its Hasse diagram, and a permuted threshold representation

We follow the idea of the constructive proof of Prop. 12.18 and first embed this semiorder into its weakorder closure obtaining the left diagram and matrix of Fig. 12.15. The link (A, F), for example, is added since (A, D) belongs to the relation S and (D, F) to its indifference, so that in total (A, F) is in $S_i I$, i.e., in the weakorder closure.

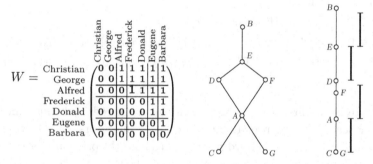

$$W =$$

	Christian	George	Alfred	Frederick	Donald	Eugene	Barbara
Christian	0	0	1	1	1	1	1
George	0	0	1	1	1	1	1
Alfred	0	0	0	1	1	1	1
Frederick	0	0	0	0	0	1	1
Donald	0	0	0	0	0	1	1
Eugene	0	0	0	0	0	0	1
Barbara	0	0	0	0	0	0	0

Fig. 12.15 Weakorder closure and threshold representation for the semiorder of Fig. 12.14

With 'yard sticks' and along a linear thin-line scale, it is indicated which of the linear order relations are considered 'too short for exceeding the threshold'; it is the semiorder of Fig. 12.14 which is left over. □

We show different weakorder closure of a semiorder in Fig. 13.6 along with the discussion of indifference.

Embedding a weakorder into a linear strictorder

The most immediate way is to use the Szpilrajn extension because a weakorder is necessarily a strictorder. There exists, however, a less costly alternative since we work with relations on basesets. The baseset has its baseset ordering, and this may be employed to define a linear extension of the weakorder.

Proposition 12.20. *We assume a weakorder* $W : X \longrightarrow X$ *together with an arbitrary linear strictorder* $C : X \longrightarrow X$ *on* X, *which may be the base order on the baseset. Then* $C_1 := W \cup (C \cap \overline{W}^{\mathsf{T}})$ *is a linear strictorder containing* W.

Proof: The construct C_1 is irreflexive since W and C are. It is semi-connex since

$$
\begin{aligned}
C_1 \cup C_1^{\mathsf{T}} &= W \cup (C \cap \overline{W}^{\mathsf{T}}) \cup \left(W \cup (C \cap \overline{W}^{\mathsf{T}})\right)^{\mathsf{T}} && \text{by definition} \\
&= W \cup (C \cap \overline{W}^{\mathsf{T}}) \cup W^{\mathsf{T}} \cup (C^{\mathsf{T}} \cap \overline{W}) && \text{transposed} \\
&= W \cup (C^{\mathsf{T}} \cap \overline{W}) \cup (C \cap \overline{W}^{\mathsf{T}}) \cup W^{\mathsf{T}} && \text{reordered} \\
&= \left[(W \cup C^{\mathsf{T}}) \cap (W \cup \overline{W})\right] \cup \left[(C \cup W^{\mathsf{T}}) \cap (\overline{W}^{\mathsf{T}} \cup W^{\mathsf{T}})\right] && \text{distributivity} \\
&= \left[(W \cup C^{\mathsf{T}}) \cap \mathbb{T}\right] \cup \left[(C \cup W^{\mathsf{T}}) \cap \mathbb{T}\right] = W \cup C^{\mathsf{T}} \cup C \cup W^{\mathsf{T}} \supseteq C^{\mathsf{T}} \cup C \\
&= \mathbb{\overline{\overline{I}}} && \text{since } C \text{ as a linear strictorder is semi-connex}
\end{aligned}
$$

Transitivity:

$$
\left(W \cup (C \cap \overline{W}^{\mathsf{T}})\right) ; \left(W \cup (C \cap \overline{W}^{\mathsf{T}})\right)
$$

is investigated term-wise:

$W ; W \subseteq W$ since W is transitive

$W ; (C \cap \overline{W}^{\mathsf{T}}) \subseteq W ; \overline{W}^{\mathsf{T}} \subseteq W$ since W is negatively transitive

$(C \cap \overline{W}^{\mathsf{T}}) ; (C \cap \overline{W}^{\mathsf{T}}) \subseteq C \cap \overline{W}^{\mathsf{T}}$ transitive C, negative transitivity of W □

Figure 12.16 shows first a weakorder, then this weakorder embedded into a linear strictorder according to Prop. 12.20, using for C the reverse of the strictorder on the interval $1, 2, \ldots, 11$. The third matrix rearranges the original relation along the strictorder C_1 thus obtained. In the latter form, the given weakorder is more easily identified as such.

$$
\begin{array}{c}
\begin{array}{c} \scriptstyle 1\;2\;3\;4\;5\;6\;7\;8\;9\;10\;11 \end{array} \\
\begin{array}{r|l}
1 & 0\;1\;0\;0\;0\;1\;0\;0\;0\;0\;0 \\
2 & 0\;0\;0\;0\;0\;0\;0\;0\;0\;0\;0 \\
3 & 1\;1\;0\;1\;1\;1\;1\;1\;0\;1\;1 \\
4 & 1\;1\;0\;0\;1\;1\;1\;0\;0\;0\;1 \\
5 & 1\;1\;0\;0\;0\;1\;0\;0\;0\;0\;0 \\
6 & 0\;0\;0\;0\;0\;0\;0\;0\;0\;0\;0 \\
7 & 1\;1\;0\;0\;0\;1\;0\;0\;0\;0\;0 \\
8 & 1\;1\;0\;0\;1\;1\;1\;0\;0\;0\;1 \\
9 & 1\;1\;0\;1\;1\;1\;1\;1\;0\;1\;1 \\
10 & 1\;1\;0\;0\;1\;1\;1\;0\;0\;0\;1 \\
11 & 1\;1\;0\;0\;0\;1\;0\;0\;0\;0\;0
\end{array}
\end{array}
\qquad
\begin{array}{c}
\begin{array}{c} \scriptstyle 1\;2\;3\;4\;5\;6\;7\;8\;9\;10\;11 \end{array} \\
\begin{array}{r|l}
1 & 0\;1\;0\;0\;0\;1\;0\;0\;0\;0\;0 \\
2 & 0\;0\;0\;0\;0\;0\;0\;0\;0\;0\;0 \\
3 & 1\;1\;0\;1\;1\;1\;1\;1\;0\;1\;1 \\
4 & 1\;1\;0\;0\;1\;1\;1\;0\;0\;0\;1 \\
5 & 1\;1\;0\;0\;0\;1\;0\;0\;0\;0\;0 \\
6 & 0\;1\;0\;0\;0\;0\;0\;0\;0\;0\;0 \\
7 & 1\;1\;0\;0\;1\;1\;0\;0\;0\;0\;0 \\
8 & 1\;1\;0\;1\;1\;1\;1\;0\;0\;0\;1 \\
9 & 1\;1\;1\;1\;1\;1\;1\;1\;0\;1\;1 \\
10 & 1\;1\;0\;1\;1\;1\;1\;1\;0\;0\;1 \\
11 & 1\;1\;0\;0\;1\;1\;1\;0\;0\;0\;0
\end{array}
\end{array}
\qquad
\begin{array}{c}
\begin{array}{c} \scriptstyle 9\;3\;10\;8\;4\;11\;7\;5\;1\;6\;2 \end{array} \\
\begin{array}{r|l}
9 & 0\;0\;1\;1\;1\;1\;1\;1\;1\;1\;1 \\
3 & 0\;0\;1\;1\;1\;1\;1\;1\;1\;1\;1 \\
10 & 0\;0\;0\;0\;0\;1\;1\;1\;1\;1\;1 \\
8 & 0\;0\;0\;0\;0\;1\;1\;1\;1\;1\;1 \\
4 & 0\;0\;0\;0\;0\;1\;1\;1\;1\;1\;1 \\
11 & 0\;0\;0\;0\;0\;0\;0\;1\;1\;1\;1 \\
7 & 0\;0\;0\;0\;0\;0\;0\;1\;1\;1\;1 \\
5 & 0\;0\;0\;0\;0\;0\;0\;0\;1\;1\;1 \\
1 & 0\;0\;0\;0\;0\;0\;0\;0\;0\;1\;1 \\
6 & 0\;0\;0\;0\;0\;0\;0\;0\;0\;0\;0 \\
2 & 0\;0\;0\;0\;0\;0\;0\;0\;0\;0\;0
\end{array}
\end{array}
$$

Fig. 12.16 Embedding a weakorder in the upper right triangle: W, C_1 and W rearranged

One will identify the pairs $(9,3), (10,8), (10,4), (8,4), (11,7), (11,5), (7,5), (6,2)$ in two ways. Firstly, precisely these are added in the step from the first to the second relation. Secondly, they are those preventing the permuted third relation being precisely the upper right triangle.

Exercises

12.6 Let an intervalorder R be given and define $\Xi := \overline{R} \cap \overline{R}^\mathsf{T}$. Prove that $\Xi; R$ as well as $R; \Xi$ are weakorders.

12.5 Relating preference and utility

Authors have often reported a basic difference between the European school of decision analysis and the American school. While the former stresses the qualitative notion of a preference relation as a basis for decision-support, the latter works via real-valued utility functions. The bridge between the two is the concept of *realizability* based on mappings into the real numbers. It is, however, just tradition to use real numbers. Some drawbacks are that monotonic transformations are allowed and that thresholds are not fixed to numbers and may be fixed almost arbitrarily. What is in fact needed is the linear order of real numbers. This indicates that an algebraic treatment should also be possible.

We start this investigation by presenting two versions of more or less the same concept of embedding into strictorders "$<$" on $\mathbb{R}$ or C. It is, of course, also possible to define realizability with respect to "$\leq$" on $\mathbb{R}$ and the order $E := \mathbb{I} \cup C$.

Definition 12.21. A (possibly heterogeneous) relation $R : X \longrightarrow Y$ is said to be

(i) $\mathbb{R}$**-realizable** if there exist mappings $f : X \longrightarrow \mathbb{R}$, $g : Y \longrightarrow \mathbb{R}$ such that
$$(x,y) \in R \quad \Longleftrightarrow \quad f(x) < g(y);$$
(ii) **realizable**, if there exists a linear strictorder C with mappings f, g such that
$$R = f; C; g^\mathsf{T}. \qquad \qquad \square$$

When following this idea over the hierarchy of orderings, the start is simple and considers linear *strict*orders, which is rather straightforward to embed into the real axis $\mathbb{R}$, at least when the relation is finite.

Theorem 12.22 (Birkhoff–Milgram Theorem). *Let $C : X \longrightarrow X$ be a linear strictorder. There exists a mapping f from the source of C into the real numbers such that*

$$(x, y) \in C \quad \Longleftrightarrow \quad f(x) < f(y)$$

if and only if C has a countable order-dense subset. ☐

A subset U is order-dense with respect to C, if $C \subseteq C \cdot (\mathbb{I} \cap U \cdot \mathbb{T}) \cdot C$; this expresses that between any pair $(x, y) \in C$ there exists a point of $u \in U$ with $(x, u) \in C$ and $(u, y) \in C$. We do not prove[7] this well-known theorem here, nor do we comment on order-density at this point. It is only relevant in non-finite cases. It is important that, once we have a mapping into a linear strictorder, we can use this theorem finally to encode the respective property in $\mathbb{R}$. According to our general approach, we will therefore only study realizations in some linear strictorder.

Realization of a weakorder

In Prop. 10.32, we have factorized Ferrers relations, and thus put them into correspondence with a strictorder. Consequently, there will exist characterizations with regard to their embeddability into $\mathbb{R}$ – or into a linear strictorder – also for weak-orders.

Proposition 12.23. *Let W be a finite relation.*

$$W \text{ weakorder} \quad \Longleftrightarrow \quad \begin{array}{l} \textit{There exists a mapping } f \textit{ and a linear} \\ \textit{strictorder } C \textit{ such that } W = f \cdot C \cdot f^{\mathsf{T}}. \end{array}$$

The constructive proof of "$\Longrightarrow$" results in f surjective.

Proof: "$\Longrightarrow$": A weakorder is a Ferrers relation, so that a linear strictorder C exists together with *two* mappings – surjective when constructed – such that $W = f \cdot C \cdot g^{\mathsf{T}}$; see Prop. 10.32. We are going to convince ourselves that row equivalence $\Xi(W) = \mathsf{syq}(W^{\mathsf{T}}, W^{\mathsf{T}})$ and column equivalence $\Psi(W) = \mathsf{syq}(W, W)$ coincide since $\overline{W} \cdot W^{\mathsf{T}} = W^{\mathsf{T}}$.

This holds as a consequence of negative transitivity and irreflexivity of W. So the corresponding natural projections are equal: $f = g$. (In addition, they are surjective by quotient construction, which is not required in the reverse direction, and satisfy $f^{\mathsf{T}} \cdot W \cdot f = C$.)

"$\Longleftarrow$": Assume a mapping f, a linear strictorder C, and consider $W := f \cdot C \cdot f^{\mathsf{T}}$.

Asymmetry propagates from C to W:

$$\begin{array}{lll} W^{\mathsf{T}} = f \cdot C^{\mathsf{T}} \cdot f^{\mathsf{T}} & \qquad & \text{definition of } W \\ \quad \subseteq f \cdot \overline{C} \cdot f^{\mathsf{T}} & & C \text{ is asymmetric} \end{array}$$

---

[7] A proof may be found, for example, as Theorem 3.1.1 in [78].

$$= \overline{f_{;} C_{;} f^{\mathsf{T}}} \qquad \text{Prop. 5.13, mapping } f \text{ slipping under negation}$$
$$= \overline{W} \qquad \text{definition of } W$$

Negatively transitive:

$$
\begin{aligned}
\overline{W}_{;}\overline{W} &= \overline{f_{;} C_{;} f^{\mathsf{T}}}_{;} \overline{f_{;} C_{;} f^{\mathsf{T}}} && \text{definition of } W \\
&= f_{;} \overline{C}_{;} f^{\mathsf{T}}_{;} f_{;} \overline{C}_{;} f^{\mathsf{T}} && \text{Prop. 5.13} \\
&\subseteq f_{;} \overline{C}_{;} \overline{C}_{;} f^{\mathsf{T}} && \text{since } f \text{ is univalent} \\
&= f_{;} \overline{C}_{;} f^{\mathsf{T}} && \text{since the linear strictorder } C \text{ is negatively transitive} \\
&= \overline{f_{;} C_{;} f^{\mathsf{T}}} && \text{Prop. 5.13} \\
&= \overline{W}
\end{aligned}
$$

It is then a weakorder according to Prop. 12.9.   □

Corollary 12.24. *The factorization of W into f, C is uniquely defined up to iso-morphism – provided that only those with f surjective are considered.*

Proof: First, we mention that in such a case

$$f_{;} f^{\mathsf{T}} = f_{;} \mathbb{I}_{;} f^{\mathsf{T}} = f_{;} (\overline{C} \cap \overline{C}^{\mathsf{T}})_{;} f^{\mathsf{T}} = \overline{f_{;} C_{;} f^{\mathsf{T}}} \cap \overline{f_{;} C^{\mathsf{T}}_{;} f^{\mathsf{T}}} = \overline{W} \cap \overline{W}^{\mathsf{T}} = \Xi(W).$$

Should a second factorization $W = f_1 {}_{;} C_1 {}_{;} f_1^{\mathsf{T}}$ be presented, there could easily be defined an isomorphism of the structure W, f, C into the structure W, f_1, C_1 with $\varphi := f^{\mathsf{T}}_{;} f_1$ (and in addition the identity on the source of W). Then, for example, φ is total and injective and the homomorphism condition is satisfied as

$$\varphi_{;} \varphi^{\mathsf{T}} = f^{\mathsf{T}}_{;} f_1 {}_{;} f_1^{\mathsf{T}}{}_{;} f = f^{\mathsf{T}}_{;} \Xi(W)_{;} f = f^{\mathsf{T}}_{;} f_{;} f^{\mathsf{T}}_{;} f = \mathbb{I}_{;} \mathbb{I} = \mathbb{I},$$
$$C_{;} \varphi = \varphi_{;} \varphi^{\mathsf{T}}_{;} C_{;} \varphi = \varphi_{;} f_1^{\mathsf{T}}{}_{;} f_{;} C_{;} f^{\mathsf{T}}_{;} f_1 = \varphi_{;} f_1^{\mathsf{T}}{}_{;} W_{;} f_1 = \varphi_{;} C_1. \qquad □$$

Assume such a weakorder realization has been found, then one has, using the Birkhoff–Milgram Theorem, finally a mapping into $\mathbb{R}$ satisfying

$$f(a) > f(b) \quad \text{if and only if} \quad (a, b) \in W.$$

Realization of a semiorder

We follow the line of weakening linear strictorders one step further and proceed from weakorders to semiorders. Also for these, there exist characterizations with regard to their embeddability into $\mathbb{R}$ via embedding first into a linear strictorder. Figure 12.17 shows a semiorder. Semi-transitivity has to be checked by testing all consecutive arrows, i.e., $(f, d, e), (b, d, e), (c, b, d), (c, a, e), (c, g, e)$. The diagram does not show a weakorder because it fails to be negatively transitive; see the triangle b, a, d with $(b, a) \notin R, (a, d) \notin R$, but $(b, d) \in R$.

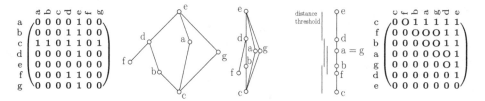

Fig. 12.17 Semiorder with Hasse diagram, squeezed, and linearized with threshold

The proposition we now aim at resembles what is referred to in the literature as the Scott–Suppes Theorem; see [131]. Traditionally it is formulated as a theorem on mapping a semiorder into the real numbers with a threshold:

Under what conditions on $R : V \longrightarrow V$ will there exist a mapping $f : V \longrightarrow \mathbb{R}$ such that

$$(x, y) \in R \iff f(x) \geq f(y) + 1.$$

Example 12.25. To find out in which way such thresholds emerge, consider a set X which has somehow been given a numeric valuation $v : X \longrightarrow \mathbb{R}$. The task is to arrange elements of X linearly according to this valuation, but to consider valuations as equal when not differing by more than some threshold number t; one does not wish to make a distinction between x and y provided $|v(x) - v(y)| < t$. We will then get a preference relation $T : X \longrightarrow X$ as well as an indifference relation $I : X \longrightarrow X$ as follows:

$$(x, y) \in T \iff v(x) > v(y) + t,$$
$$(x, y) \in I \iff |v(x) - v(y)| \leq t.$$

It turns out that then T is always a threshold in the above sense. For better reference, we take an already published example from [143]: $X = \{a, b, c, d, e, f\}$ with valuations $13, 12, 8, 5, 4, 2$, where any difference of not more than 2 shall be considered unimportant (and in which all values attached are different). Two relations will emerge as indicated in Fig. 12.18.

$$T = \begin{array}{c} a \\ b \\ c \\ d \\ e \\ f \end{array} \begin{pmatrix} 0 & 0 & 1 & 1 & 1 & 1 \\ 0 & 0 & 1 & 1 & 1 & 1 \\ 0 & 0 & 0 & 1 & 1 & 1 \\ 0 & 0 & 0 & 0 & 0 & 1 \\ 0 & 0 & 0 & 0 & 0 & 0 \\ 0 & 0 & 0 & 0 & 0 & 0 \end{pmatrix} \qquad \begin{array}{c} a \\ b \\ c \\ d \\ e \\ f \end{array} \begin{pmatrix} 0 & 1 & 0 & 0 & 0 & 0 \\ 1 & 0 & 0 & 0 & 0 & 0 \\ 0 & 0 & 0 & 0 & 0 & 0 \\ 0 & 0 & 0 & 0 & 1 & 0 \\ 0 & 0 & 0 & 1 & 0 & 1 \\ 0 & 0 & 0 & 0 & 1 & 0 \end{pmatrix} = I$$

Fig. 12.18 Preference and indifference relation stemming from a numerical valuation with threshold

One will also verify the threshold property for T. It is easily seen that the outcome

T, I is invariant when X is transformed monotonically and correspondingly t and the values of v. Therefore, assertions based on T, I alone may be considered meaningful ones. $\qquad\qquad\square$

Once this is presented, authors routinely add that every positive real number other than 1 would also do, and that the number 1 'has no meaning'. Also they stress that addition and subtraction in these formulae do not induce any particular algebraic structure. One will not spot obvious connections with our setting here. While proofs with real numbers are lengthy and usually free-hand mathematics,[8] we concentrate here on what is important, namely the strictorder aspect, and prove it point-free. So the proof may be checked – or even found – with computer help. To go from the strictorder to the reals is trivial in the finite case – in the general case it is regulated by the Birkhoff–Milgram Theorem 12.22.

Proposition 12.26 (A Scott–Suppes-type theorem). *Let R be a finite relation.*

$$R \text{ semiorder} \quad\Longleftrightarrow\quad \begin{array}{l} \textit{There exists a mapping } f, \textit{ a linear strictorder } C, \textit{ and} \\ \textit{a threshold } T \textit{ in } C, \textit{ such that } R = f \,\mathbin{;} T \,\mathbin{;} f^{\mathsf{T}}. \end{array}$$

The constructive proof of "$\Longrightarrow$" results in f surjective.

Proof: "$\Longleftarrow$" In view of Prop. 12.4, we choose to show that R is irreflexive, Ferrers, and semi-transitive. R is irreflexive since T as subrelation of C is irreflexive, $T \subseteq \overline{\mathbb{I}}$, leading to

$$R = f \,\mathbin{;} T \,\mathbin{;} f^{\mathsf{T}} \subseteq f \,\mathbin{;} \overline{\mathbb{I}} \,\mathbin{;} f^{\mathsf{T}} = \overline{f \,\mathbin{;} f^{\mathsf{T}}} \subseteq \overline{\mathbb{I}}$$

since the mapping f may slip below the negation bar and since f is total.

R is Ferrers:

$$\begin{aligned} R \,\mathbin{;} \overline{R}^{\mathsf{T}} \,\mathbin{;} R &= f \,\mathbin{;} T \,\mathbin{;} f^{\mathsf{T}} \,\mathbin{;} \overline{f \,\mathbin{;} T^{\mathsf{T}} \,\mathbin{;} f^{\mathsf{T}}} \,\mathbin{;} f \,\mathbin{;} T \,\mathbin{;} f^{\mathsf{T}} && \text{by definition and transposition} \\ &= f \,\mathbin{;} T \,\mathbin{;} f^{\mathsf{T}} \,\mathbin{;} f \,\mathbin{;} \overline{T}^{\mathsf{T}} \,\mathbin{;} f^{\mathsf{T}} \,\mathbin{;} f \,\mathbin{;} T \,\mathbin{;} f^{\mathsf{T}} && \text{mapping slips out of negation; Prop. 5.13} \\ &\subseteq f \,\mathbin{;} T \,\mathbin{;} \overline{T}^{\mathsf{T}} \,\mathbin{;} T \,\mathbin{;} f^{\mathsf{T}} && \text{univalence of } f \\ &= f \,\mathbin{;} T \,\mathbin{;} f^{\mathsf{T}} && \text{since } T \text{ is Ferrers} \\ &= R && \text{by definition} \end{aligned}$$

The proof that R is semi-transitive is nearly identical.

"$\Longrightarrow$" Negative transitivity is the key property a weakorder enjoys which a semiorder need not satisfy. So the basic idea is as follows: we enlarge the semiorder to its weakorder closure $W \supseteq R$ according to Prop. 12.18, in which R is a threshold. For this weakorder, we determine the surjective mapping f as in Prop. 12.23, so that

---

[8] Dana Scott in [131]: 'The author does not claim that the proof given here is particularly attractive.'

$W = f \,\mathbin{;} C \,\mathbin{;} f^{\mathsf{T}}$ with some linear strictorder C. Since f is surjective, we have also $C = f^{\mathsf{T}} \,\mathbin{;} W \,\mathbin{;} f$. We define $T := f^{\mathsf{T}} \,\mathbin{;} R \,\mathbin{;} f$ and show the threshold properties:

$$
\begin{aligned}
T &= f^{\mathsf{T}} \,\mathbin{;} R \,\mathbin{;} f && \text{definition of } T \\
&\subseteq f^{\mathsf{T}} \,\mathbin{;} W \,\mathbin{;} f && \text{weakorder closure } W \text{ of } R \\
&= f^{\mathsf{T}} \,\mathbin{;} f \,\mathbin{;} C \,\mathbin{;} f^{\mathsf{T}} \,\mathbin{;} f && \text{factorization of } W \\
&\subseteq C && \text{univalency}
\end{aligned}
$$

$$
\begin{aligned}
T \,\mathbin{;} C^{\mathsf{d}} &= T \,\mathbin{;} (\mathbb{I} \cup C) && \text{since } C \text{ is a linear strictorder} \\
&= T \cup T \,\mathbin{;} C && \text{distributivity of composition} \\
&= T \cup f^{\mathsf{T}} \,\mathbin{;} R \,\mathbin{;} f \,\mathbin{;} f^{\mathsf{T}} \,\mathbin{;} W \,\mathbin{;} f && \text{by definition} \\
&= T \cup f^{\mathsf{T}} \,\mathbin{;} R \,\mathbin{;} \Xi(W) \,\mathbin{;} W \,\mathbin{;} f && f \,\mathbin{;} f^{\mathsf{T}} = \Xi(W) \text{ is the row equivalence of } W \\
&\subseteq T \cup f^{\mathsf{T}} \,\mathbin{;} R \,\mathbin{;} W \,\mathbin{;} f && \text{because } \Xi(W) \,\mathbin{;} W = W \\
&\subseteq T \cup f^{\mathsf{T}} \,\mathbin{;} R \,\mathbin{;} f && R \text{ is threshold in its weakorder closure } W \\
&= T
\end{aligned}
$$

$$
\begin{aligned}
f \,\mathbin{;} T \,\mathbin{;} f^{\mathsf{T}} &= f \,\mathbin{;} f^{\mathsf{T}} \,\mathbin{;} R \,\mathbin{;} f \,\mathbin{;} f^{\mathsf{T}} && \text{definition of } T \\
&= \Xi(W) \,\mathbin{;} R \,\mathbin{;} \Xi(W) && \text{natural projection } f \\
&= (\overline{W}^{\mathsf{T}} \cap \overline{W}) \,\mathbin{;} R \,\mathbin{;} (\overline{W}^{\mathsf{T}} \cap \overline{W}) && \text{because } \overline{W} \,\mathbin{;} W^{\mathsf{T}} = W^{\mathsf{T}} \text{ for the weakorder } W \\
&\subseteq \overline{W}^{\mathsf{T}} \,\mathbin{;} R \,\mathbin{;} (\overline{W}^{\mathsf{T}} \cap \overline{W}) && \text{monotony} \\
&\subseteq R \,\mathbin{;} (\overline{W}^{\mathsf{T}} \cap \overline{W}) && \text{threshold: } \overline{W}^{\mathsf{T}} \,\mathbin{;} R \cup R \,\mathbin{;} \overline{W}^{\mathsf{T}} = R \subseteq W \\
&\subseteq R \,\mathbin{;} \overline{W}^{\mathsf{T}} && \text{monotony} \\
&\subseteq R && \text{threshold property again}
\end{aligned}
$$

In addition $\Xi(W) \supseteq \mathbb{I}$ holds trivially, so that $\ldots \supseteq R$ and, thus, $\ldots = R$. $\qquad\square$

Also in this case, we will find just one factorization according to Cor. 12.27.

Corollary 12.27. *The factorization of R into f, T, C is uniquely defined up to isomorphism – provided that only those with f surjective are considered.*

Proof: Let W be the weakorder closure of R. First, we prove

$$
f \,\mathbin{;} f^{\mathsf{T}} = f \,\mathbin{;} \mathbb{I} \,\mathbin{;} f^{\mathsf{T}} = f \,\mathbin{;} (\overline{C} \cap \overline{C}^{\mathsf{T}}) \,\mathbin{;} f^{\mathsf{T}} = \overline{f \,\mathbin{;} C \,\mathbin{;} f^{\mathsf{T}}} \cap \overline{f \,\mathbin{;} C \,\mathbin{;} f^{\mathsf{T}}}^{\mathsf{T}} = \overline{W} \cap \overline{W}^{\mathsf{T}} = \Xi(W)
$$

and remember that by construction

$$
f \,\mathbin{;} C \,\mathbin{;} f^{\mathsf{T}} = W.
$$

Now assume a second factorization $R = f_1 \,\mathbin{;} T_1 \,\mathbin{;} f_1^{\mathsf{T}}$ with T_1 threshold in the linear strictorder C_1 to be presented. With $\varphi := f^{\mathsf{T}} \,\mathbin{;} f_1$ (and in addition the identity on R), we define an isomorphism of the structure determined by R, f, T, C into the structure R, f_1, T_1, C_1; see Fig. 12.19.

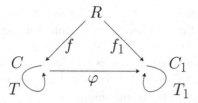

Fig. 12.19 Standard situation when studying uniqueness

$$\varphi \, \varphi^\mathsf{T} = f^\mathsf{T} \, f_1 \, f_1^\mathsf{T} \, f = f^\mathsf{T} \, \Xi(W) \, f = f^\mathsf{T} \, f \, f^\mathsf{T} \, f = \mathbb{I} \, \mathbb{I} = \mathbb{I}$$
$$C \, \varphi = \varphi \, \varphi^\mathsf{T} \, C \, \varphi = \varphi \, f_1^\mathsf{T} \, f \, C \, f^\mathsf{T} \, f_1 = \varphi \, f_1^\mathsf{T} \, W \, f_1 = \varphi \, f_1^\mathsf{T} \, f_1 \, C_1 \, f_1^\mathsf{T} \, f_1 = \varphi \, C_1$$
$$T \, \varphi = \varphi \, \varphi^\mathsf{T} \, T \, \varphi = \varphi \, f_1^\mathsf{T} \, f \, T \, f^\mathsf{T} \, f_1 = \varphi \, f_1^\mathsf{T} \, R \, f_1 = \varphi \, f_1^\mathsf{T} \, f_1 \, T_1 \, f_1^\mathsf{T} \, f_1 = \varphi \, T_1 \qquad \square$$

With Fig. 12.20, and Fig. 12.21, we present two more semiorders in original form as well as threshold-ordered form via simultaneous permutations.

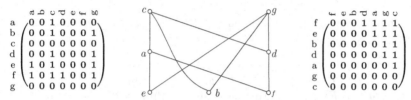

Fig. 12.20 Another semiorder

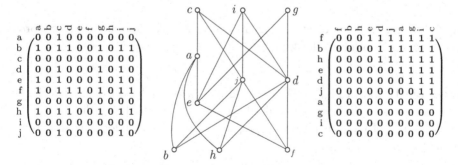

Fig. 12.21 Yet another semiorder with simultaneous rearrangement

Realization of an intervalorder

We now generalize from semiorders to intervalorders. Intervalorders also allow factorization. They have, however, their own highly elaborated theory that should also be presented so as to understand better the practical background of several examples. The purely technical part of intervalorder factorization will now be shown and it will be accepted that further theory is postponed to Section 12.6.

Although intervalorders are homogeneous relations, one will find that when factorizing them, one will have to proceed as for heterogeneous relations. This means that rows and columns require different mappings.

Proposition 12.28. *Let R be a finite relation.*

$$R \text{ intervalorder} \iff \begin{array}{l} \textit{There exist mappings } f, g \text{ and a linear strictorder} \\ C \text{ such that } R = f \,{;}\, C \,{;}\, g^{\mathsf{T}} \text{ and } g^{\mathsf{T}} \,{;}\, f \subseteq E = \mathbb{I} \cup C. \end{array}$$

The constructive proof of "$\Longrightarrow$" results in f, g surjective.

Proof: An intervalorder is by definition in particular a Ferrers relation, so that Prop. 10.32 may be applied: R is Ferrers precisely when a linear strictorder C exists together with two mappings such that $R = f \,{;}\, C \,{;}\, g^{\mathsf{T}}$.

This being established, the additional fact that R is irreflexive is now shown to be equivalent to $g^{\mathsf{T}} \,{;}\, f \subseteq E$:

$$
\begin{aligned}
& R = f \,{;}\, C \,{;}\, g^{\mathsf{T}} \subseteq \overline{\mathbb{I}} \\
\iff\ & \mathbb{I} \subseteq \overline{f \,{;}\, C \,{;}\, g^{\mathsf{T}}} = f \,{;}\, \overline{C} \,{;}\, g^{\mathsf{T}} = f \,{;}\, E^{\mathsf{T}} \,{;}\, g^{\mathsf{T}} \qquad \text{see Prop. 5.13 and Prop. 5.24.i} \\
\iff\ & \mathbb{I} \,{;}\, g \subseteq f \,{;}\, E^{\mathsf{T}} \qquad \text{shunting according to Prop. 5.12} \\
\iff\ & f^{\mathsf{T}} \,{;}\, g \subseteq E^{\mathsf{T}} \qquad \text{shunting according to Prop. 5.12} \\
\iff\ & g^{\mathsf{T}} \,{;}\, f \subseteq E \qquad \text{transposed} \qquad\qquad \square
\end{aligned}
$$

Already here, i.e., before entering into more theoretical considerations, we present two realizations of intervalorders.

Example 12.29. In Fig. 12.22, we see an example graph borrowed from the book [100]. In the first line, this intervalorder is presented with Hasse diagram and two matrices, one of which is already arranged in upper triangle form by simultaneous permutation.

In the second line, another representation is given. The intervals in a linearly ordered 5-element set are shown and then a rearrangement with independent permutations of rows and columns to staircase form, i.e., Ferrers form. The mappings f (right border) and g (left border) are also shown. Observe that their source orderings are different. $\qquad\qquad \square$

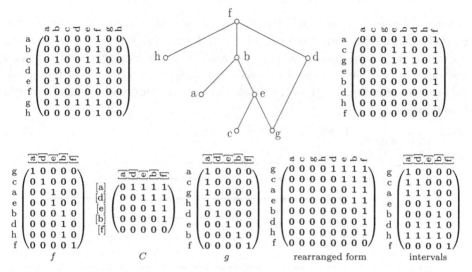

Fig. 12.22 Hasse diagram of an intervalorder together with different representations

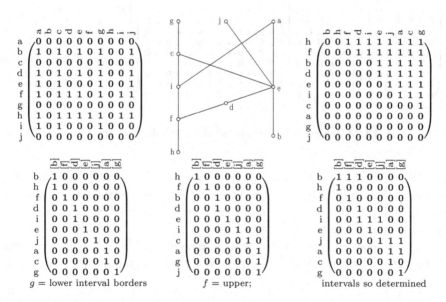

Fig. 12.23 Another intervalorder; observe different row and column arrangements

12.6 Intervalorders and interval graphs

Assume an assessment activity requiring a manager to position his employees as to their abilities on a linearly ordered scale. However, one is not interested in a linear ranking of the personnel; so in the assessment the manager need not enter just one position; he is allowed to indicate a range to which he feels the person belongs. Assume Fig. 12.24 as the result of this procedure with a linear order of qualifications $1, \ldots, 6$.

13-assessments:

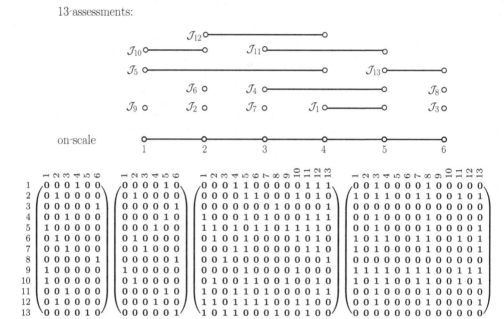

The four matrices below the diagram are labelled f, g, Γ, P with rows 1–13.

Matrix f (columns 1–6):

	1	2	3	4	5	6
1	0	0	0	1	0	0
2	0	1	0	0	0	0
3	0	0	0	0	0	1
4	0	0	1	0	0	0
5	1	0	0	0	0	0
6	0	1	0	0	0	0
7	0	0	1	0	0	0
8	0	0	0	0	0	1
9	1	0	0	0	0	0
10	1	0	0	0	0	0
11	0	0	1	0	0	0
12	0	1	0	0	0	0
13	0	0	0	0	1	0

Matrix g (columns 1–6):

	1	2	3	4	5	6
1	0	0	0	0	1	0
2	0	1	0	0	0	0
3	0	0	0	0	0	1
4	0	0	0	0	1	0
5	0	0	0	1	0	0
6	0	1	0	0	0	0
7	0	0	1	0	0	0
8	0	0	0	0	0	1
9	1	0	0	0	0	0
10	0	1	0	0	0	0
11	0	0	0	0	1	0
12	0	0	0	1	0	0
13	0	0	0	0	0	1

Matrix Γ (columns 1–13):

	1	2	3	4	5	6	7	8	9	10	11	12	13
1	0	0	0	1	1	0	0	0	0	0	1	1	1
2	0	0	0	0	1	1	0	0	0	1	0	1	0
3	0	0	0	0	0	0	0	1	0	0	0	0	1
4	1	0	0	0	1	0	1	0	0	0	1	1	1
5	1	1	0	1	0	1	1	0	1	1	1	1	0
6	0	1	0	0	1	0	0	0	0	1	0	1	0
7	0	0	0	1	1	0	0	0	0	0	1	1	0
8	0	0	1	0	0	0	0	0	0	0	0	0	1
9	0	0	0	0	1	0	0	0	0	1	0	0	0
10	0	1	0	0	1	1	0	0	1	0	0	1	0
11	1	0	0	1	1	0	1	0	0	0	0	1	1
12	1	1	0	1	1	1	1	0	0	1	1	0	0
13	1	0	1	1	0	0	0	1	0	0	1	0	0

Matrix P (columns 1–13):

	1	2	3	4	5	6	7	8	9	10	11	12	13
1	0	0	1	0	0	0	0	1	0	0	0	0	0
2	1	0	1	1	0	0	1	1	0	0	1	0	1
3	0	0	0	0	0	0	0	0	0	0	0	0	0
4	0	0	1	0	0	0	0	1	0	0	0	0	0
5	0	0	1	0	0	0	0	1	0	0	0	0	1
6	1	0	1	1	0	0	1	1	0	0	1	0	1
7	1	0	1	0	0	0	0	1	0	0	0	0	1
8	0	0	0	0	0	0	0	0	0	0	0	0	0
9	1	1	1	1	0	1	1	1	0	0	1	1	1
10	1	0	1	1	0	0	1	1	0	0	1	0	1
11	0	0	1	0	0	0	0	1	0	0	0	0	0
12	0	0	1	0	0	0	0	1	0	0	0	0	1
13	0	0	0	0	0	0	0	0	0	0	0	0	0

Fig. 12.24 Interval assessment and the intervalorder 'strictly before'

How does one come from the set of intervals attributed to persons to an intuitively understandable overall ranking of the employees? The intervals may contain one another, may just intersect, touch, or may be disjoint. When working on the real axis, we assume for simplicity always left-open and right-closed intervals. Such an assessment is, however, not detailed enough to require the use of a continuous scale. A better idea is to work with a discrete order where no topological problems will arise.

To extract a preference structure out of this setting is rather immediate. One can start from the set of intervals $\left(\mathcal{J}_i\right)_{i \in I}$ on the linearly ordered set and define a preference relation and an indifference relation as follows.

- Person i is preferred to person j if interval $\mathcal{J}_i$ is situated completely on the right of interval $\mathcal{J}_j$.
- Assessment of person i compared with person j is indifferent if intervals $\mathcal{J}_i$ and $\mathcal{J}_j$ intersect.

These ideas are now taken up in an appropriate relation-algebraic form for which we assume the following definition.

Definition 12.30. (i) A simple graph (X, Γ) is called a **represented interval graph** if it is given via a triple (f, g, E) consisting of a linear order $E : Y \longrightarrow Y$ on some set Y, with corresponding linear strictorder C, and mappings $f : X \longrightarrow Y$ and $g : X \longrightarrow Y$ satisfying $f^{\mathsf{T}} ; g \subseteq E$. Then we call

$\quad B := g ; E^{\mathsf{T}} ; f^{\mathsf{T}} \cap f ; E ; g^{\mathsf{T}}$ **intersection relation** for the intervals

$\quad \Gamma := g ; E^{\mathsf{T}} ; f^{\mathsf{T}} \cap f ; E ; g^{\mathsf{T}} \cap \overline{\mathbb{I}} = B \cap \overline{\mathbb{I}}$ **adjacency** of the interval graph

$\quad P := g ; C ; f^{\mathsf{T}}$ **'before'-strictorder** of the intervals

(ii) An **interval graph** is a simple graph (X, Γ), for which *there exist* relations (f, g, E), such that it becomes a represented interval graph. □

In Fig. 12.27 and Fig. 12.25, one will see how Def. 12.30 is supposed to be read: demanding that the mappings satisfy $f^{\mathsf{T}} ; g \subseteq E$, guarantees that to every vertex a non-empty interval according to E is assigned in the form that f, g indicate the left, respectively right, end of the interval considered. The intersection relation B expresses that, given two vertices of X, the two intervals assigned intersect, because two relationships hold together for the two vertices. One may look from the first to its right interval end and consider whatever interval starts to the left of it, but one may also go to the left end of the interval attached and consider whatever interval ends to the right of it. The adjacency Γ is nothing other than the irreflexive version thereof.

A slight variant of Def. 12.30 starts from a set of sets, represented row-wise as a relation $R : X \longrightarrow Y$ with Y ordered linearly by $E : Y \longrightarrow Y$. Then the construct $R\,R^{\mathsf{T}}$ (the **(row)intersection relation**) or, usually the irreflexive version $\overline{\mathbb{I}} \cap R\,R^{\mathsf{T}}$, is often called the **(row)intersection graph** of R. It corresponds to the edge-adjacency $K = \overline{\mathbb{I}} \cap R ; R^{\mathsf{T}}$ studied in some detail in [123, 124].

Interval graphs are frequently studied. What makes the concept interesting is the connection with intervalorders often used in operations research. Among the favorable properties of interval graphs is that important algorithms work on them in linear time; see [60].

Proposition 12.31. *The 'before'-strictorder P of a represented interval graph is an intervalorder.*

Proof: P is irreflexive according to the equivalence shown in the proof of Prop. 12.28. It is also transitive since $f^\mathsf{T}; g \subseteq E$ and C is a linear strictorder:

$$P; P = g; C; f^\mathsf{T}; g; C; f^\mathsf{T} \subseteq g; C; E; C; f^\mathsf{T} \subseteq g; C; f^\mathsf{T} = P.$$

P is Ferrers, because the linear strictorder C is Ferrers in

$$P; \overline{P}^\mathsf{T}; P = g; C; f^\mathsf{T}; \overline{g; C; f^\mathsf{T}}^\mathsf{T}; g; C; f^\mathsf{T} = g; C; f^\mathsf{T}; \overline{f; C^\mathsf{T}; g^\mathsf{T}}; g; C; f^\mathsf{T}$$
$$= g; C; f^\mathsf{T}; f; \overline{C}^\mathsf{T}; g^\mathsf{T}; g; C; f^\mathsf{T} \subseteq g; C; \overline{C}^\mathsf{T}; C; f^\mathsf{T} \subseteq g; C; f^\mathsf{T} = P.$$

Therefore, P as a Ferrers strictorder is by definition an intervalorder. □

We have proved with Prop. 12.28, that there will always exist a linear strictorder $C : Y' \longrightarrow Y'$ together with the two mappings $f_{\text{fact}}, g_{\text{fact}} : X \longrightarrow Y'$ such that for this intervalorder $P = f_{\text{fact}}; C; g_{\text{fact}}^\mathsf{T}$. Regrettably, these $f_{\text{fact}}, g_{\text{fact}}$ in Prop. 12.28 are different mappings – notationally too similar and in addition exchanged – than f, g in Def. 12.30. This is obviously incidental. The $f_{\text{fact}}, g_{\text{fact}}$ of the factorization do not point to Y but to a set Y' of equivalence classes of X as may be seen in Fig. 12.25.

Now we consider for every element $a \in X$ the pair $(g_{\text{fact}}(a), f_{\text{fact}}(a))$, which due to $g_{\text{fact}}^\mathsf{T}; f_{\text{fact}} \subseteq E$ describes a (possibly 1-point) interval in the linearly ordered set Y'. Whenever elements a, b are in relation P, their intervals $(g_{\text{fact}}(a), f_{\text{fact}}(a))$ and $(g_{\text{fact}}(b), f_{\text{fact}}(b))$ will following $(a, b) \in P = f_{\text{fact}}; C; g_{\text{fact}}^\mathsf{T}$ satisfy $\big(f_{\text{fact}}(a), g_{\text{fact}}(b)\big) \in C$, letting the interval for a reside strictly below, respectively on the left of, the interval for b. The example of Fig. 12.25 – already earlier studied as Fig. 12.11 – will make this clear.

	5	2	4	1	13	3
9	1	0	0	0	0	0
2	0	1	0	0	0	0
6	0	1	0	0	0	0
10	0	1	0	0	0	0
7	0	0	1	0	0	0
5	0	0	0	1	0	0
12	0	0	0	1	0	0
1	0	0	0	0	1	0
4	0	0	0	0	1	0
11	0	0	0	0	1	0
3	0	0	0	0	0	1
8	0	0	0	0	0	1
13	0	0	0	0	0	1

	5	2	4	1	13	3
5	1	0	0	0	0	0
9	1	0	0	0	0	0
10	1	0	0	0	0	0
2	0	1	0	0	0	0
6	0	1	0	0	0	0
12	0	1	0	0	0	0
4	0	0	1	0	0	0
7	0	0	1	0	0	0
11	0	0	1	0	0	0
1	0	0	0	1	0	0
13	0	0	0	0	1	0
3	0	0	0	0	0	1
8	0	0	0	0	0	1

	5	9	10	2	6	12	4	7	11	1	13	3	8
9	0	0	1	1	1	1	1	1	1	1	1	1	1
2	0	0	0	0	0	1	1	1	1	1	1	1	1
6	0	0	0	0	0	1	1	1	1	1	1	1	1
10	0	0	0	0	0	1	1	1	1	1	1	1	1
7	0	0	0	0	0	0	0	0	1	1	1	1	1
5	0	0	0	0	0	0	0	0	0	1	1	1	1
12	0	0	0	0	0	0	0	0	0	0	1	1	1
1	0	0	0	0	0	0	0	0	0	0	0	1	1
4	0	0	0	0	0	0	0	0	0	0	0	0	0
11	0	0	0	0	0	0	0	0	0	0	0	0	0
3	0	0	0	0	0	0	0	0	0	0	0	0	0
8	0	0	0	0	0	0	0	0	0	0	0	0	0
13	0	0	0	0	0	0	0	0	0	0	0	0	0

$f_{\text{fact}}, g_{\text{fact}}, P$ with rows and columns arranged differently to show the Ferrers aspect

Fig. 12.25 Intervalorder as earlier in Fig. 12.11 factorized with $f_{\text{fact}}, g_{\text{fact}}$

While one might say that the row sequence 3,8,13 and the column sequence 13,3,8 should be unified to 13,3,8, it will immediately become clear that a similar procedure cannot be executed for the column sequence 5,9,10 and the row sequence 9,2,6,.... What must be stressed is that being Ferrers is an essentially heterogeneous property, i.e., normally induces different permutations for rows and columns.

This is much more complicated than for semiorders. For a comparison, we consider Fig. 12.11 (the same as Fig. 12.25). It has a semiorder extension shown in Fig. 12.12. We give this the threshold form of Fig. 12.26 by *simultaneous* permutation. The seven extension entries, dashed in Fig. 12.12, are marked by fat $\mathbf{1}$ s.

$$
\begin{array}{r}
\\
9 \\ 10 \\ 2 \\ 6 \\ 7 \\ 5 \\ 12 \\ 4 \\ 11 \\ 1 \\ 13 \\ 3 \\ 8
\end{array}
\begin{array}{c}
\begin{smallmatrix} 9 & 10 & 2 & 6 & 7 & 5 & 12 & 4 & 11 & 1 & 13 & 3 & 8 \end{smallmatrix} \\
\left(\begin{array}{ccccccccccccc}
0 & 0 & 1 & 1 & 1 & \mathbf{1} & 1 & 1 & 1 & 1 & 1 & 1 & 1 \\
0 & 0 & 0 & 0 & 1 & 1 & 1 & 1 & 1 & 1 & 1 & 1 & 1 \\
0 & 0 & 0 & 0 & 1 & 1 & 1 & 1 & 1 & 1 & 1 & 1 & 1 \\
0 & 0 & 0 & 0 & 1 & 1 & 1 & 1 & 1 & 1 & 1 & 1 & 1 \\
0 & 0 & 0 & 0 & 0 & 0 & 0 & 0 & 0 & 1 & 1 & 1 & 1 \\
0 & 0 & 0 & 0 & 0 & 0 & 0 & 0 & 0 & 0 & 1 & 1 & 1 \\
0 & 0 & 0 & 0 & 0 & 0 & 0 & 0 & 0 & 0 & 1 & 1 & 1 \\
0 & 0 & 0 & 0 & 0 & 0 & 0 & 0 & 0 & 0 & 0 & 1 & 1 \\
0 & 0 & 0 & 0 & 0 & 0 & 0 & 0 & 0 & 0 & 0 & 1 & 1 \\
0 & 0 & 0 & 0 & 0 & 0 & 0 & 0 & 0 & 0 & 0 & 1 & 1 \\
0 & 0 & 0 & 0 & 0 & 0 & 0 & 0 & 0 & 0 & 0 & 0 & 0 \\
0 & 0 & 0 & 0 & 0 & 0 & 0 & 0 & 0 & 0 & 0 & 0 & 0 \\
0 & 0 & 0 & 0 & 0 & 0 & 0 & 0 & 0 & 0 & 0 & 0 & 0
\end{array}\right)
\end{array}
$$

Fig. 12.26 A semiorder extension of Fig. 12.11 permuted simultaneously

Interval graphs and semiorder

It will on some occasions be important to know whether in the given set of intervals there exists one interval strictly contained in another – or not.

Proposition 12.32. *For a represented interval graph with the added property that no interval is strictly contained in another one, P will turn out to be a semiorder; in formulae:*

$$ g\,;C\,;g^{\mathsf{T}} = f\,;C\,;f^{\mathsf{T}} \qquad \Longrightarrow \qquad P := g\,;C\,;f^{\mathsf{T}} \text{ is a semiorder.} $$

Proof: We prove semi-transitivity:

$$
\begin{aligned}
P\,;P\,;\overline{P}^{\mathsf{T}} &= g\,;C\,;f^{\mathsf{T}}\,;g\,;C\,;f^{\mathsf{T}}\,;\overline{f\,;C^{\mathsf{T}}\,;g^{\mathsf{T}}} && \text{expanded and transposed} \\
&= g\,;C\,;f^{\mathsf{T}}\,;g\,;C\,;f^{\mathsf{T}}\,;f\,;\overline{C}^{\mathsf{T}}\,;g^{\mathsf{T}} && \text{mapping slipping out of negation} \\
&\subseteq g\,;C\,;f^{\mathsf{T}}\,;g\,;C\,;\overline{C}^{\mathsf{T}}\,;g^{\mathsf{T}} && f \text{ is univalent} \\
&= g\,;C\,;f^{\mathsf{T}}\,;g\,;C\,;g^{\mathsf{T}} && C \text{ is a linear strictorder} \\
&= g\,;C\,;f^{\mathsf{T}}\,;f\,;C\,;f^{\mathsf{T}} && \text{using the assumption} \\
&\subseteq g\,;C\,;C\,;f^{\mathsf{T}} && f \text{ is univalent} \\
&\subseteq g\,;C\,;f^{\mathsf{T}} && C \text{ is transitive} \\
&= P && \text{definition of } P
\end{aligned}
$$

This completes the proof, since we know already that P is an intervalorder.   □

We are going to interpret one half $g ; C ; g^\mathsf{T} \subseteq f ; C ; f^\mathsf{T}$ of the assumption we have used: with $g ; C ; g^\mathsf{T}$, we go from an interval to all those ending strictly behind and require with $f ; C ; f^\mathsf{T} = f ; \overline{E}^\mathsf{T} ; f^\mathsf{T} - E$ is a linear order! – that these must not start strictly before or at the same position. We have, thus, given preference to the strict overlapping on the right. In the same way, we must prevent strict overlapping on the left with $f ; C^\mathsf{T} ; f^\mathsf{T} \subseteq g ; C^\mathsf{T} ; g^\mathsf{T}$. Both together result in the equality used as an assumption of the proposition.

In the literature, the relational condition for a semiorder has had other versions, not least that all intervals considered on the real axis have unit length. Of course, unit length intervals cannot be strictly included in one another. But this is not a healthy condition. The scientific texts then usually mention that the number 1 as length 'has no meaning'. This cryptic remark is nothing other than saying that this is a purely relational condition, but due to lack of other means it is expressed in real numbers.

Comparability and transitive orientation

We start this investigation with the example of Fig. 12.27.

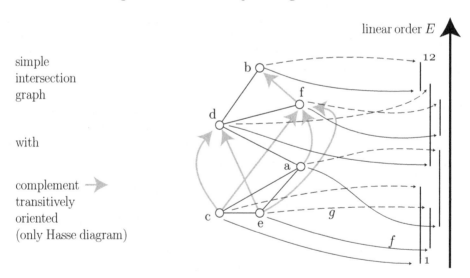

Fig. 12.27 Interval graph with transitively oriented complement

In Fig. 12.28, precisely the same as in Fig. 12.27 is expressed in terms of mappings and relations. The endpoints of the intervals may well be real numbers. Here, for simplicity, they are numbered 1 to 12 from bottom to top – with 2 to 11 hidden in order not to overcrowd the figure. The mapping f points to the lower bound, and g to the upper.

$$f = \begin{array}{cc} & \begin{array}{cccccccccccc}1&2&3&4&5&6&7&8&9&10&11&12\end{array} \\ \begin{array}{c}a\\b\\c\\d\\e\\f\end{array} & \begin{pmatrix}0&0&1&0&0&0&0&0&0&0&0&0\\0&0&0&0&0&0&0&0&0&1&0&0\\1&0&0&0&0&0&0&0&0&0&0&0\\0&0&0&0&0&1&0&0&0&0&0&0\\0&1&0&0&0&0&0&0&0&0&0&0\\0&0&0&0&0&0&0&1&0&0&0&0\end{pmatrix}\end{array} \qquad g = \begin{array}{cc} & \begin{array}{cccccccccccc}1&2&3&4&5&6&7&8&9&10&11&12\end{array} \\ \begin{array}{c}a\\b\\c\\d\\e\\f\end{array} & \begin{pmatrix}0&0&0&0&0&0&1&0&0&0&0&0\\0&0&0&0&0&0&0&0&0&0&0&1\\0&0&0&0&1&0&0&0&0&0&0&0\\0&0&0&0&0&0&0&0&0&0&1&0\\0&0&0&1&0&0&0&0&0&0&0&0\\0&0&0&0&0&0&0&0&1&0&0&0\end{pmatrix}\end{array}$$

$$\begin{array}{cc} & \begin{array}{cccccccccccc}1&2&3&4&5&6&7&8&9&10&11&12\end{array} \\ \begin{array}{c}a\\b\\c\\d\\e\\f\end{array} & \begin{pmatrix}0&0&1&1&1&1&1&0&0&0&0&0\\0&0&0&0&0&0&0&0&0&1&1&1\\1&1&1&1&1&0&0&0&0&0&0&0\\0&0&0&0&0&1&1&1&1&1&1&0\\0&1&1&1&0&0&0&0&0&0&0&0\\0&0&0&0&0&0&0&1&1&0&0&0\end{pmatrix}\end{array} \quad \begin{array}{cc} & \begin{array}{cccccc}a&b&c&d&e&f\end{array} \\ \begin{array}{c}a\\b\\c\\d\\e\\f\end{array} & \begin{pmatrix}1&0&1&1&1&0\\0&1&0&1&0&0\\1&0&1&0&1&0\\1&1&0&1&0&1\\1&0&1&0&1&0\\0&0&0&1&0&1\end{pmatrix}\end{array} \quad \begin{array}{cc} & \begin{array}{cccccc}a&b&c&d&e&f\end{array} \\ \begin{array}{c}a\\b\\c\\d\\e\\f\end{array} & \begin{pmatrix}0&1&0&0&0&1\\0&0&0&0&0&0\\0&1&0&1&0&1\\0&0&0&0&0&0\\0&1&0&1&0&1\\0&1&0&0&0&0\end{pmatrix}\end{array} \quad \begin{array}{cc} & \begin{array}{cccccc}a&b&c&d&e&f\end{array} \\ \begin{array}{r}\{a,d\}\to\\\{b,d\}\to\\\{a,c,e\}\to\\\{d,f\}\to\end{array} & \begin{pmatrix}1&0&0&1&0&0\\0&1&0&1&0&0\\1&0&1&0&1&0\\0&0&0&1&0&1\end{pmatrix}\end{array}$$

interval representation $\qquad B = \mathbb{I} \cup \Gamma \qquad P \qquad$ maxclique relation M

Fig. 12.28 Interval representation in the linearly ordered set of Fig. 12.27 by relations

The maxclique relation of the intersection relation B in Fig. 12.28 will be used only later. The complement $\overline{B}$ of B enjoys the special property[9] of being 'transitively orientable', i.e., being partitioned into a strictorder P and its converse: $\overline{B} = P \cup P^{\mathsf{T}}$.

The definition of transitive orientability does not require the interval graph to be a represented one.

Definition 12.33. Consider any symmetric and reflexive (and thus homogeneous) relation B together with the simple graph (X, Γ), where $\Gamma := \overline{\mathbb{I}} \cap B$. We say that B, respectively the simple graph (X, Γ), has the **transitive orientation property** if there exists a strictorder P such that $\overline{B} = P \cup P^{\mathsf{T}}$. The graph $(X, \overline{B})$ is then called a **comparability graph.** □

The investigations on transitive orientability we now aim at originate from a famous theorem of 1962 by Alain Ghouilà-Houri.

Proposition 12.34 (Ghouilà-Houri [58]). *The complement of an interval graph is transitively orientable.* □

This result is first studied in the following more detailed and restrictive setting of a

[9] In the literature, this usually reads 'The complement of the interval graph (X, Γ)'. This, however, requires the complement forming for a simple graph Γ to mean $\overline{\mathbb{I}} \cap \overline{\Gamma}$, where making this irreflexive again after complementation is tacitly understood.

represented interval graph and may be considered to be a corollary of the following proposition.

Proposition 12.35. *Assume the setting of Def. 12.30. Then P is a transitive orientation of the complement $\overline{B}$; i.e.,*

$$P \cap P^{\mathsf{T}} = \mathbb{L}, \quad P \cup P^{\mathsf{T}} = \overline{B}.$$

Proof: From Prop. 12.31, we know that P is an intervalorder. It is a transitive orientation because $\overline{B} = P \cup P^{\mathsf{T}}$:

$$P \cup P^{\mathsf{T}} = g_{;}C_{;}f^{\mathsf{T}} \cup f_{;}C^{\mathsf{T}}_{;}g^{\mathsf{T}} = g_{;}\overline{E}^{\mathsf{T}}_{;}f^{\mathsf{T}} \cup f_{;}\overline{E}_{;}g^{\mathsf{T}} \qquad \text{since } E \text{ is a linear order}$$
$$= \overline{g_{;}E^{\mathsf{T}}_{;}f^{\mathsf{T}}} \cup \overline{f_{;}E_{;}g^{\mathsf{T}}} = \overline{B} \qquad \text{mappings } f, g \text{ may slip below a negation}$$

Disjointness $P \cap P^{\mathsf{T}} = \mathbb{L}$ follows directly from transitivity and irreflexivity. $\square$

It was not too hard to find a transitive orientation if the interval graph is explicitly represented as (f, g, E). The definition of an interval graph, however, only assures that such a representation *exists*. It may, thus, be possible that an interval graph Γ is given, but a representation (f, g, E) has not yet been communicated. Then one will easily find $B := \mathbb{I} \cup \Gamma$, but may have no idea how to obtain P.

This raises the question as to whether a transitive orientation P may also be found when a representation is not given, i.e., is it possible to express P simply in terms of B or Γ. We will give hints how to try to obtain an orientation; of course not in a uniquely determined way. Already in Def. 12.30, we might have taken $P_1 := g_{;}C^{\mathsf{T}}_{;}f^{\mathsf{T}}$ as well, providing us with a different transitive orientation, namely the reverse one.

At this point, the maxclique relation of the intersection relation, shown already in Fig. 12.28, comes into play. We recall that every symmetric and reflexive relation B may be factorized into its maxclique relation as $B = M^{\mathsf{T}}_{;}M$; see Prop. 11.8. The fascinating idea is to relate somehow the selection of a transitive orientation P with imposing a linear strictorder C_Δ on the set of maxcliques. In the case when either one of P and C_Δ enjoys an additional property, the two are very closely related with one another. We will exhibit this first in a rather technical form and later give the classical interpretations as they emerged historically during the process of developing the theory.

Definition 12.36. We consider a maxclique relation M of an adjacency Γ (or its reflexive version $B := \mathbb{I} \cup \Gamma$). Assuming a linear strictorder C_Δ to exist on these cliques, we attribute to

M the **consecutive 1s property**, provided $C_\Delta {}_{;} M \cap C^{\mathsf{T}}_\Delta {}_{;} M \subseteq M$. $\square$

The intention of this definition may easily be understood as follows. Start from a maxclique and proceed in two ways, considering either the vertices involved in a maxclique C_Δ-strictly above or below. Then one will at most find those vertices involved in the maxclique we had been starting from. Starting from the clique $\{a, d\}$ in Fig. 12.29, one will find vertices b, d, f in cliques strictly above and a, c, e in those strictly below, getting an empty intersection. However, beginning in $\{d, f\}$, one will find b, d strictly above and a, c, d, e strictly below with intersection just d which is part of the original maxclique.

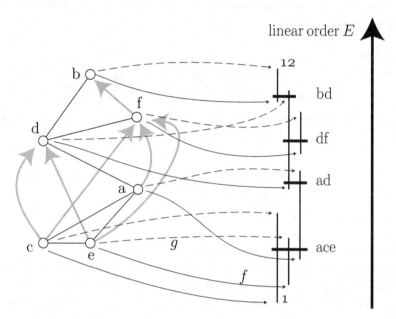

Fig. 12.29 Interval graph of Fig. 12.27 with maxclique shown along the linear order E

Now a close correspondence between transitive orientations $\overline{B} = P \cup P^\mathsf{T}$ and the strictorder C_Δ will be exhibited.

Proposition 12.37. *Let an intersection relation* B, *or an adjacency* $\Gamma := \overline{\mathbb{I}} \cap B$, *be given and consider its factorization* $B = M^\mathsf{T} \,\fatsemi\, M$ *with a maxclique relation* M.

(i) *For every transitive orientation* $\overline{B} = P \cup P^\mathsf{T}$ *with* $P \,\fatsemi\, B \,\fatsemi\, P \subseteq P$, *the construct* $C_\Delta := M \,\fatsemi\, P \,\fatsemi\, M^\mathsf{T}$ *is a linear strictorder that lets* M *have the consecutive* $\mathbb{1}s$ *property.*

(ii) *For every* M *with the consecutive* $\mathbb{1}s$ *property via the strictorder* C_Δ *on the set of maxcliques, the construct* $P := M^\mathsf{T}\,\fatsemi\,C_\Delta\,\fatsemi\,M \cap \overline{B}$ *is a transitive orientation satisfying* $P \,\fatsemi\, B \,\fatsemi\, P \subseteq P$.

(iii) *In either case,* P *is necessarily Ferrers.*

Proof: (i) C_Δ is transitive:

$$\begin{aligned}
C_\Delta \mathbin{;} C_\Delta &= M \mathbin{;} P \mathbin{;} M^\mathsf{T} \mathbin{;} M \mathbin{;} P \mathbin{;} M^\mathsf{T} && \text{by definition} \\
&= M \mathbin{;} P \mathbin{;} B \mathbin{;} P \mathbin{;} M^\mathsf{T} && \text{because } M \text{ factorizes } B \\
&\subseteq M \mathbin{;} P \mathbin{;} M^\mathsf{T} = C_\Delta && \text{using the assumption}
\end{aligned}$$

C_Δ is irreflexive and semi-connex:

$$\begin{aligned}
C_\Delta \cup C_\Delta^\mathsf{T} &= M \mathbin{;} P \mathbin{;} M^\mathsf{T} \cup M \mathbin{;} P^\mathsf{T} \mathbin{;} M^\mathsf{T} \\
&= M \mathbin{;} (P \cup P^\mathsf{T}) \mathbin{;} M^\mathsf{T} = M \mathbin{;} \overline{B} \mathbin{;} M^\mathsf{T} = M \mathbin{;} \overline{M}^\mathsf{T} = \overline{\mathbb{I}}
\end{aligned}$$

The very last equality, in transposed form, follows from the maxclique factorization with Prop. 11.8.

We have also to show the consecutive $\mathbf{1}$s property. First, we have

$$\begin{aligned}
C_\Delta \mathbin{;} M \cap C_\Delta^\mathsf{T} \mathbin{;} M &= M \mathbin{;} P \mathbin{;} M^\mathsf{T} \mathbin{;} M \cap M \mathbin{;} P^\mathsf{T} \mathbin{;} M^\mathsf{T} \mathbin{;} M = M \mathbin{;} P \mathbin{;} B \cap M \mathbin{;} P^\mathsf{T} \mathbin{;} B \\
&\subseteq \overline{M \mathbin{;} P^\mathsf{T}} \cap \overline{M \mathbin{;} P} = \overline{M \mathbin{;} (P \cup P^\mathsf{T})} = \overline{M \mathbin{;} \overline{B}} = \overline{\overline{M}} = M.
\end{aligned}$$

It remains to prove the intermediate steps using the assumption as, for example,

$$P \mathbin{;} M^\mathsf{T} \mathbin{;} M \mathbin{;} P = P \mathbin{;} B \mathbin{;} P \subseteq P \subseteq \overline{B} \quad \Longrightarrow \quad M \mathbin{;} P^\mathsf{T} \mathbin{;} B \subseteq \overline{M \mathbin{;} P}.$$

(ii) Whenever such a C_Δ has been given, we will obtain disjoint partitions as

$$\mathbb{T} = \left(C_\Delta \mathbin{;} M \cap \overline{M}\right) \cup M \cup \left(C_\Delta^\mathsf{T} \mathbin{;} M \cap \overline{M}\right) = M \mathbin{;} P \cup M \cup M \mathbin{;} P^\mathsf{T},$$

which requires several steps to prove. The idea for the first partition is intuitive: if one starts with any maxclique and proceeds to the set of all vertices, one will have those that are part of the maxclique in the middle and others that do not participate in it on the left and on the right with regard to the linear strictorder C_Δ. Disjointness of the middle term to the other two is trivial looking at $M, \overline{M}$. But also the intersection of the outer two vanishes in view of the consecutive $\mathbf{1}$s property:

$$\left(C_\Delta \mathbin{;} M \cap \overline{M}\right) \cap \left(C_\Delta^\mathsf{T} \mathbin{;} M \cap \overline{M}\right) = \left(C_\Delta \mathbin{;} M \cap C_\Delta^\mathsf{T} \mathbin{;} M\right) \cap \overline{M} \subseteq M \cap \overline{M} = \mathbb{L}$$

Finally, we prove that all three terms together decompose the universal relation, using that every vertex is contained in a maxclique, i.e., $\mathbb{T} = \mathbb{T} \mathbin{;} M$, and that C_Δ is a linear strictorder:

$$\mathbb{T} = \mathbb{T} \mathbin{;} M = (\overline{\mathbb{I}} \cup \mathbb{I}) \mathbin{;} M = \overline{\mathbb{I}} \mathbin{;} M \cup \mathbb{I} \mathbin{;} M = (C_\Delta \cup C_\Delta^\mathsf{T}) \mathbin{;} M \cup M$$

Now we prove the second partition mentioned initially where the outer terms are more easily expressed with M and P. The following is a cyclic estimation resulting in equality everywhere in between:

$$\begin{aligned}
C_\Delta \mathbin{;} M \cap \overline{M} &= M \mathbin{;} \overline{B} \cap C_\Delta \mathbin{;} M && \text{arranging differently and using } M \mathbin{;} \overline{B} = \overline{M} \\
&\subseteq (M \cap C_\Delta \mathbin{;} M \mathbin{;} \overline{B}^\mathsf{T}) \mathbin{;} (\overline{B} \cap M^\mathsf{T} \mathbin{;} C_\Delta \mathbin{;} M) && \text{Dedekind rule} \\
&\subseteq M \mathbin{;} P && \text{definition of } P \\
&= M \mathbin{;} (M^\mathsf{T} \mathbin{;} C_\Delta \mathbin{;} M \cap \overline{B}) \\
&= M \mathbin{;} \left(\left[M^\mathsf{T} \mathbin{;} C_\Delta \cap (M^\mathsf{T} \mathbin{;} C_\Delta^\mathsf{T} \cup \overline{M^\mathsf{T} \mathbin{;} C_\Delta^\mathsf{T}})\right] \mathbin{;} M \cap \overline{B}\right) && \text{intersecting } \mathbb{T}
\end{aligned}$$

$$\subseteq M \,; \left(\left[M^\mathsf{T} \cup \overline{M^\mathsf{T} \,; C_\Delta^\mathsf{T}}\right)\right] \,; M \cap \overline{B}) \qquad \text{consecutive } \mathbf{1}\text{s property}$$
$$= M \,; \left(\left[M^\mathsf{T} \,; M \cup \overline{M^\mathsf{T} \,; C_\Delta^\mathsf{T}} \,; M\right] \cap \overline{B}) \qquad \text{distributivity}$$
$$= M \,; \left(\overline{M^\mathsf{T} \,; C_\Delta^\mathsf{T}} \,; M \cap \overline{B}) \qquad \text{since } B = M^\mathsf{T} \,; M$$
$$\subseteq M \,; \overline{M^\mathsf{T} \,; C_\Delta^\mathsf{T}} \,; M \cap M \,; \overline{B} \qquad \text{subdistributive composition}$$
$$\subseteq \overline{C_\Delta}^\mathsf{T} \,; M \cap \overline{M} \qquad \text{using } M \,; \overline{B} = \overline{M}$$
$$= (\mathbb{I} \cup C_\Delta) \,; M \cap \overline{M} \qquad \text{linear strictorder}$$
$$= (M \cup C_\Delta \,; M) \cap \overline{M}$$
$$= C_\Delta \,; M \cap \overline{M}$$

In mainly the same way, we obtain

$$M^\mathsf{T} \,; C_\Delta \cap \overline{M}^\mathsf{T} = P \,; M^\mathsf{T}.$$

P provides a transitive orientation:

$$P \cup P^\mathsf{T} = (\overline{B} \cap M^\mathsf{T} \,; C_\Delta \,; M) \cup (\overline{B} \cap M^\mathsf{T} \,; C_\Delta^\mathsf{T} \,; M) \qquad \text{by definition}$$
$$= \overline{B} \cap (M^\mathsf{T} \,; C_\Delta \,; M \cup M^\mathsf{T} \,; C_\Delta^\mathsf{T} \,; M) \qquad \text{distributive}$$
$$= \overline{B} \cap M^\mathsf{T} \,; (C_\Delta \cup C_\Delta^\mathsf{T}) \,; M \qquad \text{distributive composition}$$
$$= \overline{B} \cap M^\mathsf{T} \,; \overline{\mathbb{I}} \,; M \qquad \text{because } C_\Delta \text{ is assumed to be a linear strictorder}$$
$$= \overline{B} \qquad \text{because } \overline{B} \subseteq M^\mathsf{T} \,; \overline{\mathbb{I}} \,; M, \text{ see below}$$

Obviously, M is surjective, i.e., $\mathbb{T} \,; M = \mathbb{T}$. Therefore,

$$\mathbb{T} = M^\mathsf{T} \,; \mathbb{T} \,; M = M^\mathsf{T} \,; (\mathbb{I} \cup \overline{\mathbb{I}}) \,; M = M^\mathsf{T} \,; M \cup M^\mathsf{T} \,; \overline{\mathbb{I}} \,; M = B \cup M^\mathsf{T} \,; \overline{\mathbb{I}} \,; M.$$

Transitivity, $P \,; P \subseteq P$, is a consequence of $P \,; B \,; P \subseteq P$ which we prove next:

$$P \,; B \,; P = P \,; M^\mathsf{T} \,; M \,; P = (M^\mathsf{T} \,; C_\Delta \cap \overline{M}^\mathsf{T}) \,; (C_\Delta \,; M \cap \overline{M}) \subseteq M^\mathsf{T} \,; C_\Delta \,; M \cap \overline{B} = P$$

has to be shown in order to prove transitivity of P and its postulated property. Containment in the first term follows because C_Δ is transitive. Containment in the second is a bit more tricky and requires us to make use of the fact that the strictorder C_Δ is a linear strictorder. The essential inner part of the claim

$$P \,; B \,; P = (M^\mathsf{T} \,; C_\Delta \cap \overline{M}^\mathsf{T}) \,; (C_\Delta \,; M \cap \overline{M}) \subseteq \overline{M^\mathsf{T} \,; M} = \overline{B}$$

is equivalent with

$$(C_\Delta^\mathsf{T} \,; M \cap \overline{M}) \,; M^\mathsf{T} \,; M \subseteq \overline{C_\Delta \,; M \cap \overline{M}}$$

which will be estimated in several steps

$$(C_\Delta^\mathsf{T} \,; M \cap \overline{M}) \,; M^\mathsf{T} \,; M \subseteq (C_\Delta^\mathsf{T} \,; M \cap \overline{C_\Delta \,; M \cap C_\Delta^\mathsf{T} \,; M}) \,; M^\mathsf{T} \,; M \qquad \text{consecutive } \mathbf{1}\text{s}$$
$$= (C_\Delta^\mathsf{T} \,; M \cap \overline{[C_\Delta \,; M \cup C_\Delta^\mathsf{T} \,; M]}) \,; M^\mathsf{T} \,; M \qquad \text{De Morgan rule}$$
$$= (C_\Delta^\mathsf{T} \,; M \cap \overline{C_\Delta \,; M}) \,; M^\mathsf{T} \,; M \qquad \text{for Boolean reasons}$$
$$\subseteq \overline{C_\Delta \,; M} \,; M^\mathsf{T} \,; M$$
$$\subseteq \overline{C_\Delta \,; M} \qquad \text{Schröder rule}$$
$$= \overline{(\mathbb{I} \cup C_\Delta^\mathsf{T}) \,; M} \qquad \text{linear strictorder } C_\Delta$$
$$= \overline{M \cup C_\Delta^\mathsf{T} \,; M} \qquad \text{distributive}$$
$$= \overline{M \cup (C_\Delta^\mathsf{T} \,; M \cap \overline{M})} \qquad \text{for Boolean reasons}$$
$$= \overline{C_\Delta \,; M \cap \overline{M}} \qquad \text{disjoint partition}$$

(iii) We have the disjoint union $\mathbb{T} = B \cup P \cup P^\mathsf{T}$ and, therefore, $\overline{P}^\mathsf{T} = B \cup P$. Now,

$$P_{;}\overline{P}^\mathsf{T}{}_{;}P = P_{;}(B \cup P)_{;}P = P_{;}B_{;}P \cup P_{;}P_{;}P \subseteq P$$

owing to the assumption and transitivity. $\qquad\qquad\qquad\qquad\qquad\square$

The condition on P in (i) may be replaced by another equivalent one.

Lemma 12.38. *We consider a reflexive and symmetric relation B and assume a transitive orientation $\overline{B} = P \cup P^\mathsf{T}$ of its complement. Then*

$$B_{;}P \cap P^\mathsf{T}{}_{;}B = \mathbb{\bot} \quad\Longleftrightarrow\quad P_{;}B_{;}P \subseteq P.$$

Proof: $B_{;}P \cap P^\mathsf{T}{}_{;}B = \mathbb{\bot} \quad\Longleftrightarrow\quad P^\mathsf{T}{}_{;}B \subseteq \overline{B_{;}P} \quad\Longleftrightarrow\quad P_{;}B_{;}P \subseteq \overline{B} = P \cup P^\mathsf{T},$

but we have from transitive orientation that

$$P^\mathsf{T}{}_{;}P^\mathsf{T} \subseteq P^\mathsf{T} \subseteq \overline{B} \quad\Longrightarrow\quad B_{;}P \subseteq \overline{P}^\mathsf{T} \quad\Longrightarrow\quad P_{;}B_{;}P \subseteq P_{;}\overline{P}^\mathsf{T} \subseteq \overline{P}^\mathsf{T}. \qquad\square$$

We aim at an interpretation of the conditions thus shown to be equivalent.

Proposition 12.39. *Let any two relations M, N be given that satisfy $M^\mathsf{T}{}_{;}M = N^\mathsf{T}{}_{;}N$. Then either both of them have the consecutive $\mathbf{1}$s property, or none.*

Proof: Let M, N be given. Assuming both of them not to have the consecutive $\mathbf{1}$s property, we are done. In the case, for example, that M has it, there exists some linear strictorder C_Δ satisfying the condition. From this strictorder in turn, we determine a transitive orientation $\overline{B} = P \cup P^\mathsf{T}$ with $B := M^\mathsf{T}{}_{;}M$. This is then a transitive orientation also for $B := N^\mathsf{T}{}_{;}N$, from which we obtain the linear strictorder $C'_\Delta = N_{;}P_{;}N^\mathsf{T}$. $\qquad\qquad\qquad\qquad\qquad\square$

If M and N should happen to have the same number of rows, one may even be able to determine a permutation that transforms M into N. The linear strictorders C_Δ and C'_Δ would then be relations on the same set, so that a permutation exists that transforms one into the other.

Figure 12.30 shows on the left what must not occur in order to have $B_{;}P \cap P^\mathsf{T}{}_{;}B = \mathbb{\bot}$. We start in the upper right (or left) corner and proceed in the two ways indicated by the formula to the respective other corner. We may, however, also interpret $P_{;}B_{;}P \subseteq P$, in which case we start from one of the lower two vertices: any of the diagonals marked B forbids the other. Because P is irreflexive, degenerations with coinciding vertices may only occur for B; but assuming any of the Bs contracted to one vertex, two consecutive Ps will show up making the other B impossible. We have, however, a partition of the set of pairs of vertices. It is unavoidable that the upper two vertices be linked by B, and similarly the lower two. Assuming them to belong to P leads to a contradiction with transitive orientation: one of the diagonals then must belong to P instead of B as indicated.

What has thus been shown is that the assumption in Lemma 12.38 is violated in the case of a 4-cycle in B having no chord in B.

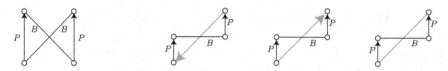

Fig. 12.30 Remaining 4-cycle

On the right side of Fig. 12.30, all three possibilities for the diagonal are discussed. The first is impossible when P is a transitive orientation, because then the horizontal line must not belong to B. The second would be fine: some sort of transitivity $P\,{;}\,B\,{;}\,P \subseteq P$. It is precisely the last that must not occur as to the conditions $B\,{;}\,P \cap P^\mathsf{T}\,{;}\,B = \mathbb{I}$ or $P\,{;}\,B\,{;}\,P \subseteq P$. Thinking a little bit ahead, the remaining diagonals have to be in B, because otherwise a contradiction with transitive orientation would happen. This gives a 4-cycle without a chord in B that we try to forbid in the next subsection.

It is this observation that leads us to investigate cycles and triangulations. We shall study the maxcliques of B in combination with simple cycles in B. The remarkable property of the set of maxcliques of an interval graph is that deciding for any two of them, and relating them by an *arrow*, can be extended to a *linear* ordering of the set of maxcliques. This then serves in a non-trivial way to find a transitive orientation directly from B. The result thus announced needs some preparation.

Triangulation

When formulating the triangulation property, we take care that it fits our relational approach. The standard form is: a simple graph is **triangulated**, provided that every simple cycle of length greater than 3 possesses a chord. Considering Fig. 12.31, the cycle on the left may turn out to have a chord as indicated in the middle, i.e., a chord linking two arbitrary vertices. We may be more specific in this regard and postulate that for some i a chord leads from a_i precisely to a_{i+2}, a '2-step chord'.

Definition 12.40. Given a simple graph with adjacency Γ, we call it a **chordal**

graph or we say that it has the **triangulation property**, provided that $\mathcal{C}^2 \cap \Gamma \neq \bot\!\!\!\bot$ for every simple circuit $\mathcal{C} \subseteq \Gamma$. $\qquad\qquad\qquad\qquad\qquad\qquad\qquad\qquad\qquad\qquad\qquad$ □

Fig. 12.31 Cycle with arbitrary chord and with '2-step chord'

It might seem that the variant definition just given is more restrictive, but it is not. If we have $\mathcal{C}^2 \cap \Gamma \neq \bot\!\!\!\bot$, this is certainly a special chord as indicated on the right of Fig. 12.31. On the other hand, given any simple cycle with a chord, this simple cycle is thus reduced to two smaller ones. Either one of the two is a simple cycle of length 3, i.e., linking some a_i with a_{i+2}, or this situation may happen only after iterating this decomposition. In any case, a simple cycle of length 3 gives rise to the situation $\mathcal{C}^2 \cap \Gamma \neq \bot\!\!\!\bot$. If the graph is chordal, i.e., triangulated, most trivially every simple 4-cycle has a chord, but not vice versa.

Proposition 12.41. *If the reflexive and symmetric relation B admits a transitive orientation $\overline{B} = P \cup P^\mathsf{T}$ of its complement satisfying $P\,\grave{,}\,B\,\grave{,}\,P \subseteq P$, the simple graph with adjacency $\Gamma := \overline{\mathbb{I}} \cap B$ is triangulated, i.e., a chordal graph.*

Proof: Assume an arbitrary simple circuit $\mathcal{C} \subseteq B$ of length greater than 3, take the second point a_2 of it and consider a_4. Then $(a_1, a_2) \in B$, $(a_2, a_3) \in B$ and $(a_3, a_4) \in B$. If there exists a 2-step chord $(a_1, a_3) \in B$ or $(a_2, a_4) \in B$, we are done; so we concentrate on the situation that none of these exists. Without loss of generality, we assume $(a_4, a_2) \in P$ – otherwise consider the reverse orientation. Now, we check the conceivable orientations of $\{a_1, a_3\}$. From $(a_1, a_3) \in P$, we deduce with $P\,\grave{,}\,B\,\grave{,}\,P \subseteq P$

$$(a_1, a_3) \in P, (a_3, a_4) \in B, (a_4, a_2) \in P \quad \Longrightarrow \quad (a_1, a_2) \in P$$

which is a contradiction, because a_1, a_2 are consecutive on the circuit. The attempt $(a_3, a_1) \in P$ leads in an analogous way to

$$(a_4, a_2) \in P, (a_2, a_3) \in B, (a_3, a_1) \in P \quad \Longrightarrow \quad (a_4, a_1) \in P.$$

In the case of the circuit $\mathcal{C}$ considered having length 4, this is already a contradiction. If there exists a next point $a_5 \neq a_1$ in the circuit, an arrow $(a_1, a_5) \in P$ is impossible because it would imply an arrow $(a_1, a_2) \in P$ as a consequence of $(a_1, a_5) \in P, (a_5, a_4) \in B, (a_4, a_1) \in P$. Therefore, $(a_5, a_1) \in P$, which would be a contradiction if the length were just 5; and so on iteratively. $\qquad$ □

This result may be traced back to the earliest in the literature concerning the triangulation property. It is here presented only as a corollary.

Corollary 12.42 (Hajós 1958 [64]). *An interval graph is always a chordal graph, i.e., a triangulated one.*

Figure 12.32 shows that Cor. 12.42 may not be reversed. The left is a triangulated graph that is not an interval graph, since there does not even exist a transitive orientation for the dashed complement. The outer 3-cycle *alone* may well be given a transitive orientation; extending this to all of the inner spikes will result in two consecutive *P*s, demanding another one by transitivity that contradicts the graph on the left.

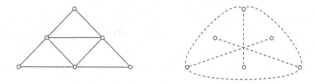

Fig. 12.32 Triangulated graph, but not an interval graph

An immediate observation shows that a pure (i.e., chordless) cycle of odd length > 3 cannot be transitively oriented, while even-length cycles can be oriented in quite a systematic way; see Fig. 12.33.

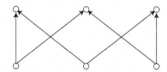

Fig. 12.33 Even-length cycle transitively oriented

Shortly after the publication of the results Cor. 12.42 and Prop. 12.34, a strengthened version appeared.

Corollary 12.43 (Gilmore and Hoffman 1964 [59]). *If an undirected graph has the triangulation property, and if in addition its complement is transitively orientable, it is necessarily an interval graph.*   □

The following diagram tries to collect all this information on cross-relationships.

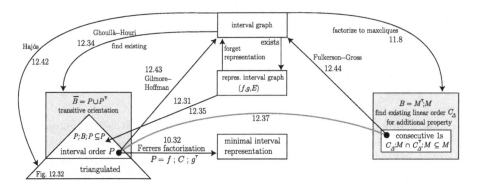

Fig. 12.34 Interdependency of results concerning interval graphs and -orders

An important property of relations shows up here, that is not easily handled algebraically. It may be visualized using the matrix representation, saying that in every column the 1s reside in consecutive positions. Then one will also see how the following integrates; originally it was discovered as an independent result.

Corollary 12.44 (Fulkerson-Gross 1965 [55, 56]). *A simple graph with adjacency* $\Gamma : X \longrightarrow X$ *is an interval graph precisely when the maxclique relation of* Γ *has the consecutive* 1s *property.* □

Example 12.45. We reconsider these rather difficult results in an extended example. Assume the same adjacency Γ to stem from two quite different interval arrangements; see the first and second lines of Fig. 12.35. While the linear orders E on 14, respectively 16, elements are trivial and need not be visualized, the respective relations f, g may easily be deduced as those pointing to the left, respectively right, end of the interval. One will observe – scanning the left interval representation from left to right – that the maxcliques are $\{b, g\}, \{b, d, f\}, \{b, e\}, \{a, c\}, \{c, h\}$. Repeating this scan for the right matrix, the same set of maxcliques will be found, however, they appear in a different sequence.

Also the transitive orientations P obtained as $P := g \,; C \,; f^\mathsf{T}$ (see Def. 12.30) look quite different, as do the maxclique factorizations $B = M^\mathsf{T} \,; M$. Both have been obtained in the first step by the same algorithm and would therefore look equal; but the linear strictorders $C_\Delta := M \,; P \,; M^\mathsf{T}$ are completely different, so that we have chosen to apply the permutation of the maxclique set to the upper right triangle. In both cases, the M is then in consecutive 1s form.   □

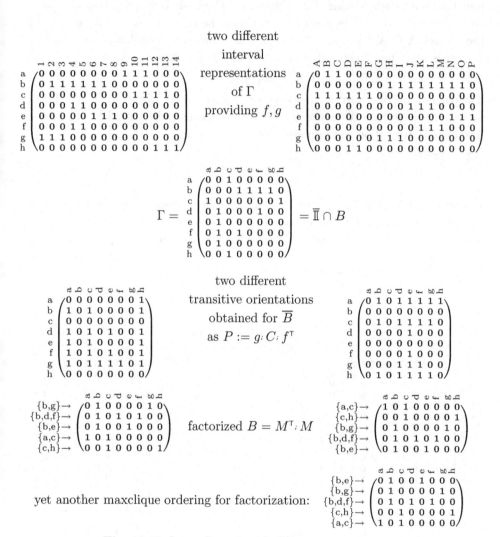

Fig. 12.35 Interval graph with different representations

The triangulation property does not lend itself to being formulated easily with relations. Even the property that every simple 4-cycle has a chord is difficult to work with relationally.

The quantification over all cycle lengths in the triangulation property has as an immediate consequence that every simple cycle may be decomposed down to 3-cycles, i.e., triangulated. It seems important to be able to describe when a graph has simple 4-cycles without chord. There are 4 elements to consider, so that we reach the borderline of representable relations, and thus, cannot hope to find a point-free formulation except when using direct products. The basic concept is that of a square with diagonal vertex pairs x, y and u, v. Given the graph $\Gamma : X \longrightarrow X$, we consider the pairset $X \times X$ with projections π, ρ as usual. The possible pairs of pairs of diagonals may then be collected in a relation on $X \times X$ as

$$W := \pi\,;\Gamma\,;\pi^{\mathsf{T}} \cap \pi\,;\Gamma\,;\rho^{\mathsf{T}} \cap \rho\,;\Gamma\,;\pi^{\mathsf{T}} \cap \rho\,;\Gamma\,;\rho^{\mathsf{T}}.$$

The pairs of diagonal vertices, however, are connected via two steps along Γ leading to a different vertex. We characterize all such vertex pairs with $\Gamma^2 \cap \overline{\Gamma}$, however lifted via vectorization to the pairset as

$$t := \left[\pi\,;(\Gamma^2 \cap \overline{\Gamma}) \cap \rho\right]\,;\mathbb{T}.$$

Proposition 12.46. *Every 4-cycle of a graph has a chord precisely when no pair of thus related vertices is in relation W to another such; expressed in a different way, when t is internally stable with respect to W, i.e., $W\,;t \subseteq \bar{t}$.*

Proof: We omit this proof here.   □

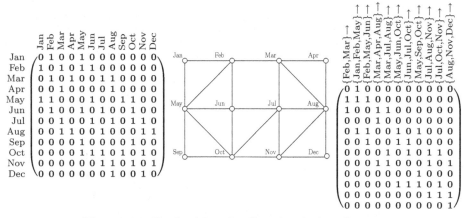

Fig. 12.36 Graph with a chordless 4-cycle, its adjacency
and its transposed maxclique relation

13

Preference and Indifference

Mankind has developed many concepts for reasoning about something that is *better than*, is *more attractive than*, is *at least as good as* something else or *similar to* something else. Such concepts lead to an enormous bulk of formulae and interdependencies. Modelling preferences means associating algebraic properties of orders/strictorders/equivalences to such colloquial qualifications of being *better than*, etc. This is a major step of abstraction that not least requires getting rid of routine-handling this in spoken language (brother of parent = uncle, taller/smaller than) – which may be different in English, German, Japanese, or Arabic, for example.

We have been starting from the concept of an *order* and a *strictorder*, defined as a transitive, antisymmetric, reflexive relation or as transitive and asymmetric, respectively. In Chapter 12, we explained the interdependency of these order concepts. Now, we will start to use them in modelling preferences.

This means, on the one hand, relating them with colloquial concepts of comparison, indifference, and incomparability, for which there exists a lot of literature. On the other, it means explaining certain nearly coincident fields of discourse that have been established, but use different notation.

We recall duality before entering into the other fields.

13.1 Duality

In order theory, one always has two variants, a reflexive and an irreflexive form, i.e., the order "$\leq$" and the strictorder "$<$". Considered algebraically, the transition is $E \mapsto \overline{\mathbb{I}} \cap E =: C$ and in reverse direction $C \mapsto \mathbb{I} \cup C =: E$. It would have been tedious always to explain everything in a twofold form; so our exposition has been based on just one of the two variants; as a consequence of Prop. 12.1, we chose the (slightly more general) irreflexive form, i.e., the strictorder.

When studying preferences, a different transition between a reflexive and an irreflexive variant is traditional, namely toggling back and forth to the dual. In trivial cases, this reduces to the aforementioned more simple form. We recall the definition of the dual: given an arbitrary relation R, in Def. 4.5 we called $R^{\mathrm{d}} := \overline{R}^{\mathsf{T}}$ its dual. Forming the dual of a relation is an involutory operation. It simply exchanges several properties.

Proposition 13.1. *When dualizing, the properties of a relation R and its dual R^{d} toggle as follows*

R^{d} *reflexive* $\iff$ R *irreflexive*
R^{d} *asymmetric* $\iff$ R *connex*
R^{d} *antisymmetric* $\iff$ R *semi-connex*
R^{d} *transitive* $\iff$ R *negatively transitive,*

while other properties prevail

R^{d} *symmetric* $\iff$ R *symmetric*
R^{d} *semi-transitive* $\iff$ R *semi-transitive*
R^{d} *Ferrers* $\iff$ R *Ferrers.*

Proof: Most of these statements are trivial. From the remark after Def. 10.28, we know that with a Ferrers relation its dual is also Ferrers. In a similar way, semi-transitivity propagates to the dual,

$$\overline{C}^{\mathsf{T}}{}_;\overline{C}^{\mathsf{T}}{}_;C \subseteq \overline{C}^{\mathsf{T}} \iff C^{\mathsf{T}}{}_;C^{\mathsf{T}}{}_;\overline{C} \subseteq C^{\mathsf{T}} \iff C{}_;C{}_;\overline{C}^{\mathsf{T}} \subseteq C. \qquad \square$$

The concept of the dual is particularly interesting when applied to orders and strictorders. With Fig. 13.1, we follow this idea over all the types of orders/strictorders of Def. 12.4. The dual of an order or strictorder need not be transitive again. To relate a (strict)order with its dual, we fix two statements in advance that are not completely symmetric for orders and strictorders:

C strictorder $\implies$ $C \subseteq C^{\mathrm{d}}$, since a strictorder is always asymmetric
E order $\implies$ $E^{\mathrm{d}} \subseteq E$, only when E should happen to be connex

So our interest will later be concentrated on the so-called indifference $I := C^{\mathrm{d}} \cap \overline{C}$ defined via the strictorder C.

We might have expected the *order* to appear somewhere in Fig. 13.1. One will see, however, that the order does not really fit in here. The order is in a sense a composite construct built from a preorder, as introduced in Def. 5.20, with quotient forming according to an equivalence immediately executed. Also the concept of a preorder proper does not show up. The dual of a semiorder need not be transitive. If it is transitive, one has started from a weakorder.

strictorder	intervalorder	semiorder	weakorder	linear strictorder	← dualization →	linear order	connex preorder	dual of semiorder	dual of intervalorder	dual of strictorder
• ○	• |	• |	•	• ○	irreflexive / reflexive	•	•	•	•	•
• |	• ○	• ○	• |	• |	asymmetric / connex	•	•	•	•	•
• | ○	•	•	•	• | ○	transitive / neg. trans.	•	•	•	•	•
—	• | ○	• | ○	•	•	Ferrers	•	•	•	•	—
—	—	• | ○	•	•	semi-transitive	•	•	•	—	—
—	—	—	• |	•	neg. trans. / transitive	•	•	—	—	—
—	—	—	—	• | ○	semi-connex / antisymm.	•	—	—	—	—

Fig. 13.1 Strictorders dualized according to Prop. 13.1

An equivalence relation can always be arranged in a block-diagonal form, putting members of classes side by side. This is possible permuting rows as well as columns simultaneously. While an ordering can always be arranged with empty subdiagonal triangle, a non-trivial preorder cannot. On the other hand, a weakorder can. Arranging an ordering to fit in the upper triangle is known as topological sorting or finding a Szpilrajn extension. With the idea of a block diagonal, we get a close analog for preorders. Observe that for preorders $R \cap R^\mathsf{T} = \Xi(R) = \Psi(R)$.

$$
\begin{pmatrix}
1 & 1 & 1 & 1 & 0 & 0 & 0 \\
0 & 1 & 0 & 1 & 0 & 0 & 0 \\
0 & 0 & 1 & 1 & 0 & 0 & 0 \\
0 & 0 & 0 & 1 & 0 & 0 & 0 \\
0 & 0 & 0 & 0 & 1 & 0 & 1 \\
0 & 0 & 0 & 0 & 0 & 1 & 1 \\
0 & 0 & 0 & 0 & 0 & 0 & 1
\end{pmatrix}
\quad
\begin{pmatrix}
1 & 1 & 0 & 0 & 0 & 1 & 1 \\
1 & 1 & 0 & 0 & 0 & 1 & 1 \\
0 & 0 & 1 & 1 & 1 & 1 & 1 \\
0 & 0 & 1 & 1 & 1 & 1 & 1 \\
0 & 0 & 1 & 1 & 1 & 1 & 1 \\
0 & 0 & 0 & 0 & 0 & 1 & 1 \\
0 & 0 & 0 & 0 & 0 & 1 & 1
\end{pmatrix}
\quad
\begin{pmatrix}
1 & 1 & 0 & 0 & 0 & 0 & 0 \\
1 & 1 & 0 & 0 & 0 & 0 & 0 \\
0 & 0 & 1 & 1 & 1 & 0 & 0 \\
0 & 0 & 1 & 1 & 1 & 0 & 0 \\
0 & 0 & 1 & 1 & 1 & 0 & 0 \\
0 & 0 & 0 & 0 & 0 & 1 & 1 \\
0 & 0 & 0 & 0 & 0 & 1 & 1
\end{pmatrix}
$$

Fig. 13.2 Arranging order, preorder, and equivalence with empty
lower left (block-)triangle

For the rightmost columns of Fig. 13.1, similar spanning subsets can be found as on the left. It is, however, more economic to use duality, so that we restrict ourselves to just one example and look at reflexive Ferrers relations.

Proposition 13.2. *Let R be a Ferrers relation.*

R *reflexive* $\implies$ R *connex and negatively transitive.*

Proof: Any Ferrers relation satisfies $R \mathbin{;} \overline{R}^{\mathsf{T}} \mathbin{;} R \subseteq R$ by definition, which implies $\overline{R}^{\mathsf{T}} \subseteq R$ in the case R is reflexive, i.e., $\mathbb{I} \subseteq R$. Therefore, $\mathbb{T} = R^{\mathsf{T}} \cup R$. Negative transitivity is equivalent to $\overline{R}^{\mathsf{T}} \mathbin{;} R \subseteq R$ with Schröders equivalence; this, however, follows immediately from the Ferrers property when R is reflexive. $\qquad\square$

A reflexive Ferrers relation is, thus, the dual of an intervalorder.

Exercises

13.1 In [48], finite relations are considered. A **quasi-series** R is then defined, postulating that it is asymmetric and transitive and that $\overline{R} \cap \overline{R}^{\mathsf{T}}$ is an equivalence. Prove that R is a quasi-series precisely when it is a weakorder.

13.2 Let R be any Ferrers relation. Then the following composite relation is a weakorder:
$$R_b := \begin{pmatrix} R \mathbin{;} \overline{R}^{\mathsf{T}} & R \\ \overline{R}^{\mathsf{T}} & \overline{R}^{\mathsf{T}} \mathbin{;} R \end{pmatrix}.$$

13.3 For any intervalorder C, the construct $\overline{C}^{\mathsf{T}} \mathbin{;} C$ is a strictorder.

13.2 Modelling indifference

There exists an amount of literature concerning the algebraic properties of relations occurring around preferences, i.e., orders and/or strictorders, etc. The very first observation is that a *preference* should be transitive – although this will not be included in the definition of a preference structure. Because it is used in the environment of financial services where optimization is omnipresent, one could easily ask a person to change the preference to a transitive one. One would ask the person to give an amount of Euros by which car a is preferred by him to car b, car b is preferred by him to car c, car c is preferred by him to car a. Once he has paid $p(a, b)$ for changing his car from a to b, $p(b, c)$ for changing his car from b to c, $p(c, a)$ for changing his car from c to a, he will find himself in the uncomfortable situation of having paid an amount of money just in order to own the same car as before.

The next point to discuss is *indifference*. For decades, papers have appeared concerning indifference with respect to some defined preference. Intuitively, people demand that $a < b$ followed by indifference between b and c, results – in a sense – in $a < c$. For a high-jump competition, this is certainly true. If *Abe* does not jump as high as *Bob*, namely 2.20 m versus 2.25 m, which *Bob* as well as *Carl* have mastered and both failed to achieve 2.30 m, then *Abe* does not jump as high as *Carl*. If "$\prec$" is used to denote this preference point-wise and if "$\approx$" is the indifference defined by it, then this means

$$a \prec b, \quad b \approx c \quad \Longrightarrow \quad a \prec c.$$

In a similar way, one will demand for attempts at preference modelling that

$$a \approx b, \quad b \prec c \quad \Longrightarrow \quad a \prec c.$$

Later in point-free form, this will show up as $P_{\,;}\,I \subseteq P$ and $I_{\,;}\,P \subseteq P$.

In this setting, however, one is traditionally unsatisfied with the concept of indifference: it should not be transitive, which it often turns out to be. Transitivity of indifference was strongly contradicted by Luce and Tversky in the 1950s. But Armstrong [3] had already mentioned that indifference must not be transitive: *. . . For this implies that indifference between alternatives can only occur if the utilities are identical, and therefore the relation of indifference must be symmetrical and transitive. But it is a well-known fact that . . . the relation of indifference is not transitive.* People are, without any doubt, 'indifferent' with respect to taking one tiny crystal of sugar more or less in their coffee; however, they would certainly try to avoid iterating throwing in tiny pieces of sugar and finally emptying a huge bag of sugar. Similarly with temperatures: heating by 0.05 degrees will hardly be recognized or measured; but unlimited iterating of such heating will certainly be catastrophic. It is, thus, a non-trivial task to characterize indifference. We start with the following definition.

Definition and **Proposition 13.3.** (i) *For any strictorder C with corresponding order $E := \mathbb{I} \cup C$, we consider its so-called*

indifference $I := I_C := C^{\mathrm{d}} \cap \overline{C} = (E^{\mathrm{d}} \cap \overline{E}) \cup \mathbb{I}$

and prove the following facts:

— $I \supseteq \mathbb{I}$,
— C^{d} *is connex and negatively transitive; in addition it satisfies*
$$C^{\mathrm{d}} = I \cup C \supseteq \mathbb{I} \cup C = E, \qquad \text{and the dual} \qquad E^{\mathrm{d}} = \overline{\mathbb{I}} \cap C^{\mathrm{d}} \supseteq \overline{I} \cap C^{\mathrm{d}} = C.$$

(ii) *If R is any connex and negatively transitive relation, $C := R^{\mathrm{d}}$ is a strictorder. Its indifference I_C satisfies $I_C = R \cap R^{\mathsf{T}}$, $\quad \overline{I_C} \cap R = C$.*

Proof: (i) It is trivial to convince oneself by Boolean evaluation that the two versions of the definition of indifference mean the same. This shows in particular that $I \supseteq \mathbb{I}$.

$$C^{\mathrm{d}} \cup C^{\mathrm{d}^{\mathsf{T}}} = \overline{C}^{\mathsf{T}} \cup \overline{C} \supseteq \overline{C}^{\mathsf{T}} \cup C^{\mathsf{T}} = \mathbb{T},$$

using asymmetry, proves connexity. Negative transitivity of C^{d} follows directly from transitivity of C. We then use that asymmetry $C^{\mathsf{T}} \subseteq \overline{C}$ of C implies $C \subseteq \overline{C}^{\mathsf{T}} = C^{\mathrm{d}}$ and prove the additional properties:

$$C^{\mathrm{d}} = C^{\mathrm{d}} \cap \mathbb{T} = C^{\mathrm{d}} \cap (\overline{C} \cup C) = (C^{\mathrm{d}} \cap \overline{C}) \cup (C^{\mathrm{d}} \cap C) = I \cup C \supseteq \mathbb{I} \cup C = E$$

and dually
$$E^d = (\mathbb{I} \cup C)^d = \overline{\mathbb{I} \cup C}^{\mathsf{T}} = \overline{\mathbb{I}} \cap \overline{C}^{\mathsf{T}} \supseteq \overline{I} \cap C^d = \overline{C^d \cap \overline{C}} \cap C^d = (\overline{C^d} \cup C) \cap C^d = C.$$

(ii) Transitivity of C follows from negative transitivity of R and asymmetry of C
from connexity of R. The statement $I_C = C^d \cap \overline{C} = R \cap R^{\mathsf{T}}$ is trivial by definition
of I_C and C
$$\overline{I_C} \cap R = \overline{I_C} \cap C^d = \overline{C^d \cap \overline{C}} \cap C^d = (C^{\mathsf{T}} \cup C) \cap \overline{C}^{\mathsf{T}} = C \cap \overline{C}^{\mathsf{T}} = C. \qquad \square$$

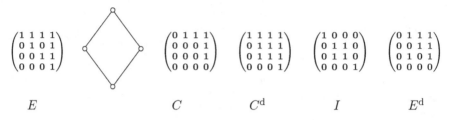

$$\qquad\qquad E \qquad\qquad\qquad\qquad\qquad C \qquad\qquad\quad C^d \qquad\qquad I \qquad\qquad E^d$$

Fig. 13.3 Example of indifference of a strictorder

If a is neither strictly inferior to b, nor the other way round, the relation shows
indifference as to their ranking, which means not least that a and b may be identical.
Proposition 13.3 has shown what holds for indifference of a strictorder in general.

Indifference in more special cases

Now we specialize to the more specific types of orders to obtain additional results.
As long as one works with linear strictorders/orders only, there is not much to say
because indifference turns out to be just the identity, as we are going to prove next.

Proposition 13.4. (i) *For a linear strictorder C,*

— *the indifference is the identity, i.e., satisfies $I = \mathbb{I}$.*
— *the dual is the corresponding linear order; it satisfies $C^d = E$.*

(ii) *If E is any linear order, then its dual $C := E^d = \overline{I_C} \cap E = \overline{\mathbb{I}} \cap E$ is the*
corresponding linear strictorder.

Proof: (i) $I = C^d \cap \overline{C} = \overline{C}^{\mathsf{T}} \cap \overline{C} = \overline{C^{\mathsf{T}} \cup C} = \mathbb{I}$ since the linear strictorder C is
semi-connex. Thus, we have equality in Prop. 13.3.i.

(ii) That $C := E^d$ is a linear strictorder follows by simply dualizing the constituent
properties. For the rest, we start from the general situation of Prop. 13.3.ii, namely
that $C = \overline{I_C} \cap E$. From (i), we already know $I_C = \mathbb{I}$, so that also $C = \overline{\mathbb{I}} \cap E$. $\qquad \square$

$$C = E^{\mathrm{d}} = \begin{array}{c} \\ 1\\2\\3\\4 \end{array}\!\!\begin{array}{c} 1\,2\,3\,4 \\ \begin{pmatrix} 0&1&1&1\\0&0&1&1\\0&0&0&1\\0&0&0&0 \end{pmatrix}\end{array} \qquad C^{\mathrm{d}} = E = \begin{array}{c} \\ 1\\2\\3\\4 \end{array}\!\!\begin{array}{c} 1\,2\,3\,4 \\ \begin{pmatrix} 1&1&1&1\\0&1&1&1\\0&0&1&1\\0&0&0&1 \end{pmatrix}\end{array} \qquad I = \begin{array}{c} \\ 1\\2\\3\\4 \end{array}\!\!\begin{array}{c} 1\,2\,3\,4 \\ \begin{pmatrix} 1&0&0&0\\0&1&0&0\\0&0&1&0\\0&0&0&1 \end{pmatrix}\end{array} = \mathbb{I}$$

Fig. 13.4 Linear order, respectively linear strictorder, with duals and indifference

Indifference will become more interesting when considered in the case of a weakorder. Abandoning semi-connexity and, thus, moving to weakorders, brings an indifference I, which is obviously transitive.[1] There exists a close relationship between weakorders and connex preorders which we will now exhibit. In the rearranged form as well as in the Hasse diagram of Fig. 12.6, it could easily be seen how weakorder and connex preorder are related by adding an equivalence, or removing it. Using duality, we now give a proof of this effect.

Proposition 13.5. (i) *For a weakorder* W,

— *indifference coincides with row and with column equivalence, i.e.,*

$$I = \Xi(W) = \Psi(W),$$

— *its dual* $W^{\mathrm{d}} = I \cup W$ *is the corresponding connex preorder.*

(ii) *For any connex preorder* P*, the dual* $W := P^{\mathrm{d}}$ *is a weakorder, the so-called corresponding weakorder.*

Proof: (i) $I = W^{\mathrm{d}} \cap \overline{W} = \overline{W}^{\mathsf{T}} \cap \overline{W}$, so that I is symmetric. From irreflexivity $W \subseteq \overline{\mathbb{I}}$, it follows that I is reflexive. Since W is negatively transitive, I is transitive. Then we have $\overline{W}^{\mathsf{T}}; W = W$. Direction "$\subseteq$" follows via the Schröder equivalence from negative transitivity. In the other direction, we have $\overline{W}^{\mathsf{T}}; W \supseteq W$ since W is irreflexive. Using this, we evaluate, for example,

$$\Psi(W) = \mathsf{syq}(W, W) = \overline{W^{\mathsf{T}}; \overline{W}} \cap \overline{\overline{W}^{\mathsf{T}}; W} = \overline{W}^{\mathsf{T}} \cap \overline{W} = W^{\mathrm{d}} \cap \overline{W} = I.$$

Using Prop. 13.1, we obtain from W weakorder, that W^{d} is a connex preorder.

(ii) Again with Prop. 13.1, $W := P^{\mathrm{d}}$ turns out to be a weakorder. □

[1] As already mentioned, transitivity of indifference has raised objections, so that one has relaxed conditions even further to arrive at semiorders. Figure 12.14 shows that indifference holds for elements *Alfred, Frederick*, as well as for *Frederick, Donald*, but not for *Alfred, Donald*, illustrating that indifference is no longer a transitive relation.

$$
W = P^{\mathrm{d}} =
\begin{array}{r}
1\\2\\3\\4\\5\\6\\7\\8\\9\\10\\11
\end{array}
\begin{pmatrix}
0&0&1&1&1&1&1&1&1&1&1\\
0&0&1&1&1&1&1&1&1&1&1\\
0&0&0&0&0&1&1&1&1&1&1\\
0&0&0&0&0&1&1&1&1&1&1\\
0&0&0&0&0&1&1&1&1&1&1\\
0&0&0&0&0&0&0&0&1&1&1\\
0&0&0&0&0&0&0&0&1&1&1\\
0&0&0&0&0&0&0&0&1&1&1\\
0&0&0&0&0&0&0&0&0&1&1\\
0&0&0&0&0&0&0&0&0&0&0\\
0&0&0&0&0&0&0&0&0&0&0
\end{pmatrix}
\qquad
W^{\mathrm{d}} = P =
\begin{pmatrix}
1&1&1&1&1&1&1&1&1&1&1\\
1&1&1&1&1&1&1&1&1&1&1\\
0&0&1&1&1&1&1&1&1&1&1\\
0&0&1&1&1&1&1&1&1&1&1\\
0&0&1&1&1&1&1&1&1&1&1\\
0&0&0&0&0&1&1&1&1&1&1\\
0&0&0&0&0&1&1&1&1&1&1\\
0&0&0&0&0&1&1&1&1&1&1\\
0&0&0&0&0&0&0&0&1&1&1\\
0&0&0&0&0&0&0&0&0&1&1\\
0&0&0&0&0&0&0&0&0&1&1
\end{pmatrix}
\qquad
I =
\begin{pmatrix}
1&1&0&0&0&0&0&0&0&0&0\\
1&1&0&0&0&0&0&0&0&0&0\\
0&0&1&1&1&0&0&0&0&0&0\\
0&0&1&1&1&0&0&0&0&0&0\\
0&0&1&1&1&0&0&0&0&0&0\\
0&0&0&0&0&1&1&1&0&0&0\\
0&0&0&0&0&1&1&1&0&0&0\\
0&0&0&0&0&1&1&1&0&0&0\\
0&0&0&0&0&0&0&0&1&0&0\\
0&0&0&0&0&0&0&0&0&1&1\\
0&0&0&0&0&0&0&0&0&1&1
\end{pmatrix}
$$

Fig. 13.5 Weakorder, respectively connex preorder, with duals and indifference

Indifference considerations for semiorders or intervalorders are more difficult to handle, but allow non-transitive indifference. They require also the study of thresholds.

13.3 Modelling thresholds

Many preferences have their origin in measuring something so that real numbers can be attached. Real numbers are of arbitrary precision while measuring is not. One will probably hesitate to say that a is better than b because $v(a) = 1.23456789$ and $v(b) = 1.23456788$, in particular when measuring is only done in a rather rough form and the digits result mostly from some sort of computing or reading from a scale. A good idea is to fix some threshold describing imprecision, say $\delta := 0.0001$. Then we consider instead of a value $v(a)$ for a an interval $v(a) - \delta \ldots v(a) + \delta$. In such a case, a is better than b only when the intervals attached do not intersect, i.e., if $v(a) - \delta > v(b) + \delta$. Altogether, we have preference with threshold (cf. Def. 12.16) and indifference so as to respect sensitivity of measuring:

$v(a) \geq v(b) + 2\delta$ means a is at least as good as b,
$v(a) > v(b) + 2\delta$ means a is better than b,
$|v(a) - v(b)| \leq 2\delta$ means a is as good as b.

But what about measuring temperatures in degrees Kelvin, for example. This would use a logarithmic scale, so that fixing a constant δ seems inadequate. The natural idea is to use a function $\delta(a)$ that is also logarithmic. Then immediately questions of monotony come up, etc.

The idea of being 'greater by more than a threshold t' in $\mathbb{R}$ shall now be transferred to a relation T. We aim at

$(x, y) \in T$ as corresponding to $v(x) > v(y) + t$

and see immediately that

$(u, y) \in T$ for every u with $(u, x) \in T$ since then by definition $v(u) > v(x) + t$

$(x, z) \in T$ for every z with $(y, z) \in T$ since then by definition $v(y) > v(z) + t$.

So the full cone above x as well as the full cone below y satisfy the condition. For this, we have already developed an algebraic machinery. A semiorder was what we finally identified as appropriate.

$S = P^{\mathrm{d}}$

	Alfred	Barbara	Christian	Donald	Eugene	Frederick	George
Alfred	0	0	1	1	1	1	1
Barbara	0	0	1	1	1	1	1
Christian	0	0	0	0	1	1	1
Donald	0	0	0	0	0	1	1
Eugene	0	0	0	0	0	1	1
Frederick	0	0	0	0	0	0	1
George	0	0	0	0	0	0	0

$S^{\mathrm{d}} = P$

	Alfred	Barbara	Christian	Donald	Eugene	Frederick	George
Alfred	1	1	1	1	1	1	1
Barbara	1	1	1	1	1	1	1
Christian	0	0	1	1	1	1	1
Donald	0	0	1	1	1	1	1
Eugene	0	0	0	1	1	1	1
Frederick	0	0	0	0	0	1	1
George	0	0	0	0	0	0	1

I_S

	Alfred	Barbara	Christian	Donald	Eugene	Frederick	George
Alfred	1	1	0	0	0	0	0
Barbara	1	1	0	0	0	0	0
Christian	0	0	1	1	0	0	0
Donald	0	0	1	1	0	0	0
Eugene	0	0	0	1	1	0	0
Frederick	0	0	0	0	0	1	0
George	0	0	0	0	0	0	1

W

	Alfred	Barbara	Christian	Donald	Eugene	Frederick	George
Alfred	0	0	1	1	1	1	1
Barbara	0	0	1	1	1	1	1
Christian	0	0	0	1	1	1	1
Donald	0	0	0	0	1	1	1
Eugene	0	0	0	0	0	1	1
Frederick	0	0	0	0	0	0	1
George	0	0	0	0	0	0	0

I_W

	Alfred	Barbara	Christian	Donald	Eugene	Frederick	George
Alfred	1	1	0	0	0	0	0
Barbara	1	1	0	0	0	0	0
Christian	0	0	1	0	0	0	0
Donald	0	0	0	1	0	0	0
Eugene	0	0	0	0	1	0	0
Frederick	0	0	0	0	0	1	0
George	0	0	0	0	0	0	1

J

	Alfred	Barbara	Christian	Donald	Eugene	Frederick	George
Alfred	0	0	0	0	0	0	0
Barbara	0	0	0	0	0	0	0
Christian	0	0	0	1	0	0	0
Donald	0	0	1	0	1	0	0
Eugene	0	0	0	1	0	0	0
Frederick	0	0	0	0	0	0	0
George	0	0	0	0	0	0	0

Fig. 13.6 Semiorder, respectively 'reflexive semiorder', with duals, indifferences, and S-incomparability

So, when given a semiorder S that is not yet a weakorder, we have already determined the smallest strict enlargement $W \supseteq S$, the weakorder closure. According to Prop. 12.18.ii, this is found by looking for additional pairs (a, b) with $(a, b') \in S$ where b' is S-indifferent to b, or else a is S-indifferent with some a' and $(a', b) \in S$. This is quite intuitive. It is then interesting to consider the two indifferences encountered, namely

$$I_S := S^{\mathrm{d}} \cap \overline{S} \quad \text{and} \quad I_W := W^{\mathrm{d}} \cap \overline{W},$$

formed from the semiorder or from its weakorder closure, respectively. From $S \subseteq W$, we conclude $I_W \subseteq I_S$. Instead of considering these two indifferences, I_W, I_S, we decide to investigate in which way the indifference I_S exceeds I_W and to introduce a name for it.

In the example of Fig. 13.6, we consider *Christian* and *Donald*. It is immediately clear that a *direct* comparison in one of the directions, either with S, or with S^{T}, will not reveal a distinction between them; both are thus S-indifferent. Things change when switching to the weakorder closure $W = I_{S^,} S \cup S^, I_S \supseteq S$; then *Christian* and *Donald* are no longer W-indifferent, but are still S-indifferent.

Definition 13.6. Let a semiorder S be given together with its weakorder closure $W := S^{\mathrm{d}} \, ; S \cup S \, ; S^{\mathrm{d}} = I_S \, ; S \cup S \, ; I_S$ according to Prop. 12.18. We define S-**incomparability** J with regard to S, W as

$$J := I_S \cap \overline{I_W} = X \cup X^{\mathsf{T}} \quad \text{where } X = W \cap \overline{S}. \qquad \square$$

Remembering $W \subseteq S^{\mathrm{d}} = \overline{S}^{\mathsf{T}}$, we convince the reader that $J = X \cup X^{\mathsf{T}}$:

$$J = S^{\mathrm{d}} \cap \overline{S} \cap \overline{W^{\mathrm{d}} \cap \overline{W}} = \overline{S}^{\mathsf{T}} \cap \overline{S} \cap (W^{\mathsf{T}} \cup W) = (W^{\mathsf{T}} \cap \overline{S}^{\mathsf{T}}) \cup (W \cap \overline{S}).$$

Proposition 13.7. (i) *Let a semiorder S be given and consider its weakorder closure $W := S^{\mathrm{d}} \, ; S \cup S \, ; S^{\mathrm{d}} = I_S \, ; S \cup S \, ; I_S$ together with the indifferences I_S and I_W as well as the S-incomparability J.*

— *The indifference I_S need not be an equivalence, but is reflexive and symmetric.*

— *S-incomparability J is symmetric and irreflexive.*

— *The dual $S^{\mathrm{d}} = I_W \cup J \cup S$ is a connex, semi-transitive, Ferrers relation.*

(ii) *For every connex, semi-transitive, Ferrers relation P, the dual $S := P^{\mathrm{d}}$ is a semiorder, the so-called corresponding semiorder. It satisfies $S = \overline{I_W \cup J} \cap P$.*

Proof: (i) I_S is reflexive and symmetric by construction. Figure 13.6 shows an example with I_S not an equivalence. Obviously, J is irreflexive and symmetric also by construction.

To prove the decomposition of S^{d} requires a simple but detailed Boolean argument. From $J = I_S \cap \overline{I_W}$ with $I_W \subseteq I_S$, we find $J \cup I_W = I_S$ where by definition $I_S = S^{\mathrm{d}} \cap \overline{S}$ with $S \subseteq S^{\mathrm{d}}$.

(ii) This follows from Def. 13.6 with $J \cup I_W = I_S$. $\qquad \square$

So a semiorder always comes together with its weakorder closure, but also with its dual that is a connex, semi-transitive, Ferrers relation. As this is not at all a simple matter, we provide yet another example with Fig. 13.7, first unordered and then with some algebraic visualization that makes perception easier.

$$
\begin{array}{c}
\begin{array}{c}
\text{Mon Tue Wed Thu Fri Sat}
\end{array}\\
\begin{array}{c}
\text{Mon}\\\text{Tue}\\\text{Wed}\\\text{Thu}\\\text{Fri}\\\text{Sat}
\end{array}
\begin{pmatrix}
0&1&0&1&1&0\\
0&0&0&0&0&0\\
0&1&0&1&1&1\\
0&1&0&0&0&0\\
0&1&0&0&0&0\\
0&1&0&0&0&0
\end{pmatrix}\\
S = P^{\mathrm d}
\end{array}
\qquad
\begin{array}{c}
\begin{pmatrix}
1&1&1&1&1&1\\
0&1&0&0&0&0\\
1&1&1&1&1&1\\
0&1&0&1&1&1\\
0&1&0&1&1&1\\
1&1&0&1&1&1
\end{pmatrix}\\
S^{\mathrm d} = P
\end{array}
\qquad
\begin{array}{c}
\begin{pmatrix}
1&0&1&0&0&1\\
0&1&0&0&0&0\\
1&0&1&0&0&0\\
0&0&0&1&1&1\\
0&0&0&1&1&1\\
1&0&0&1&1&1
\end{pmatrix}\\
I_S
\end{array}
\qquad
\begin{array}{c}
\begin{pmatrix}
1&0&0&0&0&0\\
0&1&0&0&0&0\\
0&0&1&0&0&0\\
0&0&0&1&1&0\\
0&0&0&1&1&0\\
0&0&0&0&0&1
\end{pmatrix}\\
I_W
\end{array}
\qquad
\begin{array}{c}
\begin{pmatrix}
0&0&1&0&0&1\\
0&0&0&0&0&0\\
1&0&0&0&0&0\\
0&0&0&0&0&1\\
0&0&0&0&0&1\\
1&0&0&1&1&0
\end{pmatrix}\\
J
\end{array}
$$

$$
\begin{array}{c}
\begin{array}{c}
\text{Wed Mon Sat Thu Fri Tue}
\end{array}\\
\begin{array}{c}
\text{Wed}\\\text{Mon}\\\text{Sat}\\\text{Thu}\\\text{Fri}\\\text{Tue}
\end{array}
\begin{pmatrix}
0&0&1&1&1&1\\
0&0&0&1&1&1\\
0&0&0&0&0&1\\
0&0&0&0&0&1\\
0&0&0&0&0&1\\
0&0&0&0&0&0
\end{pmatrix}\\
S = P^{\mathrm d}
\end{array}
\qquad
\begin{array}{c}
\begin{pmatrix}
1&1&1&1&1&1\\
1&1&1&1&1&1\\
0&1&1&1&1&1\\
0&0&1&1&1&1\\
0&0&1&1&1&1\\
0&0&0&0&0&1
\end{pmatrix}\\
S^{\mathrm d} = P
\end{array}
\qquad
\begin{array}{c}
\begin{pmatrix}
1&1&0&0&0&0\\
1&1&1&0&0&0\\
0&1&1&1&1&0\\
0&0&1&1&1&0\\
0&0&1&1&1&0\\
0&0&0&0&0&1
\end{pmatrix}\\
I_S
\end{array}
\qquad
\begin{array}{c}
\begin{pmatrix}
1&0&0&0&0&0\\
0&1&0&0&0&0\\
0&0&1&0&0&0\\
0&0&0&1&1&0\\
0&0&0&1&1&0\\
0&0&0&0&0&1
\end{pmatrix}\\
I_W
\end{array}
\qquad
\begin{array}{c}
\begin{pmatrix}
0&1&0&0&0&0\\
1&0&1&0&0&0\\
0&1&0&1&1&0\\
0&0&1&0&0&0\\
0&0&1&0&0&0\\
0&0&0&0&0&0
\end{pmatrix}\\
J
\end{array}
$$

Fig. 13.7 Semiorder, respectively 'reflexive semiorder', with duals, indifference, and S-incomparability; in original and in rearranged form

Exercises

13.4 Decompose the semiorder S given below into I_W, I_S and J:

$$
\begin{array}{c}
\text{1\ \ 2\ \ 3\ \ 4\ \ 5\ \ 6\ \ 7\ \ 8\ \ 9\ \ 10}\\
\begin{array}{r}
1\\2\\3\\4\\5\\6\\7\\8\\9\\10
\end{array}
\begin{pmatrix}
0&0&1&1&0&0&1&1&0&1\\
0&0&1&1&1&1&1&1&0&1\\
0&0&0&0&0&0&0&0&0&0\\
0&0&0&0&0&0&0&0&0&1\\
0&0&0&1&0&0&0&1&0&1\\
0&0&1&1&0&0&1&1&0&1\\
0&0&1&1&0&0&0&1&0&1\\
0&0&0&0&0&0&0&0&0&0\\
0&1&1&1&1&1&1&1&0&1\\
0&0&0&0&0&0&0&0&0&0
\end{pmatrix}
\end{array}
$$

13.4 Preference structures

So far, we have studied orderings in various forms. With preference structures, one tries to tackle situations where one expects orderings to occur, from another angle. One favorite approach is to partition a square universal relation into several others and to try on the basis of these parts to find some justification for arranging the items into some order. This is obviously not possible for an arbitrary partitioning; so one asks for the conditions under which it becomes possible.

Two variant forms are conceivable. It may, firstly, be assumed that somebody (an individual, a decision-maker, etc.) has been asked to provide mutually exclusive qualifications of all pairs as

- a is clearly preferred to b,
- feeling indifferent about a and b,
- having no idea how to compare a and b.

The results are then collected in three relations on the set of items, namely P, I, J. What one readily postulates is that these qualifications satisfy the rules of Def. 13.8, in order to meet the standard semantics of our language.

Definition 13.8. Any disjoint partition $\mathbb{T} = P \cup P^\mathsf{T} \cup I \cup J$ of a homogeneous universal relation into the relations denoted as P, P^T (strict) preference and its opposite, I indifference, and J incomparability shall be called a (P, I, J)-**preference-3-structure**,[2] provided

- P is asymmetric,
- I is reflexive and symmetric, and
- J is irreflexive and symmetric. ◻

From these three, one immediately arrives at the so-called underlying characteristic relation of this preference structure from which they can all be reproduced. Let a relation R be given that is considered as the outcome of asking for any two alternatives a, b whether

- a is *not worse* than b.

When meeting our colloquial use of language, this makes R immediately a reflexive relation. It is sometimes called a *weak preference relation*.

Proposition 13.9. (i) *There is a one-to-one correspondence between reflexive homogeneous relations R and preference-3-structures (P, I, J), which may be given explicitly with transitions $\alpha : R \mapsto (P, I, J)$ and $\beta : (P, I, J) \mapsto R$ as*

$$\alpha(R) := (R \cap \overline{R}^\mathsf{T}, R \cap R^\mathsf{T}, \overline{R} \cap \overline{R}^\mathsf{T}) \qquad and \qquad \beta(P, I, J) := P \cup I.$$

(ii) *There is a one-to-one correspondence between irreflexive homogeneous relations S and preference-3-structures (P, I, J), which may be given explicitly with transitions $\gamma : S \mapsto (P, I, J)$ and $\delta : (P, I, J) \mapsto S$ as*

$$\gamma(S) := (S \cap \overline{S}^\mathsf{T}, \overline{S}^\mathsf{T} \cap \overline{S}, S \cap S^\mathsf{T}) \qquad and \qquad \delta(P, I, J) := \overline{P \cup I}^\mathsf{T}.$$

Proof: (i) It is trivial but tedious to show that this is a bijection. We use that $\overline{P \cup I} = P^\mathsf{T} \cup J$ due to the partition. Let $(P', I', J') = \alpha(\beta(P, I, J))$, so that

---

[2] Preference structures with 4 constituent relations can also be found in the literature.

$$P' = (P \cup I) \cap \overline{P \cup I}^{\mathsf{T}} = (P \cup I) \cap (P^{\mathsf{T}} \cup J)^{\mathsf{T}} = P \cup (I \cap J) = P \cup \mathbb{L} = P,$$

$$I' = (P \cup I) \cap (P \cup I)^{\mathsf{T}} = (P \cup I) \cap (P^{\mathsf{T}} \cup I) = (P \cap P^{\mathsf{T}}) \cup I = \mathbb{L} \cup I = I,$$

$$J' = \overline{P \cup I} \cap \overline{P \cup I}^{\mathsf{T}} = (P^{\mathsf{T}} \cup J) \cap (P^{\mathsf{T}} \cup J)^{\mathsf{T}} = (P^{\mathsf{T}} \cap P) \cup J = \mathbb{L} \cup J = J.$$

In the other direction, we show

$$\beta(\alpha(R)) = (R \cap \overline{R}^{\mathsf{T}}) \cup (R \cap R^{\mathsf{T}}) = R \cap (\overline{R}^{\mathsf{T}} \cup R^{\mathsf{T}}) = R \cap \mathbb{T} = R.$$

So we have a bijection.

It is then trivial to see that $\beta(P,I,J)$ is reflexive. In a similar way, we convince ourselves that the second and the third components of $\alpha(R)$ are symmetric by construction. Since R is assumed to be reflexive, so is the second component of $\alpha(R)$, and the third is necessarily irreflexive. The first is asymmetric because

$$(R \cap \overline{R}^{\mathsf{T}})^{\mathsf{T}} = R^{\mathsf{T}} \cap \overline{R} \subseteq \overline{R} \cup R^{\mathsf{T}} = \overline{R \cap \overline{R}^{\mathsf{T}}}.$$

(ii) This is left as Exercise 13.6.   □

For any reflexive relation R expressing such weak preference, one may thus define

- $P := R \cap \overline{R}^{\mathsf{T}}$, *strict preference*,
- $I := R \cap R^{\mathsf{T}}$, *indifference*,
- $J := \overline{R} \cap \overline{R}^{\mathsf{T}}$, *incomparability*.

No assumptions as to transitivity are being made. It is, thus, a different approach compared with Section 13.3. Particularly irritating is the tradition of using the same name *indifference* for the concept just defined and that of Def. 13.3. One will find the example of Fig. 13.8 even more irritating; there is a semiorder at the first position, but the internal mechanics of the preference-3-structure evaluate $\beta(P,I,J)$ and $\delta(P,I,J)$ in quite an unexpected way; compare with Fig. 13.7.

$$
\begin{array}{c}
\begin{array}{cc}
& \begin{smallmatrix} 1 & 2 & 3 & 4 & 5 & 6 \end{smallmatrix} \\
\begin{smallmatrix} 1 \\ 2 \\ 3 \\ 4 \\ 5 \\ 6 \end{smallmatrix} &
\begin{pmatrix}
0 & 0 & 1 & 1 & 1 & 1 \\
0 & 0 & 0 & 1 & 1 & 1 \\
0 & 0 & 0 & 0 & 0 & 1 \\
0 & 0 & 0 & 0 & 1 & 1 \\
0 & 0 & 0 & 0 & 0 & 1 \\
0 & 0 & 0 & 0 & 0 & 0
\end{pmatrix}
\end{array}
\quad
\begin{array}{cc}
& \begin{smallmatrix} 1 & 2 & 3 & 4 & 5 & 6 \end{smallmatrix} \\
\begin{smallmatrix} 1 \\ 2 \\ 3 \\ 4 \\ 5 \\ 6 \end{smallmatrix} &
\begin{pmatrix}
1 & 0 & 0 & 0 & 0 & 0 \\
0 & 1 & 0 & 0 & 0 & 0 \\
0 & 0 & 1 & 0 & 0 & 0 \\
0 & 0 & 0 & 1 & 1 & 0 \\
0 & 0 & 0 & 1 & 1 & 0 \\
0 & 0 & 0 & 0 & 0 & 1
\end{pmatrix}
\end{array}
\quad
\begin{array}{cc}
& \begin{smallmatrix} 1 & 2 & 3 & 4 & 5 & 6 \end{smallmatrix} \\
\begin{smallmatrix} 1 \\ 2 \\ 3 \\ 4 \\ 5 \\ 6 \end{smallmatrix} &
\begin{pmatrix}
0 & 1 & 0 & 0 & 0 & 0 \\
1 & 0 & 1 & 0 & 0 & 0 \\
0 & 1 & 0 & 1 & 1 & 0 \\
0 & 0 & 1 & 0 & 0 & 0 \\
0 & 0 & 1 & 0 & 0 & 0 \\
0 & 0 & 0 & 0 & 0 & 0
\end{pmatrix}
\end{array}
\\[2em]
\begin{array}{cc}
& \begin{smallmatrix} 1 & 2 & 3 & 4 & 5 & 6 \end{smallmatrix} \\
\begin{smallmatrix} 1 \\ 2 \\ 3 \\ 4 \\ 5 \\ 6 \end{smallmatrix} &
\begin{pmatrix}
1 & 0 & 1 & 1 & 1 & 1 \\
0 & 1 & 0 & 1 & 1 & 1 \\
0 & 0 & 1 & 0 & 0 & 1 \\
0 & 0 & 0 & 1 & 1 & 1 \\
0 & 0 & 0 & 1 & 1 & 1 \\
0 & 0 & 0 & 0 & 0 & 1
\end{pmatrix}
\end{array}
\qquad
\begin{array}{cc}
& \begin{smallmatrix} 1 & 2 & 3 & 4 & 5 & 6 \end{smallmatrix} \\
\begin{smallmatrix} 1 \\ 2 \\ 3 \\ 4 \\ 5 \\ 6 \end{smallmatrix} &
\begin{pmatrix}
0 & 1 & 1 & 1 & 1 & 1 \\
1 & 0 & 1 & 1 & 1 & 1 \\
0 & 1 & 0 & 1 & 1 & 1 \\
0 & 0 & 1 & 0 & 0 & 1 \\
0 & 0 & 1 & 0 & 0 & 1 \\
0 & 0 & 0 & 0 & 0 & 0
\end{pmatrix}
\end{array}
\end{array}
$$

Fig. 13.8 Preference-3-structure (P,I,J) with corresponding $\beta(P,I,J)$ and $\delta(P,I,J)$

We are going to relate this concept step by step with our hierarchy *linear strict-order/weakorder/semiorder/intervalorder*. Another major point of concern has

always been whether there can occur $(P \cup I)$-circuits,[3] since this is considered coun-
terintuitive for concepts of preference and indifference. This will also be studied for
the special versions.

Partition axiomatics for intervalorders

Intervalorders admit a preference structure characterization.

Lemma 13.10. (i) *Let a* (P, I, J)-*preference-3-structure be given. If the additional
properties*

$$J = \mathbb{\bot} \quad and \quad P \,\!\!\,\,_\text{;}\, I \,\!\!\,\,_\text{;}\, P \subseteq P$$

are satisfied, then P *is an intervalorder.*

(ii) *If* P *is an intervalorder, then every preference-3-structure* (P, I, J) *will satisfy*

$$P \,\!\!\,\,_\text{;}\, I \,\!\!\,\,_\text{;}\, P \subseteq P;$$

among these are, for example, $(P, \overline{P \cup P^{\mathsf{T}}}, \mathbb{\bot})$ *and* $(P, \mathbb{I}, \overline{\mathbb{I}} \cap \overline{P \cup P^{\mathsf{T}}})$.

Proof: (i) With $J = \mathbb{\bot}$, we have the partition $\mathbb{T} = P \cup P^{\mathsf{T}} \cup I$. Since $\mathbb{I} \subseteq I$ holds
by definition, P is transitive and irreflexive. Because

$$P \,\!\!\,\,_\text{;}\, \overline{P}^{\mathsf{T}} \,\!\!\,\,_\text{;}\, P = P \,\!\!\,\,_\text{;}\, (P^{\mathsf{T}} \cup I)^{\mathsf{T}} \,\!\!\,\,_\text{;}\, P = P \,\!\!\,\,_\text{;}\, (P \cup I) \,\!\!\,\,_\text{;}\, P = P \,\!\!\,\,_\text{;}\, P \,\!\!\,\,_\text{;}\, P \cup P \,\!\!\,\,_\text{;}\, I \,\!\!\,\,_\text{;}\, P \subseteq P,$$

the Ferrers property is satisfied and, thus, an intervalorder established.

(ii) Let P be an intervalorder. Then due to the partition $\mathbb{T} = P \cup P^{\mathsf{T}} \cup I \cup J$
certainly

$$P \,\!\!\,\,_\text{;}\, I \,\!\!\,\,_\text{;}\, P \subseteq P \,\!\!\,\,_\text{;}\, \overline{P}^{\mathsf{T}} \,\!\!\,\,_\text{;}\, P \subseteq P.$$

Because P is asymmetric, the properties according to Def. 13.8 hold with $I :=
\overline{P \cup P^{\mathsf{T}}}$ and $J := \mathbb{\bot}$ as well as with $I' := \mathbb{I}$ and $J' := \overline{\mathbb{I}} \cap \overline{P \cup P^{\mathsf{T}}}$. □

Concerning $(P \cup I)$-circuits, i.e., picycles, the statement for intervalorders is as
follows: every circuit of length at least 2 must contain two consecutive Is. Assume
a circuit of length at least 2 without. Then, via transitivity, consecutive Ps may be
contracted to just one; sequences $P \,\!\!\,\,_\text{;}\, I \,\!\!\,\,_\text{;}\, P$ may also be contracted to just a P. What
remains is a circuit $P \,\!\!\,\,_\text{;}\, P$ or a circuit $P \,\!\!\,\,_\text{;}\, I$, both of which cannot exist since P, P^{T}, I
is a partition. A first example has already been given with Fig. 6.3.

Investigating circuits with respect to $P \cup I$ for an intervalorder P, it may, however,
happen that a proper $(P \cup I)$-circuit contains two consecutive I-arcs; see Fig. 13.9.

---

[3] In [51], this has been given the appealing denotation 'picycle'.

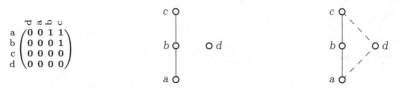

Fig. 13.9 An intervalorder with a circuit a, b, c, d, a containing two consecutive Is

We can also phrase the idea of interrelationship between intervalorders and special preference structures slightly differently.

Proposition 13.11. *Let P be an arbitrary strictorder, and consider it together with indifference $I := \overline{P \cup P^\mathsf{T}}$ according to Def. 13.3 and $J := \mathbb{1}$ as a preference-3-structure (P, I, J). Then*

$$P \text{ intervalorder} \iff P \text{ Ferrers} \iff P\,;I \text{ asymmetric} \iff I\,;P \text{ asymmetric}.$$

Proof: The first equivalence holds by definition. Now, we concentrate on the middle equivalence; the right one follows then from symmetry. The condition that $P\,;I$ is asymmetric means

$$(P\,;I)^\mathsf{T} \subseteq \overline{P\,;I} \quad \text{or, equivalently,} \quad I\,;P\,;I \subseteq \overline{P}^\mathsf{T},$$

which decomposes to

$$(\overline{P} \cap \overline{P}^\mathsf{T})\,;P\,;(\overline{P} \cap \overline{P}^\mathsf{T}) \subseteq \overline{P}^\mathsf{T}.$$

For "$\Longrightarrow$", we are allowed to use the Ferrers condition for $\overline{P}^\mathsf{T}$, since the latter is Ferrers precisely when P is:

$$\overline{P}^\mathsf{T}\,;P\,;\overline{P}^\mathsf{T} \subseteq \overline{P}^\mathsf{T}.$$

For the reverse direction "$\Longleftarrow$", one will decompose $\overline{P}^\mathsf{T} = (P \cap \overline{P}^\mathsf{T}) \cup (\overline{P} \cap \overline{P}^\mathsf{T})$ on the left side and handle the additive terms separately. One case follows from the aforementioned asymmetry condition. In the three others, one will always find a product of two non-negated Ps, which will be contracted with transitivity before applying the Schröder rule as in

$$(P \cap \overline{P}^\mathsf{T})\,;P\,;(P \cap \overline{P}^\mathsf{T}) \subseteq P\,;P\,;\overline{P}^\mathsf{T} \subseteq P\,;\overline{P}^\mathsf{T} \subseteq \overline{P}^\mathsf{T}. \qquad \square$$

Partition axiomatics for semiorders

We now relate the preference-3-structure concept with semiorders. To this end, we recall that a semiorder is irreflexive, semi-transitive, and Ferrers.

Lemma 13.12. (i) *Let a (P, I, J)-preference-3-structure be given. If the additional properties*

$$J = \mathbb{1} \qquad P \,;\, I \,;\, P \subseteq P \qquad P \,;\, P \cap I \,;\, I = \mathbb{1}$$

are satisfied, then P is a semiorder.

(ii) *If P is a semiorder, then every preference-3-structure (P, I, J) will satisfy*

$$P \,;\, I \,;\, P \subseteq P \qquad P \,;\, P \cap I \,;\, I = \mathbb{1};$$

among these are, for example, $(P, \overline{P \cup P^{\mathsf{T}}}, \mathbb{1})$ and $(P, \mathbb{I}, \overline{\mathbb{I}} \cap \overline{P \cup P^{\mathsf{T}}})$.

Proof: (i) We maintain from Lemma 13.10 that the first two conditions are satisfied precisely when an intervalorder has been presented. This leaves us with the task of proving that the third condition, in this context, implies semi-transitivity

$$P \,;\, P \,;\, \overline{P}^{\mathsf{T}} = P \,;\, P \,;\, (P \cup I) = P \,;\, P \,;\, P \cup P \,;\, P \,;\, I \subseteq P.$$

Due to transitivity, only $P \,;\, P \,;\, I$ needs further discussion. Its intersection with I is $\mathbb{1}$ since with the Dedekind rule

$$P \,;\, P \,;\, I \cap I \subseteq (P \,;\, P \cap I \,;\, I) \,;\, (\dots) = \mathbb{1}.$$

But also the intersection with P^{T} is $\mathbb{1}$:

$$P \,;\, P \,;\, I \cap P^{\mathsf{T}} \subseteq (\dots) \,;\, (I \cap P^{\mathsf{T}} \,;\, P^{\mathsf{T}} \,;\, P^{\mathsf{T}}) \subseteq (\dots) \,;\, (I \cap P^{\mathsf{T}}) = \mathbb{1}$$

(ii) Let P be a semiorder. We start from the partition $\mathbb{T} = P \cup P^{\mathsf{T}} \cup I \cup J$ and prove the second property:

$$\begin{array}{lll} P \,;\, P \,;\, \overline{P}^{\mathsf{T}} \subseteq P, & \text{i.e., starting from semi-transitivity} \\ \Longleftrightarrow \quad \overline{P} \,;\, P^{\mathsf{T}} \,;\, P^{\mathsf{T}} \subseteq P^{\mathsf{T}} & \text{transposed} \\ \Longleftrightarrow \quad \overline{P}^{\mathsf{T}} \,;\, P \,;\, P \subseteq P & \text{Schröder} \\ \Longrightarrow \quad I \,;\, P \,;\, P \subseteq P \subseteq \overline{I} & \text{since } I \subseteq \overline{P}^{\mathsf{T}} \text{ and } P \subseteq \overline{I} \text{ due to the partition} \\ \Longleftrightarrow \quad I \,;\, I \subseteq \overline{P \,;\, P} & \text{Schröder} \end{array}$$

and the latter is, up to a Boolean argument, what had to be proved. The first property follows as for intervalorders. □

Investigating circuits with respect to $P \cup I$ for a semiorder P, we observe that every $(P \cup I)$-circuit will necessarily contain more I-arcs than P-arcs. All conceivable circuits with up to three arrows are easily checked to find out that they contain strictly more Is than Ps. A circuit $a\ P\ b\ I\ c\ P\ d\ I\ a$ cannot exist because for reasons of partitioning $P \,;\, I \,;\, P \subseteq P$ does not allow the fourth arrow to be in I. A circuit $a\ P\ b\ P\ c\ I\ d\ I\ a$ cannot exist because $P \,;\, P \cap I \,;\, I = \mathbb{1}$.

Now, we conclude by induction on the number of arrows, assuming that up to n-arc circuits it has already been established that there are more I-arcs than P-arcs. Assuming that for $n + 1$ there were more P-arcs than I-arcs, there would be two of them consecutive and with transitivity these could be replaced by just one, arriving at a situation for n with at least as many P-arcs as I-arcs, contradicting the induction hypothesis. Assuming equally many P-arcs and I-arcs in a strictly alternating sequence, every sequence P_iI_iP might iteratively be cut down to P, finally arriving at a loop in the asymmetric P; a contradiction. Assuming equally many P-arcs and I-arcs non-alternating, one will identify a situation $P_iP_iI \subseteq P_iP_i\overline{P}^\top \subseteq P$, the latter by semi-transitivity', thus decreasing the number of Ps as well as the number of Is by one and arriving at a violation of the induction hypothesis.

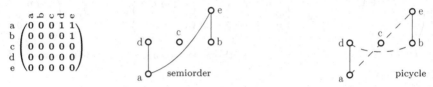

Fig. 13.10 A semiorder with a circuit a, d, b, e, c, a of 3 Is and just 2 Ps

Partition axiomatics for weakorders

There is also a very close connection between preference-3-structures and weak-orders.

Lemma 13.13. (i) *Let a (P, I, J)-preference-3-structure be given. If the additional properties*

$$J = \perp\!\!\!\perp \qquad P_iP \subseteq P \qquad I_iI \subseteq I,$$

are satisfied, then P is a weakorder.

(ii) *If P is a weakorder, then every preference-3-structure $(P, I, \perp\!\!\!\perp)$ will satisfy*

$$P_iP \subseteq P \qquad I_iI \subseteq I.$$

Proof: (i) Here again, the partition is $\mathbb{T} = P \cup P^\top \cup I$. Thus, P is irreflexive since I is postulated to be reflexive. Negative transitivity follows from

$$\overline{P}_i\overline{P} = (P^\top \cup I)_i(P^\top \cup I) \subseteq (P^2)^\top \cup P^\top_iI \cup I_iP^\top \cup I^2 \subseteq P^\top \cup \overline{P} \cup \overline{P} \cup I = \overline{P}.$$

We have used that P and I are transitive and that, for example, $P^\top_iI \subseteq \overline{P} \iff P_iP \subseteq \overline{I}$, where the latter follows from transitivity and the partitioning. Irreflexivity and negative transitivity is sufficient for a weakorder according to Prop. 12.5.

(ii) The first part $P_iP \subseteq P$ is trivially satisfied for a weakorder. We start again from the partition $\mathbb{T} = P \cup P^\top \cup I$ and prove

$$I;I = \overline{P \cup P^{\mathsf{T}}};\overline{P \cup P^{\mathsf{T}}} \subseteq \overline{P \cup P^{\mathsf{T}}} = I,$$

which follows from negative transitivity. □

$$\begin{array}{c} \\ a\\ b\\ c\\ d\\ e \end{array}\begin{array}{c} \text{a b c d e} \\ \begin{pmatrix} 0&0&1&1&1\\ 0&0&1&1&1\\ 0&0&0&0&0\\ 0&0&0&0&0\\ 0&0&0&0&0 \end{pmatrix}\end{array} \qquad \begin{array}{c} \text{a b c d e} \\ \begin{pmatrix} 1&1&0&0&0\\ 1&1&0&0&0\\ 0&0&1&1&1\\ 0&0&1&1&1\\ 0&0&1&1&1 \end{pmatrix}\end{array} = I_1 \quad J_1 = \begin{array}{c} \text{a b c d e} \\ \begin{pmatrix} 0&0&0&0&0\\ 0&0&0&0&0\\ 0&0&0&0&0\\ 0&0&0&0&0\\ 0&0&0&0&0 \end{pmatrix}\end{array}$$

$$\begin{array}{c} \\ a\\ b\\ c\\ d\\ e \end{array}\begin{array}{c} \text{a b c d e} \\ \begin{pmatrix} 1&1&0&0&0\\ 1&1&0&0&0\\ 0&0&1&0&1\\ 0&0&0&1&1\\ 0&0&1&1&1 \end{pmatrix}\end{array} = I_2 \quad J_2 = \begin{array}{c} \text{a b c d e} \\ \begin{pmatrix} 0&0&0&0&0\\ 0&0&0&0&0\\ 0&0&0&1&0\\ 0&0&1&0&0\\ 0&0&0&0&0 \end{pmatrix}\end{array} \qquad \begin{array}{c} \text{a b c d e} \\ \begin{pmatrix} 1&0&0&0&0\\ 0&1&0&0&0\\ 0&0&1&0&0\\ 0&0&0&1&0\\ 0&0&0&0&1 \end{pmatrix}\end{array} = I_3 \quad J_3 = \begin{array}{c} \text{a b c d e} \\ \begin{pmatrix} 0&1&0&0&0\\ 1&0&0&0&0\\ 0&0&0&1&1\\ 0&0&1&0&1\\ 0&0&1&1&0 \end{pmatrix}\end{array}$$

Fig. 13.11 A weakorder P amended in three ways to a preference-3-structure (P, I_i, J_i)

Investigating circuits with respect to $P \cup I$ for a weakorder P when $J = \perp\!\!\!\perp$, we observe that consecutive Is can be reduced to I in the same way as consecutive Ps can be reduced to just P. So alternating circuits of P and I remain. But also

$$P;I = P;I \cap \top\!\!\!\top = P;I \cap (P \cup P^{\mathsf{T}} \cup I) = (P;I \cap P) \cup (P;I \cap P^{\mathsf{T}}) \cup (P;I \cap I) = P;I \cap P$$

using

$$P;I \cap P^{\mathsf{T}} \subseteq (P \cap P^{\mathsf{T}};I);(I \cap P^{\mathsf{T}};P^{\mathsf{T}}) \subseteq (\dots);(I \cap P^{\mathsf{T}}) = \perp\!\!\!\perp$$
$$P;I \cap I \subseteq (P \cap I;I);(\dots) \subseteq (P \cap I);(\dots) = \perp\!\!\!\perp$$

so that $P;I \subseteq P$. This finally means that powers of $P \cup I$ are contained in $P \cup I = \overline{P}^{\mathsf{T}}$ and, thus, cannot be closed to a circuit. All this is quite intuitive: indifference iterated never leads to a case where preference holds. Preference, indifference, and preference again, will always result in preference.

Partition axiomatics for linear strictorders

We study first which additional conditions make a preference-3-structure a linear strictorder.

Lemma 13.14. *Let a (P, I, J)-preference-3-structure be given. It satisfies the additional properties*

$$J = \perp\!\!\!\perp \qquad P;P \subseteq P \qquad I = \mathbb{I},$$

if and only if P is a linear strictorder.

Proof: Under these specific circumstances, the partition is $\top\!\!\!\top = P \cup P^{\mathsf{T}} \cup \mathbb{I}$, so that P is semi-connex and irreflexive. As directly postulated, P is also transitive, i.e., a linear strictorder. The reverse direction is trivial. □

For a linear strictorder P, a proper circuit can never occur: any composition of $(P \cup I)^i$ with $i \geq 1$ can be reduced to a power of $P^i \cup \mathbb{I}^i$ since $I = \mathbb{I}$, and the first part is contained in the irreflexive P.

Exercises

13.5 Let any connex relation P be given and define

$W := P \cap \overline{P}^\top$, its asymmetric part, and $I := P \cap P^\top$, its symmetric part.

(i) Then $P = W \cup I$ as well as $\mathbb{T} = W \cup W^\top \cup I$ are partitions.

(ii) If $W \,{\scriptstyle i}\, I \,{\scriptstyle i}\, W \subseteq W$ and $I^2 \subseteq \overline{W}$, it will turn out that W is a weakorder.

13.6 Prove Prop. 13.9.ii.

14

Aggregating Preferences

Since the development of lattice theory, it became more and more evident that concepts of upper and lower bounds, suprema and infima did not require orderings to be linear. Nevertheless, fuzziness was mainly studied along the linear order of $\mathbb{R}$ and only later began to be generalized to the ordinal level: numbers indicate the relative positions of items, but no longer the magnitude of difference. Then we moved to the interval level: numbers indicate the magnitude of difference between items, but there is no absolute zero point. Examples are attitude scales and opinion scales. We proceed even further and introduce relational measures with values in a lattice. Measures traditionally provide a basis for integration. Astonishingly, this holds true for these relational measures so that it becomes possible to introduce a concept of relational integration.

With De Morgan triples, we then touch yet another closely related concept of relational methods of preference aggregation.

14.1 Modelling preferences

Anyone who is about to make important decisions will usually base these on carefully selected basic information and clean lines of reasoning. In general, it is not too difficult to apply *just one* criterion and to operate according to this criterion. If several criteria must be taken into consideration, one has also to consider the all too common situation that these provide contradictory information: 'This car looks nicer, but it is much more expensive'. Social and economic sciences have developed techniques to model what takes place when decisions have to be made in an environment with a multitude of diverging criteria. Preference is assumed to represent the degree to which one alternative is preferred to another. Often this takes the form of expressing that alternative A is considered 'not worse than' alternative B. Sometimes a linear ranking of the set of alternatives is assumed, which we avoid.

So making decisions became abstracted to a scientific task. We can observe two lines of development. The Anglo-Saxon countries, in particular, formulated *utility theory*, in which numerical values indicate the intensity of some preference. However, mainly in Continental Europe, binary relations were used to model pairwise preference; see [52], for example. While the former idea can easily be related to statistics, the latter is based on evidence via direct comparison. In earlier years, basic information was quite often statistical in nature and was expressed in real numbers. Today we have more often fuzzy, vague, rough, etc., forms of qualification.

14.2 Introductory example

We first give the example of deciding which car to buy out of several offers. It is intended to follow a set C of three criteria, namely color, price, and speed. They are, of course, not of equal importance for us; price will most certainly outweigh the color of the car, for example. Nevertheless, let the valuation with these criteria be given on an ordinal scale $\mathcal{L}$ with 5 linearly ordered values as indicated on the left side (Fig. 14.1). (Here, for simplicity, the ordering is linear, but it need not be.) We name these values 1,2,3,4,5, but do not combine this with any arithmetic; i.e., value 4 is not intended to mean twice as good as value 2. Rather they might be described with linguistic variables as *bad, not totally bad, medium, outstanding, absolutely outstanding*; these example qualifications have been chosen deliberately not to be 'equidistant'.

$$
\begin{array}{l}
\text{color} \\
\text{price} \\
\text{speed}
\end{array}
\begin{pmatrix}
0 & 0 & 0 & 1 & 0 \\
0 & 0 & 0 & 1 & 0 \\
0 & 1 & 0 & 0 & 0
\end{pmatrix}
\qquad
\begin{aligned}
4 = \mathtt{lub}\,\big[&\mathtt{glb}\,(4_{v(\text{color})}, 4_{\mu\{\text{c,p}\}}), \\
&\mathtt{glb}\,(4_{v(\text{price})}, 4_{\mu\{\text{c,p}\}}), \\
&\mathtt{glb}\,(2_{v(\text{speed})}, 5_{\mu\{\text{c,p,s}\}})\big]
\end{aligned}
$$

Fig. 14.1 A first valuation

First we concentrate on the left side of Fig. 14.1. The task is to arrive at *one* overall valuation of the car from these three criteria. In a simple-minded approach, we might indeed conceive numbers $1, 2, 3, 4, 5 \in \mathbb{R}$ and then evaluate in a classical way the average value since $\frac{1}{3}(4+4+2) = 3.3333\ldots$, which is a value not expressible on the given scale. When considering the second example, Fig. 14.2, we would arrive at the same average value although the switch from Fig. 14.1 to Fig. 14.2 between price and speed would persuade most people to decide differently.

$$
\begin{array}{l}
\text{color} \\
\text{price} \\
\text{speed}
\end{array}
\begin{pmatrix}
0 & 0 & 0 & 1 & 0 \\
0 & 1 & 0 & 0 & 0 \\
0 & 0 & 0 & 1 & 0
\end{pmatrix}
\qquad
\begin{aligned}
3 = \mathtt{lub}\,\big[&\mathtt{glb}\,(4_{v(\text{color})}, 3_{\mu\{\text{c,s}\}}), \\
&\mathtt{glb}\,(2_{v(\text{price})}, 5_{\mu\{\text{c,p,s}\}}), \\
&\mathtt{glb}\,(4_{v(\text{speed})}, 3_{\mu\{\text{c,s}\}})\big]
\end{aligned}
$$

Fig. 14.2 A second valuation

With relational integration, we learn to make explicit which set of criteria to apply with which weight. It is conceivable that criteria c_1, c_2 are given a low weight but the criteria set $\{c_1, c_2\}$ in conjunction a high one. This means that we introduce a **relational measure** assigning values in $\mathcal{L}$ to subsets of $\mathcal{C}$.

$$
\mu = \begin{matrix} \{\} \\ \{\text{color}\} \\ \{\text{price}\} \\ \{\text{color,price}\} \\ \{\text{speed}\} \\ \{\text{color,speed}\} \\ \{\text{price,speed}\} \\ \{\text{color,price,speed}\} \end{matrix} \begin{pmatrix} 1 & 0 & 0 & 0 & 0 \\ 1 & 0 & 0 & 0 & 0 \\ 0 & 0 & 1 & 0 & 0 \\ 0 & 0 & 0 & 1 & 0 \\ 0 & 1 & 0 & 0 & 0 \\ 0 & 0 & 1 & 0 & 0 \\ 0 & 0 & 1 & 0 & 0 \\ 0 & 0 & 0 & 0 & 1 \end{pmatrix}
$$

Fig. 14.3 A relational measure

For gauging purposes, we demand that the empty criteria set is assigned the least value in $\mathcal{L}$ and the full criteria set the greatest. A point to stress is that we assume the criteria themselves as well as the measuring of subsets of criteria to be *commensurable*.

The relational measure μ should obviously be monotonic with respect to the ordering Ω on the powerset of $\mathcal{C}$ and the ordering E on $\mathcal{L}$. We do not demand continuity (additivity), however. The price alone is ranked of medium importance 3, higher than speed alone, while color alone is considered completely unimportant and ranks 1. However, color and price together are ranked 4, i.e., higher than the supremum of ranks for color alone and for price alone, etc.

Since the valuations according to the criteria as well as the valuation according to the relative measuring of the criteria are now given, we can proceed as visualized on the right sides of Fig. 14.1 and Fig. 14.2. We run through the criteria and always look for two items: their corresponding value and for the value of that subset of criteria *assigning equal or higher values*. Then we determine the greatest lower bound for the two values. From the list thus obtained, the least upper bound is taken. The two examples above show how by simple evaluation along this concept, one will arrive at the overall values 4 or 3, respectively. This results from the fact that in the second case only such rather unimportant criteria as color and speed assign the higher values.

The effect is counter-running: low values of criteria as for s in Fig. 14.1 are intersected with rather high μs as many criteria give higher scores and μ is monotonic. Highest values of criteria as for color or speed in Fig. 14.2 are intersected with the μ of a small or even 1-element criteria set; i.e., with a rather small one. In total we find that here are two operations applied in a way we already know from matrix multiplication: a 'sum' operator, `lub` or "∨", following application of a 'product' operator, `glb` or "∧".

This example gave a first idea of how relational integration works and how it may be useful. Introducing a relational measure and using it for integration serves an important purpose: concerns are now separated. One may design the criteria and the measure in a design phase prior to polling. Only then will the questionnaire be filled in, or the voters polled. The procedure of coming to an overall valuation is now just computation and should no longer lead to quarrels.

14.3 Relational measures

Assume the following basic setting with a set $\mathcal{C}$ of so-called criteria and a measuring lattice $\mathcal{L}$. Depending on the application envisaged, the set $\mathcal{C}$ may also be interpreted as one of the players in a cooperative game, of attributes, of experts, or of voters in an opinion polling problem. This includes the setting with $\mathcal{L}$ the interval $[0, 1] \subseteq \mathbb{R}$ or a linear ordering for measuring. We consider a (relational) measure generalizing the concept of a fuzzy measure (or capacité in French) assigning measures in $\mathcal{L}$ for subsets of $\mathcal{C}$.

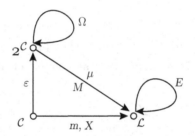

Fig. 14.4 Basic situation for relational integration

The relation ε is the membership relation between $\mathcal{C}$ and its powerset $\mathbf{2}^{\mathcal{C}}$. The measures envisaged will be called μ, other relations as M. Valuations according to the criteria will be X or m depending on the context.

For a running example assume the task of assessing members of staff according to their intellectual abilities and in addition according to the workload they can master.

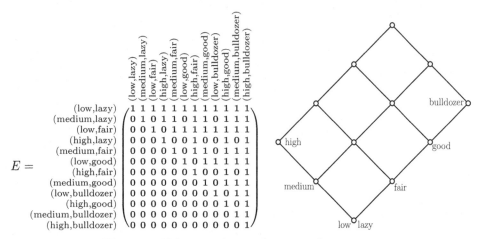

Fig. 14.5 Valuations for a 2-dimensional assessment

Definition 14.1. Suppose a set of criteria $\mathcal{C}$ given together with some lattice $\mathcal{L}$, ordered by E, in which subsets of these criteria are given a measure $\mu : 2^{\mathcal{C}} \longrightarrow \mathcal{L}$. Let Ω be the powerset ordering on $2^{\mathcal{C}}$. We call a mapping $\mu : 2^{\mathcal{C}} \longrightarrow \mathcal{L}$ a **(relational) measure** provided

— $\Omega_{;}\,\mu \subseteq \mu_{;}\,E$,   meaning that μ is isotonic with respect to orderings Ω and E,

— $\mu^{\mathsf{T}};\,0_{\Omega} = 0_{E}$,   the empty subset of $2^{\mathcal{C}}$ is mapped to the least element of $\mathcal{L}$,

— $\mu^{\mathsf{T}};\,1_{\Omega} = 1_{E}$, $\mathcal{C} \in 2^{\mathcal{C}}$ is mapped to the greatest element of $\mathcal{L}$.   □

A (relational) measure for $s \in 2^{\mathcal{C}}$, i.e., $\mu(s)$ when written as a mapping or $\mu^{\mathsf{T}};\,s$ when written in relation form, may be interpreted as the weight of importance we attribute to the combination s of criteria. It should not be confused with a probability. The latter would require the setting $\mathcal{L} = [0,1] \subseteq \mathbb{R}$ and in addition that μ be continuous.

Many ideas of this type have been collected by Glenn Shafer under the heading *theory of evidence*, calling μ a *belief function*. Using it, he explained a basis of rational behavior. We attribute certain weights to evidence, but do not explain in which way. In our case these weights are lattice-ordered. This alone gives us reason to decide rationally this or that way. Real-valued belief functions have numerous applications in artificial intelligence, expert systems, approximate reasoning, knowledge extraction from data, and Bayesian networks.

Definition 14.2. Given this setting, we call the measure μ

(i) a **Bayesian measure** if it is lattice-continuous according to Def. 9.15, i.e.,

$$\operatorname{lub}_E(\mu^{\mathsf{T}};s) = \mu^{\mathsf{T}};\operatorname{lub}_\Omega(s)$$

for a subset $s \subseteq 2^C$ or, expressed differently, a set of subsets of C,

(ii) a **simple support mapping** focused on U valued with v, if U is a non-empty subset $U \subseteq C$ and $v \in \mathcal{L}$ an element such that

$$\mu(s) = \begin{cases} 0_E & \text{if } s \not\supseteq U \\ v & \text{if } C \neq s \supseteq U \\ 1_E & \text{if } C = s. \end{cases} \qquad \square$$

In a real-valued environment, the condition for a Bayesian measure is that it is additive when non-overlapping. Lattice-continuity incorporates two concepts, namely sending 0_Ω to 0_E and additivity

$$\mu^{\mathsf{T}};(s_1 \cup s_2) = \mu^{\mathsf{T}};s_1 \ \cup_{\mathcal{L}} \ \mu^{\mathsf{T}};s_2.$$

Concerning additivity, the example given by Glenn Shafer [132] is of wondering whether a Ming vase is a genuine one or a fake. We have to put the full amount of our belief on the disjunction '*genuine* or *fake*' as one of the alternatives will certainly be the case. But the amount of trust we are willing to put on the alternatives may in both cases be very small as we have only tiny hints for the vase being genuine, but also very tiny hints for it being a fake. We do not put any trust on $0_\Omega = $ 'at the same time *genuine* and *fake*'.

In the extreme case, we have complete ignorance expressed by the so-called **vacuous belief mapping**

$$\mu_0(s) = \begin{cases} 0_E & \text{if } C \neq s \\ 1_E & \text{if } C = s. \end{cases}$$

Using the idea of probability, we could not so easily cope with ignorance. Probability does not allow one to withhold belief from a proposition without *according the withheld amount of belief to the negation*. When thinking about the Ming vase in terms of probability we would have to attribute p to *genuine* and $1 - p$ to *fake*.

On the other extreme, we may completely overspoil our trust expressed by the so-called **overcredulous belief mapping**

$$\mu_1(s) = \begin{cases} 0_E & \text{if } 0_\Omega = s \\ 1_E & \text{otherwise.} \end{cases}$$

Whenever the result for an arbitrary criterion arrives, the overcredulous belief mapping attributes it all the components of trust or belief. In particular, μ_1 is Bayesian while μ_0 is not.

$$\mu_0 = $$

	(low,lazy)	(medium,lazy)	(low,fair)	(high,lazy)	(medium,fair)	(low,good)	(high,fair)	(medium,good)	(low,bulldozer)	(high,good)	(medium,bulldozer)	(high,bulldozer)
{}	1	0	0	0	0	0	0	0	0	0	0	0
{Abe}	1	0	0	0	0	0	0	0	0	0	0	0
{Bob}	1	0	0	0	0	0	0	0	0	0	0	0
{Abe,Bob}	1	0	0	0	0	0	0	0	0	0	0	0
{Carl}	1	0	0	0	0	0	0	0	0	0	0	0
{Abe,Carl}	1	0	0	0	0	0	0	0	0	0	0	0
{Bob,Carl}	1	0	0	0	0	0	0	0	0	0	0	0
{Abe,Bob,Carl}	1	0	0	0	0	0	0	0	0	0	0	0
{Don}	1	0	0	0	0	0	0	0	0	0	0	0
{Abe,Don}	1	0	0	0	0	0	0	0	0	0	0	0
{Bob,Don}	1	0	0	0	0	0	0	0	0	0	0	0
{Abe,Bob,Don}	1	0	0	0	0	0	0	0	0	0	0	0
{Carl,Don}	1	0	0	0	0	0	0	0	0	0	0	0
{Abe,Carl,Don}	1	0	0	0	0	0	0	0	0	0	0	0
{Bob,Carl,Don}	1	0	0	0	0	0	0	0	0	0	0	0
{Abe,Bob,Carl,Don}	0	0	0	0	0	0	0	0	0	0	0	1

$$\mu_1 = $$

	(low,lazy)	(medium,lazy)	(low,fair)	(high,lazy)	(medium,fair)	(low,good)	(high,fair)	(medium,good)	(low,bulldozer)	(high,good)	(medium,bulldozer)	(high,bulldozer)
{}	1	0	0	0	0	0	0	0	0	0	0	0
{Abe}	0	0	0	0	0	0	0	0	0	0	0	1
{Bob}	0	0	0	0	0	0	0	0	0	0	0	1
{Abe,Bob}	0	0	0	0	0	0	0	0	0	0	0	1
{Carl}	0	0	0	0	0	0	0	0	0	0	0	1
{Abe,Carl}	0	0	0	0	0	0	0	0	0	0	0	1
{Bob,Carl}	0	0	0	0	0	0	0	0	0	0	0	1
{Abe,Bob,Carl}	0	0	0	0	0	0	0	0	0	0	0	1
{Don}	0	0	0	0	0	0	0	0	0	0	0	1
{Abe,Don}	0	0	0	0	0	0	0	0	0	0	0	1
{Bob,Don}	0	0	0	0	0	0	0	0	0	0	0	1
{Abe,Bob,Don}	0	0	0	0	0	0	0	0	0	0	0	1
{Carl,Don}	0	0	0	0	0	0	0	0	0	0	0	1
{Abe,Carl,Don}	0	0	0	0	0	0	0	0	0	0	0	1
{Bob,Carl,Don}	0	0	0	0	0	0	0	0	0	0	0	1
{Abe,Bob,Carl,Don}	0	0	0	0	0	0	0	0	0	0	0	1

Fig. 14.6 Vacuous belief mapping and overcredulous belief mapping

Combining measures

Dempster [45] found for the real-valued case a method of combining measures in a form closely related to conditional probability. It shows a way of adjusting opinion in the light of new evidence. We have re-modelled this for the relational case. One should be aware of how a measure behaves on upper and lower cones.

Proposition 14.3. $\mu = \text{lubR}_E(\Omega^\top; \mu)$ $\mu = \text{glbR}_E(\Omega; \mu)$.

Proof: For the measure μ we prove in advance that $\overline{E}; \mu^\top; \Omega^\top = \overline{E}; E^\top; \mu^\top = \overline{E}; \mu^\top$, which is trivial considering

$$\Omega; \mu; E \subseteq \mu; E; E = \mu; E$$

where equality holds also in between. We have applied that μ is a homomorphism $\Omega; \mu \subseteq \mu; E$, that E is transitive, and Ω reflexive.

Therefore we have $\text{glbR}_E(\mu) = \text{glbR}_E(\Omega; \mu) = \text{glbR}_E(\mu; E)$ because $\text{glbR}_E(\mu) = \left[\overline{\overline{E}; \mu^\top} \cap \overline{E^\top}; \overline{\overline{E}; \mu^\top}\right]^\top$. Finally

$$\text{glbR}_E(\mu) = \left[\overline{\overline{E}; \mu^\top} \cap \overline{E^\top}; \overline{\overline{E}; \mu^\top}\right]^\top = \overline{\mu; \overline{E}^\top} \cap \overline{\mu; \overline{E}^\top; E}$$

$$= \mu; \overline{\overline{E}}^\top \cap \mu; \overline{\overline{E}^\top; E} = \mu; E^\top \cap \mu; E = \mu; (E^\top \cap E) = \mu; \mathbb{I} = \mu \qquad \square$$

When, in addition to μ, one has got further evidence from a second measure μ', one will intersect the upper cones resulting in a possibly smaller cone positioned higher up and take its greatest lower bound

$$\mu \oplus \mu' := \mathtt{glbR}_E(\mu; E \cap \mu'; E).$$

One might, however, also see where μ and μ' agree, and thus intersect the lower cones resulting in a possibly smaller cone positioned deeper down and take its least upper bound

$$\mu \otimes \mu' := \mathtt{lubR}_E(\mu; E^{\mathsf{T}} \cap \mu'; E^{\mathsf{T}}).$$

A first observation is as follows

$$
\begin{aligned}
\mu; E \cap \mu'; E &= \overline{\overline{\mu; \overline{E}} \cap \overline{\mu'; \overline{E}}} && \text{since } \mu, \mu' \text{ are mappings}\\
&= \overline{\mu; \overline{E} \cup \mu'; \overline{E}} && \text{De Morgan}\\
&= \overline{(\mu \cup \mu'); \overline{E}} && \text{distributive}\\
&= \mathtt{ubdR}_E(\mu \cup \mu') && \text{by definition of upper bounds taken row-wise}
\end{aligned}
$$

Indeed, we have complete lattices Ω, E, so that all least upper and greatest lower bounds will exist. In this case, intersection of the cones above μ, μ' means taking their least upper bound and the cone above this. So the simpler definitions are

$$\mu \oplus \mu' := \mathtt{glbR}_E(\mu; E \cap \mu'; E) = \mathtt{glbR}_E(\mathtt{ubdR}_E(\mu \cup \mu')) = \mathtt{lubR}_E(\mu \cup \mu'),$$
$$\mu \otimes \mu' := \mathtt{lubR}_E(\mu; E^{\mathsf{T}} \cap \mu'; E^{\mathsf{T}}) = \mathtt{lubR}_E(\mathtt{lbdR}_E(\mu \cup \mu')) = \mathtt{glbR}_E(\mu \cup \mu').$$

Proposition 14.4. *If the measures μ, μ' are given, $\mu \oplus \mu'$ as well as $\mu \otimes \mu'$ are measures again. Both operations are commutative and associative. The vacuous belief mapping μ_0 is the null element while the overcredulous belief mapping is the unit element:*

$$\mu \oplus \mu_0 = \mu, \quad \mu \otimes \mu_1 = \mu, \quad and \quad \mu \otimes \mu_0 = \mu_0.$$

Proof: The gauging condition requires that the least element be sent to the least element, from which we show one part:

$$
\begin{aligned}
(\mu \oplus \mu')^{\mathsf{T}}; 0_\Omega &= \mathtt{lub}_E([\mu \cap \mu']^{\mathsf{T}}); 0_\Omega\\
&= \mathtt{syq}(E^{\mathsf{T}}, \overline{\overline{E}^{\mathsf{T}}(\mu^{\mathsf{T}} \cup \mu'^{\mathsf{T}})}); 0_\Omega && \text{Prop. 9.9}\\
&= \mathtt{syq}(E^{\mathsf{T}}, \overline{\overline{E}^{\mathsf{T}}(\mu^{\mathsf{T}} \cup \mu'^{\mathsf{T}}); 0_\Omega}) && \text{point slips into symmetric quotient}\\
&= \mathtt{syq}(E^{\mathsf{T}}, \overline{\overline{E}^{\mathsf{T}}(\mu^{\mathsf{T}} \cup \mu'^{\mathsf{T}}); 0_\Omega}) && \text{point slipping below negation}\\
&= \mathtt{syq}(E^{\mathsf{T}}, \overline{\overline{E}^{\mathsf{T}}(\mu^{\mathsf{T}}; 0_\Omega \cup \mu'^{\mathsf{T}}; 0_\Omega)}) && \text{distributive}\\
&= \mathtt{syq}(E^{\mathsf{T}}, \overline{\overline{E}^{\mathsf{T}}; 0_E}) && \text{gauging requirement}\\
&= \mathtt{syq}(E^{\mathsf{T}}, \overline{\overline{E}^{\mathsf{T}}}; 0_E) && \text{point slipping out of negation}\\
&= \mathtt{syq}(E^{\mathsf{T}}, E^{\mathsf{T}}); 0_E && \text{point slipping out of symmetric quotient}\\
&= \mathbb{I}; 0_E = 0_E && \text{property of an ordering}
\end{aligned}
$$

Because μ, μ' are measures, we have that $\mu^{\mathsf{T}} {}_{,} 1_\Omega = 1_E$ and also $\mu'^{\mathsf{T}} {}_{,} 1_\Omega = 1_E$. In both cases, the cone above the image is simply 1_E, and so their intersection as well as the greatest lower bound thereof is 1_E.

The proof of monotony of the composed measures is left for the reader.

In view of $\mu_{,} E \cap \mu_{0,} E = \mu_{,} E$ and $\mu_{,} E \cap \mu_{1,} E = \mu_{1,} E$, the 'algebraic' identities are trivial. $\qquad\qquad\square$

There exists a bulk of literature around the topic of Dempster–Shafer belief. It concentrates mostly on work with real numbers and their linear order and applies traditional free-hand mathematics. This sometimes makes it difficult to follow the basic ideas, not least as several authors all too often fall back on probability considerations.

We feel that the point-free relational reformulation of this field and the important generalization accompanying it is a clarification – at least for the strictly growing community of those who do not fear to use relations. Proofs may now be supported via proof systems. For examples where this might be applied, see [137].

14.4 Relational integration

Assume now that for all the criteria $\mathcal{C}$ a valuation has taken place resulting in a mapping $X : \mathcal{C} \longrightarrow \mathcal{L}$. The question is how to arrive at an overall valuation by rational means, for which μ shall be the guideline. See [113] for other details.

Definition 14.5. Given a relational measure μ and a mapping X indicating the values given by the criteria, we define the **relational integral**

$$(R)\!\int X \circ \mu \;:=\; \mathtt{lubR}_E(\mathbb{T}{}_{,}\; \mathtt{glbR}_E[X \cup \mathtt{syq}(X{}_{,} E^{\mathsf{T}}{}_{,} X^{\mathsf{T}}, \varepsilon){}_{,}\mu]). \qquad\square$$

The idea behind this integral is as follows. From the valuation of any criterion, proceed to all higher valuations and from these back to those criteria that assigned such higher values. With $X {}_{,} E {}_{,} X^{\mathsf{T}}$, the transition from all the criteria to the set of criteria is given. Now a symmetric quotient is needed in order to comprehend all these sets to elements of the powerset. (To this end, the converse is needed.) Once the sets are elements of the powerset, the measure μ may be applied. As already shown in the initial example, we now have the value of the respective criterion and in addition the valuation of the criteria set. From the two, we form the greatest lower bound. So in total, we have lower bounds for all the criteria. These are combined in one set multiplying the universal relation from the left side. Finally, the least upper bound is taken.

We are now in a position to understand why gauging $\mu^{\mathsf{T}} 1_\Omega = 1_E$ is necessary for μ, or 'greatest element is sent to greatest element'. Consider, for example, the special case of an X with all criteria assigning the same value. We certainly expect the relational integral to deliver this value precisely, regardless of the measure chosen. But this might not be the case if a measure is assigned too small a value to the full set.

As a running example, we provide the following highly non-continuous measure of Fig. 14.7. Here, for example,

$$\mu(\text{Abe}) = (\text{high}, \text{lazy}), \mu(\text{Bob}) = (\text{medium}, \text{fair}), \text{ with supremum (high, fair)}$$

but in excess to this, μ assigns $\mu(\text{Abe}, \text{Bob}) = (\text{high}, \text{good})$.

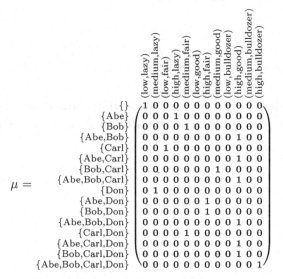

Fig. 14.7 Non-additive example measure

Assume in addition valuations X_1, X_2 to be given as shown in Fig. 14.8. We will execute the relational integrations.

Aggregating preferences

$$X_1 = \begin{matrix} \text{Abe} \\ \text{Bob} \\ \text{Carl} \\ \text{Don} \end{matrix} \begin{pmatrix} 0 & 0 & 0 & 1 & 0 & 0 & 0 & 0 & 0 & 0 & 0 & 0 \\ 0 & 0 & 0 & 0 & 0 & 0 & 1 & 0 & 0 & 0 & 0 & 0 \\ 0 & 1 & 0 & 0 & 0 & 0 & 0 & 0 & 0 & 0 & 0 & 0 \\ 0 & 0 & 0 & 0 & 1 & 0 & 0 & 0 & 0 & 0 & 0 & 0 \end{pmatrix}$$

$$X_2 = \begin{matrix} \text{Abe} \\ \text{Bob} \\ \text{Carl} \\ \text{Don} \end{matrix} \begin{pmatrix} 0 & 1 & 0 & 0 & 0 & 0 & 0 & 0 & 0 & 0 & 0 & 0 \\ 0 & 0 & 0 & 0 & 1 & 0 & 0 & 0 & 0 & 0 & 0 & 0 \\ 0 & 0 & 0 & 0 & 0 & 1 & 0 & 0 & 0 & 0 & 0 & 0 \\ 0 & 0 & 1 & 0 & 0 & 0 & 0 & 0 & 0 & 0 & 0 & 0 \end{pmatrix}$$

$$(R)\int X_1 \circ \mu = (0\ 0\ 0\ 0\ 0\ 0\ 1\ 0\ 0\ 0\ 0\ 0) \qquad (R)\int X_2 \circ \mu = (0\ 0\ 0\ 0\ 1\ 0\ 0\ 0\ 0\ 0\ 0\ 0)$$

Fig. 14.8 Two relational integrations

As already mentioned, we apply a sum operator `lub` after applying the product operator `glb`. When values are assigned with X, we look with E for those greater or equal, then with $X^\top$ for the criteria so valuated. Now comes a technically difficult step, namely proceeding to the union of the resulting sets with the symmetric quotient `syq` and the membership relation ε. The μ-score of this set is then taken. Obviously

$$\texttt{glbR}_E(X) \le (R)\int X \circ \mu \le \texttt{lubR}_E(X).$$

These considerations originate from free re-interpretation of the following concepts for work in $[0,1] \subseteq \mathbb{R}$. The **Sugeno integral** operator is defined as

$$M_{S,\mu}(x_1\ldots,x_m) = (S)\int x \circ \mu = \bigvee_{i=1}^{m} [x_i \wedge \mu(A_i)]$$

and the **Choquet integral**[1] operator as

$$M_{C,\mu}(x_1,\ldots,x_m) = (C)\int x \circ \mu = \sum_{i=1}^{m} [(x_i - x_{i-1}) \cdot \mu(A_i)].$$

In both cases, the elements of the vector $(x_1,\ldots,x_m)$ must be reordered such that

$$0 = x_0 \le x_1 \le x_2 \le \cdots \le x_m \le x_{m+1} = 1 \quad \text{and} \quad \mu(A_i) = \mu(C_i,\ldots,C_m).$$

---

[1] Named after Gustave Choquet (1915–2006), a French mathematician (Wikipedia).

The concept of Choquet integral was first introduced for a real-valued context in [34] and later used by Michio Sugeno. These two integrals are reported to have nice properties for aggregation: they are continuous, non-decreasing, and stable under certain interval preserving transformations. Not least, they reduce to the weighted arithmetic mean as soon as they become additive.

14.5 Defining relational measures

Such measures μ may be given directly, which is, however, a costly task since a powerset is involved all of whose elements need values. Therefore, measures mainly originate in some other way.

Measures originating from direct valuation of criteria

Let a **direct valuation** of the criteria be given as any relation m between $\mathcal{C}$ and $\mathcal{L}$. Although it is allowed to be contradictory and non-univalent, we provide for a way of defining a relational measure based on it. This will happen via the following constructs

$$\sigma(m) := \overline{\varepsilon^{\mathsf{T}}\,;m\,;\overline{E}} \qquad \pi(\mu) := \overline{\varepsilon\,;\mu\,;\overline{E}^{\mathsf{T}}},$$

which very obviously satisfy the Galois correspondence requirement

$$m \subseteq \pi(\mu) \iff \mu \subseteq \sigma(m).$$

They satisfy $\sigma(m\,;E^{\mathsf{T}}) = \sigma(m)$ and $\pi(\mu\,;E) = \pi(\mu)$, so that in principle only lower, respectively upper, cones occur as arguments. Applying $\overline{W\,;E} = \overline{W\,;E}\,;E^{\mathsf{T}}$, we get

$$\sigma(m)\,;E = \overline{\varepsilon^{\mathsf{T}}\,;m\,;\overline{E}}\,;E = \overline{\varepsilon^{\mathsf{T}}\,;m\,;\overline{E}\,;E^{\mathsf{T}}}\,;E = \overline{\varepsilon^{\mathsf{T}}\,;m\,;\overline{E}} = \sigma(m),$$

so that images of σ are always upper cones – and thus best described by their greatest lower bound $\mathtt{glbR}_E(\sigma(m))$.

Proposition 14.6. *Given any relation $m : \mathcal{C} \to \mathcal{L}$, the construct*

$$\mu_m := \mu_0 \oplus \mathtt{glbR}_E(\sigma(m)) = \mu_0 \oplus \mathtt{lubR}_E(\varepsilon^{\mathsf{T}}\,;m)$$

forms a relational measure, the so-called **possibility measure** *based on m.*

Proof: The gauging properties are more or less trivial to prove. For the proof of isotony we neglect gauging for simplicity:

$$\Omega\,;\mu_m \subseteq \mu_m\,;E \qquad \text{to be shown}$$
$$\iff \Omega\,;\mathsf{syq}(\overline{\overline{E}^{\mathsf{T}}\,;m^{\mathsf{T}}\,;\varepsilon},E^{\mathsf{T}}) \subseteq \mathsf{syq}(\overline{\overline{E}^{\mathsf{T}}\,;m^{\mathsf{T}}\,;\varepsilon},E^{\mathsf{T}})\,;E \qquad \text{expanding, Prop. 9.9}$$
$$\iff \overline{\varepsilon^{\mathsf{T}}\,;\overline{\varepsilon}}\,;\mathsf{syq}(\overline{\overline{E}^{\mathsf{T}}\,;m^{\mathsf{T}}\,;\varepsilon},E^{\mathsf{T}}) \subseteq \overline{\varepsilon^{\mathsf{T}}\,;m\,;\overline{E}} \qquad \text{expanding, cancelling}$$
$$\iff \overline{\overline{\varepsilon}^{\mathsf{T}}\,;\varepsilon}\,;\varepsilon^{\mathsf{T}}\,;m\,;\overline{E} \subseteq \mathsf{syq}(\overline{\overline{E}^{\mathsf{T}}\,;m^{\mathsf{T}}\,;\varepsilon},E^{\mathsf{T}}) \qquad \text{Schröder rule}$$

Remembering that $\overline{\varepsilon^{\mathsf{T}}}$ ε $\varepsilon^{\mathsf{T}} = \varepsilon^{\mathsf{T}}$, $\overline{E} = \overline{E}$ E^{T}, and expanding the symmetric quotient, the latter is certainly satisfied.

Equivalence of the two variants:

$$
\begin{aligned}
\mathtt{glbR}_E(\sigma(m)) &= \mathtt{glbR}_E\big(\overline{\varepsilon^{\mathsf{T}}\, m\, \overline{E}}\big) \\
&= \big[\mathtt{glb}_E(\overline{\varepsilon^{\mathsf{T}}\, m\, \overline{E}}\,)\big]^{\mathsf{T}} \\
&= \big[\mathtt{glb}_E(\overline{E}^{\mathsf{T}}\, m^{\mathsf{T}}\, \varepsilon)\big]^{\mathsf{T}} \\
&= \big[\mathtt{glb}_E(\mathtt{ubd}\,(m^{\mathsf{T}}\, \varepsilon))\big]^{\mathsf{T}} \\
&= \big[\mathtt{lub}_E(m^{\mathsf{T}}\, \varepsilon)\big]^{\mathsf{T}} = \mathtt{lubR}_E(\varepsilon^{\mathsf{T}}\, m)
\end{aligned}
$$
$\qquad\qquad\qquad\qquad\qquad\qquad\qquad\qquad\qquad\qquad\qquad\qquad\qquad$ □

Possibility measures need not be Bayesian. The addition of the vacuous belief mapping μ_0 is again necessary for gauging purposes. In the case that m is a mapping, the situation becomes even better. From

$$
\begin{aligned}
\pi(\sigma(m\, E^{\mathsf{T}})) &= \pi(\sigma(m)) \\
&= \varepsilon\, \overline{\varepsilon^{\mathsf{T}}\, m\, \overline{E}}\, \overline{E}^{\mathsf{T}} \qquad && \text{expanded} \\
&= m\, \overline{\overline{E}}\, \overline{E}^{\mathsf{T}} \qquad && \text{Prop. 7.14.i shows that } \varepsilon\, \overline{\varepsilon^{\mathsf{T}}\, X} = \overline{X} \text{ for all } X \\
&= m\, \overline{\overline{E}}\, \overline{E}^{\mathsf{T}} \qquad && \text{since the mapping } m \text{ may slip out of negation} \\
&= m\, E\, \overline{E}^{\mathsf{T}} \\
&= m\, E^{\mathsf{T}}
\end{aligned}
$$

we see that this is an adjunction on cones. The lower cones $m\, E^{\mathsf{T}}$ in turn are $1:1$ represented by their least upper bounds $\mathtt{lubR}_E(m\, E)$.

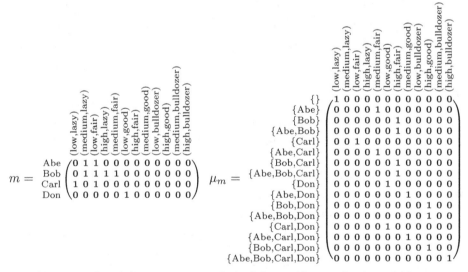

Fig. 14.9 Possibility measure μ_m derived from a direct valuation relation m

The following proposition shows that a Bayesian measure is a rather special case, namely more or less directly determined as a possibility measure for a direct valuation m that is in fact a mapping. We provide an example: one may proceed from m in Fig. 14.9 to the measure μ_m according to Prop. 14.9. However, one may also start from m_{μ_B} in Fig. 14.10, work according to Prop. 14.10, and obtain the same result.

Using direct valuations, one can also give another characterization of being Bayesian, namely that the whole measure is fully determined by the values on singleton subsets. To this end, consider $\sigma := \text{syq}(\mathbb{I}, \varepsilon)$, the mapping injecting singletons[2] into the powerset. The remarkable property of σ is that $\sigma^{\mathsf{T}}; \sigma \subseteq \mathbb{I}$ characterizes the atoms of the powerset ordering Ω (see page 145).

Proposition 14.7. *Let μ be a Bayesian measure. Then $m_\mu := \sigma; \mu$ is a mapping and in addition that direct valuation for which $\mu = \mu_{m_\mu}$.*

Proof: As a composition of two mappings, m_μ is certainly a mapping. What remains to be shown is $\mu = \mu_{m_\mu}$.

$$
\begin{aligned}
\mu^{\mathsf{T}} &= \mu^{\mathsf{T}}; \mathbb{I} \\
&= \mu^{\mathsf{T}}; \text{lub}_\Omega(\sigma^{\mathsf{T}}; \sigma; \Omega) && \text{an element in the powerset is a union of singleton sets} \\
&= \text{lub}_E(\mu^{\mathsf{T}}; \sigma^{\mathsf{T}}; \sigma; \Omega) && \text{owing to continuity of the Bayesian measure } \mu \\
&= \left[\text{lubR}_E(\Omega^{\mathsf{T}}; \sigma^{\mathsf{T}}; \sigma; \mu) \right]^{\mathsf{T}} && \text{switching to operation taken row-wise} \\
&= \left[\text{lubR}_E(\varepsilon^{\mathsf{T}}; m_\mu) \right]^{\mathsf{T}} && \text{since } \varepsilon = \sigma; \Omega, \text{ definition of } m_\mu \\
&= \mu_{m_\mu}^{\mathsf{T}} && \text{see Prop. 14.6}
\end{aligned}
$$

We need not worry about adding the vacuous belief, as we have been starting from a Bayesian measure which means that the value 1_E of the full set will be the least upper bound of all the values of the singletons. □

One will find that m_{μ_B} of Fig. 14.10 may also be obtained from the m of Fig. 14.9, taking row-wise least upper bounds according to the ordering E of Fig. 14.5. With this method just a few of the many relational measures will be found.

Measures originating from a body of evidence

We may also derive relational measures out of any relation between $\mathbf{2}^C$ and $\mathcal{L}$. Although it is allowed to be non-univalent, we provide a way of defining two measures based on it – which may coincide.

---

[2] This σ must not be mixed up with the Galois-σ.

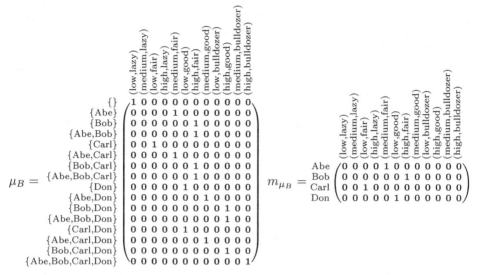

Fig. 14.10 Bayesian measure μ_B with corresponding direct valuation m_{μ_B}

Definition 14.8. Let our general setting be given.

(i) A **body of evidence** is an arbitrary relation $M : \mathbf{2}^{\mathcal{C}} \longrightarrow \mathcal{L}$, restricted only by the gauging requirement that $M^{\mathsf{T}} \,{}_{\text{!`}}\, 0_{\Omega} \subseteq 0_E$.

(ii) If the body of evidence M is in addition a mapping, we speak – following [132] – of a **basic probability assignment**.   □

We should be aware that the basic probability assignment is meant to assign something to a set regardless of what is assigned to its proper subsets. The condition $M^{\mathsf{T}} \,{}_{\text{!`}}\, 0_{\Omega} \subseteq 0_E$ expresses that M either does not assign any belief to the empty set or assigns it just 0_E.

Now a construction similar to that which led to Prop. 14.6 becomes possible, based on the following observation. If I dare to say that the occurrence of $A \subseteq \mathcal{C}$ deserves my trust to the amount $M(A)$, then $A' \subseteq A \subseteq \mathcal{C}$ deserves at least this amount of trust as it occurs whenever A occurs. I might, however, not be willing to consider that $A'' \subseteq \mathcal{C}$ with $A \subseteq A''$ deserves to be trusted by the same amount as there is a chance that it does not occur so often. We put

$$\sigma'(M) := \overline{\Omega^{\mathsf{T}} \,{}_{\text{!`}}\, M \,{}_{\text{!`}}\, \overline{E}} \qquad \pi'(\mu) := \overline{\Omega \,{}_{\text{!`}}\, \mu \,{}_{\text{!`}}\, \overline{E}^{\mathsf{T}}},$$

which again satisfies the Galois correspondence requirement

$$M \subseteq \pi'(\mu) \quad\Longleftrightarrow\quad \mu \subseteq \sigma'(M).$$

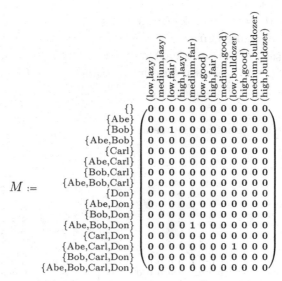

$$
M := \begin{array}{c}
\{\} \\
\{Abe\} \\
\{Bob\} \\
\{Abe,Bob\} \\
\{Carl\} \\
\{Abe,Carl\} \\
\{Bob,Carl\} \\
\{Abe,Bob,Carl\} \\
\{Don\} \\
\{Abe,Don\} \\
\{Bob,Don\} \\
\{Abe,Bob,Don\} \\
\{Carl,Don\} \\
\{Abe,Carl,Don\} \\
\{Bob,Carl,Don\} \\
\{Abe,Bob,Carl,Don\}
\end{array}
\left(
\begin{array}{cccccccccccc}
0 & 0 & 0 & 0 & 0 & 0 & 0 & 0 & 0 & 0 & 0 & 0 \\
0 & 0 & 0 & 0 & 0 & 0 & 0 & 0 & 0 & 0 & 0 & 0 \\
0 & 0 & 1 & 0 & 0 & 0 & 0 & 0 & 0 & 0 & 0 & 0 \\
0 & 0 & 0 & 0 & 0 & 0 & 0 & 0 & 0 & 0 & 0 & 0 \\
0 & 0 & 0 & 0 & 0 & 0 & 0 & 0 & 0 & 0 & 0 & 0 \\
0 & 0 & 0 & 0 & 0 & 0 & 0 & 0 & 0 & 0 & 0 & 0 \\
0 & 0 & 0 & 0 & 0 & 0 & 0 & 0 & 0 & 0 & 0 & 0 \\
0 & 0 & 0 & 0 & 0 & 0 & 0 & 0 & 0 & 0 & 0 & 0 \\
0 & 0 & 0 & 0 & 0 & 0 & 0 & 0 & 0 & 0 & 0 & 0 \\
0 & 0 & 0 & 0 & 0 & 0 & 0 & 0 & 0 & 0 & 0 & 0 \\
0 & 0 & 0 & 0 & 0 & 0 & 0 & 0 & 0 & 0 & 0 & 0 \\
0 & 0 & 0 & 1 & 0 & 0 & 0 & 0 & 0 & 0 & 0 & 0 \\
0 & 0 & 0 & 0 & 0 & 0 & 0 & 0 & 0 & 0 & 0 & 0 \\
0 & 0 & 0 & 0 & 0 & 0 & 0 & 1 & 0 & 0 & 0 & 0 \\
0 & 0 & 0 & 0 & 0 & 0 & 0 & 0 & 0 & 0 & 0 & 0 \\
0 & 0 & 0 & 0 & 0 & 0 & 0 & 0 & 0 & 0 & 0 & 0
\end{array}
\right)
$$

(columns, left to right: (low,lazy), (medium,lazy), (low,fair), (high,lazy), (medium,fair), (low,good), (high,fair), (medium,good), (low,bulldozer), (high,good), (medium,bulldozer), (high,bulldozer))

Fig. 14.11 A body of evidence – incidentally a univalent relation

Obviously $\sigma'(M_{;}E^{\mathsf{T}}) = \sigma'(M)$ and $\pi'(\mu_{;}E) = \pi'(\mu)$, so that in principle only upper (E) and lower (E^{T}), respectively, cones are set into relation. But again applying $\overline{W_{;}E} = \overline{W_{;}E}_{;}E^{\mathsf{T}}$, we get

$$
\begin{aligned}
\sigma'(M)_{;}E &= \overline{\Omega^{\mathsf{T}}_{;}M_{;}\overline{E}}_{;}E \\
&= \overline{\Omega^{\mathsf{T}}_{;}M_{;}\overline{E}}_{;}E^{\mathsf{T}}_{;}E \\
&= \overline{\Omega^{\mathsf{T}}_{;}M_{;}\overline{E}} \\
&= \sigma'(M),
\end{aligned}
$$

so that images of σ' are always upper cones – and thus best described by their greatest lower bound $\mathtt{glbR}_E(\sigma'(M))$.

$$
\begin{aligned}
\mathtt{glbR}_E(\sigma'(M)) &= \mathtt{glbR}_E(\overline{\Omega^{\mathsf{T}}_{;}M_{;}\overline{E}}) \\
&= \left[\mathtt{glb}_E(\overline{\overline{\Omega^{\mathsf{T}}_{;}M_{;}\overline{E}}}^{\mathsf{T}}) \right]^{\mathsf{T}} \\
&= \left[\mathtt{glb}_E(\overline{\overline{E}^{\mathsf{T}}_{;}M^{\mathsf{T}}_{;}\Omega}) \right]^{\mathsf{T}} \\
&= \left[\mathtt{glb}_E(\mathtt{ubd}(M^{\mathsf{T}}_{;}\Omega)) \right]^{\mathsf{T}} \\
&= \left[\mathtt{lub}_E(M^{\mathsf{T}}_{;}\Omega) \right]^{\mathsf{T}} \\
&= \mathtt{lubR}_E(\Omega^{\mathsf{T}}_{;}M)
\end{aligned}
$$

which – up to gauging by adding μ_0 – leads us to the following proposition.

Proposition 14.9. *Should some body of evidence M be given, there exist two relational measures closely resembling M,*

(i) *the* **belief measure**

$$\mu_{\text{belief}}(M) := \mu_0 \oplus \mathtt{lubR}_E(\Omega^{\mathsf{T}} \cdot M),$$

(ii) *the* **plausibility measure**

$$\mu_{\text{plausibility}}(M) := \mu_0 \oplus \mathtt{lubR}_E(\Omega^{\mathsf{T}} \cdot (\Omega \cap \overline{\Omega} \cdot \mathbb{T}) \cdot M),$$

(iii) *the belief measure assigns values not exceeding those of the plausibility measure, i.e.,* $\mu_{\text{belief}}(M) \subseteq \mu_{\text{plausibility}}(M) \cdot E^{\mathsf{T}}.$

Proof: We leave this proof for the reader. □

$\mu_{\text{belief}}(M)$

	(low,lazy)	(medium,lazy)	(low,fair)	(high,lazy)	(medium,fair)	(low,good)	(high,fair)	(medium,good)	(low,bulldozer)	(high,good)	(medium,bulldozer)	(high,bulldozer)
{}	1	0	0	0	0	0	0	0	0	0	0	0
{Abe}	1	0	0	0	0	0	0	0	0	0	0	0
{Bob}	0	0	1	0	0	0	0	0	0	0	0	0
{Abe,Bob}	0	0	1	0	0	0	0	0	0	0	0	0
{Carl}	1	0	0	0	0	0	0	0	0	0	0	0
{Abe,Carl}	1	0	0	0	0	0	0	0	0	0	0	0
{Bob,Carl}	0	0	1	0	0	0	0	0	0	0	0	0
{Abe,Bob,Carl}	0	0	1	0	0	0	0	0	0	0	0	0
{Don}	1	0	0	0	0	0	0	0	0	0	0	0
{Abe,Don}	1	0	0	0	0	0	0	0	0	0	0	0
{Bob,Don}	0	0	1	0	0	0	0	0	0	0	0	0
{Abe,Bob,Don}	0	0	0	0	1	0	0	0	0	0	0	0
{Carl,Don}	1	0	0	0	0	0	0	0	0	0	0	0
{Abe,Carl,Don}	0	0	0	0	0	0	0	0	1	0	0	0
{Bob,Carl,Don}	0	0	1	0	0	0	0	0	0	0	0	0
{Abe,Bob,Carl,Don}	0	0	0	0	0	0	0	0	0	0	0	1

$\mu_{\text{plausibility}}(M)$

	(low,lazy)	(medium,lazy)	(low,fair)	(high,lazy)	(medium,fair)	(low,good)	(high,fair)	(medium,good)	(low,bulldozer)	(high,good)	(medium,bulldozer)	(high,bulldozer)
{}	1	0	0	0	0	0	0	0	0	0	0	0
{Abe}	0	0	0	0	0	0	0	0	0	0	1	0
{Bob}	0	0	0	1	0	0	0	0	0	0	0	0
{Abe,Bob}	0	0	0	0	0	0	0	0	0	0	1	0
{Carl}	0	0	0	0	0	0	0	0	1	0	0	0
{Abe,Carl}	0	0	0	0	0	0	0	0	0	0	1	0
{Bob,Carl}	0	0	0	0	0	0	0	0	0	0	1	0
{Abe,Bob,Carl}	0	0	0	0	0	0	0	0	0	0	1	0
{Don}	0	0	0	0	0	0	0	0	0	0	1	0
{Abe,Don}	0	0	0	0	0	0	0	0	0	0	1	0
{Bob,Don}	0	0	0	0	0	0	0	0	0	0	1	0
{Abe,Bob,Don}	0	0	0	0	0	0	0	0	0	0	1	0
{Carl,Don}	0	0	0	0	0	0	0	0	0	0	1	0
{Abe,Carl,Don}	0	0	0	0	0	0	0	0	0	0	1	0
{Bob,Carl,Don}	0	0	0	0	0	0	0	0	0	0	1	0
{Abe,Bob,Carl,Don}	0	0	0	0	0	0	0	0	0	0	0	1

Fig. 14.12 Belief measure and plausibility measure for M of Fig. 14.11

The belief measure adds information to the extent that all evidence of subsets with evidence attached is incorporated.

Another idea is followed by the plausibility measure. One asks which sets have a non-empty intersection with the set one is about to determine the measure for. Assuming the evidence to be capable of floating freely around the set – and possibly flowing fully into the non-empty intersection mentioned – one determines the least upper bound of all these. The plausibility measure collects those pieces of evidence that do *not* indicate trust against occurrence of the event or non-void parts of it as they might all together convene in the set considered.

A belief measure need not be Bayesian as can be seen from row {Abe, Bob, Don} in Fig. 14.12. The belief as well as the plausibility measure more or less precisely determine their original body of evidence.

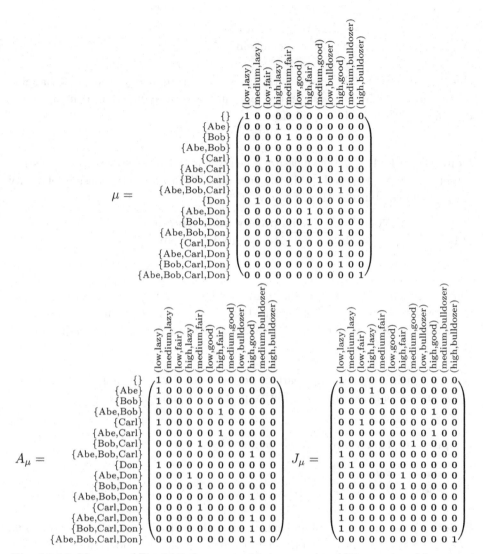

Fig. 14.13 Measure of Fig. 14.7, decomposed into a purely additive part and jump part

Proposition 14.10. *Should the body of evidence be concentrated on singleton sets only, the belief and the plausibility measure will coincide.*

Proof: That M is concentrated on arguments which are singleton sets means that $M = a \, ; M$ with a the partial diagonal relation describing the atoms of the ordering Ω; see page 145. For Ω and a one can prove $(\Omega \cap \overline{\Omega} \, ; \, \mathbb{T}) \, ; a = a$ as the only other element less than or equal to an atom, namely the least one, is cut out via $\overline{\Omega}$. Then

$$\Omega^{\mathsf{T}}{}_{;}(\Omega \cap \overline{\Omega}{}_{;}\mathbb{T})_{;}M = \Omega^{\mathsf{T}}{}_{;}(\Omega \cap \overline{\Omega}{}_{;}\mathbb{T})_{;}a_{;}M \qquad M = a_{;}M$$
$$= \Omega^{\mathsf{T}}{}_{;}a_{;}M \qquad \text{see above}$$
$$= \Omega^{\mathsf{T}}{}_{;}M \qquad \text{again since } M = a_{;}M \qquad \qquad \square$$

One should compare this result with the former one assuming m to be a mapping putting $m := \varepsilon_{;}M$. One can also try to go in the reverse direction, namely from a measure back to a body of evidence.

Definition 14.11. Let some measure μ be given and define the corresponding strict subset containment $C := \overline{\mathbb{I}} \cap \Omega$ in the powerset. We introduce two basic probability assignments combined with μ, namely

(i) $A_\mu := \mathtt{lubR}_E(C^{\mathsf{T}}{}_{;}\mu)$,   its **purely additive part**,

(ii) $J_\mu := \mu_1 \otimes (\mu \cap \overline{\mathtt{lubR}_E(C^{\mathsf{T}}{}_{;}\mu)})$,   its **jump part**. $\square$

The purely additive part is 0_E for atoms and for 0_Ω. It is not a measure. The pure jump part first shows what is assigned to atoms; in addition, it identifies where more than the least upper bound of assignments to proper subsets is assigned. It is not a measure.

Now some arithmetic on these parts is possible, not least providing the insight that a measure decomposes into an additive part and a jump part.

Proposition 14.12. *Given the present setting, we have*

(i) $A_\mu \oplus J_\mu = \mu$,

(ii) $\mu_{\text{belief}}(J_\mu) = \mu$.

Proof: (i) We can disregard multiplication with μ_1. It is introduced only for some technical reason: it converts empty rows to rows with 0_E assigned. This is necessary when adding, i.e., intersecting two upper cones and determining their greatest lower bound. Now, they will not be empty. In total, we have obviously

$$\mu_{;}E = A_{\mu;}E \cap J_{\mu;}E$$

so that the greatest lower bounds will coincide.

(ii) is again omitted and, thus, left for the reader. $\square$

In the real-valued case, this result is not surprising at all as one can always decompose into a part continuous from the left and a jump part.

In view of these results it seems promising to investigate how concepts such as commonality, consonance, necessity measures, focal sets, and cores may also be found using the relational approach. This seems particularly interesting as the concepts of De Morgan triples have also been transferred to the point-free relational side.

14.6 De Morgan triples

The introduction of triangular norms in fuzzy sets has strongly influenced the fuzzy set community and made it an accepted part of mathematics.

As long as the set C of criteria is comparatively small, it seems possible to work with the powerset 2^C and, thus, to take into consideration specific combinations of criteria. If the size of C increases so as to handle a voting-type or polling-type problem, one will soon handle voters on an equal basis – at least in democracies. This means that the measure applied must not attribute different values to differently chosen n-element sets, for example. That the value for an n-element set is different from the value attached to an $(n + 1)$-element set, will probably be accepted.

As a result, the technique used to define the measure will be based on *operations* in $\mathcal{L}$ alone. In total, instead of a measure on 2^C we work with a unary or binary operation on $\mathcal{L}$.

Norms and conorms

Researchers have frequently studied fuzzy valuations in the interval $[0, 1]$ and then asked for methods of negation, conjunction, disjunction, and subjunction (implication). What they discovered in the fuzzy environment was finally the concept of a De Morgan triple, a combination of three, or even four, real-valued functions that resemble more or less the concepts just mentioned. We here introduce corresponding functions on the relational side. To this end, we start by recalling the basic situation for a De Morgan triple.

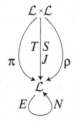

Fig. 14.14 Basic situation for a De Morgan triple

The set of fuzzy values is $\mathcal{L}$ with ordering E; it corresponds, thus, to the interval $[0, 1]$ traditionally used. With T, two values are combined in one in the sense of a conjunction, with S as a disjunction. While N is intended to model negation, J is derived so as to resemble subjunction (implication).

By now it is traditional to axiomatize what is intended as follows. We will later give the corresponding relational versions.

Definition 14.13 (*t*-norm). Given a set $\mathcal{L}$ of values, ordered by E with top element 1 and bottom element 0, one will ask for a *t*-norm to work as a **conjunction operator** T and demand it to be

normalized	$T(1, x) = x$	
commutative	$T(x, y) = T(y, x)$	
monotonic	$T(x, y) \leq T(u, v)$	whenever $0 \leq x \leq u \leq 1,\ 0 \leq y \leq v \leq 1$
associative	$T(x, T(y, z)) = T(T(x, y), z)$.	$\square$

While this may be considered a substitute for conjunction, the following represents disjunction.

Definition 14.14 (*t*-conorm). Given a set $\mathcal{L}$ of values, ordered by E with top element 1 and bottom element 0, one will ask for a *t*-conorm to work as a **disjunction operator** S and demand it to be

normalized	$S(0, x) = x$	
commutative	$S(x, y) = S(y, x)$	
monotonic	$S(x, y) \leq S(u, v)$	whenever $0 \leq x \leq u \leq 1,\ 0 \leq y \leq v \leq 1$
associative	$S(x, S(y, z)) = S(S(x, y), z)$.	$\square$

Once this is available, researchers traditionally look for negation and subjunction. There are several versions of negation in the fuzzy community. Negation will not be available in every lattice $\mathcal{L}$; weaker versions, however, show up sufficiently often. They are postulated as follows.

Definition 14.15 (Strict and strong negation). One postulates that for negation to work as a **complement operator** $N : \mathcal{L} \longrightarrow \mathcal{L}$ it is a bijective mapping that is

normalized	$N(0) = 1$	$N(1) = 0$
anti-monotonic	$x \leq y \implies$	$N(x) \geq N(y)$.

Two stronger versions are traditionally studied, namely

strictly antitonic $x < y \implies N(x) > N(y)$
strongly antitonic $N(N(x)) = x$.   □

Subjunction is then modelled over the interval $[0,1]$ along $a \to b = \neg a \vee b$.

Definition 14.16 (Subjunction). One postulates that a **subjunction operator** J is a binary mapping $J : \mathcal{L} \times \mathcal{L} \longrightarrow \mathcal{L}$ that is

normalized $J(0, x) = 1$ $J(x, 1) = 1$ $J(1, 0) = 0$
left-antitonic $x \le z \implies J(x, y) \ge J(z, y)$
right-isotonic $y \le t \implies J(x, y) \le J(x, t)$.   □

Two possibly different subjunctions may turn out to exist. The first is defined for a given negation N and t-conorm S

$$J_{S,N}(x, y) := S(N(x), y).$$

Instead of starting with a t-conorm S, we could also have started with a t-norm T. The result may well be another subjunction when obtained as

$$J_T(x, y) := \sup \{ z \mid T(x, z) \le y \}.$$

Interlude on binary mappings

We have reported on t-norms and subjunctions although they were not yet in the point-free form preferred in the present book. So we will start translating, which requires some further technicalities concerning binary mappings. The easiest transition is for commutativity.

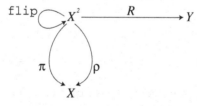

Fig. 14.15 Typing around commutativity

Commutative means that the result of R must not change when the arguments are flipped via the construct $\texttt{flip} := \rho_{\cdot}\pi^{\mathsf{T}} \cap \pi_{\cdot}\rho^{\mathsf{T}} = (\rho \otimes \pi)$.

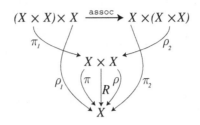

Fig. 14.16 Associativity formulated point-free

It is rather difficult to formulate associativity in a point-free fashion. While we have mentioned on many occasions that the point-free version will be shorter (often just one sixth in length), we find here a situation where this may not be the case.

Proposition and **Definition 14.17** (Relational conditions for associativity and commutativity).

(i) *For a direct product $X \times X$, the construct*
$$\texttt{flip} := (\rho \otimes \pi)$$
is a bijective mapping; it will also satisfy
$$\texttt{flip} = \texttt{flip}^{\mathsf{T}}$$
$$\texttt{flip} \, ; \pi = \rho \qquad \texttt{flip} \, ; \rho = \pi$$
$$\texttt{flip} \, ; \texttt{flip} = \mathbb{I}.$$

(ii) *A relation $R : X \times X \longrightarrow Y$ will be called* **commutative** *if $R = \texttt{flip} \, ; R$.*

(iii) *The construct*
$$\texttt{assoc} := (\pi_1 \, ; \pi \otimes (\pi_1 \, ; \rho \otimes \rho_1))$$
is a bijective mapping for any direct product configuration as in Fig. 14.16; it will also satisfy
$$\texttt{assoc} \, ; \texttt{flip} \, ; \texttt{assoc} = \texttt{flip} \, ; \texttt{assoc}^{\mathsf{T}} \, ; \texttt{flip}.$$

(iv) *A relation $R : X \times X \longrightarrow X$ will be called* **associative** *if*
$$(\pi_1 \, ; R \otimes \rho_1) \, ; R = \texttt{assoc} \, ; (\pi_2 \otimes \rho_2 \, ; R) \, ; R.$$

Proof: (i) The property of being self-converse is trivial:
$$\texttt{flip} \, ; \pi = (\rho \, ; \pi^{\mathsf{T}} \cap \pi \, ; \rho^{\mathsf{T}}) \, ; \pi = \rho \cap \pi \, ; \rho^{\mathsf{T}} \, ; \pi = \rho \cap \pi \, ; \mathbb{T} = \rho \cap \mathbb{T} = \rho.$$

Obviously, $\texttt{flip}$ is a mapping, and so is $\texttt{flip} \, ; \texttt{flip}$. When the latter is contained in the identity, it will, thus, be identical to it:

$\mathtt{flip}\,;\mathtt{flip} \subseteq \rho\,;\pi^{\mathsf{T}}\,;\pi\,;\rho^{\mathsf{T}} \subseteq \rho\,;\rho^{\mathsf{T}}$ since π is univalent

$\mathtt{flip}\,;\mathtt{flip} \subseteq \pi\,;\rho^{\mathsf{T}}\,;\rho\,;\pi^{\mathsf{T}} = \pi\,;\pi^{\mathsf{T}}$ since ρ is univalent

$\mathtt{flip}\,;\mathtt{flip} \subseteq \pi\,;\pi^{\mathsf{T}} \cap \rho\,;\rho^{\mathsf{T}} = \mathbb{I}$ intersecting the latter two

$\mathtt{flip}\,;\mathbb{T} = \mathtt{flip}\,;\pi\,;\mathbb{T} = \rho\,;\mathbb{T} = \mathbb{T}$ see above

(iii) This proof is left for the reader. $\qquad\qquad\qquad\qquad\qquad\qquad$ □

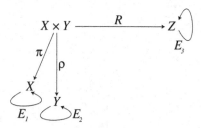

Fig. 14.17 Right-monotonicity formulated point-free

We call a relation R

right-monotonic if $(\pi\,;\pi^{\mathsf{T}} \cap \rho\,;E_2\,;\rho^{\mathsf{T}})\,;R \subseteq R\,;E_3$.

Demanding right-monotonicity is, however, just a way of writing this down economically; using that t-norms as well as t-conorms are assumed to be commutative, we need not propagate this restriction to the point-free level; so we may simply say that R is

monotonic if $(E_1 \otimes E_2)\,;R = (\pi\,;E_1\,;\pi^{\mathsf{T}} \cap \rho\,;E_2\,;\rho^{\mathsf{T}})\,;R \subseteq R\,;E_3$.

De Morgan triples point-free

We use these properties of binary mappings in the relational and, thus, point-free definition. The general setting is always that a set $\mathcal{L}$ of values is given, lattice-ordered by E. Let $1 := \mathtt{lub}_E(\mathbb{T})$ be the top element and $0 := \mathtt{glb}_E(\mathbb{T})$ the bottom element.

Definition 14.18 (Relational norm and conorm). Considering the two relations $T : \mathcal{L} \times \mathcal{L} \longrightarrow \mathcal{L}$ and $S : \mathcal{L} \times \mathcal{L} \longrightarrow \mathcal{L}$, we say that

(i) T is a **relational norm** if the following holds

$\quad (E \otimes E)\,;T \subseteq T\,;E$ monotony for both sides,

$\quad (\mathbb{T}\,;1^{\mathsf{T}} \otimes \mathbb{I})\,;T = \mathbb{I}$ $(\mathbb{T}\,;0^{\mathsf{T}} \otimes \mathbb{I})\,;T = \mathbb{T}\,;0^{\mathsf{T}}$ for gauging,

$\quad$ commutativity and associativity,

(ii) S is a **relational conorm** if the following holds

$(E \otimes E) ; S \subseteq S ; E$ monotony for both sides,

$(\mathbb{T} ; 0^{\mathsf{T}} \otimes \mathbb{I}) ; S = \mathbb{I}$ $(\mathbb{T} ; 1^{\mathsf{T}} \otimes \mathbb{I}) ; S = \mathbb{T} ; 1^{\mathsf{T}}$ for gauging,

commutativity and associativity.   □

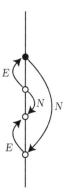

Fig. 14.18 De Morgan negation formulated point-free; start at the filled vertex

In an analogous way, the relational case also admits considering several concepts to model negation.

Definition 14.19 (Relational negation of t-norms). One postulates that a relation $N : \mathcal{L} \longrightarrow \mathcal{L}$ is a **relational negation** if it is a bijective mapping satisfying

$N^{\mathsf{T}} ; 0 = 1$ $N^{\mathsf{T}} ; 1 = 0$ for normalization,

$E^{\mathsf{T}} ; N \subseteq N ; E$ for being antitonic.

Two stronger versions are

$C ; N \subseteq N ; C^{\mathsf{T}}$ for being 'strictly' antitonic,

$N ; N = \mathbb{I}$ for being a 'strong' negation.   □

When talking about conjunction, disjunction, negation, and subjunction, one has certain expectations as to how these behave relatively to one another. However, we do not remain in Boolean algebra, so things may have changed. Not least, we will find ourselves in a situation where more than one subjunction operator is conceivable. What we will do, nonetheless, is to specify the properties the subjunction operators should satisfy. We will avoid the word *implication* which seems to insist too strongly on one thing implying another; we will talk rather about *subjunction*.

Definition 14.20 (Relational subjunction). For a relation $J : \mathcal{L} \times \mathcal{L} \longrightarrow \mathcal{L}$ to be a **relational subjunction** one postulates that it be a mapping satisfying

$$(\mathbb{T}_i 0^\mathsf{T} \otimes \mathbb{I})_i J = \mathbb{T}_i 1^\mathsf{T} \qquad (\mathbb{I} \otimes \mathbb{T}_i 1^\mathsf{T})_i J = \mathbb{T}_i 1^\mathsf{T} \qquad (\mathbb{T}_i 1^\mathsf{T} \otimes \mathbb{T}_i 0^\mathsf{T})_i J = \mathbb{T}_i 0^\mathsf{T}$$
$$(E^\mathsf{T} \otimes E)_i J \subseteq J_i E \qquad \text{anti-/isotonic.} \qquad\qquad \Box$$

So far, this is just postulational, resembling somehow $a \to b \iff \neg a \vee b$. For two cases, we show that the specification may be satisfied. The two cases are based on $J_{S,N}$ and J_T above.

Proposition 14.21. *Let E, S, N be given with the properties postulated; i.e., a relational conorm S and a relational negation N, the relation*

$$J_{S,N} := (N \otimes \mathbb{I})_i S$$

can be defined, which turns out to be a subjunction, the so-called S-N-subjunction.

Proof: We start with the main property and prove gauging additions afterwards.

$$
\begin{aligned}
(E^\mathsf{T} \otimes E)_i J &= (E^\mathsf{T} \otimes E)_i (N \otimes \mathbb{I})_i S && \text{by definition} \\
&= ((E^\mathsf{T}_i N) \otimes E)_i S && \text{due to Prop. 7.2.ii with the Point Axiom} \\
&\subseteq ((N_i E) \otimes E)_i S && \text{Def. 14.19 for a relational negation} \\
&= (N \otimes \mathbb{I})_i (E \otimes E)_i S && \text{Prop. 7.2.ii with the Point Axiom} \\
&\subseteq (N \otimes \mathbb{I})_i S_i E && \text{monotony of Def. 14.18.ii} \\
&= J_i E && \text{by definition}
\end{aligned}
$$

Only one of the gauging properties is shown; the others are left as an exercise.

$$
\begin{aligned}
((\mathbb{T}_i 0^\mathsf{T}) \otimes \mathbb{I})_i J &= ((\mathbb{T}_i 0^\mathsf{T}) \otimes \mathbb{I})_i (N \otimes \mathbb{I})_i S && \text{expanding} \\
&= ((\mathbb{T}_i 0^\mathsf{T}_i N) \otimes \mathbb{I})_i S && \text{according to Prop. 7.5.ii} \\
&= ((\mathbb{T}_i 1^\mathsf{T}) \otimes \mathbb{I})_i S && \text{according to Def. 14.19} \\
&= \mathbb{T}_i 1^\mathsf{T} && \text{according to Def. 14.18.ii} \qquad \Box
\end{aligned}
$$

The following more difficult construct also satisfies the requirements of a subjunction. In this case, we try to explain the construction in more detail. With the first ρ, we proceed to y, E-below which we consider an arbitrary result $T(u, z)$ of T. The latter is then traced back to its argument pair (u, z). Intersecting, the first component will stay x. Projecting with ρ, we pick out the result z:

$$_{(x,y)}(\pi_i \pi^\mathsf{T}_{(x,?)} \cap \rho_{y_i} E^\mathsf{T}_{T(u,z)_i} T^\mathsf{T}_{(u,z)})_{(x,z)_i} \rho_z.$$

Therefore we consider

$$(\pi_i \pi^\mathsf{T} \cap \rho_i E^\mathsf{T}_i T^\mathsf{T})_i \rho$$

to which we apply the `lub` operator row-wise.

Proposition 14.22. *Let E, T be given with the properties postulated above. Then the relation*

$$J_T = \text{lubR}_E \left[(\pi_i \pi^\mathsf{T} \cap \rho_i E^\mathsf{T}_i T^\mathsf{T})_i \rho \right]$$

*can be defined, which turns out to be a subjunction, the so-called T-**subjunction**.*

Proof: The non-trivial proof is left to the reader. □

In the next two examples, we try to give intuition for these propositions.

Example 14.23. Consider the attempt to define something similar to Boolean operations on the three element qualifications {low, medium, high}. Based on E, one will try to define conjunction and disjunction as well as negation.

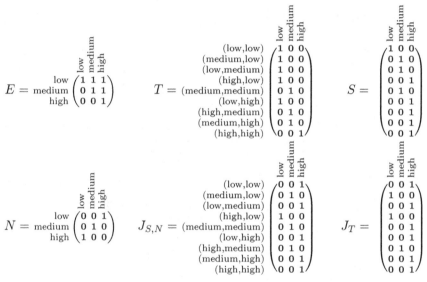

Fig. 14.19 Subjunctions $J_{S,N}$ and J_T derived from E, T, S, N

According to Prop. 14.21 and Prop. 14.22, we have evaluated the relations so as to obtain the two different forms of subjunction. □

Example 14.24. In the following example we use a very specific ordering, namely the powerset ordering of a 3-element set. It will turn out that in this case both subjunctions coincide. The powerset ordering is, of course, highly regular so that a special result could be expected.

One will have noticed that the result is given here in a different style: not as a very sparse but voluminous Boolean matrix, but as a table for the function. □

$$E = \begin{array}{c} \\ 1 \\ 2 \\ 3 \\ 4 \\ 5 \\ 6 \\ 7 \\ 8 \end{array}\begin{array}{c} {\scriptstyle 1\ 2\ 3\ 4\ 5\ 6\ 7\ 8} \\ \left(\begin{array}{cccccccc} 1 & 1 & 1 & 1 & 1 & 1 & 1 & 1 \\ 0 & 1 & 0 & 1 & 0 & 1 & 0 & 1 \\ 0 & 0 & 1 & 1 & 0 & 0 & 1 & 1 \\ 0 & 0 & 0 & 1 & 0 & 0 & 0 & 1 \\ 0 & 0 & 0 & 0 & 1 & 1 & 1 & 1 \\ 0 & 0 & 0 & 0 & 0 & 1 & 0 & 1 \\ 0 & 0 & 0 & 0 & 0 & 0 & 1 & 1 \\ 0 & 0 & 0 & 0 & 0 & 0 & 0 & 1 \end{array}\right)\end{array}$$

$$T = \begin{array}{l} [[1,1,1,1,1,1,1,1], \\ {}[1,2,1,2,1,2,1,2], \\ {}[1,1,3,3,1,1,3,3], \\ {}[1,2,3,4,1,2,3,4], \\ {}[1,1,1,1,5,5,5,5], \\ {}[1,2,1,2,5,6,5,6], \\ {}[1,1,3,3,5,5,7,7], \\ {}[1,2,3,4,5,6,7,8]] \end{array}$$

$$S = \begin{array}{l} [[1,2,3,4,5,6,7,8], \\ {}[2,2,4,4,6,6,8,8], \\ {}[3,4,3,4,7,8,7,8], \\ {}[4,4,4,4,8,8,8,8], \\ {}[5,6,7,8,5,6,7,8], \\ {}[6,6,8,8,6,6,8,8], \\ {}[7,8,7,8,7,8,7,8], \\ {}[8,8,8,8,8,8,8,8]] \end{array}$$

$$N = \begin{array}{c} \\ 1 \\ 2 \\ 3 \\ 4 \\ 5 \\ 6 \\ 7 \\ 8 \end{array}\begin{array}{c} {\scriptstyle 1\ 2\ 3\ 4\ 5\ 6\ 7\ 8} \\ \left(\begin{array}{cccccccc} 0 & 0 & 0 & 0 & 0 & 0 & 0 & 1 \\ 0 & 0 & 0 & 0 & 0 & 0 & 1 & 0 \\ 0 & 0 & 0 & 0 & 0 & 1 & 0 & 0 \\ 0 & 0 & 0 & 0 & 1 & 0 & 0 & 0 \\ 0 & 0 & 0 & 1 & 0 & 0 & 0 & 0 \\ 0 & 0 & 1 & 0 & 0 & 0 & 0 & 0 \\ 0 & 1 & 0 & 0 & 0 & 0 & 0 & 0 \\ 1 & 0 & 0 & 0 & 0 & 0 & 0 & 0 \end{array}\right)\end{array}$$

$$J_{S,N} = \begin{array}{l} [[8,8,8,8,8,8,8,8], \\ {}[7,8,7,8,7,8,7,8], \\ {}[6,6,8,8,6,6,8,8], \\ {}[5,6,7,8,5,6,7,8], \\ {}[4,4,4,4,8,8,8,8], \\ {}[3,4,3,4,7,8,7,8], \\ {}[2,2,4,4,6,6,8,8], \\ {}[1,2,3,4,5,6,7,8]] \end{array} = J_T$$

Fig. 14.20 Subjunctions $J_{S,N} = J_T$ derived from E, T, S, N

We now have sufficient experience of working with relational qualifications to be able to replace the traditionally used fuzzy qualifications. The main point of criticism concerning fuzzy methods is that every valuation is coded in the interval $[0,1]$, and that therefore diverging, antagonistic, and counter-running criteria cannot be kept sufficiently separate. Using the concept of relational norms and conorms, we are now in a position to try more realistic tasks.

Exercises

14.1 Prove or disprove providing a counter example: R is commutative precisely when $\pi^{\mathsf{T}}; R = \rho^{\mathsf{T}}; R$.

15

Relational Graph Theory

Many of the problems handled in applications are traditionally formulated in terms of graphs. This means that graphs will often be drawn and the reasoning will be pictorial. On the one hand, this is nice and intuitive when executed with chalk on a blackboard. On the other hand there is a considerable gap from this point to treating the problem on a computer. Often ad hoc programs are written in which more time is spent on I/O handling than on precision of the algorithm. Graphs are well suited to visualization of a result, even with the possibility of generating the graph via a graph drawing program. What is nearly impossible is the *input* of a problem given by means of a graph – when not using RELVIEW's interactive graph input (see [18], for example.). In such cases some sort of relational interpretation of the respective graph is usually generated and input in some way.

We will treat reducibility and irreducibility first, mentioning also partial decomposability. Then difunctional relations are studied in the homogeneous context which provides additional results. The main aim of this chapter is to provide algorithms to determine relationally specified subsets of a graph or relation in a declarative way.

15.1 Reducibility and irreducibility

We are now going to study in more detail the reducibility and irreducibility introduced in a phenomenological form as Def. 6.12. Many of these results for relations stem from Georg Frobenius [54] and his study of eigenvalues of non-negative real-valued matrices; a comprehensive presentation is given in [92].

Definition 15.1. We call a homogeneous relation

$\quad$ A **reducible** $\quad :\Longleftrightarrow \quad$ There exists a vector $\mathbb{L} \neq r \neq \mathbb{T}$ with $A ; r \subseteq r$

otherwise A is called **irreducible**. $\hspace{2cm}$ $\square$

A relation A on a set X is, thus, called reducible if there exists an $⊥ \neq r \neq ⊤$, which *reduces* A, i.e., $A ; r \subseteq r$ in the sense of Def. 6.12. Arrows of the graph according to A ending in the subset r will always start in r. Figure 15.1 indicates this with the dashed arrow convention. It is easy to see that the reducing vectors r, in this case including the trivial ones $r = ⊥$ and $r = ⊤$, form a lattice.

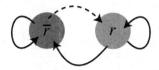

Fig. 15.1 Schema of a reducible relation using the dashed arrow convention

Using Schröder's rule, a relation A is reduced by a set r precisely when its transpose A^T is reduced by $\bar{r}$: $A ; r \subseteq r \iff A^\mathsf{T} ; \bar{r} \subseteq \bar{r}$. Therefore, a relation is reducible precisely when its transpose is.

If such a (non-trivial, i.e., $⊥ \neq r \neq ⊤$) vector r does not exist, A is called irreducible. Irreducible means that A is reduced by precisely two vectors, namely by $r = ⊥$ and $r = ⊤$. In particular, we have for an irreducible A that $A ; A^* \subseteq A^*$, meaning column-wise irreducible, so that $A^* = ⊤$, since obviously $A^* \neq ⊥$. Therefore, one can translate this into the language of graph theory:

A irreducible $\iff$ 1-graph with associated relation A is strongly connected.

An irreducible relation A will necessarily be total: A certainly contracts $A ; ⊤$ since $A ; A ; ⊤ \subseteq A ; ⊤$. From irreducibility we obtain that $A ; ⊤ = ⊥$ or $A ; ⊤ = ⊤$. The former would mean $A = ⊥$, so that A would contract every relation x, and, thus, violate irreducibility. Therefore, only the latter is possible, i.e., A is total.

For a reducible relation A and arbitrary k, the powers A^k are also reducible because $A^k ; x \subseteq A^{k-1} ; x \subseteq \ldots \subseteq A ; x \subseteq x$. However, Fig. 15.2 provides an example of an irreducible relation A with A^2 reducible. Therefore, the property of being irreducible is not multiplicative.

$$A = \begin{pmatrix} 0 & 0 & 0 & 1 \\ 0 & 0 & 0 & 1 \\ 0 & 0 & 0 & 1 \\ 1 & 1 & 1 & 0 \end{pmatrix} \quad \begin{pmatrix} 1 & 1 & 1 & 0 \\ 1 & 1 & 1 & 0 \\ 1 & 1 & 1 & 0 \\ 0 & 0 & 0 & 1 \end{pmatrix} = A^2$$

Fig. 15.2 Irreducible relation with reducible square

The following applies to all finite $n \times n$-relations; later, we will use it mainly for irreducible relations.

Proposition 15.2. *If A is any finite $n \times n$-relation, then*

$$A^n \subseteq (\mathbb{I} \cup A)^{n-1} \qquad and \qquad (\mathbb{I} \cup A)^{n-1} = A^*.$$

Proof: It is trivial that $A^i \subseteq A^*$ for all $i \geq 0$. By the pigeon hole principle, i.e., using finiteness, always $A^n \subseteq (\mathbb{I} \cup A)^{n-1}$ as otherwise $n + 1$ vertices would be needed to delimit a non-selfcrossing path of length n while only n distinct vertices are available.   □

Irreducible relations satisfy further important formulae as in Theorem 1.1.2 of [6], for example.

Theorem 15.3. *For any finite $n \times n$-relation A the following hold:*

(i)   *A irreducible* $\quad \Longleftrightarrow \quad (\mathbb{I} \cup A)^{n-1} = \mathbb{T} \quad \Longleftrightarrow \quad A_{;}(\mathbb{I} \cup A)^{n-1} = \mathbb{T}$,*

(ii) *A irreducible* $\quad \Longrightarrow \quad$ *There exists an exponent k such that $\mathbb{I} \subseteq A^k$.*

Proof: (i) We start with the first equivalence. By definition, A is irreducible, if we have for all vectors $x \neq \mathbb{\bot}$ that $A_{;}x \subseteq x$ implies $x = \mathbb{T}$. Now, by the preceding proposition

$$A_{;}(\mathbb{I} \cup A)^{n-1} \subseteq (\mathbb{I} \cup A)^{n-1}$$

so that indeed $(\mathbb{I} \cup A)^{n-1} = \mathbb{T}$.

For the other direction assume A reducible, so that $\mathbb{\bot} \neq x \neq \mathbb{T}$ exists with $A_{;}x \subseteq x$. Then also $A^k_{;}x \subseteq x$ for arbitrary k. This is a contradiction, since it would follow that also $(\mathbb{I} \cup A)^{n-1}_{;}x \subseteq x$ resulting in $\mathbb{T}_{;}x \subseteq x$ and, in the case of Boolean matrices, i.e., with Tarski's rule satisfied, in $x = \mathbb{T}$, a contradiction.

Now we prove the second equivalence of which "$\Longleftarrow$" is trivial when remembering Prop. 15.2. For "$\Longrightarrow$", we also use Prop. 15.2, $(\mathbb{I} \cup A)^{n-1} \supseteq A_{;}(\mathbb{I} \cup A)^{n-1}$, so that also $A_{;}(\mathbb{I} \cup A)^{n-1} \supseteq A_{;}A_{;}(\mathbb{I} \cup A)^{n-1}$. Since A is irreducible, this leads to $A_{;}(\mathbb{I} \cup A)^{n-1}$ being equal to $\mathbb{\bot}$ or $\mathbb{T}$, from which only the latter is possible.

(ii) Consider the irreducible $n \times n$-relation A and its powers. According to (i), there exists for every row number j a least power $1 \leq p_j \leq n$ with position $(j, j) \in A^{p_j}$. For the least common multiple p of all these p_j we have $\mathbb{I} \subseteq A^p$, and p is the smallest positive number with this property.   □

For irreducible relations yet a further distinction can be made; see, for example, Theorem 2.5 of [142].

Definition 15.4. An irreducible relation A is called **primitive** if there exists some integer $k \geq 1$ such that $A^k = \mathbb{T}$. If this is not the case, the irreducible relation may be called **cyclic of order** k, indicating that the (infinitely many) powers $A, A^2, A^3 \ldots$ reproduce cyclically and k is the greatest common divisor of all the periods occurring. $\qquad \square$

We first observe powers of primitive relations in Fig. 15.3.

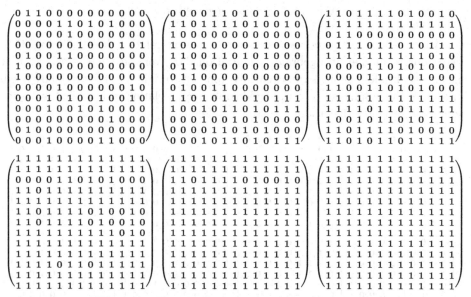

Fig. 15.3 Powers of a primitive irreducible relation

Proposition 15.5. *A relation R is primitive precisely when its powers R^k are irreducible for all $k \geq 1$.*

Proof: "$\Longrightarrow$": Assume $R^k \mathbin{;} x \subseteq x$ with $x \neq \mathbb{\bot}$ and $x \neq \mathbb{T}$ for some $k \geq 1$. Then we have also $R^{nk} \mathbin{;} x \subseteq R^{(n-1)k} \mathbin{;} x \ldots \subseteq R^k \mathbin{;} x \subseteq x$ for all $n \geq 1$. This contradicts primitivity of R because from some index on all powers of a primitive R should have been $\mathbb{T}$.

"$\Longleftarrow$": Assume R is not primitive, i.e., $R^k \neq \mathbb{T}$ for all k. It is impossible for any R^k to have a column $\mathbb{\bot}$ since this would directly show reducibility of R^k. It follows from finiteness that there will exist identical powers $R^l = R^k \neq \mathbb{T}$ with $l > k$, for example. This results in $R^{l-k} \mathbin{;} R^k = R^k$. Power R^{l-k}, therefore, is reduced by all columns of R^k – and at least one column which is unequal $\mathbb{T}$ (see, for example Theorem 1.8.2 of [6]). $\qquad \square$

Now we observe how powers of a cyclic relation behave.

$$
\begin{array}{c}
\begin{array}{cccccccccc}
 & 1\;2\;3\;4\;5\;6\;7\;8\;9
\end{array}\\
\begin{array}{c}1\\2\\3\\4\\5\\6\\7\\8\\9\end{array}
\begin{pmatrix}
0\;0\;1\;1\;0\;0\;0\;0\;0\\
0\;0\;0\;0\;0\;1\;0\;0\;1\\
0\;0\;0\;0\;0\;0\;0\;1\;0\\
0\;0\;0\;0\;1\;0\;0\;0\;0\\
1\;0\;0\;0\;0\;0\;1\;0\;0\\
1\;0\;0\;0\;0\;0\;0\;0\;0\\
0\;1\;0\;0\;0\;0\;0\;0\;0\\
0\;0\;0\;0\;0\;0\;1\;0\;0\\
1\;0\;0\;0\;0\;0\;0\;0\;0
\end{pmatrix}
\end{array}
$$

original relation with
rearranged and
subdivided version

$$
\begin{array}{c}
\begin{array}{cccccccc}
 & 5\;6\;8\;9\;1\;7\;2\;3\;4
\end{array}\\
\begin{array}{c}5\\6\\8\\9\\1\\7\\2\\3\\4\end{array}
\left(\begin{array}{cccc|cc|ccc}
0\;0\;0\;0 & 1\;1 & 0\;0\;0\\
0\;0\;0\;0 & 1\;0 & 0\;0\;0\\
0\;0\;0\;0 & 0\;1 & 0\;0\;0\\
0\;0\;0\;0 & 1\;0 & 0\;0\;0\\
\hline
0\;0\;0\;0 & 0\;0 & 0\;1\;1\\
0\;0\;0\;0 & 0\;0 & 1\;0\;0\\
\hline
0\;1\;0\;1 & 0\;0 & 0\;0\;0\\
0\;0\;1\;0 & 0\;0 & 0\;0\;0\\
1\;0\;0\;0 & 0\;0 & 0\;0\;0
\end{array}\right)
\end{array}
$$

Fig. 15.4 A cyclic irreducible relation

One will not be able to observe the essence of a cyclic relation directly in Fig. 15.5. Using our technique of algebraic visualization, however, one may grasp what it means: block side-diagonals run cyclically from left to right and are gradually filled. In a sense this applies also to the irreducible cyclic relation of Fig. 15.3, with the additional property that there is just a 1-cycle.

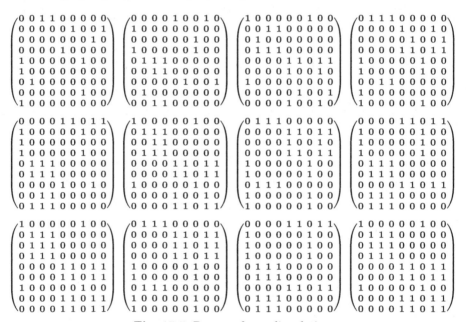

Fig. 15.5 Powers of a cyclic relation

Reducible permutations

It is interesting to look for irreducibility of permutations P. We observe $P_; x = x$ for $x := \overline{(P^k \cap \mathbb{I})_; \mathbb{T}}$ and arbitrary k because obviously

$$
P_; x = P_; \overline{(P^k \cap \mathbb{I})_; \mathbb{T}} = \overline{P_; (P^k \cap \mathbb{I})_; \mathbb{T}} = \overline{(P_; P^k \cap P)_; \mathbb{T}} = \overline{(P^k \cap \mathbb{I})_; P_; \mathbb{T}}
$$

$$= \overline{(P^k \cap \mathbb{I}) ; \mathbb{T}} = x$$

owing to the doublesided mapping properties of P. For $k = 0$, for example, this means $x = \mathbb{\perp\!\!\!\perp}$ and is rather trivial. Also cases with $P^k \cap \mathbb{I} = \mathbb{\perp\!\!\!\perp}$ resulting in $x = \mathbb{T}$ are trivial.

Top row of matrices:

```
/0 0 0 0|1 1|0 0 0\   /0 0 0 0|0 0|1 1 1\   /1 1 1 1|0 0|0 0 0\   /0 0 0 0|1 1|0 0 0\
 0 0 0 0|1 0|0 0 0     0 0 0 0|0 0|0 1 1     1 0 1 0|0 0|0 0 0     0 0 0 0|1 1|0 0 0
 0 0 0 0|0 1|0 0 0     0 0 0 0|0 0|1 0 0     0 1 0 1|0 0|0 0 0     0 0 0 0|1 0|0 0 0
 0 0 0 0|1 0|0 0 0     0 0 0 0|0 0|0 1 1     1 0 1 0|0 0|0 0 0     0 0 0 0|1 1|0 0 0
 0 0 0 0|0 0|0 1 1     1 0 1 0|0 0|0 0 0     0 0 0 0|1 1|0 0 0     0 0 0 0|0 0|1 1 1
 0 0 0 0|0 0|1 0 0     0 1 0 1|0 0|0 0 0     0 0 0 0|1 0|0 0 0     0 0 0 0|0 0|0 1 1
 0 1 0 1|0 0|0 0 0     0 0 0 0|1 0|0 0 0     0 0 0 0|0 0|0 1 1     1 0 1 0|0 0|0 0 0
 0 0 1 0|0 0|0 0 0     0 0 0 0|0 1|0 0 0     0 0 0 0|0 0|1 0 0     0 1 0 1|0 0|0 0 0
\1 0 0 0|0 0|0 0 0/   \0 0 0 0|1 1|0 0 0/   \0 0 0 0|0 0|1 1 1/   \1 1 1 1|0 0|0 0 0/
```

Middle row of matrices:

```
/0 0 0 0|0 0|1 1 1\   /1 1 1 1|0 0|0 0 0\   /0 0 0 0|1 1|0 0 0\   /0 0 0 0|0 0|1 1 1\
 0 0 0 0|0 0|1 1 1     1 1 1 1|0 0|0 0 0     0 0 0 0|1 1|0 0 0     0 0 0 0|0 0|1 1 1
 0 0 0 0|0 0|0 1 1     1 0 1 0|0 0|0 0 0     0 0 0 0|1 1|0 0 0     0 0 0 0|0 0|1 1 1
 0 0 0 0|0 0|1 1 1     1 1 1 1|0 0|0 0 0     0 0 0 0|1 1|0 0 0     0 0 0 0|0 0|1 1 1
 1 1 1 1|0 0|0 0 0     0 0 0 0|1 1|0 0 0     0 0 0 0|0 0|1 1 1     1 1 1 1|0 0|0 0 0
 1 0 1 0|0 0|0 0 0     0 0 0 0|1 1|0 0 0     0 0 0 0|0 0|1 1 1     1 1 1 1|0 0|0 0 0
 0 0 0 0|1 1|0 0 0     0 0 0 0|0 0|1 1 1     1 1 1 1|0 0|0 0 0     0 0 0 0|1 1|0 0 0
 0 0 0 0|1 0|0 0 0     0 0 0 0|0 0|0 1 1     1 0 1 0|0 0|0 0 0     0 0 0 0|0 0|1 1 1
\0 0 0 0|1 1|0 0 0/   \0 0 0 0|0 0|1 1 1/   \1 1 1 1|0 0|0 0 0/   \0 0 0 0|1 1|0 0 0/
```

Bottom row of matrices:

```
/1 1 1 1|0 0|0 0 0\   /0 0 0 0|1 1|0 0 0\   /0 0 0 0|0 0|1 1 1\   /1 1 1 1|0 0|0 0 0\
 1 1 1 1|0 0|0 0 0     0 0 0 0|1 1|0 0 0     0 0 0 0|0 0|1 1 1     1 1 1 1|0 0|0 0 0
 1 1 1 1|0 0|0 0 0     0 0 0 0|1 1|0 0 0     0 0 0 0|0 0|1 1 1     1 1 1 1|0 0|0 0 0
 1 1 1 1|0 0|0 0 0     0 0 0 0|1 1|0 0 0     0 0 0 0|0 0|1 1 1     1 1 1 1|0 0|0 0 0
 0 0 0 0|1 1|0 0 0     0 0 0 0|0 0|1 1 1     1 1 1 1|0 0|0 0 0     0 0 0 0|1 1|0 0 0
 0 0 0 0|1 1|0 0 0     0 0 0 0|0 0|1 1 1     1 1 1 1|0 0|0 0 0     0 0 0 0|1 1|0 0 0
 0 0 0 0|0 0|1 1 1     1 1 1 1|0 0|0 0 0     0 0 0 0|1 1|0 0 0     0 0 0 0|0 0|1 1 1
 0 0 0 0|0 0|1 1 1     0 0 0 0|0 0|1 1 1     0 0 0 0|1 1|0 0 0     0 0 0 0|0 0|1 1 1
\0 0 0 0|0 0|1 1 1/   \1 1 1 1|0 0|0 0 0/   \0 0 0 0|1 1|0 0 0/   \0 0 0 0|0 0|1 1 1/
```

Fig. 15.6 Powers of the cyclic relation of Fig. 15.4 in nicely arranged form

The permutation P is reducible, namely reduced by x, when neither $P^k \cap \mathbb{I} = \mathbb{\perp\!\!\!\perp}$ nor $P^k \cap \mathbb{I} = \mathbb{I}$. In recalling permutations, every cycle of P of length c will lead to $P^c \cap \mathbb{I} \neq \mathbb{\perp\!\!\!\perp}$. If k is the least common multiple of all cycle lengths occurring in P, obviously $P^k = \mathbb{I}$. If there is just one cycle – as for the cyclic successor relation for $n > 1$ – the permutation P is irreducible. Other permutations with more than one cycle are reducible.

$$\begin{pmatrix} 0 & 1 & 0 & 0 & 0 \\ 0 & 0 & 1 & 0 & 0 \\ 0 & 0 & 0 & 1 & 0 \\ 0 & 0 & 0 & 0 & 1 \\ 1 & 0 & 0 & 0 & 0 \end{pmatrix} \qquad \begin{pmatrix} 0 & 1 & 0 & 0 & 0 \\ 1 & 0 & 0 & 0 & 0 \\ 0 & 0 & 0 & 1 & 0 \\ 0 & 0 & 0 & 0 & 1 \\ 0 & 0 & 1 & 0 & 0 \end{pmatrix}$$

Fig. 15.7 Irreducible and reducible permutation

Partly decomposable relations

We now investigate a property that is not so commonly known, but is also related to finding smaller parts of a relation. In a sense, we switch to permuting independently.

Definition 15.6. Let a homogeneous relation A be given. We call

$$A \text{ \textbf{partly decomposable}} \quad :\Longleftrightarrow \quad \begin{array}{l} \text{There exists a vector } \mathbb{L} \neq x \neq \mathbb{T} \text{ and a} \\ \text{permutation } P \text{ with } A_{;}\, x \subseteq P_{;}\, x. \end{array}$$

If not partly decomposable, it is called **fully** or **totally indecomposable.** □

The most prominent property of partly decomposable relations is that, given x and P, it becomes possible to rearrange the relation into the following schema

$$P'_{;}\, x = \begin{pmatrix} \mathbb{L} \\ \mathbb{T} \end{pmatrix} \qquad P''_{;}\, P_{;}\, x = \begin{pmatrix} \mathbb{L} \\ \mathbb{T} \end{pmatrix}.$$

Rows and columns are permuted independently so as to bring $\mathbf{0}$s of x to the initial part of the vector with some permutation P' and afterwards in addition to bring $\mathbf{0}$s of $P_{;}\, x$ to the front with some permutation P''. This being achieved, the structure of $A_{;}\, x \subseteq P_{;}\, x$ is now

$$P''_{;}\, A_{;}\, P'^{\mathsf{T}}_{;} \begin{pmatrix} \mathbb{L} \\ \mathbb{T} \end{pmatrix} \subseteq \begin{pmatrix} \mathbb{L} \\ \mathbb{T} \end{pmatrix} \qquad \Longrightarrow \qquad P''_{;}\, A_{;}\, P'^{\mathsf{T}} = \begin{pmatrix} * & \mathbb{L} \\ * & * \end{pmatrix}.$$

In other words: there exists a vector $\mathbb{L} \neq x \neq \mathbb{T}$ such that x reduces $P''_{;}\, A_{;}\, P'^{\mathsf{T}}$; formally, $P''_{;}\, A_{;}\, P'^{\mathsf{T}}_{;}\, x \subseteq x$. This in turn can only be the case with the matrix above, i.e., with an empty zone in the upper right. The permutation P in the definition is not really related to A. All permutations P are tested, and it suffices that one exists. This makes it a counting argument: for a fully indecomposable relation, the number of entries $\mathbf{1}$ in x does not suffice to cover the number of $\mathbf{1}$s in $A_{;}\, x$.

$$\begin{pmatrix} 0 & 1 & 0 & 1 \\ 1 & 0 & 1 & 0 \\ 0 & 1 & 1 & 0 \\ 1 & 0 & 0 & 1 \end{pmatrix} \qquad \begin{pmatrix} 1 & 1 & 0 & 0 \\ 0 & 1 & 1 & 0 \\ 0 & 0 & 1 & 1 \\ 1 & 0 & 0 & 1 \end{pmatrix}$$

Fig. 15.8 A fully indecomposable relation with rearrangement to side diagonal

With rows and columns permuted independently, the left relation can be transferred into the right one. The latter shows a permutation which, when removed, leaves another permutation.

In numerical mathematics, a fully indecomposable shape of a matrix has quite unpleasant properties. If it is used for iteration purposes, full indecomposability would mean that any tiny alteration is strongly distributed during this iteration and not confined in its implied changes to a small region.

Proposition 15.7.

(i) *A fully indecomposable* $\quad\Longleftrightarrow\quad$ *Except for $x = \mathbb{L}$ and $x = \mathbb{T}$, the product $A_{;}\, x$ has strictly more $\mathbf{1}$s than x.*

(ii) *A fully indecomposable* $\quad\Longrightarrow\quad$ *A satisfies the Hall condition.*

(iii) A *fully indecomposable* $\implies$ A *contains a permutation.*
(iv) A *fully indecomposable* $\implies$ A *is chainable.*
(v) A, B *fully indecomposable* $\implies$ $A; B$ *fully indecomposable.*

Proof: (i) We write the definition of being partly decomposable more formally:

$$\exists x : x \neq \mathbb{L} \wedge x \neq \mathbb{T} \wedge \left[\exists P : A; x \subseteq P; x\right].$$

This is now negated to obtain the case of being fully indecomposable

$$\forall x : x = \mathbb{L} \vee x = \mathbb{T} \vee \left[\forall P : A; x \not\subseteq P; x\right].$$

We observe that the permutation P is not really connected with the problem, so that it is nothing more than a counting argument saying $|A; x| > |x|$.

(ii) The Hall condition Def. 10.8 follows from what has just been shown.

(iii) The famous assignment theorem asserts existence of a matching when the Hall condition is satisfied according to (ii).

(iv) Using (iii), we may find permutation $P \subseteq A$. Assume A is not chainable, then there exists, according to Prop. 10.26, a pair s, t that is non-trivial, i.e., neither $s = \mathbb{L}$ nor $t = \mathbb{L}$ satisfying

$$A; t \subseteq \overline{s} \quad \text{and} \quad A; \overline{t} \subseteq s.$$

In such a case, $A; t \subseteq \overline{s} \subseteq \overline{A; \overline{t}} \subseteq \overline{P; \overline{t}} = P; \overline{\overline{t}} = P; t$, a contradiction to full indecomposability.

(v) With (iii), A as well as B contain a permutation, i.e., $A \supseteq P_A$ and $B \supseteq P_B$. Assuming $A; B$ partly decomposable, $P_A; B; x \subseteq A; B; x \subseteq P; x$ and $B; x \subseteq P_A^\mathsf{T}; P; x$, meaning that also B is partly decomposable, violating the assumption. $\qquad\square$

There exists an interesting link to real-valued matrices that leads to many important applications. Let us call a non-negative matrix $B \in \mathbb{R}^{n \times n}$, equivalent with a mapping $B : X \times X \to \mathbb{R}$, with row- as well as column-sums always equal to 1 a **doubly stochastic** matrix. We say that a relation $R : X \longrightarrow X$ has **doubly stochastic pattern** provided that there exists a doubly stochastic matrix B such that $R_{ik} = 1$ if and only if $B_{ik} > 0$. It is a remarkable fact that every fully indecomposable relation has doubly stochastic pattern.

15.2 Homogeneous difunctional relations

Difunctional relations have already been mentioned at several occasions; in particular, they were defined in a phenomenological way in Def. 5.30, and their rectangle-based properties were investigated in Section 10.4. Here, the difunctional relation will have to be homogeneous in addition, and permutations will always be executed simultaneously. For this case, we also relate difunctionality with irreducibility.

Proposition 15.8. *Let an arbitrary finite homogeneous relation R be given. Then in addition to the left and right equivalences Ω, Ω' of Prop. 5.36 $\Theta := (\Omega \cup \Omega')^{*}$, $G := \Theta \mathbin{;} R$, and $G' := R \mathbin{;} \Theta$ may also be formed.*

  (i) Θ *is an equivalence.*
  (ii) $G \mathbin{;} G^{\mathsf{T}} \subseteq \Theta$ *and* $G'^{\mathsf{T}} \mathbin{;} G' \subseteq \Theta$.
  (iii) G *as well as* G' *are difunctional.*

Proof: (i) is trivial since Ω, Ω' are symmetric by construction.

(ii) $G \mathbin{;} G^{\mathsf{T}} = \Theta \mathbin{;} R \mathbin{;} (\Theta \mathbin{;} R)^{\mathsf{T}} = \Theta \mathbin{;} R \mathbin{;} R^{\mathsf{T}} \mathbin{;} \Theta \subseteq \Theta \mathbin{;} \Omega \mathbin{;} \Theta \subseteq \Theta \mathbin{;} \Theta \mathbin{;} \Theta = \Theta.$

(iii) $G \mathbin{;} G^{\mathsf{T}} \mathbin{;} G \subseteq \Theta \mathbin{;} G = \Theta \mathbin{;} \Theta \mathbin{;} R = \Theta \mathbin{;} R = G$, using (ii).   □

It need not be that $G^{\mathsf{T}} \mathbin{;} G \subseteq \Theta$; see the example $R = \begin{pmatrix} 0&0&0 \\ 0&1&1 \\ 1&0&0 \end{pmatrix}$ with $G = \begin{pmatrix} 0&0&0 \\ 1&1&1 \\ 1&1&1 \end{pmatrix}$. Nor need the pair (Θ, Θ) be an R-congruence as Fig. 15.9 shows, where also $G \neq G'$.

$$R = \begin{array}{c|ccccccccccccc}
 & 1&2&3&4&5&6&7&8&9&10&11&12&13 \\ \hline
1 & 0&0&0&0&0&1&0&0&0&0&0&0&0 \\
2 & 0&0&0&0&1&0&1&1&0&0&0&0&0 \\
3 & 0&0&0&0&0&0&0&0&0&0&0&0&1 \\
4 & 0&0&0&0&0&0&0&0&0&0&0&0&0 \\
5 & 0&0&0&0&0&0&0&0&0&0&0&0&0 \\
6 & 0&0&0&0&0&0&0&0&0&1&0&0&0 \\
7 & 1&0&0&0&0&0&0&0&0&0&0&0&0 \\
8 & 0&0&0&0&0&0&0&0&0&0&0&0&0 \\
9 & 0&0&0&0&0&0&0&0&0&1&0&1&0 \\
10 & 0&0&0&0&0&0&0&0&0&0&0&0&0 \\
11 & 0&1&0&0&0&0&0&0&0&0&0&0&0 \\
12 & 0&0&1&0&0&0&0&0&0&0&0&0&0 \\
13 & 0&0&0&0&0&1&0&0&0&0&0&1&0
\end{array}$$

$$\Theta = \begin{array}{c|ccccccccccccc}
 & 1&11&13&2&3&6&9&12&4&5&7&8&10 \\ \hline
1 & 1&1&1&0&0&0&0&0&0&0&0&0&0 \\
11 & 1&1&1&0&0&0&0&0&0&0&0&0&0 \\
13 & 1&1&1&0&0&0&0&0&0&0&0&0&0 \\
2 & 0&0&0&1&0&0&0&0&0&0&0&0&0 \\
3 & 0&0&0&0&1&1&1&1&0&0&0&0&0 \\
6 & 0&0&0&0&1&1&1&1&0&0&0&0&0 \\
9 & 0&0&0&0&1&1&1&1&0&0&0&0&0 \\
12 & 0&0&0&0&1&1&1&1&0&0&0&0&0 \\
4 & 0&0&0&0&0&0&0&0&1&0&0&0&0 \\
5 & 0&0&0&0&0&0&0&0&0&1&1&1&0 \\
7 & 0&0&0&0&0&0&0&0&0&1&1&1&0 \\
8 & 0&0&0&0&0&0&0&0&0&1&1&1&0 \\
10 & 0&0&0&0&0&0&0&0&0&0&0&0&1
\end{array}$$

$$R_{\text{rearr}} = \begin{array}{c|ccccccccccccc}
 & 1&11&13&2&3&6&9&12&4&5&7&8&10 \\ \hline
1 & 0&0&0&0&1&0&0&0&0&0&0&0&0 \\
2 & 0&0&1&0&0&0&0&0&0&0&0&0&0 \\
3 & 0&0&0&0&0&1&0&1&0&0&0&0&0 \\
4 & 0&0&0&0&0&0&0&0&0&1&1&1&0 \\
5 & 0&0&1&0&0&0&0&0&0&0&0&0&0 \\
6 & 0&1&0&0&0&0&0&0&0&0&0&0&0 \\
7 & 0&1&1&0&0&0&0&0&0&0&0&0&0 \\
8 & 0&0&0&0&0&0&0&0&1&0&0&0&0 \\
9 & 0&0&0&0&0&0&0&0&0&0&0&0&0 \\
10 & 0&0&0&0&0&0&0&0&0&0&0&0&0 \\
11 & 1&0&0&0&0&0&0&0&0&0&0&0&0 \\
12 & 0&0&0&0&0&0&0&0&0&0&0&0&0 \\
13 & 0&0&0&0&0&0&0&0&0&0&0&0&0
\end{array}$$

$$G = \begin{array}{c|ccccccccccccc}
 & 1&11&13&2&3&6&9&12&4&5&7&8&10 \\ \hline
1 & 0&0&0&1&0&1&0&1&0&0&0&0&0 \\
11 & 0&0&0&1&0&1&0&1&0&0&0&0&0 \\
13 & 0&0&0&1&0&1&0&1&0&0&0&0&0 \\
2 & 0&0&0&0&0&0&0&0&0&1&1&1&0 \\
3 & 0&1&1&0&0&0&0&0&1&0&0&0&0 \\
6 & 0&1&1&0&0&0&0&0&1&0&0&0&0 \\
9 & 0&1&1&0&0&0&0&0&1&0&0&0&0 \\
12 & 0&1&1&0&0&0&0&0&1&0&0&0&0 \\
4 & 0&0&0&0&0&0&0&0&0&0&0&0&0 \\
5 & 1&0&0&0&0&0&0&0&0&0&0&0&0 \\
7 & 1&0&0&0&0&0&0&0&0&0&0&0&0 \\
8 & 1&0&0&0&0&0&0&0&0&0&0&0&0 \\
10 & 0&0&0&0&0&0&0&0&0&0&0&0&0
\end{array}$$

$$G' = \begin{array}{c|ccccccccccccc}
 & 1&11&13&2&3&6&9&12&4&5&7&8&10 \\ \hline
1 & 0&0&0&0&1&1&1&1&0&0&0&0&0 \\
11 & 0&0&0&1&0&0&0&0&0&0&0&0&0 \\
13 & 0&0&0&0&1&1&1&1&0&0&0&0&0 \\
2 & 0&0&0&0&0&0&0&0&0&1&1&1&0 \\
3 & 1&1&1&0&0&0&0&0&0&0&0&0&0 \\
6 & 1&1&1&0&0&0&0&0&0&0&0&0&0 \\
9 & 1&1&1&0&0&0&0&0&0&0&0&0&0 \\
12 & 0&0&0&0&0&0&0&1&0&0&0&0&0 \\
4 & 0&0&0&0&0&0&0&0&0&0&0&0&0 \\
5 & 0&0&0&0&0&0&0&0&0&0&0&0&0 \\
7 & 1&1&1&0&0&0&0&0&0&0&0&0&0 \\
8 & 0&0&0&0&0&0&0&0&0&0&0&0&0 \\
10 & 0&0&0&0&0&0&0&0&0&0&0&0&0
\end{array}$$

Fig. 15.9 A relation R with Θ, then R rearranged according to it, and G, G'

One easily observes that the relation is not yet block-wise injective, nor need it be block-wise univalent. Also, G, G' are difunctional, but with differing block-schemes. The blocks of G, G' are not yet completely filled, but only filled row- or column-wise, respectively. So by applying the permutations simultaneously, we have lost some of the properties the relations enjoyed when permuting independently in the heterogeneous case. In the next theorem, we define a larger congruence where we get back what has just been lost.

Proposition 15.9. *Let a finite and homogeneous relation R be given, and consider the constructs Ω, Ω' of Prop. 5.36 and $\Theta := (\Omega \cup \Omega')^*$ as in Prop. 15.8. Define $\mathcal{O}$ as the stationary value of the iteration*

$$X \mapsto \tau(X) := (X \cup R; X; R^{\mathsf{T}} \cup R^{\mathsf{T}}; X; R)^*$$

started with $X_0 := \mathbb{I}$.

(i) *$\mathcal{O}$ is an equivalence containing Ω, Ω', and Θ.*
(ii) *'Considered modulo $\mathcal{O}$', the relation R is*
$$\text{univalent } R^{\mathsf{T}}; \mathcal{O}; R \subseteq \mathcal{O}, \text{ and}$$
$$\text{injective } R; \mathcal{O}; R^{\mathsf{T}} \subseteq \mathcal{O}.$$
(iii) *$H := \mathcal{O}; R; \mathcal{O}$ is difunctional and commutes with $\mathcal{O}$, i.e., $\mathcal{O}; H = H; \mathcal{O}$, so that the pair $(\mathcal{O}, \mathcal{O})$ constitutes an H-congruence.*

Proof: (i) The isotone iteration $X \mapsto \tau(X)$ will be stationary after a finite number of steps with a relation $\mathcal{O}$ satisfying $\mathbb{I} \subseteq \mathcal{O} = \tau(\mathcal{O}) = (\mathcal{O} \cup R; \mathcal{O}; R^{\mathsf{T}} \cup R^{\mathsf{T}}; \mathcal{O}; R)^*$. Thus, $\mathcal{O}$ is reflexive, symmetric, and transitive by construction, i.e., it is an equivalence. This equivalence certainly contains $R; R^{\mathsf{T}}$, and therefore, Ω. Also $\Omega' \subseteq \mathcal{O}$ in an analogous way, so that it also contains Θ.

(ii) is trivial.

(iii) $H; H^{\mathsf{T}}; H = \mathcal{O}; R; \mathcal{O}; (\mathcal{O}; R; \mathcal{O})^{\mathsf{T}}; \mathcal{O}; R; \mathcal{O} = \mathcal{O}; R; \mathcal{O}; R^{\mathsf{T}}; \mathcal{O}; R; \mathcal{O} \subseteq \mathcal{O}; R; \mathcal{O} = H$
$\mathcal{O}; H = \mathcal{O}; \mathcal{O}; R; \mathcal{O} = \mathcal{O}; R; \mathcal{O} = \mathcal{O}; R; \mathcal{O}; \mathcal{O} = H; \mathcal{O}.$ □

Another characterization is
$$\mathcal{O} = \inf \{Q \mid Q \text{ equivalence}, \quad Q; R \subseteq R; Q, \quad R^{\mathsf{T}}; R \subseteq Q, \quad R; R^{\mathsf{T}} \subseteq Q\}.$$
So $(\mathcal{O}, \mathcal{O})$ is the smallest R-congruence above (Ω, Ω').

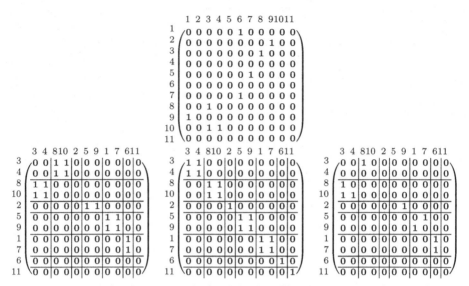

Fig. 15.10 A relation R with H, $\mathcal{O}$, and itself rearranged according to $\mathcal{O}$

In the second relation H of Fig. 15.10, one will observe a block-successor form with a 2-cycle first and then a terminating strand of 4.

Very often $\mathcal{O}$ will be much too big an equivalence, close to $\mathbb{T}$, to be interesting. There are special cases, however, where we encounter the well-known Moore–Penrose configuration again.

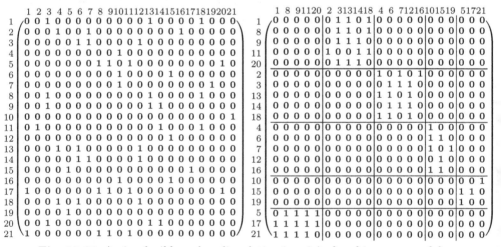

Fig. 15.11 An irreducible and cyclic relation in original and in rearranged form

Irreducibility of the difunctional closure

In the present homogeneous scenario where we permute rows and columns simultaneously, we may also study difunctionality together with irreducibility.

Proposition 15.10. *Assume the settings of Prop. 15.9, and assume that R in addition be irreducible. Then the following hold.*

(i) *H is irreducible.*
(ii) *H is total and surjective making it a $1:1$-correspondence of $\mathcal{O}$-classes.*
(iii) *H^T acts as an 'inverse' of H in as far as*
$$H^\mathsf{T}; H = \mathcal{O} \qquad H^\mathsf{T}; H^2 = H \qquad H^\mathsf{T}; H^3 = H^2 \qquad \ldots$$
(iv) *There exists a power k such that $R^k = \mathcal{O}$ and $R^{k+1} = H$.*

Proof: (i) Assuming $H; x \subseteq x$ to hold, then $R; x \subseteq x$. If it were the case that $\mathbb{\bot} \neq x \neq \mathbb{T}$, we would obtain that R were reducible, i.e., a contradiction.

(ii) As proved shortly after Def. 15.1, an irreducible relation R is total. This holds for $\mathcal{O}; R; \mathcal{O}$ as well. Surjectivity is shown analogously. Dividing out equivalences in Prop. 15.9.ii, we obtain the $1:1$-correspondence.

(iii) We now deduce that $H^\mathsf{T}; H = \mathcal{O}$, since with surjectivity and definition of $\mathcal{O}$
$$H^\mathsf{T}; H = \mathcal{O}; R^\mathsf{T}; \mathcal{O}; \mathcal{O}; R; \mathcal{O} = \mathcal{O}; R^\mathsf{T}; \mathcal{O}; R; \mathcal{O} = \mathcal{O}; \mathcal{O}; \mathcal{O} = \mathcal{O}.$$
Then also by induction
$$H^\mathsf{T}; H^{k+1} = H^\mathsf{T}; H; H^k = \mathcal{O}; H^k = H^k.$$

(iv) We prove a property we had already on <u>page 401</u> for permutations proper, i.e., with all block-widths 1: the construct $x := \overline{(H^k \cap \mathcal{O}); \mathbb{T}}$ satisfies $H; x \subseteq x$ for all powers k since

$$
\begin{aligned}
H; x &= H; \overline{(H^k \cap \mathcal{O}); \mathbb{T}} && \text{by definition} \\
&= \overline{H; (H^k \cap \mathcal{O}); \mathbb{T}} && \text{see } (*) \text{ below} \\
&= \overline{(H^{k+1} \cap H); \mathbb{T}} && \text{see } (\dagger) \text{ below} \\
&= \overline{(H^k \cap \mathcal{O}); H; \mathbb{T}} && \text{see } (\dagger) \text{ again} \\
&= \overline{(H^k \cap \mathcal{O}); \mathbb{T}} = x
\end{aligned}
$$

First, we concentrate on proving $(*)$. From the totality of H we have $\mathbb{T} = H; \mathbb{T} = H; X \cup H; \overline{X}$ so that always $\overline{H; X} \subseteq H; \overline{X}$. The opposite inclusion is satisfied for X satisfying $\mathcal{O}; X = X$, since $H^\mathsf{T}; H; X \subseteq \mathcal{O}; X = X$.

Secondly, we convince ourselves concerning $(\dagger)$. With the Dedekind rule
$$H; (H^k \cap \mathcal{O}) \subseteq H; H^k \cap H \subseteq (H \cap H; H^{k^\mathsf{T}}); (H^k \cap H^\mathsf{T}; H)$$

$$\subseteq H_{\,;}(H^k \cap H^{\mathsf{T}}_{\,;}H) = H_{\,;}(H^k \cap \mathcal{O})$$

giving equality everywhere in between.

We have, after all, that $R_{\,;}x \subseteq H_{\,;}x = x$, regardless of how k has been chosen. However, R is irreducible, which means that no x unequal to $\bot, \top$ is allowed to occur. This restricts H^k to be either $\mathcal{O}$ or to be disjoint therefrom. $\qquad\square$

An example is provided with Fig. 15.11. The relation H is obtained on the right with the respective boxes of the 5-cycle filled. It is certainly difunctional and has the Moore–Penrose property. The relation H is, however, not the difunctional closure of R. One may observe this at the rightmost non-empty 3×3-block which is not chainable. What has been obtained in Fig. 15.11 *by simultaneous permutation* is a block structure that is very close to the difunctional structure with a block diagonal. When *permuting rows and columns independently* for $h_{\mathrm{difu}}(R)$, one more block would show up, namely the rightmost split into two: $\{10\} \times \{21\}$ and $\{15, 19\} \times \{5, 17\}$.

Exercises

15.1 Assume a finite irreducible relation $A : X \longrightarrow X$ and a vector $y \subseteq X$ satisfying $\bot \not\subsetneq y \not\subsetneq \top$. Prove that $(\mathbb{I} \cup A)_{\,;}y$ has strictly more $\mathbf{1}$s than y.

15.2 Assume a finite relation

$$A = \begin{pmatrix} A_1 & B_1 & \bot & \cdots & & \bot \\ \bot & A_2 & B_2 & \cdots & & \vdots \\ \vdots & & & \ddots & & \bot \\ \bot & \bot & & & A_{r-1} & B_{r-1} \\ B_r & \bot & & \cdots & \bot & A_r \end{pmatrix}$$

with A_i square and fully indecomposable and $B_i \neq \bot$. Prove that in this case A will be fully indecomposable.

15.3 Subsets characterized by relational properties

Often in this book, we have encountered applications where subsets have been characterized by some property: chains, antichains, kernels, cliques, to mention just a few. It may, however, also be a stable set in a simple graph or an independent set of vertices in a hypergraph. Sometimes one is also interested in finding *all* the sets with the respective property. It is this point we address here in a synchronizing way. Sometimes this is a rather trivial task while in other cases even the existence of some set may not be evident in the first place, and one would be happy to find at least one.

There is an additional case to be mentioned. Sometimes one has to consider sets of subsets, but is mainly interested in their supremum, for example, as for the initial part of a relation. Such cases are also included.

A big portion of what is presented here rests on independent practical work by Rudolf Berghammer and his RELVIEW group. This research was heavily application driven. Methods sometimes scale up to spectacular size. As an example, the determination of kernels may be considered – an $\mathcal{NP}$-complete and thus inherently difficult problem. For a broad class of graphs, kernels could be computed exactly for up to 300 vertices, making this a selection out of a set of size 2^{300}.

What we present here is some sort of 'declarative programming' as experienced at other occasions with functional programming. Not least, it serves the purpose of separating the concerns of proof and correctness as opposed to 'algorithmics' dealing with efficiency in the first place.

The homogeneous case

We are going to determine subsets of vertices of a graph satisfying certain properties, and strive for a standard treatment of such tasks so that the respective requirement occurs as a parameter.

To explain the method, let us take the stable sets, the first case of the following list, as an example. Stable sets s with respect to a homogeneous relation B, i.e., a simple graph, are characterized by $B \, ; s \subseteq \overline{s}$. How can one determine all stable sets? First, the condition is applied to all sets simultaneously, using the membership relation ε. Whenever a set, that is a column of ε, satisfies the condition, the corresponding column of $\overline{B \, ; \varepsilon} \cup \overline{\varepsilon}$ will be full of $\mathbb{1}$s – simply recalling that $a \to b \iff \neg a \vee b$. When a universal relation is multiplied from the left to the negative of this relation, the columns corresponding to the condition will be zero. To mark them positively, this result is negated. It is also transposed because we work mainly with column vectors to represent subsets. Therefore we obtain

$$v := \overline{(B \, ; \varepsilon \cap \varepsilon)^{\mathsf{T}} \, ; \mathbb{T}}$$

which characterizes the stable sets along the powerset.

There is another interpretation for this result that goes back to the definition of residuals in Def. 4.6, where we have pointed out that the left residual $R \backslash S$ precisely marks which column of R is contained in which column of S. Having this in mind, one may identify v also as a left residual, namely as $v = (B \, ; \varepsilon \cap \varepsilon) \backslash \mathbb{L}$.

Most of the following list explains itself in this way. Sometimes, however, two conditions have to be satisfied. Then of course an intersection has to be taken, which is easy to observe in most cases.

For all the different set characterizations in the table, the first line describes the task; the second line shows the definition for just one set first, then the vector of all, and finally refers to the location where this has been treated.

---

stable subset s of a 1-graph $B : X \longrightarrow X$

$B_{;}\, s \subseteq \overline{s}$ $v := \overline{(B_{;}\, \varepsilon \cap \varepsilon)^{\mathsf{T}}_{;}\, \mathbb{T}}$ Def. 6.14

---

absorbant subset x of a 1-graph $B : X \longrightarrow X$

$\overline{x} \subseteq B_{;}\, x$ $v := \overline{(\overline{\varepsilon} \cap \overline{B_{;}\, \varepsilon})^{\mathsf{T}}_{;}\, \mathbb{T}}$ Def. 6.15

---

kernel s in a 1-graph $B : X \longrightarrow X$

$B_{;}\, s = \overline{s}$ $v := \overline{\left[(\overline{\varepsilon} \cap \overline{B_{;}\, \varepsilon}) \cup (\varepsilon \cap B_{;}\, \varepsilon)\right]^{\mathsf{T}}_{;}\, \mathbb{T}}$ Def. 6.16

---

covering point-set v in a 1-graph $B : X \longrightarrow X$

$B_{;}\, \overline{v} \subseteq v$ $v := \overline{(B_{;}\, \overline{\varepsilon} \cap \overline{\varepsilon})^{\mathsf{T}}_{;}\, \mathbb{T}}$ Def. 6.17

---

clique u in a reflexive and symmetric 1-graph $R : X \longrightarrow X$

$\overline{R}_{;}\, u \subseteq \overline{u}$ $v := \overline{(\overline{R}_{;}\, \varepsilon \cap \varepsilon)^{\mathsf{T}}_{;}\, \mathbb{T}}$ Def. 6.9

---

maxclique u in a reflexive and symmetric 1-graph

$\overline{R}_{;}\, u = \overline{u}$ $v := \overline{\left[(\overline{\varepsilon} \cap \overline{\overline{R}_{;}\, \varepsilon}) \cup (\varepsilon \cap \overline{R}_{;}\, \varepsilon)\right]^{\mathsf{T}}_{;}\, \mathbb{T}}$ Prop. 10.5

---

reducing vector r for a homogeneous relation $A : X \longrightarrow X$

$A_{;}\, r \subseteq r$ $v := \overline{(A_{;}\, \varepsilon \cap \overline{\varepsilon})^{\mathsf{T}}_{;}\, \mathbb{T}}$ Def. 6.12.i

---

contracting vector q for a homogeneous relation $A : X \longrightarrow X$

$A^{\mathsf{T}}_{;}\, q \subseteq q$ $v := \overline{(A^{\mathsf{T}}_{;}\, \varepsilon \cap \overline{\varepsilon})^{\mathsf{T}}_{;}\, \mathbb{T}}$ Def. 6.12.ii

---

all chains v in an order $E : X \longrightarrow X$

$\overline{E \cup E^{\mathsf{T}}}_{;}\, v \subseteq \overline{v}$ $c_{\mathrm{all}} := \overline{(\overline{E \cup E^{\mathsf{T}}}_{;}\, \varepsilon \cap \varepsilon)^{\mathsf{T}}_{;}\, \mathbb{T}}$ Def. 9.17.i

---

all longest chains v in an order $E : X \longrightarrow X$

$\qquad$ $\mathbf{max}_\Omega(c_{\mathrm{all}})$ with Ω the powerset ordering on $\mathbf{2}^X$ $\qquad$ Def. 9.17.i

all antichains v in a strictorder $C : X \longrightarrow X$

$C_i v \subseteq \overline{v}$ $\qquad$ $a_{\mathrm{all}} := \overline{(C_i \varepsilon \cap \varepsilon)^{\mathsf{T}}_i \mathbb{T}}$ $\qquad$ Def. 9.17.ii

all cardinality-maximum antichains v in a strictorder $C : X \longrightarrow X$

$\qquad$ $\mathbf{max}_{O_{||}}(a_{\mathrm{all}})$ with $O_{||}$ the cardinality-preorder on $\mathbf{2}^X$ $\quad$ Def. 9.17.ii

all progressively infinite subsets y for a homogeneous relation $A : X \longrightarrow X$

$y \subseteq A_i y$ $\qquad$ $v := \overline{(\varepsilon \cap \overline{A_i \varepsilon})^{\mathsf{T}}_i \mathbb{T}}$ $\qquad$ Def. 6.13.i

all complement-expanded subsets v for a homogeneous relation $A : X \longrightarrow X$

$\overline{v} \subseteq A_i \overline{v}$ $\qquad$ $c := \overline{(\overline{\varepsilon} \cap \overline{A_i \overline{\varepsilon}})^{\mathsf{T}}_i \mathbb{T}}$ $\qquad$ Def. 6.13.ii

the initial part of a homogeneous relation $A : X \longrightarrow X$

$\qquad$ $\mathbf{max}_\Omega(c)$ with Ω the powerset ordering on $\mathbf{2}^X$ $\qquad$ Def. 16.3

A closer look at these terms reveals many similarities. An antichain is a clique in the negation of the respective ordering, for example.

The heterogeneous case

Pairs of sets have also frequently been used as, for example, independent pairs or covering pairs of sets. It is not that simple to characterize pairs in a similar way. But sometimes the elements of these pairs are heavily interrelated as, for example, the non-enlargeable rectangles in a relation. In such cases mainly a condition for *one* of the two has to be observed, and these cases are included in the following table. Then, instead of a definition, a proposition is mentioned where the latter is stated.

An overview shows that the following entries mean – up to minor modifications – mainly the same. It is sometimes even more difficult to distinguish between the variants.

all vertical parts s of a minimal covering pair of sets of a relation $A : X \longrightarrow Y$

$$A ; \overline{A^{\mathsf{T}} ; \overline{s}} = s \qquad v := \left[\overline{(A ; \overline{\overline{A^{\mathsf{T}} ; \overline{\varepsilon}} \cap \overline{\varepsilon}}) \cup (A ; \overline{\overline{A^{\mathsf{T}} ; \overline{\varepsilon}} \cap \varepsilon})} \right]^{\mathsf{T}} ; \mathbb{T} \qquad \text{Prop. 10.6.i}$$

all horizontal parts t of a minimal covering pair of sets of a relation $R : X \longrightarrow Y$

$$A^{\mathsf{T}} ; \overline{A ; \overline{t}} = t \qquad v := \left[\overline{(A^{\mathsf{T}} ; \overline{\overline{A ; \overline{\varepsilon}} \cap \overline{\varepsilon}}) \cup (A^{\mathsf{T}} ; \overline{\overline{A ; \overline{\varepsilon}} \cap \varepsilon})} \right]^{\mathsf{T}} ; \mathbb{T} \qquad \text{Prop. 10.6.i}$$

all vertical parts u of a diclique in a relation $R : X \longrightarrow Y$

$$R ; \overline{R^{\mathsf{T}} ; u} = \overline{u} \qquad v := \left[\overline{(R ; \overline{\overline{R^{\mathsf{T}} ; \varepsilon} \cap \varepsilon}) \cup (R ; \overline{\overline{R^{\mathsf{T}} ; \varepsilon} \cap \overline{\varepsilon}})} \right]^{\mathsf{T}} ; \mathbb{T} \qquad \text{Prop. 10.2.ii}$$

all horizontal parts v of a diclique in a relation $R : X \longrightarrow Y$

$$R^{\mathsf{T}} ; \overline{R ; v} = \overline{v} \qquad v := \left[\overline{(R^{\mathsf{T}} ; \overline{\overline{R ; \varepsilon} \cap \varepsilon}) \cup (R^{\mathsf{T}} ; \overline{\overline{R ; \varepsilon} \cap \overline{\varepsilon}})} \right]^{\mathsf{T}} ; \mathbb{T} \qquad \text{Prop. 10.2.ii}$$

all vertical parts s of a maximal independent pair of sets of a relation $A : X \longrightarrow Y$

$$A ; \overline{A^{\mathsf{T}} ; s} = \overline{s} \qquad v := \left[\overline{(A ; \overline{\overline{A^{\mathsf{T}} ; \varepsilon} \cap \varepsilon}) \cup (A ; \overline{\overline{A^{\mathsf{T}} ; \varepsilon} \cap \overline{\varepsilon}})} \right]^{\mathsf{T}} ; \mathbb{T} \qquad \text{Prop. 10.6.ii}$$

all horizontal parts t of a maximal independent pair of sets of a relation $A : X \longrightarrow Y$

$$A^{\mathsf{T}} ; \overline{A ; t} = \overline{t} \qquad v := \left[\overline{(A^{\mathsf{T}} ; \overline{\overline{A ; \varepsilon} \cap \varepsilon}) \cup (A^{\mathsf{T}} ; \overline{\overline{A ; \varepsilon} \cap \overline{\varepsilon}})} \right]^{\mathsf{T}} ; \mathbb{T} \qquad \text{Prop. 10.6.ii}$$

The heterogeneous case with two argument relations

Among the many sets one is interested in are deadlocks and traps in Petri nets. At a sufficiently abstract level, a Petri net is nothing other then a bipartitioned graph, i.e., a pair $R : X \longrightarrow Y \quad S : Y \longrightarrow X$ of counter-running relations with X usually called the set of places and Y correspondingly the set of transitions. During modelling, it is important that the processes do not terminate unexpectedly, for example. So liveness of the processes should guaranteed by proof. From this requirement comes the interest in finding possible traps or deadlocks in Petri nets to which we will now contribute.

To this end, one says that a set $v \subseteq X$ is a **deadlock** when its S-predecessor set s is contained in the R-successor set. Further, a set $w \subseteq Y$ is called a **trap** when its R-successor set is contained in the S-predecessor set. In relational form, this reads

$$
\begin{aligned}
v \text{ deadlock} &\iff S ; v \subseteq R^{\mathsf{T}} ; v \\
v \text{ trap} &\iff R^{\mathsf{T}} ; v \subseteq S ; v.
\end{aligned}
$$

Example 15.11. We provide deadlock examples where only the smallest non-empty deadlocks are computed. It is easy to convince oneself that unions of deadlocks are deadlocks again.

$$
R = \begin{array}{c}
\\1\\2\\3\\4\\5\\6\\7\\8\\9
\end{array}
\begin{array}{c}
\begin{smallmatrix} a&b&c&d&e&f&g \end{smallmatrix}\\
\left(\begin{array}{ccccccc}
0&0&1&0&0&0&0\\
0&0&0&0&1&0&0\\
0&0&0&1&0&1&1\\
0&0&0&0&0&1&0\\
0&0&0&0&0&0&0\\
0&1&0&0&0&0&0\\
0&0&0&0&0&0&0\\
0&0&0&0&0&1&1\\
1&0&0&0&0&1&0
\end{array}\right)
\end{array}
\qquad
S = \begin{array}{c}
\\a\\b\\c\\d\\e\\f\\g
\end{array}
\begin{array}{c}
\begin{smallmatrix} 1&2&3&4&5&6&7&8&9 \end{smallmatrix}\\
\left(\begin{array}{ccccccccc}
0&0&1&0&0&0&0&1&0\\
0&1&1&1&0&1&0&0&0\\
0&0&1&0&1&0&0&1&0\\
1&0&1&0&0&0&1&1&0\\
1&0&1&1&0&1&0&1&1\\
0&1&0&0&1&0&1&0&0\\
0&1&0&0&1&0&0&1&0
\end{array}\right)
\end{array}
\qquad
\begin{array}{c}
\\a\\b\\c\\d\\e\\f\\g
\end{array}
\begin{array}{c}
\begin{smallmatrix} \{b\}&\{d\}&\{e,f\}&\{e,g\} \end{smallmatrix}\\
\left(\begin{array}{cccc}
0&0&0&0\\
1&0&0&0\\
0&0&0&0\\
0&1&0&0\\
0&0&1&1\\
0&0&1&0\\
0&0&0&1
\end{array}\right)
\end{array}
$$

Fig. 15.12 All minimal non-empty deadlocks of a random Petri net

Also in this case, a single term allows us to determine all deadlocks or traps, respectively:

---

deadlocks of the Petri net $R : X \longrightarrow Y \quad S : Y \longrightarrow X$

$S\,{}_;v \subseteq R^{\mathsf{T}}\,{}_;v \qquad vs := \left(\overline{\varepsilon_X^{\mathsf{T}}\,{}_;S^{\mathsf{T}}} \cup \varepsilon_X^{\mathsf{T}}\,{}_;R\right)/\mathbb{T}$

---

traps in the Petri net $R : X \longrightarrow Y \quad S : Y \longrightarrow X$

$R^{\mathsf{T}}\,{}_;v \subseteq S\,{}_;v \qquad vs := \left(\overline{\varepsilon_X^{\mathsf{T}}\,{}_;R} \cup \varepsilon_X^{\mathsf{T}}\,{}_;S^{\mathsf{T}}\right)/\mathbb{T}$

Hammocks

While one usually knows what a hammock is in everyday life, one may have problems identifying them in graph theory or even in programming. Pictorially, however, it is the same, namely something that hangs fastened at two points.

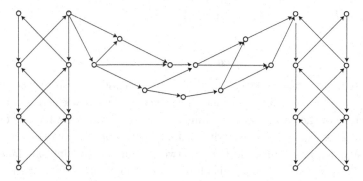

Fig. 15.13 A hammock in a graph

When one wishes to decompose a given large graph in order to produce pieces of manageable size, one will sometimes look for such hammocks. Since legacy code of early **go to**-style programming has to be reworked so as to arrive at modern clearly specified code, one checks for labels and **go to** statements and usually constructs a very large graph. When one manages to find a hammock, there is a chance that this hammock may by abstracted to a function. This function will then be handled separately and only the calls will remain in the graph, which are, thus, reduced to just an arrow.

Even when modelling with a graph, it is often helpful to identify a hammock. One is then in a position to draw the hammock separately and to insert an arrow instead, which may be marked with reference to the separately drawn picture. This is, thus, a means of decomposing and reducing complexity.

According to, for example, [12], a hammock is determined by two vertices, namely the ingoing and the outgoing vertices. So we begin to ask which formula describes the proper successors of subset u – first simply conceived as a column vector – according to a relation R; obviously

$$s_R(u) = \overline{u} \cap R^{\mathsf{T}} ; u.$$

We apply this mapping simultaneously to all subsets. The subsets conceived as elements are the row entries, so we form

$$[s_R(\varepsilon)]^{\mathsf{T}}.$$

For our current purpose, the interesting cases are those which constitute the source and the sink of a hammock, i.e., where successors or predecessors, respectively, are uniquely determined. For this we employ the construct of a univalent part of a relation

$$\mathbf{upa}(R) := R \cap \overline{R ; \overline{\mathbb{I}}} = \mathbf{syq}\,(R^{\mathsf{T}}, \mathbb{I})$$

as defined in [123, 124]. Intersecting this for successors (R) and predecessors (R^{T}) and demanding the source and sink to be different, we obtain the vector characterizing the hammocks as

$$\mathbf{hammocks}(R) := [\mathbf{upa}(s_R(\varepsilon)^{\mathsf{T}}) \cap \mathbf{upa}(s_{R^{\mathsf{T}}}(\varepsilon)^{\mathsf{T}}) ; \overline{\mathbb{I}}] ; \mathbb{T}.$$

Feedback vertex sets

There exist many more tasks of this type, of which we mention another one. Let a directed graph be given and consider all its cycles. A feedback vertex set is defined to be a subset of the vertex set containing at least one vertex from every cycle. Such feedback vertex sets have a diversity of important applications such as switching circuits, signal flow graphs, electrical networks, and constraint satisfaction problems. Feedback vertex sets have also been determined successfully with RELVIEW. One is mainly interested in minimal sets; they are, however, difficult to compute.

16

Standard Galois Mechanisms

The work in this chapter – although stemming from various application fields – is characterized by *two antitone mappings* leading in opposite directions that co-operate in a certain way. In most cases they are related to one or more relations which are often heterogeneous. An iteration leads to a fixed point of a Galois correspondence. Important classes of applications lead to these investigations. Trying to find out where a program terminates, and thus correctness considerations, also invoke such iterations. Looking for the solution of games is accompanied by these iterations. Applying the Hungarian alternating chain method to find maximum matchings or to solve assignment problems subsumes to these iterations. All this is done in structurally the same way, and deserves to be studied separately.

16.1 Galois iteration

When Evariste Galois wrote down his last notes, in preparation for the duel in 1832, in which he expected to die, he probably could not have imagined to what extent these notes would later influence mathematics and applications. What he had observed may basically be presented with the correspondence of permutations of a set and their fixed points. Consider the 5-element sequence $\{1, 2, 3, 4, 5\}$ for which there exist in total $5! = 120$ permutations. The idea is now to observe which set of permutations leaves which set of elements fixed. Demanding more elements to be untouched by a permutation results, of course, in fewer permutations. Having elements $2, 4$ fixed allows only 6 permutations, namely

$\quad$ 1,2,3,4,5 $\quad$ 1,2,5,4,3 $\quad$ 3,2,1,4,5 $\quad$ 3,2,5,4,1 $\quad$ 5,2,1,4,3 $\quad$ 5,2,3,4,1.

If, for example, $2, 3, 4$ are fixed, only permutations $1, 2, 3, 4, 5$ and $5, 2, 3, 4, 1$ remain. On the other hand, when we increase the set of permutations adding a last one to obtain

$\quad$ 1,2,3,4,5 $\quad$ 1,2,5,4,3 $\quad$ 3,2,1,4,5 $\quad$ 3,2,5,4,1 $\quad$ 5,2,1,4,3 $\quad$ 5,2,3,4,1 $\quad$ 5,3,2,4,1

the set of fixed points reduces to 4, i.e., to just one point. It is this counterplay of antitone functions that is put to work in what follows. In our finite case, it is more or less immaterial which antitone functions we work with. The schema stays the same. This shall now be demonstrated with two examples.

Example 16.1. Assume a set V and consider all of its subsets, i.e., the powerset 2^V. Then assume the following obviously antitone mappings which depend on some given relation $R : V \longrightarrow V$

$$\sigma : 2^V \longrightarrow 2^V, \quad \text{here } v \mapsto \sigma(v) = R_i\,\overline{v}$$

$$\pi : 2^V \longrightarrow 2^V, \quad \text{here } w \mapsto \pi(w) = \overline{w}.$$

In such a setting it is a standard technique to proceed as follows, starting with the empty set in the upper row of Fig. 16.1 and the full set in the lower. Then the mappings are applied from the upper to the next lower subset as well as from the lower to the next upper subset as shown in Fig. 16.1.

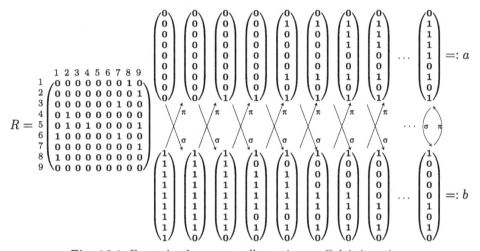

Fig. 16.1 Example of an eventually stationary Galois iteration

The first important observation is that the elements of both sequences are ordered monotonically, the upper increasing and the lower decreasing. They will thus become stationary in the finite case. When looking at the sequences in detail, it seems that there are two different sequences, but somehow alternating and respecting the order of the others. Even more important is the second observation, namely that the upper and the lower stationary values a, b will satisfy

$$b = \sigma(a) = R_i\,\overline{a} \quad \text{and} \quad a = \pi(b) = \overline{b}.$$

One has, thus, a strong algebraic property for the final result of the iteration and one should try to interpret and to make use of it.   □

Example 16.2. Consider now a completely different situation with two heterogeneous relations Q, λ and the obviously antitone mappings

$$\sigma : \mathbf{2}^V \longrightarrow \mathbf{2}^W, \quad \text{here } v \mapsto \sigma(v) = \overline{Q^{\mathsf{T}}; v}$$
$$\pi : \mathbf{2}^W \longrightarrow \mathbf{2}^V, \quad \text{here } w \mapsto \pi(w) = \overline{\lambda; w}.$$

Given this setting, we again iterate with the two start vectors and obtain what is shown in Fig. 16.2.

$$Q = \begin{array}{c} \\ 1 \\ 2 \\ 3 \\ 4 \\ 5 \\ 6 \\ 7 \end{array} \begin{array}{c} \begin{array}{ccccc} a & b & c & d & e \end{array} \\ \begin{pmatrix} 1&0&0&1&0 \\ 0&0&0&0&0 \\ 1&0&0&1&0 \\ 0&0&0&1&0 \\ 0&1&1&1&1 \\ 1&0&0&1&0 \\ 0&0&1&0&1 \end{pmatrix} \end{array} \qquad \lambda = \begin{array}{c} \\ 1 \\ 2 \\ 3 \\ 4 \\ 5 \\ 6 \\ 7 \end{array} \begin{array}{c} \begin{array}{ccccc} a & b & c & d & e \end{array} \\ \begin{pmatrix} 1&0&0&0&0 \\ 0&0&0&0&0 \\ 0&0&0&1&0 \\ 0&0&0&0&0 \\ 0&0&0&0&1 \\ 0&0&0&0&0 \\ 0&0&1&0&0 \end{pmatrix} \end{array}$$

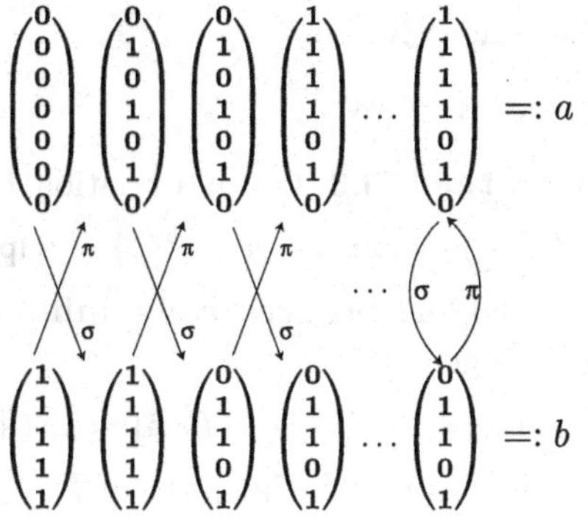

Fig. 16.2 Stationary heterogeneous Galois iteration

Also here, both sequences are monotonic, increasing, respectively decreasing, and they eventually become stationary. In addition, the upper and the lower stationary values a, b satisfy

$$b = \sigma(a) = \overline{Q^{\mathsf{T}}; b} \quad \text{and} \quad a = \pi(b) = \overline{\lambda; a},$$

i.e., schematically the same formulae as above. □

In the sections to follow, different applications will be traced back to this type of iteration. It will turn out that several well-known problems have a (relation-)algebraic flavor. In several cases, in addition, we will be able to deduce subdivisions of the sets in question from a and b. When by permutation of the rows and columns of the matrices representing the relation the sets come together, additional visual properties will show up.

16.2 Termination

The first group of considerations is usually located in graph theory, where one often looks for loops in the graph. The task arises to characterize the point set y of an infinite path of a graph in an algebraic fashion. This is then complementary to looking for sets of points from which only non-infinite sequences start in the execution of programs, i.e., terminating sequences. This is thus directly related to program semantics. We recall the basics from Section 6.5. There are two rather obvious facts:

- $B_{;}y \subseteq y$ expresses that all predecessors of y also belong to y,
- $y \subseteq B_{;}y$ expresses that every point of y precedes a point of y.

The first statement is certainly trivial, but so is the second. Because there is an infinite path assumed to exist from every point of y, at least one successor of each of the points of y will be the starting point of an infinite path. Based on this, we called a subset y progressively infinite if $y \subseteq B_{;}y$ which we now extend with the following definition.

Definition 16.3. Given a relation B, the construct

$$PI(B) := \sup\{y \mid y \subseteq B_{;}y\} = \sup\{y \mid y = B_{;}y\}$$

is called the **progressively infinite part** of B. More often, one speaks of its complement

$$J(B) := \inf\{x \mid \overline{x} \subseteq B_{;}\overline{x}\} = \inf\{x \mid \overline{x} = B_{;}\overline{x}\},$$

calling this the **initial part** of B.   □

The initial part characterizes the set of all those points from which only finite paths emerge.[1] There is already a minor statement included in this definition, namely that one may use the "="-version as well as the "$\subseteq$"-version. The set selected for the "="-version is contained in the other, so that its supremum will not exceed the other supremum. On the other hand, with every y satisfying $y \subseteq B_{;}y$, the larger construct $y' := B^{*}_{;}y$ will rather obviously satisfy $y' = B_{;}y'$, thus giving the reverse inclusion

$$y' = B^{*}_{;}y = y \cup B^{+}_{;}y \subseteq B_{;}y \cup B^{+}_{;}y = B^{+}_{;}y = B_{;}B^{*}_{;}y = B_{;}y' = B^{+}_{;}y \subseteq B^{*}_{;}y = y'.$$

The initial part is easily computed using the Galois mechanism with the antitone functions from left to right and from right to left given as

$$\sigma : 2^{V} \longrightarrow 2^{W}, \quad \text{here } v \mapsto B_{;}\overline{v}$$
$$\pi : 2^{W} \longrightarrow 2^{V}, \quad \text{here } w \mapsto \overline{w},$$

which has been shown already in Fig. 16.1. It necessarily ends with two vectors a, b satisfying

$$a = B_{;}\overline{b} \quad \text{and} \quad b = \overline{a}.$$

When started from $\perp\!\!\!\perp$ (meaning $\emptyset$) and $\top\!\!\!\top$ (meaning 2^{V}), we have in general

$$\perp\!\!\!\perp \subseteq \pi(\top\!\!\!\top) \subseteq \pi(\sigma(\perp\!\!\!\perp)) \subseteq \pi(\sigma(\pi(\top\!\!\!\top))) \subseteq \ldots$$
$$\ldots \subseteq \sigma(\pi(\sigma(\perp\!\!\!\perp))) \subseteq \sigma(\pi(\top\!\!\!\top)) \subseteq \sigma(\perp\!\!\!\perp) \subseteq \top\!\!\!\top.$$

---

[1] Be aware, however, that this includes the case of a relation which is not finitely branching, in which all paths are finite but there exist paths that exceed any prescribed length.

The outermost inclusions are obvious, the more inner inclusions follow by induction because σ, π are antitonic. The situation is a bit more specific in the present case. Since $\pi(w) = \overline{w}$, we can go one step further and write this down with σ, π expanded, i.e., $w \mapsto \sigma(\pi(w)) = B_i w$ and $v \mapsto \pi(\sigma(v)) = \overline{B_i \overline{v}}$, thereby eliminating all the duplicates:

$$\mathbb{I} \subseteq \overline{B_i \mathbb{T}} \subseteq \overline{B_i B_i \mathbb{T}} \subseteq \overline{B_i B_i B_i \mathbb{T}} \subseteq \ldots \subseteq a$$

$$b \subseteq \ldots \subseteq B_i B_i B_i \mathbb{T} \subseteq B_i B_i \mathbb{T} \subseteq B_i \mathbb{T} \subseteq \mathbb{T}.$$

Both sequences will end in the progressively infinite part a; they are actually the same except for the negation of the right one. The situation is captured more easily when Fig. 16.1 is presented in the following permuted form.

$$B = \begin{array}{c} \\ 2 \\ 3 \\ 4 \\ 5 \\ 7 \\ 9 \\ 1 \\ 6 \\ 8 \end{array} \begin{array}{c} 2\ 3\ 4\ 5\ 7\ 9\ \ 1\ 6\ 8 \\ \left(\begin{array}{cccccc|ccc} 0&0&0&0&0&1&0&0&0 \\ 0&0&0&0&1&0&0&0&0 \\ 1&0&0&0&0&0&0&0&0 \\ 1&0&1&0&0&1&0&0&0 \\ 0&0&0&0&0&1&0&0&0 \\ 0&0&0&0&0&0&0&0&0 \\ \hline 0&0&0&0&0&0&0&0&1 \\ 0&0&0&0&1&0&1&0&0 \\ 0&0&0&0&0&0&1&0&0 \end{array}\right) \end{array} \quad a = \begin{pmatrix} 0 \\ 0 \\ 0 \\ 0 \\ 0 \\ 0 \\ 1 \\ 1 \\ 1 \end{pmatrix} \quad b = \begin{pmatrix} 1 \\ 1 \\ 1 \\ 1 \\ 1 \\ 1 \\ 0 \\ 0 \\ 0 \end{pmatrix}$$

Fig. 16.3 The relation with its initial part in permuted and partitioned form

Between 1 and 8 the relation will oscillate infinitely often, as could already be seen in Fig. 6.11. From 6, one can follow the arc leading to 1 – and will then oscillate forever. One has, however, also the choice of going from 6 to 7, and then to be stopped finally at 9.

A careful reader may ask to what extent the decision to start the iteration with $\mathbb{I}, \mathbb{T}$ was arbitrary. When starting the other way round, i.e., with $\mathbb{T}, \mathbb{I}$, the result will be the initial part for the transposed graph.

Figure 16.4 shows yet another computation of the initial part of a relation. Duplicates in the iteration have been eliminated for reasons of space. It is possible to bring the subset a of months, namely Aug, Jul, Jun, Mar, Nov, to the front by simultaneous permutation. Then the upper right rectangle will again turn out to be empty.

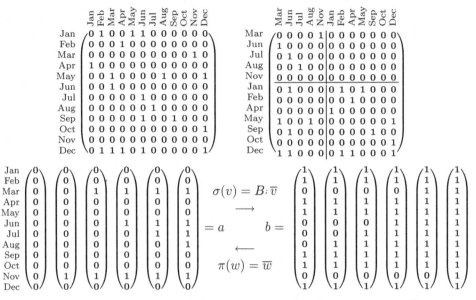

Fig. 16.4 Iterative computation of the initial part and rearrangement according to it

The initial part may be studied in more detail by exhaustion from the terminal vertices which leads us to progressive boundedness. Further investigations of the progressively infinite part are possible, recognizing basins, attractors etc. as in the field of system dynamics; see Section 19.3.

Definition 16.4. Let B be a homogeneous relation. We call

B **progressively bounded** $\;:\Longleftrightarrow\; \mathbb{T} = \sup_{h \geq 0} \overline{B^h ; \mathbb{T}}$

$\;:\Longleftrightarrow\; B$ may be exhausted by $\;z_0 := \overline{B ; \mathbb{T}},\; z_{n+1} := \overline{B ; \overline{z_n}}.$ □

Here again, we have two versions in the definition that should be shown to be equal. Firstly, we convince ourselves by induction starting with the obvious "z_{-1}" $= \perp\!\!\!\perp \subseteq z_0 \subseteq z_1$ that the sequence $z_0 \subseteq z_1 \subseteq z_2 \subseteq \ldots$ is monotonic. But we see also by induction that $z_n = \overline{B^{n+1} ; \mathbb{T}}$. For $n = 0$, it is given by definition; $n \mapsto n+1$ follows from

$$z_n = \overline{B^{n+1} ; \mathbb{T}} \;\Longleftrightarrow\; \overline{z_n} = B^{n+1} ; \mathbb{T}$$
$$\Longrightarrow\; B ; \overline{z_n} = B^{n+2} ; \mathbb{T} \;\Longleftrightarrow\; z_{n+1} = \overline{B ; \overline{z_n}} = \overline{B^{n+2} ; \mathbb{T}}.$$

The situation between progressive boundedness and progressive finiteness is shown in Theorem 16.5, but without the proof which has been fully elaborated already in [123, 124].

Theorem 16.5 (Termination formulae).

(i)
$$\sup\nolimits_{h\geq 0}\overline{B^{h};\mathbb{T}} \subseteq J(B) \subseteq B^{*};\overline{B;\mathbb{T}}$$

(ii) B *univalent* $\implies$
$$\sup\nolimits_{h\geq 0}\overline{B^{h};\mathbb{T}} = J(B) = B^{*};\overline{B;\mathbb{T}}$$

(iii) B *progressively bounded* $\implies$
$$\mathbb{T} = \sup\nolimits_{h\geq 0}\overline{B^{h};\mathbb{T}} = J(B) = B^{*};\overline{B;\mathbb{T}}$$

(iv) B *progressively finite* $\implies$
$$\sup\nolimits_{h\geq 0}\overline{B^{h};\mathbb{T}} \subseteq J(B) = B^{*};\overline{B;\mathbb{T}} = \mathbb{T}$$

(v) B *finite* $\implies$
$$\sup\nolimits_{h\geq 0}\overline{B^{h};\mathbb{T}} = J(B) \subseteq B^{*};\overline{B;\mathbb{T}} \qquad \square$$

If a relation is progressively bounded, all paths are finite and in addition have a finite common upper bound. This means that starting from the terminal vertices it is possible to exhaust all the vertices by iteratively going one step backwards. Expressed differently, the transitive closure B^{+} of the given relation B will turn out to be a strictorder which then in turn may be arranged to the upper right triangle of the matrix. This together with the exhaustion is visualized in Fig. 16.5.

```
        a b c d e f g h i j k l m n o p q              i p d n b m a c o e f h j g k l q
    a  /0 0 0 0 0 0 0 0 1 0 0 0 0 0 0 0 0\        i  /0 0|1 0|0 0|0 0 0|0 0 0 0|0 0 0|0\
    b  |0 0 0 0 0 0 0 0 0 0 1 0 0 1 0 0 |        p  |0 0|1 0|1 0|0 0 0|0 0 0 0|0 0 0|0 |
    c  |0 0 0 0 0 1 1 0 0 0 0 0 0 0 0 0 0 |       d  |0 0|0 0|1 0|0 0 0|0 0 0 0|0 1 0|0 |
    d  |0 1 0 0 0 0 0 0 0 0 1 0 0 0 0 0 0 |       n  |0 0|0 0|1 0|0 0 0|0 0 0 0|0 0 0|0 |
    e  |0 0 0 0 0 0 0 0 0 0 1 0 0 0 0 1   |       b  |0 0|0 0|0 0|0 1 0|0 0 0 0|0 0 1|0 |
    f  |0 0 0 0 0 0 1 0 0 0 0 0 0 0 0 0 0 |       m  |0 0|0 0|0 0|0 1 0|0 0 0 0|0 1 0|1 |
    g  |0 0 0 0 0 0 0 0 0 0 0 0 0 0 0 0 0 |       a  |0 0|0 0|0 0|0 0 0|0 0 1 0|0 0 0|0 |
    h  |0 0 0 0 0 1 0 0 1 0 1 0 0 0 0 0 0 |       c  |0 0|0 0|0 0|0 0 0|1 0 0 1|0 0 0|0 |
    i  |0 0 0 1 0 0 0 0 0 0 0 0 0 0 0 0 0 |       o  |0 0|0 0|0 0|0 0 0|1 0 0 0|0 0 0|0 |
    j  |0 0 0 0 0 0 0 0 0 0 1 0 0 0 0 0 0 |       e  |0 0|0 0|0 0|0 0 0|0 0 0 0|0 0 1|1 |
    k  |0 0 0 0 0 0 0 0 0 0 0 0 0 0 0 0 0 |       f  |0 0|0 0|0 0|0 0 0|0 0 0 0|1 0 0|0 |
    l  |0 0 0 0 0 0 0 0 0 0 0 0 0 0 0 0 0 |       h  |0 0|0 0|0 0|0 0 0|0 0 0 0|1 1 0|0 |
    m  |0 0 0 0 0 0 0 0 0 1 0 0 0 1 0 1   |       j  |0 0|0 0|0 0|0 0 0|0 0 0 0|0 1 0|0 |
    n  |0 1 0 0 0 0 0 0 0 0 0 0 0 0 0 0 0 |       g  |0 0|0 0|0 0|0 0 0|0 0 0 0|0 0 0|0 |
    o  |0 0 0 0 0 1 0 0 0 0 0 0 0 0 0 0 0 |       k  |0 0|0 0|0 0|0 0 0|0 0 0 0|0 0 0|0 |
    p  |0 1 0 1 0 0 0 0 0 0 0 0 0 0 0 0 0 |       l  |0 0|0 0|0 0|0 0 0|0 0 0 0|0 0 0|0 |
    q  \0 0 0 0 0 0 0 0 0 0 0 0 0 0 0 0 0/        q  \0 0|0 0|0 0|0 0 0|0 0 0 0|0 0 0|0/
```

Fig. 16.5 Exhaustion of a progressively bounded relation starting with the terminal vertices

We may even go one step further and join the latter two aspects. Then the progressively infinite vertices go to the end of the row as well as column arrangement. Then the progressively bounded rest of the finite relation is handled as in Fig. 16.5. What we have explained so far can be collected without further proof in the following proposition.

Proposition 16.6. *Any finite homogeneous relation may by simultaneously permuting rows and columns be transformed into a matrix satisfying the following basic structure with square diagonal entries:*

$$\left(\begin{array}{cc} \text{progressively bounded} & \underline{\mathbb{1}} \\ * & \text{total} \end{array}\right).$$ □

Many more aspects of terminality and foundedness can be found in our general reference [123, 124]. In particular, it is explained how principles of induction and transfinite induction are treated relationally. Confluence in connection with progressive boundedness is also presented in this book and concepts of discreteness are discussed. It did not seem appropriate to recall all this here.

Exercises

16.1 Determine initial parts of the following relations and execute the exhaustion:

$$
\begin{array}{c|ccccccccccc}
 & a & b & c & d & e & f & g & h & i & j & k \\
\hline
a & 0&0&0&0&0&0&0&0&0&0&0 \\
b & 0&0&0&0&0&0&0&0&1&1&0 \\
c & 0&0&0&0&1&1&0&0&1&0&1 \\
d & 0&0&0&0&0&0&1&0&0&1&0 \\
e & 0&0&0&0&0&0&0&1&0&0&\\
f & 0&0&0&1&0&0&0&0&0&0&\\
g & 0&0&0&1&1&0&0&0&0&0&\\
h & 0&0&0&0&0&0&0&0&0&0&\\
i & 0&0&0&0&0&1&0&0&0&0&\\
j & 0&0&0&0&0&0&0&1&0&0&0\\
k & 0&1&0&0&0&0&0&0&0&0&\\
\end{array}
\qquad
\begin{array}{c|cccccccccc}
 & 1 & 2 & 3 & 4 & 5 & 6 & 7 & 8 & 9 & 10 \\
\hline
1 & 0&0&0&0&0&0&0&0&0&0 \\
2 & 0&1&0&0&0&0&0&1&0&0 \\
3 & 0&0&0&0&0&0&0&0&0&0 \\
4 & 0&0&0&1&0&0&1&0&0&0 \\
5 & 0&0&0&1&0&0&0&1&0&0 \\
6 & 0&0&0&0&0&0&0&0&0&0 \\
7 & 1&0&0&0&0&0&0&0&0&0 \\
8 & 1&0&0&0&0&0&0&1&1&0 \\
9 & 0&0&0&0&0&0&0&0&0&1 \\
10 & 0&0&0&0&0&0&0&1&0&0 \\
\end{array}
$$

16.3 Games

For another application, we look at solutions of relational games. Let an arbitrary homogeneous relation $B : V \longrightarrow V$ be given. Two players are supposed to make alternating moves according to B, choosing a consecutive arrow to follow. The player who has no further move, i.e., who is about to move and finds an empty row in the relation B, or a terminal vertex in the graph, has lost.

Such a game is easily visualized taking a relation B represented by a graph, on which players have to determine a path in an alternating way. We study it for the NIM-type game starting with 6 matches from which we are allowed to take 1 or 2.

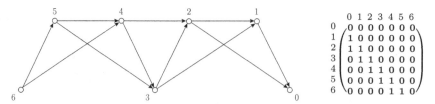

Fig. 16.6 A NIM-type game in graph and in matrix form

The two levels of Fig. 16.6 already anticipate the subdivision of the positions into win (upper level) and loss positions (lower level); since the graph is progressively

finite, there are no draw positions. The antitone functionals based on this relation
are formed in a manner quite similar to termination.

There exists a well-known technique for solving such games. Solving means to decompose the set of positions into three (in general non-empty) subsets of loss-
positions, draw-positions, and win-positions, always qualified from the point of
view of the player who is about to move. The basic aspect is to formulate the
rather obvious game conditions in point-free form.

- From a draw position there cannot exist a move to a losing position
  $$\forall\, x:\ \mathcal{D}_x \longrightarrow (\forall\, y:\ B_{xy} \longrightarrow \overline{\mathcal{L}_y}).$$
  In point-free relational notation: $\mathcal{D} \subseteq \overline{B\,\raise1pt\hbox{$\scriptscriptstyle;$}\,\mathcal{L}}$.
- From every position of win there exists at least one move to a losing position
  $$\forall\, x:\ \mathcal{W}_x \longrightarrow (\exists\, y:\ B_{xy} \wedge \mathcal{L}_y).$$
  In point-free relational notation: $\mathcal{W} \subseteq B\,\raise1pt\hbox{$\scriptscriptstyle;$}\,\mathcal{L}$ or $\overline{B\,\raise1pt\hbox{$\scriptscriptstyle;$}\,\mathcal{L}} \subseteq \overline{\mathcal{W}}$.

In $\mathcal{L} \mapsto \overline{B\,\raise1pt\hbox{$\scriptscriptstyle;$}\,\mathcal{L}}$ we recognize the same antitone mapping two times and conceive, thus,
just one function ζ based on the game relation B, namely

$$\zeta : 2^V \to 2^V, \qquad v \mapsto \zeta(v) = \overline{B\,\raise1pt\hbox{$\scriptscriptstyle;$}\,v}.$$

We use it twice in our standard iteration, $\sigma := \zeta$ and $\pi := \zeta$ and obtain the
stationary iteration of Fig. 16.7, resulting in a, b, which in turn give rise to the
rearrangement of the original relation.

$$
\begin{array}{c}
0 \\ 1 \\ 2 \\ 3 \\ 4 \\ 5 \\ 6
\end{array}
\begin{pmatrix}0\\0\\0\\0\\0\\0\\0\end{pmatrix}
\subseteq
\begin{pmatrix}1\\0\\0\\0\\0\\0\\0\end{pmatrix}
\subseteq
\begin{pmatrix}1\\0\\0\\0\\0\\0\\0\end{pmatrix}
\subseteq
\begin{pmatrix}1\\0\\0\\1\\0\\0\\0\end{pmatrix}
\subseteq
\begin{pmatrix}1\\0\\0\\1\\0\\0\\0\end{pmatrix}
\subseteq
\begin{pmatrix}1\\0\\0\\1\\0\\0\\1\end{pmatrix}
= a
$$

$$
\begin{array}{c}
0 \\ 1 \\ 2 \\ 3 \\ 4 \\ 5 \\ 6
\end{array}
\begin{pmatrix}1\\1\\1\\1\\1\\1\\1\end{pmatrix}
\supseteq
\begin{pmatrix}1\\1\\1\\1\\1\\1\\1\end{pmatrix}
\supseteq
\begin{pmatrix}1\\0\\0\\1\\1\\1\\1\end{pmatrix}
\supseteq
\begin{pmatrix}1\\0\\0\\1\\1\\1\\1\end{pmatrix}
\supseteq
\begin{pmatrix}1\\0\\0\\1\\0\\0\\1\end{pmatrix}
\supseteq
\begin{pmatrix}1\\0\\0\\1\\0\\0\\1\end{pmatrix}
= b
$$

$$
\begin{array}{c}
0 \\ 3 \\ 6 \\ 1 \\ 2 \\ 4 \\ 5
\end{array}
\begin{array}{c}
{\scriptstyle 0\ \ 3\ \ 6}\ {\scriptstyle 1\ \ 2\ \ 4\ \ 5} \\
\left(\begin{array}{ccc|cccc}
0 & 0 & 0 & 0 & 0 & 0 & 0 \\
0 & 0 & 0 & 1 & 1 & 0 & 0 \\
0 & 0 & 0 & 0 & 0 & 1 & 1 \\ \hline
1 & 0 & 0 & 0 & 0 & 0 & 0 \\
1 & 0 & 0 & 1 & 0 & 0 & 0 \\
0 & 1 & 0 & 0 & 1 & 0 & 0 \\
0 & 1 & 0 & 0 & 0 & 1 & 0
\end{array}\right)
\end{array}
$$

Fig. 16.7 The game of Fig. 16.6 with iteration and rearranged
with zones loss $= a$ and win $= \overline{b}$

Loss, draw, and win may turn out to be empty after iteration. If loss is empty, then
draw cannot be – provided the point set of the graph is not. There is one further
point to mention concerning the result. This time, we have a homogeneous relation,
and we easily observe that the two sequences from page 417 reduce using monotony
to just one

$$\mathbb{1} \subseteq \pi(\mathbb{T}) \subseteq \pi(\sigma(\mathbb{1})) \subseteq \pi(\sigma(\pi(\mathbb{T}))) \subseteq \dots$$
$$\dots \subseteq \sigma(\pi(\sigma(\mathbb{1}))) \subseteq \sigma(\pi(\mathbb{T})) \subseteq \sigma(\mathbb{1}) \subseteq \mathbb{T}.$$

The combined sequence is given here explicitly, and we observe equalities in an alternating pattern:

$$\mathbb{1} \subseteq \overline{B;\mathbb{T}} = \overline{B;\overline{B;\mathbb{1}}} \subseteq \overline{B;\overline{B;\overline{B;\mathbb{T}}}} = \ldots \subseteq \ldots \subseteq \overline{B;\overline{B;\overline{B;\mathbb{1}}}} = \overline{B;\overline{B;\mathbb{T}}} \subseteq \overline{B;\mathbb{1}} = \mathbb{T}.$$

Again, the final situation is characterized by the formulae $a = \pi(b)$ and $\sigma(a) = b$, which this time turn out to be $a = \overline{B;b}$ and $\overline{B;a} = b$. In addition, we will always have $a \subseteq b$. The smaller set a gives loss positions, while the larger one then indicates win positions as $\overline{b}$ and draw positions as $b \cap \overline{a}$. This is visualized by the following diagram for sets of win, loss, and draw, the arrows of which indicate moves that must exist, may exist, or are not allowed to exist.

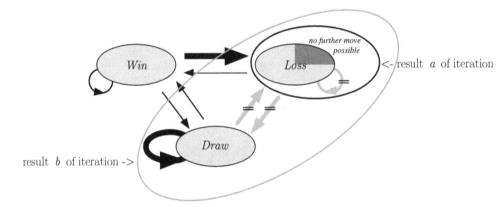

Fig. 16.8 Schema of a game solution: the iteration results a, b
in relation to win, draw, loss

A result will be found for all homogeneous relations. Often, however, all vertices will be qualified as draw. The set of draw positions may also be empty as in the solution of our initial game example.

Figure 16.9 visualizes another game solution, again concentrating on the subdivision of the matrix B and the vectors a as a first zone resembling *Loss*, b as first plus second zone – the latter resembling *Draw* – and finally the complement of b resembling *Win*.

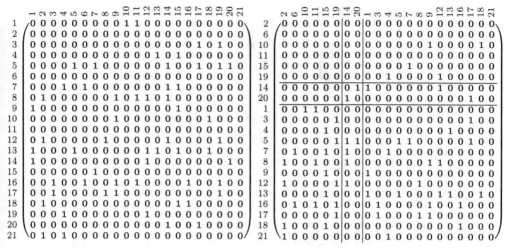

Fig. 16.9 A random relation and its game solution rearranged to zones loss–draw–win

Proposition 16.7 (Rearranging relations with respect to a game). *By simultaneously permuting rows and columns, an arbitrary finite homogeneous relation can be transformed into a matrix satisfying the following structure with square diagonal entries:*

$$\begin{pmatrix} \bot\!\!\bot & \bot\!\!\bot & * \\ \bot\!\!\bot & \text{total} & * \\ \text{total} & * & * \end{pmatrix}.$$

Proof: The subdivision with the iteration results a, b into groups loss/draw/win is uniquely determined, and the termination conditions of the iteration are written down: $a = \overline{B\,;\,b}$ and $\overline{B\,;\,a} = b$:

$$a = \begin{pmatrix} \top\!\!\top \\ \bot\!\!\bot \\ \bot\!\!\bot \end{pmatrix} = \overline{\begin{pmatrix} B_{11} & B_{12} & B_{13} \\ B_{21} & B_{22} & B_{23} \\ B_{31} & B_{32} & B_{33} \end{pmatrix} ; \begin{pmatrix} \top\!\!\top \\ \top\!\!\top \\ \bot\!\!\bot \end{pmatrix}} = \overline{B\,;\,b} \qquad b = \begin{pmatrix} \top\!\!\top \\ \top\!\!\top \\ \bot\!\!\bot \end{pmatrix} = \overline{\begin{pmatrix} B_{11} & B_{12} & B_{13} \\ B_{21} & B_{22} & B_{23} \\ B_{31} & B_{32} & B_{33} \end{pmatrix} ; \begin{pmatrix} \top\!\!\top \\ \bot\!\!\bot \\ \bot\!\!\bot \end{pmatrix}} = \overline{B\,;\,a}.$$

From this, we calculate one after the other
 — position b_1 : $\top\!\!\top = \overline{B_{11}\,;\,\top\!\!\top}$, i.e., $B_{11} = \bot\!\!\bot$, similarly $B_{21} = \bot\!\!\bot$,
 — position a_1 : $\top\!\!\top = \overline{B_{12}\,;\,\top\!\!\top}$, i.e., $B_{12} = \bot\!\!\bot$,
 — position b_3 : $\bot\!\!\bot = \overline{B_{31}\,;\,\top\!\!\top}$, i.e., $\top\!\!\top = B_{31}\,;\,\top\!\!\top$, so that B_{31} is total. □

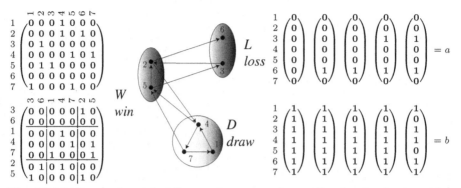

Fig. 16.10 A game in original form and rearranged according to win, draw, and loss

It seems extremely interesting to find out how these standard iterations behave if matrices are taken with coefficients which are drawn from other relation algebras. For example, do matrices over an interval algebra lead to steering algorithms? Will game algorithms over matrices with pairs (interval, compass) give hints to escape games? Will there be targeting games?

The full power of this approach, however, will only be seen when we assign the two players different and heterogeneous relations $B : V \longrightarrow W$ and $B' : W \longrightarrow V$ to follow.

Exercises

16.2 Let the matrix M represent the game relation. Determine the sets of win, loss, and draw.

$$M = \begin{array}{c} \\ a \\ b \\ c \\ d \\ e \\ f \\ g \\ h \\ i \\ j \\ k \end{array} \begin{array}{c} \begin{array}{cccccccccccc} a & b & c & d & e & f & g & h & i & j & k \end{array} \\ \begin{pmatrix} 0 & 0 & 0 & 0 & 0 & 0 & 0 & 0 & 0 & 0 & 1 \\ 0 & 0 & 0 & 0 & 0 & 0 & 0 & 0 & 0 & 1 & 0 \\ 0 & 0 & 0 & 0 & 1 & 0 & 0 & 0 & 0 & 0 & 1 \\ 0 & 0 & 0 & 0 & 0 & 0 & 0 & 0 & 0 & 0 & 0 \\ 1 & 1 & 0 & 0 & 0 & 0 & 0 & 0 & 0 & 0 & 0 \\ 1 & 0 & 0 & 0 & 0 & 0 & 0 & 0 & 0 & 0 & 0 \\ 1 & 0 & 1 & 0 & 0 & 0 & 0 & 0 & 0 & 0 & 0 \\ 0 & 0 & 1 & 0 & 0 & 0 & 1 & 1 & 0 & 0 & 0 \\ 0 & 0 & 0 & 0 & 0 & 0 & 0 & 0 & 0 & 0 & 0 \\ 0 & 0 & 0 & 0 & 1 & 0 & 0 & 0 & 1 & 0 & 0 \\ 0 & 0 & 0 & 0 & 0 & 0 & 1 & 0 & 0 & 0 & 0 \end{pmatrix} \end{array}$$

16.4 Specialization to kernels

The study of kernels in a graph has a long tradition. Even today interest in finding kernels and computing them is alive. The interest comes from playing games and trying to win, or at least characterizing winning positions, losing positions, and draw positions. In modern multi-criteria decision analysis kernels are also being investigated with much intensity.

The game aspect has already been studied in the previous section. Here we simply give the definition of a kernel and try to visualize the effect. The main point is that we look for a subset s of points of the graph with a specific property which can be described as follows. There is no arrow leading from one vertex of the subset s to any other in the subset. On the other hand, from every vertex outside one will find an arrow leading into the subset s. This is then captured by the following definition.

Definition 16.8. Let a graph be given with associated relation B. In the graph described by B, we call (as already in Def. 6.16)

$$s \text{ a } \textbf{kernel} \quad :\Longleftrightarrow \quad B\,;\,s = \bar{s}$$

$$\Longleftrightarrow \quad \text{no arrow will begin } and \text{ end in } s; \text{ from every}$$
$$\text{vertex outside } s \text{ an arrow leads into } s.$$

There is another way of expressing this saying that s must at the same time be stable $B\,;\,s \subseteq \bar{s}$ and absorbant $B\,;\,s \supseteq \bar{s}$. ◻

```
       1 2 3 4 5 6 7 8 9
  1  / 0 1 1 0 0 0 0 0 1 \
  2  | 1 1 1 0 1 1 0 1 0 |
  3  | 1 1 0 1 0 1 1 1 0 |
  4  | 0 0 0 0 1 1 1 0 1 |
  5. | 0 1 0 1 0 1 0 0 0 |
  6  | 0 0 1 1 1 1 0 1 1 |
  7  | 0 1 1 1 0 0 0 0 1 |
  8  | 0 1 1 0 0 1 0 0 1 |
  9  \ 1 1 0 1 0 0 1 1 0 /
```

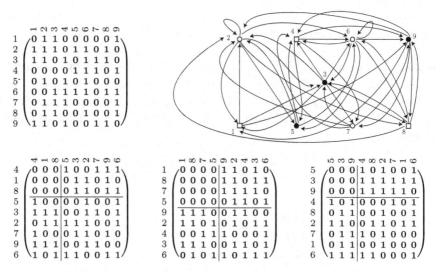

```
       4 1 8 | 5 3 2 7 9 6
  4  / 0 0 0 | 1 0 0 1 1 1 \
  1  | 0 0 0 | 0 1 1 0 1 0 |
  8  | 0 0 0 | 0 1 1 0 1 1 |
  5  | 1 0 0 | 0 0 1 0 0 1 |
  3  | 1 1 1 | 0 0 1 1 0 1 |
  2  | 0 1 1 | 1 1 1 0 0 1 |
  7  | 1 0 0 | 0 1 1 0 1 0 |
  9  | 1 1 1 | 0 0 1 1 0 0 |
  6  \ 1 0 1 | 1 1 0 0 1 1 /
```

```
       1 8 7 5 | 9 2 4 3 6
  1  / 0 0 0 0 | 1 1 0 1 0 \
  8  | 0 0 0 0 | 1 1 0 1 1 |
  7  | 0 0 0 0 | 1 1 1 1 0 |
  5  | 0 0 0 0 | 0 1 1 0 1 |
  9  | 1 1 1 0 | 0 1 1 0 0 |
  2  | 1 1 0 1 | 0 1 0 1 1 |
  4  | 0 0 1 1 | 1 0 0 0 1 |
  3  | 1 1 1 0 | 0 1 1 0 1 |
  6  \ 0 1 0 1 | 1 0 1 1 1 /
```

```
       5 3 9 | 4 8 2 7 1 6
  5  / 0 0 0 | 1 0 1 0 0 1 \
  3  | 0 0 0 | 1 1 1 1 1 1 |
  9  | 0 0 0 | 1 1 1 1 1 0 |
  4  | 1 0 1 | 0 0 0 1 0 1 |
  8  | 0 1 1 | 0 0 1 0 0 1 |
  2  | 1 1 0 | 0 1 1 0 1 1 |
  7  | 0 1 1 | 1 0 1 0 0 0 |
  1  | 0 1 1 | 0 0 1 0 0 0 |
  6  \ 1 1 1 | 1 1 0 0 0 1 /
```

Fig. 16.11 All three kernels of a graph in a rather independent position

Looking at the graph of Fig. 16.11, one will see that finding kernels may not be an easy task; it is in fact $\mathcal{NP}$-complete. The graph shown has exactly the three kernels $\{4,1,8\}, \{1,8,7,5\}, \{5,3,9\}$, for which the decomposition and permutation has also been executed.

Proposition 16.9 (Rearranging relations with respect to a kernel). *An arbitrary finite homogeneous relation may have a kernel s, in which case, by simultaneously permuting rows and columns, it can be transformed into a matrix satisfying the following basic structure with square diagonal entries:*

$$\begin{pmatrix} \mathbb{l} & * \\ \text{total} & * \end{pmatrix}.$$

Proof: The subdivision with the kernel s into groups s in front and then $\overline{s}$ is written down

$$B_{;} s = \begin{pmatrix} B_{11} & B_{12} \\ B_{21} & B_{22} \end{pmatrix}_{;} \begin{pmatrix} \mathbb{T} \\ \mathbb{l} \end{pmatrix} = \overline{\begin{pmatrix} \mathbb{T} \\ \mathbb{l} \end{pmatrix}} = \overline{s}.$$

Now we investigate components s_1 and s_2 of s. Obviously, $B_{11;} \mathbb{T} = \mathbb{l}$ implies that $B_{11} = \mathbb{l}$ and $B_{21;} \mathbb{T} = \mathbb{T}$ means that B_{21} has to be total. $\square$

16.5 Matching and assignment

Matching and assignment problems are well-known topics of operations research. Highly efficient algorithms have been developed, often presented in the language of graph theory. Here, we try to add a relational aspect and derive some results that may help in understanding the effects. In particular, we will switch from an operational point of view to a declarative form with algebraic visualization.

The tasks may easily be formulated in an environment of sympathy and marriage. Let, therefore, a sympathy relation Q be given between the set of boys V and the set of girls W of a village. This relation is typically neither univalent, nor total, surjective, or injective. The relation Q of Fig. 16.12 visualizes such a distribution of sympathy.

$$
Q =
\begin{matrix}
 & \begin{matrix} a & b & c & d & e \end{matrix} \\
\begin{matrix} 1 \\ 2 \\ 3 \\ 4 \\ 5 \\ 6 \\ 7 \end{matrix} &
\begin{pmatrix}
1 & 0 & 0 & 1 & 0 \\
0 & 0 & 0 & 0 & 0 \\
1 & 0 & 0 & 1 & 0 \\
0 & 0 & 0 & 1 & 0 \\
0 & 1 & 1 & 1 & 1 \\
1 & 0 & 0 & 1 & 0 \\
0 & 0 & 1 & 0 & 1
\end{pmatrix}
\end{matrix}
\supseteq
\begin{matrix}
 & \begin{matrix} a & b & c & d & e \end{matrix} \\
\begin{matrix} 1 \\ 2 \\ 3 \\ 4 \\ 5 \\ 6 \\ 7 \end{matrix} &
\begin{pmatrix}
1 & 0 & 0 & 0 & 0 \\
0 & 0 & 0 & 0 & 0 \\
0 & 0 & 0 & 1 & 0 \\
0 & 0 & 0 & 0 & 0 \\
0 & 0 & 0 & 0 & 1 \\
0 & 0 & 0 & 0 & 0 \\
0 & 0 & 1 & 0 & 0
\end{pmatrix}
\end{matrix}
= \lambda
$$

Fig. 16.12 Sympathy and matching

Assume now that a mass wedding (or simply a round of pairwise dancing) is going to be organized, i.e., that a univalent and injective relation $\lambda \subseteq Q$ inside the given sympathy is sought. This is the basis of the following definition that extends Def. 5.7.

Definition 16.10. Given a (possibly heterogeneous) relation $Q : V \longrightarrow W$, we call λ a Q-**matching** provided it is a matching that is contained in Q, i.e., if it satisfies altogether

$$\lambda \subseteq Q \qquad \lambda_; \lambda^\mathsf{T} \subseteq \mathbb{I} \qquad \lambda^\mathsf{T}_; \lambda \subseteq \mathbb{I}. \qquad\qquad\qquad \square$$

Of course, $\lambda := \mathbb{L}$ is a Q-matching, but certainly the most uninteresting one. An assignment that gives the maximum number of weddings (or the largest number of pairs of dancers sympathetic to one another) is not easily reached. When trying to find such a match, one will first make some assignments that later turn out to be a hindrance for achieving the maximum. It is then required that some tentative assignments be loosened again and replaced by others in order to provide further boys with girls. A set x of boys is saturated, according to Def. 10.8, when it can be assigned totally for marriage or dancing.

Algorithms to solve such problems have frequently been published and have been studied in some detail. They are known to be of quite acceptable polynomial complexity $\mathcal{O}(n^{2.5})$, theoretically even faster. These algorithms do not lend themselves readily to be formulated relationally. The static situation of a sympathy relation Q together with an arbitrary matching λ, analyzed iteratively as in Section 16.1, leads to very helpful relational identities. We consider sets of young lads $v \subseteq V$ or ladies $w \subseteq W$, respectively, and design two antitone mappings as in Section 16.1:

$$v \mapsto \sigma(v) := \overline{Q^\mathsf{T}_; v}$$
$$w \mapsto \pi(w) := \overline{\lambda_; w}.$$

The first relates a set of boys to those girls not sympathetic to any one of them. The second presents the set of boys not assigned to some set of girls. Using these antitone mappings, we execute the standard Galois iteration. It can be started in two ways namely from $\mathbb{L}, \mathbb{T}$ as well as from $\mathbb{T}, \mathbb{L}$. In the iteration of Fig. 16.2, the decision was in favor of the first of these cases. We will discuss the second later.

We recall that we have in general

$$\mathbb{L} \subseteq \pi(\mathbb{T}) \subseteq \pi(\sigma(\mathbb{L})) \subseteq \pi(\sigma(\pi(\mathbb{T}))) \subseteq \dots$$

$$\dots \subseteq \sigma(\pi(\sigma(\mathbb{L}))) \subseteq \sigma(\pi(\mathbb{T})) \subseteq \sigma(\mathbb{L}) \subseteq \mathbb{T}.$$

The outermost inclusions are obvious, the more inner inclusions follow by induction because σ, π are antitonic. The iteration will terminate after a finite number of steps with two vectors (a, b) satisfying $a = \pi(b)$ and $\sigma(a) = b$ as before. Here, this means (in negated form)

$$\overline{b} = Q^\mathsf{T}_; a \qquad\qquad \overline{a} = Q_; b$$
$$\overline{a} = \lambda_; b.$$

We have in addition $\overline{a} = Q_; b$. This follows from the chain $\overline{a} = \lambda_; b \subseteq Q_; b \subseteq \overline{a}$, which implies equality at every intermediate state. Only the resulting equalities for a, b have been used, together with monotony and the Schröder rule.

Remembering Prop. 10.6, we learn that the pair a, b is an inclusion-maximal independent pair of sets for Q, or else $\bar{a}, \bar{b}$ is an inclusion-minimal line covering for the relation Q.

Maximum matchings

Remembering Prop. 10.6 again, we learn that the pair a, b may not be an inclusion-maximal independent pair of sets for λ, or else $\bar{a}, \bar{b}$ may not be an inclusion-minimal line covering for λ. This will only be the case when in addition $\bar{b} = \lambda^{\mathsf{T}} {}_{;} a$, which is indeed a very important case that we are going to study in advance:

$$\bar{b} = Q^{\mathsf{T}} {}_{;} a \qquad \bar{a} = Q {}_{;} b$$
$$\bar{b} = \lambda^{\mathsf{T}} {}_{;} a \qquad \bar{a} = \lambda {}_{;} b.$$

A first look detects further symmetry. One may certainly ask whether we had been right in deciding for the variant of the iteration procedure starting with $\mathbb{L}, \mathbb{T}$ as opposed to $\mathbb{T}, \mathbb{L}$. Assume now we had decided the other way round. This would obviously mean the same as starting as before for Q^{T} and λ^{T}. One will observe easily that again four conditions would be valid at the end of the iteration with Q^{T} for Q and λ^{T} for λ as well as, say a', b'. Then a' corresponds to b and b' corresponds to a. This means that the resulting decomposition of the matrices does *not* depend on the choice – as long as all four equations are satisfied.

Figure 16.13 is intended to symbolize the interrelationship between λ and Q as expressed by the iteration result (a, b). The light gray zones are all filled with zeros. We deduce this from the matrix equations rewriting $\bar{a} = Q {}_{;} b$ and $\bar{b} = Q^{\mathsf{T}} {}_{;} a$ as matrices:

$$\begin{pmatrix} \mathbb{L} \\ \mathbb{L} \\ \mathbb{L} \\ \mathbb{T} \end{pmatrix} = \begin{pmatrix} Q_{11} & Q_{12} & Q_{13} & Q_{14} \\ Q_{21} & Q_{22} & Q_{23} & Q_{24} \\ Q_{31} & Q_{32} & Q_{33} & Q_{34} \\ Q_{41} & Q_{42} & Q_{43} & Q_{44} \end{pmatrix} {}_{;} \begin{pmatrix} \mathbb{L} \\ \mathbb{T} \\ \mathbb{T} \\ \mathbb{T} \end{pmatrix} \qquad \begin{pmatrix} \mathbb{T} \\ \mathbb{L} \\ \mathbb{L} \\ \mathbb{L} \end{pmatrix} = \begin{pmatrix} Q_{11}^{\mathsf{T}} & Q_{21}^{\mathsf{T}} & Q_{31}^{\mathsf{T}} & Q_{41}^{\mathsf{T}} \\ Q_{12}^{\mathsf{T}} & Q_{22}^{\mathsf{T}} & Q_{32}^{\mathsf{T}} & Q_{42}^{\mathsf{T}} \\ Q_{13}^{\mathsf{T}} & Q_{23}^{\mathsf{T}} & Q_{33}^{\mathsf{T}} & Q_{43}^{\mathsf{T}} \\ Q_{14}^{\mathsf{T}} & Q_{24}^{\mathsf{T}} & Q_{34}^{\mathsf{T}} & Q_{44}^{\mathsf{T}} \end{pmatrix} {}_{;} \begin{pmatrix} \mathbb{T} \\ \mathbb{T} \\ \mathbb{T} \\ \mathbb{L} \end{pmatrix}.$$

Then immediately $Q_{12} = \mathbb{L}, Q_{13} = \mathbb{L}, Q_{14} = \mathbb{L}, Q_{22} = \mathbb{L}, Q_{23} = \mathbb{L}, Q_{24} = \mathbb{L}, Q_{32} = \mathbb{L}, Q_{33} = \mathbb{L}, Q_{34} = \mathbb{L}$. One may wish to exclude the first horizontal zone and the last vertical zone from the beginning, i.e., demand that Q be total and surjective. If one leaves them out, $Q_{11} = \mathbb{L}$ simply by sorting empty rows to the front, which also makes Q_{21} total. In the same way, these results follow also from the second equation. Sorting empty columns to the end makes $Q_{44} = \mathbb{L}$ and leaves Q_{43} surjective. The same may then correspondingly be executed for λ. Because λ is a matching, it may be sorted so as to have it as a diagonal in the respective zones.

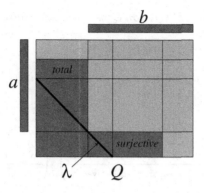

Fig. 16.13 The schema of a decomposition according to a maximum matching

In the dark zone, relation entries of Q may occur almost arbitrarily with the exception that the first formula $\overline{a} = Q\,\dot{,}\,b$ means that from outside a one will always find an image in b, guaranteed not least by part of the diagonal λ. The second formula correspondingly says that outside b one will always find an inverse image in a, which is certainly achieved with part of the diagonal λ. The other two formulae underline that this depends heavily on λ. Figure 16.14 shows this in a concrete example.

$$
\begin{array}{c}
\;\;\text{a b c d e f g h i j k l m n o p q} \\
\begin{array}{r}
1 \\ 2 \\ 3 \\ 4 \\ 5 \\ 6 \\ 7 \\ 8 \\ 9 \\ 10 \\ 11 \\ 12 \\ 13 \\ 14 \\ 15 \\ 16 \\ 17 \\ 18 \\ 19
\end{array}
\left(
\begin{array}{ccccccccccccccccc}
0&0&1&0&0&0&0&0&0&0&0&0&0&0&0&0&1 \\
0&0&1&0&0&0&0&0&0&0&0&0&0&1&0&0& \\
0&0&1&0&0&0&0&1&0&1&1&0&0&0&0&0& \\
0&0&0&0&0&0&0&0&0&0&0&1&0&0&0&1& \\
0&0&0&0&0&0&0&0&0&0&0&0&0&0&0&1& \\
0&0&1&0&0&0&0&0&0&0&1&0&0&0&0&0& \\
0&0&0&0&0&0&0&0&0&0&0&0&0&0&0&0& \\
0&0&0&0&0&0&0&0&0&0&0&1&0&0&0& \\
1&0&1&0&0&0&0&0&0&0&0&0&0&0&0&0& \\
0&0&0&1&0&0&0&0&0&0&0&0&0&0&0&0& \\
0&0&0&0&1&0&0&0&0&0&0&0&0&0&0&0& \\
0&0&0&1&0&0&0&0&0&0&0&0&0&0&0&0& \\
0&0&0&0&0&0&1&0&0&0&0&0&0&0&0&0& \\
0&0&1&0&0&0&0&0&0&0&0&0&1&0&0& \\
1&0&0&1&0&0&0&0&0&0&1&0&0&0&0&0& \\
0&0&0&0&0&0&0&0&0&0&0&0&0&0&0&0& \\
1&0&1&1&0&0&0&1&0&0&0&0&0&0&0&0& \\
0&0&0&0&0&0&0&0&0&0&0&0&0&0&1&0& \\
0&1&0&1&1&0&0&0&0&1&1&0&0&1&0&0&
\end{array}
\right)
\end{array}
$$

$$
\begin{array}{c}
\;\;\text{c d f m q d i n a e g l h p b k f j} \\
\begin{array}{r}
7 \\ 16 \\ 6 \\ 12 \\ 14 \\ 1 \\ 2 \\ 4 \\ 5 \\ 10 \\ 3 \\ 8 \\ 9 \\ 11 \\ 13 \\ 15 \\ 17 \\ 18 \\ 19
\end{array}
\left(
\begin{array}{ccccc|cccccccc|cc|cc}
0&0&0&0&0&0&0&0&0&0&0&0&0&0&0&0&0 \\
0&0&0&0&0&0&0&0&0&0&0&0&0&0&0&0&0 \\
1&0&1&0&0&0&0&0&0&0&0&0&0&0&0&0&0 \\
0&0&0&0&1&0&0&0&0&0&0&0&0&0&0&0&0 \\
1&1&0&0&0&0&0&0&0&0&0&0&0&0&0&0&0 \\
1&0&0&1&0&0&0&0&0&0&0&0&0&0&0&0&0 \\
1&1&0&0&0&0&0&0&0&0&0&0&0&0&0&0&0 \\
0&0&1&1&0&0&0&0&0&0&0&0&0&0&0&0&0 \\
0&0&0&1&0&0&0&0&0&0&0&0&0&0&0&0&0 \\
0&0&0&0&1&0&0&0&0&0&0&0&0&0&0&0&0 \\
1&0&0&0&0&1&0&0&0&0&1&0&0&0&1&0&0 \\
0&0&0&0&0&0&1&0&0&0&0&0&0&0&0&0&0 \\
1&0&0&0&0&0&0&1&0&0&0&0&0&0&0&0&0 \\
0&0&0&0&0&0&0&1&0&0&0&0&0&0&0&0&0 \\
0&0&0&0&0&0&0&0&1&0&0&0&0&0&0&0&0 \\
0&0&0&0&1&0&0&1&0&0&1&0&0&0&0&0&0 \\
1&0&0&0&1&0&0&1&0&0&0&1&0&0&0&0&0 \\
0&0&0&0&0&0&0&0&0&0&0&0&1&0&0&0&0 \\
0&1&0&0&1&0&0&0&1&0&1&0&0&1&1&0&0
\end{array}
\right)
\end{array}
$$

Fig. 16.14 Arbitrary relation with a rearrangement according to a
cardinality-maximum matching – the diagonals

We can give the present result two different flavors. The first remembers the Hall condition of Def. 10.8 guaranteeing for every subset that there are at least as many counterparts available via the relation in question. When we use 'Hall$^\mathsf{T}$', we mean the same in reverse direction. In (ii), the first three rows are put together and also the last three columns.

Proposition 16.11 (Decomposing according to matching and assignment). *Let any (possibly heterogeneous) relation Q be given and consider it together with some maximum Q-matching λ.*

(i) *By independently permuting rows and columns it will allow the following subdivision into a 4×4 schema:*

$$
\left(
\begin{array}{c|ccc}
\begin{array}{c} \text{⫫} \\ \text{total} \\ \text{Hall}^{\mathsf{T}} + \text{square} \end{array} & \begin{array}{c} \text{⫫} \\ \text{⫫} \\ \text{⫫} \end{array} & \begin{array}{c} \text{⫫} \\ \text{⫫} \\ \text{⫫} \end{array} & \begin{array}{c} \text{⫫} \\ \text{⫫} \end{array} \\
\hline
* & \text{Hall} + \text{square} & \text{surjective} & \text{⫫}
\end{array}
\right)
\quad
\left(
\begin{array}{c|ccc}
\begin{array}{c} \text{⫫} \\ \text{⫫} \\ \text{permutation} \end{array} & \begin{array}{c} \text{⫫} \\ \text{⫫} \\ \text{⫫} \end{array} & \begin{array}{cc} \text{⫫} & \text{⫫} \\ \text{⫫} & \text{⫫} \\ \text{⫫} & \text{⫫} \end{array} \\
\hline
\text{⫫} & \text{permutation} & \text{⫫} \quad \text{⫫}
\end{array}
\right)
$$

(ii) *By independently permuting rows and columns it can be transformed into the following 2×2 pattern with not necessarily square diagonal blocks:*

$$
\begin{pmatrix} \text{Hall}^{\mathsf{T}} & \text{⫫} \\ * & \text{Hall} \end{pmatrix}
\quad
\begin{pmatrix} \text{univalent}+\text{surjective}+\text{injective} & \text{⫫} \\ \text{⫫} & \text{univalent}+\text{total}+\text{injective} \end{pmatrix}
$$

Proof: We let the discussion above stand for a proof. □

Not yet maximum matchings

Thus, it was interesting to introduce the condition $\overline{b} = \lambda^{\mathsf{T}}{}_{;}\, a$. Now, we drop it and accept that λ may not yet be a maximal matching. Therefore, a, b need not be an inclusion-maximal independent pair of sets for λ (as opposed to Q), nor need $\overline{a}, \overline{b}$ be an inclusion-minimal line covering for λ. This would only be the case when, in addition, $\overline{b} = \lambda^{\mathsf{T}}{}_{;}\, a$. We visualize the outcome of the iteration with Fig. 16.15. The light gray zones are still filled with **0**-entries. In the dark gray zone, relation entries may occur arbitrarily with the exception that Q restricted to the dark zone be total and surjective. The matching λ is by permuting rows and columns arranged so as to resemble a partial diagonal of **1**s.

The first resulting equation, $\overline{a} = \lambda_{;}\, b$, indicates for Fig. 16.15 that in the broader dark horizontal zone below a the matching λ will definitely assign an element of b. The second equation, $\overline{b} = Q^{\mathsf{T}}{}_{;}\, a$, expresses that for any column y outside b a Q-sympathetic element always exists inside a.

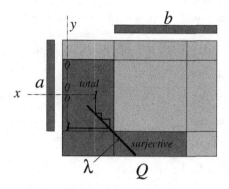

Fig. 16.15 Schema of a decomposition according to a not yet maximum matching

This element may be used immediately to increment λ provided that no other assignment has been given to it so far. It may, however, already have been assigned otherwise. This would occur if there were only **0**-entries above λ. It is this case where the famous Hungarian method of alternating chains starts, for which the equations give justification. From any element of $\overline{b} \cap \overline{\lambda^\mathsf{T}; a}$ such an alternating chain can be started. Because $\overline{b} = Q^\mathsf{T}; a$, there must be a **1** in the y-column inside the a-area. This is then assigned and the old λ-assignment in the same row removed. In the trivial case the y' thereby encountered has a Q-sympathetic x above the λ-area, which is then taken, thus having increased the matching by 1.

This will also become clear from Fig. 16.16. Dashed arrows[2] symbolize sympathy Q being possible/impossible, and also that every element of $\overline{b}$ definitely has Q-inverse images in a. On the other hand, there may be marriages from a to $\overline{b}$, but must exist from $\overline{a}$ to b.

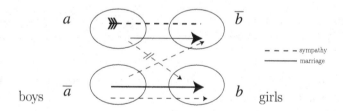

Fig. 16.16 Schema of an assignment iteration

We assume, thus, to have determined a matching and executed the iteration obtaining a, b. We further assume $\overline{b} = \lambda^\mathsf{T}; a$ *not* to hold, which means that $\overline{b} = Q^\mathsf{T}; a \not\supseteq \lambda^\mathsf{T}; a$.

---

[2] This time, it has nothing to do with the dashed arrow convention.

We make use of the formula $\lambda_{;} \overline{S} = \lambda_{;} \mathbb{T} \cap \overline{\lambda_{;} S}$, which holds since λ is univalent; see Prop. 5.6.i. The iteration ends with $\overline{b} = Q^{\mathsf{T}}{}_{;} a$ and $\overline{a} = \lambda_{;} b$. This easily expands to

$$\overline{b} = Q^{\mathsf{T}}{}_{;} a = Q^{\mathsf{T}}{}_{;} \overline{\lambda_{;} b} = Q^{\mathsf{T}}{}_{;} \overline{\lambda_{;} \overline{Q^{\mathsf{T}}{}_{;} a}} = Q^{\mathsf{T}}{}_{;} \overline{\lambda_{;} Q^{\mathsf{T}}{}_{;} \overline{\lambda_{;} \overline{Q^{\mathsf{T}}{}_{;} a}}} \ \ldots$$

from which the last but one becomes

$$\overline{b} = Q^{\mathsf{T}}{}_{;} a = Q^{\mathsf{T}}{}_{;} \overline{\lambda_{;} b} = Q^{\mathsf{T}}{}_{;} \overline{\lambda_{;} \mathbb{T} \cap \overline{\lambda_{;} \overline{Q^{\mathsf{T}}{}_{;} a}}} = Q^{\mathsf{T}}{}_{;} (\overline{\lambda_{;} \mathbb{T}} \cup \lambda_{;} Q^{\mathsf{T}}{}_{;} (\overline{\lambda_{;} \mathbb{T}} \cup \lambda_{;} Q^{\mathsf{T}}{}_{;} a))$$

indicating how to prove that

$$\overline{b} = (Q^{\mathsf{T}} \cup Q^{\mathsf{T}}{}_{;} \lambda_{;} Q^{\mathsf{T}} \cup Q^{\mathsf{T}}{}_{;} \lambda_{;} Q^{\mathsf{T}}{}_{;} \lambda_{;} Q^{\mathsf{T}} \cup \ \ldots)_{;} \overline{\lambda_{;} \mathbb{T}}.$$

If $\lambda^{\mathsf{T}}{}_{;} a \subsetneqq \overline{b}$, we may thus find a point in

$$\overline{\lambda^{\mathsf{T}}{}_{;} a} \cap (Q^{\mathsf{T}} \cup Q^{\mathsf{T}}{}_{;} \lambda_{;} Q^{\mathsf{T}} \cup Q^{\mathsf{T}}{}_{;} \lambda_{;} Q^{\mathsf{T}}{}_{;} \lambda_{;} Q^{\mathsf{T}} \cup \ \ldots)_{;} \overline{\lambda_{;} \mathbb{T}}$$

which leads to the following alternating chain algorithm. We start by choosing

$$y \subseteq \overline{\lambda^{\mathsf{T}}{}_{;} a} \cap (Q^{\mathsf{T}} \cup Q^{\mathsf{T}}{}_{;} \lambda_{;} Q^{\mathsf{T}} \cup Q^{\mathsf{T}}{}_{;} \lambda_{;} Q^{\mathsf{T}}{}_{;} \lambda_{;} Q^{\mathsf{T}} \cup \ \ldots)_{;} \overline{\lambda_{;} \mathbb{T}}$$

which is guaranteed to exist. Because the point y is an atom, we have

$$y \subseteq (Q^{\mathsf{T}}{}_{;} \lambda)^{i}{}_{;} Q^{\mathsf{T}}{}_{;} \overline{\lambda_{;} \mathbb{T}} \cap \overline{\lambda^{\mathsf{T}}{}_{;} a} \quad \text{or else} \quad y \subseteq (Q^{\mathsf{T}}{}_{;} \lambda)^{i}{}_{;} Q^{\mathsf{T}}{}_{;} \overline{\lambda_{;} \mathbb{T}} \quad \text{and} \quad y \subseteq \overline{\lambda^{\mathsf{T}}{}_{;} a}$$

for some minimal $i \in \mathbb{N}$. If i is 0, we obtain with the Dedekind rule that

$$y = Q^{\mathsf{T}}{}_{;} \overline{\lambda_{;} \mathbb{T}} \cap y \subseteq (Q^{\mathsf{T}} \cap y_{;} \overline{\lambda_{;} \mathbb{T}}^{\mathsf{T}})_{;} (\overline{\lambda_{;} \mathbb{T}} \cap Q_{;} y)$$

so that there exists an intermediate point $x \subseteq \overline{\lambda_{;} \mathbb{T}} \cap Q_{;} y$. This means that $x_{;} y^{\mathsf{T}} \subseteq \overline{\lambda}$ as well as $\subseteq Q$, so that it may simply be added to the current matching obtaining $\lambda' := x_{;} y^{\mathsf{T}} \cup \lambda$ as an enlarged matching.

Affairs are more difficult when $i > 0$. Purposefully we then call the point chosen y_i. Proceeding in a similar way with the Dedekind rule, we obtain that

$$y_i = (Q^{\mathsf{T}}{}_{;} \lambda)^{i}{}_{;} Q^{\mathsf{T}}{}_{;} \overline{\lambda_{;} \mathbb{T}} \cap y_i \subseteq ((Q^{\mathsf{T}}{}_{;} \lambda)^{i}{}_{;} Q^{\mathsf{T}} \cap y_{i;} \overline{\lambda_{;} \mathbb{T}}^{\mathsf{T}})_{;} (\overline{\lambda_{;} \mathbb{T}} \cap Q_{;} (\lambda^{\mathsf{T}}{}_{;} Q)^{i}{}_{;} y_i)$$

so that there exists an intermediate point $x_i \subseteq \overline{\lambda_{;} \mathbb{T}} \cap Q_{;} (\lambda^{\mathsf{T}}{}_{;} Q)^{i}{}_{;} y_i$. The first part means $x_i \subseteq \overline{\lambda_{;} \mathbb{T}}$ and, therefore, that $\lambda_{;} \mathbb{T} \subseteq \overline{x_i}$ or $\lambda_{;} y_i \subseteq \lambda_{;} \mathbb{T} \subseteq \overline{x_i}$ or $x_{i;} y_i^{\mathsf{T}} \subseteq \overline{\lambda}$.

When eventually a maximum matching with respect to cardinality is reached, the following situation will hold:

$$a = \sup \{ \mathbb{L}, \lambda_{;} \overline{Q^{\mathsf{T}}{}_{;} \mathbb{L}}, \lambda_{;} Q^{\mathsf{T}}{}_{;} \overline{\lambda_{;} \overline{Q^{\mathsf{T}}{}_{;} \mathbb{L}}}, \ldots \}$$

$$b = \inf \{ \mathbb{T}, \overline{Q^{\mathsf{T}}{}_{;} \overline{\lambda_{;} \mathbb{T}}}, \overline{Q^{\mathsf{T}}{}_{;} \lambda_{;} \overline{Q^{\mathsf{T}}{}_{;} \overline{\lambda_{;} \mathbb{T}}}}, \ldots \}.$$

We visualize the results of this matching iteration by concentrating on the subdivision of the matrices Q, λ initially considered by the resulting vectors $a = \{2, 4, 6, 1, 3\}$ and $b = \{e, c, b\}$.

$$
\begin{array}{c}
\begin{array}{cccccc}
 & a & d & e & c & b
\end{array}\\
\begin{array}{c}2\\6\\4\\1\\3\\5\\7\end{array}
\left(\begin{array}{cc|ccc}
0 & 0 & 0 & 0 & 0\\
1 & 1 & 0 & 0 & 0\\
0 & 1 & 0 & 0 & 0\\
1 & 1 & 0 & 0 & 0\\
1 & 1 & 0 & 0 & 0\\
0 & 1 & 1 & 1 & 1\\
0 & 0 & 1 & 1 & 0
\end{array}\right)
\end{array}
\qquad
\begin{array}{c}
\begin{array}{cccccc}
 & a & d & e & c & b
\end{array}\\
\begin{array}{c}2\\6\\4\\1\\3\\5\\7\end{array}
\left(\begin{array}{cc|ccc}
0 & 0 & 0 & 0 & 0\\
0 & 0 & 0 & 0 & 0\\
0 & 0 & 0 & 0 & 0\\
1 & 0 & 0 & 0 & 0\\
0 & 1 & 0 & 0 & 0\\
0 & 0 & 1 & 0 & 0\\
0 & 0 & 0 & 1 & 0
\end{array}\right)
\end{array}
$$

Fig. 16.17 Sympathy and matching rearranged

Further remarks

These investigations have also been presented in a completely different setting and terminology. Assume a set V together with a family W of subsets taken therefrom, which means nothing other than that a relation $Q : V \longrightarrow W$ is considered. In [93], for example, so-called partial transversals are defined as nothing other than a univalent and injective relation $\lambda \subseteq Q$. So the partial transversal is a matching. Of course, interest concentrates on maximum matchings. Then one is immediately confronted with systems of distinct representatives and the like.

Exercises

16.3 Consider the relation

$$
Q =
\begin{array}{c}
a\\b\\c\\d\\e\\f\\g\\h\\i\\j\\k
\end{array}
\begin{array}{c}
\begin{array}{ccccccccccccccccc}
1&2&3&4&5&6&7&8&9&10&11&12&13&14&15&16&17
\end{array}\\
\left(\begin{array}{ccccccccccccccccc}
0&0&0&0&0&1&0&0&0&0&0&0&1&1&1&0&0\\
0&0&0&0&0&1&0&0&1&0&1&1&0&0&1&0&0\\
0&0&0&0&0&0&0&0&0&0&1&0&0&0&0&0&0\\
0&0&0&0&0&0&0&0&0&0&0&0&0&0&0&0&0\\
0&0&1&0&1&0&0&0&0&0&0&0&0&0&1&0&0\\
0&0&0&0&0&0&0&0&0&0&0&1&0&0&0&1&0\\
1&0&0&0&0&0&1&0&0&1&0&0&0&0&0&0&0\\
0&0&0&0&1&0&0&0&0&0&0&0&0&0&1&0&0\\
0&0&0&1&0&0&0&0&0&0&0&0&0&0&0&1&0\\
0&0&0&0&0&0&0&0&0&0&0&0&0&0&0&0&0\\
0&0&0&0&1&0&0&0&0&0&0&1&1&0&0&0&0
\end{array}\right)
\end{array}
$$

and determine some maximal matching λ for it as well as the vectors a, b. Rearrange following the pattern of Fig. 16.13.

16.6 König's Theorems

We will now put the concepts together and relate them to combinatorial results. The first of these are line-coverings and assignments. Some sort of counting comes into play, however, here in its algebraic form. Permutations allow $1 : 1$-comparison of subsets. Often this means transferring heterogeneous concepts to the $n \times n$-case.

We first consider a matching (or an assignment) which is maximal with respect to cardinality, i.e., satisfying all four equations:

$$\overline{b} = Q^{\mathsf{T}};a \qquad\qquad \overline{a} = Q;b$$
$$\overline{b} = \lambda^{\mathsf{T}};a \qquad\qquad \overline{a} = \lambda;b.$$

An easy observation leads to the following result.

Proposition 16.12. *Let some relation Q be given together with a matching $\lambda \subseteq Q$ and the results a, b of the iteration. Then the following hold:*

(i) *$(\overline{a}, \overline{b})$ forms a line-covering and $Q \subseteq \overline{a};\mathbb{T} \cup \mathbb{T};\overline{b}^{\mathsf{T}}$,*

(ii) *term $\operatorname{rank}(Q) \leq |\overline{a}| + |\overline{b}|$,*

(iii) *$|\lambda| \leq$ term $\operatorname{rank}(Q)$,*

(iv) *if $\overline{b} = \lambda^{\mathsf{T}};a$, then term $\operatorname{rank}(Q) = |\overline{a}| + |\overline{b}| = |\lambda|$.*

Proof: (i) We recall Def. 6.10 and consider two parts of Q separately, starting with $\overline{a};\mathbb{T} \cap Q \subseteq \overline{a};\mathbb{T}$. Then, we have $\overline{b} = Q^{\mathsf{T}};a$ as a result of the iteration, so that

$$a;\mathbb{T} \cap Q \subseteq (a \cap Q;\mathbb{T}^{\mathsf{T}});(\mathbb{T} \cap a^{\mathsf{T}};Q) \subseteq \mathbb{T};a^{\mathsf{T}};Q = \mathbb{T};\overline{b}^{\mathsf{T}}.$$

(ii) According to (i), the rows of $\overline{a}$ together with the columns of $\overline{b}$ cover all of Q, so that the term rank cannot be strictly above the sum of the cardinalities.

(iii) A line-covering of a matching λ can obviously not be achieved with less than $|\lambda|$ lines. The matching properties $\lambda^{\mathsf{T}};\lambda \subseteq \mathbb{I}$, $\lambda;\lambda^{\mathsf{T}} \subseteq \mathbb{I}$ of univalency and injectivity require that every entry of λ be covered by a separate line.

(iv) Condition $\overline{b} = \lambda^{\mathsf{T}};a$ together with $\overline{a} = \lambda;b$ shows that $|\overline{b}|$ entries of λ are needed to end in $\overline{b}$ and $|\overline{a}|$ to start in $\overline{a}$. According to injectivity no entry of λ will start in $\overline{a}$ and end in $\overline{b}$ since $\lambda;\overline{b} = \lambda;\lambda^{\mathsf{T}};a \subseteq a$. Therefore, $|\overline{a}| + |\overline{b}| \leq |\lambda|$, which in combination with (ii,iii) leads to equality. $\qquad\square$

The following is an easy consequence which sometimes is formulated directly as a result.

Corollary 16.13 (König–Egerváry Theorem). *For an arbitrary heterogeneous relation we have that the maximum cardinality of a matching equals the minimum cardinality of a line-covering.* $\qquad\square$

We now specialize to the homogeneous case and investigate what happens when the term rank of a homogeneous relation is less than n.

Proposition 16.14 (Frobenius–König;[4] see [6], Section 2.1.4). *For a finite homogeneous relation* Q *on an* n*-element set* X *the following are equivalent.*

(i) *None of the permutations* P *on the elements of* X *satisfies* $P \subseteq Q$.

(ii) *There exists a pair of sets* a, b *such that*

$\quad\quad \bar{a}, \bar{b}$ *is a line-covering with* $|\bar{a}| + |\bar{b}| < n$, $\quad\quad$ *or equivalently,*

$\quad\quad a, b$ *is an independent pair of sets with* $n < |a| + |b|$, $\quad\quad\quad$ *or equivalently*

$\quad\quad$ *term rank* $< n$.

(iii) *There exists a vector* z *together with a permutation* P *such that* $Q^{\mathsf{T}}\!\raisebox{-0.3ex}{\tiny,}\, z \not\supseteq P\raisebox{-0.3ex}{\tiny,}\, z$. *In other words: there exists a subset* z *which is mapped by* Q *onto a set with strictly fewer elements than* z.

Proof: (i) $\Longrightarrow$ (ii): Find a maximum cardinality matching $\lambda \subseteq Q$ and execute the assignment iteration on page 429. It will end in a, b with $|\bar{a}| + |\bar{b}| = |\lambda| < n$ since λ can by assumption not be a permutation.

(ii) $\Longrightarrow$ (iii): Take $z := a$, which satisfies $Q^{\mathsf{T}}\!\raisebox{-0.3ex}{\tiny,}\, a = \bar{b}$ according to the iteration. Then $|\bar{b}| = n - |b| < |a|$.

(iii) $\Longrightarrow$ (i): When (iii) holds, Q violates the Hall condition for the subset z. If there were a permutation P contained in Q, we would have $Q^{\mathsf{T}}\!\raisebox{-0.3ex}{\tiny,}\, z \supseteq P^{\mathsf{T}}\!\raisebox{-0.3ex}{\tiny,}\, z$, and thus $|Q^{\mathsf{T}}\!\raisebox{-0.3ex}{\tiny,}\, z| \geq |P^{\mathsf{T}}\!\raisebox{-0.3ex}{\tiny,}\, z| = |z|$, a contradiction. $\quad\quad\quad\quad$ □

Exercises

16.4 The permanent of an $m \times n$-matrix A is defined as $\mathbf{per}(A) := \sum_\sigma \prod_{i=1}^m a_{i\sigma(i)}$, where σ ranges over all mappings of $\{1 \dots m\}$ to $\{1 \dots n\}$. Without loss of generality, we assume $m \leq n$. (Recall that this resembles the definition of a determinant, but without multiplying with powers of -1 in the case of $m = n$.) Prove that the permanent vanishes when there exists an $s \times t$-submatrix of $\mathbf{0}$s such that $s + t = n + 1$.

---

[4] While the scientific community has agreed to put the two names peacefully one next to the other, both Frobenius and König included harsh deprecatory remarks concerning priority in their publications, not least in [82, 83].

PART V
ADVANCED TOPICS

Beyond what has been shown so far, there exist further fascinating areas. One of these is concerned with relational specification. While standard relation algebra formulates what has to be related, here only a region is circumscribed within which the relation envisaged is confined. Although this area has attracted considerable interest from researchers, we omit the presentation of demonic operators used for it.

We continue by mentioning what we cannot present here. With relational grammar applied to natural languages, it has been shown that the translation of relations of time, possession, or modality, for example, can be handled when translating to another natural language. Spatial reasoning has long used relation algebra in order to reason about the relative situatedness of items. This is directed at real-time scanning of TV scenes.

One of the application areas that we present may not be seen as an application in the first place. We also use relations to review parts of standard mathematics, the homomorphism and isomorphism theorems. Additional results may be reported and deeper insights gained when using relations. This area seems particularly promising because many other such reconsiderations may be hoped for. We show possible directions to encourage further research with the Farkas Lemma, topology, projective geometry, etc.

Implication structures study long-range implications, preferably with only a few alternatives; they are helpful in investigating dependency situations. This makes the examples of Sudoku and timetables interesting.

A remarkable amount of time has been spent by functional programmers on investigating power operations such as the existential image and the power transpose. These are also helpful when dealing with reactive systems via simulation and bisimulation.

17

Mathematical Applications

Homomorphism and isomorphism theorems will now serve as an application area for relational methods. We review, thus, parts of standard mathematics. When using relations, minor generalizations become possible. The traditional theorems aim at algebraic structures only, i.e., with mappings satisfying certain algebraic laws. When allowing relational structures, there may not always exist images, nor need these be uniquely defined. In spite of such shortcomings, many aspects of the theorems remain valid.

17.1 Multi-coverings

The concept of congruences discussed earlier is very closely related to the concept of a multi-covering we are going to introduce now. We recall from Def. 5.38 that, given B, the pair Ξ, Θ is called a B-congruence provided $\Xi; B \subseteq B; \Theta$. This means that when varying the argument of B restricted to classes according to Ξ, the related elements may vary also, but restricted to classes according to Θ. Often several structural relations B_i are necessary to model some situation. Then, of course, the following considerations have to be applied to all of these.

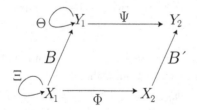

Fig. 17.1 Basic concept of a multi-covering

A multi-covering is something half-way in between relational and algebraic structures; in fact, it is the adequate generalization of a surjective homomorphism of an algebraic structure as we will see in Prop. 17.3. In a general setting, a homomorphism from a structure B into a structure B' is a pair of mappings Φ, Ψ, mapping the source side as well as the target side so that $B \,;\, \Psi \subseteq \Phi \,;\, B'$. One may, thus, expect specific formulae when dividing out the equivalences Ξ, Θ. This is the basic idea of the results we are going to present.

Definition 17.1. A homomorphism (Φ, Ψ) from B to B' is called a **multi-covering**, if the mappings are in addition surjective and satisfy $\Phi \,;\, B' \subseteq B \,;\, \Psi$.   □

So in fact, $\Phi \,;\, B' = B \,;\, \Psi$. The relationship between congruences and multi-coverings is very close.

Theorem 17.2.

  (i) *If (Φ, Ψ) is a multi-covering from B to B', then $(\Xi, \Theta) := (\Phi \,;\, \Phi^{\mathsf{T}}, \Psi \,;\, \Psi^{\mathsf{T}})$ is a B-congruence.*

  (ii) *If the pair (Ξ, Θ) is a B-congruence, then there can exist – up to isomorphism – at most one multi-covering (Φ, Ψ) satisfying $\Xi = \Phi \,;\, \Phi^{\mathsf{T}}$ and $\Theta = \Psi \,;\, \Psi^{\mathsf{T}}$.*

Proof: (i) Ξ is certainly reflexive and transitive, since Φ is total and univalent. In the same way, Θ is reflexive and transitive. The relation $\Xi = \Phi \,;\, \Phi^{\mathsf{T}}$ is symmetric by construction and so is Θ. Now we prove

$$\Xi \,;\, B = \Phi \,;\, \Phi^{\mathsf{T}} \,;\, B \subseteq \Phi \,;\, B' \,;\, \Psi^{\mathsf{T}} \subseteq B \,;\, \Psi \,;\, \Psi^{\mathsf{T}} = B \,;\, \Theta$$

applying one after the other the definition of Ξ, one of the homomorphism definitions of Prop. 5.45, the multi-covering condition, and the definition of Θ.

(ii) Let (Φ_i, Ψ_i) be multi-coverings from B to B_i, $i = 1, 2$. Then

$$B_i \subseteq \Phi_i^{\mathsf{T}} \,;\, \Phi_i \,;\, B_i \subseteq \Phi_i^{\mathsf{T}} \,;\, B \,;\, \Psi_i \subseteq B_i,$$

and therefore "=" everywhere in between, applying surjectivity, the multi-covering property and one of the homomorphism conditions of Prop. 5.45.

Now we show that $(\xi, \vartheta) := (\Phi_1^{\mathsf{T}} \,;\, \Phi_2, \Psi_1^{\mathsf{T}} \,;\, \Psi_2)$ is a homomorphism from B_1 onto B_2 – which is then of course also an isomorphism:

$$\xi^{\mathsf{T}} \,;\, \xi = \Phi_2^{\mathsf{T}} \,;\, \Phi_1 \,;\, \Phi_1^{\mathsf{T}} \,;\, \Phi_2 = \Phi_2^{\mathsf{T}} \,;\, \Xi \,;\, \Phi_2 = \Phi_2^{\mathsf{T}} \,;\, \Phi_2 \,;\, \Phi_2^{\mathsf{T}} \,;\, \Phi_2 = \mathbb{I} \,;\, \mathbb{I} = \mathbb{I}$$
$$B_1 \,;\, \vartheta = \Phi_1^{\mathsf{T}} \,;\, B \,;\, \Psi_1 \,;\, \Psi_1^{\mathsf{T}} \,;\, \Psi_2 = \Phi_1^{\mathsf{T}} \,;\, B \,;\, \Theta \,;\, \Psi_2 = \Phi_1^{\mathsf{T}} \,;\, B \,;\, \Psi_2 \,;\, \Psi_2^{\mathsf{T}} \,;\, \Psi_2 \subseteq \Phi_1^{\mathsf{T}} \,;\, \Phi_2 \,;\, B_2 \,;\, \mathbb{I} = \xi \,;\, B_2. \quad □$$

The multi-covering (Φ, Ψ) for some given congruences Ξ, Θ according to (ii) need not exist in the given relation algebra. It may, however, be constructed by setting Φ, Ψ to be the quotient mappings according to the equivalences Ξ, Θ together with

$B' := \Phi^{\mathsf{T}} \, ; B \, ; \Psi$. In particular, one can formulate this construction simply using the language TITUREL:

$$\Phi := \text{Project } \Xi, \qquad \Psi := \text{Project } \Theta, \qquad B' := \Phi^{\mathsf{T}} \, ; B \, ; \Psi.$$

The interpretation provides some standard realization and the fact that it is unique up to isomorphism then guarantees that one cannot make a mistake in using it. Compare also quotient forming in Prop. 10.20, 10.32, and 10.38.

The following provides a first example of a multi-covering.

Proposition 17.3. *Surjective homomorphisms of an algebraic structure onto another one are always multi-coverings.*

Proof: Assume the homomorphism (Φ, Ψ) from the mapping B to the mapping B', so that $B \, ; \Psi \subseteq \Phi \, ; B'$. The relation $\Phi \, ; B'$ is univalent, since Φ and B' are mappings. The sources $B \, ; \Psi \, ; \mathbb{T} = \mathbb{T} = \Phi \, ; B' \, ; \mathbb{T}$ of $B \, ; \Psi$ and $\Phi \, ; B'$ coincide, because all the relations are mappings and, therefore, total. So we can use Prop. 5.2.iii and obtain $B \, ; \Psi = \Phi \, ; B'$. $\qquad\qquad\square$

We have another trivial standard example of a multi-covering. In an arbitrary relation it may happen that several rows and/or columns coincide. We are then accustomed to consider classes of rows, for example. This is for economy of thinking, but also often for reasons of efficient memory management. In the following proposition, we write down what holds in algebraic formulae.

Proposition 17.4. *Consider a (possibly heterogeneous) relation R and its*

row equivalence	$\Xi := \mathsf{syq}(R^{\mathsf{T}}, R^{\mathsf{T}})$ *and its*
column equivalence	$\Psi := \mathsf{syq}(R, R)$

according to Def. 5.28 as well as the corresponding natural projections η_Ξ, η_Ψ (i.e., satisfying $\eta_\Xi^{\mathsf{T}} \, ; \eta_\Xi = \mathbb{I}$, $\eta_\Xi \, ; \eta_\Xi^{\mathsf{T}} = \Xi$, $\eta_\Psi^{\mathsf{T}} \, ; \eta_\Psi = \mathbb{I}$, $\eta_\Psi \, ; \eta_\Psi^{\mathsf{T}} = \Psi$), and define

$$Q := \eta_\Xi^{\mathsf{T}} \, ; R \, ; \eta_\Psi.$$

Then the following assertions hold:

(i) $\Xi \, ; R = R = R \, ; \Psi$,
(ii) $\mathsf{syq}(Q^{\mathsf{T}}, Q^{\mathsf{T}}) = \mathbb{I}$, $\mathsf{syq}(Q, Q) = \mathbb{I}$, $R = \eta_\Xi \, ; Q \, ; \eta_\Psi^{\mathsf{T}}$,
(iii) η_Ξ, η_Ψ *form a multi-covering from R onto Q as do η_Ψ, η_Ξ from R^{T} onto Q^{T}.*

Proof: For (i), we apply several rules concerning symmetric quotients

$$\Xi \, ; R = \mathsf{syq}(R^{\mathsf{T}}, R^{\mathsf{T}}) \, ; R = \left\{ R^{\mathsf{T}} \, ; [\mathsf{syq}(R^{\mathsf{T}}, R^{\mathsf{T}})]^{\mathsf{T}} \right\}^{\mathsf{T}} = R = R \, ; \mathsf{syq}(R, R) = R \, ; \Psi.$$

From (ii), the symmetric quotient formulae simply state that there are no more duplicate rows and/or columns. To prove them formally is lengthy, so we propose to execute them as Exercise 17.1:

$$\eta_{\Xi};Q;\eta_{\Psi}^{\mathsf{T}} = \eta_{\Xi};\eta_{\Xi}^{\mathsf{T}};R;\eta_{\Psi};\eta_{\Psi}^{\mathsf{T}} = \Xi;R;\Psi = R;\Psi = R.$$

(iii) $\eta_{\Xi};Q = \eta_{\Xi};\eta_{\Xi}^{\mathsf{T}};R;\eta_{\Psi} = \Xi;R;\eta_{\Psi} = R\eta_{\Psi}.$ □

A clarifying remark is in order: there exist also covering *sets* – as opposed to independent sets – which are not easily mixed up with the multi-coverings discussed presently. There also exists, however, another concept of a *graph* covering that is closely related. It will be found in the next section.

Exercises

17.1 Execute the proof of $\mathsf{syq}\,(Q^{\mathsf{T}}, Q^{\mathsf{T}}) = \mathbb{I}$ in Prop. 17.4.ii.

17.2 Covering of graphs and path equivalence

Graph coverings are only sketched here; they may be found in more detail in [123, 124]. We start with a more or less immediate lifting property of a multi-covering that we present without proof.

Proposition 17.5 (Lifting property). *Let a homogeneous relation B be given together with a multi-covering (Φ, Φ) on the relation B'. Furthermore, let some rooted graph B_0 with root a_0, i.e., satisfying $B_0;B_0^{\mathsf{T}} \subseteq \mathbb{I}$ and $B_0^{\mathsf{T}*};a_0 = \mathbb{T}$, be given together with a homomorphism Φ_0 that sends the root a_0 to $a' := \Phi_0^{\mathsf{T}};a_0$. If $a \subseteq \Phi;a'$ is any point mapped by Φ to a', there exists always a relation Ψ – not necessarily a mapping – satisfying the properties*

$$\Psi^{\mathsf{T}};a_0 = a \quad and \quad B_0;\Psi \subseteq \Psi;B.$$

Idea of **proof**: Define $\Psi := \mathsf{inf}\,\{X \mid a_0;a^{\mathsf{T}} \cup (B_0^{\mathsf{T}};X;B \cap \Phi_0;\Phi^{\mathsf{T}}) \subseteq X\}.$ □

The relation Ψ enjoys the homomorphism property but fails to be a mapping in general. In order to make it a mapping, one chooses one of the following two possibilities.

- Firstly, one might follow the recursive definition starting from a_0 and at every stage make an arbitrary choice among the relational images offered, thus choosing a fiber.

- Secondly, one might further restrict the multi-covering condition to 'locally univalent' fans in Φ, requiring $B_0^\mathsf{T}; \Psi; B \cap \Phi_0; \Phi^\mathsf{T} \subseteq \mathbb{I}$ to hold for it, which leads to a well-developed theory, see [109, 110, 111].

We will end, thus, with a homomorphism from B_0 to B in both cases, but discuss only the second, namely graph coverings, defined as multi-coverings with the additional property that the 'outgoing fan' is always isomorphic to the corresponding one in the underlying image graph. The formal definition reads as follows.

Definition 17.6. A surjective homomorphism $\Phi: G \longrightarrow G'$ is called a **graph covering**, provided that it is a multi-covering satisfying

$$B^\mathsf{T}; B \cap \Phi; \Phi^\mathsf{T} \subseteq \mathbb{I}. \qquad \square$$

The multi-covering Φ compares, thus, the two relations between the points of G and of G' and ensures that for any inverse image point x of some point x' and successor y' of x' there is *at least* one successor y of x which is mapped onto y'. The new condition guarantees that there is *at most* one such y since it requires that the relation 'have a common predecessor according to B, and have a common image under Φ' is contained in the identity.

This is an important concept that allows us to define the semantics of recursive programs; it can be applied in uniformization theory of Riemann surfaces as well as in other topological contexts. The added condition, namely, allows lifting paths in G' in a unique way to paths in G, provided any inverse image in G' of the starting point of the path in G is given.

We have again reworked mathematical topics from a relational perspective. First the step from an algebraic to a relational structure has been made. This is so serious a generalization, that one would not expect much of the idea of homomorphism and isomorphism theorems to survive. With the concept of a multi-covering, however, a new and adequate concept seems to have been found. Proposition 17.3 shows that it reduces completely to homomorphisms when going back to the algebraic case. For relational structures, a multi-covering behaves nicely with respect to quotient forming. This relates to earlier papers (see [109, 110, 111]) where the semantics of programs (partial correctness, total correctness, and flow equivalence, even for systems of recursive procedures) were first given a point-free relational form.

17.3 Homomorphism and Isomorphism Theorems

Now we study the homomorphism and isomorphism theorems traditionally offered in a course on group theory or on universal algebra, however generalized from the relational point of view. In the courses mentioned, R, S are often n-ary mappings such as addition and multiplication. Here, we allow them more generally to be relations, i.e., not necessarily mappings. The algebraic laws they satisfy in the algebra are completely irrelevant. These theorems have for a long time been considered general concepts of universal algebra – not of group theory, for example. Here, we go even further and identify them as relational properties. Study of these theorems does not require the concept of an algebra in the classical sense as consisting of mappings.[1]

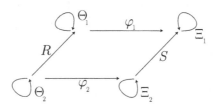

Fig. 17.2 Basic situation of the homomorphism theorem

Proposition 17.7 (Homomorphism Theorem). *Let a relation R be given with an R-congruence (Θ_2, Θ_1) as well as a relation S together with an S-congruence (Ξ_2, Ξ_1) as in Fig. 17.2. Assume a multi-covering (φ_2, φ_1) from R to S such that $\Theta_i = \varphi_i \cdot \Xi_i \cdot \varphi_i^{\mathsf{T}}$ for $i = 1, 2$. Introducing the natural projections η_i for Θ_i and δ_i for Ξ_i, see Fig. 17.3, one has that $\psi_i := \eta_i^{\mathsf{T}} \cdot \varphi_i \cdot \delta_i$, $i = 1, 2$, establish an isomorphism from $R' := \eta_2^{\mathsf{T}} \cdot R \cdot \eta_1$ to $S' := \delta_2^{\mathsf{T}} \cdot S \cdot \delta_1$.*

Proof: The equivalences (Θ_2, Θ_1) satisfy $\Theta_2 \cdot R \subseteq R \cdot \Theta_1$ while (Ξ_2, Ξ_1) satisfy $\Xi_2 \cdot S \subseteq S \cdot \Xi_1$ because they are assumed to be congruences. Furthermore, we have that (φ_2, φ_1) are surjective mappings satisfying $R \cdot \varphi_1 \subseteq \varphi_2 \cdot S$ for homomorphism and $R \cdot \varphi_1 \supseteq \varphi_2 \cdot S$ for multi-covering.

The ψ_i are bijective mappings, which we prove omitting indices:

$$\begin{aligned}
\psi^{\mathsf{T}} \cdot \psi &= \delta^{\mathsf{T}} \cdot \varphi^{\mathsf{T}} \cdot \eta \cdot \eta^{\mathsf{T}} \cdot \varphi \cdot \delta && \text{by definition and executing transposition} \\
&= \delta^{\mathsf{T}} \cdot \varphi^{\mathsf{T}} \cdot \Theta \cdot \varphi \cdot \delta && \text{natural projection } \eta \\
&= \delta^{\mathsf{T}} \cdot \varphi^{\mathsf{T}} \cdot \varphi \cdot \Xi \cdot \varphi^{\mathsf{T}} \cdot \varphi \cdot \delta && \text{condition relating } \Theta, \varphi, \Xi \\
&= \delta^{\mathsf{T}} \cdot \Xi \cdot \delta && \text{since } \varphi \text{ is surjective and univalent}
\end{aligned}$$

---

[1] In George Grätzer's *Universal Algebra*, for example, reported as:
Homomorphism Theorem: Let $\mathcal{A}$ and $\mathcal{B}$ be algebras, and $\varphi : A \to B$ a homomorphism of A onto B. Let Θ denote the congruence relation induced by φ. Then we have that $\mathcal{A}/\Theta$ is isomorphic to $\mathcal{B}$.
First Isomorphism Theorem: Let $\mathcal{A}$ be an algebra, $\mathcal{B}$ a subalgebra of $\mathcal{A}$, and Θ a congruence relation of $\mathcal{A}$. Then $\langle [B]\Theta/\Theta_{[B]\Theta}; F \rangle \approx \langle B/\Theta_B; F \rangle$.
Second Isomorphism Theorem: Let $\mathcal{A}$ be an algebra, let Θ, Φ be congruence relations of $\mathcal{A}$, and assume that $\Theta \subseteq \Phi$. Then $\mathcal{A}/\Phi \approx (\mathcal{A}/\Theta)/(\Phi/\Theta)$.

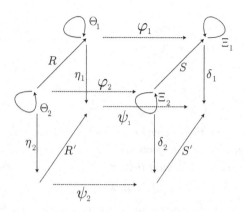

Fig. 17.3 Natural projections added to Fig. 17.2

$$= \delta^{\mathsf{T}}; \delta; \delta^{\mathsf{T}}; \delta = \mathbb{I}; \mathbb{I} = \mathbb{I} \qquad \text{natural projection } \delta$$

The other property, $\psi; \psi^{\mathsf{T}} = \mathbb{I}$, is proved analogously. Proof of the isomorphism property:

$$
\begin{aligned}
R'; \psi_1 &= \eta_2^{\mathsf{T}}; R; \eta_1; \eta_1^{\mathsf{T}}; \varphi_1; \delta_1 && \text{by definition} \\
&= \eta_2^{\mathsf{T}}; R; \Theta_1; \varphi_1; \delta_1 && \text{natural projection } \eta_1 \\
&= \eta_2^{\mathsf{T}}; R; \varphi_1; \Xi_1; \varphi_1^{\mathsf{T}}; \varphi_1; \delta_1 && \text{property of } \varphi_1 \text{ with respect to } \Theta_1, \Xi_1 \\
&= \eta_2^{\mathsf{T}}; R; \varphi_1; \Xi_1; \delta_1 && \text{since } \varphi_1 \text{ is surjective and univalent} \\
&= \eta_2^{\mathsf{T}}; \varphi_2; S; \Xi_1; \delta_1 && \text{multi-covering} \\
&= \eta_2^{\mathsf{T}}; \varphi_2; \Xi_2; S; \Xi_1; \delta_1 && S; \Xi_1 \subseteq \Xi_2; S; \Xi_1 \subseteq S; \Xi_1; \Xi_1 = S; \Xi_1 \\
&= \eta_2^{\mathsf{T}}; \varphi_2; \delta_2; \delta_2^{\mathsf{T}}; S; \delta_1; \delta_1^{\mathsf{T}}; \delta_1 && \text{natural projections} \\
&= \eta_2^{\mathsf{T}}; \varphi_2; \delta_2; S'; \delta_1^{\mathsf{T}}; \delta_1 && \text{definition of } S' \\
&= \eta_2^{\mathsf{T}}; \varphi_2; \delta_2; S' && \text{since } \delta_1 \text{ is surjective and univalent} \\
&= \psi_2; S' && \text{definition of } \psi_2
\end{aligned}
$$

According to Lemma 5.48, this suffices to establish an isomorphism. □

One should bear in mind that this proposition was in several respects slightly more general than the classical homomorphism theorem: R, S need not be mappings, nor need they be homogeneous relations, Ξ was not confined to be the identity congruence, and, not least, relation algebra admits non-standard models.

Proposition 17.8 (First Isomorphism Theorem). *Let a homogeneous relation R on X be given and consider it in connection with an equivalence Ξ and a subset U. Assume that U is contracted by R and that Ξ is an R-congruence*

$R^\mathsf{T}{}_;U \subseteq U$ *and* $\Xi{}_;R \subseteq R{}_;\Xi$.

Now extrude both U and its Ξ-saturation $\Xi{}_;U$, so as to obtain natural injections

$$\iota : Y \longrightarrow X \quad and \quad \lambda : Z \longrightarrow X,$$

universally characterized by

$$\iota^\mathsf{T}{}_;\iota = \mathbb{I}_X \cap U{}_;\mathbb{T}, \qquad \iota{}_;\iota^\mathsf{T} = \mathbb{I}_Y,$$
$$\lambda^\mathsf{T}{}_;\lambda = \mathbb{I}_X \cap \Xi{}_;U{}_;\mathbb{T}, \quad \lambda{}_;\lambda^\mathsf{T} = \mathbb{I}_Z.$$

On Y and Z, we consider the derived equivalences $\Xi_Y := \iota{}_;\Xi{}_;\iota^\mathsf{T}$ and $\Xi_Z := \lambda{}_;\Xi{}_;\lambda^\mathsf{T}$ and in addition their natural projections $\eta : Y \longrightarrow Y_\Xi$ and $\delta : Z \longrightarrow Z_\Xi$. In a standard way, restrictions of R may then be defined on both sides, namely

$$S := \eta^\mathsf{T}{}_;\iota{}_;R{}_;\iota^\mathsf{T}{}_;\eta \quad and \quad T := \delta^\mathsf{T}{}_;\lambda{}_;R{}_;\lambda^\mathsf{T}{}_;\delta.$$

In this setting, $\varphi := \delta^\mathsf{T}{}_;\lambda{}_;\iota^\mathsf{T}{}_;\eta$ gives an isomorphism between S and T.

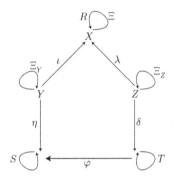

Fig. 17.4 Situation of the First Isomorphism Theorem

Proof: We omit this proof and refer to the next proposition that gives an even more general form.   □

This theorem allows a generalization to a heterogeneous version.

Proposition 17.9 (First Isomorphism Theorem, generalized version). *Let a possibly heterogeneous relation $R : X_1 \longrightarrow X_2$ be given. Consider it in connection with subsets U_1 on the source side as well as U_2 on the target side and assume that U_1 is sent completely into U_2 by R. Furthermore assume an R-congruence (Ξ_1, Ξ_2) to be given:*

$$R^\mathsf{T}{}_;U_1 \subseteq U_2 \quad and \quad \Xi_1{}_;R \subseteq R{}_;\Xi_2$$

Now extrude the U_i as well as their saturations $\Xi_i\mathbin{;}U_i$, so as to obtain natural injections

$$\iota_1 : Y_1 \longrightarrow X_1, \qquad \lambda_1 : Z_1 \longrightarrow X_1, \qquad \iota_2 : Y_2 \longrightarrow X_2, \qquad \lambda_2 : Z_2 \longrightarrow X_2,$$

universally characterized by

$$\iota_i^{\mathsf{T}}\mathbin{;}\iota_i = \mathbb{I}_{X_i}\cap U_i\mathbb{T}, \qquad \iota_i\mathbin{;}\iota_i^{\mathsf{T}} = \mathbb{I}_{Y_i}, \qquad \lambda_i^{\mathsf{T}}\mathbin{;}\lambda_i = \mathbb{I}_{X_i}\cap\Xi_i\mathbin{;}U_i\mathbb{T}, \qquad \lambda_i\mathbin{;}\lambda_i^{\mathsf{T}} = \mathbb{I}_{Z_i}.$$

On Y_i and Z_i, we consider the equivalences $\Xi_{i_Y} := \iota_i\mathbin{;}\Xi_i\mathbin{;}\iota_i^{\mathsf{T}}$ and $\Xi_{i_Z} := \lambda_i\mathbin{;}\Xi_i\mathbin{;}\lambda_i^{\mathsf{T}}$ and in addition their natural projections $\eta_i : Y_i \longrightarrow Y_{\Xi_i}$ and $\delta_i : Z_i \longrightarrow Z_{\Xi_i}$. In a standard way, restrictions of R may be defined on both sides, namely

$$S := \eta_1^{\mathsf{T}}\mathbin{;}\iota_1\mathbin{;}R\mathbin{;}\iota_2^{\mathsf{T}}\mathbin{;}\eta_2 \qquad and \qquad T := \delta_1^{\mathsf{T}}\mathbin{;}\lambda_1\mathbin{;}R\mathbin{;}\lambda_2^{\mathsf{T}}\mathbin{;}\delta_2.$$

In this setting, $\varphi_1 := \delta_1^{\mathsf{T}}\mathbin{;}\lambda_1\mathbin{;}\iota_1^{\mathsf{T}}\mathbin{;}\eta_1$ and $\varphi_2 := \delta_2^{\mathsf{T}}\mathbin{;}\lambda_2\mathbin{;}\iota_2^{\mathsf{T}}\mathbin{;}\eta_2$ give an isomorphism (φ_1,φ_2) between T and S.

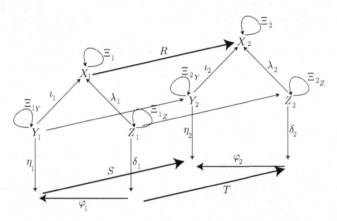

Fig. 17.5 Situation of the generalized version of the First Isomorphism Theorem

Proof: We prove several results in advance, namely

$$\Xi_i\mathbin{;}\iota_i^{\mathsf{T}}\mathbin{;}\iota_i\mathbin{;}\Xi_i = \Xi_i\mathbin{;}\lambda_i^{\mathsf{T}}\mathbin{;}\lambda_i\mathbin{;}\Xi_i, \qquad\qquad (17.1)$$

proved using rules for composition of equivalences:

$$
\begin{aligned}
\Xi_i\mathbin{;}\iota_i^{\mathsf{T}}\mathbin{;}\iota_i\mathbin{;}\Xi_i &= \Xi_i\mathbin{;}(\mathbb{I}\cap U_i\mathbin{;}\mathbb{T})\mathbin{;}\Xi_i && \text{extrusion of } U_i\\
&= \Xi_i\mathbin{;}\Xi_i\mathbin{;}(\mathbb{I}\cap U_i\mathbin{;}\mathbb{T}\mathbin{;}\Xi_i)\mathbin{;}\Xi_i\mathbin{;}\Xi_i && \text{since } \Xi_i \text{ is an equivalence}\\
&= \Xi_i\mathbin{;}\Xi_i\mathbin{;}(\Xi_i\cap U_i\mathbin{;}\mathbb{T}\mathbin{;}\Xi_i)\mathbin{;}\Xi_i && \text{Prop. 5.27}\\
&= \Xi_i\mathbin{;}(\Xi_i\cap\Xi_i\mathbin{;}U_i\mathbin{;}\mathbb{T}\mathbin{;}\Xi_i)\mathbin{;}\Xi_i && \text{Prop. 5.27 again}\\
&= \Xi_i\mathbin{;}(\mathbb{I}\cap\Xi_i\mathbin{;}U_i\mathbin{;}\mathbb{T})\mathbin{;}\Xi_i\mathbin{;}\Xi_i && \text{Prop. 5.27 again}\\
&= \Xi_i\mathbin{;}\lambda_i^{\mathsf{T}}\mathbin{;}\lambda_i\mathbin{;}\Xi_i && \text{extrusion of } \Xi_i\mathbin{;}U_i
\end{aligned}
$$

In a similar way

$$\iota_i\mathbin{;}\lambda_i^{\mathsf{T}}\mathbin{;}\lambda_i = \iota_i \qquad and \qquad \iota_1\mathbin{;}R\mathbin{;}\iota_2^{\mathsf{T}}\mathbin{;}\iota_2 = \iota_1\mathbin{;}R. \qquad\qquad (17.2)$$

The left identity is proved with

$$
\begin{aligned}
\iota_i{}_i \lambda_i^\mathsf{T}{}_i \lambda_i &= \iota_i{}_i \iota_i^\mathsf{T}{}_i \iota_i{}_i \lambda_i^\mathsf{T}{}_i \lambda_i && \iota_i \text{ is injective and total} \\
&= \iota_i{}_i (\mathbb{I} \cap U_i{}_i \mathbb{T}){}_i (\mathbb{I} \cap \Xi_i{}_i U_i{}_i \mathbb{T}) && \text{definition of natural injection} \\
&= \iota_i{}_i (\mathbb{I} \cap U_i{}_i \mathbb{T} \cap \Xi_i{}_i U_i{}_i \mathbb{T}) && \text{intersecting partial identities} \\
&= \iota_i{}_i (\mathbb{I} \cap U_i{}_i \mathbb{T}) = \iota_i{}_i \iota_i^\mathsf{T}{}_i \iota_i = \iota_i && \text{since } \Xi_i \supseteq \mathbb{I}
\end{aligned}
$$

The conditions $R^\mathsf{T}{}_i U_1 \subseteq U_2$ and $\Xi_1{}_i R \subseteq R{}_i \Xi_2$ allow us to prove the right identity of which "$\subseteq$" is obvious. For "$\supseteq$", we apply $\iota_i{}_i \iota_i^\mathsf{T} = \mathbb{I}$ after having shown

$$
\begin{aligned}
\iota_1^\mathsf{T}{}_i \iota_1{}_i R &= (\mathbb{I} \cap U_1{}_i \mathbb{T}){}_i R = U_1{}_i \mathbb{T}{}_i \mathbb{I} \cap R && \text{masking, Prop. 8.5} \\
&\subseteq (U_1{}_i \mathbb{T} \cap R{}_i \mathbb{I}^\mathsf{T}){}_i (\mathbb{I} \cap (U_1{}_i \mathbb{T})^\mathsf{T}{}_i R) && \text{Dedekind} \\
&\subseteq (R \cap U_1{}_i \mathbb{T}){}_i (\mathbb{I} \cap \mathbb{T}{}_i U_2^\mathsf{T}) && \text{since } R^\mathsf{T}{}_i U_1 \subseteq U_2 \\
&= (R \cap U_1{}_i \mathbb{T}){}_i (\mathbb{I} \cap U_2{}_i \mathbb{T}) && \text{since } Q \subseteq \mathbb{I} \text{ implies } Q = Q^\mathsf{T} \\
&= (\mathbb{I} \cap U_1{}_i \mathbb{T}){}_i R{}_i (\mathbb{I} \cap U_2{}_i \mathbb{T}) && \text{according to Prop. 8.5 again} \\
&= \iota_1^\mathsf{T}{}_i \iota_1{}_i R{}_i \iota_2^\mathsf{T}{}_i \iota_2 && \text{definition of natural injection}
\end{aligned}
$$

With $R^\mathsf{T}{}_i \Xi_1{}_i U_1 \subseteq \Xi_2{}_i R^\mathsf{T}{}_i U_1 \subseteq \Xi_2{}_i U_2$, in a similar way we obtain

$$
\lambda_1{}_i R{}_i \lambda_2^\mathsf{T}{}_i \lambda_2 = \lambda_1{}_i R. \tag{17.3}
$$

We show that φ_i is univalent and surjective:

$$
\begin{aligned}
\varphi_i^\mathsf{T}{}_i \varphi_i &= \eta_i^\mathsf{T}{}_i \iota_i{}_i \lambda_i^\mathsf{T}{}_i \delta_i{}_i \delta_i^\mathsf{T}{}_i \lambda_i{}_i \iota_i^\mathsf{T}{}_i \eta_i && \text{by definition} \\
&= \eta_i^\mathsf{T}{}_i \iota_i{}_i \lambda_i^\mathsf{T}{}_i \Xi_{i_Z}{}_i \lambda_i{}_i \iota_i^\mathsf{T}{}_i \eta_i && \text{natural projection} \\
&= \eta_i^\mathsf{T}{}_i \iota_i{}_i \lambda_i^\mathsf{T}{}_i \lambda_i{}_i \Xi_i{}_i \lambda_i^\mathsf{T}{}_i \lambda_i{}_i \iota_i^\mathsf{T}{}_i \eta_i && \text{definition of } \Xi_{i_Z} \\
&= \eta_i^\mathsf{T}{}_i \iota_i{}_i \Xi_i{}_i \iota_i^\mathsf{T}{}_i \eta_i && \text{as proved initially} \\
&= \eta_i^\mathsf{T}{}_i \Xi_{i_Y}{}_i \eta_i && \text{definition of } \Xi_{i_Y} \\
&= \eta_i^\mathsf{T}{}_i \eta_i{}_i \eta_i^\mathsf{T}{}_i \eta_i = \mathbb{I}{}_i \mathbb{I} = \mathbb{I} && \text{natural projection}
\end{aligned}
$$

To show that φ_i is injective and total, we start

$$
\begin{aligned}
\delta_i{}_i \varphi_i{}_i \varphi_i^\mathsf{T}{}_i \delta_i^\mathsf{T} &= \delta_i{}_i \delta_i^\mathsf{T}{}_i \lambda_i{}_i \iota_i^\mathsf{T}{}_i \eta_i{}_i \eta_i^\mathsf{T}{}_i \iota_i{}_i \lambda_i^\mathsf{T}{}_i \delta_i{}_i \delta_i^\mathsf{T} && \text{by definition} \\
&= \Xi_{i_Z}{}_i \lambda_i{}_i \iota_i^\mathsf{T}{}_i \Xi_{i_Y}{}_i \iota_i{}_i \lambda_i^\mathsf{T}{}_i \Xi_{i_Z} && \text{natural projections} \\
&= \lambda_i{}_i \Xi_i{}_i \lambda_i^\mathsf{T}{}_i \lambda_i{}_i \iota_i^\mathsf{T}{}_i \iota_i{}_i \Xi_i{}_i \iota_i^\mathsf{T}{}_i \iota_i{}_i \lambda_i^\mathsf{T}{}_i \lambda_i{}_i \Xi_i{}_i \lambda_i^\mathsf{T} && \text{by definition of } \Xi_{i_Y}, \Xi_{i_Z} \\
&= \lambda_i{}_i \Xi_i{}_i \iota_i^\mathsf{T}{}_i \iota_i{}_i \Xi_i{}_i \iota_i^\mathsf{T}{}_i \iota_i{}_i \Xi_i{}_i \lambda_i^\mathsf{T} && \text{since } \iota_i{}_i \lambda_i^\mathsf{T}{}_i \lambda_i = \iota_i \\
&= \lambda_i{}_i \Xi_i{}_i \lambda_i^\mathsf{T}{}_i \lambda_i{}_i \Xi_i{}_i \lambda_i^\mathsf{T}{}_i \lambda_i{}_i \Xi_i{}_i \lambda_i^\mathsf{T} && \text{following Eqn. (17.1)} \\
&= \Xi_{i_Z}{}_i \Xi_{i_Z}{}_i \Xi_{i_Z} = \Xi_{i_Z} && \text{by definition of } \Xi_{i_Z}
\end{aligned}
$$

so that we may go on with

$$
\begin{aligned}
\varphi_i{}_i \varphi_i^\mathsf{T} &= \delta_i^\mathsf{T}{}_i \delta_i{}_i \varphi_i{}_i \varphi_i^\mathsf{T}{}_i \delta_i^\mathsf{T}{}_i \delta_i && \text{by definition} \\
&= \delta_i^\mathsf{T}{}_i \Xi_{i_Z}{}_i \delta_i && \text{as shown before} \\
&= \delta_i^\mathsf{T}{}_i \delta_i{}_i \delta_i^\mathsf{T}{}_i \delta_i = \mathbb{I}{}_i \mathbb{I} = \mathbb{I} && \text{natural projection}
\end{aligned}
$$

The interplay of subset forming and equivalence classes is shown schematically in Fig. 17.6.

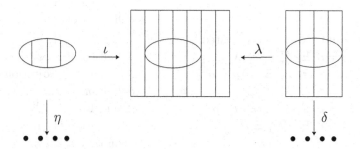

Fig. 17.6 Idea of the interconnection between subsets and classes in the First Isomorphism Theorem

It turns out that Ξ_{1_Y}, Ξ_{2_Y} is an R_Y-congruence for the intermediate construct $R_Y := \iota_1 \cdot R \cdot \iota_2^\mathsf{T}$:

$$
\begin{aligned}
\Xi_{1_Y} \cdot R_Y &= \iota_1 \cdot \Xi_1 \cdot \iota_1^\mathsf{T} \cdot \iota_1 \cdot R \cdot \iota_2^\mathsf{T} && \text{by definition} \\
&\subseteq \iota_1 \cdot \Xi_1 \cdot R \cdot \iota_2^\mathsf{T} && \iota_1 \text{ is univalent} \\
&\subseteq \iota_1 \cdot R \cdot \Xi_2 \cdot \iota_2^\mathsf{T} && \text{congruence} \\
&\subseteq \iota_1 \cdot R \cdot \iota_2^\mathsf{T} \cdot \iota_2 \cdot \Xi_2 \cdot \iota_2^\mathsf{T} && \text{due to Eqn. (17.2)} \\
&\subseteq R_Y \cdot \Xi_{2_Y} && \text{definitions of } R_Y, \Xi_{2_Y}
\end{aligned}
$$

The constructs $\alpha_i := \iota_i \cdot \Xi_i \cdot \lambda_i^\mathsf{T} \cdot \delta_i$ are surjective mappings:

$$
\begin{aligned}
\alpha_i^\mathsf{T} \cdot \alpha_i &= \delta_i^\mathsf{T} \cdot \lambda_i \cdot \Xi_i \cdot \iota_i^\mathsf{T} \cdot \iota_i \cdot \Xi_i \cdot \lambda_i^\mathsf{T} \cdot \delta_i && \text{by the definition just given} \\
&= \delta_i^\mathsf{T} \cdot \lambda_i \cdot \Xi_i \cdot \lambda_i^\mathsf{T} \cdot \lambda_i \cdot \Xi_i \cdot \lambda_i^\mathsf{T} \cdot \delta_i && \text{due to Eqn. (17.1)} \\
&= \delta_i^\mathsf{T} \cdot \Xi_{i_Z} \cdot \Xi_{i_Z} \cdot \delta_i && \text{definition of } \Xi_{i_Z} \\
&= \delta_i^\mathsf{T} \cdot \Xi_{i_Z} \cdot \delta_i && \Xi_{i_Z} \text{ is an equivalence} \\
&= \delta_i^\mathsf{T} \cdot \delta_i \cdot \delta_i^\mathsf{T} \cdot \delta_i = \mathbb{I} \cdot \mathbb{I} = \mathbb{I} && \delta_i \text{ is natural projection for } \Xi_{i_Z} \\
\alpha_i \cdot \alpha_i^\mathsf{T} &= \iota_i \cdot \Xi_i \cdot \lambda_i^\mathsf{T} \cdot \delta_i \cdot \delta_i^\mathsf{T} \cdot \lambda_i \cdot \Xi_i \cdot \iota_i^\mathsf{T} && \text{by definition} \\
&= \iota_i \cdot \Xi_i \cdot \lambda_i^\mathsf{T} \cdot \Xi_{i_Z} \cdot \lambda_i \cdot \Xi_i \cdot \iota_i^\mathsf{T} && \delta_i \text{ is natural projection for } \Xi_{i_Z} \\
&= \iota_i \cdot \Xi_i \cdot \lambda_i^\mathsf{T} \cdot \lambda_i \cdot \Xi_i \cdot \lambda_i^\mathsf{T} \cdot \lambda_i \cdot \Xi_i \cdot \iota_i^\mathsf{T} && \text{definition of } \Xi_{i_Z} \\
&= \iota_i \cdot \Xi_i \cdot \iota_i^\mathsf{T} \cdot \iota_i \cdot \Xi_i \cdot \iota_i^\mathsf{T} \cdot \iota_i \cdot \Xi_i \cdot \iota_i^\mathsf{T} && \text{due to Eqn. (17.1)} \\
&= \Xi_{i_Y} \cdot \Xi_{i_Y} \cdot \Xi_{i_Y} = \Xi_{i_Y} \supseteq \mathbb{I} && \text{definition of equivalence } \Xi_{i_Y}
\end{aligned}
$$

With the α_i, we may express S, T in a shorter way:

$$
\begin{aligned}
\alpha_1^\mathsf{T} \cdot R_Y \cdot \alpha_2 &= \delta_1^\mathsf{T} \cdot \lambda_1 \cdot \Xi_1 \cdot \iota_1^\mathsf{T} \cdot R_Y \cdot \iota_2 \cdot \Xi_2 \cdot \lambda_2^\mathsf{T} \cdot \delta_2 && \text{definition of } \alpha_1, \alpha_2 \\
&= \delta_1^\mathsf{T} \cdot \lambda_1 \cdot \Xi_1 \cdot \iota_1^\mathsf{T} \cdot \iota_1 \cdot R \cdot \iota_2^\mathsf{T} \cdot \iota_2 \cdot \Xi_2 \cdot \lambda_2^\mathsf{T} \cdot \delta_2 && \text{definition of } R_Y \\
&= \delta_1^\mathsf{T} \cdot \lambda_1 \cdot \Xi_1 \cdot \iota_1^\mathsf{T} \cdot \iota_1 \cdot R \cdot \Xi_2 \cdot \lambda_2^\mathsf{T} \cdot \delta_2 && \text{due to Eqn. (17.2)} \\
&= \delta_1^\mathsf{T} \cdot \lambda_1 \cdot \Xi_1 \cdot \iota_1^\mathsf{T} \cdot \iota_1 \cdot \Xi_1 \cdot R \cdot \Xi_2 \cdot \lambda_2^\mathsf{T} \cdot \delta_2 && \Xi_1 \cdot R \cdot \Xi_2 \subseteq R \cdot \Xi_2 \cdot \Xi_2 = R \cdot \Xi_2 \subseteq \Xi_1 \cdot R \cdot \Xi_2 \\
&= \delta_1^\mathsf{T} \cdot \lambda_1 \cdot \Xi_1 \cdot \lambda_1^\mathsf{T} \cdot \lambda_1 \cdot \Xi_1 \cdot R \cdot \Xi_2 \cdot \lambda_2^\mathsf{T} \cdot \delta_2 && \text{due to Eqn. (17.1)} \\
&= \delta_1^\mathsf{T} \cdot \Xi_{1_Z} \cdot \lambda_1 \cdot R \cdot \Xi_2 \cdot \lambda_2^\mathsf{T} \cdot \delta_2 && \text{as before, definition of } \Xi_{1_Z}, \Xi_{2_Z} \\
&= \delta_1^\mathsf{T} \cdot \Xi_{1_Z} \cdot \lambda_1 \cdot R \cdot \lambda_2^\mathsf{T} \cdot \lambda_2 \cdot \Xi_2 \cdot \lambda_2^\mathsf{T} \cdot \delta_2 && \text{due to Eqn. (17.3)} \\
&= \delta_1^\mathsf{T} \cdot \Xi_{1_Z} \cdot \lambda_1 \cdot R \cdot \lambda_2^\mathsf{T} \cdot \Xi_{2_Z} \cdot \delta_2 && \text{definition of } \Xi_{2_Z} \\
&= \delta_1^\mathsf{T} \cdot \delta_1 \cdot \delta_1^\mathsf{T} \cdot \lambda_1 \cdot R \cdot \lambda_2^\mathsf{T} \cdot \delta_2 \cdot \delta_2^\mathsf{T} \cdot \delta_2 && \delta_1, \delta_2 \text{ natural projections for } \Xi_{1_Z}, \Xi_{2_Z} \\
&= \delta_1^\mathsf{T} \cdot \lambda_1 \cdot R \cdot \lambda_2^\mathsf{T} \cdot \delta_2 = T && \delta_1, \delta_2 \text{ surjective mappings}
\end{aligned}
$$

$$\eta_1^\mathsf{T} R_Y \eta_2 = \eta_1^\mathsf{T} \iota_1 R \iota_2^\mathsf{T} \eta_2 \qquad\qquad \text{definition of } R_Y$$
$$= S \qquad\qquad\qquad\qquad\quad \text{definition of } S$$

Relations α and φ are closely related:

$$\alpha_i \varphi_i = \iota_i \Xi_i \lambda_i^\mathsf{T} \delta_i \delta_i^\mathsf{T} \lambda_i \iota_i^\mathsf{T} \eta_i \qquad\qquad \text{definition of } \alpha_i, \varphi_i$$
$$= \iota_i \Xi_i \lambda_i^\mathsf{T} \Xi_{i_Z} \lambda_i \iota_i^\mathsf{T} \eta_i \qquad\qquad \delta_i \text{ is natural projection for } \Xi_{i_Z}$$
$$= \iota_i \Xi_i \lambda_i^\mathsf{T} \lambda_i \Xi_i \lambda_i^\mathsf{T} \lambda_i \iota_i^\mathsf{T} \eta_i \qquad\qquad \text{definition of } \Xi_{i_Z}$$
$$= \iota_i \Xi_i \lambda_i^\mathsf{T} \lambda_i \Xi_i \iota_i^\mathsf{T} \eta_i \qquad\qquad \text{due to Eqn. (17.2)}$$
$$= \iota_i \Xi_i \iota_i^\mathsf{T} \iota_i \Xi_i \iota_i^\mathsf{T} \eta_i \qquad\qquad \text{due to Eqn. (17.1)}$$
$$= \Xi_{i_Y} \Xi_{i_Y} \eta_i \qquad\qquad \text{definition of } \Xi_{i_Y}$$
$$= \eta_i \eta_i^\mathsf{T} \eta_i \eta_i^\mathsf{T} \eta_i = \eta_i \qquad\qquad \eta_i \text{ is natural projection for } \Xi_{i_Y}$$
$$\alpha_i^\mathsf{T} \eta_i = \alpha_i^\mathsf{T} \alpha_i \varphi_i \qquad\qquad \text{see before}$$
$$= \varphi_i \qquad\qquad\qquad\quad \alpha_i \text{ are univalent and surjective}$$

This enables us already to prove the homomorphism condition:

$$T \varphi_2 = \alpha_1^\mathsf{T} R_Y \alpha_2 \alpha_2^\mathsf{T} \eta_2 \qquad\qquad \text{above results on } T, \varphi_2$$
$$= \alpha_1^\mathsf{T} R_Y \Xi_{2_Y} \eta_2 \qquad\qquad \alpha_2 \alpha_2^\mathsf{T} = \Xi_{2_Y}, \text{ see above}$$
$$= \alpha_1^\mathsf{T} \Xi_{1_Y} R_Y \Xi_{2_Y} \eta_2 \qquad\qquad \Xi_{1_Y}, \Xi_{2_Y} \text{ is an } R_Y\text{-congruence}$$
$$= \alpha_1^\mathsf{T} \eta_1 \eta_1^\mathsf{T} R_Y \eta_2 \eta_2^\mathsf{T} \eta_2 \qquad\qquad \eta_i \text{ natural projection for } \Xi_{i_Y}$$
$$= \varphi_1 \eta_1^\mathsf{T} R_Y \eta_2 \qquad\qquad \eta_2 \text{ univalent and surjective}$$
$$= \varphi_1 S \qquad\qquad\qquad\quad \text{see above}$$

This was an equality, so that it suffices according to Lemma 5.48. □

An attempt to generalize the Second Isomorphism Theorem in a similar way immediately suggests itself. The reader should, however, check that this leads to no other result as for the Homomorphism Theorem.

17.4 Further mathematical snippets

The following is intended to show that relations may help also to formulate other concepts in mathematics. We give a glimpse of projective geometry, topology, and the Farkas Lemma.

Discrete geometry

Some nice relational applications are possible in finite geometry. The basic setting is that of a set $\mathcal{P}$ of **points** and set $\mathcal{L}$ of **lines** related by incidence $\mathcal{I} : \mathcal{P} \longrightarrow \mathcal{L}$. Lines may also be related by orthogonality, which in turn may be defined starting from the incidences. One then easily arrives at Desargues, Pappus, or other properties.

Definition 17.10. A relation $\mathcal{I} : \mathcal{P} \longrightarrow \mathcal{L}$ between points and lines is called an **affine incidence plane**, provided

(i) for every two different points there exists precisely one line incident with both of them,

(ii) for every line g and every point P not incident to it, there exists precisely one line h through P without common point with g (i.e., that is 'parallel' to g),

(iii) for every line g there exists a point not incident to it. □

Figure 17.7 shows an example of an affine incidence plane. It is an easy relational exercise to determine all classes of mutually parallel lines and to add them as points of a new kind together with a new line containing all these so as to obtain the projective geometry of Fig. 17.8. Lines $1, 2, 3$ are mutually parallel and form the class $[1]$.

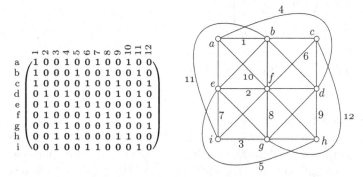

Fig. 17.7 Incidence relation $\mathcal{I}$ of an affine incidence plane

Closely related to these – and more symmetric – are the projective planes.

Definition 17.11. A relation $\mathcal{I} : \mathcal{P} \longrightarrow \mathcal{L}$ is called a **projective incidence plane**, provided

(i) for every two different points there exists precisely one line to which both are incident,

(ii) for every two different lines there exists precisely one point incident with both of them,

(iii) there exist four points and no group consists of three of them all on a line. □

We have already mentioned the well-known transition from an affine to a projective plane. From a projective incidence plane in turn, we can delete an arbitrary line together with the points incident to it in order to obtain an affine incidence plane.

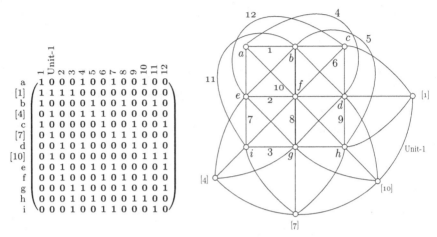

$$
\begin{array}{c|ccccccccccccc}
 & 1 & \text{Unit-1} & 2 & 3 & 4 & 5 & 6 & 7 & 8 & 9 & 10 & 11 & 12 \\
\hline
a & 1 & 0 & 0 & 0 & 1 & 0 & 0 & 1 & 0 & 0 & 1 & 0 & 0 \\
{[1]} & 1 & 1 & 1 & 1 & 0 & 0 & 0 & 0 & 0 & 0 & 0 & 0 & 0 \\
b & 1 & 0 & 0 & 0 & 0 & 1 & 0 & 0 & 1 & 0 & 0 & 1 & 0 \\
{[4]} & 0 & 1 & 0 & 0 & 1 & 1 & 1 & 0 & 0 & 0 & 0 & 0 & 0 \\
c & 1 & 0 & 0 & 0 & 0 & 0 & 1 & 0 & 0 & 1 & 0 & 0 & 1 \\
{[7]} & 0 & 1 & 0 & 0 & 0 & 0 & 0 & 1 & 1 & 1 & 0 & 0 & 0 \\
d & 0 & 0 & 1 & 0 & 1 & 0 & 0 & 0 & 0 & 1 & 0 & 1 & 0 \\
{[10]} & 0 & 1 & 0 & 0 & 0 & 0 & 0 & 0 & 0 & 0 & 1 & 1 & 1 \\
e & 0 & 0 & 1 & 0 & 0 & 1 & 0 & 1 & 0 & 0 & 0 & 0 & 1 \\
f & 0 & 0 & 1 & 0 & 0 & 0 & 1 & 0 & 1 & 0 & 1 & 0 & 0 \\
g & 0 & 0 & 0 & 1 & 1 & 0 & 0 & 0 & 1 & 0 & 0 & 0 & 1 \\
h & 0 & 0 & 0 & 1 & 0 & 1 & 0 & 0 & 0 & 1 & 1 & 0 & 0 \\
i & 0 & 0 & 0 & 1 & 0 & 0 & 1 & 1 & 0 & 0 & 0 & 1 & 0 \\
\end{array}
$$

Fig. 17.8 A projective incidence plane obtained from Fig. 17.7

We transfer the definition of a projective plane into relation-algebraic form (disregarding the condition on degeneration). From the baseorder of points, respectively lines, we take the duals $\mathsf{Base0}_{\mathcal{P}}^{\mathrm{d}}$ and $\mathsf{Base0}_{\mathcal{L}}^{\mathrm{d}}$, thus restricting ourselves to two-element sets – as represented by pairs strictly below the diagonal – when vectorizing as

$$v_{\mathcal{P}} := \mathsf{vec}(\mathsf{Base0}_{\mathcal{P}}^{\mathrm{d}}), \quad v_{\mathcal{L}} := \mathsf{vec}(\mathsf{Base0}_{\mathcal{L}}^{\mathrm{d}}).$$

Both are then extruded resulting in the natural injections $\iota_{\mathcal{P}}, \iota_{\mathcal{L}}$, from which the conditions for a projective geometry are immediately written down demanding that

$$\mathsf{LineForPointPair} := \iota_{\mathcal{P}}; (\mathcal{I} \otimes \mathcal{I}) \quad \text{and} \quad \mathsf{PointForLinePair} := \iota_{\mathcal{L}}; (\mathcal{I}^{\mathsf{T}} \otimes \mathcal{I}^{\mathsf{T}})$$

be mappings. In Fig. 17.10 (below), for example, the incidence on the left has been used in this way to compute the binary mapping given in two different representations.

The relation in Fig. 17.9 shows a second projective plane. It has a close connection with the complete quadrangle presented as an example in Fig. 3.5: one will find just one line added, namely line 7 connecting the diagonal points D_1, D_2, D_3, shown in the graph via node marking.

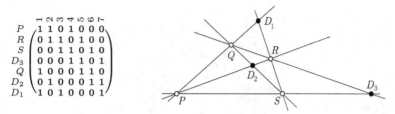

$$
\begin{array}{c}
\begin{array}{ccccccc} 1 & 2 & 3 & 4 & 5 & 6 & 7 \end{array}\\
\begin{array}{c} P \\ R \\ S \\ D_3 \\ Q \\ D_2 \\ D_1 \end{array}
\begin{pmatrix}
1 & 1 & 0 & 1 & 0 & 0 & 0\\
0 & 1 & 1 & 0 & 1 & 0 & 0\\
0 & 0 & 1 & 1 & 0 & 1 & 0\\
0 & 0 & 0 & 1 & 1 & 0 & 1\\
1 & 0 & 0 & 0 & 1 & 1 & 0\\
0 & 1 & 0 & 0 & 0 & 1 & 1\\
1 & 0 & 1 & 0 & 0 & 0 & 1
\end{pmatrix}
\end{array}
$$

Fig. 17.9 Another projective incidence plane

Mathematical folklore tells us that every projective plane has $m^2 + m + 1$ lines as well as points. Those for a given m are all isomorphic.[2] However, a projective plane does not exist for every m. A rather recent result, obtained only with massive computer help, says that it does not exist for $m = 10$, for example.

Given an $\mathcal{I}$ as in Fig. 17.10, the terms above evaluate to the binary mapping in one of the two representations shown there in the middle and on the right.

A hint is given of the possibility of using projective geometry techniques in cryptography. To this end assume the projective plane $\mathcal{I} : \mathcal{P} \longrightarrow \mathcal{L}$ from Fig. 17.10 and determine the mapping that assigns to any set of two different points their linking line as

    `decode := PointForLinePair.`

The cryptographic effect is now as follows. Assume words over the alphabet $\mathcal{P}$ are to be communicated. Coding means to produce via

    `encode := decode`$^\mathsf{T}$

some 2-element set over $\mathcal{L}$, with the corresponding lines crossing at the respective point. Then two words are sent over the communication line, perhaps even over two different communication channels. The recipient will then put these two words side by side and zip the binary mapping over the sequence of pairs so as to obtain back the original word. One will observe that the choice of the pair is to a certain extent arbitrary, so that additional care may be taken to approach an equal distribution of characters sent. (One may, however, also decide to encode an alphabet of 42 characters, i.e., every matrix entry with a different interpretation.)

---

[2] Although of different origin, the projective planes of Fig. 17.8 and Fig. 17.11 are, thus, isomorphic.

Incidence $\mathcal{I}$ (left):

	V_1	V_2	V_3	V_4	V_5	V_6	V_7
L_1	0	1	0	1	1	0	0
L_2	1	0	1	0	1	0	0
L_3	1	1	0	0	0	1	0
L_4	0	0	0	0	1	1	1
L_5	1	0	0	1	0	0	1
L_6	0	1	1	0	0	0	1
L_7	0	0	1	1	0	1	0

decode relation (middle):

	V_1	V_2	V_3	V_4	V_5	V_6	V_7
$(L_2,L_1)\to$	0	0	0	0	1	0	0
$(L_3,L_1)\to$	0	1	0	0	0	0	0
$(L_4,L_1)\to$	0	0	0	0	1	0	0
$(L_3,L_2)\to$	1	0	0	0	0	0	0
$(L_5,L_1)\to$	0	0	0	1	0	0	0
$(L_4,L_2)\to$	0	0	0	0	1	0	0
$(L_6,L_1)\to$	0	1	0	0	0	0	0
$(L_5,L_2)\to$	1	0	0	0	0	0	0
$(L_4,L_3)\to$	0	0	0	0	0	1	0
$(L_7,L_1)\to$	0	0	0	1	0	0	0
$(L_6,L_2)\to$	0	0	1	0	0	0	0
$(L_5,L_3)\to$	1	0	0	0	0	0	0
$(L_7,L_2)\to$	0	0	1	0	0	0	0
$(L_6,L_3)\to$	0	1	0	0	0	0	0
$(L_5,L_4)\to$	0	0	0	0	0	0	1
$(L_7,L_3)\to$	0	0	0	0	0	1	0
$(L_6,L_4)\to$	0	0	0	0	0	0	1
$(L_7,L_4)\to$	0	0	0	0	0	1	0
$(L_6,L_5)\to$	0	0	0	0	0	0	1
$(L_7,L_5)\to$	0	0	0	1	0	0	0
$(L_7,L_6)\to$	0	0	1	0	0	0	0

decode as symmetric table (right):

	L_1	L_2	L_3	L_4	L_5	L_6	L_7
L_1	–	5	2	5	4	2	4
L_2	5	–	1	5	1	3	3
L_3	2	1	–	6	1	2	6
L_4	5	5	6	–	7	7	6
L_5	4	1	1	7	–	7	4
L_6	2	3	2	7	7	–	3
L_7	4	3	6	6	4	3	–

to be communicated: $V_7V_3V_3V_1V_2V_6$ sent independently along two channels: $L_5L_6L_6L_5L_6L_7$ / $L_4L_2L_7L_2L_3L_4$ received: $V_7V_3V_3V_1V_2V_6$

Fig. 17.10 Projective cryptography

Figure 17.10 shows the original incidence $\mathcal{I}$ and derived therefrom the relation *decode* as a relation that is a mapping. On the right also another representation of *decode* is given as a symmetric table for a binary function – defined only when arguments are different.

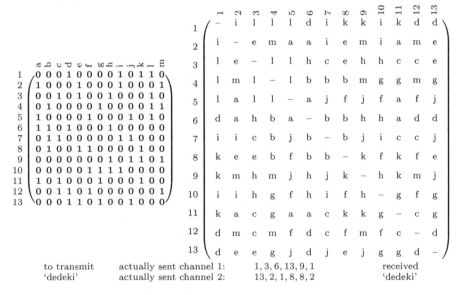

Incidence matrix:

	a	b	c	d	e	f	g	h	i	j	k	l	m
1	0	0	1	0	0	0	0	0	1	0	1	1	0
2	1	0	0	0	1	0	0	0	1	0	0	0	1
3	0	0	1	0	1	0	0	1	0	0	0	1	0
4	0	1	0	0	0	0	1	0	0	0	0	1	1
5	1	0	0	0	0	1	0	0	0	1	0	1	0
6	1	1	0	1	0	0	0	1	0	0	0	0	0
7	0	1	1	0	0	0	0	0	1	1	0	0	0
8	0	1	0	0	1	1	0	0	0	0	1	0	0
9	0	0	0	0	0	0	0	1	0	1	1	0	1
10	0	0	0	0	0	1	1	1	1	0	0	0	0
11	1	0	1	0	0	0	1	0	0	0	1	0	0
12	0	0	1	1	0	1	0	0	0	0	0	0	1
13	0	0	0	1	1	0	1	0	0	1	0	0	0

Decoding as binary table:

	1	2	3	4	5	6	7	8	9	10	11	12	13
1	–	i	l	l	l	d	i	k	k	i	k	d	d
2	i	–	e	m	a	a	i	e	m	i	a	m	e
3	l	e	–	l	l	h	c	e	h	h	c	c	e
4	l	m	l	–	l	b	b	b	m	g	g	m	g
5	l	a	l	l	–	a	j	f	j	f	a	f	j
6	d	a	h	b	a	–	b	b	h	h	a	d	d
7	i	i	c	b	j	b	–	b	j	i	c	c	j
8	k	e	e	b	f	b	b	–	k	f	k	f	e
9	k	m	h	m	j	h	j	k	–	h	k	m	j
10	i	i	h	g	f	h	i	f	h	–	g	f	g
11	k	a	c	g	a	a	c	k	k	g	–	c	g
12	d	m	c	m	f	d	c	f	m	f	c	–	d
13	d	e	e	g	j	d	j	e	j	g	g	d	–

to transmit 'dedeki'   actually sent channel 1: 1, 3, 6, 13, 9, 1   actually sent channel 2: 13, 2, 1, 8, 8, 2   received 'dedeki'

Fig. 17.11 Second example of projective cryptography: incidence and decoding as binary table

If in the example of Fig. 17.11 one decided on less redundancy, at least $78 = (169 - 13)/2$ characters could be encoded – more than sufficient to transmit the full 'Dedekind'.

Topology

Topology can be defined via open or closed sets, neighborhoods, etc. We show that at least the latter – in the form given to it by Felix Hausdorff[3] – is an inherently 'linear' configuration.

Definition 17.12. A set X endowed with a system $\mathcal{U}(p)$ of subsets for every $p \in X$ is called a **topological structure**, provided

(i) $p \in U$ for every neighborhood $U \in \mathcal{U}(p)$,
(ii) if $U \in \mathcal{U}(p)$ and $V \supseteq U$, then $V \in \mathcal{U}(p)$,
(iii) if $U_1, U_2 \in \mathcal{U}(p)$, then $U_1 \cap U_2 \in \mathcal{U}(p)$,
(iv) for every $U \in \mathcal{U}(p)$ there exists $V \in \mathcal{U}(p)$ so that $U \in \mathcal{U}(y)$ for all $y \in V$. $\qquad\square$

We can here give only a sketch of the idea. To this end, we express the same as (iv) with ε and $\mathcal{U}$ conceived as relations

$$\varepsilon : X \longrightarrow 2^X \quad \text{and} \quad \mathcal{U} : X \longrightarrow 2^X$$

and derive from this a relation-algebraic formula as in many former cases.

'For every $U \in \mathcal{U}(p)$ there exists a $V \in \mathcal{U}(p)$ such that $U \in \mathcal{U}(y)$ for all $y \in V$'

transition to slightly more formal notation

$\Longleftrightarrow \quad \forall p, U : U \in \mathcal{U}(p) \ \rightarrow \ \left[\exists V : V \in \mathcal{U}(p) \wedge \left\{\forall y : y \in V \rightarrow U \in \mathcal{U}(y)\right\}\right]$

interpreting "$\in$" as "ε"

$\Longleftrightarrow \quad \forall p, U : \mathcal{U}_{pU} \ \rightarrow \ \left[\exists V : \mathcal{U}_{pV} \wedge \left\{\forall y : \varepsilon_{yV} \rightarrow \mathcal{U}_{yU}\right\}\right]$

$\forall x : p(x) = \overline{\exists x : \overline{p(x)}}$

$\Longleftrightarrow \quad \forall p, U : \mathcal{U}_{pU} \ \rightarrow \ \left[\exists V : \mathcal{U}_{pV} \wedge \overline{\exists y : \varepsilon_{yV} \wedge \overline{\mathcal{U}_{yU}}}\right]$

composition and transposition

$\Longleftrightarrow \quad \forall p, U : \mathcal{U}_{pU} \ \rightarrow \ \left[\exists V : \mathcal{U}_{pV} \wedge \overline{\varepsilon^\mathsf{T}; \overline{\mathcal{U}}}_{VU}\right]$

definition of composition

$\Longleftrightarrow \quad \forall p, U : \mathcal{U}_{pU} \ \rightarrow \ \left(\mathcal{U}; \overline{\varepsilon^\mathsf{T}; \overline{\mathcal{U}}}\right)_{pU}$

transition to point-free form

$\Longleftrightarrow \quad \mathcal{U} \subseteq \mathcal{U}; \overline{\varepsilon^\mathsf{T}; \overline{\mathcal{U}}}$

---

[3] Felix Hausdorff (1868–1942), a famous mathematician, remained in Germany even though he was Jewish. He committed suicide in 1942, after having been imprisoned.

We may, thus, lift the definition of a topological structure to a point-free version.

Definition 17.13. Consider any relation $\mathcal{U} : X \longrightarrow 2^X$ between some set and its powerset. One will then automatically also have the membership relation, the powerset ordering $\varepsilon, \Omega : 2^X \longrightarrow 2^X$, and the powerset meet $\mathcal{M} : 2^X \times 2^X \longrightarrow 2^X$. The relation $\mathcal{U}$ will be called a **topological structure** if the following properties are satisfied.

(i) $\mathcal{U}$ is total and contained in the respective membership relation:
$$\mathcal{U}_{;} \mathbb{T} = \mathbb{T} \quad \text{and} \quad \mathcal{U} \subseteq \varepsilon.$$

(ii) $\mathcal{U}$ is an upper cone, i.e., $\mathcal{U}_{;} \Omega \subseteq \mathcal{U}$.

(iii) Finite meets of neighborhoods are neighborhoods again, i.e.,
$$(\mathcal{U} \otimes \mathcal{U})_{;} \mathcal{M} \subseteq \mathcal{U}.$$

(iv) $\mathcal{U} \subseteq \mathcal{U}_{;} \overline{\varepsilon^{\mathsf{T}}_{;} \overline{\mathcal{U}}}.$ □

In total: $\mathcal{U}_{;} \Omega = \mathcal{U} = \mathcal{U}_{;} \overline{\varepsilon^{\mathsf{T}}_{;} \overline{\mathcal{U}}}$. A lot of the basics of topology follows easily from this definition, not least the interconnection with the definition of a topological space via open sets.

Definition 17.14. A system $\mathcal{O}$ of subsets of a set is an **open set topology** if

(i) $\perp\!\!\!\perp, \mathbb{T} \in \mathcal{O}$,

(ii) arbitrary unions of elements of $\mathcal{O}$ belong to $\mathcal{O}$,

(iii) finite intersections of elements of $\mathcal{O}$ belong to $\mathcal{O}$.   □

It is relatively easy to lift this definition to the relation-algebraic level so as to obtain the following definition.

Definition 17.15. Given a set X with membership relation ε and powerset ordering $\Omega : 2^X \longrightarrow 2^X$, we call a vector $\mathcal{O}$ along 2^X an **open set topology** provided

(i) $\overline{\varepsilon^{\mathsf{T}}_{;} \mathbb{T}} \subseteq \mathcal{O} \qquad \overline{\overline{\varepsilon}^{\mathsf{T}}_{;} \mathbb{T}} \subseteq \mathcal{O}$,

(ii) $v \subseteq \mathcal{O} \qquad \Longrightarrow \qquad \mathrm{syq}(\varepsilon, \varepsilon_{;} v) \subseteq \mathcal{O}$,

(iii) $\mathcal{M}^{\mathsf{T}}_{;} (\mathcal{O} \otimes \mathcal{O}) \subseteq \mathcal{O}$.   □

It is possible (and well known) how to define neighborhoods via open sets and vice versa. This is also possible in a point-free relation-algebraic form. There is also a close interrelationship with Aumann contacts and what we have called the topography around a relation. Recalling the discussion of contact and closure in Section 11.4, there exist other interesting examples of topologies. We have, however, to confine ourselves here to this remark.

Farkas's Lemma

The following is a highly difficult result which is related to the primal/dual transition for linear programming tasks. Its relational formulation here is new. The closest correspondence to a version dealing with real-valued matrices and (possibly) non-negative vectors may be found as Theorem 1 in Chapter 7 of [51]. The theorem is more or less devoid of any assumptions, and may, thus, be considered extremely basic/general. Its origins date back as far as to [49].

Proposition 17.16 (A result resembling Farkas's Lemma for relations). *Let any relation* $A : X \longrightarrow Y$ *be given. In addition assume an arbitrary vector* $\mathbb{L} \neq v \subseteq Y$ *along the target side. Defining the union* $s := A^{\mathsf{T}} {}_; \mathbb{T}$ *of all rows, exactly one of the following two statements will hold:*

(i) $v \subseteq s$,
(ii) *there exists a vector* $r \subseteq Y$ *satisfying* $A {}_; r = \mathbb{L}$ *and* $v \cap r \neq \mathbb{L}$.

Proof: First, we deduce a contradiction from the assumption that both (i) and (ii) hold true:

$$s^{\mathsf{T}} {}_; r = \mathbb{T}^{\mathsf{T}} {}_; A {}_; r = \mathbb{L}$$

implies $v \subseteq s = s {}_; \mathbb{T} \subseteq \overline{r}$, or else $v \cap r = \mathbb{L}$, thus violating the assumption (ii).

Secondly, a contradiction is deduced from the assumption that both (i) and (ii) are *not* satisfied. Negation of (ii), slightly modified, means

$$\forall r \subseteq Y : A {}_; r = \mathbb{L} \quad \rightarrow \quad v \cap r = \mathbb{L}.$$

From $v \not\subseteq s$, we define the non-empty $r := v \cap \overline{s}$. Then $r = v \cap \overline{s} = v \cap \overline{A^{\mathsf{T}} {}_; \mathbb{T}} \subseteq \overline{A^{\mathsf{T}} {}_; \mathbb{T}}$, or equivalently $A^{\mathsf{T}} {}_; \mathbb{T} \subseteq \overline{r}$. This means $A {}_; r \subseteq \mathbb{L}$. The r so defined contradicts the requirement $v \cap r = \mathbb{L}$. $\qquad\square$

This was certainly a result in quite a different style compared with the others reported previously. When considering relations as matrices it is close to trivial, as may be seen in Fig. 17.12.

Putting $v := \mathbb{T}$, there is a certain relationship with a variant of the Tarski rule, for which Tarski in [138] gave the equivalent formulation that for all relations R

$$R_{;}\mathbb{T} = \mathbb{T} \qquad \text{or} \qquad \mathbb{T}_{;}\overline{R} = \mathbb{T}.$$

The left formula means that R is total, the right statement may stem from a row of $\mathbf{0}$s.

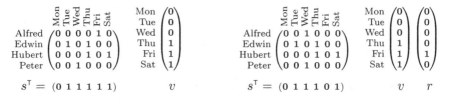

$$s^{\mathsf{T}} = (0\ 1\ 1\ 1\ 1\ 1) \qquad\qquad v \qquad\qquad s^{\mathsf{T}} = (0\ 1\ 1\ 1\ 0\ 1) \qquad\qquad v \qquad r$$

Fig. 17.12 The alternative with regard to the relational Farkas Lemma

Proposition 17.17 (A result resembling Farkas's Lemma for relations, generalized). *Let any relation $A : X \longrightarrow Y$ be given. In addition assume an arbitrary vector $u \subseteq X$ along the source side and an arbitrary vector $v \subseteq Y$ along the target side. Then exactly one of the following two statements will hold:*

(i) $v \subseteq A^{\mathsf{T}}_{;} u$,

(ii) *there exists a vector $w \not\supseteq v$ that satisfies $A^{\mathsf{T}}_{;} u \subseteq w$.*

Proof: First, we deduce a contradiction from the assumption that both (i) and (ii) hold true. Assuming such a w with $w \not\supseteq v$ in (ii), we have the contradiction $v \subseteq A^{\mathsf{T}}_{;} u \subseteq w$.

Secondly, a contradiction is deduced from the assumption that both (i) and (ii) are *not* satisfied. We start, thus, from

$$v \not\subseteq A^{\mathsf{T}}_{;} u$$
$$\forall w : A^{\mathsf{T}}_{;} u \subseteq w \quad \rightarrow \quad v \subseteq w$$

and choose $w := \overline{v} \cup A^{\mathsf{T}}_{;} u$. Then the statement left of the subjunction arrow is satisfied and $v \subseteq w$ has to hold true. This cannot be, since there exists an element x in v that does not reside in $A^{\mathsf{T}}_{;} u$ and, thus, is not in w. $\qquad\square$

When writing this down as

$$v \subseteq A^{\mathsf{T}}_{;} u \qquad \text{and} \qquad A_{;}\overline{w} \subseteq \overline{u},$$

it looks so astonishingly similar to the traditional Schröder equivalence, that one may suspect a deeper relationship. This would certainly be important for the area of linear programming, where the Farkas Lemma is presented in ever new formulations, adapted to the application side. Some texts include five or more versions of the Farkas Lemma.

18

Implication Structures

What is here called an *implication structure* is an abstraction of situations that occur frequently. One is given a set of conflicting alternatives and the task is to choose among them as many non-conflicting ones as possible. This is also called *attribute dependency*.

The criteria to observe include choices that forbid others to be chosen. If someone is planned to participate in an event, s/he cannot participate in a different event at the same time. Sometimes a choice, or a non-choice, implies that another choice is decided for. Situations thus described occur not least when constructing a timetable or solving a Sudoku. They also describe, however, how one tries to learn from a chunk of raw data in machine learning.

18.1 Attribute dependency

We assume an implication situation where we have a set of items that may *imply (enforce), forbid,* or *counter-imply (counter-enforce)* one another. The task is simply to select a subset such that all the given postulates are satisfied.

In order to model this, let a set N be given. We are looking for subsets $s \subseteq N$ satisfying whatever has been demanded as implication concerning two elements $i, k \in N$

$$s_i \rightarrow s_k, \qquad s_i \rightarrow \neg s_k, \qquad \neg s_i \rightarrow s_k.$$

Subsets s are here conceived as Boolean vectors $s \in 2^N$. Therefore, s_i is shorthand for $i \in s$. Enforcing, forbidding, and counter-enforcing are conceived as relations $E, F, C : N \longrightarrow N$.

An arbitrary subset $s \subseteq N$ may either satisfy the implicational requirements imposed by E, F, C, or may not. We are usually not interested in *all* solutions, in much the same way as *one* timetable satisfying formulated requirements will suffice for a school. For a theoretical investigation, nevertheless, we consider the set

$S \subseteq \mathbf{2}^N$ of all subsets fulfilling the given postulates, i.e., the possible solutions for the postulated set of implications. They satisfy, thus,

$$\forall s \in S: \quad s_i \rightarrow \quad s_k \qquad \text{if } (i,k) \in E, \tag{$*$}$$

$$\forall s \in S: \quad s_i \rightarrow \neg s_k \qquad \text{if } (i,k) \in F, \tag{$\dagger$}$$

$$\forall s \in S: \neg s_i \rightarrow \quad s_k \qquad \text{if } (i,k) \in C. \tag{$\ddagger$}$$

We aim at a relational formulation of these predicate-logic versions of the requirements. As already at several other occasions, we indicate loosely how the respective transition is justified.

$$\forall i \in N : \forall k \in N : E_{i,k} \rightarrow \left(\forall s \in S : s_i \rightarrow s_k \right)$$

$$a \rightarrow b = \neg a \vee b$$

$$\Longleftrightarrow \quad \forall i \in N : \forall k \in N : \overline{E_{i,k}} \vee \left(\forall s \in S : s_k \vee \overline{s_i} \right)$$

$$\forall k : p(k) = \neg \exists k : \neg p(k)$$

$$\Longleftrightarrow \quad \forall i \in N : \neg \Big\{ \exists k \in N : E_{i,k} \wedge \big[\exists s \in S : \overline{s_k} \wedge s_i \big] \Big\}$$

$$a \wedge \big[\exists s \in S : q(s) \big] = \exists s \in S : [a \wedge q(s)]$$

$$\Longleftrightarrow \quad \forall i \in N : \neg \Big\{ \exists k \in N : \big[\exists s \in S : E_{i,k} \wedge \overline{s_k} \wedge s_i \big] \Big\}$$

exchanging quantifiers

$$\Longleftrightarrow \quad \forall i \in N : \neg \Big\{ \exists s \in S : \big[\exists k \in N : E_{i,k} \wedge \overline{s_k} \wedge s_i \big] \Big\}$$

$$\exists k \in N : \big[q(k) \wedge a \big] = \big[\exists k \in N : q(k) \big] \wedge a$$

$$\Longleftrightarrow \quad \forall i \in N : \neg \Big\{ \exists s \in S : \big[\exists k \in N : E_{i,k} \wedge \overline{s_k} \big] \wedge s_i \Big\}$$

definition of composition

$$\Longleftrightarrow \quad \forall i \in N : \neg \big\{ \exists s \in S : (E \,\mathbin{;}\, \overline{s})_i \wedge s_i \big\}$$

$$\neg \exists s : p(s) = \forall s : \neg p(s)$$

$$\Longleftrightarrow \quad \forall i \in N : \forall s \in S : \overline{E \,\mathbin{;}\, \overline{s}_i} \vee \overline{s_i}$$

$$\neg a \vee b = a \rightarrow b$$

$$\Longleftrightarrow \quad \forall i \in N : \forall s \in S : (E \,\mathbin{;}\, \overline{s})_i \rightarrow \overline{s_i}$$

exchanging quantifiers

$$\Longleftrightarrow \quad \forall s \in S : \forall i \in N : (E \,\mathbin{;}\, \overline{s})_i \rightarrow \overline{s_i}$$

transition to point-free version

$$\Longleftrightarrow \quad \forall s \in S : E \,\mathbin{;}\, \overline{s} \subseteq \overline{s}$$

A Galois connection

Assume for the moment three relations E, F, C to be given arbitrarily. Our aim is to conceive the relations as some implication structure and to look for the underlying set of solutions. To this end, we define

$$(E,F,C) \mapsto \sigma(E,F,C) := \big\{ v \mid E \,\mathbin{;}\, \overline{v} \subseteq \overline{v}, \quad F \,\mathbin{;}\, v \subseteq \overline{v}, \quad C \,\mathbin{;}\, \overline{v} \subseteq v \big\}$$

as the transition to the set of solutions of the triple E, F, C. The versions for F, C are deduced in quite a similar way. In Fig. 18.1, such a situation is shown: three arbitrary relations E, F, C for which all solutions $s \in \mathbf{2}^N$ have been determined and for simplicity been collected as columns into one relation.

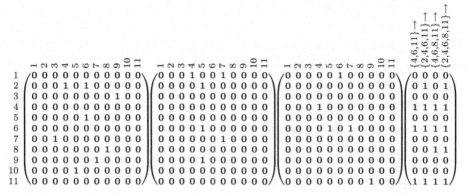

Fig. 18.1 Prescribed enforcing E, forbidding F, counter-enforcing C with all solutions

This was a step from a relation triple E, F, C to a set of solutions. We may, however, also start with any set $S \subseteq \mathbf{2}^N$ of subsets of N and ask whether it is a solution of some triple of implication relations. Then we define the transition to the triple of implication relations of S

$$S \mapsto \pi(S) := \left(\inf{}_{s \in S}\{\overline{s_i\, \overline{s}^{\mathsf{T}}}\},\ \inf{}_{s \in S}\{\overline{s_i\, s^{\mathsf{T}}}\},\ \inf{}_{s \in S}\{\overline{\overline{s}_i\, \overline{s}^{\mathsf{T}}}\} \right).$$

This will in particular result in $\pi(\emptyset) = (\mathbb{T}, \mathbb{T}, \mathbb{T})$. Altogether, this gives a situation we are already acquainted with, namely – according to the following proposition – a Galois correspondence.

Proposition 18.1. *The functionals σ, π just defined form a Galois correspondence from subsets $S \subseteq \mathbf{2}^N$ to relation triples E, F, C on N.*

Proof: We have to show that for the three components of the triple

$$\left. \begin{array}{l} E \subseteq \pi(S)_1 \text{ and} \\ F \subseteq \pi(S)_2 \text{ and} \\ C \subseteq \pi(S)_3 \end{array} \right\} \quad \Longleftrightarrow \quad S \subseteq \sigma(E, F, C).$$

We start from $E \subseteq \pi(S)_1 = \inf{}_{s \in S} \overline{s_i\, \overline{s}^{\mathsf{T}}}$, which implies that we have $E \subseteq \overline{s_i\, \overline{s}^{\mathsf{T}}}$ for all $s \in S$. Negating results in $s_i\, \overline{s}^{\mathsf{T}} \subseteq \overline{E}$. Using Schröder's rule, we get $E_i\, \overline{s} \subseteq \overline{s}$ for all s and, thus, the first condition in forming $\sigma(E, F, C)$. The other two cases are handled in a largely analogous way.

Now, we work in the reverse direction, assuming

$$S \subseteq \sigma(E, F, C) = \{ s \mid E_i\, \overline{s} \subseteq \overline{s},\quad F_i\, s \subseteq \overline{s},\quad C_i\, \overline{s} \subseteq s \}.$$

This means that we have $E_i\, \overline{s} \subseteq \overline{s}$, $F_i\, s \subseteq \overline{s}$, $C_i\, \overline{s} \subseteq s$ for every $s \in S$. The negations and the Schröder steps taken before were equivalences, and may, thus, be reversed. This means that for all $s \in S$ we have $E \subseteq \overline{s_{;}\overline{s^{\mathsf{T}}}}$, $F \subseteq \overline{\overline{s}_{;}s^{\mathsf{T}}}$, $C \subseteq \overline{\overline{s}_{;}\overline{s^{\mathsf{T}}}}$. In this way, we see that E, F, C stay below the infima. □

Given the Galois correspondence, it is straightforward to prove all the results that follow simply by Galois folklore. In particular we study

$$\varphi(S) := \sigma(\pi(S)) \qquad \text{and} \qquad \rho(E, F, C) := \pi(\sigma(E, F, C))$$

with properties as follows:

- σ, π are antitone mappings,
- ρ and φ are expanding, i.e.,
  $E \subseteq \pi_1(\sigma(E, F, C))$ $F \subseteq \pi_2(\sigma(E, F, C))$ $C \subseteq \pi_3(\sigma(E, F, C))$ for all E, F, C
  $S \subseteq \sigma(\pi(S))$ for all S,
- ρ and φ are idempotent, i.e.,
  $\rho(E, F, C) = \rho(\rho(E, F, C))$ for all E, F, C
  $\varphi(S) = \varphi(\varphi(S))$ for all S,
- ρ, φ are monotonic and, thus, closure operations,
- there exist fixed points for ρ, φ,
- the fixed point sets with respect to ρ, φ are mapped antitonely onto one another.

Proofs of these results may not least be found in [112, 118, 120, 135].

In Fig. 18.2, we are going to start from a set of vectors (it is in fact the solution $S \subseteq \mathbf{2}^N$ as obtained in Fig. 18.1).

$$\begin{array}{r} \\ 1\\2\\3\\4\\5\\6\\7\\8\\9\\10\\11 \end{array}
\begin{array}{c} \begin{array}{cccc} A&B&C&D \end{array} \\ \begin{pmatrix} 0&0&0&0\\0&1&0&1\\0&0&0&0\\1&1&1&1\\0&0&0&0\\1&1&1&1\\0&0&0&0\\0&0&1&1\\0&0&0&0\\0&0&0&0\\1&1&1&1 \end{pmatrix} \end{array}
\begin{array}{c} \begin{array}{ccccccccccc} 1&2&3&4&5&6&7&8&9&10&11 \end{array} \\ \begin{pmatrix} 1&1&1&1&1&1&1&1&1&1&1\\0&1&0&1&0&1&0&0&0&0&1\\1&1&1&1&1&1&1&1&1&1&1\\0&0&0&1&0&1&0&0&0&0&1\\1&1&1&1&1&1&1&1&1&1&1\\0&0&0&1&0&1&0&0&0&0&1\\1&1&1&1&1&1&1&1&1&1&1\\0&0&0&1&0&1&0&1&0&0&1\\1&1&1&1&1&1&1&1&1&1&1\\1&1&1&1&1&1&1&1&1&1&1\\0&0&0&1&0&1&0&0&0&0&1 \end{pmatrix} \end{array}
\begin{array}{c} \begin{array}{ccccccccccc} 1&2&3&4&5&6&7&8&9&10&11 \end{array} \\ \begin{pmatrix} 1&1&1&1&1&1&1&1&1&1&1\\1&0&1&0&1&0&1&0&1&1&0\\1&1&1&1&1&1&1&1&1&1&1\\1&0&1&0&1&0&1&0&1&1&0\\1&1&1&1&1&1&1&1&1&1&1\\1&0&1&0&1&0&1&0&1&1&0\\1&1&1&1&1&1&1&1&1&1&1\\1&0&1&0&1&0&1&0&1&1&0\\1&1&1&1&1&1&1&1&1&1&1\\1&1&1&1&1&1&1&1&1&1&1\\1&0&1&0&1&0&1&0&1&1&0 \end{pmatrix} \end{array}
\begin{array}{c} \begin{array}{ccccccccccc} 1&2&3&4&5&6&7&8&9&10&11 \end{array} \\ \begin{pmatrix} 0&0&0&1&0&1&0&0&0&0&1\\0&0&0&1&0&1&0&0&0&0&1\\0&0&0&1&0&1&0&0&0&0&1\\1&1&1&1&1&1&1&1&1&1&1\\0&0&0&1&0&1&0&0&0&0&1\\1&1&1&1&1&1&1&1&1&1&1\\0&0&0&1&0&1&0&0&0&0&1\\0&0&0&1&0&1&0&0&0&0&1\\0&0&0&1&0&1&0&0&0&0&1\\0&0&0&1&0&1&0&0&0&0&1\\1&1&1&1&1&1&1&1&1&1&1 \end{pmatrix} \end{array}$$

Fig. 18.2 A set S of vectors with implication triple $\pi(S)$ obtained from them

Fixed point closure for an implication structure

In order to have a clear distinction, we decide to denote the fixed points of closure forming with calligraphic letters. We need not stress the status of closure forming as an operation, because we will always start with the same E, F, C so that we have

$$E \subseteq \mathcal{E} \quad F \subseteq \mathcal{F} \quad C \subseteq \mathcal{C} \qquad\qquad\qquad S \subseteq \mathcal{S},$$
$$S := \sigma(E, F, C) \qquad\qquad\qquad\qquad (\mathcal{E}, \mathcal{F}, \mathcal{C}) := \pi(S),$$
$$(\mathcal{E}, \mathcal{F}, \mathcal{C}) = \rho(\mathcal{E}, \mathcal{F}, \mathcal{C}) = \rho(\rho(\mathcal{E}, \mathcal{F}, \mathcal{C})) \qquad \mathcal{S} = \varphi(\mathcal{S}) = \varphi(\varphi(\mathcal{S})).$$

It should be noted that $E \subseteq \mathcal{E}$ means containment for relations while $S \subseteq \mathcal{S}$ denotes a set of vectors contained in another vector set. In no way does this mean that we are able to execute these computations efficiently – although the tasks are finite. The calligraphic versions are, thus, just theoretically existing relations together with their properties. With tiny examples, however, we are in a position to show everything explicitly. In Fig. 18.2, for example, we find the relation triple $\mathcal{E}, \mathcal{F}, \mathcal{C}$ for the three relations E, F, C of Fig. 18.1, since there we started from the solution set of the latter.

In what follows, we investigate the much more stringent properties that such fixed points $(\mathcal{E}, \mathcal{F}, \mathcal{C})$ of an implication structure obviously have. We restrict ourselves to the finite case and, thus, avoid difficult continuity considerations.

Proposition 18.2. *All the fixed points of the Galois correspondence satisfy*

  (i) $\mathcal{F} = \mathcal{F}^\mathsf{T}, \quad \mathcal{C} = \mathcal{C}^\mathsf{T},$
  (ii) $\mathbb{I} \subseteq \mathcal{E} = \mathcal{E}^2,$
  (iii) $\mathcal{E} \,{}_\mathrm{;}\, \mathcal{F} = \mathcal{F}, \quad \mathcal{C} \,{}_\mathrm{;}\, \mathcal{E} = \mathcal{C},$
  (iv) $\mathcal{F} \,{}_\mathrm{;}\, \mathcal{C} \subseteq \mathcal{E}.$

Proof: We see immediately that the second as well as the third components of $\pi(S)$ are symmetric by definition. $\mathbb{I} \subseteq \mathcal{E}$ since obviously $\mathbb{I} \subseteq \overline{s \,{}_\mathrm{;}\, s^\mathsf{T}}$ for all s.

Transitivity of $\mathcal{E}$ follows since $\overline{s \,{}_\mathrm{;}\, s^\mathsf{T}} \,{}_\mathrm{;}\, \overline{s \,{}_\mathrm{;}\, s^\mathsf{T}} \subseteq \overline{s \,{}_\mathrm{;}\, s^\mathsf{T}}$ and transitivity is '$\cap$'-hereditary. Similarly, for $\mathcal{F}$ and $\mathcal{C}$ in (iii) and for (iv). □

The enforcing relation $\mathcal{E}$ of a fixed point is, thus, a preorder. While these properties concerned non-negated implication relations, there are others including also negated relations.

Corollary 18.3. *The fixed points satisfy in addition*

  (i) $\mathcal{E} \,{}_\mathrm{;}\, \overline{\mathcal{C}} = \overline{\mathcal{C}}, \quad \overline{\mathcal{F}} \,{}_\mathrm{;}\, \mathcal{E} = \overline{\mathcal{F}},$
  (ii) $\mathcal{C} \,{}_\mathrm{;}\, \overline{\mathcal{C}} \subseteq \overline{\mathcal{E}}, \quad \overline{\mathcal{E}} \,{}_\mathrm{;}\, \mathcal{C} \subseteq \overline{\mathcal{F}},$
  (iii) $\overline{\mathcal{F}} \,{}_\mathrm{;}\, \mathcal{F} \subseteq \overline{\mathcal{E}}, \quad \mathcal{F} \,{}_\mathrm{;}\, \overline{\mathcal{E}} \subseteq \overline{\mathcal{C}}.$

Proof: In (i), "⊇" is obvious since $\mathcal{E}$ is reflexive. "⊆" follows with the Schröder rule from $\mathcal{C} \,;\, \mathcal{E} = \mathcal{C}$; the others in a similar way. □

Considered from a computational point of view, the formulae of Cor. 18.3 do not give any additional information, because they all follow from the earlier ones. Another example illustrates these results further, showing the closure forming on the S-side.

Example 18.4. In Fig. 18.3, we start with an arbitrary set S of vectors. Then we determine the relation triple $(\mathcal{E}, \mathcal{F}, \mathcal{C}) := \pi(S)$, which are already fixed points. Finally, the set $\mathcal{S} := \varphi(S)$ of all vectors satisfying $\mathcal{E}, \mathcal{F}, \mathcal{C}$ is determined, which, as its closure, contains S. Here, therefore, every column of S again shows up as a column of $\mathcal{S}$.

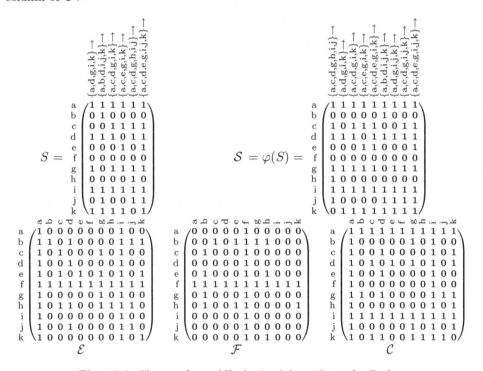

Fig. 18.3 Closure $\mathcal{S} = \varphi(S)$ obtained from S via $\mathcal{E}, \mathcal{F}, \mathcal{C}$

Computing implication closures

With the theory presented so far, the logical mechanism of implication structures has been described. Implication structures often arise in practice and can be defined in several ways. The practical tasks typically start with requirements E, F, C that are formulated in an application specific way, or given more or less explicitly in predicate-logic form. In any case, the task is to look for – at least one – solution vector s of the closure $\mathcal{S} = \sigma(E, F, C)$ in order to reflect the application dependent requirements.

As this is now settled, we consider implication structures from a computational point of view. While it seems possible to handle three $n \times n$-matrices E, F, C with regard to composition or transitive closure forming for really big numbers n in a relation-algebraic way with RELVIEW (see [18] for example), it will soon become extremely difficult to determine vectors of length n satisfying certain given implications. While we do not have enough mathematical structure on the S-side, the formulae proved earlier for the (E, F, C)-side may be helpful. To find a solution s is normally $\mathcal{NP}$-complete, as for the timetable problem or for satisfiability problems.

One, therefore, tries to compute bigger and bigger approximations

$$E \subseteq E_{\text{approx}} \subseteq \mathcal{E} \qquad F \subseteq F_{\text{approx}} \subseteq \mathcal{F} \qquad C \subseteq C_{\text{approx}} \subseteq \mathcal{C}$$

for the closure with the hope that these bigger relations will make it easier to find one of the solutions s. One should bear in mind that this means working largely heuristically. In the case that there exist very long chains of implications, the approximation of the closure may well make it easier to determine a solution s by applying the following concepts.

Considering Prop. 18.2, one will easily suggest starting with E, F, C and applying the following steps round-robinwise until a stable situation is reached:

- determine the reflexive-transitive closure of E,

- determine the symmetric closure of F and C,

- expand F to $E_; F$ and C to $C_; E$,

- add $F_; C$ to E.

We provide a fully elaborated tiny example with Fig. 18.4.

$$E = \begin{array}{c|ccccc} & 1&2&3&4&5\\ \hline 1&1&0&0&0&0\\ 2&0&0&1&0&0\\ 3&0&0&0&1&0\\ 4&0&0&0&0&0\\ 5&1&0&0&0&0 \end{array} \qquad F = \begin{array}{c|ccccc} & 1&2&3&4&5\\ \hline 1&0&0&0&0&0\\ 2&1&0&0&0&0\\ 3&0&1&0&0&0\\ 4&0&0&1&0&0\\ 5&0&0&0&0&0 \end{array} \qquad C = \begin{array}{c|ccccc} & 1&2&3&4&5\\ \hline 1&0&0&0&0&0\\ 2&0&0&0&0&0\\ 3&0&0&0&0&0\\ 4&0&0&0&0&0\\ 5&0&0&0&0&1 \end{array}$$

$$E_{\text{approx}} = \begin{array}{c|ccccc} & 1&2&3&4&5\\ \hline 1&1&0&0&0&0\\ 2&1&1&1&1&1\\ 3&0&0&1&1&0\\ 4&0&0&0&1&0\\ 5&1&0&0&0&1 \end{array} \qquad F_{\text{approx}} = \begin{array}{c|ccccc} & 1&2&3&4&5\\ \hline 1&0&1&0&0&0\\ 2&1&1&1&1&1\\ 3&0&1&1&1&0\\ 4&0&1&1&0&0\\ 5&0&1&0&0&0 \end{array} \qquad C_{\text{approx}} = \begin{array}{c|ccccc} & 1&2&3&4&5\\ \hline 1&1&0&0&0&1\\ 2&0&0&0&0&0\\ 3&0&0&0&0&0\\ 4&0&0&0&0&0\\ 5&1&0&0&0&1 \end{array}$$

$$\mathcal{E} = \begin{array}{c|ccccc} & 1&2&3&4&5\\ \hline 1&1&0&0&0&1\\ 2&1&1&1&1&1\\ 3&1&1&1&1&1\\ 4&1&0&0&1&1\\ 5&1&0&0&0&1 \end{array} \qquad \mathcal{F} = \begin{array}{c|ccccc} & 1&2&3&4&5\\ \hline 1&0&1&1&0&0\\ 2&1&1&1&1&1\\ 3&1&1&1&1&1\\ 4&0&1&1&0&0\\ 5&0&1&1&0&0 \end{array} \qquad \mathcal{C} = \begin{array}{c|ccccc} & 1&2&3&4&5\\ \hline 1&1&1&1&1&1\\ 2&1&0&0&0&1\\ 3&1&0&0&0&1\\ 4&1&0&0&0&1\\ 5&1&1&1&1&1 \end{array}$$

$$\mathcal{E}_{\text{perm}} = \begin{array}{c|ccccc} & 2&3&4&1&5\\ \hline 2&1&1&1&1&1\\ 3&1&1&1&1&1\\ 4&0&0&1&1&1\\ 1&0&0&0&1&1\\ 5&0&0&0&1&1 \end{array} \qquad \mathcal{F}_{\text{perm}} = \begin{array}{c|ccccc} & 2&3&4&1&5\\ \hline 2&1&1&1&1&1\\ 3&1&1&1&1&1\\ 4&1&1&0&0&0\\ 1&1&1&0&0&0\\ 5&1&1&0&0&0 \end{array} \qquad \mathcal{C}_{\text{perm}} = \begin{array}{c|ccccc} & 2&3&4&1&5\\ \hline 2&0&0&0&1&1\\ 3&0&0&0&1&1\\ 4&0&0&0&1&1\\ 1&1&1&1&1&1\\ 5&1&1&1&1&1 \end{array}$$

$$\mathcal{S} = \begin{array}{c|cc} & \{1,5\}\!\rightarrow & \{1,4,5\}\!\rightarrow\\ \hline 1&1&1\\ 2&0&0\\ 3&0&0\\ 4&0&1\\ 5&1&1 \end{array} \qquad \mathcal{S}_{\text{perm}} = \begin{array}{c|cc} & \{1,5\}\!\rightarrow & \{1,4,5\}\!\rightarrow\\ \hline 2&0&0\\ 3&0&0\\ 4&0&1\\ 1&1&1\\ 5&1&1 \end{array}$$

Fig. 18.4 Approximating solutions of implication relations

One will observe that the approximations satisfy all the requirements of Prop. 18.2, but are not yet closures.

We provide yet another example and show the approximations for the E, F, C of Fig. 18.1. The relations are filled considerably but are far from reaching the closure already shown in Fig. 18.2.

$$\begin{array}{c|ccccccccccc} & 1&2&3&4&5&6&7&8&9&10&11\\ \hline 1&1&0&0&1&0&1&0&0&0&0&1\\ 2&0&1&0&1&0&1&0&0&0&0&0\\ 3&0&0&1&0&0&1&1&0&1&0&1\\ 4&0&0&0&1&0&1&0&0&0&0&0\\ 5&1&0&1&1&1&1&1&0&1&0&1\\ 6&0&0&0&0&0&1&0&0&0&0&0\\ 7&0&0&1&0&0&1&1&0&1&0&1\\ 8&0&0&0&0&0&0&0&1&0&0&0\\ 9&0&0&1&0&0&1&1&0&1&0&1\\ 10&1&0&1&1&1&1&1&0&1&1&1\\ 11&0&0&0&0&0&0&0&0&0&0&1 \end{array} \quad \begin{pmatrix} 1&1&1&1&1&0&1&0&1&1&0\\ 1&0&0&0&1&0&0&0&0&1&0\\ 1&0&1&0&1&0&1&0&1&1&0\\ 1&0&0&0&1&0&0&0&0&1&0\\ 1&1&1&1&1&1&1&0&1&1&0\\ 0&0&0&0&1&0&0&0&0&1&0\\ 1&0&1&0&1&0&1&0&1&1&0\\ 0&0&0&0&0&0&0&0&0&0&0\\ 1&0&1&0&1&0&1&0&1&1&0\\ 1&1&1&1&1&1&1&0&1&1&0\\ 0&0&0&0&0&0&0&0&0&0&0 \end{pmatrix} \quad \begin{pmatrix} 0&0&0&0&0&1&0&0&0&0&0\\ 0&0&0&0&0&0&0&0&0&0&0\\ 0&0&0&0&0&1&0&0&0&0&1\\ 0&0&0&1&0&1&0&0&0&0&0\\ 0&0&0&0&0&1&0&0&0&0&0\\ 1&0&1&1&1&1&1&0&1&0&1\\ 0&0&0&0&0&1&0&0&0&0&1\\ 0&0&0&0&0&0&0&0&0&0&0\\ 0&0&0&0&0&1&0&0&0&0&1\\ 0&0&0&0&0&0&0&0&0&0&0\\ 0&0&1&0&0&1&1&0&1&0&1 \end{pmatrix}$$

Fig. 18.5 Approximations E_{approx}, F_{approx}, and C_{approx} for Fig. 18.1

Tight and pseudo elements

In addition to the computational attempts discussed so far, another interesting observation can be made.

Proposition 18.5. *Let any three relations* $E, F, C : N \longrightarrow N$ *be given and consider an arbitrary element* $i \in N$.

  (i) $F(i, i) = 1 \implies i \notin s$ *for any solution s.*
  (ii) $C(i, i) = 1 \implies i \in s$ *for any solution s.*
 (iii) *If there exists an* i *with* $F(i, i) = 1 = C(i, i)$, *then the solution set is empty.*

Proof: (i,ii) We instantiate $(\dagger, \ddagger)$ of page 462 for $k := i$:

$$\forall s \in S : \quad s_i \to \neg s_i \quad \text{if } (i, i) \in F, \quad \text{meaning} \quad \forall s \in S : \neg s_i,$$
$$\forall s \in S : \neg s_i \to \quad s_i \quad \text{if } (i, i) \in C, \quad \text{meaning} \quad \forall s \in S : s_i.$$

(iii) This is simply a consequence of two non-satisfiable conditions. $\qquad\square$

Given this setting, the definition of implication relations leads us to call $i \in N$ with respect to E, F, C

- an (E, F, C)-**pseudo** element if $F(i, i) = 1$,
- an (E, F, C)-**tight** element if $C(i, i) = 1$,
- an (E, F, C)-**flexible** element otherwise.

So, for every 1 in the diagonal of C, the corresponding element *must* and for every 1 in the diagonal of F, it *cannot* belong to *any* solution $s \in \mathcal{S}$. These facts *together with all their implications* according to E, F, C may also be helpful in looking for solutions S.

This raises the question as to the interrelationship of being (E, F, C)-tight, (E, F, C)-pseudo, or (E, F, C)-flexible and of being $(\mathcal{E}, \mathcal{F}, \mathcal{C})$-tight, $(\mathcal{E}, \mathcal{F}, \mathcal{C})$-pseudo, or $(\mathcal{E}, \mathcal{F}, \mathcal{C})$-flexible. For the latter, we will always omit the qualification $(\mathcal{E}, \mathcal{F}, \mathcal{C})$ and simply speak of being **tight**, **pseudo**, and **flexible**.

A first rather trivial statement says that

$$\begin{aligned}
(E, F, C)\text{-tight} &\implies \text{tight,} \\
(E, F, C)\text{-pseudo} &\implies \text{pseudo,} \quad \text{but} \\
(E, F, C)\text{-flexible} &\impliedby \text{flexible.}
\end{aligned}$$

Proposition 18.6. *For the closure* $\mathcal{E}, \mathcal{F}, \mathcal{C}$ *of an implication structure* E, F, C *together with the set* $\mathcal{S} \subseteq 2^N$ *of solution subsets, the following holds.*

(i) *The vector* $v := (\mathcal{F} \cap \mathbb{I}); \mathbb{T}$ *satisfies*

$$v = \inf\nolimits_{s \in S} \overline{s} \qquad\qquad v; \mathbb{T} \subseteq \mathcal{E} \qquad\qquad v_i = 1 \implies i \in N \text{ is pseudo.}$$

(ii) *The vector* $w := (\mathcal{C} \cap \mathbb{I}); \mathbb{T}$ *satisfies*

$$w = \inf\nolimits_{s \in S} s \qquad\qquad \mathbb{T}; w^\mathsf{T} \subseteq \mathcal{E} \qquad\qquad w_i = 1 \implies i \in N \text{ is tight.}$$

Proof: We restrict to (ii) and start the proof, which is unexpectedly difficult on the technical side, with two principally trivial steps that are completely obvious when looking at the column representation of a vector $s \subseteq N$. Then

$$\overline{s; s^\mathsf{T}} \cap \mathbb{I}_{N,N} = \overline{s; \mathbb{T}_N^\mathsf{T}} \cap \mathbb{I}_{N,N} = \overline{s}; \mathbb{T}_N^\mathsf{T} \cap \mathbb{I}_{N,N},$$
$$\overline{\overline{s}; \overline{s}^\mathsf{T}} \cap \mathbb{I}_{N,N} = \overline{\overline{s}; \mathbb{T}_N^\mathsf{T}} \cap \mathbb{I}_{N,N} = s; \mathbb{T}_N^\mathsf{T} \cap \mathbb{I}_{N,N}.$$

For the first equations above, the direction "$\supseteq$" is trivial; furthermore

$$\overline{s; s^\mathsf{T}} \cap \mathbb{I}_{N,N} \subseteq \overline{s; \mathbb{T}_N^\mathsf{T}} \quad\Longleftrightarrow\quad s; \mathbb{T}_N^\mathsf{T} \subseteq s; s^\mathsf{T} \cup \overline{\mathbb{I}}$$

where the latter follows from $s; \mathbb{T}_N^\mathsf{T} = s; (s \cup \overline{s})^\mathsf{T} = s; s^\mathsf{T} \cup s; \overline{s}^\mathsf{T} \subseteq s; s^\mathsf{T} \cup \overline{\mathbb{I}}$. The second line above means the same, but formulated for $\overline{s}$ instead of s.

Using this:

$$\begin{aligned}
w := (\mathcal{C} \cap \mathbb{I}); \mathbb{T} &= \big(\pi_3(\mathcal{S}) \cap \mathbb{I} \big); \mathbb{T} \\
&= \big(\inf \{ \overline{\overline{s}; \overline{s}^\mathsf{T}} \mid s \in \mathcal{S} \} \cap \mathbb{I} \big); \mathbb{T} \\
&= \big(\inf \{ \overline{\overline{s}; \overline{s}^\mathsf{T}} \cap \mathbb{I} \mid s \in \mathcal{S} \} \big); \mathbb{T} \\
&= \big(\inf \{ s; \mathbb{T}_N^\mathsf{T} \cap \mathbb{I} \mid s \in \mathcal{S} \} \big); \mathbb{T} \\
&= \inf \{ s \mid s \in \mathcal{S} \} \qquad\qquad\qquad\qquad \square
\end{aligned}$$

In addition to the former approximation operations, one should, therefore, also increment as a consequence of tightness and pseudo availability adding the corresponding rows and columns as follows

$$\begin{aligned}
E &\mapsto E \cup (F \cap \mathbb{I}); \mathbb{T} \cup \mathbb{T}; (C \cap \mathbb{I}), \\
F &\mapsto F \cup (F \cap \mathbb{I}); \mathbb{T} \cup \mathbb{T}; (F \cap \mathbb{I}), \\
C &\mapsto C \cup (C \cap \mathbb{I}); \mathbb{T} \cup \mathbb{T}; (C \cap \mathbb{I}).
\end{aligned}$$

The justification follows directly from $(*, \dagger, \ddagger)$ mentioned initially on page 462.

Rearranging implication structures

Closed implication matrices are only interesting modulo the equivalence $\mathcal{E} \cap \mathcal{E}^\mathsf{T}$ derived from the preorder $\mathcal{E}$. It is an easy consequence that by simultaneous permutation of rows and columns every triple of implication matrices can be arranged in the following standard form of row and column groups. An example may be found in the illustrating closure matrices of Fig. 18.2, which are here given schematically in reordered form:

$$\mathcal{E} = \begin{pmatrix} \top & \top & \top \\ \bot & E_0 & \top \\ \bot & \bot & \top \end{pmatrix}, \qquad \mathcal{F} = \begin{pmatrix} \top & \top & \top \\ \top & F_0 & \bot \\ \top & \bot & \bot \end{pmatrix}, \qquad \mathcal{C} = \begin{pmatrix} \bot & \bot & \top \\ \bot & C_0 & \top \\ \top & \top & \top \end{pmatrix}.$$

All rows corresponding to pseudo elements are positioned as the first group of rows, followed by the group of rows corresponding to flexible elements, and those for tight elements. E_0, F_0, C_0 satisfy some additional rules. This will now be demonstrated with an example.

```
 (1)  (1)  (1)  (1)        1 2 3 4 5 6 7        1 2 3 4 5 6 7        1 2 3 4 5 6 7
 (0)  (0)  (0)  (0)     1 /1 0 0 0 0 0 1\    1 /0 1 0 0 0 1 0\    1 /1 1 1 1 1 1 1\
 (1)  (0)  (1)  (0)     2 |1 1 1 1 1 1 1|    2 |1 1 1 1 1 1 1|    2 |1 0 0 0 0 0 1|
 (0)  (1)  (0)  (1)     3 |1 0 1 0 0 0 1|    3 |0 1 0 1 0 1 0|    3 |1 0 0 0 1 0 1|
 (0)  (1)  (1)  (1)     4 |1 0 0 1 1 0 1|    4 |0 1 1 0 1 0 0|    4 |1 0 0 0 0 0 1|
 (0)  (0)  (0)  (0)     5 |1 0 0 0 1 0 1|    5 |0 1 0 0 0 1 0|    5 |1 0 1 0 0 0 1|
 (1)  (1)  (1)  (1)     6 |1 1 1 1 1 1 1|    6 |1 1 1 1 1 1 1|    6 |1 0 0 0 0 0 1|
                        7 \1 0 0 0 0 0 1/    7 \0 1 0 0 0 1 0/    7 \1 1 1 1 1 1 1/
```

Fig. 18.6 Solution set S and implication relations $\mathcal{E}, \mathcal{F}, \mathcal{C}$

These solutions together with the resulting relations E, F, C are now arranged so as to get an easier overview.

```
(0) (0) (0) (0)  pseudo        2 6 3 4 5 1 7        2 6 3 4 5 1 7        2 6 3 4 5 1 7
(0) (0) (0) (0)            2 /1 1|1 1 1|1 1\    2 /1 1|1 1 1|1 1\    2 /0 0|0 0 0|1 1\
(1) (0) (1) (0)            6 |1 1|1 1 1|1 1|    6 |1 1|1 1 1|1 1|    6 |0 0|0 0 0|1 1|
(0) (0) (0) (1) flexible    3 |0 0|1 0 0|1 1|    3 |1 1|0 1 0|0 0|    3 |0 0|1 0 0|1 1|
(0) (1) (1) (1)            4 |0 0|0 1 1|1 1|    4 |1 1|1 0 0|0 0|    4 |0 0|0 0 0|1 1|
(1) (1) (1) (1)            5 |0 0|0 0 1|1 1|    5 |1 1|0 0 0|0 0|    5 |0 0|1 0 0|1 1|
(1) (1) (1) (1)  tight      1 |0 0|0 0 0|1 1|    1 |1 1|0 0 0|0 0|    1 \1 1|1 1 1|1 1|
                            7 \0 0|0 0 0|1 1/    7 \1 1|0 0 0|0 0/    7 \1 1|1 1 1|1 1/
```

Fig. 18.7 Matrices of Fig. 18.6 with $\mathcal{E}, \mathcal{F}, \mathcal{C}$ rearranged

We can execute this rearrangement correspondingly for Fig. 18.3, obtaining Fig. 18.8.

```
      6 2 3 4 5 7 8 10 11 1 9          6 2 3 4 5 7 8 10 11 1 9          6 2 3 4 5 7 8 10 11 1 9
  6 /1 1 1 1 1 1 1 1 1 1 1\       6 /1 1 1 1 1 1 1 1 1 1 1\       6 /0 0 0 0 0 0 0 0 0 1 1\
  2 |0 1 0 1 0 0 0 1 1 1 1|       2 |1 0 1 0 1 1 1 0 0 0 0|       2 |0 0 0 0 0 1 0 0 0 1 1|
  3 |0 0 1 0 0 1 0 0 0 1 1|       3 |1 1 0 0 0 0 0 0 0 0 0|       3 |0 0 0 1 0 0 0 0 0 1 1|
  4 |0 0 0 1 0 0 0 0 0 1 1|       4 |1 0 0 0 0 0 0 0 0 0 0|       4 |0 0 1 0 1 1 0 0 1 1 1|
  5 |0 0 1 0 1 1 0 0 1 1 1|       5 |1 1 0 0 0 0 1 0 0 0 0|       5 |0 0 0 1 0 0 0 0 0 1 1|
  7 |0 0 0 0 0 1 0 0 0 1 1|       7 |1 1 0 0 0 0 0 0 0 0 0|       7 |0 1 0 1 0 0 0 1 1 1 1|
  8 |0 0 1 1 0 1 1 1 0 1 1|       8 |1 1 0 0 1 0 0 0 1 0 0|       8 |0 0 0 0 0 0 0 0 1 1 1|
 10 |0 0 0 1 0 0 1 0 1 1 1|      10 |1 0 0 0 0 0 0 0 0 0 0|      10 |0 0 0 0 0 0 1 0 0 1 1|
 11 |0 0 0 0 0 0 0 1 1 1 1|      11 |1 0 0 0 0 0 1 0 0 0 0|      11 |0 0 1 1 0 1 1 1 0 1 1|
  1 |0 0 0 0 0 0 0 0 0 1 1|       1 |1 0 0 0 0 0 0 0 0 0 0|       1 |1 1 1 1 1 1 1 1 1 1 1|
  9 \0 0 0 0 0 0 0 0 0 1 1/       9 \1 0 0 0 0 0 0 0 0 0 0/       9 \1 1 1 1 1 1 1 1 1 1 1/
```

Fig. 18.8 Matrices of Fig. 18.3 with $\mathcal{E}, \mathcal{F}, \mathcal{C}$ rearranged

18.2 Examples of implication closures

We will now provide two extended examples: timetabling and Sudoku solving. In both cases we are going to write down terms that denote the implication structure. These are then used for interpretation in TITUREL so as to obtain the example relations.

Timetables

Implication structures have been studied not least in the context of timetables. For a timetable problem, there exists a set of hours or time slots. To these time slots, lessons or meetings are assigned to which a set of participants has to convene. The problem is that the participants may not be available at all the given hours.

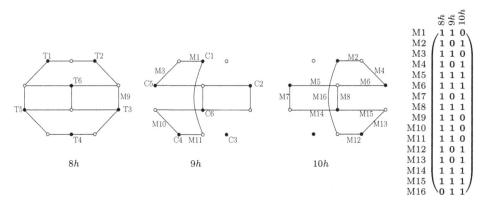

	$8h$	$9h$	$10h$
M1	1	1	0
M2	1	0	1
M3	1	1	0
M4	1	0	1
M5	1	1	1
M6	1	1	1
M7	1	0	1
M8	1	1	1
M9	1	1	0
M10	1	1	0
M11	1	1	0
M12	1	0	1
M13	1	0	1
M14	1	1	1
M15	1	1	1
M16	0	1	1

Fig. 18.9 Timetable problem with 3 hours. The availability of the meetings to be scheduled is indicated via graph and relation

A simple version of such an investigation provides an extended example. There are given a set T of teachers and a set C of school classes. Using the direct sum construct $P := T + C$, we bind these together in one set P of participants. Then there is a set H of school hours. The lessons or meetings to be scheduled are each composed of a class and a teacher and a set of available hours. The availability may be restricted because a teacher or class is not available for some reason or another. It may, however, also be the case that a lesson must not be set during some specific hour or time slot, since, for example, a day must not start with a gymnastics lesson.

Figure 18.9 shows all the lessons of our example[1] together with their availability in two different ways; the graph, respectively the relation, have already been determined in an earlier phase of analysis. We have only 6 teachers $T1, \ldots, T6$, 6 classes $C1, \ldots, C6$ and the possible hours $H = \{8^h, 9^h, 10^h\}$. The necessary meetings (meets for short) have been numbered M1,...,M16; participation is shown in

---

[1] One may find this example in [107].

the graph. The relation on the right expresses information that can immediately be read off the graph: in lesson $M13$, teacher $T3$ is involved as well as class $C3$.

We now present a relational formulation referring to Fig. 18.10. It must be prescribed who has to convene for a lesson or meeting. Here, we list which set of teachers has to meet with relations $\mu_T : M \longrightarrow T$ and the classes with $\mu_C : M \longrightarrow C$. We obtain the participants (teachers or classes) via the direct sum construct with natural injections ι, κ by

$$\mu_P := \mu_T{}_; \iota \cup \mu_C{}_; \kappa.$$

Every teacher has an availability $\alpha_T : T \longrightarrow H$ in a similar way as every class has $\alpha_C : C \longrightarrow H$. Via the direct sum, then immediately the availability

$$\alpha_P := \iota^{\mathsf{T}}{}_; \alpha_T \cup \kappa^{\mathsf{T}}{}_; \alpha_C$$

of every participant is obtained. The a priori availability of lessons is given as α_M. So far, it may be that a teacher is available for many hours but because of restrictions by other participants in his meetings this is completely misleading, as all the others are very tight. The reduction to availability of meetings eliminates in a straightforward manner the availabilities of participants which can never be used. From this and the joint availability of the participants the availability of the lesson is determined as

$$\gamma := \alpha_M \cap \overline{\mu_P{}_; \overline{\alpha_P}}.$$

In γ, the availability α_M of the lesson itself has been taken care of; in addition, it is *not* the case that there exists a participant of the lesson in question that is *not* available. The relation γ is shown on the right of Fig. 18.9.

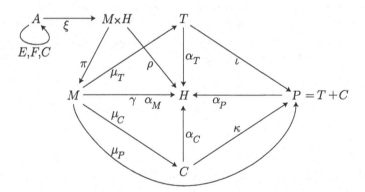

Fig. 18.10 The relations involved in a timetable problem

To solve this timetable problem we must assign lessons or meetings to hours, i.e., find a mapping $\vartheta \subseteq \gamma$ satisfying the availability restrictions – assuming in this example precisely one hour is necessary per lesson.

However, the timetable problem is known to be $\mathcal{NP}$-complete, i.e., inherently difficult. It needs a little bit of construction to configure it as an implication structure. To this end, we introduce the possible associations of meetings and hours as a direct product $M \times H$ together with the two projections $\pi : M \times H \longrightarrow M$ and $\rho : M \times H \longrightarrow H$. Then we convert γ with vectorization according to Section 7.9 into a vector

$$v := (\pi \, \dot{} \, \gamma \cap \rho) \, \dot{} \, \mathbb{T},$$

defined along $M \times H$; this is then extruded with $\xi := \mathtt{Extrude}\ v$ in order to cut out from the beginning what corresponds to non-availabilities. Due to this extrusion, the row and column labels in the following figures always end with $\rightarrow$. A solution is now a vector along $M \times H$ constrained with certain relations E, F, C that we are now going to construct from the timetable requirements.

The criteria are a conjunction of three properties:

- a meet is assigned only if all of its participants are available;
- every meet is assigned precisely once, and
- no participant is assigned two different meets in the same time slot.

The elementary forbidding relation may thus be formulated as

$$F_H := \pi \, \dot{} \, \pi^{\mathsf{T}} \cap \rho \, \dot{} \, \overline{\mathbb{I}} \, \dot{} \, \rho^{\mathsf{T}}$$

saying that a lesson being assigned to a time slot is forbidden for any *other* time slot. Also, having assigned a lesson to some hour, there cannot be assigned *another* lesson to this hour in case they overlap in participants, as shown algebraically by forbidding restricted to one participant

$$F_P := \pi \, \dot{} \, \left(\overline{\mathbb{I}} \cap \mu_P \, \dot{} \, \mu_P^{\mathsf{T}} \right) \, \dot{} \, \pi^{\mathsf{T}} \cap \rho \, \dot{} \, \rho^{\mathsf{T}}.$$

Both these forbidding requirements are then joined and extruded with the aforementioned ξ so as to obtain

$$F := \xi \, \dot{} \, (F_H \cup F_P) \, \dot{} \, \xi^{\mathsf{T}}.$$

Writing this term down in TITUREL instead of TeX, one can interpret it and obtain the result of Fig. 18.11.

For example, $M1$ assigned to $8h$ forbids assigning it at $9h$, but also assigning $M2$ at $8h$. In addition to F, we need the elementary counter-enforcing relation C. We

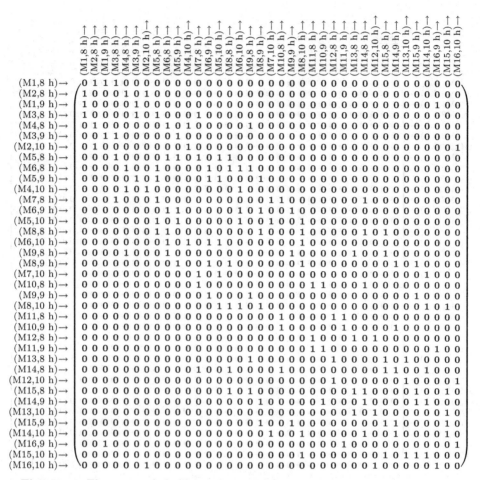

Fig. 18.11 Elementary forbidding relation F of the timetable problem of Fig. 18.9

will observe here that C and F overlap. The basic idea is that when there are just two lessons for a participant p at some time slot, the not-assigning of one of these enforces the assignment of the other to be made. So, we design the functional

$$c(p) := \xi \cdot \mathsf{upa}(\pi \cdot \mu_P \cdot (\mathbb{I} \cap p \cdot \mathbb{T}) \cdot \mu_P^\mathsf{T} \cdot \pi^\mathsf{T} \cap \rho \cdot \rho^\mathsf{T} \cap \overline{\mathbb{I}}) \cdot \xi^\mathsf{T}$$

and join the result of the application to all participants

$$C := \sup_{p \in P} c(p) \cup \xi \cdot \mathsf{upa}(\pi \cdot \pi^\mathsf{T} \cap \rho \cdot \overline{\mathbb{I}} \cdot \rho^\mathsf{T}) \cdot \xi^\mathsf{T}.$$

In both cases, the univalent part functional

$$\mathsf{upa}(R) := R \cap \overline{R \cdot \overline{\mathbb{I}}} = \mathsf{syq}(R^\mathsf{T}, \mathbb{I})$$

has been used. The result of an evaluation of this term is shown in Fig. 18.12.

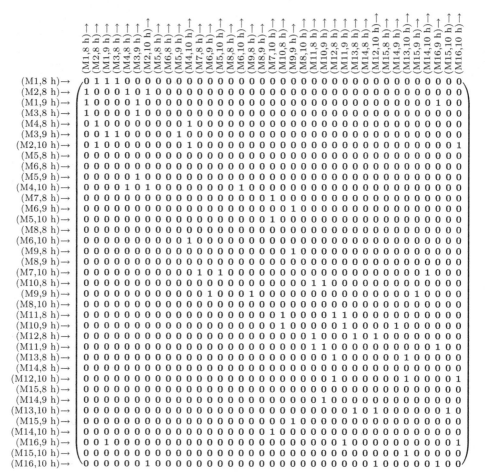

Fig. 18.12 Elementary counter-enforcing relation C of the timetable problem of Fig. 18.9

As an example consider lesson $M3$; if it is not scheduled for $8h$, it must be assigned to $9h$ since there is no other possibility. In addition, considering participant $C2$, the lesson $M6$ not assigned to $10h$ implies lesson $M4$ to be assigned at $10h$. Now we investigate what the elementary enforcing E should look like. For the first time, we now use a forbidding relation F as a parameter – it need not be the elementary forbidding:

$$e_F(p) := F / \big(\iota; \big(\pi; \mu_P; (\mathbb{I} \cap p; \mathbb{T}); \mu_P^\mathsf{T}; \pi^\mathsf{T}\big) \cap \rho; \rho^\mathsf{T} \cap \overline{\mathbb{I}}\big); \iota^\mathsf{T}\big)\big)$$

$$E := \big(\sup_{p \in P} e_F(p)\big)^*.$$

The teacher appointed to produce the timetable will most certainly *not* establish

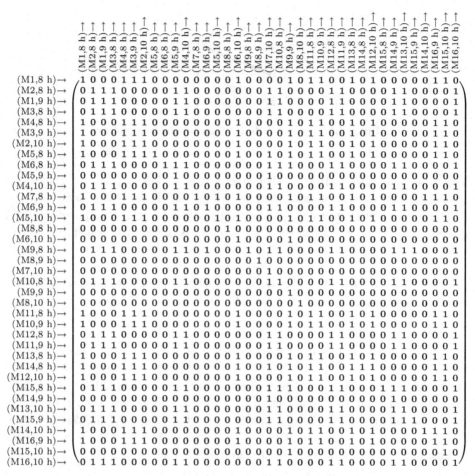

Fig. 18.13 Elementary enforcing relation E of the timetable problem of Fig. 18.9

such matrices. He will probably start the process of solving simply by trying to assign meetings to hours. In order to solve this $\mathcal{NP}$-complete problem, he will mostly apply heuristics. These heuristics will mainly consist of some sort of accounting. That is, he will try immediately to update availabilities which are reduced when an assignment has been made. He may also immediately apply the highly efficient mincut analysis. This means that for every participant all the meetings he is involved in will be assigned testwise regardless of the other participants. This will detect efficiently, for example, that the bold assignment of c in Fig. 18.14 must not be used, as then not all of $\{a, b, c, d\}$ can be accommodated.

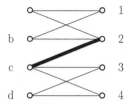

Fig. 18.14 Mincut tightness in a timetabling problem

The result may also be that a full assignment is no longer possible. Then back-tracking is necessary. The result may finally be that from mincut analysis alone, it follows that some meeting is already tight with regard to the possibilities for assigning it. This meeting will then be chosen to be formally assigned.

In this way, one decision to assign a meeting may result in some sequence of imme-diate consequences. These are fixed before the next choice is made.

In spite of this flexibility, there exist pseudo-availabilities, i.e., availabilities that cannot be used in a solution. We give an example with meeting M7 scheduled at 8^h: this immediately forbids assignment of (3,8),(5,8),(14,8),(10,8),(7,10). These in turn counter-enforce assignment of (1,8) (because otherwise teacher T1 cannot be accommodated at 8^h),(11,8). These then forbid (2,8),(12,8),(11,9),(1,9), thus counter-enforcing (4,8),(13,8),(16,9), etc. In total, we find that this is just pseudo, so that we have to decide for the only alternative as shown in Fig. 18.15.

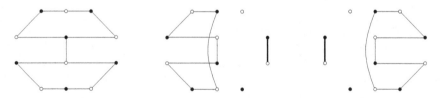

Fig. 18.15 Timetable problem after assigning tight assignments
and deleting forbidden ones

In the timetable problem of Fig. 18.15, it will turn out that mincut analysis for any participant (i.e., vertex) will not provide tight assignments in the sense just indicated. There exist, however, long-range implications to be derived by iteration of enforce, forbid, and counter-enforce.

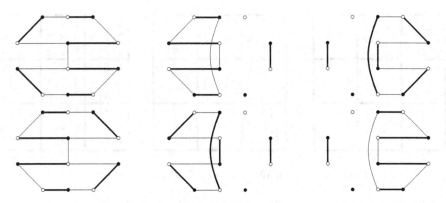

Fig. 18.16 Two solutions of the timetable problem: bold lines indicate assignment

There is something else to be considered carefully: according to Prop. 18.2, E is a preorder. One may ask what influence it has on the necessarily heuristic algorithm when one chooses minimal/maximal elements to be handled first. Selecting minimal elements with regard to E first makes fundamental decisions early in a backtracking algorithm. It may, however, also be wise to assign maximal elements first. Then some freedom is still left to assign the others – and to fit to criteria one has not yet formalized.

Sudoku

Another nice example of an implication structure is given by one of the now popular *Sudoku* puzzles. A table of 9×9 squares is given, partially filled with natural numbers $1, 2, 3, 4, 5, 6, 7, 8, 9$, which is to be filled completely. The problem is that one is not allowed to enter the same number twice in a row or column and one is not allowed to enter duplicates in the nine 3×3 subzones. There exist variant forms with 12×12 or 15×15 squares, or others.

Sudokus published in a newspaper typically allow for just one solution. In general, there may be none or many solutions. Often the first steps are rather trivial; looking from the empty position i, k to row i and also column k as well as to subzone s, one may find out that already 8 of the nine numbers have been used. It is then obvious that the ninth has to be entered in the so far empty field i, k.

From the first Sudoku to the second in Fig. 18.17, at position $(2, 3)$ one easily observes that the corresponding row already contains 2,9,5,6, the corresponding column 2,3,1,8, and the upper left sub-square 2,7. So all the digits except 4 are present and 4 must, therefore, be assigned to position $(2, 3)$.

Fig. 18.17 A Sudoku puzzle with immediate insertions

Another trivial step is to look for some field whether all *other* non-filled places in a sub-square (or row, or column) are already forbidden for some number. Of course, then this number has to be entered in the field in question. This is what leads us to insert a 7 at position (4,3) since the first two columns are already blocked with a 7. Also, we have to put a 1 at (7,4): the last two rows already contain a 1 as well as column 6.

These two steps follow different ideas but are in principle trivial. With a grandson on one's knees, one will easily make mistakes; nevertheless, it is a trivial accounting job and, thus, not really interesting. The Sudoku of Fig. 18.17 may be considered trivial, as already the trivial operations iterated lead to the full solution on the right.

What interests us more is the task beyond this. Other Sudokus will not be filled after such operations. This is the point when anyone trying to solve the Sudoku becomes in trouble and may be tempted to start case analyses or branch and bound techniques. So we assume a new example, as on the left of Fig. 18.18, and immediately execute all these trivial – but error-prone – steps to arrive at the diagram on the right.

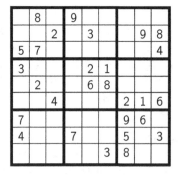

Fig. 18.18 A less trivial Sudoku puzzle that is maximally filled by trivial insertions

With a computer, one is in a position to avoid such procedures of search style, applying implication structures to the not yet completed Sudoku on the right. We recall that after the steps executed so far the following two statements hold true.

- It is not the case that any entry must be made as *all other numbers* in row, column, or subzone are already used and, thus, no longer available.
- It is not the case, that *all other entries* in the respective row, column, or subzone are already forbidden for some digit.

One may check and find out that this indeed holds true for the right Sudoku of Fig. 18.18. But there is plenty of sources of local implications and similarly counter-enforcing.

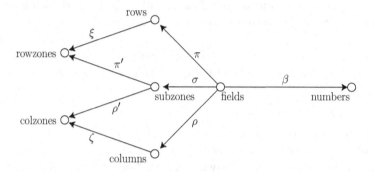

Fig. 18.19 Relations around a Sudoku

We now study this with implication structures, i.e., from a relational point of view. It would be extremely tedious and error-prone to write down such a relation by hand. Other techniques are, however, available. We start from the very basics introducing the sets in question, namely

    `rowSet,colSet,digSet,rowZonSet,colZonSet`

of which the first three stand for rows, columns, and digits $1, 2, 3, 4, 5, 6, 7, 8, 9$ – or others in variant forms. The last denotes the three zones *low, medium, high* so that the upper right subzone is represented, for example, by the pair (*low,high*). The entries of the Sudoku are then given as elements of the direct product

    `fields = DirPro rowSet colSet`

with projections

    $\pi = $ `Pi   rowSet colSet`,
    $\rho = $ `Rho rowSet colSet`.

The squares are heavily based on the classifications of rows and columns as $\xi(1) = \xi(2) = \xi(3) = low$, $\xi(4) = \xi(5) = \xi(6) = medium$, $\xi(7) = \xi(8) = \xi(9) = high$ and analogously with ζ for columns. Obviously, the 'projection' σ to the subzones may be derived from this setting as

$$\sigma := \pi_{;} \xi_{;} {\pi'}^{\mathsf{T}} \cap \rho_{;} \zeta_{;} {\rho'}^{\mathsf{T}}.$$

Several interconnections between these projections are immediate, the row equivalence $\pi_{;} \pi^{\mathsf{T}}$, the column equivalence $\rho_{;} \rho^{\mathsf{T}}$, the subzone equivalence $\sigma_{;} \sigma^{\mathsf{T}}$, the identity on fields $(1,1), (1,2), \ldots, (9,8), (9,9)$, etc. Most importantly, the univalent relation β indicates the current accommodation of numbers in fields, be it from the start or after insertions already executed.

The union of the equivalences by projections π, ρ, σ is obviously important although not an equivalence itself; we give them a name

$$\Xi_{\text{row}} := \pi_{;} \pi^{\mathsf{T}} \qquad \Xi_{\text{col}} := \rho_{;} \rho^{\mathsf{T}} \qquad \Xi_{\text{sub}} := \sigma_{;} \sigma^{\mathsf{T}} \qquad \Xi := \Xi_{\text{row}} \cup \Xi_{\text{col}} \cup \Xi_{\text{sub}}.$$

Using this, $\Xi_{;} \beta$ sums up which valuations are attached to all the fields related to a given one by having the same row, column, or subzone.

If in some intermediate state β of the insertions there should be just one digit missing among these, this *must* be inserted. All these cases are computed in one step and added to the actual β:

$$\beta \mapsto \beta \cup \left\{ \mathsf{upa}(\Xi_{;} \beta) \cap \overline{\beta_{;} \mathbb{T}} \right\}.$$

This term immediately computes at least a part of the solution. It goes from some field to all that are with it in the same row, column, or subzone, looks with β for the numbers entered and cuts out with $\overline{\beta_{;} \mathbb{T}}$ all fields that are already filled. The main point is the application of the univalent-part function $\mathsf{upa}(R) = R \cap \overline{R_{;} \overline{\mathbb{I}}}$; see Def. 6.2.

We have, thus, indicated how a Sudoku solving program might be based on implication structures. Several versions have been written in TITUREL.

Power Operations

There exist many other areas of possibly fruitful application of relational mathematics from which we select three topics – and thus omit many others. In the first section, we will again study the lifting into the powerset. Credit is mainly due to theoretical computer scientists (e.g. [22, 39]) who have used and propagated such constructs as an existential image or a power transpose. When relations (and not just mappings as for homomorphisms) are employed to compare state transitions, we will correspondingly need relators as a substitute for functors. We will introduce the power relator. While rules concerning power operations have often been only postulated, we go further here and deduce such rules in our general axiomatic setting.

In Section 19.2, we treat questions of simulation using relational means. When trying to compare the actions of two black box state transition systems, one uses the concept of a bisimulation to model the idea of behavioral equivalence. In Section 19.3, a glimpse of state transition systems and system dynamics is offered. The key concept is that of an orbit, the sequence of sets of states that result when continuously executing transitions of the system starting from some given set of states. There exist many orbits in a delicately related relative position. Here the application of relations seems particularly promising.

19.1 Existential image and power transpose

The first two power operations are at least known to a small community. The idea is as follows. Relations, in contrast to functions, are mainly introduced in order to assign not precisely one result but a – possibly empty – *set* of results, so that they are often conceived as set-valued functions. Since we have formally constructed our domains, we know that there are different possibilities for handling a subset, namely as a vector, or as an element in the powerset. It is mainly the latter transition that leads us to introduce the following two power operations.

Definition 19.1. Given any relation $R : X \longrightarrow Y$ together with membership relations $\varepsilon : X \longrightarrow \mathbf{2}^X$ and $\varepsilon' : Y \longrightarrow \mathbf{2}^Y$ on the source, respectively, target side, we call

(i) $\Lambda := \Lambda_R := \mathsf{syq}(R^\mathsf{T}, \varepsilon')$ its **power transpose**,

(ii) $\vartheta := \vartheta_R := \mathsf{syq}(R^\mathsf{T}{}_{;}\, \varepsilon, \varepsilon')$ its **existential image**. $\square$

We have included in this definition our decision how to abbreviate the lengthy denotations – as long as there does not arise any confusion as to which R is meant. With Fig. 19.1, we show the typing of these two operations, and of the result ζ of applying the power relator yet to be introduced. However, from now on we will be a bit sloppy, talking simply of the power relator ζ.

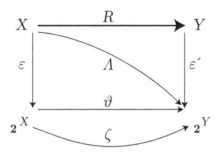

Fig. 19.1 Typing of existential image, power transpose, and power relator ζ

It may be observed directly in Fig. 19.2, where a first example of such operators is shown, that Λ and ϑ are mappings. This will now also be proved formally. Usually the power relator is not a mapping.

Proposition 19.2. *Power transposes as well as existential images of relations are mappings.*

Proof: The proof is analogous for power transpose and existential image. Using Prop. 8.13, ϑ is univalent since

$$\vartheta^\mathsf{T}{}_{;}\,\vartheta = \mathsf{syq}(\varepsilon', R^\mathsf{T}{}_{;}\,\varepsilon){}_{;}\,\mathsf{syq}(R^\mathsf{T}{}_{;}\,\varepsilon, \varepsilon') \qquad \text{by definition and transposition}$$
$$\subseteq \mathsf{syq}(\varepsilon', \varepsilon') \qquad \text{cancelling according to Prop. 8.13.i}$$
$$= \mathbb{I} \qquad \text{Def. 7.13.i for a membership relation}$$

It is in addition total because of Def. 7.13.ii for a membership relation. $\square$

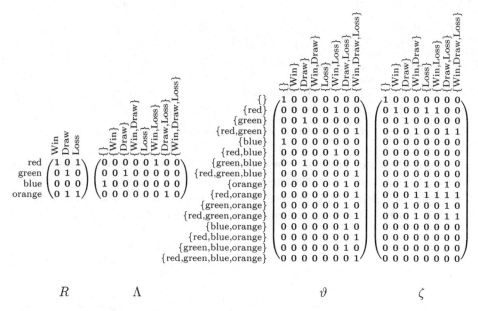

Fig. 19.2 R together with its power transpose, existential image, and power relator

One may have observed that Λ means nothing other than set comprehension concerning the set assigned to an argument. The element orange in R of Fig. 19.2 is assigned the two elements Draw and Loss which is reflected in Λ by assigning it the element corresponding to the set {Draw,Loss}.

The existential image ϑ is a little bit more detailed compared with Λ as it takes into account *sets* of arguments and assigns them the union of the sets assigned to its elements. Some people, therefore, say that ϑ is additive. For example, the set {red,orange} is assigned {Win,Draw,Loss} because this is the union of the images of the sets assigned to elements of the argument set. This is expressed in the name: for every constituent of the result, there *exists* an argument for which it is the *image*. Rows for singleton sets correspond directly to the specific rows of Λ.

Proposition 19.3. *Let any relation* $R : X \longrightarrow Y$ *be given together with the two membership relations* $\varepsilon : X \longrightarrow 2^X$ *and* $\varepsilon' : Y \longrightarrow 2^Y$ *and consider the two operations*

$$\sigma(R) := \overline{\varepsilon^{\mathsf{T}} ; R ; \overline{\varepsilon'}} \qquad and \qquad \pi(W) := \overline{\varepsilon ; W ; \overline{\varepsilon'}^{\mathsf{T}}}.$$

(i) *These two operations form a so-called Galois correspondence, i.e.,*
$$R \subseteq \pi(W) \qquad \Longleftrightarrow \qquad W \subseteq \sigma(R).$$

(ii) *The operation* $\sigma(R)$ *is injective, i.e.,* $\sigma(R) = \sigma(S)$ *implies* $R = S$.

Proof: (i) We use the Schröder rule several times in combination with negation.

$$R \subseteq \overline{\varepsilon \,;\, W \,;\, \overline{\varepsilon'}^{\mathsf{T}}} \iff \varepsilon \,;\, W \,;\, \overline{\varepsilon'}^{\mathsf{T}} \subseteq \overline{R} \iff R \,;\, \overline{\varepsilon'} \,;\, W^{\mathsf{T}} \subseteq \overline{\varepsilon}$$
$$\iff \overline{\varepsilon'}^{\mathsf{T}} \,;\, R^{\mathsf{T}} \,;\, \varepsilon \subseteq \overline{W}^{\mathsf{T}} \iff W \subseteq \overline{\varepsilon^{\mathsf{T}} \,;\, R \,;\, \overline{\varepsilon'}}$$

(ii) Injectivity follows from

$$\pi(\sigma(R)) = \overline{\varepsilon \,;\, \overline{\varepsilon^{\mathsf{T}} \,;\, R \,;\, \overline{\varepsilon'}} \,;\, \overline{\varepsilon'}^{\mathsf{T}}} \qquad \text{by definition}$$
$$= \overline{\overline{R \,;\, \overline{\varepsilon'}} \,;\, \overline{\varepsilon'}^{\mathsf{T}}} \qquad \text{Prop. 7.14 of membership relation } \varepsilon$$
$$= \overline{\overline{R}} = R \qquad \text{Prop. 7.14 of membership relation again} \qquad \square$$

It is evident that we did not use the specific properties of the membership relations in proving (i), so that other relations (A, B, for example, instead of $\varepsilon, \varepsilon'$) would also lead to such a Galois correspondence. Here, however, it is important that according to (ii) we have an embedding via $\sigma(R)$.

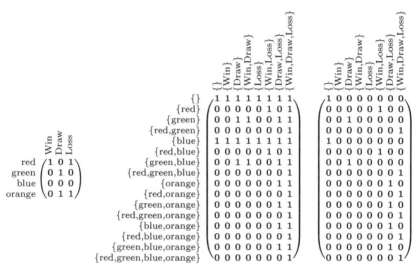

Fig. 19.3 Relation R with the cone $\sigma(R)$ and its existential image ϑ_R as row-wise `glb` thereof

With Fig. 19.3, we show that the image of $\sigma(R)$ is always an upper cone, taken row-wise. For this cone, there always exists a greatest lower bound, which – in the next proposition – turns out to be the existential image.

Proposition 19.4. *Assume the same setting as in Prop. 19.3 together with the powerset orderings* $\Omega := \overline{\varepsilon^{\mathsf{T}} \,;\, \overline{\varepsilon}}$, Ω' *on the source and on the target side. Then the relation R and its existential image* $\vartheta := \texttt{ExImag}(R) := \texttt{syq}(R^{\mathsf{T}} \,;\, \varepsilon, \varepsilon')$ *are $1:1$ related by*

$$\pi(\vartheta) = R \qquad and \qquad \vartheta = \texttt{glbR}_{\Omega'}(\sigma(R)).$$

Proof: The first identity is proved as follows:

$$\pi(\vartheta) = \overline{\varepsilon_{;} \vartheta_{;} \overline{\varepsilon'}^{\mathsf{T}}} \qquad\qquad \text{by definition}$$

$$= \overline{\varepsilon_{;} \mathsf{syq}\left(R^{\mathsf{T}}_{;} \varepsilon, \varepsilon'\right)_{;} \overline{\varepsilon'}^{\mathsf{T}}} \qquad \text{expanding } \vartheta$$

$$= \overline{\varepsilon_{;} \overline{\mathsf{syq}\left(\overline{R^{\mathsf{T}}_{;} \varepsilon}, \varepsilon'\right)_{;} \overline{\varepsilon'}}^{\mathsf{T}}} \qquad \text{Prop. 8.10.i}$$

$$= \overline{\varepsilon_{;} \overline{R^{\mathsf{T}}_{;} \varepsilon}}^{\mathsf{T}} \qquad\qquad \text{Prop. 7.14}$$

$$= \overline{\varepsilon_{;} \overline{\varepsilon^{\mathsf{T}}_{;} R}} \qquad\qquad \text{transposing}$$

$$= \overline{\overline{R}} = R \qquad\qquad \text{standard property Prop. 7.14 of membership again}$$

We evaluate the lower bounds $\mathsf{lbd}_{\Omega'}\left(\sigma(R)^{\mathsf{T}}\right)$ first

$$\mathsf{lbd}_{\Omega'}\left(\sigma(R)^{\mathsf{T}}\right) = \overline{\overline{\Omega'}_{;} \sigma(R)^{\mathsf{T}}} \qquad\qquad \text{by definition of lower bound } \mathsf{lbd}$$

$$= \overline{\varepsilon'^{\mathsf{T}}_{;} \overline{\varepsilon'}_{;} \overline{\varepsilon'^{\mathsf{T}}_{;} R^{\mathsf{T}}_{;} \varepsilon}} \qquad\qquad \text{expanding } \Omega', \text{ definition of } \sigma(R)$$

$$= \overline{\varepsilon'^{\mathsf{T}}_{;} \overline{R^{\mathsf{T}}_{;} \varepsilon}} \qquad\qquad\qquad \text{Prop. 7.14}$$

and then work on the greatest lower bound:

$$\mathsf{glb}_{\Omega'}\left(\sigma(R)^{\mathsf{T}}\right) = \overline{\varepsilon'^{\mathsf{T}}_{;} \overline{R^{\mathsf{T}}_{;} \varepsilon}} \cap \overline{\overline{\varepsilon'}^{\mathsf{T}}_{;} \varepsilon'_{;} \overline{\varepsilon'^{\mathsf{T}}_{;} \overline{R^{\mathsf{T}}_{;} \varepsilon}}} \qquad \text{definition of } \mathsf{glb}, \text{ expanding } \Omega'$$

$$= \overline{\varepsilon'^{\mathsf{T}}_{;} \overline{R^{\mathsf{T}}_{;} \varepsilon}} \cap \overline{\overline{\varepsilon'}^{\mathsf{T}}_{;} R^{\mathsf{T}}_{;} \varepsilon} \qquad\qquad \text{Prop. 7.14}$$

$$= \mathsf{syq}\left(\varepsilon', R^{\mathsf{T}}_{;} \varepsilon\right) \qquad\qquad\qquad \text{definition of } \mathsf{syq}$$

$$= \vartheta^{\mathsf{T}} \qquad\qquad\qquad\qquad \text{definition of } \vartheta \qquad\qquad \square$$

Now the existential image is considered in more detail. We have already seen that the existential image is in a sense additive. When applied in a non-finite environment one can better express this as lattice-theoretic continuity.

Proposition 19.5. $\vartheta := \mathsf{syq}\left(R^{\mathsf{T}}_{;} \varepsilon, \varepsilon'\right)$ *is a (lattice-)continuous mapping.*

Proof: We prove the continuity condition starting from $\mathsf{lub}_{\Omega}(X) = \mathsf{syq}(\varepsilon, \varepsilon_{;} X)$ according to Prop. 9.10 and temporarily abbreviating $g := \left[\mathsf{lub}_{\Omega}(X)\right]^{\mathsf{T}}$:

$$\vartheta^{\mathsf{T}}_{;} \mathsf{lub}_{\Omega}(X) = \left[\mathsf{syq}\left(R^{\mathsf{T}}_{;} \varepsilon, \varepsilon'\right)\right]^{\mathsf{T}}_{;} \mathsf{lub}_{\Omega}(X) \qquad \text{by definition}$$

$$= \mathsf{syq}\left(\varepsilon', R^{\mathsf{T}}_{;} \varepsilon\right)_{;} g^{\mathsf{T}} \qquad\qquad \text{transposing } \mathsf{syq}, \text{ definition of } g$$

$$= \mathsf{syq}\left(\varepsilon', R^{\mathsf{T}}_{;} \varepsilon_{;} g^{\mathsf{T}}\right) \qquad\qquad \text{Prop. 8.16.iii}$$

$$= \mathsf{syq}\left(\varepsilon', R^{\mathsf{T}}_{;} \varepsilon_{;} \mathsf{syq}(\varepsilon, \varepsilon_{;} X)\right) \qquad \text{expanding and transposing } g$$

$$= \mathsf{syq}\left(\varepsilon', R^{\mathsf{T}}_{;} \varepsilon_{;} X\right) \qquad\qquad \text{property of the membership}$$

$$= \mathsf{syq}\left(\varepsilon', \varepsilon'_{;} \mathsf{syq}(\varepsilon', R^{\mathsf{T}}_{;} \varepsilon)_{;} X\right) \qquad \text{Prop. 7.14 for membership } \varepsilon$$

$$= \mathsf{lub}_{\Omega'}\left(\mathsf{syq}(\varepsilon', R^{\mathsf{T}}_{;} \varepsilon)_{;} X\right) \qquad\qquad \text{Prop. 9.10}$$

$$= \mathsf{lub}_{\Omega'}\left(\vartheta^{\mathsf{T}}_{;} X\right) \qquad\qquad\qquad \text{definition of } \vartheta \qquad\qquad \square$$

The existential image behaves nicely with respect to relational composition; it is multiplicative and respects identities as is shown in the next proposition.

Proposition 19.6. *We consider the existential image* $\vartheta_Q := \text{syq}(Q^{\mathsf{T}} \,;\, \varepsilon, \varepsilon')$:

(i) $\vartheta_{Q\,;\,R} = \vartheta_Q \,;\, \vartheta_R$,

(ii) $\vartheta_{\mathbb{I}_X} = \mathbb{I}_{\mathbf{2}^X}$.

Proof: (i) ϑ_Q is a mapping, so that we may reason as follows

$$\vartheta_Q \,;\, \vartheta_R = \vartheta_Q \,;\, \text{syq}(R^{\mathsf{T}} \,;\, \varepsilon', \varepsilon'')$$

$$= \text{syq}(R^{\mathsf{T}} \,;\, \varepsilon' \,;\, \vartheta_Q^{\mathsf{T}}, \varepsilon'') \qquad \text{because of Prop. 8.16.ii}$$

$$= \text{syq}(R^{\mathsf{T}} \,;\, \varepsilon' \,;\, \text{syq}(\varepsilon', Q^{\mathsf{T}} \,;\, \varepsilon), \varepsilon'') \qquad \text{definition and transposition of } \vartheta_Q$$

$$= \text{syq}(R^{\mathsf{T}} \,;\, Q^{\mathsf{T}} \,;\, \varepsilon, \varepsilon'') \qquad \text{cancelling the symmetric quotient}$$

$$= \text{syq}((Q\,;\,R)^{\mathsf{T}} \,;\, \varepsilon, \varepsilon'')$$

$$= \vartheta_{Q\,;\,R}$$

(ii) $\vartheta_{\mathbb{I}_X} = \text{syq}(\mathbb{I}_X^{\mathsf{T}} \,;\, \varepsilon, \varepsilon) = \text{syq}(\varepsilon, \varepsilon) = \mathbb{I}_{\mathbf{2}^X}.$ □

Interpreting this, we may say: forming the existential image is a multiplicative operation that in addition preserves identities. Rather obviously, however, conversion is not preserved.

Simulation via power transpose and existential image

Looking at Fig. 19.2, one will easily realize that the existential image and power transpose together with singleton injection σ (not to be mixed up with $\sigma(R)$) always satisfy

$$\sigma \,;\, \vartheta = \Lambda,$$

which may be proved with

$$\sigma \,;\, \vartheta = \sigma \,;\, \text{syq}(R^{\mathsf{T}} \,;\, \varepsilon, \varepsilon')$$

$$= \text{syq}(R^{\mathsf{T}} \,;\, \varepsilon \,;\, \sigma^{\mathsf{T}}, \varepsilon') \qquad \text{Prop. 8.16.ii since } \sigma \text{ is a mapping}$$

$$= \text{syq}(R^{\mathsf{T}} \,;\, \varepsilon \,;\, \text{syq}(\varepsilon, \mathbb{I}), \varepsilon') \qquad \text{by definition of } \sigma \text{ and transposition}$$

$$= \text{syq}(R^{\mathsf{T}} \,;\, \mathbb{I}, \varepsilon') \qquad \text{standard rule for the membership relation}$$

$$= \text{syq}(R^{\mathsf{T}}, \varepsilon') = \Lambda$$

We will first see that a relation R and its power transpose can replace one another to a certain extent. In most cases, however, it is a better idea to let the relation R be simulated by its existential image.

Proposition 19.7. *Let any relation* $R : X \longrightarrow Y$ *be given together with the membership relation* $\varepsilon' : Y \longrightarrow \mathbf{2}^Y$ *on its target side. Then*

$$f = \Lambda_R \qquad \Longleftrightarrow \qquad R = f \mathbin{;} \varepsilon'^{\mathsf{T}}.$$

Proof: "$\Longleftarrow$" $\Lambda_R = \mathsf{syq}\,(R^{\mathsf{T}}, \varepsilon')$ by definition
$= \mathsf{syq}\,((f \mathbin{;} \varepsilon'^{\mathsf{T}})^{\mathsf{T}}, \varepsilon')$ by assumption
$= \mathsf{syq}\,(\varepsilon' \mathbin{;} f^{\mathsf{T}}, \varepsilon')$ transposing
$= f \mathbin{;} \mathsf{syq}\,(\varepsilon', \varepsilon')$ Prop. 8.16.ii
$= f \mathbin{;} \mathbb{I} = f$ Def. 7.13 of membership relation

"$\Longrightarrow$" $f \mathbin{;} \varepsilon'^{\mathsf{T}} = \Lambda_R \mathbin{;} \varepsilon'^{\mathsf{T}}$ by assumption
$= \mathsf{syq}\,(R^{\mathsf{T}}, \varepsilon') \mathbin{;} \varepsilon'^{\mathsf{T}}$ definition of power transpose
$= \left[\varepsilon' \mathbin{;} \mathsf{syq}\,(\varepsilon', R^{\mathsf{T}}) \right]^{\mathsf{T}}$ transposing
$= \left(R^{\mathsf{T}} \right)^{\mathsf{T}} = R$ since $\varepsilon \mathbin{;} \mathsf{syq}\,(\varepsilon, X) = X$ for any membership relation $\qquad \square$

The power transpose, for typing reasons, cannot be multiplicative as the existential image is. Nevertheless, it has a similar property.

Proposition 19.8. *For any two relations* $R : X \longrightarrow Y$ *and* $S : Y \longrightarrow Z$

$$\Lambda_R \mathbin{;} \vartheta_S = \Lambda_{R \mathbin{;} S}.$$

Proof: We denote the membership relations as $\varepsilon' : Y \longrightarrow 2^Y$ and $\varepsilon'' : Z \longrightarrow 2^Z$. Then the power transpose is a mapping $f := \Lambda_R$ by construction, so that

$$\begin{aligned}
\Lambda_R \mathbin{;} \vartheta_S &= f \mathbin{;} \mathsf{syq}\,(S^{\mathsf{T}} \mathbin{;} \varepsilon', \varepsilon'') &&\text{by definition}\\
&= \mathsf{syq}\,(S^{\mathsf{T}} \mathbin{;} \varepsilon' \mathbin{;} f^{\mathsf{T}}, \varepsilon'') &&\text{Prop. 8.16.ii}\\
&= \mathsf{syq}\,(S^{\mathsf{T}} \mathbin{;} R^{\mathsf{T}}, \varepsilon'') &&\text{Prop. 19.7}\\
&= \mathsf{syq}\,((R \mathbin{;} S)^{\mathsf{T}}, \varepsilon'') &&\text{transposing}\\
&= \Lambda_{R \mathbin{;} S} &&\text{by definition} \qquad \square
\end{aligned}$$

We have already seen that every relation R is $1:1$ related with its existential image; but even more, the existential image $\vartheta := \vartheta_R$ may in many respects represent R. For this investigation, we refer to the concept of an L- or U-simulation of Section 19.2.

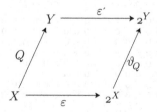

Fig. 19.4 A relation and its existential image simulating each other

Proposition 19.9. *We consider the existential image* $\vartheta_Q := \mathtt{syq}(Q^{\mathsf{T}}; \varepsilon, \varepsilon')$ *of some relation* Q *together with the membership relations on its source and target side as shown in Fig. 19.4.*

(i) *Via* $\varepsilon, \varepsilon'$, *the relation* Q *is an* L*-simulation of its existential image* ϑ_Q,
$$\varepsilon^{\mathsf{T}}; Q \subseteq \vartheta_Q; \varepsilon'^{\mathsf{T}}.$$
(ii) *Via* $\varepsilon^{\mathsf{T}}, \varepsilon'^{\mathsf{T}}$, *the existential image* ϑ_Q *is an* L^{T}*-simulation of* Q,
$$\vartheta_Q; \varepsilon'^{\mathsf{T}} \subseteq \varepsilon^{\mathsf{T}}; Q.$$
(iii) *In total, we always have*
$$\varepsilon^{\mathsf{T}}; Q = \vartheta_Q; \varepsilon'^{\mathsf{T}}.$$

Proof: $\vartheta_Q; \varepsilon'^{\mathsf{T}} = (\varepsilon'; \vartheta_Q^{\mathsf{T}})^{\mathsf{T}} = (\varepsilon'; [\mathtt{syq}(Q^{\mathsf{T}}; \varepsilon, \varepsilon')]^{\mathsf{T}})^{\mathsf{T}}$
$= [\varepsilon'; \mathtt{syq}(\varepsilon', Q^{\mathsf{T}}; \varepsilon)]^{\mathsf{T}} = (Q^{\mathsf{T}}; \varepsilon)^{\mathsf{T}} = \varepsilon^{\mathsf{T}}; Q$ □

There is a special situation in the case when a relation Q is simulated by its existential image ϑ_Q. The relation Q itself is always an L-simulation via the membership relations on the source as well as on the target side. But in the reverse direction, ϑ_Q is an L^{T}-simulation of Q; of course via the reversed membership relations. It is normally neither a U- nor a U^{T}-simulation. We provide an example with Fig. 19.5.

$Q =$

	Mon	Tue	Wed	Thu	Fri
Win	0	1	1	0	1
Draw	0	1	1	1	0
Loss	1	0	0	1	0

	Mon	Tue	Wed	Thu	Fri
{}	0	0	0	0	0
{Win}	0	1	1	0	1
{Draw}	0	1	1	1	0
{Win,Draw}	0	1	1	1	1
{Loss}	1	0	0	1	0
{Win,Loss}	1	1	1	1	1
{Draw,Loss}	1	1	1	1	0
{Win,Draw,Loss}	1	1	1	1	1

$= \vartheta_Q; \varepsilon'^{\mathsf{T}} = \varepsilon^{\mathsf{T}}; Q$

$\vartheta_Q =$

	{}	{Mon}	{Tue}	{Mon,Tue}	{Wed}	{Mon,Wed}	{Tue,Wed}	{Mon,Tue,Wed}	{Thu}	{Mon,Thu}	{Tue,Thu}	{Mon,Tue,Thu}	{Wed,Thu}	{Mon,Wed,Thu}	{Tue,Wed,Thu}	{Mon,Tue,Wed,Thu}	{Fri}	{Mon,Fri}	{Tue,Fri}	{Mon,Tue,Fri}	{Wed,Fri}	{Mon,Wed,Fri}	{Tue,Wed,Fri}	{Mon,Tue,Wed,Fri}	{Thu,Fri}	{Mon,Thu,Fri}	{Tue,Thu,Fri}	{Mon,Tue,Thu,Fri}	{Wed,Thu,Fri}	{Mon,Wed,Thu,Fri}	{Tue,Wed,Thu,Fri}	{Mon,Tue,Wed,Thu,Fri}
{}	1	0	0	0	0	0	0	0	0	0	0	0	0	0	0	0	0	0	0	0	0	0	0	0	0	0	0	0	0	0	0	0
{Win}	0	0	0	0	0	0	0	0	0	0	0	0	0	0	0	0	0	0	0	0	0	0	1	0	0	0	0	0	0	0	0	0
{Draw}	0	0	0	0	0	0	0	0	0	0	0	0	0	0	1	0	0	0	0	0	0	0	0	0	0	0	0	0	0	0	0	0
{Win,Draw}	0	0	0	0	0	0	0	0	0	0	0	0	0	0	0	0	0	0	0	0	0	0	0	0	0	0	0	0	0	0	1	0
{Loss}	0	0	0	0	0	0	0	0	0	1	0	0	0	0	0	0	0	0	0	0	0	0	0	0	0	0	0	0	0	0	0	0
{Win,Loss}	0	0	0	0	0	0	0	0	0	0	0	0	0	0	0	0	0	0	0	0	0	0	0	0	0	0	0	0	0	0	0	1
{Draw,Loss}	0	0	0	0	0	0	0	0	0	0	0	0	0	0	0	1	0	0	0	0	0	0	0	0	0	0	0	0	0	0	0	0
{Win,Draw,Loss}	0	0	0	0	0	0	0	0	0	0	0	0	0	0	0	0	0	0	0	0	0	0	0	0	0	0	0	0	0	0	0	1

Fig. 19.5 Crosswise simulation of a relation and its existential image

Power relators

Since conversion is not preserved by the existential image, researchers kept looking for a construct for which this also holds, and invented the power relator we are now going to define. It is different in nature; in particular it is not a mapping as are the power transpose and the existential image.

Definition 19.10. Given any relation $R : X \longrightarrow Y$ together with membership relations $\varepsilon : X \longrightarrow 2^X$ and $\varepsilon' : Y \longrightarrow 2^Y$ on the source, respectively, target side, we call

$$\mathtt{PowRlt}(R) := \big(\varepsilon \backslash (R \,\varsemi\, \varepsilon')\big) \cap \big((\varepsilon^\mathsf{T} \,\varsemi\, R)/\varepsilon'^\mathsf{T}\big) = \overline{\varepsilon^\mathsf{T} \,\varsemi\, \overline{R \,\varsemi\, \varepsilon'}} \cap \overline{\varepsilon^\mathsf{T} \,\varsemi\, \overline{R \,\varsemi\, \varepsilon'}}$$

its **power relator**.　　　□

We will abbreviate this notation when appropriate as

$$\zeta := \zeta_R := \mathtt{PowRlt}(R).$$

It is some sort of symmetrized existential image since it satisfies

$$\zeta_R \supseteq \vartheta_R \qquad \text{and} \qquad \zeta_R \supseteq \big[\vartheta_{R^\mathsf{T}}\big]^\mathsf{T}.$$

In many cases, we have deduced the point-free version from a predicate-logic version in order to justify the shorter form. This time, we proceed in a similar way starting from two points e_x, e_y in the powerset. Then we expand the relational form so as to obtain the corresponding non-lifted expression for x, y.

$$e_x \,\varsemi\, e_y^\mathsf{T} \subseteq \zeta_R = \overline{\varepsilon^\mathsf{T} \,\varsemi\, \overline{R \,\varsemi\, \varepsilon'}} \cap \overline{\varepsilon^\mathsf{T} \,\varsemi\, \overline{R \,\varsemi\, \varepsilon'}}$$

$\Longleftrightarrow \quad e_x \,\varsemi\, e_y^\mathsf{T} \subseteq \overline{\varepsilon^\mathsf{T} \,\varsemi\, \overline{R \,\varsemi\, \varepsilon'}}$ and $e_x \,\varsemi\, e_y^\mathsf{T} \subseteq \overline{\varepsilon^\mathsf{T} \,\varsemi\, \overline{R \,\varsemi\, \varepsilon'}}$ ⠀splitting conjunction

$\Longleftrightarrow \quad e_x \subseteq \overline{\varepsilon^\mathsf{T} \,\varsemi\, \overline{R \,\varsemi\, \varepsilon'}} \,\varsemi\, e_y$ and $e_y^\mathsf{T} \subseteq e_x^\mathsf{T} \,\varsemi\, \overline{\varepsilon^\mathsf{T} \,\varsemi\, \overline{R \,\varsemi\, \varepsilon'}}$ ⠀Prop. 5.12.ii

$\Longleftrightarrow \quad e_x \subseteq \overline{\varepsilon^\mathsf{T} \,\varsemi\, \overline{R \,\varsemi\, \varepsilon'} \,\varsemi\, e_y}$ and $e_y^\mathsf{T} \subseteq \overline{e_x^\mathsf{T} \,\varsemi\, \varepsilon^\mathsf{T} \,\varsemi\, \overline{R \,\varsemi\, \varepsilon'}}$ ⠀points slipping

$\Longleftrightarrow \quad e_x \subseteq \overline{\varepsilon^\mathsf{T} \,\varsemi\, \overline{R \,\varsemi\, y}}$ and $e_y^\mathsf{T} \subseteq \overline{x^\mathsf{T} \,\varsemi\, \overline{R \,\varsemi\, \varepsilon'}}$ ⠀$x = \varepsilon \,\varsemi\, e_x$ and $y = \varepsilon' \,\varsemi\, e_y$

$\Longleftrightarrow \quad \varepsilon^\mathsf{T} \,\varsemi\, \overline{R \,\varsemi\, y} \subseteq \overline{e_x}$ and $\varepsilon'^\mathsf{T} \,\varsemi\, \overline{R^\mathsf{T} \,\varsemi\, x} \subseteq \overline{e_y}$ ⠀negated, transposed

$\Longleftrightarrow \quad \varepsilon \,\varsemi\, e_x \subseteq \overline{R \,\varsemi\, y}$ and $\varepsilon' \,\varsemi\, e_y \subseteq \overline{R^\mathsf{T} \,\varsemi\, x}$ ⠀Schröder equivalence

$\Longleftrightarrow \quad x \subseteq \overline{R \,\varsemi\, y}$ and $y \subseteq \overline{R^\mathsf{T} \,\varsemi\, x}$ ⠀because $x = \varepsilon \,\varsemi\, e_x$, $y = \varepsilon' \,\varsemi\, e_y$

We have, thus, identified what the power relator expresses for pairs x, y consisting of a vector x on the source side and a vector y on the target side. When considering R restricted to the rectangle made up by x and y, it is at the same time total and surjective. This means that every point of x is an inverse image of some point of y and in addition that every point in y is an image of some point in x.

Proposition 19.11. *A power relator is multiplicative, and reproduces identity and conversion:*

$$\zeta_R \,\mathring{,}\, \zeta_S = \zeta_{R\,\mathring{,}\, S}, \qquad \zeta_{\mathbb{I}_X} = \mathbb{I}_{2^X}, \qquad and \qquad \zeta_{R^{\mathsf{T}}} = \left[\zeta_R\right]^{\mathsf{T}}.$$

Proof: We start with the simple cases:

$$\zeta_{\mathbb{I}_X} = \left(\varepsilon\backslash(\mathbb{I}\,\mathring{,}\,\varepsilon)\right) \cap \left((\varepsilon^{\mathsf{T}}\,\mathring{,}\,\mathbb{I})/\varepsilon^{\mathsf{T}}\right) = \varepsilon\backslash\varepsilon \cap \varepsilon^{\mathsf{T}}/\varepsilon^{\mathsf{T}} = \mathbf{syq}\,(\varepsilon,\varepsilon) = \mathbb{I}_{2^X}$$

$$
\begin{aligned}
\zeta_{R^{\mathsf{T}}} &= \overline{\varepsilon'^{\mathsf{T}}\,\mathring{,}\,\overline{R^{\mathsf{T}}\,\mathring{,}\,\varepsilon}} \cap \overline{\varepsilon'^{\mathsf{T}}\,\mathring{,}\,\overline{R^{\mathsf{T}}}\,\mathring{,}\,\varepsilon} \qquad && \text{expanded}\\
&= \left[\overline{\varepsilon^{\mathsf{T}}\,\mathring{,}\,\overline{R\,\mathring{,}\,\varepsilon'}} \cap \overline{\varepsilon^{\mathsf{T}}\,\mathring{,}\,\overline{R}\,\mathring{,}\,\varepsilon'}\right]^{\mathsf{T}} && \text{transposed and exchanged}\\
&= \left[\zeta_R\right]^{\mathsf{T}} && \text{by definition}
\end{aligned}
$$

$$
\begin{aligned}
\zeta_R \,\mathring{,}\, \zeta_S \subseteq \zeta_{R\,\mathring{,}\, S} \quad &\Longleftrightarrow\quad \overline{\zeta_{R\,\mathring{,}\, S}}\,\mathring{,}\,\zeta_S^{\mathsf{T}} \subseteq \overline{\zeta_R} \qquad && \text{Schröder}\\
&\Longleftrightarrow\quad \left(\overline{\varepsilon^{\mathsf{T}}\,\mathring{,}\,\overline{R\,\mathring{,}\, S\,\mathring{,}\,\varepsilon''}} \cup \overline{\varepsilon^{\mathsf{T}}\,\mathring{,}\,\overline{R}\,\mathring{,}\, S\,\mathring{,}\,\varepsilon''}\right) \mathring{,} \left(\overline{\varepsilon'^{\mathsf{T}}\,\mathring{,}\,\overline{S\,\mathring{,}\,\varepsilon''}} \cap \overline{\varepsilon'^{\mathsf{T}}\,\mathring{,}\,\overline{S}\,\mathring{,}\,\varepsilon''}\right)^{\mathsf{T}} \subseteq \overline{\varepsilon^{\mathsf{T}}\,\mathring{,}\,\overline{R\,\mathring{,}\,\varepsilon'}} \cup \overline{\varepsilon^{\mathsf{T}}\,\mathring{,}\,\overline{R}\,\mathring{,}\,\varepsilon'}\\
&\Longleftarrow\quad \overline{\varepsilon^{\mathsf{T}}\,\mathring{,}\,\overline{R\,\mathring{,}\, S\,\mathring{,}\,\varepsilon''}}\,\mathring{,}\,\overline{\varepsilon'^{\mathsf{T}}\,\mathring{,}\,\overline{S\,\mathring{,}\,\varepsilon''}}^{\mathsf{T}} \subseteq \overline{\varepsilon^{\mathsf{T}}\,\mathring{,}\,\overline{R\,\mathring{,}\,\varepsilon'}} \quad && \text{and}\quad \overline{\varepsilon^{\mathsf{T}}\,\mathring{,}\,\overline{R\,\mathring{,}\, S\,\mathring{,}\,\varepsilon''}}\,\mathring{,}\,\overline{\varepsilon'^{\mathsf{T}}\,\mathring{,}\,\overline{S}\,\mathring{,}\,\varepsilon''}^{\mathsf{T}} \subseteq \overline{\varepsilon^{\mathsf{T}}\,\mathring{,}\,\overline{R}\,\mathring{,}\,\varepsilon'}\\
&\Longleftarrow\quad \overline{\overline{R\,\mathring{,}\, S\,\mathring{,}\,\varepsilon''}}\,\mathring{,}\,\overline{\varepsilon'^{\mathsf{T}}\,\mathring{,}\,\overline{S\,\mathring{,}\,\varepsilon''}}^{\mathsf{T}} \subseteq \overline{R\,\mathring{,}\,\varepsilon'} \quad && \text{and}\quad \overline{\varepsilon^{\mathsf{T}}\,\mathring{,}\,\overline{R\,\mathring{,}\, S}\,\mathring{,}\,\varepsilon''\,\mathring{,}\,\varepsilon''^{\mathsf{T}}\,\mathring{,}\,\overline{\varepsilon'^{\mathsf{T}}\,\mathring{,}\,\overline{S}}}^{\mathsf{T}} \subseteq \overline{\varepsilon^{\mathsf{T}}\,\mathring{,}\,\overline{R}\,\mathring{,}\,\varepsilon'}\\
&\Longleftrightarrow\quad \overline{R\,\mathring{,}\,\varepsilon'}\,\mathring{,}\,\overline{\varepsilon'^{\mathsf{T}}\,\mathring{,}\,\overline{S\,\mathring{,}\,\varepsilon''}} \subseteq \overline{R\,\mathring{,}\, S\,\mathring{,}\,\varepsilon''} \quad && \text{and}\quad \overline{\varepsilon^{\mathsf{T}}\,\mathring{,}\,\overline{R\,\mathring{,}\, S}\,\mathring{,}\,\left(\varepsilon'^{\mathsf{T}}\,\mathring{,}\,\overline{S}\right)^{\mathsf{T}}} \subseteq \overline{\varepsilon^{\mathsf{T}}\,\mathring{,}\,\overline{R}\,\mathring{,}\,\varepsilon'}
\end{aligned}
$$

where the latter are more or less immediately satisfied according to Prop. 7.14.

The reverse containment requires the Point Axiom. Therefore, we assume points $e_x, e_y \subseteq \zeta_{R\,\mathring{,}\, S}$ and start with shunting a point and letting it slip below negation:

$$
\begin{aligned}
e_x\,\mathring{,}\, e_y^{\mathsf{T}} \subseteq \zeta_{R\,\mathring{,}\, S} &= \overline{\varepsilon^{\mathsf{T}}\,\mathring{,}\,\overline{R\,\mathring{,}\, S\,\mathring{,}\,\varepsilon''}} \cap \overline{\varepsilon^{\mathsf{T}}\,\mathring{,}\,\overline{R}\,\mathring{,}\, S\,\mathring{,}\,\varepsilon''}\\
\Longleftrightarrow\quad e_x &\subseteq \left[\overline{\varepsilon^{\mathsf{T}}\,\mathring{,}\,\overline{R\,\mathring{,}\, S\,\mathring{,}\,\varepsilon''}} \cap \overline{\varepsilon^{\mathsf{T}}\,\mathring{,}\,\overline{R}\,\mathring{,}\, S\,\mathring{,}\,\varepsilon''}\right]\mathring{,}\, e_y = \overline{\varepsilon^{\mathsf{T}}\,\mathring{,}\,\overline{R\,\mathring{,}\, S\,\mathring{,}\,\varepsilon''}\,\mathring{,}\, e_y} \cap \overline{\varepsilon^{\mathsf{T}}\,\mathring{,}\,\overline{R}\,\mathring{,}\, S\,\mathring{,}\,\varepsilon''\,\mathring{,}\, e_y}\\
&= \overline{\varepsilon^{\mathsf{T}}\,\mathring{,}\,\overline{R\,\mathring{,}\, S\,\mathring{,}\, y}} \cap \overline{\varepsilon^{\mathsf{T}}\,\mathring{,}\,\overline{R}\,\mathring{,}\, S\,\mathring{,}\, y} \subseteq \overline{\varepsilon^{\mathsf{T}}\,\mathring{,}\,\overline{R}\,\mathring{,}\, S\,\mathring{,}\, y}\\
\Longleftrightarrow\quad \varepsilon^{\mathsf{T}}\,\mathring{,}\,&\overline{R}\,\mathring{,}\, S\,\mathring{,}\, y \subseteq \overline{e_x} \quad \Longleftrightarrow\quad x = \varepsilon\,\mathring{,}\, e_x \subseteq \overline{R}\,\mathring{,}\, S\,\mathring{,}\, y
\end{aligned}
$$

Here, we take the opportunity to define an intermediate vector as $z := R^{\mathsf{T}}\,\mathring{,}\, x \cap S\,\mathring{,}\, y$, together with the point $e_z := \mathbf{syq}\,(\varepsilon', z)$ and proceed as follows:

$$x = \varepsilon\,\mathring{,}\, e_x \subseteq \overline{R}\,\mathring{,}\, S\,\mathring{,}\, y \cap x \subseteq (R \cap x\,\mathring{,}\, y^{\mathsf{T}}\,\mathring{,}\, S^{\mathsf{T}})\,\mathring{,}\,(S\,\mathring{,}\, y \cap R^{\mathsf{T}}\,\mathring{,}\, x) \subseteq \overline{R}\,\mathring{,}\, z \quad \Longleftrightarrow\quad \varepsilon^{\mathsf{T}}\,\mathring{,}\,\overline{R}\,\mathring{,}\, z \subseteq \overline{e_x}$$

We prove $e_x\,\mathring{,}\, e_z^{\mathsf{T}} \subseteq \zeta_R$, and symmetrically $e_z\,\mathring{,}\, e_y^{\mathsf{T}} \subseteq \zeta_S$, so that multiplying the latter two gives $e_x\,\mathring{,}\, e_z^{\mathsf{T}}\,\mathring{,}\, e_z\,\mathring{,}\, e_y^{\mathsf{T}} = e_x\,\mathring{,}\, e_y^{\mathsf{T}} \subseteq \zeta_R\,\mathring{,}\, \zeta_S$:

$$
\begin{aligned}
\varepsilon^{\mathsf{T}}\,\mathring{,}\,\overline{R}\,\mathring{,}\, z = \varepsilon^{\mathsf{T}}\,\mathring{,}\,\overline{R}\,\mathring{,}\,\varepsilon'\,\mathring{,}\, e_z &= \varepsilon^{\mathsf{T}}\,\mathring{,}\,\overline{R\,\mathring{,}\,\varepsilon'}\,\mathring{,}\, e_z \subseteq \overline{e_x} \quad && \Longleftrightarrow\quad e_x\,\mathring{,}\, e_z^{\mathsf{T}} \subseteq \overline{\varepsilon^{\mathsf{T}}\,\mathring{,}\,\overline{R\,\mathring{,}\,\varepsilon'}}\\
z = \varepsilon'\,\mathring{,}\, e_z \subseteq R^{\mathsf{T}}\,\mathring{,}\, x \quad &\Longleftrightarrow\quad \varepsilon'^{\mathsf{T}}\,\mathring{,}\,\overline{R^{\mathsf{T}}\,\mathring{,}\,\varepsilon}\,\mathring{,}\, e_x = \varepsilon'^{\mathsf{T}}\,\mathring{,}\,\overline{R^{\mathsf{T}}}\,\mathring{,}\,\varepsilon\,\mathring{,}\, e_x = \varepsilon'^{\mathsf{T}}\,\mathring{,}\,\overline{R^{\mathsf{T}}\,\mathring{,}\, x} \subseteq \overline{e_z}\\
&\Longleftrightarrow\quad e_z\,\mathring{,}\, e_x^{\mathsf{T}} \subseteq \overline{\varepsilon'^{\mathsf{T}}\,\mathring{,}\,\overline{R^{\mathsf{T}}\,\mathring{,}\,\varepsilon}} \quad && \Longleftrightarrow\quad e_x\,\mathring{,}\, e_z^{\mathsf{T}} \subseteq \overline{\varepsilon^{\mathsf{T}}\,\mathring{,}\,\overline{R}\,\mathring{,}\,\varepsilon'} \qquad\qquad \square
\end{aligned}
$$

This result allows a byproduct that one might not have thought of in the first place.

Proposition 19.12 (Intermediate Vector Theorem). *Let any relations R, S be given and a vector x along the source side of R together with a vector y along the target side of S. Assuming the Intermediate Point Theorem Prop. 8.24 to hold, and thus the Point Axiom, the following are equivalent:*

- $x \subseteq R \,\fatsemi\, S \,\fatsemi\, y$ *and* $y \subseteq (R \,\fatsemi\, S)^{\mathsf{T}} \,\fatsemi\, x$
- $x \subseteq R \,\fatsemi\, z$ *and* $z \subseteq R^{\mathsf{T}} \,\fatsemi\, x$ *and* $z \subseteq S \,\fatsemi\, y$ *and* $y \subseteq S^{\mathsf{T}} \,\fatsemi\, z$ *for some vector z.*

Proof: With the remarks immediately after Def. 19.10, we reduce the problem so as to be able to apply the Intermediate Point Theorem.

$$
\begin{aligned}
& x \subseteq R \,\fatsemi\, S \,\fatsemi\, y \qquad\qquad \text{and} \qquad\qquad y \subseteq (R \,\fatsemi\, S)^{\mathsf{T}} \,\fatsemi\, x \\
\Longleftrightarrow\quad & e_x \,\fatsemi\, e_y^{\mathsf{T}} \subseteq \zeta_{R \,\fatsemi\, S} = \zeta_R \,\fatsemi\, \zeta_S \qquad \text{by the remark just mentioned and Prop. 19.11} \\
\Longleftrightarrow\quad & e_x \,\fatsemi\, e_z^{\mathsf{T}} \subseteq \zeta_R \quad \text{and} \quad e_z \,\fatsemi\, e_y^{\mathsf{T}} \subseteq \zeta_S \qquad \text{Intermediate Point Theorem 8.24} \\
\Longleftrightarrow\quad & x \subseteq R \,\fatsemi\, z \quad z \subseteq R^{\mathsf{T}} \,\fatsemi\, x \quad z \subseteq S \,\fatsemi\, y \quad y \subseteq S^{\mathsf{T}} \,\fatsemi\, z \qquad \text{re-interpreting} \qquad \square
\end{aligned}
$$

This result may be rephrased in lifted form as

$$
e_x \,\fatsemi\, e_y^{\mathsf{T}} \subseteq \zeta_{R \,\fatsemi\, S} \qquad \Longleftrightarrow \qquad e_x \,\fatsemi\, e_z^{\mathsf{T}} \subseteq \zeta_R \text{ and } e_z \,\fatsemi\, e_y^{\mathsf{T}} \subseteq \zeta_S \text{ for some point } e_z.
$$

Also the power relator is to a certain extent connected via simulation with the original relation.

Proposition 19.13. *Let a relation R together with its power relator ζ_R and the two membership relations $\varepsilon, \varepsilon'$ be given. Then ζ_R is an $\varepsilon^{\mathsf{T}}, \varepsilon'^{\mathsf{T}}$-L-simulation of R.*

Proof: $\varepsilon \,\fatsemi\, \zeta_R \subseteq \varepsilon \,\fatsemi\, \overline{\varepsilon^{\mathsf{T}} \,\fatsemi\, \overline{R \,\fatsemi\, \varepsilon'}} = R \,\fatsemi\, \varepsilon'$, from the definition and using Prop 7.14 $\square$

Besides the powerset ordering, there exists a preorder on the direct power that stems from an ordering on the underlying set. It is usually called an Egli–Milner[1] preorder.

Definition 19.14. For any given ordering relation $E : X \longrightarrow X$ and membership relation $\varepsilon : X \longrightarrow 2^X$, one may define the **Egli–Milner preorder**[2] on the powerset 2^X as

$$
\mathsf{EM}_E := \left(\varepsilon \backslash (E \,\fatsemi\, \varepsilon) \right) \cap \left((\varepsilon^{\mathsf{T}} \,\fatsemi\, E) / \varepsilon^{\mathsf{T}} \right) = \overline{\varepsilon^{\mathsf{T}} \,\fatsemi\, \overline{E \,\fatsemi\, \varepsilon}} \cap \overline{\overline{\varepsilon^{\mathsf{T}} \,\fatsemi\, E} \,\fatsemi\, \varepsilon} = \mathsf{PowRlt}(E). \qquad \square
$$

A first example of an Egli–Milner preorder is provided by the identity relation on X that produces the identity on 2^X since $\mathsf{EM}_{\mathbb{I}} := \overline{\varepsilon^{\mathsf{T}} \,\fatsemi\, \overline{\varepsilon}} \cap \overline{\overline{\varepsilon^{\mathsf{T}}} \,\fatsemi\, \varepsilon} = \Omega \cap \Omega^{\mathsf{T}} = \mathbb{I}_{2^X}$. A less trivial Egli–Milner example is provided by the linear 4-element-order of Fig. 19.6.

---

[1] Robin Milner (more precisely according to Wikipedia: Arthur John Robin Gorell Milner, born 1934) is a British computer scientist; in 1994 at TU München, he was awarded the F.-L.-Bauer Prize.

[2] Many people traditionally cut out the element $\emptyset$ because it is not related to any other element in the Egli–Milner preorder. We have chosen not to do so because it does not disturb too much. Cutting out would lead to a more complicated term.

When working in semantics of non-deterministic programs and domain construction, one often uses the Egli–Milner preorder but would like to use an order instead.

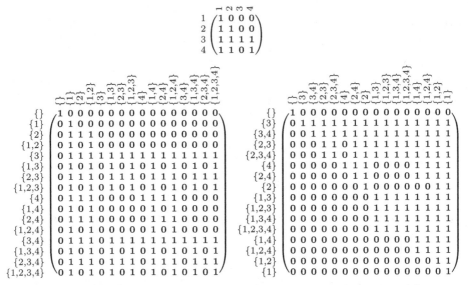

Fig. 19.6 Egli-Milner preorder of an ordering, rearranged to upper right

In such cases, one typically starts from a so-called flat ordering as in Fig. 19.7. The order on 4 elements is flat, i.e., there are 3 elements unordered and the 4th positioned beneath. Such a situation will always produce an Egli–Milner *ordering*, which may be seen here easily considering the rearranged version.

Exercises

19.1 Let a relation $R : X \longrightarrow Y$ be given together with the membership relation $\varepsilon' : Y \longrightarrow 2^Y$. Prove that $\Lambda_{\varepsilon'^{\mathsf{T}}} = \mathbb{I}$.

19.2 Provide a point-free proof that the Egli–Milner preorder of a flat ordering is in fact an ordering itself.

19.3 Prove that the Egli–Milner preorder as defined in Def. 19.14 is indeed a preorder.

19.2 Simulation and bisimulation

As announced, we will no longer have *mappings* when we are about to compare and have just *relations*. The main problem is that we will lose the ability to 'roll' the conditions so as to have four different but equivalent forms according to Prop. 5.45.

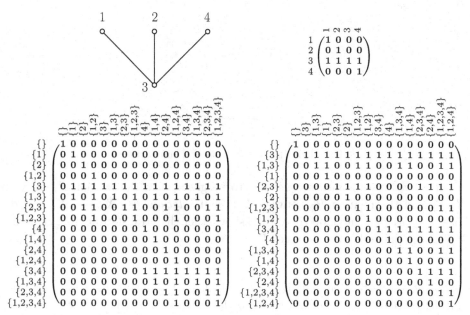

Fig. 19.7 Egli–Milner preorder of a flat ordering, also rearranged to upper right triangle

An important concept of process algebra is the state transition model; a state transition system is an abstract machine consisting of a set of states and (labelled) transitions between these. In a state transition system the set of states is not necessarily finite or countable, as for finite state automata; nor need the set of transitions be finite or countable. If finite, however, the state transition system can be represented by (labelled) directed graphs.

We first give the standard definition and proceed to a relational one afterwards.

Definition 19.15. A so-called **state transition system** consists of a set X of states and a set Λ of labels. These labels serve as indexes or names for relations $(\lambda : X \longrightarrow X)_{\lambda \in \Lambda}$ on X. □

In the case $|\Lambda| = 1$, the system is often called an *unlabelled* system since then there is no longer a need to mention the label explicitly; one will agree upon this label prior to starting investigations. If there is more than one label, one can consider the conditions that follow as combined by conjunction. As this is the case, we will mostly consider just one relation as a state transition.

We will begin more concretely with a heterogeneous situation.

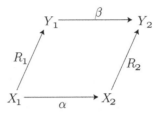

Fig. 19.8 Basic situation of a simulation

Assume two 'state transition systems' to be given, here the relation R_1 between X_1, Y_1 and the relation R_2 between X_2, Y_2. If *relations* $\alpha : X_1 \longrightarrow X_2$ and $\beta : Y_1 \longrightarrow Y_2$ are presented, one may still ask whether they transfer the 'first structure sufficiently precisely into the second'.

Definition 19.16. Relations α, β are called an *L*-**simulation**, sometimes also called **forward simulation**, of the second structure R_2 by the first structure R_1, provided the following holds. If an element x_2 is simulated via α by some element in X_1 which is in relation R_1 with y_1, then y_1 simulates via β some element in Y_2 to which y_1 is in relation R_2.

$$\forall x_2 \in X_2 : \forall y_1 \in Y_1 :$$
$$\big[\exists x_1 \in X_1 : (x_1, x_2) \in \alpha \wedge (x_1, y_1) \in R_1\big] \quad \longrightarrow$$
$$\big[\exists y_2 \in Y_2 : (x_2, y_2) \in R_2 \wedge (y_1, y_2) \in \beta\big] \qquad \qquad \square$$

The first observation is that the 'rolling' of Prop. 5.45 is no longer possible, since it depends heavily on univalence and totality. When bringing it to a point-free form in the same way as for homomorphisms, we will, therefore, arrive at several possibly non-equivalent forms. Anyway, a considerable amount of literature has emerged dealing with these questions.

Definition 19.17. Assume two state transition systems R_1, R_2. Given two relations α, β, we call

(i) R_1 an α, β-*L*-**simulation** of R_2 provided that $\alpha^\mathsf{T} {\,}_\mathsf{;} R_1 \subseteq R_2 {\,}_\mathsf{;} \beta^\mathsf{T}$,
(ii) R_1 an α, β-*L*$^\mathsf{T}$-**simulation** of R_2 provided that $R_1 {\,}_\mathsf{;} \beta \subseteq \alpha {\,}_\mathsf{;} R_2$,
(iii) R_1 an α, β-*U*-**simulation** of R_2 provided that $\alpha^\mathsf{T} {\,}_\mathsf{;} R_1 {\,}_\mathsf{;} \beta \subseteq R_2$,
(iv) R_1 an α, β-*U*$^\mathsf{T}$-**simulation** of R_2 provided that $R_1 \subseteq \alpha {\,}_\mathsf{;} R_2 {\,}_\mathsf{;} \beta^\mathsf{T}$. $\qquad \square$

Immediately, some technical questions arise. For example, do these simulations compose in some way similar to homomorphisms? Does any (or do any two) of these simulations imply another one? There exist a lot of interdependencies between these concepts which are explained at length in [42]. Two main directions may be found. First, one is interested in simulating something that is already a simulation, thus building larger simulations from tiny steps. This follows the line of Prop. 19.18. One may, however, also build simulations of large Rs that are composed of smaller parts via relational composition. For this, other results are adequate. Putting it geometrically, the first aligns rectangles of Fig. 19.9 horizontally while the second idea aligns them vertically; see Fig. 19.10.

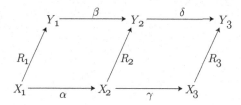

Fig. 19.9 Simulation of an already simulated relation

Proposition 19.18. *Let relations $\alpha, \beta, \gamma, \delta$ be given. Then one has – sometimes under conditions on functionality or totality – that the simulations considered compose horizontally:*

(i) $\begin{array}{l} R_1 \text{ is a } \alpha, \beta\text{-}L\text{-simulation of } R_2 \\ R_2 \text{ is a } \gamma, \delta\text{-}L\text{-simulation of } R_3 \end{array} \implies R_1 \text{ is a } (\alpha\,;\gamma), (\beta\,;\delta)\text{-}L\text{-simulation of } R_3.$

(ii) $\begin{array}{l} R_1 \text{ is a } \alpha, \beta\text{-}U\text{-simulation of } R_2 \\ R_2 \text{ is a } \gamma, \delta\text{-}U\text{-simulation of } R_3 \end{array} \implies R_1 \text{ is a } (\alpha\,;\gamma), (\beta\,;\delta)\text{-}U\text{-simulation of } R_3.$

(iii) $\begin{array}{l} R_1 \text{ is a } \alpha, \beta\text{-}U^{\mathsf{T}}\text{-simulation of } R_2 \\ R_2 \text{ is a } \gamma, \delta\text{-}L\text{-simulation of } R_3 \\ \alpha \text{ is univalent} \end{array} \implies R_1 \text{ is a } (\alpha\,;\gamma), (\beta\,;\delta)\text{-}L\text{-simulation of } R_3.$

Proof: (i) $\gamma^{\mathsf{T}}\,;\alpha^{\mathsf{T}}\,;R_1 \subseteq \gamma^{\mathsf{T}}\,;R_2\,;\beta^{\mathsf{T}} \subseteq R_3\,;\delta^{\mathsf{T}}\,;\beta^{\mathsf{T}}.$

(ii) $\gamma^{\mathsf{T}}\,;\alpha^{\mathsf{T}}\,;R_1\,;\beta\,;\delta \subseteq \gamma^{\mathsf{T}}\,;R_2\,;\delta \subseteq R_3.$

(iii) $\gamma^{\mathsf{T}}\,;\alpha^{\mathsf{T}}\,;R_1 \subseteq \gamma^{\mathsf{T}}\,;\alpha^{\mathsf{T}}\,;\alpha\,;R_2\,;\beta^{\mathsf{T}} \subseteq \gamma^{\mathsf{T}}\,;R_2\,;\beta^{\mathsf{T}} \subseteq R_3\,;\delta^{\mathsf{T}}\,;\beta^{\mathsf{T}},$ using that α is univalent. □

Already in Prop. 19.9, we have found simulations that are very interesting from the theoretical point of view.

Proposition 19.19. *Let relations α, β, γ be given. Then one has – maybe under* *conditions on functionality or totality – that the simulations considered compose vertically:*

(i) R_1 *is a* α, β-*L-simulation of* R_2
     S_1 *is a* β, γ-*L-simulation of* S_2 $\implies$ $R_1 \, ; S_1$ *is a* α, γ-*L-simulation of* $R_2 \, ; S_2$,

(ii) R_1 *is a* α, β-*U-simulation of* R_2
     S_1 *is a* β, γ-*U-simulation of* S_2 $\implies$ $R_1 \, ; S_1$ *is a* α, γ-*U-simulation of* $R_2 \, ; S_2$.
     β *is total*

Proof: (i) $\alpha^{\mathsf{T}} \, ; R_1 \, ; S_1 \subseteq R_2 \, ; \beta^{\mathsf{T}} \, ; S_1 \subseteq R_2 \, ; S_2 \, ; \gamma^{\mathsf{T}}$.

(ii) $\alpha^{\mathsf{T}} \, ; R_1 \, ; S_1 \, ; \gamma \subseteq \alpha^{\mathsf{T}} \, ; R_1 \, ; \beta \, ; \beta^{\mathsf{T}} \, ; S_1 \, ; \gamma \subseteq R_2 \, ; S_2$. $\square$

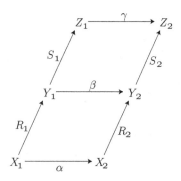

Fig. 19.10 Simulating composed steps

Bisimulation

Simulation has an obvious direction: R simulates S with some relations α, β from their sources to their targets. We will now try to get rid of a constituent of this concept. We look for a far more symmetric situation bringing simulation in forward and backward directions at the same time. In one respect it will become simpler; we will study only homogeneous relations.

Definition 19.20. Let two homogeneous relations $R : X \longrightarrow X$ and $S : Y \longrightarrow Y$ be given and consider them as *processes* working on the *state sets* X, respectively Y. We call a relation, not necessarily a mapping,

φ a **bisimulation** of S by R $:\Longleftrightarrow$ $\varphi^{\mathsf{T}} \, ; R \subseteq S \, ; \varphi^{\mathsf{T}}$ and $\varphi \, ; S \subseteq R \, ; \varphi$. $\square$

It is an easy exercise to express bisimulation by just one condition as

$$\varphi \subseteq \overline{R\!:\!\varphi\!:\!S^{\mathsf{T}}} \cap \overline{R\!:\!\varphi\!:\!S^{\mathsf{T}}} = R^{\mathsf{T}}\backslash(\varphi\!:\!S^{\mathsf{T}}) \cap (R\!:\!\varphi)/S$$

which looks quite similar to, but is not, a symmetric quotient. A first example of a bisimulation is provided in Fig. 19.11.

Bisimulation expresses equivalence of the flow of execution as represented by sequences of labels. When either one of the processes 'makes a move' according to its relation, then the other is able to measure up to it. When each of the processes is assembled in a black box that communicates the transitions executed only to the extent of giving the graphical label ● ○ ■ □ ☆ of the respective state, the systems cannot be distinguished from one another by an outside observer.

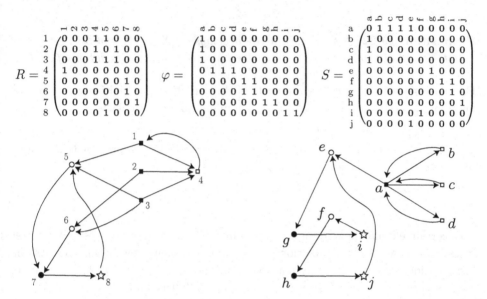

Fig. 19.11 A bisimulation φ of S by R

Often one studies the **simulation preorder**, saying nothing more than that *there exists* a simulating relation $\varphi : X \longrightarrow Y$ so that it simulates S by R. Note that it need not be the case that if R simulates S and S simulates R then they are bisimilar. Of course not; there may exist $\alpha : X \longrightarrow Y$ and $\beta : Y \longrightarrow X$, but one is not guaranteed that $\beta = \alpha^{\mathsf{T}}$. The simulation preorder is the largest simulation relation over a given transition system.

Not least in model-checking, one is very interested in an overview of the class of all (bi)simulations. We give an overview of all possible bisimulations with the following proposition, the proof of which is so trivial that it may be omitted.

Proposition 19.21. *Let the processes $R : X \longrightarrow X$ and $S : Y \longrightarrow Y$ be given.*

(i) *The empty relation* $\perp\!\!\!\perp : X \longrightarrow Y$ *is always a bisimulation.*
(ii) *The identity relation* $\mathbb{I} : X \longrightarrow X$ *is a bisimulation provided* $R = S$.
(iii) *The converse of a bisimulation is a bisimulation in the opposite direction.*
(iv) *The composition of bisimulations is a bisimulation again.*
(v) *Unions of bisimulations are bisimulations again.*   □

The class of all bisimulations has a nice structure, but contains very uninteresting elements such as $\perp\!\!\!\perp$. One may ask whether there exists a good method of comparison. We recall for this purpose the concept of a multi-covering.

Proposition 19.22. *Assume two multi-coverings* $\alpha : X \longrightarrow Z$ *and* $\beta : Y \longrightarrow Z$ *of* R *onto* T *and of* S *onto* T *to be given. Then* $\varphi := \alpha \,\dot{;}\, \beta^{\mathsf{T}}$ *is a bisimulation of* S *by* R.

Proof: First we recall the requirements of the definition of a multi-covering on the mappings α and β, namely that they be homomorphisms and satisfy

$$R \,\dot{;}\, \alpha = \alpha \,\dot{;}\, T \qquad S \,\dot{;}\, \beta = \beta \,\dot{;}\, T.$$

With homomorphisms α, β rolled according to Prop. 5.45, the proof is now simple:

$$\varphi^{\mathsf{T}} \,\dot{;}\, R = \beta \,\dot{;}\, \alpha^{\mathsf{T}} \,\dot{;}\, R \subseteq \beta \,\dot{;}\, T \,\dot{;}\, \alpha^{\mathsf{T}} = S \,\dot{;}\, \beta \,\dot{;}\, \alpha^{\mathsf{T}} = S \,\dot{;}\, \varphi^{\mathsf{T}}$$
$$\varphi \,\dot{;}\, S = \alpha \,\dot{;}\, \beta^{\mathsf{T}} \,\dot{;}\, S \subseteq \alpha \,\dot{;}\, T \,\dot{;}\, \beta^{\mathsf{T}} = R \,\dot{;}\, \alpha \,\dot{;}\, \beta^{\mathsf{T}} = R \,\dot{;}\, \varphi.$$ □

To a certain extent, the opposite direction may also be proved. To this end, we first investigate how close a simulation comes to a congruence on either side. As the bisimulation $\perp\!\!\!\perp$ shows, this may not always be the case. In Fig. 19.12, for example, $\varphi \,\dot{;}\, \varphi^{\mathsf{T}}$ is not transitive: it relates 6 to 9 and 9 to 1, but not 6 to 1.

Proposition 19.23. *For every bisimulation* φ *between the processes* R *and* S, *the reflexive-transitive closures give rise to the* R-*congruence* $\Xi := (\varphi \,\dot{;}\, \varphi^{\mathsf{T}})^{*}$ *and the* S-*congruence* $\Psi := (\varphi^{\mathsf{T}} \,\dot{;}\, \varphi)^{*}$.

Proof: Ξ and Ψ are symmetric; transitivity and reflexivity have been added by construction.

$$\varphi \,\dot{;}\, \varphi^{\mathsf{T}} \,\dot{;}\, R \subseteq \varphi \,\dot{;}\, S \,\dot{;}\, \varphi^{\mathsf{T}} \subseteq R \,\dot{;}\, \varphi \,\dot{;}\, \varphi^{\mathsf{T}}$$

so that this extends to the powers, and finally to the closure; similarly for S.   □

It is an appealing idea to relate the concepts of simulation and bisimulation in the following way.

Proposition 19.24. *Let the processes* $R : X \longrightarrow X$ *and* $S : Y \longrightarrow Y$ *be given and consider the injections* $\iota : X \longrightarrow X + Y$ *and* $\kappa : Y \longrightarrow X + Y$ *into the direct sum of the state spaces. The relation* φ *is a bisimulation of* S *by* R *precisely when* $\alpha := \varphi_{;} \kappa \cup \varphi^{\mathsf{T}}{}_{;}\iota$ *is an* L-*simulation of the sum process* $\Sigma := \iota^{\mathsf{T}}{}_{;} R_{;} \iota \cup \kappa^{\mathsf{T}}{}_{;} S_{;} \kappa$ *of itself.*

Proof: We explain what happens with a matrix of relations. This makes the proof of $\alpha^{\mathsf{T}}{}_{;} \Sigma \subseteq \Sigma_{;} \alpha^{\mathsf{T}}$ evident:

$$\Sigma := \begin{pmatrix} R & \mathbb{1} \\ \mathbb{1} & S \end{pmatrix} \qquad \alpha = \alpha^{\mathsf{T}} := \begin{pmatrix} \mathbb{1} & \varphi \\ \varphi^{\mathsf{T}} & \mathbb{1} \end{pmatrix}. \qquad \square$$

In many respects, these relational investigations go back to [75, 76, 148, 149].

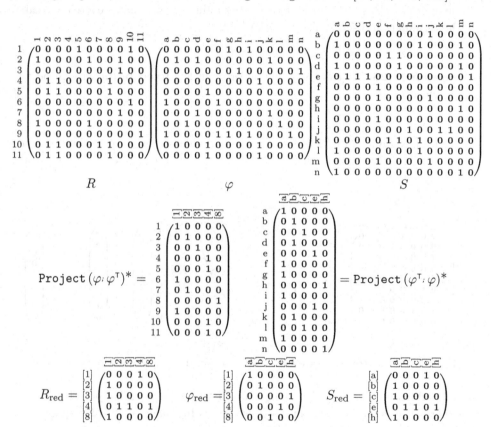

Fig. 19.12 A bisimulation with multi-coverings on a common quotient

19.3 State transition dynamics

Dynamical systems are traditionally studied with dynamics given over space and time. Usually, one uses differential equations that regulate the local situations in an infinitesimal manner and one expects to find answers concerning the global behavior. Our attempt here is a bit different since it starts in a discrete situation and tries to stay there. Nonetheless, we will get important results. Our initial reference here is [130], according to which we introduce some application-oriented mode of speaking.

Definition 19.25. In the present context, we call the total relation $Q : S \longrightarrow S$, a **state transition dynamics**, while the set S is called a **state space**. □

We assume Q to be total in order to exclude handling topics that are already treated in the termination discussions of Section 16.2.

Definition 19.26. We call the point set or point x a **fixed point** with respect to Q, provided $Q^{\mathsf{T}} {\,;\,} x = x$. □

We give an example in Fig. 19.13. A relation Q is given for state transition dynamics together with two vectors, each describing a set that is reproduced under application $v \mapsto Q^{\mathsf{T}} {\,;\,} v$.

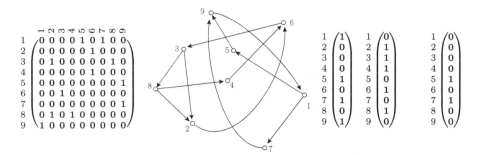

Fig. 19.13 State transition dynamics with two fixed point sets and a 3-periodic one

Fixed points are not really interesting since it will more often happen that a given starting set v changes when mapped to $Q^{\mathsf{T}} {\,;\,} v$. In many practical respects it is necessary to keep track of how v changes.

Definition 19.27. We call the infinite sequence $v = (v_i)_{i \in \mathbb{N}}$ of vectors an **orbit** with respect to Q starting with the **origin** v_0, if $v_{i+1} = Q^{\mathsf{T}} {\,;\,} v_i$ for all $i \geq 0$. □

$$
\begin{array}{c}
\;\; \scriptstyle 1\;2\;3\;4\;5\;6\;7\;8\;9 \\
\begin{array}{c} 1\\2\\3\\4\\5\\6\\7\\8\\9 \end{array}
\left(\begin{array}{ccccccccc}
0&0&0&0&1&0&1&0&0\\
0&0&0&0&0&1&0&0&0\\
0&1&0&0&0&0&0&1&0\\
0&0&0&0&0&1&0&0&0\\
0&0&0&0&0&0&0&0&1\\
0&0&1&0&0&0&0&0&0\\
0&0&0&0&0&0&0&0&1\\
0&1&0&1&0&0&0&0&0\\
1&0&0&0&0&0&0&0&0
\end{array}\right)
\end{array}
\;
\begin{pmatrix}0\\1\\1\\0\\0\\0\\0\\0\\0\end{pmatrix}
\begin{pmatrix}0\\1\\0\\0\\0\\1\\0\\1\\0\end{pmatrix}
\begin{pmatrix}0\\1\\1\\1\\0\\1\\0\\0\\0\end{pmatrix}
\begin{pmatrix}0\\1\\1\\0\\0\\1\\0\\1\\0\end{pmatrix}
\begin{pmatrix}0\\1\\1\\1\\0\\1\\0\\1\\0\end{pmatrix}
\begin{pmatrix}0\\1\\1\\1\\0\\1\\0\\1\\0\end{pmatrix}
$$

Fig. 19.14 Eventually periodic orbit of the state transition dynamics of Fig. 19.13

System dynamics is mainly concerned with such orbits and handles questions of the type: is this an orbit that will have some guaranteed behavior, at least eventually? Such behavior is characterized by facts such as 'is periodic', 'will later definitely be caught by some narrower orbit'. An example is the behavior of a system in the case of an error: will it after successful or unsuccessful reaction *definitely* fall back into some safe situation.

Definition 19.28. We call the orbit $(v_i)_{i \in \mathbb{N}}$ **periodic** if there exists an $i > 0$ with $v_i = v_0$. The orbit is **eventually periodic** if there exists some $0 \leq k \in \mathbb{N}$ such that the orbit starting with $v_0' := v_k = Q^{k^\mathsf{T}} \cdot v_0$ is periodic. □

Of course, this will then hold in the same way for every k' exceeding k. In Fig. 19.14, we see that the subset $\{2, 3\}$ is eventually periodic. The set $\{2, 3, 4, 6, 8\}$ has already been exhibited as a fixed point in Fig. 19.13, i.e., 1-periodic.

Instead of becoming periodic as in this example, an orbit may stay completely inside another orbit over all time, a situation for which we now provide notation.

Definition 19.29. We say that the orbit v is **contained** in the orbit w if for all $i \in \mathbb{N}$ $v_i \subseteq w_i$. The orbit starting at origin v will **fully discharge** into the orbit starting at origin w if a natural number j exists such that $v_j \subseteq w$. □

One should observe that being contained has been defined as a relation between orbits. To discharge fully, in contrast, is a relation between states, i.e., point sets. How difficult it is to get an overview on the relative situation of orbits may be estimated from the following two orbits

$\{2\}, \{6\}, \{3\}, \{2, 8\}, \{2, 4, 6\}, \{3, 6\}, \{2, 3, 8\}, \{2, 4, 6, 8\},$
$\qquad\{2, 3, 4, 6\}, \{2, 3, 6, 8\}, \{2, 3, 4, 6, 8\}, \{2, 3, 4, 6, 8\}, \{2, 3, 4, 6, 8\}, \ldots$

$\{4\}, \{6\}, \{3\}, \{2, 8\}, \{2, 4, 6\}, \{3, 6\}, \{2, 3, 8\}, \{2, 4, 6, 8\},$
$\qquad\{2, 3, 4, 6\}, \{2, 3, 6, 8\}, \{2, 3, 4, 6, 8\}, \{2, 3, 4, 6, 8\}, \{2, 3, 4, 6, 8\}, \ldots$

of our standard example. The orbit starting at $\{2\}$ is contained in the orbit starting at $\{2,3\}$, while the one starting at $\{2,8\}$ will fully discharge into the orbit starting with $\{3,6\}$. So comparing orbits, seeing whether one is contained in the other, is not too easy by just looking at the matrix. An algebraic criterion should be developed. To facilitate an algebraic characterization the view is restricted to relations between point sets in the concept of fully discharging.

For the first time we will now unavoidably have to use the element representation of a subset in the powerset, namely $e_v := \mathsf{syq}(\varepsilon, v)$ instead of just v. Parallel to this, we will also lift Q so as to obtain the existential image $\vartheta := \vartheta_Q := \mathsf{syq}(Q^{\mathsf{T}}; \varepsilon, \varepsilon)$. It is a mapping of the powerset of the state set into itself; see Fig. 19.5, where already examples were given showing how the relation Q and its existential image mutually simulate each other. The setting is slightly simplified here because Q is assumed to be homogeneous; an example is given in Fig. 19.15.

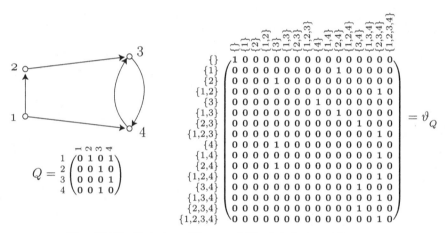

Fig. 19.15 State transitions and lifted state transitions

We recall from Prop. 19.9 that a relation Q – in the present context homogeneous – and its existential image ϑ simulate one another as follows

$$\varepsilon^{\mathsf{T}}; Q = \vartheta; \varepsilon^{\mathsf{T}}.$$

Furthermore, we distinguish always a subset or vector v and its corresponding point e_v in the powerset with interrelationship

$$e_v := \mathsf{syq}(\varepsilon, v) \quad v = \varepsilon; e_v.$$

The next formulae compare the operation of Q on v and the operation of ϑ on e_v:

$$\vartheta^{\mathsf{T}}; e_v = \mathsf{syq}(\varepsilon, Q^{\mathsf{T}}; \varepsilon); e_v \qquad \text{expanded with Def. 19.1.ii and transposed}$$
$$= \mathsf{syq}(\varepsilon, Q^{\mathsf{T}}; \varepsilon; e_v) \qquad \text{a point is a transposed mapping, Prop. 8.16.iii}$$
$$= \mathsf{syq}(\varepsilon, Q^{\mathsf{T}}; v) \qquad \text{see above}$$

Thus, applying Q to v and 'lifting' the result gives the same result as lifting to e_v

first and applying ϑ. In an analogous way $Q^{\mathsf{T}} ; v = \varepsilon ; \vartheta^{\mathsf{T}} ; e_v$, i.e., applying Q to the vector v is the same as lifting v, looking for its image via ϑ and 'flattening' back to the vector.

Given a graph or relation Q, one will almost immediately think of its reflexive-transitive closure Q^* representing reachability. Once we have the lifted relation ϑ for Q, we may also look for reachability ϑ^*. Already a short look at Fig. 19.16 shows that the lifted version offers a far more detailed view. While with Q^*, the iteration starting from v must be executed operationally, ϑ^* shows directly all vectors that can occur when starting from e_v. However, it does not give any indication as to the sequence in which these occur; for example, $\{1,3\}, \{2,4\}, \{3\}, \{4\}, \{3\}, \{4\} \ldots$

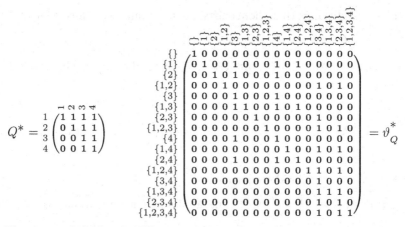

Fig. 19.16 Lifted and unlifted reachability of state transitions of Fig. 19.15

A vertex y is reachable via Q from some vertex x, provided

$$x ; y^{\mathsf{T}} \subseteq Q^* \qquad \text{or after shunting if} \qquad x \subseteq Q^* ; y.$$

A vector w is reachable via ϑ from some vector v, provided the points e_v, e_w satisfy

$$e_v ; e_w^{\mathsf{T}} \subseteq \vartheta^* \qquad \text{or after shunting if} \qquad e_v \subseteq \vartheta^* ; e_w.$$

It is, however, more difficult to reason with Q about *vectors* v, w – as opposed to *points*. To find out in what way, we look for 'reachability in n or more steps'. While the lifted version keeps the sets passed sufficiently separate, the non-lifted version does not. In the following construction, one must not mix up these two aspects: work in the lifted source and work in the original one.

In view of Fig. 19.16, one may wonder in particular why vectors $\{3\}, \{4\}, \{3,4\}$ are not in one equivalence class modulo $\vartheta^* \cap \vartheta^{*\mathsf{T}}$. But when starting from 1, for example, one will pass $\{1\}, \{2,4\}$ and finally toggle $\{3\}, \{4\}, \{3\} \ldots$ infinitely. In contrast, starting from $\{1,2\}$ proceeds with $\{2,3,4\}, \{3,4\}, \{3,4\} \ldots$ Only the starting point makes the difference between the orbits starting at $\{3\}$ and $\{4\}$. All three, i.e., those starting from $\{3\}$, $\{4\}$, and $\{1,2\}$ are eventually periodic, but only $\{3,4\}$ and $\{4\}$ are periodic.

Proposition 19.30. *The orbit starting at origin v will fully discharge into the orbit starting at origin w precisely when the lifted versions of v, w are in the 'fully discharges into' preorder $\vartheta^{*}{}_{;}\Omega$, i.e., when*

$$e_{v}{}_{;}e_{w}^{\mathsf{T}} \subseteq \vartheta^{*}{}_{;}\Omega.$$

Proof: The condition is that there exists a natural number k such that $Q^{k^{\mathsf{T}}}{}_{;}v \subseteq w$:

$$\varepsilon_{;}\vartheta^{k^{\mathsf{T}}}{}_{;}e_{v} = Q^{k^{\mathsf{T}}}{}_{;}\varepsilon_{;}e_{v} = Q^{k^{\mathsf{T}}}{}_{;}v \subseteq w = \varepsilon_{;}e_{w}$$

is obtained using the simulation property and the condition initially given. Shunting results in

$$\varepsilon_{;}\vartheta^{k^{\mathsf{T}}}{}_{;}e_{v}{}_{;}e_{w}^{\mathsf{T}} \subseteq \varepsilon$$

$$\overline{\Omega} = \varepsilon^{\mathsf{T}}{}_{;}\overline{\varepsilon} \subseteq \overline{\vartheta^{k^{\mathsf{T}}}{}_{;}e_{v}{}_{;}e_{w}^{\mathsf{T}}} \qquad \text{Schröder equivalence; definition of powerset order } \Omega$$

$$\vartheta^{k^{\mathsf{T}}}{}_{;}e_{v}{}_{;}e_{w}^{\mathsf{T}} \subseteq \Omega \qquad \text{negated}$$

$$\vartheta^{k}{}_{;}\overline{\Omega} \subseteq \overline{e_{v}{}_{;}e_{w}^{\mathsf{T}}} \qquad \text{Schröder equivalence}$$

$$e_{v}{}_{;}e_{w}^{\mathsf{T}} \subseteq \overline{\vartheta^{k}{}_{;}\overline{\Omega}} = \vartheta^{k}{}_{;}\overline{\overline{\Omega}} = \vartheta^{k}{}_{;}\Omega \qquad \text{negated and since } \vartheta \text{ is a mapping}$$

Now, we may sum up over all k so as to obtain $\vartheta^{*}{}_{;}\Omega$. $\qquad\qquad\square$

It is obvious that Q^{*} does not allow such a detailed analysis as ϑ^{*}_{Q} does. For the example of Fig. 19.15, the 'fully discharges into' preorder is shown in Fig. 19.17. The first observation concerns the bars of $\mathsf{0}$s. They result from the fact that, for example,

$$\{1,2\}, \{2,3,4\}, \{3,4\}, \{3,4\} \ldots \quad \text{but} \quad \{1,3\}, \{2,4\}, \{3\}, \{4\}, \{3\}, \{4\} \ldots$$

Effects of this type make it difficult to reason formally about orbits being eventually caught by others.

We have seen so far that quite difficult situations may occur when looking for orbits and when one will eventually be fully discharged into another. It often depends on another concept we are going to develop next.

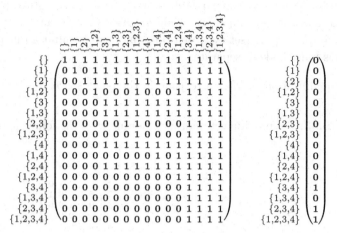

Fig. 19.17 'Fully discharges into' preorder for state transitions of Fig. 19.15 and all its basins

Definition 19.31. We call a non-empty set b of states a **basin** if it contracts Q, i.e., if $Q^{\mathsf{T}};b \subseteq b$. □

Non-emptiness has been postulated only in order to avoid degeneration. The so-called basins constitute orbits that are easier to describe as they will never reach new states, but rather are contractions of a given one. A basin is closed under Q-transitions. The state transition relation Q has basins $\{3,4\}$, $\{2,3,4\}$ and $\{1,2,3,4\}$. Figure 19.17 shows it on the right side.

Once we have provided the algebraic definition of the relation between sets that one eventually fully discharges into the other, one can start further investigations that are beyond the scope of this text. It was intended with this chapter to show the use of the existential image in this context which seems a new idea. Systems analysts, and logicians who help them, will then approach concepts of recurrence, for example. They will also be confronted with situations that have already been studied here concerning irreducibility, cyclicity, etc. We simply continue with two concepts without going into too much detail.

Definition 19.32. We consider a basin given by a vector b together with one of its subsets $a \subseteq b$, for which we also consider the point $e_a := \mathsf{syq}(\varepsilon, a)$.

(i) a is an **unavoidably attracting set** of b provided that $b \subseteq \varepsilon; \vartheta_Q^*; e_a$.

(ii) a is a **potentially attracting set** of b if $b \subseteq Q^*; \varepsilon; \vartheta_Q^*; e_a$. □

Power operations

The set a inside basin b will unavoidably attract if every orbit starting somewhere inside b eventually stays completely inside a. One cannot formulate this properly without going to the powerset because orbits may enter a after a different number of steps. This must be kept separate.

Both these qualifications lead to a relation in the powerspace.

Example 19.33. We provide yet another example with the following relations.

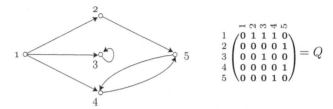

Fig. 19.18 Another example of state transition dynamics

	{}	{1}	{2}	{1,2}	{3}	{1,3}	{2,3}	{1,2,3}	{4}	{1,4}	{2,4}	{1,2,4}	{3,4}	{1,3,4}	{2,3,4}	{1,2,3,4}	{5}	{1,5}	{2,5}	{1,2,5}	{3,5}	{1,3,5}	{2,3,5}	{1,2,3,5}	{4,5}	{1,4,5}	{2,4,5}	{1,2,4,5}	{3,4,5}	{1,3,4,5}	{2,3,4,5}	{1,2,3,4,5}	b
{}	1	0	0	0	0	0	0	0	0	0	0	0	0	0	0	0	0	0	0	0	0	0	0	0	0	0	0	0	0	0	0	0	0
{1}	0	1	0	0	0	0	0	0	0	0	0	0	1	0	1	0	0	0	0	0	1	0	0	0	0	0	0	0	0	0	0	0	0
{2}	0	0	1	0	0	0	0	0	1	0	0	0	0	0	0	1	0	0	0	0	0	0	0	0	0	0	0	0	0	0	0	0	0
{1,2}	0	0	0	1	0	0	0	0	0	0	0	0	0	0	0	0	0	0	0	0	0	0	0	0	0	0	0	1	0	1	0	0	0
{3}	0	0	0	0	1	0	0	0	0	0	0	0	0	0	0	0	0	0	0	0	0	0	0	0	0	0	0	0	0	0	0	0	1
{1,3}	0	0	0	0	0	1	0	0	0	0	0	0	1	0	1	0	0	0	0	0	1	0	0	0	0	0	0	0	0	0	0	0	0
{2,3}	0	0	0	0	0	0	1	0	0	0	0	0	1	0	0	0	0	0	0	0	1	0	0	0	0	0	0	0	0	0	0	0	0
{1,2,3}	0	0	0	0	0	0	0	1	0	0	0	0	0	0	0	0	0	0	0	0	0	0	0	0	0	0	0	1	0	1	0	0	0
{4}	0	0	0	0	0	0	0	0	1	0	0	0	0	0	0	1	0	0	0	0	0	0	0	0	0	0	0	0	0	0	0	0	0
{1,4}	0	0	0	0	0	0	0	0	0	1	0	0	0	0	0	0	0	0	0	0	0	0	0	0	0	0	0	1	0	1	0	0	0
{2,4}	0	0	0	0	0	0	0	0	1	0	1	0	0	0	0	1	0	0	0	0	0	0	0	0	0	0	0	0	0	0	0	0	0
{1,2,4}	0	0	0	0	0	0	0	0	0	0	0	1	0	0	0	0	0	0	0	0	0	0	0	0	0	0	0	1	0	1	0	0	0
{3,4}	0	0	0	0	0	0	0	0	0	0	0	0	1	0	0	0	0	0	0	0	1	0	0	0	0	0	0	0	0	0	0	0	0
{1,3,4}	0	0	0	0	0	0	0	0	0	0	0	0	0	1	0	0	0	0	0	0	0	0	0	0	0	0	0	1	0	1	0	0	0
{2,3,4}	0	0	0	0	0	0	0	0	0	0	0	0	1	0	1	0	0	0	0	0	1	0	0	0	0	0	0	0	0	0	0	0	0
{1,2,3,4}	0	0	0	0	0	0	0	0	0	0	0	0	0	0	0	1	0	0	0	0	0	0	0	0	0	0	0	1	0	1	0	0	0
{5}	0	0	0	0	0	0	0	1	0	0	0	0	0	0	0	1	0	0	0	0	0	0	0	0	0	0	0	0	0	0	0	0	0
{1,5}	0	0	0	0	0	0	0	0	0	0	0	1	0	1	0	0	1	0	0	1	0	0	0	0	0	0	0	0	0	0	0	0	0
{2,5}	0	0	0	0	0	0	0	0	0	0	0	0	0	0	0	0	0	0	1	0	0	0	0	1	0	0	0	0	0	0	0	0	0
{1,2,5}	0	0	0	0	0	0	0	0	0	0	0	0	0	0	0	0	0	0	0	1	0	0	0	0	0	0	0	1	0	1	0	0	0
{3,5}	0	0	0	0	0	0	0	0	0	0	0	0	1	0	0	0	0	0	0	0	1	0	0	0	0	0	0	0	0	0	0	0	0
{1,3,5}	0	0	0	0	0	0	0	0	0	0	0	1	0	1	0	0	0	0	0	1	1	0	0	0	0	0	0	0	0	0	0	0	0
{2,3,5}	0	0	0	0	0	0	0	0	0	0	0	0	0	0	0	0	0	0	0	0	0	0	1	0	0	0	0	1	0	0	0	0	0
{1,2,3,5}	0	0	0	0	0	0	0	0	0	0	0	0	0	0	0	0	0	0	0	0	0	0	0	1	0	0	0	1	0	1	0	0	0
{4,5}	0	0	0	0	0	0	0	0	0	0	0	0	0	0	0	0	0	0	0	0	0	0	0	0	1	0	0	0	0	0	0	0	1
{1,4,5}	0	0	0	0	0	0	0	0	0	0	0	0	0	0	0	0	0	0	0	0	0	0	0	0	0	1	0	0	1	0	1	0	0
{2,4,5}	0	0	0	0	0	0	0	0	0	0	0	0	0	0	0	0	0	0	0	0	0	0	0	0	0	0	1	0	1	0	0	0	1
{1,2,4,5}	0	0	0	0	0	0	0	0	0	0	0	0	0	0	0	0	0	0	0	0	0	0	0	0	0	0	0	1	1	0	1	0	0
{3,4,5}	0	0	0	0	0	0	0	0	0	0	0	0	0	0	0	0	0	0	0	0	0	0	0	0	0	0	0	0	1	0	0	0	1
{1,3,4,5}	0	0	0	0	0	0	0	0	0	0	0	0	0	0	0	0	0	0	0	0	0	0	0	0	0	0	0	1	1	1	0	0	0
{2,3,4,5}	0	0	0	0	0	0	0	0	0	0	0	0	0	0	0	0	0	0	0	0	0	0	0	0	0	0	0	1	0	1	0	0	1
{1,2,3,4,5}	0	0	0	0	0	0	0	0	0	0	0	0	0	0	0	0	0	0	0	0	0	0	0	0	0	0	0	1	0	1	1	1	1

Fig. 19.19 Reflexive-transitive closure of the existential image ϑ^* and all basins b

In addition, we present the 'fully discharges into' preorder for this example which can easily be computed.

Column headers (left to right):
{}, {1}, {2}, {1,2}, {3}, {1,3}, {2,3}, {1,2,3}, {4}, {1,4}, {2,4}, {1,2,4}, {3,4}, {1,3,4}, {2,3,4}, {1,2,3,4}, {5}, {1,5}, {2,5}, {1,2,5}, {3,5}, {1,3,5}, {2,3,5}, {1,2,3,5}, {4,5}, {1,4,5}, {2,4,5}, {1,2,4,5}, {3,4,5}, {1,3,4,5}, {2,3,4,5}, {1,2,3,4,5}

```
{}          1 1 1 1 1 1 1 1 1 1 1 1 1 1 1 1 1 1 1 1 1 1 1 1 1 1 1 1 1 1 1 1
{1}         0 1 0 1 0 1 0 1 0 1 0 1 1 1 1 1 1 0 1 0 1 1 1 1 0 1 0 1 1 1 1 1
{2}         0 0 1 1 0 0 1 1 1 1 1 1 1 1 1 1 1 1 1 1 1 1 1 1 1 1 1 1 1 1 1 1
{1,2}       0 0 0 1 0 0 0 1 0 0 0 1 0 0 0 1 0 0 0 1 0 0 0 1 0 0 0 1 1 1 1 1
{3}         0 0 0 0 1 1 1 1 0 0 0 0 1 1 1 1 0 0 0 0 1 1 1 1 0 0 0 0 1 1 1 1
{1,3}       0 0 0 0 0 1 0 1 0 0 0 0 0 1 0 1 0 0 0 0 0 1 0 1 0 0 0 0 0 1 1 1
{2,3}       0 0 0 0 0 0 1 1 0 0 0 0 0 1 1 1 0 0 0 0 0 1 1 1 0 0 0 0 1 1 1 1
{1,2,3}     0 0 0 0 0 0 0 1 0 0 0 0 0 0 0 1 0 0 0 0 0 0 0 1 0 0 0 0 1 1 1 1
{4}         0 0 0 0 0 0 0 0 1 1 1 1 1 1 1 1 1 1 1 1 1 1 1 1 1 1 1 1 1 1 1 1
{1,4}       0 0 0 0 0 0 0 0 0 1 0 1 0 1 0 1 0 0 0 0 0 0 0 0 0 1 0 1 1 1 1 1
{2,4}       0 0 0 0 0 0 0 0 1 1 1 1 1 1 1 1 1 1 1 1 1 1 1 1 1 1 1 1 1 1 1 1
{1,2,4}     0 0 0 0 0 0 0 0 0 0 0 1 0 0 1 0 0 0 0 0 0 0 0 0 0 0 0 1 1 1 1 1
{3,4}       0 0 0 0 0 0 0 0 0 0 0 0 1 1 1 1 0 0 0 0 1 1 1 1 0 0 0 0 1 1 1 1
{1,3,4}     0 0 0 0 0 0 0 0 0 0 0 0 0 1 0 1 0 0 0 0 0 0 0 0 0 0 0 0 0 1 1 1
{2,3,4}     0 0 0 0 0 0 0 0 0 0 0 0 0 1 1 1 0 0 0 0 0 1 1 1 0 0 0 0 1 1 1 1
{1,2,3,4}   0 0 0 0 0 0 0 0 0 0 0 0 0 0 0 1 0 0 0 0 0 0 0 0 0 0 0 0 1 1 1 1
{5}         0 0 0 0 0 0 0 0 1 1 1 1 1 1 1 1 1 1 1 1 1 1 1 1 1 1 1 1 1 1 1 1
{1,5}       0 0 0 0 0 0 0 0 0 0 0 0 1 1 1 1 0 1 0 1 1 1 1 0 1 0 1 1 1 1 1 1
{2,5}       0 0 0 0 0 0 0 0 0 0 0 0 0 0 0 1 1 0 0 1 1 1 1 1 1 1 1 1 1 1 1 1
{1,2,5}     0 0 0 0 0 0 0 0 0 0 0 0 0 0 0 0 0 0 0 1 0 0 1 0 0 0 1 1 1 1 1 1
{3,5}       0 0 0 0 0 0 0 0 0 0 0 0 1 1 1 1 0 0 0 0 1 1 1 1 0 0 0 0 1 1 1 1
{1,3,5}     0 0 0 0 0 0 0 0 0 0 0 0 0 0 1 1 0 0 0 0 1 1 1 1 0 0 0 0 0 1 1 1
{2,3,5}     0 0 0 0 0 0 0 0 0 0 0 0 0 0 0 1 0 0 0 0 0 1 1 1 0 0 0 0 1 1 1 1
{1,2,3,5}   0 0 0 0 0 0 0 0 0 0 0 0 0 0 0 0 0 0 0 0 0 0 0 1 0 0 0 0 1 1 1 1
{4,5}       0 0 0 0 0 0 0 0 0 0 0 0 0 0 0 0 0 0 0 0 0 0 0 0 1 1 1 1 1 1 1 1
{1,4,5}     0 0 0 0 0 0 0 0 0 0 0 0 0 0 0 0 0 0 0 0 0 0 0 0 0 1 0 1 1 1 1 1
{2,4,5}     0 0 0 0 0 0 0 0 0 0 0 0 0 0 0 0 0 0 0 0 0 0 0 0 1 1 1 1 1 1 1 1
{1,2,4,5}   0 0 0 0 0 0 0 0 0 0 0 0 0 0 0 0 0 0 0 0 0 0 0 0 0 0 0 1 1 1 1 1
{3,4,5}     0 0 0 0 0 0 0 0 0 0 0 0 0 0 0 0 0 0 0 0 0 0 0 0 0 0 0 0 1 1 1 1
{1,3,4,5}   0 0 0 0 0 0 0 0 0 0 0 0 0 0 0 0 0 0 0 0 0 0 0 0 0 0 0 0 0 1 1 1
{2,3,4,5}   0 0 0 0 0 0 0 0 0 0 0 0 0 0 0 0 0 0 0 0 0 0 0 0 0 0 0 0 1 1 1 1
{1,2,3,4,5} 0 0 0 0 0 0 0 0 0 0 0 0 0 0 0 0 0 0 0 0 0 0 0 0 0 0 0 0 1 1 1 1
```

Fig. 19.20 The 'fully discharges into' preorder for Fig. 19.18

Appendix A
Notation

Our topic has different roots and, thus, diverging notation. But there is a second source of differences: in mathematical expositions, usually written in TEX, one often uses abbreviations when items are concerned that are clear from the context. Much of the background of this book, however, rests on programming work. For use on a computer via the relational language TITUREL, such contexts must be given in a more detailed form. The following tables show notational correspondences.

One should recall that names starting with a capital letter and also infix operators starting with a colon ":" indicate so-called constructors in HASKELL, that may be *matched* against one another. Furthermore, variables must start with lower case letters, so that we have to tolerate the frequently occurring transition from R, X, Y to r,x,y.

A.1 Handling heterogeneous relations

Handling heterogeneous relations in computer programs means not least keeping track of sources and targets of relations. This may be achieved by implementing the category of types and, above that, a Boolean and a monoid part.

Description	TEX form	TITUREL version
Category part relation from to	$R : X \longrightarrow Y$	r with x = src r, y = tgt r
Boolean lattice part union, intersection negation null relation universal relation	$R \cup S,\ R \cap S$ $\overline{R}$ $⊥_{X,Y}$ (abbreviated $⊥$) $⊤_{X,Y}$ (abbreviated $⊤$)	r :|||: s, r :&&&: s NegaR r NullR x y UnivR x y
Monoid part converse composition identity	R^T $R\,\mathbf{;}\,S$ $\mathbb{I}_X$ (abbreviated $\mathbb{I}$)	Convs r r :***: s Ident x

Fig. A.1 Correspondence of notation in TEX and in TITUREL

A.2 Constructing new domains

TITUREL also offers the possibility of generic domain construction. In all cases, a new domain is generated and from that point on treated solely with the generic means created in the course of this construction. For reasons of space, we have always abbreviated the newly constructed source in the right column of Fig. A.2 as d. In every case, the first line gives the newly constructed domain in its full denotation, for example, as d = DirPro x y.

Description	TEX form	TITUREL version
Direct product		
product domain	$X \times Y$	d = DirPro x y
project to the left	$\pi_{X,Y} : X \times Y \longrightarrow X$	Pi x y d -> x
project to the right	$\rho_{X,Y} : X \times Y \longrightarrow Y$	Rho x y d -> y
definable vectors	$\pi_{X,Y}{}^{\mathsf{:}}v_X$, v_X a vector on X	Pi x y :***: vX,
	$\rho_{X,Y}{}^{\mathsf{:}}v_Y$, v_Y a vector on Y	Rho x y :***: vY
definable elements	$\pi_{X,Y}{}^{\mathsf{:}}e_X \cap \rho_{X,Y}{}^{\mathsf{:}}e_Y$	(Pi x y :***: eX) :&&&:
	e_X, e_Y elements of X and Y	(Rho x y :***: eY)
Direct sum		
sum domain	$X + Y$	d = DirSum x y
inject left variant	$\iota_{X,Y} : X \longrightarrow X + Y$	Iota x y x -> d
inject right variant	$\kappa_{X,Y} : Y \longrightarrow X + Y$	Kappa x y y -> d
definable vectors	$\iota_{X,Y}^{\mathsf{T}}{}^{\mathsf{:}}v_X$, v_X a vector on X	Convs (Iota x y) :***: vX,
	$\kappa_{X,Y}^{\mathsf{T}}{}^{\mathsf{:}}v_Y$, v_Y a vector on Y	Convs (Kappa x y) :***: vY
definable elements	$\iota_{X,Y}^{\mathsf{T}}{}^{\mathsf{:}}e_X$ or $\kappa_{X,Y}^{\mathsf{T}}{}^{\mathsf{:}}e_Y$	Convs (Iota x y) :***: eX,
	e_X, e_Y elements of X or Y	Convs (Kappa x y) :***: eY
Direct power		
power domain	2^X	d = DirPow x
membership	$\varepsilon : X \longrightarrow 2^X$	Member x x -> d
definable vectors	$\sup_{i \in I}[\mathsf{syq}(\varepsilon, v_i)]$	SupVect (Syq (Member x) vI)
	v_i vectors on X	
definable elements	$\mathsf{syq}(\varepsilon, v)$, v vector of X	Syq (Member x) v
Quotient		
quotient domain	X_Ξ, Ξ an equivalence on X	d = QuotMod xi
natural projection	$\eta_\Xi : X \longrightarrow X_\Xi$	Project xi src xi -> d
definable vectors	$\eta_\Xi^{\mathsf{T}}{}^{\mathsf{:}}v_X$, v_X vector of X	Convs (Project xi) :***: vX
definable elements	$\eta_\Xi^{\mathsf{T}}{}^{\mathsf{:}}e_X$, e_X element of X	Convs (Project xi) :***: eX
Extrusion		
extruded domain	$E(V)$, V vector of X	d = Extrude v
natural injection	$\iota_V : E(V) \longrightarrow X$	Inject v d -> src v
definable vectors	$\iota_V{}^{\mathsf{:}}v_X$, v_X vector of $V \subseteq X$	Inject v :***: vX
definable elements	$\iota_V{}^{\mathsf{:}}e_X$, e_X element of $V \subseteq X$	Inject v :***: eX
Permutation		
permuted target	Y^ξ, $\xi : X \longrightarrow Y$ bijective map	d = PermTgt xi
rearrangement	$\rho_\xi : Y \longrightarrow Y^\xi$	ReArrTo xi tgt xi -> d
definable vectors	$\rho_\xi^{\mathsf{T}}{}^{\mathsf{:}}v_Y$, v_Y vector of Y	Convs (ReArrTo xi) :***: vY
definable elements	$\rho_\xi^{\mathsf{T}}{}^{\mathsf{:}}e_Y$, e_Y element of Y	Convs (ReArrTo xi) :***: eY

Fig. A.2 Generically available relations in domain construction

To the right of the generic relations such as `Pi`, `Iota`, `Member`, their typing is mentioned. Since TITUREL is fully typed,[1] one can always find the source of a relation as here with `src xi`. In the same way, v carries inside itself information on source x which need not be mentioned.

These domain construction operations make up a big part of the language. On top of this, terms may be formed, starting with constants and variables of category object terms, element terms, vector terms, and relational terms. This is roughly indicated with the definitions of Fig. A.2.

A.3 Substrate

In Fig. A.3, we show the layer structure of the language. All this makes up the left syntactical side. The aims in designing TITUREL were that it would allow all of the problems tackled so far to be formulated using relational methods, thereby offering syntax- and type-controls to reduce the likelihood of running into errors.

- It allows relational terms and formulae to be *transformed* in order to optimize them so that later they can be handled efficiently with the help of some system. In particular, a distinction is made between the matchable denotation of an operation and its execution.
- There exists the possibility of *interpreting* the relational language. For this, mainly three methods are possible. In the most simple method, one is able to attach Boolean matrices to the relational constants, for example, and evaluate terms built from these. In a second more sophisticated form, one is able to interpret using the RELVIEW system. In a third variant, interpretation is possible using the RATH-system. RATH is a HASKELL-based system with which non-representable relation algebras may also be studied; see [73].
- It is also intended to be able to *prove* relational formulae. Again, several forms are possible. In a first variant, a system will allow proofs in the style of RALF, a former interactive proof assistant for executing relational proofs [66, 67]. Already now, however, a variant has been initiated that allows proofs in Rasiowa–Sikorski style [98].
- In order to support people in their work with relations, it is possible to *translate* relational formulae into TEX-representation or into some pure ASCII-form. It is also possible to translate relational terms and formulae automatically from the point-free form into a form of first-order predicate logic.

In addition, there is a semantic side as indicated on the right of Fig. A.3. Interpretation of the language constructs is intended to take place in sets, elements and

---

[1] On top of the always underlying HASKELL typing.

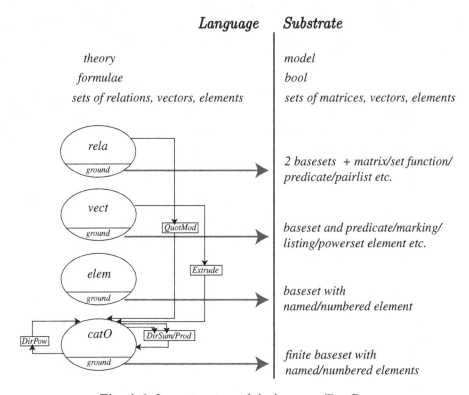

Fig. A.3 Layer structure of the language TITUREL

subsets of these, relations, and unary or binary functions between sets. We restrict ourselves to mentioning some basic facts.

- Sets are cared for using a data structure `BaseSet` mentioned in Section 2.1 that shows a name for the finite set and a namestring for all the elements – even in the case that they are just numbered.
- Elements may appear using the data structure `ElemInBaseSet` presented in Section 2.2 mentioning the baseset the element is assumed to belong to and in addition the number of the element position, or a Boolean vector marking the element along the baseset, or the aforementioned name of the element.
- Subsets may appear using the data structure `SubSet` presented in Section 2.3 mentioning the baseset they are assumed to belong to and in addition the list of numbers of the elements in the baseset, or the list of names of the elements in the baseset, or a Boolean vector marking the subset along the baseset, or a unary predicate, or an element in the corresponding powerset.

- Relations may appear using the data structure `Rel` mentioned in Section 3.1 with two basesets to indicate row and column type, amended by one of the following: a set of pairs of elements, a binary predicate, a Boolean matrix, a Boolean vector along the product space, or a set-valued function.
- Unary and binary functions are represented using the data structures indicated in Section 3.5: `FuncOnBaseSets`, `Fct2OnBaseSets`, i.e., mentioning the basesets between which they are defined first and then enabling the Boolean matrix or a list of lists of element numbers to be given.

Appendix B
Proofs postponed from Part II

Here follow the proofs we decided not to present in Part II, i.e., in Chapters 4 to 7. At that early stage of the book, these proofs would have interrupted visual understanding. Of course, they should not be omitted completely, so they are enclosed here. For better reference the respective proposition is also repeated. The original numbering is attached directly.

B.1 Proofs for Chapter 5

Proposition B.1 (5.2). (i) $Q_i R$ is univalent, whenever Q and R are.

(ii) R univalent $\iff$ $R_i \overline{\mathbb{I}} \subseteq \overline{R}$.

(iii) $R \subseteq Q$, Q univalent, $R_i \mathbb{T} \supseteq Q_i \mathbb{T}$ $\implies$ $R = Q$.

Proof: (i) $(Q_i R)^{\mathsf{T}} {}_i Q_i R = R^{\mathsf{T}} {}_i Q^{\mathsf{T}} {}_i Q_i R \subseteq R^{\mathsf{T}} {}_i \mathbb{I}_i R = R^{\mathsf{T}} {}_i R \subseteq \mathbb{I}$

(ii) $R^{\mathsf{T}} {}_i R \subseteq \mathbb{I}$ $\iff$ $R_i \overline{\mathbb{I}} \subseteq \overline{R}$ using the Schröder equivalence

(iii) $Q = Q_i \mathbb{T} \cap Q \subseteq R_i \mathbb{T} \cap Q \subseteq (R \cap Q_i \mathbb{T}^{\mathsf{T}})_i (\mathbb{T} \cap R^{\mathsf{T}} {}_i Q) \subseteq R_i R^{\mathsf{T}} {}_i Q \subseteq R_i Q^{\mathsf{T}} {}_i Q \subseteq R$ using Dedekind's rule $\qquad\square$

Proposition B.2 (5.3). Q univalent $\implies$ $Q_i (A \cap B) = Q_i A \cap Q_i B$.

Proof: Direction "$\subseteq$" is trivial; the other follows from
$$Q_i A \cap (Q_i B) \subseteq (Q \cap (Q_i B)_i A^{\mathsf{T}})_i (A \cap Q^{\mathsf{T}} {}_i (Q_i B)) \subseteq Q_i (A \cap Q^{\mathsf{T}} {}_i Q_i B) \subseteq Q_i (A \cap B). \qquad\square$$

Proposition B.3 (5.4). Q univalent $\implies$ $A \cap B_i Q = (A_i Q^{\mathsf{T}} \cap B)_i Q$.

Proof: $B_i Q \cap A \subseteq (B \cap A_i Q^{\mathsf{T}})_i (Q \cap B^{\mathsf{T}} {}_i A) \subseteq (A_i Q^{\mathsf{T}} \cap B)_i Q$
$\subseteq (A \cap B_i Q)_i (Q^{\mathsf{T}} \cap A^{\mathsf{T}} {}_i B)_i Q \subseteq (A \cap B_i Q)_i Q^{\mathsf{T}} {}_i Q = A \cap B_i Q \qquad\square$

Proposition B.4 (5.6).

(i) Q *univalent* $\implies$ $Q \overline{A} = Q \mathbb{T} \cap \overline{Q A}$.

(ii) Q *univalent* $\implies$ $\overline{Q A} = Q \overline{A} \cup \overline{Q \mathbb{T}}$.

Proof: (i) "$\subseteq$" consists of two parts, the trivial $Q \overline{A} \subseteq Q \mathbb{T}$ and $Q \overline{A} \subseteq \overline{Q A}$, which follows with the Schröder rule from univalence. Direction "$\supseteq$" is obtained via Boolean reasoning from $\mathbb{T} \subseteq \overline{Q \mathbb{T}} \cup Q A \cup Q \overline{A}$.

(ii) $\overline{Q A} \supseteq \overline{Q \mathbb{T}}$ is trivial and $\overline{Q A} \supseteq Q \overline{A}$ follows from univalency with the Schröder equivalence. The other direction is obtained as for (i). □

Proposition B.5 (5.12). *Let R, S be arbitrary relations for which the following constructs in connection with x, y and f exist:*

(i) *if f is a mapping,* $\quad R \subseteq S f^{\mathsf{T}} \iff R f \subseteq S$,

(ii) *if x is a point,* $\quad R \subseteq S x \iff R x^{\mathsf{T}} \subseteq S$,

(iii) *if x, y are points,* $\quad y \subseteq S x \iff x \subseteq S^{\mathsf{T}} y$. □

Proof: (i) "$\implies$" is immediate multiplying f from the right and using $f^{\mathsf{T}} f \subseteq \mathbb{I}$.

"$\impliedby$" is also immediate multiplying f^{T} from the right and using $\mathbb{I} \subseteq f f^{\mathsf{T}}$.

We mention here also how the non-trivial direction of the equivalence of the definitions of totality, $\mathbb{T} = R \mathbb{T}$ as opposed to $\mathbb{I} \subseteq R R^{\mathsf{T}}$, is proved applying the Dedekind rule:

$$\mathbb{I} = \mathbb{T} \cap \mathbb{I} = R \mathbb{T} \cap \mathbb{I} \subseteq (R \cap \mathbb{I} \mathbb{T}^{\mathsf{T}}) (\mathbb{T} \cap R^{\mathsf{T}} \mathbb{I}) \subseteq R R^{\mathsf{T}}.$$

(ii) Same as (i), remembering that the converse of a point is always a mapping.

(iii) From (ii) we have $y \subseteq S x \iff y x^{\mathsf{T}} \subseteq S$ which, transposed, gives $x y^{\mathsf{T}} \subseteq S^{\mathsf{T}}$. The proof is completed by employing (ii) again. □

Proposition B.6 (5.24). (i) *A linear order E and its associated strictorder C satisfy $\overline{E}^{\mathsf{T}} = C$.*

(ii) *A linear order E satisfies $E \overline{E}^{\mathsf{T}} E = C \subseteq E$.*

(iii) *A linear strictorder C satisfies $C C \overline{C}^{\mathsf{T}} = C \overline{C}^{\mathsf{T}} C = C^2 \subseteq C$.*

(iv) *E is a linear order precisely when E^{d} is a linear strictorder.*

(v) *C is a linear strictorder precisely when C^{d} is a linear order.*

Proof: (i) By definition, a linear order is connex, i.e., satisfies $\mathbb{T} = E \cup E^{\mathsf{T}}$, so that $\overline{E}^{\mathsf{T}} \subseteq E$. We have in addition $\overline{E}^{\mathsf{T}} \subseteq \overline{\mathbb{I}}$ from reflexivity. Both together result in $\overline{E}^{\mathsf{T}} \subseteq C$. The reverse containment $C = \overline{\mathbb{I}} \cap E \subseteq \overline{E}^{\mathsf{T}}$ follows from antisymmetry.

(ii) $E \,\bar{E}^{\mathsf{T}} = E \,C = C.$

(iii)–(v) These proofs are immediate. $\square$

Proposition B.7 (5.27). *Let Ξ be an equivalence and let A, B be arbitrary relations.*

(i) $\Xi \,(\Xi \,A \cap B) = \Xi \,A \cap \Xi \,B = \Xi \,(A \cap \Xi \,B).$

(ii) $\Xi \,\overline{\Xi \,R} = \overline{\Xi \,R}.$

Proof: (i) $\Xi \,(\Xi \,A \cap B) \subseteq \Xi^2 \,A \cap \Xi \,B = \Xi \,A \cap \Xi \,B$
$\subseteq (\Xi \cap \Xi \,B \,A^{\mathsf{T}}) \,(A \cap \Xi^{\mathsf{T}} \,\Xi \,B) \subseteq \Xi \,(A \cap \Xi \,B) \subseteq \Xi \,A \cap \Xi^2 \,B$
$= \Xi \,B \cap \Xi \,A \subseteq (\Xi \cap \Xi \,A \,B^{\mathsf{T}}) \,(B \cap \Xi^{\mathsf{T}} \,\Xi \,A) \subseteq \Xi \,(B \cap \Xi \,A) = \Xi \,(\Xi \,A \cap B)$

(ii) $\Xi \,\overline{\Xi \,R} \supseteq \overline{\Xi \,R}$ is trivial since Ξ is reflexive

$\Xi \,\overline{\Xi \,R} \subseteq \overline{\Xi \,R} \quad \Longleftrightarrow \quad \Xi^{\mathsf{T}} \,\overline{\Xi \,R} \subseteq \overline{\Xi \,R}$ via the Schröder equivalence $\square$

Proposition B.8 (5.29). *For an arbitrary relation R and its row and column equivalence, always*

(i) $\Xi(R) \,R = R = R \,\Psi(R),$

(ii) $\mathcal{R}(\overline{R}) = \big(\mathcal{R}(R)\big)^{\mathsf{T}} \qquad \Xi(\overline{R}) = \Xi(R) \qquad \mathcal{R}(R^{\mathsf{T}}) = \mathcal{C}(\overline{R}),$

(iii) $\Xi(R) = \Xi(\mathcal{R}(R)) \quad$ *or, equivalently,* $\quad \mathrm{syq}(R^{\mathsf{T}}, R^{\mathsf{T}}) = \mathrm{syq}(\overline{\overline{R} \,R^{\mathsf{T}}}^{\mathsf{T}}, \overline{\overline{R} \,R^{\mathsf{T}}}^{\mathsf{T}}).$

Proof: (i) $\Xi(R) \,R = \mathrm{syq}(R^{\mathsf{T}}, R^{\mathsf{T}}) \,R = R$ with an application of Prop. 8.11.

(ii) These proofs are trivial.

(iii) The following starts expanding the definition, reducing double negations, and executing transpositions.

$$\Xi(\mathcal{R}(R)) = \mathrm{syq}(\overline{\overline{R} \,R^{\mathsf{T}}}^{\mathsf{T}}, \overline{\overline{R} \,R^{\mathsf{T}}}^{\mathsf{T}}) = \overline{\overline{R} \,R^{\mathsf{T}} \,\overline{R} \,\overline{R}^{\mathsf{T}}} \cap \overline{\overline{\overline{R} \,R^{\mathsf{T}}} \,\overline{R} \,\overline{R}^{\mathsf{T}}}$$
$$= \overline{\overline{R} \,R^{\mathsf{T}}} \cap \overline{R} \,\overline{R}^{\mathsf{T}} \text{ using Prop. 4.7}$$
$$= \mathrm{syq}(R^{\mathsf{T}}, R^{\mathsf{T}}) = \Xi(R)$$

                                                                                        $\square$

Proposition B.9 (5.31). *If Q is a difunctional relation, the following holds:*

$$Q \,(A \cap Q^{\mathsf{T}} \,B) = Q \,A \cap Q \,Q^{\mathsf{T}} \,B.$$

Proof: "$\subseteq$" is trivially satisfied. "$\supseteq$" is proved using the Dedekind formula

$$Q \,A \cap (Q \,Q^{\mathsf{T}} \,B) \subseteq (Q \cap (Q \,Q^{\mathsf{T}} \,B) \,A^{\mathsf{T}}) \,(A \cap Q^{\mathsf{T}} \,(Q \,Q^{\mathsf{T}} \,B))$$

$\subseteq Q_i (A \cap Q^\mathsf{T}_i Q_i Q^\mathsf{T}_i B) \subseteq Q_i (A \cap Q^\mathsf{T}_i B).$ □

Proposition B.10 (5.32). *If Q is a difunctional relation, the following holds:*

(i) $\quad Q_i \overline{Q^\mathsf{T}_i A} = Q_i \mathbb{T} \cap \overline{Q_i Q^\mathsf{T}_i A} \qquad$ *and* $\qquad \overline{Q_i Q^\mathsf{T}_i A} = Q_i \overline{Q^\mathsf{T}_i A} \cup \overline{Q_i \mathbb{T}},$

(ii) $\quad Q_i \overline{Q^\mathsf{T}_i A} = \overline{Q_i Q^\mathsf{T}_i A} \quad$ *in the case that Q is in addition total.*

Proof: (i) "$\supseteq$" is trivial by a Boolean argument. The first part of "$\subseteq$" is again trivial, while the second is deduced applying the Schröder equivalence and using the difunctionality condition. (ii) is a special case of (i). □

Proposition B.11 (5.34). *The following holds for an arbitrary finite homogeneous relation R on a set of n elements:*

(i) $R^n \subseteq (\mathbb{I} \cup R)^{n-1}$,

(ii) $R^* = \sup_{0 \leq i < n} R^i$,

(iii) $R^+ = \sup_{0 < i \leq n} R^i$,

(iv) $(\mathbb{I} \cup R)_i (\mathbb{I} \cup R^2)_i (\mathbb{I} \cup R^4)_i (\mathbb{I} \cup R^8)_i \ldots_i (\mathbb{I} \cup R^{2^{\lfloor \log n \rfloor}}) = R^*$.

Proof: (i) We use the pigeon-hole principle to interpret $(\mathbb{I} \cup R)^{n-1} = \sup_{0 \leq i < n} R^i$:

$$R^n_{xz} = \exists y_1 : \exists y_2 : \ldots \exists y_{n-1} : R_{xy_1} \cap R_{y_1 y_2} \cap \ldots \cap R_{y_{n-1} z}.$$

This means $n + 1$ indices $y_0 := x, y_1, \ldots, y_{n-1}, y_n := z$ of which at least two will coincide, for example, $y_r = y_s$, $0 \leq r < s \leq n$. From $R^n_{xz} = 1$ it follows that $R^{n-(s-r)}_{xz} = 1$.

(ii), (iii), and (iv) are, thus, obvious. □

Proposition B.12 (5.36). *Let some (possibly heterogeneous) relation R be given and consider Ω, Ω', its left and right equivalence:*

(i) Ω and Ω' are equivalences,

(ii) $\Omega_i R = R_i \Omega'$,

(iii) $\mathbb{I} \cup \overline{R^\mathsf{T}_i \Omega_i R} = \Omega'$,

(iv) $\mathbb{I} \cup \overline{R_i \Omega'_i R^\mathsf{T}} = \Omega$.

Proof: (i) The two are reflexive, symmetric, and transitive by construction. (ii), (iii), and (iv) are all proved with the same mechanism of regular algebra:

$$\Omega_; R = \big[\sup_{0\leq i}(R_; R^{\mathsf{T}})^i\big]_; R = R_; \big[\sup_{0\leq i}(R^{\mathsf{T}}_; R)^i\big] = R_; \Omega' \qquad \Box$$

Proposition B.13 (5.50). *Let any relation R between sets V and W be given and assume that Ξ, Ω is an R-congruence. Denoting the natural projections as η_Ξ, η_Ω, respectively, we form the quotient sets and consider the relation $S := \eta_\Xi^{\mathsf{T}}_; R_; \eta_\Omega$ between V_Ξ and W_Ω. Then η_Ξ, η_Ω is a homomorphism of the structure R into the structure S that satisfies $R_; \eta_\Omega = \eta_\Xi_; S$.*

Proof: η_Ξ and η_Ω are mappings by construction.

$$
\begin{aligned}
R_; \eta_\Omega \;&\subseteq\; \eta_\Xi_; \eta_\Xi^{\mathsf{T}}_; R_; \eta_\Omega = \eta_\Xi_; S &&\text{since } \eta_\Xi \text{ is total and by definition of } S\\
&=\; \Xi_; R_; \eta_\Omega &&\text{since } \Xi = \eta_\Xi_; \eta_\Xi^{\mathsf{T}}\\
&\subseteq\; R_; \Omega_; \eta_\Omega &&\text{since } \Xi_; R \subseteq R_; \Omega\\
&=\; R_; \eta_\Omega_; \eta_\Omega^{\mathsf{T}}_; \eta_\Omega &&\text{since } \Omega = \eta_\Omega_; \eta_\Omega^{\mathsf{T}}\\
&=\; R_; \eta_\Omega_; \mathbb{I}_{W_\Omega} &&\text{since } \mathbb{I}_{W_\Omega} = \eta_\Omega^{\mathsf{T}}_; \eta_\Omega\\
&=\; R_; \eta_\Omega &&\qquad\qquad\qquad\qquad\Box
\end{aligned}
$$

B.2 Proofs for Chapter 6

Definition B.14 (6.4). Given two vectors $u \subseteq X$ and $v \subseteq Y$, together with (possibly heterogeneous) universal relations $\mathbb{T}$,

$$u_; v^{\mathsf{T}} = u_; \mathbb{T} \cap (v_; \mathbb{T})^{\mathsf{T}}.$$

Proof: "$\subseteq$" is trivial. With the Dedekind rule, we prove "$\supseteq$"

$$u_; \mathbb{T} \cap (v_; \mathbb{T})^{\mathsf{T}} \subseteq (u \cap (v_; \mathbb{T})^{\mathsf{T}}_; \mathbb{T})_; (\mathbb{T} \cap u^{\mathsf{T}}_; (v_; \mathbb{T})^{\mathsf{T}}) \subseteq u_; u^{\mathsf{T}}_; \mathbb{T}v^{\mathsf{T}} \subseteq u_; \mathbb{T}_; v^{\mathsf{T}} = u_; v^{\mathsf{T}} \qquad \Box$$

Proposition B.15 (6.5). *For a relation R the following are equivalent:*

(i) R *is a rectangle,*

(ii) $R_; \mathbb{T}_; R \subseteq R$,

(iii) $R_; \mathbb{T}_; R = R$,

(iv) $R_; \overline{R}^{\mathsf{T}}_; R = \mathbb{L}$,

(v) *For any fitting pair A, B, the Dedekind rule becomes an equality,*
$$A_; B \cap R = (A \cap R_; B^{\mathsf{T}})_; (B \cap A^{\mathsf{T}}_; R).$$

Proof: (i) $\Longrightarrow$ (ii): Since R is a rectangle, there exist vectors u, v with $u_i v^\mathsf{T} = R$. Therefore,

$$R_i \mathbb{T}_i R = u_i v^\mathsf{T}_i \mathbb{T}_i u_i v^\mathsf{T} \subseteq u_i \mathbb{T}_i v^\mathsf{T} = u_i v^\mathsf{T} = R$$

(ii) $\Longleftrightarrow$ (iii): The inclusion "$\subseteq$" is given; "$\supseteq$" holds for arbitrary relations since

$$R = R_i \mathbb{I} \cap R \subseteq (R \cap R_i \mathbb{I}^\mathsf{T})_i (\mathbb{I} \cap R^\mathsf{T}_i R) \subseteq R_i R^\mathsf{T}_i R \subseteq R_i \mathbb{T}_i R.$$

(ii) $\Longleftrightarrow$ (iv): We use the Schröder rule two times:

$$R_i \mathbb{T}_i R \subseteq R \iff \mathbb{T}_i R^\mathsf{T}_i \overline{R} \subseteq \overline{R} \iff R_i \overline{R}^\mathsf{T}_i R \subseteq \mathbb{L}.$$

(ii) $\Longrightarrow$ (v): Direction "$\subseteq$" is given by the Dedekind rule. From "$\supseteq$", containment in $A_i B$ is trivial, while containment in R follows from:

$$(A \cap R_i B^\mathsf{T})_i (B \cap A^\mathsf{T}_i R) \subseteq R_i B^\mathsf{T}_i A^\mathsf{T}_i R \subseteq R_i \mathbb{T}_i R \subseteq R$$

(v) $\Longrightarrow$ (i): Take $A := \mathbb{T}_{\mathtt{src}(R),\mathtt{tgt}(R)}$, $B := \mathbb{T}_{\mathtt{tgt}(R),\mathtt{tgt}(R)}$ as a fitting pair to get

$$R = A_i B \cap R = (A \cap R_i B^\mathsf{T})_i (B \cap A^\mathsf{T}_i R) = R_i B^\mathsf{T}_i A^\mathsf{T}_i R = R_i \mathbb{T}_i (R^\mathsf{T}_i \mathbb{T})^\mathsf{T}$$

We have, thus, $R = u_i v^\mathsf{T}$ with vectors $u := R_i \mathbb{T}$ and $v := R^\mathsf{T}_i \mathbb{T}$. $\qquad\square$

Proposition B.16 (6.8). *Given a relation R, the subsets $u := \mathtt{dom}(R), v := \mathtt{cod}(R)$ together constitute the smallest rectangle containing R, i.e.,*

$$h_{\mathrm{rect}}(R) = u_i v^\mathsf{T} = R_i \mathbb{T}_i R = R_i \mathbb{T} \cap \mathbb{T}_i R.$$

Proof: Let u', v' be an arbitrary rectangle containing R. Then the containment $R \subseteq u'_i v'^\mathsf{T}$ implies $R_i \mathbb{T} \subseteq u'_i v'^\mathsf{T}_i \mathbb{T} \subseteq u'_i \mathbb{T} = u'$ and similarly for v'. Therefore, $u = \mathtt{dom}(R) \subseteq u'$ and $v = \mathtt{cod}(R) \subseteq v'$. The rest follows with Prop. 6.5. $\qquad\square$

B.3 Proofs for Chapter 7

Proposition B.17 (7.9). *The natural projection η onto the quotient domain modulo an equivalence Ξ is defined in an essentially unique way.*

Proof: The natural projection η is uniquely determined up to isomorphism: should a second natural projection χ be presented, i.e., we assume two such projections $V_\Xi \xleftarrow{\ \eta\ } V \xrightarrow{\ \chi\ } W_\Xi$, for which therefore

$$\Xi = \eta_i \eta^\mathsf{T}, \qquad \eta^\mathsf{T}_i \eta = \mathbb{I}_{V_\Xi}, \qquad \text{but also}$$
$$\Xi = \chi_i \chi^\mathsf{T}, \qquad \chi^\mathsf{T}_i \chi = \mathbb{I}_{W_\Xi}.$$

Looking at this setting, the only way to relate V_Ξ with W_Ξ is to define $\Phi := \eta^\mathsf{T}_i \chi$ and proceed showing

$$\Phi^\mathsf{T}_{\,;}\Phi = (\chi^\mathsf{T}_{\,;}\eta)_{;}(\eta^\mathsf{T}_{\,;}\chi) \qquad \text{by definition of } \Phi$$
$$= \chi^\mathsf{T}_{\,;}(\eta_{;}\eta^\mathsf{T})_{;}\chi \qquad \text{associative}$$
$$= \chi^\mathsf{T}_{\,;}\Xi_{;}\chi \qquad \text{since } \Xi = \eta_{;}\eta^\mathsf{T}$$
$$= \chi^\mathsf{T}_{\,;}(\chi_{;}\chi^\mathsf{T})_{;}\chi \qquad \text{since } \Xi = \chi_{;}\chi^\mathsf{T}$$
$$= (\chi^\mathsf{T}_{\,;}\chi)_{;}(\chi^\mathsf{T}_{\,;}\chi) \qquad \text{associative}$$
$$= \mathbb{I}_{W_\Xi\,;}\mathbb{I}_{W_\Xi} \qquad \text{since } \chi^\mathsf{T}_{\,;}\chi = \mathbb{I}_{W_\Xi}$$
$$= \mathbb{I}_{W_\Xi} \qquad \text{since } \mathbb{I}_{W_\Xi\,;}\mathbb{I}_{W_\Xi} = \mathbb{I}_{W_\Xi}$$

$\Phi_{;}\Phi^\mathsf{T} = \mathbb{I}_{V_\Xi}$ is shown analogously. Furthermore, $(\mathbb{I}, \Phi)$ satisfies the property of an isomorphism between η and χ following Lemma 5.48:

$$\eta_{;}\Phi = \eta_{;}\eta^\mathsf{T}_{\,;}\chi = \Xi_{;}\chi = \chi_{;}\chi^\mathsf{T}_{\,;}\chi = \chi_{;}\mathbb{I}_{W_\Xi} = \chi \qquad\qquad \square$$

Proposition B.18 (7.10). *Let an equivalence Ξ be given and consider its natural projection η. If any two relations A, B are presented, one of which satisfies $\Xi A = A$, the following holds,*

$$\eta^\mathsf{T}_{\,;}(A \cap B) = \eta^\mathsf{T}_{\,;}A \cap \eta^\mathsf{T}_{\,;}B.$$

Proof: $\eta^\mathsf{T}_{\,;}(A \cap B) = \eta^\mathsf{T}_{\,;}\eta_{;}\eta^\mathsf{T}_{\,;}(A \cap B) \qquad$ because $\eta^\mathsf{T}_{\,;}\eta = \mathbb{I}$
$$= \eta^\mathsf{T}_{\,;}\Xi_{;}(\Xi_{;}A \cap B) \qquad \text{using } \Xi = \eta_{;}\eta^\mathsf{T} \text{ and } \Xi_{;}A = A$$
$$= \eta^\mathsf{T}_{\,;}(\Xi_{;}A \cap \Xi_{;}B) \qquad \text{Prop. 5.27.i}$$
$$= \eta^\mathsf{T}_{\,;}(\eta_{;}\eta^\mathsf{T}_{\,;}A \cap \eta_{;}\eta^\mathsf{T}_{\,;}B) \qquad \text{expanding } \Xi = \eta_{;}\eta^\mathsf{T}$$
$$= \eta^\mathsf{T}_{\,;}\eta_{;}(\eta^\mathsf{T}_{\,;}A \cap \eta^\mathsf{T}_{\,;}B) \qquad \text{because } \eta \text{ is univalent}$$
$$= \eta^\mathsf{T}_{\,;}A \cap \eta^\mathsf{T}_{\,;}B \qquad \text{since } \eta^\mathsf{T}_{\,;}\eta = \mathbb{I} \qquad\qquad \square$$

Proposition B.19 (7.11). *Let any subset $\mathbb{L} \neq U \subseteq V$ of some baseset V be given. Then the natural injection $\iota_U : D_U \longrightarrow V$, with the properties*

$$\iota_{U\,;}\iota_U^\mathsf{T} = \mathbb{I}_{D_U}, \qquad \iota_U^\mathsf{T}_{\,;}\iota_U = \mathbb{I}_V \cap U_{;}\mathbb{T}_{V,V},$$

which thereby introduces the new domain D_U, is defined in an essentially unique form.

Proof: Assume $D_U \overset{\iota_U}{\longrightarrow} V \overset{\chi}{\longleftarrow} D$, i.e., another injection $\chi : D \longrightarrow V$ with the corresponding properties

$$\chi_{;}\chi^\mathsf{T} = \mathbb{I}_D, \qquad \chi^\mathsf{T}_{\,;}\chi = \mathbb{I} \cap U_{;}\mathbb{T}_{V,V}$$

to be given. We define $\Phi := \iota_{U\,;}\chi^\mathsf{T}$ and show

$$\Phi^\mathsf{T}_{\,;}\Phi = \chi_{;}\iota_U^\mathsf{T}_{\,;}\iota_{U\,;}\chi^\mathsf{T} = \chi_{;}(\mathbb{I}_V \cap U_{;}\mathbb{T})_{;}\chi^\mathsf{T} = \chi_{;}\chi^\mathsf{T}_{\,;}\chi_{;}\chi^\mathsf{T} = \mathbb{I}_{D\,;}\mathbb{I}_D = \mathbb{I}_D$$

and also $\Phi_{;}\Phi^\mathsf{T} = \mathbb{I}_{D_U}$ as well as

$$\Phi_{;}\chi = \iota_{U\,;}\chi^\mathsf{T}_{\,;}\chi = \iota_{U\,;}(\mathbb{I}_V \cap U_{;}\mathbb{T}) = \iota_{U\,;}\iota_U^\mathsf{T}_{\,;}\iota_U = \mathbb{I}_{D_U\,;}\iota_U = \iota_U. \qquad\qquad \square$$

Proposition B.20 (7.12). *Let any relation* $R : X \longrightarrow Y$ *be given with* $R \neq \perp\!\!\!\perp$ *and consider the extrusion* $S := \iota_{;} R_{;} \iota'^{\mathsf{T}}$ *according to its domain* $\mathrm{dom}(R) = R_{;} \mathbb{T}$ *and codomain* $\mathrm{cod}(R) = R^{\mathsf{T}}_{;} \mathbb{T}$, *i.e., the relations*

$$\iota := \mathtt{Inject}\,(R_{;}\mathbb{T}) \quad and \quad \iota' := \mathtt{Inject}\,(R^{\mathsf{T}}_{;}\mathbb{T}).$$

Then the following hold:

 (i) $\iota^{\mathsf{T}}_{;} \mathbb{T} = R_{;} \mathbb{T} \qquad \iota'^{\mathsf{T}}_{;} \mathbb{T} = R^{\mathsf{T}}_{;} \mathbb{T}$,
 (ii) S *is total and surjective,*
 (iii) $\iota^{\mathsf{T}}_{;} \iota_{;} R = R = R_{;} \iota'^{\mathsf{T}}_{;} \iota'$,
 (iv) $\iota^{\mathsf{T}}_{;} S_{;} \iota' = R$,
 (v) $\iota^{\mathsf{T}}_{;} \iota = \mathbb{I} \cap R_{;} \mathbb{T}$.

Proof: (i) $\iota^{\mathsf{T}}_{;} \mathbb{T} = \iota^{\mathsf{T}}_{;} \iota_{;} \mathbb{T} \qquad$ because ι is total
 $= (\mathbb{I} \cap R_{;} \mathbb{T}) \mathbb{T} \qquad$ by definition of an extrusion
 $= \mathbb{T} \cap R_{;} \mathbb{T} \qquad$ masking according to Prop. 8.5
 $= R_{;} \mathbb{T}$

(ii) needs that for every relation $A_{;} A^{\mathsf{T}}_{;} \mathbb{T} = A_{;} \mathbb{T}$ together with totality of ι:
$$S_{;} \mathbb{T} = \iota_{;} R_{;} \iota'^{\mathsf{T}}_{;} \mathbb{T} = \iota_{;} R_{;} R^{\mathsf{T}}_{;} \mathbb{T} = \iota_{;} R_{;} \mathbb{T} = \iota_{;} \iota^{\mathsf{T}}_{;} \mathbb{T} = \mathbb{I}_{;} \mathbb{T} = \mathbb{T}$$

(iii) is shown via a cyclic estimation using the Dedekind rule to obtain the first half:
$$R = R_{;} \mathbb{T} \cap R = \iota^{\mathsf{T}}_{;} \mathbb{T} \cap R \subseteq (\iota^{\mathsf{T}} \cap R_{;} \mathbb{T})_{;} (\mathbb{T} \cap \iota_{;} R) = \iota^{\mathsf{T}}_{;} \iota_{;} R \subseteq R$$

(iv) is then trivial.

(v) $R_{;} \mathbb{T} \cap \mathbb{I} = \iota^{\mathsf{T}}_{;} \mathbb{T} \cap \mathbb{I} \subseteq (\iota^{\mathsf{T}} \cap \mathbb{I}_{;} \mathbb{T}^{\mathsf{T}})_{;} (\mathbb{T} \cap \iota_{;} \mathbb{I}) \subseteq \iota^{\mathsf{T}}_{;} \iota$ $\qquad\qquad\qquad$ $\square$

Proposition B.21 (7.14). *Let a membership relation* $\varepsilon : A \longrightarrow 2^A$ *be given. Then*

 (i) $X = \varepsilon_{;} \mathsf{syq}(\varepsilon, X) = \overline{\varepsilon_{;} \overline{\varepsilon^{\mathsf{T}}_{;} X}}$,
 (ii) $\overline{X} = \overline{\varepsilon}_{;} \mathsf{syq}(\varepsilon, X) = \overline{\overline{\varepsilon}_{;} \overline{\overline{\varepsilon}^{\mathsf{T}}_{;} X}}$.

Proof: (i) We use Prop. 8.12.iii together with the properties of the symmetric quotient as explained in Section 8.5. Then we have $X = \varepsilon_{;} \mathsf{syq}(\varepsilon, X)$ because ε is surjective as a membership relation. But also

$X = \underline{\varepsilon_{;} \mathsf{syq}(\varepsilon, X)} \qquad$ because ε is a membership relation
 $= \varepsilon_{;} (\overline{\overline{\varepsilon^{\mathsf{T}}_{;} X}} \cap \overline{\varepsilon^{\mathsf{T}}_{;} \overline{X}}) \qquad$ expanded
 $\subseteq \overline{\varepsilon_{;} \overline{\varepsilon^{\mathsf{T}}_{;} X}} \qquad$ monotony
 $\subseteq X \qquad$ Schröder rule

so that equality holds everywhere in between. Similar reasoning is possible for (ii), recalling the two natures of negation $N = \mathsf{syq}(\varepsilon, \overline{\varepsilon})$ and $\varepsilon_{;} N = \overline{\varepsilon}$. $\qquad\qquad$ $\square$

Appendix C
Algebraic Visualization

Quite frequently in this book, we have tried to get additional intuition using rearranged relations. This follows the tradition of numerical mathematics where eigenvalue considerations for matrices, for example, are supported by figures with diagonal blocks. The intuition obtained will often help one to understand better and even to find a proof. A proof based on just such a rearrangement is usually not acceptable unless additional justification is given. It is this point we concentrate on here.

Our rearrangement algorithms are all based on pure relation algebra, supported by the fact that we assume basesets always to be equipped with an ordering. In this appendix, we provide those relational terms that have been written down in the relational language TITUREL to achieve appropriate rearrangements for the basic cases. More detailed information may be found in [17].

C.1 Rearranging a bijective mapping

We assume a bijective mapping $\xi : V \longrightarrow W$ to start with, i.e., a possibly heterogeneous relation. The aim is to arrange this bijective mapping ξ via a permutation $\rho_\xi : W \longrightarrow W^\xi$ of its target, so that $\xi_; \rho_\xi : V \longrightarrow W^\xi$ 'looks like a diagonal matrix', see Fig. C.1.

It is not least at this point where our decision to use *basesets* instead of *sets* has an important effect. To denote row entries of a matrix, or column entries respectively, we use basesets. The basic idea is therefore rather trivial: the matrix underlying the bijective mapping looks more or less like a permutation, but it will often not be a permutation as it is possibly heterogeneous. In Fig. C.1, source and target are $V = \{1, 2, 3, 4, 5\}$ and $W = \{US, French, German, British, Spanish\}$. What we need in order to obtain the diagonal shape for the matrix is a new target, namely $W^\xi = \{French, US, Spanish, German, British\}$.

	US	French	German	British	Spanish
1	0	1	0	0	0
2	1	0	0	0	0
3	0	0	0	1	0
4	0	0	0	0	1
5	0	0	1	0	0

$\xi : V \longrightarrow W$
original relation

	French	US	British	Spanish	German
US	0	1	0	0	0
French	1	0	0	0	0
German	0	0	0	0	1
British	0	0	1	0	0
Spanish	0	0	0	1	0

$\rho_\xi : W \longrightarrow W^\xi$
rearrangement transition

	French	US	British	Spanish	German
1	1	0	0	0	0
2	0	1	0	0	0
3	0	0	1	0	0
4	0	0	0	1	0
5	0	0	0	0	1

$\xi; \rho_\xi : V \longrightarrow W^\xi$
rearranged relation

Fig. C.1 Rearranging a bijective mapping

We remember the domain construction for a new baseset to denote a permuted codomain from Section 7.7 as `PermTgt` ξ (or shorter W^ξ). Then the relation `ReArrTo` ξ (or shorter ρ_ξ) gives the generic transition from W to W^ξ. Considered as a matrix, it is – up to the source and target – the converse of ξ. The two constructs `PermTgt` and `ReArrTo` have been incorporated into the relational language TITUREL and proved to be sufficient to denote in a consistent way.

C.2 Rearranging a linear order

Let $E : V \longrightarrow V$ be a linear order on a set V and assume Ω to be its baseorder in the case that V is considered a baseset. How can one permute the baseset V, obtaining π_E, so as to see the permuted E as the upper right triangle? This is a completely trivial task – but tedious when one actually has to execute it. Our considerations here aim at the finite case only. With $0_E := \mathtt{lea}(E)$ and $0_\Omega := \mathtt{lea}(\Omega)$, we determine the respective least elements. Then the Hasse relations

$$H_E := C \cap \overline{C; C} \qquad \text{with } C := \overline{\mathbb{I}} \cap E \text{ the respective linear strictorder,}$$

$$H_\Omega := C_\Omega \cap \overline{C_\Omega; C_\Omega} \qquad \text{with } C_\Omega := \overline{\mathbb{I}} \cap \Omega \text{ the respective linear strictorder,}$$

are computed. The permutation is defined recursively, starting with

$$P_0 := 0_E; 0_\Omega^\mathsf{T}$$

to send the least element to the least element, followed by successive application of the functional

$$\tau(X) := X \cup H_E^\mathsf{T}; X; H_\Omega,$$

becoming stationary when X is a total relation. Then

$$\pi_E := \mathsf{sup} \left[P_0, \tau(P_0), \tau(\tau(P_0)), \dots \right].$$

With the permutation thus obtained, the otherwise unfamiliar domain construction `PermTgt` π_E – meaning nothing other than rearranging – is executed, in Fig. C.2 from baseset $V = \{1, 2, 3, 4\}$ to baseset $V^{\pi_E} = \{3, 2, 4, 1\}$.

$$
E : V \longrightarrow V \quad
\begin{array}{c}
 & \begin{smallmatrix}1 & 2 & 3 & 4\end{smallmatrix} \\
\begin{smallmatrix}1\\2\\3\\4\end{smallmatrix} &
\begin{pmatrix}
1 & 0 & 0 & 0 \\
1 & 1 & 0 & 1 \\
1 & 1 & 1 & 1 \\
1 & 0 & 0 & 1
\end{pmatrix}
\end{array}
\qquad
\Omega : V \longrightarrow V \quad
\begin{array}{c}
 & \begin{smallmatrix}1 & 2 & 3 & 4\end{smallmatrix} \\
\begin{smallmatrix}1\\2\\3\\4\end{smallmatrix} &
\begin{pmatrix}
1 & 1 & 1 & 1 \\
0 & 1 & 1 & 1 \\
0 & 0 & 1 & 1 \\
0 & 0 & 0 & 1
\end{pmatrix}
\end{array}
$$

E : V ⟶ V   Ω : V ⟶ V

original relation   its base order

$$
\pi_E : V \longrightarrow V^{\pi_E} \quad
\begin{array}{c}
 & \begin{smallmatrix}3 & 2 & 4 & 1\end{smallmatrix} \\
\begin{smallmatrix}1\\2\\3\\4\end{smallmatrix} &
\begin{pmatrix}
0 & 0 & 0 & 1 \\
0 & 1 & 0 & 0 \\
1 & 0 & 0 & 0 \\
0 & 0 & 1 & 0
\end{pmatrix}
\end{array}
\qquad
\pi_E^{\mathsf{T}} {;}\, E {;}\, \pi_E : V^{\pi_E} \longrightarrow V^{\pi_E} \quad
\begin{array}{c}
 & \begin{smallmatrix}3 & 2 & 4 & 1\end{smallmatrix} \\
\begin{smallmatrix}3\\2\\4\\1\end{smallmatrix} &
\begin{pmatrix}
1 & 1 & 1 & 1 \\
0 & 1 & 1 & 1 \\
0 & 0 & 1 & 1 \\
0 & 0 & 0 & 1
\end{pmatrix}
\end{array}
$$

$\pi_E : V \longrightarrow V^{\pi_E}$ $\pi_E^{\mathsf{T}} {;}\, E {;}\, \pi_E : V^{\pi_E} \longrightarrow V^{\pi_E}$

rearrangement transition   rearranged relation

Fig. C.2 Rearranging a linear order to upper triangular form

Note that we have here a simultaneous permutation of source and target side – in contrast to the last section where only the target was permuted. It is then easy to generalize this to strictorders.

C.3 Composite rearrangements

The preceding two rearrangements are basic steps on which we build to handle more complicated cases.

Rearranging a weakorder

Any weakorder $W : X \longrightarrow X$ can be transformed into an upper right block triangle form. To obtain a permutation relation on X that rearranges W this way, we perform three steps. First, W is joined with the identity $\mathbb{I}$. The resulting reflexive closure $E = W \cup \mathbb{I}$ of W is an order on X. Next, a linear extension E' of E might be determined as a Szpilrajn extension according to Prop. 12.14; however, we apply Prop. 12.20 to achieve this in a simpler way. And finally, a permutation $P : X \longrightarrow X$ is computed that rearranges the linear order E' into the full upper right triangle $P^{\mathsf{T}} {;}\, E' {;}\, P$. A little reflection shows that the same permutation also transforms the original weakorder relation W into the desired upper right block triangle form $P^{\mathsf{T}} {;}\, W {;}\, P$.

$$
\begin{array}{c}
 & \begin{smallmatrix}1 & 2 & 3 & 4 & 5 & 6 & 7 & 8 & 9 & 10 & 11\end{smallmatrix} \\
\begin{smallmatrix}1\\2\\3\\4\\5\\6\\7\\8\\9\\10\\11\end{smallmatrix} &
\begin{pmatrix}
0 & 1 & 0 & 1 & 1 & 1 & 1 & 0 & 1 & 0 & 1 \\
0 & 0 & 0 & 0 & 0 & 0 & 0 & 0 & 0 & 0 & 0 \\
1 & 1 & 0 & 1 & 1 & 1 & 1 & 1 & 1 & 1 & 0 \\
0 & 1 & 0 & 0 & 1 & 1 & 0 & 0 & 0 & 0 & 0 \\
0 & 1 & 0 & 0 & 0 & 1 & 0 & 0 & 0 & 0 & 0 \\
0 & 0 & 0 & 0 & 0 & 0 & 0 & 0 & 0 & 0 & 0 \\
0 & 1 & 0 & 1 & 1 & 1 & 0 & 1 & 0 & 1 & 0 \\
0 & 1 & 0 & 0 & 1 & 1 & 0 & 0 & 0 & 0 & 0 \\
0 & 1 & 0 & 1 & 1 & 1 & 0 & 1 & 0 & 1 & 0 \\
0 & 1 & 0 & 0 & 1 & 1 & 0 & 0 & 0 & 0 & 0 \\
1 & 1 & 0 & 1 & 1 & 1 & 1 & 1 & 1 & 1 & 0
\end{pmatrix}
\end{array}
$$

$$
\begin{array}{c}
 & \begin{smallmatrix}3 & 11 & 1 & 7 & 9 & 4 & 8 & 10 & 5 & 2 & 6\end{smallmatrix} \\
\begin{smallmatrix}1\\2\\3\\4\\5\\6\\7\\8\\9\\10\\11\end{smallmatrix} &
\begin{pmatrix}
0 & 0 & 1 & 0 & 0 & 0 & 0 & 0 & 0 & 0 & 0 \\
0 & 0 & 0 & 0 & 0 & 0 & 0 & 0 & 0 & 1 & 0 \\
1 & 0 & 0 & 0 & 0 & 0 & 0 & 0 & 0 & 0 & 0 \\
0 & 0 & 0 & 0 & 0 & 1 & 0 & 0 & 0 & 0 & 0 \\
0 & 0 & 0 & 0 & 0 & 0 & 0 & 0 & 1 & 0 & 0 \\
0 & 0 & 0 & 0 & 0 & 0 & 0 & 0 & 0 & 0 & 1 \\
0 & 0 & 0 & 1 & 0 & 0 & 0 & 0 & 0 & 0 & 0 \\
0 & 0 & 0 & 0 & 0 & 0 & 1 & 0 & 0 & 0 & 0 \\
0 & 0 & 0 & 0 & 1 & 0 & 0 & 0 & 0 & 0 & 0 \\
0 & 0 & 0 & 0 & 0 & 0 & 0 & 1 & 0 & 0 & 0 \\
0 & 1 & 0 & 0 & 0 & 0 & 0 & 0 & 0 & 0 & 0
\end{pmatrix}
\end{array}
$$

$$
\begin{array}{c}
 & \begin{smallmatrix}3 & 11 & 1 & 7 & 9 & 4 & 8 & 10 & 5 & 2 & 6\end{smallmatrix} \\
\begin{smallmatrix}3\\11\\1\\7\\9\\4\\8\\10\\5\\2\\6\end{smallmatrix} &
\begin{pmatrix}
0 & 0 & 1 & 1 & 1 & 1 & 1 & 1 & 1 & 1 & 1 \\
0 & 0 & 1 & 1 & 1 & 1 & 1 & 1 & 1 & 1 & 1 \\
0 & 0 & 0 & 0 & 0 & 1 & 1 & 1 & 1 & 1 & 1 \\
0 & 0 & 0 & 0 & 0 & 1 & 1 & 1 & 1 & 1 & 1 \\
0 & 0 & 0 & 0 & 0 & 1 & 1 & 1 & 1 & 1 & 1 \\
0 & 0 & 0 & 0 & 0 & 0 & 0 & 0 & 1 & 1 & 1 \\
0 & 0 & 0 & 0 & 0 & 0 & 0 & 0 & 1 & 1 & 1 \\
0 & 0 & 0 & 0 & 0 & 0 & 0 & 0 & 1 & 1 & 1 \\
0 & 0 & 0 & 0 & 0 & 0 & 0 & 0 & 0 & 1 & 1 \\
0 & 0 & 0 & 0 & 0 & 0 & 0 & 0 & 0 & 0 & 0 \\
0 & 0 & 0 & 0 & 0 & 0 & 0 & 0 & 0 & 0 & 0
\end{pmatrix}
\end{array}
$$

Fig. C.3 Rearranging a weakorder to upper block triangular form

When a connex preorder is presented, we are now also able to rearrange it. To this end, the corresponding weakorder is determined first. Then the permutation is obtained for this weakorder.

Rearranging a not yet linear (strict-)order

First we consider a not necessarily linear order E. In this case, it is advisable first to complete it with a Szpilrajn extension according to Prop. 12.14 to a linear order. It may then be handled as such using the method of Section C.2, for which procedure we give an example in Fig. C.4.

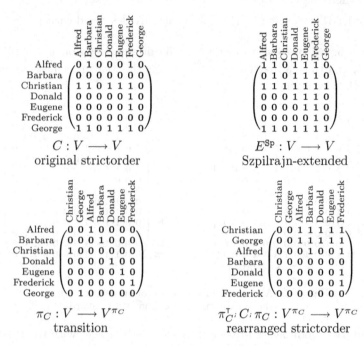

$$C : V \longrightarrow V$$
original strictorder

$$E^{\mathrm{Sp}} : V \longrightarrow V$$
Szpilrajn-extended

$$\pi_C : V \longrightarrow V^{\pi_C}$$
transition

$$\pi_C^{\mathsf{T}} {}_{;} C {}_{;} \pi_C : V^{\pi_C} \longrightarrow V^{\pi_C}$$
rearranged strictorder

Fig. C.4 Rearranging a not yet linear order to upper triangular form

The rearrangement obtained did not take into consideration that this linear strict-order C was in fact a semiorder. Recognizing this additional property, we obtain a much nicer upper triangular form respecting this in the style of Fig. 12.17.

Rearranging symmetric idempotents

Symmetric idempotents are sometimes also sloppily called 'partial' equivalence relations and abbreviated as PERs. They satisfy $R \mathbin{;} R \subseteq R$ – implying $R \mathbin{;} R = R$ – and $R^{\mathsf{T}} \subseteq R$, but may have empty rows and columns, i.e., may not be reflexive, in contrast to an equivalence. A nice presentation would be correspondingly similar to an equivalence with its diagonal blocks. Our decision is to present empty rows and columns collected at the end. That is, we construct the row equivalence $\Xi := \Xi(R)$ of R. Then the base strictorder C of the quotient source `QuotMod` Ξ would already give an indication how to arrange blocks coherently with the weakorder

$$\Omega := \pi \,;\, C \,;\, \pi^{\mathsf{T}}$$

where we let $\pi := $ `Project` Ξ denote the natural projection. With a little bit of manipulation on Ω we achieve the block of empty rows positioned last. To this end we decompose the underlying set of rows, respectively columns, into $\mathbb{T} = $ `empty` $\cup$ `non-empty` and form the weakorder

$$\Omega' := \Omega \cup \text{non-empty} \,;\, \text{empty}^{\mathsf{T}}.$$

The weakorder Ω' may then be handled as already described earlier.

With Fig. C.5, we provide an example. The original relation is symmetric and idempotent, and thus, a 'partial' equivalence. As the weakorder just developed, we obtain the right relation.

```
         1 2 3 4 5 6 7 8 9 10 11 12 13              1 2 3 4 5 6 7 8 9 10 11 12 13
    1  / 1 1 0 0 1 0 1 1 1 1  0  0  1 \        1  / 0 0 1 1 0 1 0 0 0 0  1  1  0 \
    2  | 1 1 0 0 1 0 1 1 1 1  0  0  1 |        2  | 0 0 1 1 0 1 0 0 0 0  1  1  0 |
    3  | 0 0 1 0 0 0 0 0 0 0  0  1  0 |        3  | 0 0 0 1 0 1 0 0 0 0  1  0  0 |
    4  | 0 0 0 0 0 0 0 0 0 0  0  0  0 |        4  | 0 0 0 0 0 0 0 0 0 0  0  0  0 |
    5  | 1 1 0 0 1 0 1 1 1 1  0  0  1 |        5  | 0 0 1 1 0 1 0 0 0 0  1  1  0 |
    6  | 0 0 0 0 0 0 0 0 0 0  0  0  0 |        6  | 0 0 0 0 0 0 0 0 0 0  0  0  0 |
    7  | 1 1 0 0 1 0 1 1 1 1  0  0  1 |        7  | 0 0 1 1 0 1 0 0 0 0  1  1  0 |
    8  | 1 1 0 0 1 0 1 1 1 1  0  0  1 |        8  | 0 0 1 1 0 1 0 0 0 0  1  1  0 |
    9  | 1 1 0 0 1 0 1 1 1 1  0  0  1 |        9  | 0 0 1 1 0 1 0 0 0 0  1  1  0 |
   10  | 1 1 0 0 1 0 1 1 1 1  0  0  1 |       10  | 0 0 1 1 0 1 0 0 0 0  1  1  0 |
   11  | 0 0 0 0 0 0 0 0 0 0  0  0  0 |       11  | 0 0 0 0 0 0 0 0 0 0  0  0  0 |
   12  | 0 0 1 0 0 0 0 0 0 0  0  1  0 |       12  | 0 0 0 1 0 1 0 0 0 0  1  0  0 |
   13  \ 1 1 0 0 1 0 1 1 1 1  0  0  1 /       13  \ 0 0 1 1 0 1 0 0 0 0  1  1  0 /
            original relation                            weakorder
```

Fig. C.5 A symmetric and idempotent relation with
the constructed corresponding weakorder

The permutation obtained is presented in Fig. C.6 together with the result of the rearrangement.

```
         1 2 5 7 8 9 10 13 3 12 4 6 11            1 2 5 7 8 9 10 13 3 12 4 6 11
    1  / 1 0 0 0 0 0 0 0 0 0 0 0 0 \        1  / 1 1 1 1 1 1 1 1 0 0 0 0 0 \
    2  | 0 1 0 0 0 0 0 0 0 0 0 0 0 |        2  | 1 1 1 1 1 1 1 1 0 0 0 0 0 |
    3  | 0 0 0 0 0 0 0 0 1 0 0 0 0 |        5  | 1 1 1 1 1 1 1 1 0 0 0 0 0 |
    4  | 0 0 0 0 0 0 0 0 0 0 1 0 0 |        7  | 1 1 1 1 1 1 1 1 0 0 0 0 0 |
    5  | 0 0 1 0 0 0 0 0 0 0 0 0 0 |        8  | 1 1 1 1 1 1 1 1 0 0 0 0 0 |
    6  | 0 0 0 0 0 0 0 0 0 0 0 1 0 |        9  | 1 1 1 1 1 1 1 1 0 0 0 0 0 |
    7  | 0 0 0 1 0 0 0 0 0 0 0 0 0 |       10  | 1 1 1 1 1 1 1 1 0 0 0 0 0 |
    8  | 0 0 0 0 1 0 0 0 0 0 0 0 0 |       13  | 1 1 1 1 1 1 1 1 0 0 0 0 0 |
    9  | 0 0 0 0 0 1 0 0 0 0 0 0 0 |        3  | 0 0 0 0 0 0 0 0 1 1 0 0 0 |
   10  | 0 0 0 0 0 0 1 0 0 0 0 0 0 |       12  | 0 0 0 0 0 0 0 0 1 1 0 0 0 |
   11  | 0 0 0 0 0 0 0 0 0 0 0 0 1 |        4  | 0 0 0 0 0 0 0 0 0 0 0 0 0 |
   12  | 0 0 0 0 0 0 0 0 0 1 0 0 0 |        6  | 0 0 0 0 0 0 0 0 0 0 0 0 0 |
   13  \ 0 0 0 0 0 0 0 1 0 0 0 0 0 /       11  \ 0 0 0 0 0 0 0 0 0 0 0 0 0 /
            permutation        and             rearranged original
```

Fig. C.6 Permutation that transforms the relation of Fig. C.5 to block-diagonal form

While the arrangement of the other diagonal blocks may give a more or less arbitrary sequence, we have been careful to assemble the empty rows and columns at the end.

Rearranging difunctional relations

When given a difunctional relation R, one might attempt to consider the symmetric idempotents $R_i R^\mathsf{T}$ and $R^\mathsf{T}_i R$ for the rows and columns, respectively, determine the corresponding permutations P_R, P_C for the rows as well as for the columns, and then form $R_{\mathbf{rearr}} := P_R^\mathsf{T}_i R_i P_C$. This is, however, not sufficient as it rearranges independently and does not take into account that a (partial) block diagonal should finally appear.

So we execute the arrangement as to the symmetric idempotent in a first step only for the rows and obtain $R_{\mathbf{rearr},R} := P_R^\mathsf{T}_i R$. Here equal rows are adjacent and the empty rows, if any, reside at the end. Remembering row-to-column difunctional, it is then possible to relate this with a corresponding investigation on the target side. Figure C.7 shows an example.

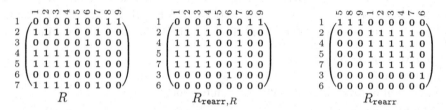

Fig. C.7 Rearranging a difunctional relation

Rearranging independence and covering

In Section 10.2, we learned about independent and covering sets. Not least, conditions were presented that are satisfied when such a pair is maximal/minimal with respect to set inclusion. With the techniques mentioned so far, it is relatively easy to configure the respective permutations.

Let a relation R be given together with an independent pair of sets u, v, so that it will satisfy $R_i v \subseteq \overline{u}$. Then form the weakorder $W_R := u_i \overline{u}^\mathsf{T}$ on the source side as well as the weakorder $W_C := \overline{v}v^\mathsf{T}$ on the target side. We have already learned how to obtain the necessary permutation P_R, P_C to arrange such weakorders in the upper right form. Using these, $R_{\mathbf{rearr}} := P_R^\mathsf{T}_i R_i P_C$ shows a rearrangement $R_{u,v \text{ independent}}$ of the relation R so as to have the $\mathbf{0}$-rectangle in the upper right.

One may also start an analogous procedure with a covering pair s, t of sets of R and obtain $R_{s,t \text{ covering}} := R_{\overline{s},\overline{t} \text{ independent}}$.

Rearranging around matching and assignment

We recall the discussion in Section 16.5 concerning (maximum) matchings and the assignment problem. Our aim is to arrive at a decomposition according to Fig. 16.13 for an arbitrary relation with a given cardinality-maximum matching. First, the Galois iteration is performed that produces for relation $Q : V \longrightarrow W$ and matching $\lambda \subseteq Q$ the sets $a \subseteq V$ and $b \subseteq W$. Then a, b serve to build rearranged relations as shown earlier.

Once this is understood, one can treat any given relation in this way, i.e., first determine a cardinality-maximum matching and then iterate so as to obtain the corresponding non-enlargeable rectangle outside the relation. Once this is known, one may indeed get an upper right rectangle of zeros and a diagonal in the lower left.

Rearranging others

These sketches will have provided enough evidence that a lot of other concepts can also be visualized in this way. Among those examples where this is conceivable – and has already been worked out and used in this book – are the following

- block versions of orderings,
- implication structure rearrangement,
- exhaustion of progressively bounded points/blocks,
- game rearrangement.

Appendix D
Historical Annotations

A very brief account shall be given of those persons and developments that advanced relational mathematics. This will, of course, turn out to be a rather personal choice and include secondary reporting from several historical sources. In particular we will ignore most of what has been reported over the centuries on 'Logics in general'. In view of the Schröder rule

$$A_{;}B \subseteq C \quad \Longleftrightarrow \quad A^{\mathsf{T}}{}_{;}\overline{C} \subseteq \overline{B} \quad \approx \quad \forall i,k : \left(\exists j : A_{ij}^{\mathsf{T}} \wedge \overline{C}_{jk} \right) \to \overline{B}_{i,k}$$

we look for what makes up relations proper; that is, liberating conversion and composition from being expressed only in natural language and 'quantifying over the predicate'.

D.1 From Aristotle to the Scholastics

Although he was the very first to treat relations, Aristotle (384–322 BC) did not advance very far. On several occasions, it has been reported that his method of syllogisms was in fact not suited to reasoning that the head of a horse is the head of an animal, given the statement that the horse is an animal. His works on what we today call Logic were reorganized by his pupils after his death to form the *Organon*. One of its ten categories[1] is called *relatio* (in Latin, Greek πρός τι), a topic discussed and passed on over the centuries.

According to [24], Aristotle had already dealt with an early form of quantification: *Some things are universal, others individual. By the term 'universal' I mean that which is of such a nature as to be predicated of many objects, by 'individual' that which is not thus predicated. Thus 'man' is a universal, 'Callias' an individual... If, then, a man states a positive and a negative proposition of universal character with regard to a universal, these two propositions are 'contrary'.*

The scholastic tradition concerning relations appears to have been started with the adoption of the Organon (after Plato's work) by one of the greatest teachers

---

[1] The 10 categories are: substance, quantity, quality, relation, place, time, situation, state, action, and passion.

Fig. D.1 Aristotle (384–322 BC) in the Louvre

and philosophers[2] of his time, known for having had thousands of students, Peter Abelard[3] (1079–1142).

Much of his unpublished work stayed unknown and was rediscovered only late in 1836 by Cousin. In [57], it is reported that Aristotle's natural science was forbidden in Paris from 1210 because divine revelation was preferred to formal reasoning; this lasted until about 1255.

Another person to be mentioned is John Duns, the Scot (1266–1308), more frequently referred to as Duns Scotus. Several achievements have been attributed to him, in particular that $p \rightarrow (\overline{p} \rightarrow q)$ will always hold. But already 200 years after his death his personal authorship concerning logic (besides his important work in theology) began to be mistrusted.

Scholastic understanding of the quantifications 'every' and 'some' was already close to the modern interpretation, but was expressed in a verbose – not yet formal – style: ... *if one says: 'every man runs' it follows formally: 'therefore this man runs, and that man runs, etc.' But of the particular sign I have said that it signifies that a universal term to which it is adjoined stands disjunctively for all its supposita.*

---

[2] In [24] an epitaph is recalled qualifying Abelard as 'the Aristotle of our time, the equal or superior of all logicians there have been'.

[3] Also known as Abaelard(us). He is most famous for having taught Héloise in the home of her uncle. They fell in love and produced a child. Peter even married Héloise, but secretly so that he could maintain his clerical status. However, this did not prevent him from being castrated one night in the dormitory of his abbey, as revenge on the part of Héloise's uncle.

That is evident since if one says: 'some man runs' it follows that Socrates or Plato runs, or Cicero runs, and so of each (de singulis). ... since it is sufficient for the truth of a disjunctive that one of its parts be true.

Fig. D.2 Duns Scotus (1266–1308)

The great tradition of the schoolmen (i.e., scholastics) concerning Logics was carried forward by William of Ockham (1287–1347). According to [79], page 295, the De Morgan rule was already known to William of Ockham and 'occurs explictly' in his *Summa Totius Logicae*.

D.2 Anticipations by Leibniz

Gottfried Wilhelm Leibniz (1646–1716) was one of the first to calculate with bit values as can be seen in Fig. D.3 taken from one of his manuscripts [69]. In [136], he is called the founder of symbolic logic. In [79] he is ranked *among the greatest of all logicians. ... most of what he had written remained unpublished in the library at Hanover, where he had served the Elector as a historian, scientific adviser, and expert on international law.*

In his work, Leibniz often already used the word 'binary'. In addition, he came close to something like a relational calculus in handling congruences of geometrical objects; in the many historical remarks of the voluminous thesis [134] Leibniz is cited as: *Characteres sunt res quaedam, quibus aliarum rerum inter se relationes exprimuntur, et quarum facilior est quam illarum tractatio.*

In 1970, the logic historian Bocheński [24] harshly criticized ... *modern philosophers such as Spinoza, ..., Kant, Hegel etc. could have no interest for the historian of formal logic. When compared with the logicians of the 4th century* B.C., *the 13th and 20th centuries* A.D. *they were simply ignorant of what pertains logic ... But there*

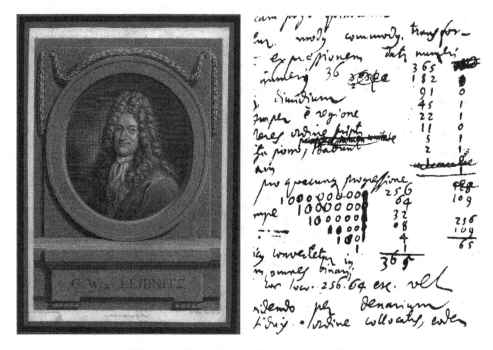

Fig. D.3 Gottfried Wilhelm Leibniz (1646–1716) and one of his manuscripts with **0, 1**

is one exception, Leibniz (1646–1716). So far from being an ignorabimus,[4] *he was one of the greatest logicians of all time, which is the more remarkable in that his historical knowledge* (The present author adds: concerning earlier work in logics!) *was rather limited. His place in the history of logic is unique. On the one hand his achievement constitutes a peak in the treatment of a part of the Aristotelian syllogistic, where he introduced many new, or newly developed features, such as the completion of the combinatorial method, the exact working out of various methods of reduction, the method of substitution, the so-called 'Eulerian' diagrams, etc. On the other hand he is the founder of mathematical logic. ... his real achievements in the realm of mathematical logic are little relevant to the history of problems, since they remained for long unpublished and were first discovered at the end of the 19th century when the problems he had dealt with had already been raised independently.*

It was mainly Louis Couturat who made public the enormous achievements hidden in the unpublished papers of Leibniz. According to [36]: *It was in 1676 that he* (i.e., Leibniz) *first dreamed of a kind of algebra of thought ....*

---

[4] ignoramus et ignorabimus: 'We do not know' and 'We will not know'. A famous statement causing controversy by Emil du Bois-Reymond, from his two talks to the *Versammlung Deutscher Naturforscher und Ärzte* in Leipzig, 1880.

D.3 Quantifying the predicate

One main point to resolve in order to arrive at relations as we conceive them today was quantification which is needed to introduce relation composition. 'Quantifying the predicate' became a major issue in the first half of the nineteenth century. How difficult it was to introduce quantification is hard to imagine today; it can be estimated from the attempts of Giuseppe Peano as late as 1888: $a \supset_{x,y...} b$ meant 'whatever $x, y. . .$ may be, b is deduced from a'. He is the man to whom we owe the foundations of natural and real numbers!

Fig. D.4 The botanist George Bentham (1800–1884)

The first to achieve certain progress was George Bentham (1800–1884), see [8, 9]. He has been called 'the premier systematic botanist of the nineteenth century'. As a byproduct of his classification work he published in 1827 the *Outline of a New System of Logic, with a Critical Examination of Dr. Whately's 'Elements of Logic'.* He 'quantified over predicates', which looked like

X in toto, Y ex parte,

or abbreviated by him as

$t X$, $p Y$.

We write this down in a fictitious form closer to our notation: assuming predicates $X, Y : U \longrightarrow \mathbb{B}$, what he formulated was

$\forall X$ or $\exists Y$.

Nowadays, we would prefer to have a quantification over an individual variable that may satisfy the predicate

$\forall x : X(x)$, $\exists y : Y(y)$.

It seems that Bentham did not quite achieve this. However, he reasoned routinely in natural language that 'every man is' is merely the substitution of 'no man is not', again not denoting with a bound variable the individuals over which he was about to quantify.

Fig. D.5 Sir William Hamilton (1788–1856)

The two people to be mentioned next are Augustus De Morgan (1806–1871) and Sir William Hamilton[5] (1788–1856). They fought for decades on quantification over predicates and their respective priorities, developing a considerable degree of personal animosity.[6] Since 1839, Hamilton had taught as an Edinburgh professor of logic and metaphysics on the principle of a quantified predicate; a topic which De Morgan began to study only in 1846 – in more depth and with greater and longer-lasting success. It may well be that these really harsh quarrels led De Morgan to think about relations and relational composition in more depth than he might have done otherwise.

Prior [102] reports on page 148 concerning quantification over predicates: *In the nineteenth century, the 'quantification of the predicate' was advocated again, but with a quite fantastic incompetence, by the Edinburgh logician Sir William Hamilton*. Hamilton favored

All X is all Y

with all the other conceivable versions *Some, Any, is-not*, in contrast to the more familiar

All X is Y.

---

[5] Not to be confused with Sir William Rowan Hamilton (1805–1865), see Fig. D.8 and [96], the great Irish mathematician and proven postal correspondent of De Morgan.
[6] De Morgan called Hamilton disrespectfully the 'Edinburgh Aristotle'.

Fig. D.6 Augustus De Morgan (1806–1871)

As Peter Heath put it in his introduction to [41], *... whatever may have been done by European logicians, Bentham was the first writer in English to quantify the predicate, Hamilton the first to make any extensive use of it, and De Morgan the first to grasp what it was all about.*

Around 1873, Bentham's claims to priority were finally settled by William Stanley Jevons in [70] and termed *the most fruitful discovery made in abstract logical science since the time of Aristotle.* He made it clear that priority in quantifying the predicate had to be attributed to George Bentham – not to Sir William Hamilton. Jevons also reported that Boolean algebra had already been fully anticipated by Leibniz. He mentioned that Leibniz was not 'afraid' of $xx = x$ for disjunction. This was judged almost sensational by readers of Boole's ideas and probably made Boole concentrate mainly on the 'exclusive or'.

According to [41], De Morgan *was among the first to discern the genius of George Boole, then an unknown schoolmaster, and that he not only encouraged and obtained a hearing for his work, but was instrumental in securing for Boole his first and only academic post, as professor at Queen's College, Cork.*

The Boole–De Morgan letters, ranging over the period 1842–1864, have been edited in [133]. In his letter to George Boole of 1860 (no. 64 according to the numbering in [133]), De Morgan referred to Hamilton (at that time already deceased) thus: *He is a monster of capability because so unequally balanced that some parts are of gigantic development and others only rudimentary.*

In a sense, George Boole's 1847 pamphlet *The mathematical analysis of logic, being an essay towards a calculus of deductive reasoning* initiated the study of relations, [25, 27]. Boole[7] still used unary predicates in his work as opposed to binary ones which are closer to relations. Later, his work was summarized in [26, 28].

Fig. D.7 George Boole (1815–1864)

An example of his many syllogisms:

All Xs are Ys,
Some Zs are not Ys $\implies$ Some Zs are not Xs (memorial word: Baroko)

In 1859, De Morgan published *On the Syllogism: IV; and on the Logic of Relations* [40, 41], which contains both

$\overline{a \vee b} = \overline{a} \wedge \overline{b}$ and 'Theorem K'

the famous De Morgan rule and Theorem K (roughly corresponding to what we have called the Schröder rule), known far less well than the De Morgan rule but of at least equal importance. In the running text of [40], it reads as follows:

162. If two relations combine into what is contained in a third relation, then the converse of either of the two combined with the contrary of the third, in the same

---

[7] The right picture of Fig. D.7 is the frontispiece of [28].

order, is contained in the contrary *of the other of the two. Thus the following three assertions are* identically the same, ...

To this a footnote is attached saying:

This theorem ought to be called theorem K, being in fact the theorem on which depends the process ... indicated by the letter K in the old memorial verses.

In his symbolic notation, it is

'If $LM))N,$ then $L^{-1}n))m$ and $nM^{-1}))l$',

which is obviously 'isomorphic' to our present version

'If $A \,; B \subseteq C,$ then $A^{\mathsf{T}}\,; \overline{C} \subseteq \overline{B}$ and $\overline{C}\,; B^{\mathsf{T}} \subseteq \overline{A}$'.

Heath [41] also reports that Charles Sanders Peirce *in a moment of enthusiasm* called De Morgan *the greatest formal logician that ever lived.* In spite of this – recall the verbose textual description of Theorem K above – Alfred Tarski wrote in [138]: *Nevertheless, De Morgan cannot be regarded as the creator of the modern theory of relations, since he did not possess an adequate apparatus for treating the subject in which he was interested, and was apparently unable to create such an apparatus. His investigations on relations show a lack of clarity and rigor which perhaps accounts for the neglect into which they fell in the following years.*

It is highly unsatisfactory that as late as in 1914, and after De Morgan's, Peirce's, and Schröder's work, Couturat's booklet on *The Algebra of Logic* [36] does not mention relational composition, or conversion, or quantification. However, it gives a lot of credit to the achievements of Leibniz.

D.4 Getting rid of bindings to natural languages

Already in 1827, George Bentham complained of the shortcomings of natural languages: *Because a figure is used in one langauge, it does not follow that it must be translated into another by the same figure. Should we have to translate the French expression, 'Ces hommes sont utiles dans le rapport de leur puissance,' should we say, 'These man are useful in the report of their power?' or ...*

In [79], William of Ockham's reasonings read as follows:

Socrates is running, so something is running.
A white man is running, therefore a man is running.
I have an imitation pearl, therefore I have a pearl.

The last statement is marked as obviously questionable. It shows the problem of relying too much on natural languages and their irregular structure. It is much too simplistic a point of view to assume that *white man* restricts the concept of *man* in the same way as *imitation pearl* restricts *pearl*. Over a very long period of time, Scholastics and Logic were confined to such disputes. This was what George Bentham harshly criticized.[8]

Other deficiencies stemming from natural language were shown by Boole in [25, 27]:

No virtuous man is a tyrant,   *is converted into*
No tyrant is a virtuous man.

All birds are animals,   *is converted into*
Some animals are birds.

Every poet is a man of genius,   *is converted into*
He who is not a man of genius is not a poet.

As one easily sees, Boole quantifies over predicates, but not yet in the form of today, i.e., not with a bound variable. This bound variable now makes it much simpler to talk about the predicates to be satisfied and/or to satisfy further predicates. In Boole's time, one had to look for appropriate quantifying particles, *some, all*, for example, and had to struggle with their insufficiencies and dependence on language.

De Morgan added the following footnote:

> *Though I take the following only from a newspaper, yet I feel confident it really happened: there is the truth of nature about it, and the enormity of the cases is not incredible to those who have taught beginners in reasoning. The scene is a ragged school.* TEACHER. *Now, boys, Shem, Ham, and Japheth were Noah's sons; who was the father of Shem, Ham, and Japheth? No answer.* TEACHER. *Boys, you know Mr Smith, the carpenter, opposite; has he any sons?* BOYS. *Oh! yes, Sir! there's Bill and Ben.* TEACHER. *And who is the father of Bill and Ben Smith?* BOYS. *Why, Mr Smith, to be sure.* TEACHER. *Well, then, once more, Shem, Ham, and Japheth were Noah's sons; who was the father of Shem, Ham, and Japheth? A long pause; at last a boy, indignant at what he thought the attempted trick, cried out* It couldn't *have been Mr Smith! These boys had never converted the relation of father and son, except under the material aid of a common surname: if Shem Arkwright, &c., had been described as the sons of Noah Arkwright, part of the difficulty, not all, would have been removed.*

Schröder discusses

> *Dass man im Deutschen nichts geschlechtslos, ohne ein bestimmtes genus sagen kann ist für unsre Disziplin sehr hinderlich und begründet einen grossen Vorsprung des Englischen, wo einfach „lover" eintritt für „der, die oder das Lieben-*

---

[8] ... *that the author had done much towards divesting the science of that useless jargon, of those unmeaning puerilities, with which it had been loaded by the schoolmen; and which, being the only apparent result of their efforts, have cast so much opprobrium and ridicule on the very name of Logic.*

Fig. D.8 Sir William Rowan Hamilton (1805–1865)

de". Auch können wir im Deutschen für B$^\mathsf{T}$ [using our notation of the present book!] *(benefitted by-) nicht „bewohlthatet von-" sagen sondern müssen zu der Umschreibung „Empfänger oder Empfängerin von Wohlthaten seitens-" unsre Zuflucht nehmen, u.s.w. ... Wir werden in unserer Disziplin fein unterscheiden müssen zwischen den Partikeln „ausser" (englisch:* but, save?, besides?*) und „ausgenommen" (englisch:* excepting).

In [63], Hans Hahn,[9] spiritus rector of the famous Vienna Circle, is recalled with

... Die Sätze unserer Sprache sind im wesentlichen so gebaut: sie sagen von einem Subjekt, eventuell mehreren Subjekten, ein Prädikat aus. 'Dieses Kreidestück ist weiß.' 'Dieses Kreidestück und dieses Kreidestück sind weiß.' So weit wäre alles in Ordnung. Die Sprache sagt aber auch: 'Dieses Kreidestück und dieses Kreidestück sind gleichfarbig.' Die Sprache tut also so, als ob – wie früher jedem der beiden Kreidestücke die Eigenschaft 'weiß' – so jetzt jedem von beiden die Eigenschaft 'gleichfarbig' zugeschrieben würde. Das ist aber offenbar Unsinn: 'Gleichfarbig' ist nicht eine Eigenschaft eines Individuums, wie etwa 'weiß', sondern ist eine Beziehung, eine Relation zwischen zwei Individuen.

---

[9] Hans Hahn (1879–1934), to whom Kurt Gödel submitted his spectacular dissertation in 1929; also famous for the Hahn–Banach Theorem, for example.

D.5 Relations as Boolean matrices

We know for a fact that the concept of a matrix was not yet available when relations were first studied. The term 'matrix' is reported to have been coined in 1850 by James Joseph Sylvester as the name of an array of numbers. Matrices then were used by Arthur Cayley (1821–1895) in 1858 [31] for a rectangular array of elements of a field. Rules for multiplication were given, but the transposition for a product, $(AB)^\mathsf{T} = B^\mathsf{T} A^\mathsf{T}$, was not yet mentioned among the rules [62]. Sir William Rowan Hamilton (1805–1865) was one of the first to use them. They were later intensively studied by Otto Toeplitz (1881–1940), as circulant matrices and as theory of operators conceived as infinite matrices [68].

The origins of matrices and determinants (intimately related to the former), however, date back some time. It is again Leibniz to whom [80] ascribes the first treatment of determinants: *Gottfried Wilhelm Leibniz was the first mathematician to elaborate a determinant theory. His contributions included coining the term 'resultant' (*resultans sc. aequatio*) to denote certain combinatorial sums of the terms in the determinant; inventing a symbol for this resultant; formulating (though not proving) some general theorems about resultants; and deducing important results in the theory of systems of linear equations and in elimination theory, formulated by means of determinants. His numerous relevant manuscripts, dating from 1678 to 1713, remained completely unpublished until recently* ... (meaning 1980!).

Fig. D.9 Takakazu Seki (1637/1642–1708) on Japanese stamp

Only very recently in [145], the following appeared in the *Mathematical Intelligencer*: 'Takakazu Seki, also known as 'Seki Kowa', was the first mathematician to investigate determinants, a few years before Leibniz (who is usually given priority) contributes to the subject. In 1683 Seki explained how to calculate determinants up to size 5×5, and a Japanese stamp shows his diagram for calculating the products that arise in the evaluation of 4×4 determinants.'

In [62], it is reported that *Matrix theory is today one of the staples of higher-level mathematics education; so it is surprising to find that its history is fragmentary, and that only fairly recently did it acquire its present status.* According to this text, one of the origins of matrix theory is bilinear forms and quadratic forms, now written down as $x^\mathsf{T} A y$ and $x^\mathsf{T} A x$. Quadratic forms were studied by Gottfried Wilhelm Leibniz from the late seventeenth century onwards. ...

Carl Friedrich Gauss's Disquisitiones arithmeticae *(1801), his masterpiece in number theory, contains a superb passage on the treatment of quadratic forms in which the coefficients are laid out as a rectangular array, and matrix inversion and multiplication and reduction to special forms are described; but nothing came out of it.*

 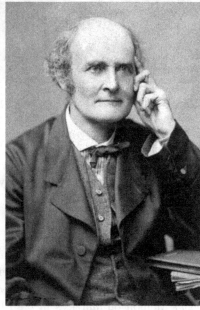

Fig. D.10 James Joseph Sylvester (1814–1897) and Arthur Cayley (1821–1895)

Sylvester as well as Cayley had been students of Augustus De Morgan; Lord Byron's daughter Ada[10] – after whom the programming language is named – was his private pupil.

In 1876 J. J. Sylvester was called on to help start Mathematics at the newly founded Johns Hopkins University in Baltimore. According to [23], Christine Ladd-Franklin resumed her studies with him after several years of working as a teacher. Her dissertation *On the Algebra of Logic* was, however, submitted to Charles Sanders Peirce who included it in a volume [84] he edited. Since, as a woman, she was not

---

[10] More formally, The Right Honorable Augusta Ada, Countess of Lovelace (1815–1852).

Fig. D.11 James Joseph Sylvester (1814–1897) and Arthur Cayley (1821–1895)

formally enrolled at the university, she could not obtain a doctorate. But 44 years later in 1926, and after she had already obtained an honorary doctorate elsewhere, she was awarded the formal degree.

In about 1870, Charles S. Peirce was *... looking for a good general algebra of logic.* He was initially unaware of Cayley's matrix concept, but on becoming informed of it, he immediately realized it to be useful for relations as well; see [35], where he is cited as saying *I have this day had the delight of reading for the first time Professor Cayley's* Memoir on matrices *in the* Philosophical transactions *for 1858 ....*

In [138], Tarski wrote: *The title of creator of the theory of relations was reserved for C. S. Peirce. In several papers published between 1870 and 1882, he introduced and made precise all the fundamental concepts of the theory of relations and formulated and established its fundamental laws.*

D.6 From equality to containment

After 1895, Ernst Schröder (see, for example [90]) wrote his voluminous three part *Algebra der Logik*; most pertinent to the present topic is Part III [127, 128] following earlier publications such as [129]. He himself seemed overwhelmed by his topic,

Fig. D.12 Charles Sanders Peirce (1839–1914)

praising it as *grandiose, overabundant, full of beauty and harmony.*[11] Schröder was the first to distinguish between the *algebra* and the *logic* of relatives, thus beginning to open a view on the diversity of models.

In the preface of [36], it is mentioned that '. . . Schröder departed from the custom of BOOLE, JEVONS, and himself (1877), which consisted in the making fundamental of the notion of *equality*, and adopted the notion of *subordination* or *inclusion* as a primitive notion'. The symbol he used looks like a forerunner of our present €-symbol.

It is obvious to logicians, that it is more or less immaterial – except for highly sophisticated investigations – whether one uses

$$A \subseteq B \qquad \text{or} \qquad A \cap \overline{B} = \bot \qquad \text{or} \qquad \top = \overline{A} \cup B,$$

i.e., an implicational as opposed to an equational style. The former is more adapted to engineers who are accustomed to estimations with regard to real numbers.

---

[11] *Es ist eine grossartige Disziplin, reich an Ausdrucksmitteln und mächtigen Schlussmethoden, fast überreich an Sätzen, wenn auch von unvergleichlichem Ebenmaasse, in welche ich versuchen will den Leser hiermit einzuführen.*

Fig. D.13 Ernst Schröder (1841–1902)

D.7 Final establishment of relations

It was the appearance of the by now well-known antinomies in combination with the brilliance of the *Principia Mathematica* [144] of Bertrand Russell and Alfred North Whitehead that made relational methods dormant for a long time. It is an irony that Russell had originally intended to lay a foundation of mathematics with the help of relations. It is reported that Russell expressed his assessment in an essay on metamathematics and metaphysics as: *The nineteenth century, which prided itself upon the invention of steam and evolution might have derived a more legitimate title to fame from the discovery of pure mathematics.*[12]

---

[12] This came to the author's attention via the novella *Alan Turing* by Rolf Hochhuth: 'Bertrand Russell hat 1901 in einem Essay über Mathematik und Metaphysik geschrieben: *Das 19. Jahrhundert, dessen ganzer Stolz die Entdeckung von Dampfkraft und Evolution war, hätte seinen Anspruch auf Nachruhm noch eher auf die Entdeckung der reinen Mathematik gründen können, also auf Booles Laws of Thought, 1854.*'

SECTION D]   THE RELATIVE PRODUCT OF TWO RELATIONS

*34·3. $\vdash : \dot{\exists}!(P\mid Q) . \equiv . \exists!(\mathrm{D}'P \cap \mathrm{D}'Q)$

Dem.

$\vdash . *25·5 . \supset$

$\vdash :: \dot{\exists}!(P\mid Q) . \equiv :. (\exists x,y) . x(P\mid Q)y :.$

[*34·1] $\equiv :. (\exists x,y) : (\exists z) . xPz . zQy :.$

[*11·27] $\equiv :. (\exists x,y,z) . xPz . zQy :.$

[*11·24] $\equiv :. (\exists z,x,y) . xPz . zQy :.$

[*11·27] $\equiv :. (\exists z) :. (\exists x,y) . xPz . zQy :.$

[*11·54] $\equiv :. (\exists z) :. (\exists x) . xPz : (\exists y) . zQy :.$

[*33·13·131] $\equiv :. (\exists z) :. z \in \mathrm{D}'P . z \in \mathrm{D}'Q :.$

[*22·33] $\equiv :. (\exists z) :. z \in \mathrm{D}'P \cap \mathrm{D}'Q :.$

[*24·5] $\equiv :. \exists!(\mathrm{D}'P \cap \mathrm{D}'Q) :: \supset \vdash . \mathrm{Prop}$

*34·301. $\vdash : \mathrm{D}'P \cap \mathrm{D}'Q = \Lambda . \equiv . P\mid Q = \dot{\Lambda}$ [*34·3 . Transp]

*34·302. $\vdash : C'P \cap C'Q = \Lambda . \supset . P\mid Q = \dot{\Lambda} . Q\mid P = \dot{\Lambda}$

Dem.

$\vdash . *33·16 . \supset \vdash : \mathrm{Hp} . \supset . \mathrm{D}'P \cap \mathrm{D}'Q = \Lambda . \mathrm{D}'Q \cap \mathrm{D}'P = \Lambda .$

[*34·301] $\supset . P\mid Q = \dot{\Lambda} . Q\mid P = \dot{\Lambda} : \supset \vdash . \mathrm{Prop}$

*34·31. $\vdash : \dot{\exists}!(P\mid Q) . \supset . \dot{\exists}!P . \dot{\exists}!Q$

Dem.

$\vdash . *34·3 . \supset \vdash : \mathrm{Hp} . \supset . \exists!(\mathrm{D}'P \cap \mathrm{D}'Q) .$

[*24·561] $\supset . \exists!\mathrm{D}'P . \exists!\mathrm{D}'Q .$

[*33·24] $\supset . \dot{\exists}!P . \dot{\exists}!Q : \supset \vdash . \mathrm{Prop}$

*34·32. $\vdash :. P = \dot{\Lambda} . \mathbf{v} . Q = \dot{\Lambda} : \supset . P\mid Q = \dot{\Lambda}$ [*34·31 . Transp . *25·51]

Fig. D.14 Bertrand Russell (1872–1970) with page 259 of the *Principia Mathematica*

In his 1915 paper, Leopold Löwenheim initiated finite model theory, heavily using relations. He demanded with emphasis to 'Schröderize' all of mathematics [86, 87].

Fig. D.15 Leopold Löwenheim (1878–1957)

Only after decades, in 1941, Alfred Tarski (1902–1983, who until just before his doctorate published as Teitelbaum and as Tajtelbaum) revitalized relational methods with his prominent paper *On the calculus of relations* [138]. In August 1939 Tarski travelled – planned for three weeks – to attend a conference on *Unity of Science* held in the United States. He was equipped with a rather small suitcase – which was to serve him for the next six years of war.

Fig. D.16 Alfred Tarski (1902–1983)

With his paper *Relations binaires, fermetures, correspondances de Galois* of 1948, [105], Jacques Riguet used relations in a variety of fields, thus founding relational methods in applications as a research topic.

Fig. D.17 Jacques Riguet

The group of co-workers, friends, and colleagues of the present author have also contributed to relational mathematics, mainly with respect to the use of heterogeneous relations, introducing the symmetric quotient, typing and domain construction, stressing the matrix interpretation,[13] and in looking for many applications.

---

[13] When using typed matrices, there is no longer reason to be hesitant about applying negation; for a long time this had been avoided owing to the unrestricted universe in which the complement had to be formed; see the remark by George Boole reported on page 36.

References

[1] Fuad Aleskerov and Bernard Monjardet. *Utility Maximization, Choice and Preference.* Studies in Economic Theory. Springer-Verlag, 2002.

[2] Masahiko Aoki, John S. Chipman, and Peter C. Fishburn. A selected bibliography of works relating to the theory of preferences, utility, and demand. In Chipman et al. [33], pages 437–492.

[3] W. E. Armstrong. The determinateness of the utility function. *Economic Journal,* 49:453–467, 1939.

[4] Georg Aumann. *Kontakt-Relationen.* Sitzungsberichte der Bayer. Akademie der Wissenschaften, Math.-Nat. Klasse, 1970.

[5] Georg Aumann. *Ad Artem Ultimam – Eine Einführung in die Gedankenwelt der Mathematik.* München: R. Oldenbourg, 1974. ISBN 3-486-34481-1.

[6] R. B. Bapat and T. E. S. Raghavan. *Nonnegative Matrices and Applications,* volume 64 of *Encyclopaedia of Mathematics and its Applications.* Cambridge University Press, 1996.

[7] Alexander R. Bednarek and Stanislaw M. Ulam. Projective algebra and the calculus of relations. *Journal of Symbolic Logic,* 43:56–64, 1978.

[8] George Bentham. *Outline of a New System of Logic – With a Critical Examination of Dr. Whately's 'Elements of Logic'.* London: Hunt and Clarke, 1827.

[9] George Bentham. *Outline of a New System of Logic – With a Critical Examination of Dr Whately's 'Elements of Logic'.* In *Nineteenth-Century British Philosophy – Thoemmes Reprints* [8], 1990. Reprint of the edition of 1827. ISBN 1-85506-029-9

[10] Claude Berge. *Graphs and Hypergraphs.* North-Holland, 1973.

[11] Rudolf Berghammer. *Ordnungen, Verbände, und Relationen mit Anwendungen.* Vieweg + Teubner, 2008.

[12] Rudolf Berghammer, Thomas Gritzner, and Gunther Schmidt. Prototyping relational specifications using higher-order objects. Technical Report 1993/04, Fakultät für Informatik, Universität der Bundeswehr München, 33 p., 1993.

[13] Rudolf Berghammer, Thomas Gritzner, and Gunther Schmidt. Prototyping relational specifications using higher-order objects. In Jan Heering, Kurt Meinke, Bernhard Möller, and Tobias Nipkow, editors, *Higher-Order Algebra, Logic, and Term Rewriting, 1. Int. Workshop, HOA '93, Amsterdam,* volume 816 of *Lecture Notes in Computer Science,* pages 56–75. Springer-Verlag, 1994. 1. Int. Workshop, HOA '93, Amsterdam.

[14] Rudolf Berghammer, Armando Martín Haeberer, Gunther Schmidt, and Paulo A. S. Veloso. Comparing two different approaches to products in abstract relation algebra. In Nivat et al. [95], pages 167–176.

[15] Rudolf Berghammer, Thorsten Hoffmann, Barbara Leoniuk, and Ulf Milanese. Prototyping and programming with relations. *Electronic Notes in Theoretical Computer Science*, 44(3):27–50, 2003.

[16] Rudolf Berghammer and Gunther Schmidt. Discrete ordering relations. *Discrete Mathematics*, 43:1–7, 1983.

[17] Rudolf Berghammer and Gunther Schmidt. Algebraic visualization of relations using RELVIEW. In Victor G. Ganzha, Ernst W. Mayr, and Evgenii V. Vorozhtsov, editors, *Computer Algebra in Scientific Computing, 10th Int. Workshop, CASC 2007, Bonn, Germany, September 16–20, 2007, Proceedings*, volume 4770 of *Lecture Notes in Computer Science*, pages 58–72. Springer-Verlag, 2007.

[18] Rudolf Berghammer, Gunther Schmidt, and Michael Winter. RELVIEW and RATH – two systems for dealing with relations. In de Swart et al. [43], pages 1–16.

[19] Rudolf Berghammer, Gunther Schmidt, and Hans Zierer. Symmetric quotients. Technical Report TUM-INFO 8620, Technische Universität München, Institut für Informatik, 1986.

[20] Rudolf Berghammer, Gunther Schmidt, and Hans Zierer. Symmetric quotients and domain constructions. *Information Processing Letters*, 33(3):163–168, 1989/90.

[21] Rudolf Berghammer and Hans Zierer. Relational algebraic semantics of deterministic and nondeterministic programs. *Theoretical Computer Science*, 43:123–147, 1986.

[22] Richard S. Bird and Oege de Moor. *Algebra of Programming*. Prentice-Hall International, 1996.

[23] Andrea Blunck. Frauen in der Geschichte der Mathematik: zum Beispiel Christine Ladd-Franklin (1847–1930). In Gudrun Wolfschmidt, editor, *'Es gibt für Könige keinen besonderen Weg zur Geometrie' – Festschrift für Karin Reich*, volume 60 of *Algorismus: Studien zur Geschichte der Mathematik und der Naturwissenschaften*, chapter 16, pages 199–206. Dr. Erwin Rauner Verlag, 2007.

[24] Józef Maria Bocheński. *A History of Formal Logic*. New York: Chelsea Publishing, 1970. Translation of the German original of 1956; translated by Ivo Thomas.

[25] George Boole. *The Mathematical Analysis of Logic – Being an Essay Towards a Calculus of Deductive Reasoning*. Cambridge: Macmillan, Barclay, & Macmillan; London: George Bell, 1847.

[26] George Boole. *An Investigation of the Laws of Thought, on Which are Founded the Mathematical Theories of Logic and Probabilities*. Macmillan, 1854.

[27] George Boole. *The Mathematical Analysis of Logic – Being an Essay Towards a Calculus of Deductive Reasoning*. In a reprint of [25], 1948. Oxford: Basil Blackwell.

[28] George Boole. *An Investigation of the Laws of Thought, on Which are*

Founded the Mathematical Theories of Logic and Probabilities. Dover Publications, 1951. First American Printing of the 1854 edition with all corrections made within the text.

[29] Chris Brink, Wolfram Kahl, and Gunther Schmidt (editors). *Relational Methods in Computer Science.* Advances in Computing Science. Vienna: Springer-Verlag, 1997. ISBN 3-211-82971-7.

[30] Ilja N. Bronstein and Konstantin A. Semendjajew. *Taschenbuch der Mathematik.* Thun und Frankfurt/Main: Verlag Harri Deutsch, 1979. ISBN 3-87144-492-8.

[31] Arthur Cayley. A memoir on the theory of matrices. *Philosophical Transactions of the Royal Society of London*, 148 (Part 1):17–37, 1858.

[32] John S. Chipman. Consumption theory without transitive indifference. In Chipman et al. [33], pages 224–253.

[33] John S. Chipman, Leonid Hurwicz, Marcel K. Richter, and Hugo F. Sonnenschein, editors. *Preferences, Utility and Demand – A Minnesota Symposium.* New York: Harcourt Brace Jovanovich, 1971.

[34] Gustave Choquet. Theory of capacities. *Annales de l'institut Fourier*, 5:131–295, 1954.

[35] Irving M. Copilowish. Matrix development of the calculus of relations. *Journal of Symbolic Logic*, 13(4):193–203, 1948.

[36] Louis Couturat. *The Algebra of Logic.* Chicago, IL: Open Court Publishing Company, 1914. Authorized English Translation by Lydia Gillingham Robinson.

[37] B. A. Davey and H. A. Priestley. *Introduction to Lattices and Order.* Cambridge University Press, 1990.

[38] Philip J. Davis and Reuben Hersh. *The Mathematical Experience.* Boston, MA: Houghton Mifflin Company, 1981.

[39] Oege de Moor. *Categories, Relations and Dynamic Programming.* PhD Thesis, Oxford University, 1992.

[40] Augustus De Morgan. On the Syllogism, No. IV, and on the Logic of Relations. *Transactions of the Cambridge Philosophical Society*, X:331–358, 1864. Marked as 'Read April 23, 1860'.

[41] Augustus De Morgan. *On the Syllogism – and Other Logical Writings.* Rare Masterpieces of Philosophy and Science. Routledge & Kegan Paul, 1966. Reprinted in part. Edited, with an Introduction by Peter Heath.

[42] Willem-Paul de Roever and Kai Engelhardt. *Data Refinement: Model-Oriented Proof Methods and their Comparison*, volume 47 of *Cambridge Tracts in Theoretical Computer Science.* Cambridge University Press, 1998.

[43] Harrie de Swart, Ewa S. Orłowska, Gunther Schmidt, and Marc Roubens, editors. *Theory and Applications of Relational Structures as Knowledge Instruments. COST Action 274: TARSKI*, volume 2929 of *Lecture Notes in Computer Science.* Springer-Verlag, 2003. ISBN 3-540-20780-5

[44] Harrie de Swart, Ewa S. Orłowska, Gunther Schmidt, and Marc Roubens, editors. *Theory and Applications of Relational Structures as Knowledge Instruments II. COST Action 274: TARSKI*, volume 4342 of *Lecture Notes in Computer Science.* Springer-Verlag, 2006. ISBN-10: 3-540-69223-1, ISBN-13: 978-3-540-69223-2.

[45] Arthur Pentland Dempster. Upper and lower probabilities induced by a multivalued mapping. *Annals of Mathematical Statistics*, 38:325–339, 1967.

[46] Jean-Paul Doignon and Jean-Claude Falmagne. Matching relations and the dimensional structure of social sciences. *Mathematical Social Sciences*, 7:211–229, 1984.

[47] J. L. E. Dreyer and H. H. Turner, editors. *History of the Royal Astronomical Society, 1820–1920*. Royal Astronomical Society, sold by Wheldon & Wesley, 1923. Reprinted by Blackwell Scientific Publications, 1987.

[48] André Ducamp and Jean-Claude Falmagne. Composite measurement. *Journal of Mathematical Psychology*, 6:359–390, 1969.

[49] Julius (Gyula) Farkas. Über die Theorie der einfachen Ungleichungen. *Journal für die Reine und Angewandte Mathematik*, 124:1–27, 1902.

[50] L. M. G. Feijs and R. C. van Ommering. Relation partition algebra – mathematical aspects of uses and part-of relations. *Science of Computer Programming*, 33:163–212, 1999.

[51] Peter C. Fishburn. *Interval Orders and Interval Graphs – A Study of Partially Ordered Sets*. Wiley-Interscience Series in Discrete Mathematics. John Wiley and sons, 1985.

[52] János Fodor and Marc Roubens. *Fuzzy Preference Modelling and Multicriteria Decision Support*, volume 14 of *Theory and Decision Library, Series D: System Theory, Knowledge Engineering and Problem Solving*. Kluwer Academic Publishers, 1994.

[53] Peter J. Freyd and Andre Scedrov. *Categories, Allegories*, volume 39 of *North-Holland Mathematical Library*. Amsterdam: North-Holland, 1990. ISBN 0-444-70368-3 and 0-444-70367-5.

[54] Georg Frobenius. Über Matrizen aus nicht negativen Elementen. *Sitzungsberichte der Preussischen Akademie der Wissenschaften zu Berlin*, Math.-Phys. Klasse: 456–477, 1912.

[55] Delbert Ray Fulkerson and O. A. Gross. Incidence matrices with the consecutive 1s property. *Bulletin of the American Mathematical Society*, 70:681–684, 1964.

[56] Delbert Ray Fulkerson and O. A. Gross. Incidence matrices and interval graphs. *Pacific Journal of Mathematics*, 15(3):835–855, 1965.

[57] Helmuth Gericke. *Mathematik in Antike, Orient und Abendland*. Wiesbaden: fourierverlag, 2003. Reprint of two earlier parts *Mathematik in Antike und Orient* and *Mathematik im Abendland*.

[58] Alain Ghouilà-Houri. Caractérization des graphes non orientés dont on peut orienter les arrêtes de manière à obtenir le graphe d'une relation d'ordre. *Comptes Rendus de l'Académie des Sciences de Paris*, 254:1370–1371, 1962.

[59] Paul C. Gilmore and Alan J. Hoffman. A characterization of comparability graphs and of interval graphs. *Canadian Journal of Mathematics*, 16:539–548, 1964.

[60] Martin Charles Golumbic. *Algorithmic Graph Theory and Perfect Graphs*, volume 57 of *Annals of Discrete Mathematics*, edited by Peter L. Hammer. Elsevier, 2nd edition, 2004.

[61] Ivor Grattan-Guinness, editor. *Companion Encyclopedia of the History and Philosophy of the Mathematical Sciences*, volume 1. London: Routledge, 1994.

[62] Ivor Grattan-Guinness and W. Ledermann. Matrix theory. In Grattan-Guinness [61], chapter 6.7, pages 775–786.

[63] Hans Hahn. *Empirismus, Logik, Mathematik*. Number 645 in Suhrkamp Taschenbuch Wissenschaft. Suhrkamp, 1988.

[64] G. Hajós. Über eine Art von Graphen. *Intern. Math. Nachr.*, 11:Problem 65, 1957.

[65] Robert M. Haralick. The diclique representation and decomposition of binary relations. *Journal of the ACM*, 21:356–366, 1974.

[66] Claudia Hattensperger. *Rechnergestütztes Beweisen in heterogenen Relationenalgebren.* PhD Thesis, Universität der Bundeswehr München, 1997. München: Dissertationsverlag NG Kopierladen, ISBN 3-928536-99-0.

[67] Claudia Hattensperger, Rudolf Berghammer, and Gunther Schmidt. RALF – a relation-algebraic formula manipulation system and proof checker (Notes to a system demonstration). In Nivat et al. [95], pages 405–406.

[68] Stefan Hildebrandt and Peter D. Lax. *Otto Toeplitz*, volume 319 of *Bonner Mathematische Schriften*. Mathematisches Institut der Universität Bonn, 1999.

[69] Erich Hochstetter, Hermann-Josef Greve, and Heinz Gumin. *Herrn von Leibniz' Rechnung mit Null und Eins. Dritte Auflage.* Siemens AG, 1979. ISBN 3-8009-1279-1.

[70] William Stanley Jevons. *The Principles of Science – a Treatise on Logic and Scientific Method.* London: MacMillan, 1887. Second edition.

[71] Peter Jipsen, Chris Brink, and Gunther Schmidt. Background material. In Brink et al. [29], chapter 1, pages 1–21.

[72] Wolfram Kahl. A Relation-Algebraic Approach to Graph Structure Transformation. Technical Report 2002/03, Fakultät für Informatik, Universität der Bundeswehr München, June 2002. http://ist.unibw-muenchen.de/Publications/TR/2002-03/.

[73] Wolfram Kahl and Gunther Schmidt. Exploring (Finite) Relation Algebras With Tools Written in Haskell. Technical Report 2000/02, Fakultät für Informatik, Universität der Bundeswehr München, October 2000. http://ist.unibw-muenchen.de/Publications/TR/2000-02/.

[74] Britta Kehden. *Vektoren und Vektorprädikate und ihre Verwendung bei der Entwicklung relationaler Algorithmen.* PhD Thesis, Christian-Albrechts-Universität zu Kiel, 2008.

[75] Peter Kempf and Michael Winter. Relational semantics for processes. Technical Report 1998/05, Fakultät für Informatik, Universität der Bundeswehr München, 1998.

[76] Peter Kempf and Michael Winter. Relational Semantics for Processes. *Studies in Fuzziness and Soft Computing*, pages 59–73, 2001.

[77] Ki Hang Kim. *Boolean Matrix Theory and Applications*, volume 70 of *Monographs and Textbooks in Pure and Applied Mathematics*. New York: Marcel Dekker, 1982.

[78] Ki Hang Kim and Fred W. Roush. *Introduction to Mathematical Consensus Theory*, volume 59 of *Lecture Notes in Pure and Applied Mathematics*. Marcel Dekker, 1980.

[79] William Kneale and Martha Kneale. *The Development of Logic*. Oxford: Clarendon Press, 1962, reprinted with corrections as paperback 1984.

[80] Eberhard Knobloch. Determinants. In Grattan-Guinness [61], chapter 6.6, pages 766–774.

[81] Janusz Konieczny. On cardinalities of row spaces of Boolean matrices. *Semigroup Forum*, 44:393–402, 1992.

[82] Dénes König. *Theorie der endlichen und unendlichen Graphen – Kombinatorische Topologie der Streckenkomplexe.* Leipzig: Akademische Verlagsgesellschaft, 1936.

[83] Dénes König. *Theorie der endlichen und unendlichen Graphen – Kombinatorische Topologie der Streckenkomplexe.* New York: Chelsea, Post war reprint of the 1936 edition.

[84] Christine Ladd-Franklin. The Algebra of Logic. In Charles Sanders Peirce, editor, *Studies in Logic by Members of the Johns Hopkins University*, pages 17–71. Little, Brown & Co., 1882.

[85] Barbara Leoniuk. *ROBDD-basierte Implementierung von Relationen und relationalen Operationen mit Anwendungen.* PhD Thesis, Christian-Albrechts-Universität zu Kiel, 2001.

[86] Leopold Löwenheim. Über Möglichkeiten im Relativkalkül. *Mathematische Annalen*, 76:447–470, 1915.

[87] Leopold Löwenheim. Einkleidung der Mathematik in Schröderschen Relativkalkul. *Journal of Symbolic Logic*, 5:1–15, 1940.

[88] Robert Duncan Luce. A note on boolean matrix theory. *Proceedings of the American Mathematical Society*, 3:382–388, 1952.

[89] Robert Duncan Luce. Semi-orders and a theory of utility discrimination. *Econometrica*, 24:178–191, 1956.

[90] Jakob Lüroth. Ernst Schröder †. *Jahresberichte der Deutschen Mathematiker-Vereinigung*, 12:249–265, 1903.

[91] Roger Duncan Maddux. On the derivation of identities involving projection functions. In Lásló Csirmaz, Dov Gabbay, and Maarten de Rijke, editors, *Logic Colloquium'92. Center for the Study of Language and Information Publications, Stanford 1995*, Studies in Logic, Language, and Information, pages 145–163, 1995.

[92] Henryk Minc. *Nonnegative Matrices.* Wiley, 1988.

[93] Leonid Mirsky. *Transversal Theory – An Account of Some Aspects of Combinatorial Mathematics*, volume 75 of *Mathematics in Science and Engineering.* Academic Press, 1971.

[94] Bernard Monjardet. Axiomatiques et propriétés des quasi-ordres. *Mathematiques et Sciences Humaines*, 16(63):51–82, 1978.

[95] Maurice Nivat, Charles Rattray, Teodore Rus, and Giuseppe Scollo, editors. *Algebraic Methodology and Software Technology, Proc. 3rd Int. Conf. (AMAST '93), University of Twente, Enschede, The Netherlands, June 21–25, 1993*, Workshops in Computing. Springer-Verlag, 1994.

[96] Seán O'Donnell. *William Rowan Hamilton – Portrait of a Prodigy.* Dublin: Boole Press, 1983.

[97] Jean-Pierre Olivier and Dany Serrato. Squares and rectangles in relation categories. Three cases: semilattice, distributive lattice and boolean non-unitary. *Fuzzy Sets and Systems*, 72:167–187, 1995.

[98] Ewa S. Orłowska and Gunther Schmidt. Rasiowa-Sikorski Proof Systems in Relation Algebra, 2004. Internal notes.

[99] Marc Pirlot. Synthetic description of a semiorder. *Discrete and Applied Mathematics*, 31:299–308, 1991.

[100] Marc Pirlot and Philippe Vincke. *Semiorders – Properties, Representations, Applications*, volume 36 of *Theory and Decision Library, Series B: Mathematical and Statistical Methods.* Kluwer Academic Publishers, 1997.

[101] Robert J. Plemmons and M. T. West. On the semigroup of binary relations. *Pacific Journal of Mathematics*, 35(3):743–753, 1970.

[102] A. N. Prior. *Formal Logic*. Oxford: Clarendon Press, Second edition, 1962, reprinted 1963.

[103] Ingrid Rewitzky. Binary multirelations. In de Swart et al. [43], pages 259–274.

[104] Ingrid Rewitzky and Chris Brink. Monotone predicate transformers as up-closed multirelations. In Schmidt [126], pages 311–327.

[105] Jacques Riguet. Relations binaires, fermetures, correspondances de Galois. *Bulletin de la Société Mathématique de France*, 76:114–155, 1948.

[106] Jacques Riguet. Les relations de Ferrers. *Comptes Rendus de l'Académie des Sciences de Paris*, 231:936–937, 1950.

[107] Gunther Schmidt. Wenn der Lehrer halbiert werden soll. *Bild der Wissenschaft*, 7:100–102, 1974. Mathematisches Kabinett.

[108] Gunther Schmidt. Eine relationenalgebraische Auffassung der Graphentheorie. Technical Report 7619, Fachbereich Mathematik der Technischen Universität München, 1976.

[109] Gunther Schmidt. Programme als partielle Graphen. Habil. Thesis 1977 and report 7813, Fachbereich Mathematik der Technischen Universität München, 1977. In English as [110, 111].

[110] Gunther Schmidt. Programs as partial graphs I: Flow equivalence and correctness. *Theoretical Computer Science*, 15:1–25, 1981.

[111] Gunther Schmidt. Programs as partial graphs II: Recursion. *Theoretical Computer Science*, 15:159–179, 1981.

[112] Gunther Schmidt. Implication structures. In Ivo Düntsch and Michael Winter, editors, *8th Int. Seminar RelMiCS*, pages 227–237, 2005.

[113] Gunther Schmidt and Rudolf Berghammer. Relational measures and integration in preference modeling. *Journal of Logic and Algebraic Programming*, 76/1:112–129, 2008. Special Issue edited by Georg Struth; DOI: 10.1016/j.jlap.2007.10.001; ISSN 1567-8326.

[114] Gunther Schmidt and Rudolf Berghammer. Contact relations with applications. In Rudolf Berghammer, Ali Jaoua, and Bernhard Möller, editors, *RelMiCS*, volume 5827 of *Lecture Notes in Computer Science*, pages 306–321. Springer, 2009.

[115] Gunther Schmidt, Rudolf Berghammer, and Hans Zierer. Beschreibung semantischer Bereiche mit Keimen. Technical Report TUM-I8611, Institut für Informatik, Technische Universität München, 1986. 33 p.

[116] Gunther Schmidt, Rudolf Berghammer, and Hans Zierer. Describing semantic domains with sprouts. In Franz-Josef Brandenburg, Guy Vidal-Naquet, and Martin Wirsing, editors, *Proc. 4th Symposium on Theoretical Aspects of Computer Science (STACS '87)*, volume 247 of *Lecture Notes in Computer Science*, pages 299–310. Springer-Verlag, 1987. Shortened version of [115].

[117] Gunther Schmidt, Claudia Hattensperger, and Michael Winter. Heterogeneous relation algebra. In Brink et al. [29], chapter 3, pages 40–54.

[118] Gunther Schmidt and Thomas Ströhlein. A boolean matrix iteration in timetable construction. Technical Report 7406, Abteilung Mathematik der Technischen Universität München, 1974.

[119] Gunther Schmidt and Thomas Ströhlein. Relationen, Graphen und Strukturen, 1975. Internal Report at Technische Universität München.

[120] Gunther Schmidt and Thomas Ströhlein. A boolean matrix iteration in timetable construction. *Linear Algebra and Its Applications*, 15:27–51, 1976.

[121] Gunther Schmidt and Thomas Ströhlein. Relationen und Graphen, 1981. Internal Report at Technische Universität München.

[122] Gunther Schmidt and Thomas Ströhlein. Diskrete Mathematik – Relationen, Graphen und Programme I. Technical report, Institut für Informatik der Technischen Universität München, 1985.

[123] Gunther Schmidt and Thomas Ströhlein. *Relationen und Graphen*. Mathematik für Informatiker. Springer-Verlag, 1989. ISBN 3-540-50304-8, ISBN 0-387-50304-8.

[124] Gunther Schmidt and Thomas Ströhlein. *Relations and Graphs – Discrete Mathematics for Computer Scientists*. EATCS Monographs on Theoretical Computer Science. Springer-Verlag, 1993. ISBN 3-540-56254-0, ISBN 0-387-56254-0.

[125] Jürgen Schmidt, editor. *Mengenlehre – Einführung in die axiomatische Mengenlehre; Band 1: Grundbegriffe*. BI-Hochschultaschenbücher 56. Bibliographisches Institut, 1974.

[126] Renate A. Schmidt, editor. *RelMiCS '9 – Relations and Kleene-Algebra in Computer Science*, volume 4136 of *Lecture Notes in Computer Science*. Springer-Verlag, 2006.

[127] Ernst Schröder. *Vorlesungen über die Algebra der Logik (exacte Logik)*, volume 3, Algebra und Logik der Relative, Part I; no Part II was ever published. Leipzig: Teubner, 1895.

[128] Ernst Schröder. *Vorlesungen über die Algebra der Logik (exacte Logik)*, volume 3 of a reprint of [127], 1966. Algebra und Logik der Relative; 2nd edition.

[129] Ernst Schröder. Note über die Algebra der binaren Relative. *Mathematische Annalen*, 46:144–158, 1895.

[130] Giuseppe Scollo, Giuditta Franco, and Vincenzo Manca. A relational view of recurrence and attractors in state transititon systems. In Schmidt [126], pages 358–372.

[131] Dana Scott. Measurement Structures and Linear Inequalities. *Journal of Mathematical Psychology*, 1:233–247, 1964.

[132] Glenn Shafer. *A Mathematical Theory of Evidence*. Princeton University Press, 1976.

[133] G. C. Smith. *The Boole-De Morgan Correspondence 1842–1864*. Oxford Logic Guides; General Editor: Dana Scott. Oxford: Clarendon Press, 1982.

[134] Roland Soltysiak. *Die Projektion affiner Strukturen über Fastkörpern mit Hilfe relationentheoretischer Methoden*. PhD Thesis, Universität Duisburg, 1980.

[135] Thomas Ströhlein. Tightness and flexibility in timetable problems. In Manfred Nagl and Hans-J. Schneider, editors, *Graphs, Data Structures, Algorithms*, pages 299–306. Hanser, 1978.

[136] Nicholai Ivanovich Styazhkin. *History of Mathematical Logic from Leibniz to Peano*. MIT Press, 1969. Translation of the Russian original of 1964.

[137] Michio Sugeno. *Industrial Applications of Fuzzy Control*. North-Holland, 1985.

[138] Alfred Tarski. On the calculus of relations. *Journal of Symbolic Logic*, 6:73–89, 1941.

[139] Alfred Tarski. Some metalogical results concerning the calculus of relations. *Journal of Symbolic Logic*, 18:188–189, 1953. Abstract.

[140] Alfred Tarski and Steven R. Givant. *A Formalization of Set Theory without Variables*, volume 41 of *Colloquium Publications*. American Mathematical Society, 1987.

[141] Gottfried Tinhofer and Gunther Schmidt, editors. *Graph-Theoretic Concepts in Computer Science*, volume 246 of *Lecture Notes in Computer Science*. Springer-Verlag, 1987. Proc. 12th Int. Workshop WG '86, Kloster Bernried, June 17–19, 1986. ISBN 3-540-17218-1, ISBN 0-387-17218-1.

[142] Richard S. Varga. *Matrix Iterative Analysis*. Englewood Cliffs, NJ: Prentice-Hall, 1962.

[143] Philippe Vincke. Preferences and numbers. In A. Colorni, M. Paruccini, and B. Roy, editors, *A-MCD-A – Aide Multi Critère à la Décision – Multiple Criteria Decision Aiding*, pages 343–354. Joint Research Centre of the European Commission, 2001.

[144] Alfred North Whitehead and Bertrand Russell. *Principia Mathematica*. Cambridge University Press, 1910. 3 volumes, second edition 1962. ISBN 978-0521067911.

[145] Robin Wilson. The Philamath's Alphabet – S. *The Mathematical Intelligencer*, 30(4):88, 2008.

[146] Michael Winter. Decomposing relations into orderings. In Rudolf Berghammer and Bernhard Möller, editors, *Participants Proc. of the Int. Workshop RelMiCS '7 Relational Methods in Computer Science and 2nd Int. Workshop on Applications of Kleene Algebra, in combination with a Workshop of the* COST *Action 274:* TARSKI, *Malente, May 12–17, 2003*, pages 190–196, 2003.

[147] Michael Winter. Decomposing relations into orderings. In Rudolf Berghammer, Bernhard Möller, and Georg Struth, editors, *Proc. of the Internat. Workshop RelMiCS '7 and 2nd Internat. Workshop on Applications of Kleene Algebra, in combination with a workshop of the COST Action 274: TARSKI. Revised Selected Papers*, volume 3051 of *Lecture Notes in Computer Science*, pages 261–272. Springer-Verlag, 2004. ISBN 3-540-22145-X

[148] Michael Winter. A Relation Algebraic Theory of Bisimulation. Technical Report # CS-06-04, Brock University, St. Catharines, Ontario, Canada, 2006.

[149] Michael Winter. A relation algebraic theory of bisimulations. *Fundamenta Informaticae*, 83(4):429–449, 2008.

[150] Hans Zierer. Programmierung mit Funktionsobjekten: Konstruktive Erzeugung semantischer Bereiche und Anwendung auf die partielle Auswertung. Dissertation, 1988.

Symbols

Sets

It is widely known that $\{\, x \mid E(x)\,\}$ is used for defining sets descriptively, with $x \in M$ for being an element and $M \subseteq X$ for being a subset of a set. We use 2^X or $\mathcal{P}(X)$ for the powerset and $\emptyset$ or $\{\}$ for the empty set. Union and intersection are denoted as $M \cup N$ and $M \cap N$. The complement is $\overline{M}$, provided the ground set is tacitly given. The Cartesian product is $M \times N$.

Logic

For metalanguage consequence, equivalence, and definition, "$\Longrightarrow$", "$\Longleftrightarrow$", and "$:\Longleftrightarrow$" are used. Definitional equality is denoted as "$:=$". The set of Boolean truth-values is $\mathbb{B} = \{\, \mathbf{0}, \mathbf{1}\,\}$. In the context of propositional logic, "$\wedge$", "$\vee$" are used for 'and' respectively 'or', together with "$\rightarrow$" for 'if... then' and "$\leftrightarrow$" for 'precisely when'. In the context of predicate logic, "$\exists$" and "$\forall$" denote the existential quantifier and the universal quantifier.

Relations

$R : X \longrightarrow Y$	relation with source and target		15
$\mathrm{src}(R)$	source of a relation		36
$\mathrm{tgt}(R)$	target of a relation		36
$\mathrm{dom}(R)$	domain of a relation		91
$\mathrm{cod}(R)$	codomain of a relation		91
$R \cup S$	union		37
$\sup \mathcal{A}$, $\sup \{\, R \mid R \in \mathcal{A}\,\}$	supremum		38
$R \cap S$	intersection		37
$\inf \mathcal{A}$, $\inf \{\, R \mid R \in \mathcal{A}\,\}$	infimum		38
$\bot\!\!\bot$	empty relation	provided the	20
$\top\!\!\top$	universal relation	ground sets	20
$\mathbb{I}$	identity	are tacitly given	20
R^{T}	transposed relation, converse		39
R^{d}	dual of a relation		43
$R \,\mathring{;}\, S$	product, composition		40

Domains

DirPro	t1	t2	direct product of t1 and t2	108
Pi	t1	t2	first projection from direct product	108
Rho	t1	t2	second projection from direct product	108
DirSum	t1	t2	direct sum of t1 and t2	128
Iota	t1	t2	first injection into direct sum	128
Kappa	t1	t2	second injection into direct sum	128
QuotMod	xi		quotient domain modulo equivalence $\xi \approx$ xi	132
Project	xi		natural projection for ξ onto quotient domain	132
Extrude	u		extruded subset u conceived as separate domain	136
Inject	u		natural injection onto subset u	136
DirPow	t		power domain for domain t	141
Member	t		membership relation	141
PermTgt	p		permuted target for bijective mapping P	149
ReArrTo	p		permutation relation for bijective mapping P	149

For further details of the notation used in domain construction see Appendix A.

Index